SCHAUM'S 3000 SOLVED PROBLEMS IN

CALCULUS

by

Elliott Mendelson, Ph.D.
Queens College

McGRAW-HILL BOOK COMPANY
New York   St. Louis   San Francisco   Auckland   Bogotá
Hamburg   London   Madrid   Mexico   Milan   Montreal   New Delhi
Panama   Paris   Saõ Paulo   Singapore   Sydney   Tokyo   Toronto

/ Elliott Mendelson, Ph.D., *Professor of Mathematics at Queens College.*

Dr. Mendelson taught at the University of Chicago and Columbia University before his tenure at Queens College. He has authored Schaum's Outlines of BEGINNING CALCULUS and BOOLEAN ALGEBRA AND SWITCHING CIRCUITS.

Other Contributors to This Volume

/ Frank Ayres, Jr., Ph.D., *Dickinson College*

/ George Simmons, Ph.D., *Colorado College*

/ Sherman Stein, Ph.D., *University of California, Davis*

/ David Beckwith, Ph.D., *McGraw-Hill Book Company*

/ Joseph H. Kindle, Ph.D., *University of Cincinnati*

Project supervision by The Total Book.

Library of Congress Cataloging-in-Publication Data

Schaum's 3000 solved problems in calculus / edited by Elliott
  Mendelson.
        p.   cm.
  ISBN 0-07-041480-7
  1. Calculus—Problems, exercises, etc.   I. Mendelson, Elliott.
  II. Title: Schaum's three thousand solved problems in calculus.
  QA303.S288   1988
  515′.076—dc19   87-23265
                                                                          CIP

  2 3 4 5 6 7 8 9 0 SHP/SHP 8 9 2 1 0 9 8

ISBN 0-07-041480-7

CONTENTS

To the Student

This collection of solved problems covers elementary and intermediate calculus, and much of advanced calculus. We have aimed at presenting the broadest range of problems that you are likely to encounter—the old chestnuts, all the current standard types, and some not so standard.

Each chapter begins with very elementary problems. Their difficulty usually increases as the chapter progresses, but there is no uniform pattern.

It is assumed that you have available a calculus textbook, including tables for the trigonometric, logarithmic, and exponential functions. Our ordering of the chapters follows the customary order found in many textbooks, but as no two textbooks have exactly the same sequence of topics, you must expect an occasional discrepancy from the order followed in your course.

The printed solution that immediately follows a problem statement gives you all the details of *one way* to solve the problem. You might wish to delay consulting that solution until you have outlined an attack in your own mind. You might even disdain to read it until, with pencil and paper, you have solved the problem yourself (or failed gloriously). Used thus, *3000 Solved Problems in Calculus* can almost serve as a supplement to any course in calculus, or even as an independent refresher course.

1.1   Solve $3 + 2x < 7$.

▌ $2x < 4$ [Subtract 3 from both sides.   This is equivalent to adding -3 to both sides.]
Answer $x < 2$ [Divide both sides by 2.   This is equivalent to multiplying by $\frac{1}{2}$.]   In interval notation, the solution is the set $(-\infty, 2)$.

1.2   Solve $5 - 3x < 5x + 2$.

▌ $5 - 3x < 5x + 2$, $5 < 8x + 2$ [Add $3x$ to both sides.], $3 < 8x$ [Subtract 2 from both sides.]
Answer $\frac{3}{8} < x$ [Divide both sides by 8.]   In interval notation, the solution is the set $(\frac{3}{8}, \infty)$.

1.3   Solve $-7 < 2x + 5 < 9$.

▌ $-7 < 2x + 5 < 9$, $-12 < 2x < 4$ [Subtract 5 from all terms.]
Answer $-6 < x < 2$ [Divide by 2.]   In interval notation, the solution is the set $(-6, 2)$.

1.4   Solve $3 \le 4x - 1 < 5$.

▌ $3 \le 4x - 1 < 5$, $4 \le 4x < 6$ [Add 1 to all terms.]
Answer $1 \le x < \frac{3}{2}$ [Divide by 4.]   In interval notation, the solution is the set $[1, \frac{3}{2})$.

1.5   Solve $4 < -2x + 5 \le 7$.

▌ $4 < -2x + 5 \le 7$, $-1 < -2x \le 2$ [Subtract 5.]
Answer $\frac{1}{2} > x \ge -1$ [Divide by -2.   Since -2 is negative, we must reverse the inequalities.]   In interval notation, the solution is the set $[-1, \frac{1}{2})$.

1.6   Solve $5 \le \frac{1}{3}x + 1 \le 6$.

▌ $5 \le \frac{1}{3}x + 1 \le 6$, $4 \le \frac{1}{3}x \le 5$ [Subtract 1.]
Answer $12 \le x \le 15$ [Multiply by 3.]   In interval notation, the solution is the set $[12, 15]$.

1.7   Solve $2/x < 3$.

▌ x may be positive or negative.   **Case 1.** $x > 0$. $2/x < 3$. $2 < 3x$ [Multiply by x.], $\frac{2}{3} < x$ [Divide by 3.]
Case 2. $x < 0$. $2/x < 3$. $2 > 3x$ [Multiply by x.   Reverse the inequality.], $\frac{2}{3} > x$ [Divide by 3.]   Notice that this condition $\frac{2}{3} > x$ is satisfied whenever $x < 0$.   Hence, in the case where $x < 0$,   the inequality is satisfied by all such x.
Answer $\frac{2}{3} < x$ or $x < 0$.   As shown in Fig. 1-1, the solution is the union of the intervals $(\frac{2}{3}, \infty)$ and $(-\infty, 0)$.

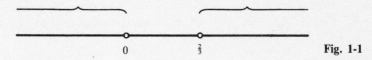

0            $\frac{2}{3}$ **Fig. 1-1**

1.8   Solve

$$\frac{x + 4}{x - 3} < 2 \qquad\qquad (1)$$

▌ We cannot simply multiply both sides by $x - 3$,   because we do not know whether $x - 3$ is positive or negative.   **Case 1.** $x - 3 > 0$ [This is equivalent to $x > 3$.]   Multiplying the given inequality (1) by the positive quantity $x - 3$ preserves the inequality: $x + 4 < 2x - 6$, $4 < x - 6$ [Subtract x.], $10 < x$ [Add 6.]   Thus, when $x > 3$,   the given inequality holds when and only when $x > 10$.   **Case 2.** $x - 3 < 0$ [This is equivalent to $x < 3$].   Multiplying the given inequality (1) by the negative quantity $x - 3$ reverses the inequality: $x + 4 > 2x - 6$, $4 > x - 6$ [Subtract x.], $10 > x$ [Add 6.]   Thus, when $x < 3$,   the inequality

1

(*1*) holds when and only when $x < 10$. But $x < 3$ implies $x < 10$, and, therefore, the inequality (*1*) holds for all $x < 3$.

Answer $x > 10$ or $x < 3$. As shown in Fig. 1-2, the solution is the union of the intervals $(10, \infty)$ and $(-\infty, 3)$.

$$0 \qquad 3 \qquad\qquad 10 \qquad\qquad \textbf{Fig. 1-2}$$

1.9 Solve

$$\frac{x}{x+5} < 1 \qquad\qquad (1)$$

❚ **Case 1.** $x + 5 > 0$ [This is equivalent to $x > -5$.]. We multiply the inequality (*1*) by $x + 5$. $x < x + 5$, $0 < 5$ [Subtract x.] This is always true. So, (*1*) holds throughout this case, that is, whenever $x > -5$. **Case 2.** $x + 5 < 0$ [This is equivalent to $x < -5$.]. We multiply the inequality (*1*) by $x + 5$. The inequality is reversed, since we are multiplying by a negative number. $x > x + 5$, $0 > 5$ [Subtract x.] But $0 > 5$ is false. Hence, the inequality (*1*) does not hold at all in this case.

Answer $x > -5$. In interval notation, the solution is the set $(-5, \infty)$.

1.10 Solve

$$\frac{x-7}{x+3} > 2 \qquad\qquad (1)$$

❚ **Case 1.** $x + 3 > 0$ [This is equivalent to $x > -3$.]. Multiply the inequality (*1*) by $x + 3$. $x - 7 > 2x + 6$, $-7 > x + 6$ [Subtract x.], $-13 > x$ [Subtract 6.] But $x < -13$ is always false when $x > -3$. Hence, this case yields no solutions. **Case 2.** $x + 3 < 0$ [This is equivalent to $x < -3$.]. Multiply the inequality (*1*) by $x + 3$. Since $x + 3$ is negative, the inequality is reversed. $x - 7 < 2x + 6$, $-7 < x + 6$ [Subtract x.] $-13 < x$ [Subtract 6.] Thus, when $x < -3$, the inequality (*1*) holds when and only when $x > -13$.

Answer $-13 < x < -3$. In interval notation, the solution is the set $(-13, -3)$.

1.11 Solve $(2x - 3)/(3x - 5) \geq 3$.

❚ **Case 1.** $3x - 5 > 0$ [This is equivalent to $x > \frac{5}{3}$.]. $2x - 3 \geq 9x - 15$ [Multiply by $3x - 5$.], $-3 \geq 7x - 15$ [Subtract $2x$.], $12 \geq 7x$ [Add 15.], $\frac{12}{7} \geq x$ [Divide by 7.] So, when $x > \frac{5}{3}$, the solutions must satisfy $x \leq \frac{12}{7}$. **Case 2.** $3x - 5 < 0$ [This is equivalent to $x < \frac{5}{3}$.]. $2x - 3 \leq 9x - 15$ [Multiply by $3x - 5$. Reverse the inequality.], $-3 \leq 7x - 15$ [Subtract $2x$.], $12 \leq 7x$ [Add 15.], $\frac{12}{7} \leq x$ [Divide by 7.] Thus, when $x < \frac{5}{3}$, the solutions must satisfy $x \geq \frac{12}{7}$. This is impossible. Hence, this case yields no solutions.

Answer $\frac{5}{3} < x \leq \frac{12}{7}$. In interval notation, the solution is the set $(\frac{5}{3}, \frac{12}{7}]$.

1.12 Solve $(2x - 3)/(3x - 5) \geq 3$.

❚ Remember that a product is positive when and only when both factors have the same sign. **Case 1.** $x - 2 > 0$ and $x + 3 > 0$. Then $x > 2$ and $x > -3$. But these are equivalent to $x > 2$ alone, since $x > 2$ implies $x > -3$. **Case 2.** $x - 2 < 0$ and $x + 3 < 0$. Then $x < 2$ and $x < -3$, which are equivalent to $x < -3$, since $x < -3$ implies $x < 2$.

Answer $x > 2$ or $x < -3$. In interval notation, this is the union of $(2, \infty)$ and $(-\infty, -3)$.

1.13 Solve Problem 1.12 by considering the sign of the function $f(x) = (x - 2)(x + 3)$.

❚ Refer to Fig. 1-3. To the left of $x = -3$, both $x - 2$ and $x + 3$ are negative and $f(x)$ is positive. As one passes through $x = -3$, the factor $x - 3$ changes sign and, therefore, $f(x)$ becomes negative. $f(x)$ remains negative until we pass through $x = 2$, where the factor $x - 2$ changes sign and $f(x)$ becomes and then remains positive. Thus, $f(x)$ is positive for $x < -3$ and for $x > 2$. **Answer**

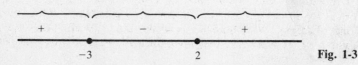

$$+ \qquad\qquad - \qquad\qquad + $$
$$-3 \qquad\qquad 2 \qquad\qquad \textbf{Fig. 1-3}$$

1.14   Solve $(x-1)(x+4)<0$.

❚ The key points of the function $g(x)=(x-1)(x+4)$ are $x=-4$ and $x=1$ (see Fig. 1-4).   To the left of $x=-4$,   both $x-1$ and $x+4$ are negative and, therefore, $g(x)$ is positive.   As we pass through $x=-4$, $x+4$ changes sign and $g(x)$ becomes negative.   When we pass through $x=1$, $x-1$ changes sign and $g(x)$ becomes and then remains positive.   Thus, $(x-1)(x+4)$ is negative for $-4<x<1$.   **Answer**

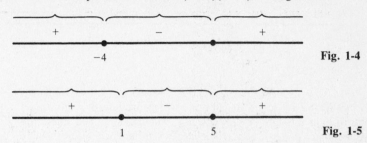

Fig. 1-4

Fig. 1-5

1.15   Solve $x^2-6x+5>0$.

❚ Factor: $x^2-6x+5=(x-1)(x-5)$.   Let $h(x)=(x-1)(x-5)$.   To the left of $x=1$ (see Fig. 1-5), both $x-1$ and $x-5$ are negative and, therefore, $h(x)$ is positive.   When we pass through $x=1$, $x-1$ changes sign and $h(x)$ becomes negative.   When we move further to the right and pass through $x=5$, $x-5$ changes sign and $h(x)$ becomes positive again.   Thus, $h(x)$ is positive for $x<1$ and for $x>5$.   **Answer** $x>5$ or $x<1$.   This is the union of the intervals $(5,\infty)$ and $(-\infty,1)$.

1.16   Solve $x^2+7x-8<0$.

❚ Factor: $x^2+7x-8=(x+8)(x-1)$,   and   refer   to   Fig. 1-6.   For $x<-8$,   both $x+8$ and $x-1$ are negative and, therefore, $F(x)=(x+8)(x-1)$ is positive.   When we pass through $x=-8$, $x+8$ changes sign and, therefore, so does $F(x)$.   But when we later pass through $x=1$, $x-1$ changes sign and $F(x)$ changes back to being positive.   Thus, $F(x)$ is negative for $-8<x<1$.   **Answer**

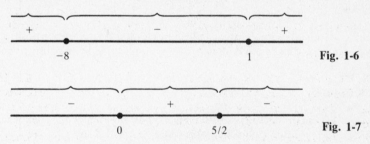

Fig. 1-6

Fig. 1-7

1.17   Solve $5x-2x^2>0$.

❚ Factor: $5x-2x^2=x(5-2x)$,   and refer to Fig. 1-7.   The key points for the function $G(x)=x(5-2x)$ are $x=0$ and $x=\frac{5}{2}$.   For $x<0$, $5-2x$ is positive and, therefore, $G(x)$ is negative.   As we pass through $x=0$, x changes sign and, therefore, $G(x)$ becomes positive.   When we pass through $x=\frac{5}{2}$, $5-2x$ changes sign and, therefore, $G(x)$ changes back to being negative.   Thus, $G(x)$ is positive when and only when $0<x<\frac{5}{2}$.   **Answer**

1.18   Solve $(x-1)^2(x+4)<0$.

❚ $(x-1)^2$ is always positive except when $x=1$ (when it is 0).   So, the only solutions occur when $x+4<0$ and $x\neq1$.
Answer $x<-4$ [In interval notation, $(-\infty,-4)$.]

1.19   Solve $x(x-1)(x+1)>0$.

❚ The key points for $H(x)=x(x-1)(x+1)$ are $x=0$, $x=1$,   and $x=-1$ (see Fig. 1-8).   For x to the left of -1, x, $x-1$,   and $x+1$ all are negative and, therefore, $H(x)$ is negative.   As we pass through $x=-1$, $x+1$ changes sign and, therefore, so does $H(x)$.   When we later pass through $x=0$, x changes sign and, therefore, $H(x)$ becomes negative again.   Finally, when we pass through $x=1$, $x-1$ changes sign and $H(x)$ becomes and remains positive.   Therefore, $H(x)$ is positive when and only when $-1<x<0$ or $x>1$.   **Answer**

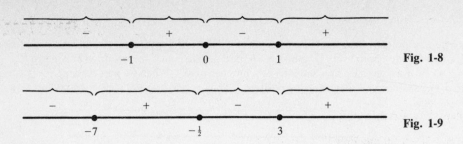

Fig. 1-8

Fig. 1-9

1.20 Solve $(2x + 1)(x - 3)(x + 7) < 0$.

▎ See Fig. 1-9. The key points for the function $K(x) = (2x + 1)(x - 3)(x + 7)$ are $x = -7$, $x = -\frac{1}{2}$, and $x = 3$. For x to the left of $x = -7$, all three factors are negative and, therefore, $K(x)$ is negative. When we pass from left to right through $x = -7$, $x + 7$ changes sign, and, therefore, $K(x)$ becomes positive. When we later pass through $x = -\frac{1}{2}$, $2x + 1$ changes sign, and, therefore, $K(x)$ becomes negative again. Finally, as we pass through $x = 3$, $x - 3$ changes sign and $K(x)$ becomes and remains positive. Hence, $K(x)$ is negative when and only when $x < -7$ or $3 < x < 7$. **Answer**

1.21 Does $\sqrt{a^2} < \sqrt{b^2}$ always imply $a < b$?

▎ No. Let $a = 1$ and $b = -2$.

1.22 Solve $x > x^2$.

▎ $x > x^2$ is equivalent to $x^2 - x < 0$, $x(x - 1) < 0$, $0 < x < 1$.

1.23 Solve $x^2 > x^3$.

▎ $x^2 > x^3$ is equivalent to $x^3 - x^2 < 0$, $x^2(x - 1) < 0$, $x < 1$, and $x \neq 0$.

1.24 Find all solutions of $\dfrac{1}{x} < \dfrac{1}{y}$.

▎ This is clearly true when x is negative and y positive, and false when x is positive and y negative. When x and y are both positive, or x and y are both negative, multiplication by the positive quantity xy yields the equivalent inequality $y < x$.

1.25 Solve $(x - 1)(x - 2)(x - 3)(x - 4) < 0$.

▎ When $x > 4$, the product is positive. Figure 1-10 shows how the sign changes as one passes through the points $4, 3, 2, 1$. Hence, the inequality holds when $1 < x < 2$ or $3 < x < 4$.

Fig. 1-10

CHAPTER 2
Absolute Value

2.1 Solve $|x + 3| \le 5$.

$\blacksquare$ $|x + 3| \le 5$ if and only if $-5 \le x + 3 \le 5$.
Answer $-8 \le x \le 2$ [Subtract 3.] In interval notation, the solution is the set $[-8, 2]$.

2.2 Solve $|3x + 2| < 1$.

$\blacksquare$ $|3x + 2| < 1$ if and only if $-1 < 3x + 2 < 1$, $-3 < 3x < -1$ [Subtract 2.]
Answer $-1 < x < -\frac{1}{3}$ [Divide by 3.] In interval notation, the solution is the set $(-1, -\frac{1}{3})$.

2.3 Solve $|5 - 3x| < 2$.

$\blacksquare$ $|5 - 3x| < 2$ if and only if $-2 < 5 - 3x < 2$, $-7 < -3x < -3$ [Subtract 5.]
Answer $\frac{7}{3} > x > 1$ [Divide by -3 and reverse the inequalities.] In interval notation, the solution is the set $(1, \frac{7}{3})$.

2.4 Solve $|3x - 2| \ge 1$.

$\blacksquare$ Let us solve the negation of the given relation: $|3x - 2| < 1$. This is equivalent to $-1 < 3x - 2 < 1$, $1 < 3x < 3$ [Add 2.], $\frac{1}{3} < x < 1$ [Divide by 3.]
The points not satisfying this condition correspond to x such that $x \le \frac{1}{3}$ or $x \ge 1$. **Answer**

2.5 Solve $|3 - x| = x - 3$.

$\blacksquare$ $|u| = -u$ when and only when $u \le 0$. So, $|3 - x| = x - 3$ when and only when $3 - x \le 0$; that is, $3 \le x$. **Answer**

2.6 Solve $|3 - x| = 3 - x$.

$\blacksquare$ $|u| = u$ when and only when $u \ge 0$. So, $|3 - x| = 3 - x$ when and only when $3 - x \ge 0$; that is, $3 \ge x$. **Answer**

2.7 Solve $|2x + 3| = 4$.

$\blacksquare$ If $c > 0$, $|u| = c$ if and only if $u = \pm c$. So, $|2x + 3| = 4$ when and only when $2x + 3 = \pm 4$. There are two cases: **Case 1.** $2x + 3 = 4$. $2x = 1$, $x = \frac{1}{2}$. **Case 2.** $2x + 3 = -4$. $2x = -7$, $x = -\frac{7}{2}$.
So, either $x = \frac{1}{2}$ or $x = -\frac{7}{2}$. **Answer**

2.8 Solve $|7 - 5x| = 1$.

$\blacksquare$ $|7 - 5x| = |5x - 7|$. So, there are two cases: **Case 1.** $5x - 7 = 1$. $5x = 8$, $x = \frac{8}{5}$. **Case 2.** $5x - 7 = -1$.
$5x = 6$, $x = \frac{6}{5}$.
So, either $x = \frac{8}{5}$ or $x = \frac{6}{5}$. **Answer**

2.9 Solve $|x/2 + 3| < 1$.

$\blacksquare$ This inequality is equivalent to $-1 < x/2 + 3 < 1$, $-4 < x/2 < -2$ [Subtract 3.], $-8 < x < -4$ [Multiply by 2.] **Answer**

2.10 Solve $|1/x - 2| < 4$.

$\blacksquare$ This inequality is equivalent to $-4 < 1/x - 2 < 4$, $-2 < 1/x < 6$ [Add 2.] When we multiply by x, there are two cases: **Case 1.** $x > 0$. $-2x < 1 < 6x$, $x > -\frac{1}{2}$ and $\frac{1}{6} < x$, $\frac{1}{6} < x$. **Case 2.** $x < 0$. $-2x > 1 > 6x$, $x < -\frac{1}{2}$ and $\frac{1}{6} > x$, $x < -\frac{1}{2}$.
So, either $x < -\frac{1}{2}$ or $\frac{1}{6} < x$. **Answer**

2.11 Solve $|1 + 3/x| > 2$.

▊ This breaks up into two cases: **Case 1.** $1 + 3/x > 2$. $3/x > 1$ [Hence, $x > 0$.], $3 > x$. **Case 2.** $1 + 3/x < -2$. $3/x < -3$ [Hence, $x < 0$.], $3 > -3x$ [Reverse $<$ to $>$.], $-1 < x$ [Reverse $>$ to $<$.]. So, either $0 < x < 3$ or $-1 < x < 0$. **Answer**

2.12 Solve $|x^2 - 10| \le 6$.

▊ This is equivalent to $-6 \le x^2 - 10 \le 6$, $4 \le x^2 \le 16$, $2 \le |x| \le 4$. So, either $2 \le x \le 4$ or $-4 \le x \le -2$. **Answer**

2.13 Solve $|2x - 3| = |x + 2|$.

▊ There are two cases: **Case 1.** $2x - 3 = x + 2$. $x - 3 = 2$, $x = 5$. **Case 2.** $2x - 3 = -(x + 2)$. $2x - 3 = -x - 2$, $3x - 3 = -2$, $3x = 1$, $x = \frac{1}{3}$. So, either $x = 5$ or $x = \frac{1}{3}$. **Answer**

2.14 Solve $2x - 1 = |x + 7|$.

▊ Since an absolute value is never negative, $2x - 1 \ge 0$. There are two cases: **Case 1.** $x + 7 \ge 0$. $2x - 1 = x + 7$, $x - 1 = 7$, $x = 8$. **Case 2.** $x + 7 < 0$. $2x - 1 = -(x + 7)$, $2x - 1 = -x - 7$, $3x - 1 = -7$, $3x = -6$, $x = -2$. But then, $2x - 1 = -5 < 0$. So, the only solution is $x = 8$. **Answer**

2.15 Solve $|2x - 3| < |x + 2|$.

▊ This is equivalent to $-|x + 2| < 2x - 3 < |x + 2|$. There are two cases: **Case 1.** $x + 2 \ge 0$. $-(x + 2) < 2x - 3 < x + 2$, $-x - 2 < 2x - 3 < x + 2$, $1 < 3x$ and $x < 5$, $\frac{1}{3} < x < 5$. **Case 2:** $x + 2 < 0$. $-(x + 2) > 2x - 3 > x + 2$, $-x - 2 > 2x - 3 > x + 2$, $1 > 3x$ and $x > 5$, $\frac{1}{3} > x$ and $x > 5$ [impossible]. So, $\frac{1}{3} < x < 5$ is the solution.

2.16 Solve $|2x - 5| = -4$.

▊ There is no solution since an absolute value cannot be negative.

2.17 Solve $0 < |3x + 1| < \frac{1}{3}$.

▊ First solve $|3x + 1| < \frac{1}{3}$. This is equivalent to $-\frac{1}{3} < 3x + 1 < \frac{1}{3}$, $-\frac{4}{3} < 3x < -\frac{2}{3}$ [Subtract 1.], $-\frac{4}{9} < x < -\frac{2}{9}$ [Divide by 3.] The inequality $0 < |3x + 1|$ excludes the case where $0 = |3x + 1|$, that is, where $x = -\frac{1}{3}$.
Answer All x for which $-\frac{4}{9} < x < -\frac{2}{9}$ except $x = -\frac{1}{3}$.

2.18 The well-known triangle inequality asserts that $|u + v| \le |u| + |v|$. Prove by mathematical induction that, for $n \ge 2$, $|u_1 + u_2 + \cdots + u_n| \le |u_1| + |u_2| + \cdots + |u_n|$.

▊ The case $n = 2$ is the triangle inequality. Assume the result true for some n. By the triangle inequality and the inductive hypothesis,

$$|u_1 + u_2 + \cdots + u_n + u_{n+1}| \le |u_1 + u_2 + \cdots + u_n| + |n_{n+1}| \le (|u_1| + |u_2| + \cdots + |u_n|) + |u_{n+1}|$$

and, therefore, the result also holds for $n + 1$.

2.19 Prove $|u - v| \ge ||u| - |v||$.

▊ $|u| = |v + (u - v)| \le |v| + |u - v|$ [Triangle inequality.] Hence, $|u - v| \ge |u| - |v|$. Similarly, $|v - u| \ge |v| - |u|$. But, $|v - u| = |u - v|$. So, $|u - v| \ge$ (maximum of $|u| - |v|$ and $|v| - |u|$) $= ||u| - |v||$.

2.20 Solve $|x - 1| < |x - 2|$.

▊ **Analytic solution.** The given equation is equivalent to $-|x - 2| < x - 1 < |x - 2|$. **Case 1.** $x - 2 \ge 0$. $-(x - 2) < x - 1 < x - 2$. Then, $-1 < -2$, which is impossible. **Case 2.** $x - 2 < 0$. $-(x - 2) > x - 1 > x - 2$, $-x + 2 > x - 1 > x - 2$, $3 > 2x$, $\frac{3}{2} > x$. Thus, the solution consists of all x such that $x < \frac{3}{2}$.

Geometric solution. $|u - v|$ is the distance between u and v. So, the solution consists of all points x that are closer to 1 than to 2. Figure 2-1 shows that these are all points x such that $x < \frac{3}{2}$.

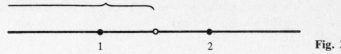

Fig. 2-1

2.21 Solve $|x + 1/x| > 2$.

▮ This is equivalent to $\left|\dfrac{x^2 + 1}{x}\right| > 2$, $\dfrac{x^2 + 1}{|x|} > 2$ [Since $x^2 + 1 > 0$.], $x^2 + 1 > 2|x|$, $x^2 - 2|x| + 1 > 0$, $|x|^2 - 2|x| + 1 > 0$ [Since $x^2 = |x|^2$.], $(|x| - 1)^2 > 0$, $|x| \neq 1$.
Answer All x except $x = +1$ and $x = -1$.

2.22 Solve $|x + 1/x| < 4$.

▮ This is equivalent to $\left|\dfrac{x^2 + 1}{x}\right| < 4$, $x^2 + 1 < 4|x|$, $x^2 - 4|x| + 1 < 0$, $|x|^2 - 4|x| + 1 < 0$, $(|x| - 2)^2 < 3$ [Completing the square], $||x| - 2| < \sqrt{3}$, $-\sqrt{3} < |x| - 2 < \sqrt{3}$, $2 - \sqrt{3} < |x| < 2 + \sqrt{3}$.
When $x > 0$, $2 - \sqrt{3} < x < 2 + \sqrt{3}$, and, when $x < 0$, $-2 - \sqrt{3} < x < -2 + \sqrt{3}$. **Answer**

2.23 Solve $x + 1 < |x|$.

▮ When $x \geq 0$, this reduces to $x + 1 < x$, which is impossible. When $x < 0$, the inequality becomes $x + 1 < -x$, which is equivalent to $2x + 1 < 0$, or $2x < -1$, or $x < -\frac{1}{2}$. **Answer**

2.24 Prove $|ab| = |a| \cdot |b|$.

▮ From the definition of absolute value, $|a| = \pm a$ and $|b| = \pm b$. Hence, $|a| \cdot |b| = (\pm a) \cdot (\pm b) = \pm (ab)$. Since $|a| \cdot |b|$ is nonnegative, $|a| \cdot |b|$ must be $|ab|$.

2.25 Solve $|2(x - 4)| < 10$.

▮ $|2| \cdot |x - 4| = |2(x - 4)| < 10$, $2|x - 4| < 10$, $|x - 4| < 5$, $-5 < x - 4 < 5$, $-1 < x < 9$. **Answer**

2.26 Solve $|x^2 - 17| = 8$.

▮ There are two cases. **Case 1.** $x^2 - 17 = 8$. $x^2 = 25$, $x = \pm 5$. **Case 2.** $x^2 - 17 = -8$. $x^2 = 9$, $x = \pm 3$. So, there are four solutions: ± 3, ± 5. **Answer**

2.27 Solve $|x - 1| < 1$.

▮ $-1 < x - 1 < 1$, $0 < x < 2$.

2.28 Solve $|3x + 5| \leq 4$.

▮ $-4 \leq 3x + 5 \leq 4$, $-9 \leq 3x \leq -1$, $-3 \leq x \leq -\frac{1}{3}$.

2.29 Solve $|x + 4| > 2$.

▮ First solve the negation, $|x + 4| \leq 2$: $-2 \leq x + 4 \leq 2$, $-6 \leq x \leq -2$. Hence, the solution of the original inequality is $x < -6$ or $x > -2$.

2.30 Solve $|2x - 5| \geq 3$.

▮ First solve the negation $|2x - 5| < 3$: $-3 < 2x - 5 < 3$, $2 < 2x < 8$, $1 < x < 4$. Hence, the solution of the original inequality is $x \leq 1$ or $x \geq 4$.

2.31 Solve $|7x - 5| = |3x + 4|$.

▮ **Case 1.** $7x - 5 = 3x + 4$. Then $4x = 9$, $x = \frac{9}{4}$. **Case 2.** $7x - 5 = -(3x + 4)$. Then $7x - 5 = -3x - 4$, $10x = 1$, $x = \frac{1}{10}$. Thus, the solutions are $\frac{9}{4}$ and $\frac{1}{10}$.

2.32   Solve $|3x - 2| \le |x - 1|$.

▮ This is equivalent to $-|x - 1| \le 3x - 2 \le |x - 1|$. **Case 1.** $x - 1 \ge 0$. Then $-(x - 1) \le 3x - 2 \le x - 1$, $-x + 1 \le 3x - 2 \le x - 1$; the first inequality is equivalent to $\frac{3}{4} \le x$ and the second to $x \le \frac{1}{2}$. But this is impossible. **Case 2.** $x - 1 < 0$. $-x + 1 \ge 3x - 2 \ge x - 1$; the first inequality is equivalent to $x \le \frac{3}{4}$ and the second to $x \ge \frac{1}{2}$. Hence, we have $\frac{1}{2} \le x \le \frac{3}{4}$. **Answer**

2.33   Solve $|x - 2| + |x - 5| = 9$.

▮ **Case 1.** $x \ge 5$. Then $x - 2 + x - 5 = 9$, $2x - 7 = 9$, $2x = 16$, $x = 8$. **Case 2.** $2 \le x < 5$. Then $x - 2 + 5 - x = 9$, $3 = 9$, which is impossible. **Case 3.** $x < 2$. Then $2 - x + 5 - x = 9$, $7 - 2x = 9$, $2x = -2$, $x = -1$. So, the solutions are 8 and -1.

2.34   Solve $4 - x \ge |5x + 1|$.

▮ **Case 1.** $5x + 1 \ge 0$, that is, $x \ge -\frac{1}{5}$. Then $4 - x \ge 5x + 1$, $3 \ge 6x$, $\frac{1}{2} \ge x$. Thus, we obtain the solutions $-\frac{1}{5} \le x \le \frac{1}{2}$. **Case 2.** $5x + 1 < 0$, that is, $x < -\frac{1}{5}$. Then $4 - x \ge -5x - 1$, $4x \ge -5$, $x \ge -\frac{5}{4}$. Thus, we obtain the solutions $-\frac{5}{4} \le x < -\frac{1}{5}$. Hence, the set of solutions is $[-\frac{1}{5}, \frac{1}{2}] \cup [-\frac{5}{4}, -\frac{1}{5}) = [-\frac{5}{4}, \frac{1}{2}]$.

2.35   Prove $|a - b| \le |a| + |b|$.

▮ By the triangle inequality, $|a - b| = |a + (-b)| \le |a| + |-b| = |a| + |b|$.

2.36   Solve the inequality $|x - 1| \ge |x - 3|$.

▮ We argue geometrically from Fig. 2-2. $|x - 1|$ is the distance of x from 1, and $|x - 3|$ is the distance of x from 3. The point $x = 2$ is equidistant from 1 and 3. Hence, the solutions consist of all $x \ge 2$.

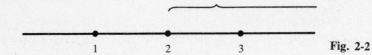

$$1 \qquad 2 \qquad 3 \qquad\qquad \textbf{Fig. 2-2}$$

3.1 Find the slope of the line through the points $(-2, 5)$ and $(7, 1)$.

❚ Remember that the slope m of the line through two points (x_1, y_1) and (x_2, y_2) is given by the equation $m = \dfrac{y_2 - y_1}{x_2 - x_1}$. Hence, the slope of the given line is $\dfrac{1-5}{7-(-2)} = \dfrac{-4}{9} = -\dfrac{4}{9}$.

3.2 Find a point-slope equation of the line through the points $(1, 3)$ and $(3, 6)$.

❚ The slope m of the given line is $(6-3)/(3-1) = \frac{3}{2}$. Recall that the point-slope equation of the line through point (x_1, y_1) and with slope m is $y - y_1 = m(x - x_1)$. Hence, one point-slope equation of the given line, using the point $(1, 3)$, is $y - 3 = \frac{3}{2}(x - 1)$. **Answer**
Another point-slope equation, using the point $(3, 6)$, is $y - 6 = \frac{3}{2}(x - 3)$. **Answer**

3.3 Write a point-slope equation of the line through the points $(1, 2)$ and $(1, 3)$.

❚ The line through $(1, 2)$ and $(1, 3)$ is vertical and, therefore, does not have a slope. Thus, there is no point-slope equation of the line.

3.4 Find a point-slope equation of the line going through the point $(1, 3)$ with slope 5.

❚ $y - 3 = 5(x - 1)$. **Answer**

3.5 Find the slope of the line having the equation $y - 7 = 2(x - 3)$ and find a point on the line.

❚ $y - 7 = 2(x - 3)$ is a point-slope equation of the line. Hence, the slope $m = 2$, and $(3, 7)$ is a point on the line.

3.6 Find the slope-intercept equation of the line through the points $(2, 4)$ and $(4, 8)$.

❚ Remember that the slope-intercept equation of a line is $y = mx + b$, where m is the slope and b is the y-intercept (that is, the y-coordinate of the point where the line cuts the y-axis). In this case, the slope $m = (8 - 4)/(4 - 2) = \frac{4}{2} = 2$.
Method 1. A point-slope equation of the line is $y - 8 = 2(x - 4)$. This is equivalent to $y - 8 = 2x - 8$, or, finally, to $y = 2x$. **Answer**
Method 2. The slope-intercept equation has the form $y = 2x + b$. Since $(2, 4)$ lies on the line, we may substitute 2 for x and 4 for y. So, $4 = 2 \cdot 2 + b$, and, therefore, $b = 0$. Hence, the equation is $y = 2x$. **Answer**

3.7 Find the slope-intercept equation of the line through the points $(-1, 6)$ and $(2, 15)$.

❚ The slope $m = (15 - 6)/[2 - (-1)] = \frac{9}{3} = 3$. Hence, the slope-intercept equation looks like $y = 3x + b$. Since $(-1, 6)$ is on the line, $6 = 3 \cdot (-1) + b$, and therefore, $b = 9$. Hence, the slope-intercept equation is $y = 3x + 9$.

3.8 Find the slope-intercept equation of the line through $(2, -6)$ and the origin.

❚ The origin has coordinates $(0, 0)$. So, the slope $m = (-6 - 0)/(2 - 0) = -\frac{6}{2} = -3$. Since the line cuts the y-axis at $(0, 0)$, the y-intercept b is 0. Hence, the slope-intercept equation is $y = -3x$.

3.9 Find the slope-intercept equation of the line through $(2, 5)$ and $(-1, 5)$.

❚ The line is horizontal. Since it passes through $(2, 5)$, an equation for it is $y = 5$. But, this is the slope-intercept equation, since the slope $m = 0$ and the y-intercept b is 5.

3.10 Find the slope and y-intercept of the line given by the equation $7x + 4y = 8$.

▌ If we solve the equation $7x + 4y = 8$ for y, we obtain the equation $y = -\frac{7}{4}x + 2$, which is the slope-intercept equation. Hence, the slope $m = -\frac{7}{4}$ and the y-intercept $b = 2$.

3.11 Show that every line has an equation of the form $Ax + By = C$, where A and B are not both 0, and that, conversely, every such equation is the equation of a line.

▌ If a given line is vertical, it has an equation $x = C$. In this case, we can let $A = 1$ and $B = 0$. If the given line is not vertical, it has a slope-intercept equation $y = mx + b$, or, equivalently, $-mx + y = b$. So, let $A = -m$, $B = 1$, and $C = b$. Conversely, assume that we are given an equation $Ax + By = C$, with A and B not both 0. If $B = 0$, the equation is equivalent to $x = C/A$, which is the equation of a vertical line. If $B \ne 0$, solve the equation for y: $y = -\dfrac{A}{B}x + \dfrac{C}{B}$. This is the slope-intercept equation of the line with slope $-\dfrac{A}{B}$ and y-intercept $\dfrac{C}{B}$.

3.12 Find an equation of the line L through $(-1, 4)$ and parallel to the line M with the equation $3x + 4y = 2$.

▌ Remember that two lines are parallel if and only if their slopes are equal. If we solve $3x + 4y = 2$ for y, namely, $y = -\frac{3}{4}x + \frac{1}{2}$, we obtain the slope-intercept equation for M. Hence, the slope of M is $-\frac{3}{4}$ and, therefore, the slope of the parallel line L also is $-\frac{3}{4}$. So, L has a slope-intercept equation of the form $y = -\frac{3}{4}x + b$. Since L goes through $(-1, 4)$, $4 = -\frac{3}{4} \cdot (-1) + b$, and, therefore, $b = 4 - \frac{3}{4} = \frac{13}{4}$. Thus, the equation of L is $y = -\frac{3}{4}x + \frac{13}{4}$.

3.13 Show that the lines parallel to a line $Ax + By = C$ are those lines having equations of the form $Ax + By = E$ for some E. (Assume that $B \ne 0$.)

▌ If we solve $Ax + By = C$ for y, we obtain the slope-intercept equation $y = -\dfrac{A}{B}x + \dfrac{C}{B}$. So, the slope is $-A/B$. Given a parallel line, it must also have slope $-A/B$ and, therefore, has a slope-intercept equation $y = -\dfrac{A}{B}x + b$, which is equivalent to $\dfrac{A}{B}x + y = b$, and, thence to $Ax + By = bB$. Conversely, a line with equation $Ax + By = E$ must have slope $-A/B$ (obtained by putting the equation in slope-intercept form) and is, therefore, parallel to the line with equation $Ax + By = C$.

3.14 Find an equation of the line through $(2, 3)$ and parallel to the line with the equation $4x - 2y = 7$.

▌ By Problem 3.13, the required line must have an equation of the form $4x - 2y = E$. Since $(2, 3)$ lies on the line, $4(2) - 2(3) = E$. So, $E = 8 - 6 = 2$. Hence, the desired equation is $4x - 2y = 2$.

3.15 Find an equation of the line through $(2, 3)$ and parallel to the line with the equation $y = 5$.

▌ Since $y = 5$ is the equation of a horizontal line, the required parallel line is horizontal. Since it passes through $(2, 3)$, an equation for it is $y = 3$.

3.16 Show that any line that is neither vertical nor horizontal and does not pass through the origin has an equation of the form $\dfrac{x}{a} + \dfrac{y}{b} = 1$, where b is the y-intercept and a is the x-intercept (Fig. 3-1).

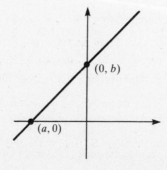

Fig. 3-1

▌ In Problem 3.11, set $C/A = a$ and $C/B = b$. Notice that, when $y = 0$, the equation yields the value $x = a$, and, therefore, a is the x-intercept of the line. Similarly for the y-intercept.

3.17 Find an equation of the line through the points $(0, 2)$ and $(3, 0)$.

▮ The y-intercept is $b = 2$ and the x-intercept is $a = 3$. So, by Problem 3.16, an equation of the line is $\frac{x}{3} + \frac{y}{2} = 1$.

3.18 If the point $(2, k)$ lies on the line with slope $m = 3$ passing through the point $(1, 6)$, find k.

▮ A point-slope equation of the line is $y - 6 = 3(x - 1)$. Since $(2, k)$ lies on the line, $k - 6 = 3(2 - 1)$. Hence, $k = 9$.

3.19 Does the point $(-1, -2)$ lie on the line L through the points $(4, 7)$ and $(5, 9)$?

▮ The slope of L is $(9 - 7)/(5 - 4) = 2$. Hence, a point-slope equation of L is $y - 7 = 2(x - 4)$. If we substitute -1 for x and -2 for y in this equation, we obtain $-2 - 7 = 2(-1 - 4)$, or $-9 = -10$, which is false. Hence, $(-1, -2)$ does not lie on L.

3.20 Find the slope-intercept equation of the line M through $(1, 4)$ that is perpendicular to the line L with equation $2x - 6y = 5$.

▮ Solve $2x - 6y = 5$ for y, obtaining $y = \frac{1}{3}x - \frac{5}{6}$. So, the slope of L is $\frac{1}{3}$. Recall that two lines with slopes m_1 and m_2 are perpendicular if and only if $m_1 m_2 = -1$, or, equivalently, $m_1 = -1/m_2$. Hence, the slope of M is the negative reciprocal of $\frac{1}{3}$, that is, -3. The slope-intercept equation of M has the form $y = -3x + b$. Since $(1, 4)$ is on M, $4 = -3 \cdot 1 + b$. Hence, $b = 7$, and the required equation is $y = -3x + 7$.

3.21 Show that, if a line L has the equation $Ax + By = C$, then a line M perpendicular to L has an equation of the form $-Bx + Ay = E$.

▮ Assume first that L is not a vertical line. Hence, $B \neq 0$. So, $Ax + By = C$ is equivalent to $y = -\frac{A}{B}x + \frac{C}{B}$, and, therefore, the slope of L is $-(A/B)$. Hence, the slope of M is the negative reciprocal B/A; its slope-intercept equation has the form $y = \frac{B}{A}x + b$, which is equivalent to $-\frac{B}{A}x + y = b$ and thence to $-Bx + Ay = Ab$. In this case, $E = Ab$. (In the special case when $A = 0$, L is horizontal and M is vertical. Then M has an equation $x = a$, which is equivalent to $-Bx = -Ba$. Here, $E = -Ba$ and $A = 0$.) If L is vertical (in which case, $B = 0$), M is horizontal and has an equation of the form $y = b$, which is equivalent to $Ay = Ab$. In this case, $E = Ab$ and $B = 0$.

3.22 Find an equation of the line through the point $(2, -3)$ and perpendicular to the line $4x - 5y = 7$.

▮ The required equation has the form $5x + 4y = E$ (see Problem 3.21). Since $(2, -3)$ lies on the line, $5(2) + 4(-3) = E$. Hence, $E = -2$, and the desired equation is $5x + 4y = -2$.

3.23 Show that two lines, L with equation $A_1x + B_1y = C_1$ and M with equation $A_2x + B_2y = C_2$, are parallel if and only if their coefficients of x and y are proportional, that is, there is a nonzero number r such that $A_2 = rA_1$ and $B_2 = rB_1$.

▮ Assume that $A_2 = rA_1$ and $B_2 = rB_1$, with $r \neq 0$. Then the equation of M is $rA_1x + rB_1y = C_2$, which is equivalent to $A_1x + B_1y = \frac{1}{r} \cdot C_2$. Then, by Problem 3.13, M is parallel to L. Conversely, assume M is parallel to L. By solving the equations of L and M for y, we see that the slope of L is $-(A_1/B_1)$ and the slope of M is $-(A_2/B_2)$. Since M and L are parallel, their slopes are equal:

$$-\frac{A_1}{B_1} = -\frac{A_2}{B_2} \qquad \text{or} \qquad \frac{A_2}{A_1} = \frac{B_2}{B_1} = r$$

(In the special case where the lines are vertical, $B_1 = B_2 = 0$, and we can set $r = A_2/A_1$.)

3.24 Determine whether the lines $3x + 6y = 7$ and $2x + 4y = 5$ are parallel.

▮ The coefficients of x and y are proportional: $\frac{2}{3} = \frac{4}{6}$. Hence, by Problem 3.23, the lines are parallel.

3.25 Use slopes to determine whether the points $A(4, 1)$, $B(7, 3)$, and $C(3, 9)$ are the vertices of a right triangle.

▌ The slope m_1 of line $\overleftrightarrow{AB}$ is $(3-1)/(7-4) = \frac{2}{3}$. The slope m_2 of line $\overleftrightarrow{BC}$ is $(9-3)/(3-7) = -\frac{6}{4} = -\frac{3}{2}$. Since m_2 is the negative reciprocal of m_1, the lines $\overleftrightarrow{AB}$ and $\overleftrightarrow{BC}$ are perpendicular. Hence, $\triangle ABC$ has a right angle at B.

3.26 Determine k so that the points $A(7, 5)$, $B(-1, 2)$, and $C(k, 0)$ are the vertices of a right triangle with right angle at B.

▌ The slope of line $\overleftrightarrow{AB}$ is $(5-2)/[7-(-1)] = \frac{3}{8}$. The slope of line $\overleftrightarrow{BC}$ is $(2-0)/(-1-k) = -2/(1+k)$. The condition for $\triangle ABC$ to have a right angle at B is that lines $\overleftrightarrow{AB}$ and $\overleftrightarrow{BC}$ are perpendicular, which holds when and only when the product of their slopes is -1, that is $(\frac{3}{8})[-2/(1+k)] = -1$. This is equivalent to $6 = 8(1+k)$, or $8k = -2$, or $k = -\frac{1}{4}$.

3.27 Find the slope-intercept equation of the line through $(1, 4)$ and rising 5 units for each unit increase in x.

▌ Since the line rises 5 units for each unit increase in x, its slope must be 5. Hence, its slope-intercept equation has the form $y = 5x + b$. Since $(1, 4)$ lies on the line, $4 = 5(1) + b$. So, $b = -1$. Thus, the equation is $y = 5x - 1$.

3.28 Use slopes to show that the points $A(5, 4)$, $B(-4, 2)$, $C(-3, -3)$, and $D(6, -1)$ are vertices of a parallelogram.

▌ The slope of $\overleftrightarrow{AB}$ is $(4-2)/[5-(-4)] = \frac{2}{9}$ and the slope of $\overleftrightarrow{CD}$ is $[-3-(-1)]/(-3-6) = \frac{2}{9}$; hence, $\overleftrightarrow{AB}$ and $\overleftrightarrow{CD}$ are parallel. The slope of $\overleftrightarrow{BC}$ is $(-3-2)/[-3-(-4)] = -5$ and the slope of $\overleftrightarrow{AD}$ is $(-1-4)/(6-5) = -5$, and, therefore, $\overleftrightarrow{BC}$ and $\overleftrightarrow{AD}$ are parallel. Thus, $ABCD$ is a parallelogram.

3.29 For what value of k will the line $kx + 5y = 2k$ have y-intercept 4?

▌ When $x = 0$, $y = 4$. Hence, $5(4) = 2k$. So, $k = 10$.

3.30 For what value of k will the line $kx + 5y = 2k$ have slope 3?

▌ Solve for y: $y = -\dfrac{k}{5}x + \dfrac{2k}{5}$. This is the slope-intercept equation. Hence, the slope $m = -k/5 = 3$. So, $k = -15$.

3.31 For what value of k will the line $kx + 5y = 2k$ be perpendicular to the line $2x - 3y = 1$?

▌ By the solution to Problem 3.30, the slope of $kx + 5y = 2k$ is $-k/5$. By solving for y, the slope of $2x - 3y = 1$ is found to be $\frac{2}{3}$. For perpendicularity, the product of the slopes must be -1. Hence, $(-k/5) \cdot \frac{2}{3} = -1$. So, $2k = 15$, and, therefore, $k = \frac{15}{2}$.

3.32 Find the midpoint of the line segment between $(2, 5)$ and $(-1, 3)$.

▌ By the midpoint formula, the coordinates of the midpoint are the averages of the coordinates of the endpoints. In this case, the midpoint (x, y) is given by $([2+(-1)]/2, (5+3)/2) = (\frac{1}{2}, 4)$.

3.33 A triangle has vertices $A(1, 2)$, $B(8, 1)$, $C(2, 3)$. Find the equation of the median from A to the midpoint M of the opposite side.

▌ The midpoint M of segment $\overline{BC}$ is $((8+2)/2, (1+3)/2) = (5, 2)$. So, $\overleftrightarrow{AM}$ is horizontal, with equation $y = 2$.

3.34 For the triangle of Problem 3.33, find an equation of the altitude from B to the opposite side $\overleftrightarrow{AC}$.

▌ The slope of $\overleftrightarrow{AC}$ is $(3-2)/(2-1) = 1$. Hence, the slope of the altitude is the negative reciprocal of 1, namely, -1. Thus, its slope-intercept equation has the form $y = -x + b$. Since $B(8, 1)$ is on the altitude, $1 = -8 + b$, and, so, $b = 9$. Hence, the equation is $y = -x + 9$.

3.35 For the triangle of Problem 3.33, find an equation of the perpendicular bisector of side $\overline{AB}$.

▮ The midpoint N of $\overline{AB}$ is $((1+8)/2, (2+1)/2) = (9/2, 3/2)$. The slope of $\overleftrightarrow{AB}$ is $(2-1)/(1-8) = -\frac{1}{7}$. Hence, the slope of the desired line is the negative reciprocal of $-\frac{1}{7}$, that is, 7. Thus, the slope-intercept equation of the perpendicular bisector has the form $y = 7x + b$. Since $(9/2, 3/2)$ lies on the line, $\frac{3}{2} = 7(\frac{9}{2}) + b$. So, $b = \frac{3}{2} - \frac{63}{2} = -30$. Thus, the desired equation is $y = 7x - 30$.

3.36 If a line L has the equation $3x + 2y = 4$, prove that a point $P(x, y)$ is above L if and only if $3x + 2y > 4$.

▮ Solving for y, we obtain the equation $y = -\frac{3}{2}x + 2$. For any fixed x, the vertical line with that x-coordinate cuts the line at the point Q where the y-coordinate is $-\frac{3}{2}x + 2$ (see Fig. 3-2). The points along that vertical line and above Q have y-coordinates $y > -\frac{3}{2}x + 2$. This is equivalent to $2y > -3x + 4$, and thence to $3x + 2y > 4$.

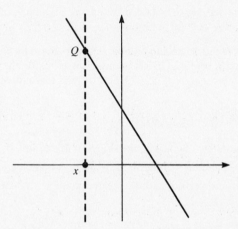

Fig. 3-2

3.37 Generalize Problem 3.36 to the case of any line $Ax + By = C$ $(B \neq 0)$.

▮ **Case 1.** $B > 0$. As in the solution of Problem 3.36, a point $P(x, y)$ is above this line if and only if $Ax + By > C$. **Case 2.** $B < 0$. Then a procedure similar to that in the solution of Problem 3.36 shows that a point $P(x, y)$ is above this line if and only if $Ax + By < C$.

3.38 Use two inequalities to describe the set of all points above the line L: $4x + 3y = 9$ and below the line M: $2x + y = 1$.

▮ By Problem 3.37, to be above L, we must have $4x + 3y > 9$. To be below M, we must have $2x + y < 1$.

3.39 Describe geometrically the family of lines $y = mx + 2$.

▮ The set of all nonvertical lines through $(0, 2)$.

3.40 Describe geometrically the family of lines $y = 3x + b$.

▮ The family of mutually parallel lines of slope 3.

3.41 Prove by use of coordinates that the altitudes of any triangle meet at a common point.

▮ Given $\triangle ABC$, choose the coordinate system so that A and B lie on the x-axis and C lies on the y-axis (Fig. 3-3). Let the coordinates of A, B, and C be $(u, 0)$, $(v, 0)$, and $(0, w)$. *(i)* The altitude from C to $\overleftrightarrow{AB}$ is the y-axis. *(ii)* The slope of $\overleftrightarrow{BC}$ is $-w/v$. So, the altitude from A to $\overleftrightarrow{BC}$ has slope v/w. Its slope-intercept equation has the form $y = (v/w)x + b$. Since $(u, 0)$ lies on the line, $0 = (v/w)(u) + b$; hence, its y-intercept $b = -vu/w$. Thus, this altitude intersects the altitude from C (the y-axis) at the point $(0, -vu/w)$. *(iii)* The slope of $\overleftrightarrow{AC}$ is $-w/u$. So, the altitude from B to $\overleftrightarrow{AC}$ has slope u/w, and its slope-intercept equation is $y = (u/w)x + b$. Since $(v, 0)$ lies on the altitude, $0 = (u/w)(v) + b$, and its y-intercept $b = -uv/w$. Thus, this altitude also goes through the point $(0, -vu/w)$.

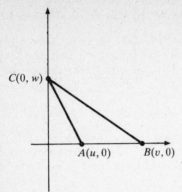

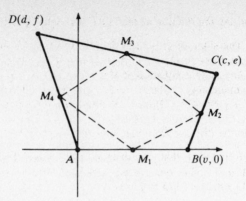

Fig. 3-3   Fig. 3-4

3.42 Using coordinates, prove that the figure obtained by joining midpoints of consecutive sides of a quadrilateral *ABCD* is a parallelogram.

❚ Refer to Fig. 3-4. Let *A* be the origin and let *B* be on the *x*-axis, with coordinates $(v, 0)$. Let *C* be (c, e) and let *D* be (d, f). The midpoint M_1 of $\overline{AB}$ has coordinates $(v/2, 0)$, the midpoint M_2 of $\overline{BC}$ is $((c + v)/2, e/2)$, the midpoint M_3 of $\overline{CD}$ is $((c + d)/2, (e + f)/2)$, and the midpoint M_4 of $\overline{AD}$ is $(d/2, f/2)$.

$$\text{Slope of line } \overleftrightarrow{M_1M_2} = \frac{e/2 - 0}{(c + v)/2 - v/2} = \frac{e}{c} \qquad \text{Slope of line } \overleftrightarrow{M_3M_4} = \frac{(e + f)/2 - f/2}{(c + d)/2 - d/2} = \frac{e}{c}$$

Thus, $\overleftrightarrow{M_1M_2}$ and $\overleftrightarrow{M_3M_4}$ are parallel. Similarly, the slopes of $\overleftrightarrow{M_2M_3}$ and $\overleftrightarrow{M_1M_4}$ both turn out to be $f/(d - v)$, and therefore $\overleftrightarrow{M_2M_3}$ and $\overleftrightarrow{M_1M_4}$ are parallel. Thus, $M_1M_2M_3M_4$ is a parallelogram. (Note two special cases. When $c = 0$, both $\overleftrightarrow{M_1M_2}$ and $\overleftrightarrow{M_3M_4}$ are vertical and, therefore, parallel. When $d = v$, both $\overleftrightarrow{M_1M_4}$ and $\overleftrightarrow{M_2M_3}$ are vertical and, therefore, parallel.)

3.43 Using coordinates, prove that, if the medians $\overline{AM_1}$ and $\overline{BM_2}$ of $\triangle ABC$ are equal, then $\overline{CA} = \overline{CB}$.

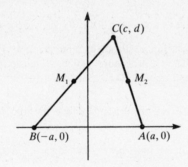

Fig. 3-5

❚ Choose the *x*-axis so that it goes through *A* and *B* and let the origin be halfway between *A* and *B*. Let *A* be $(a, 0)$. Then *B* is $(-a, 0)$. Let *C* be (c, d). Then M_1 is $((c - a)/2, d/2)$ and M_2 is $((c + a)/2, d/2)$. By the distance formula,

$$\overline{AM_1} = \sqrt{\left(a - \frac{c - a}{2}\right)^2 + \left(0 - \frac{d}{2}\right)^2} = \sqrt{\left(\frac{3a - c}{2}\right)^2 + \left(\frac{d}{2}\right)^2}$$

and

$$BM_2 = \sqrt{\left[\frac{c + a}{2} - (-a)\right]^2 + \left(\frac{d}{2} - 0\right)^2} = \sqrt{\left(\frac{3a + c}{2}\right)^2 + \left(\frac{d}{2}\right)^2}$$

Setting $AM_1 = BM_2$ and squaring both sides, we obtain $[(3a - c)/2]^2 + (d/2)^2 = [(3a + c)/2]^2 + (d/2)^2$, and, simplifying, $(3a - c)^2 = (3a + c)^2$. So, $(3a + c)^2 - (3a - c)^2 = 0$, and, factoring the left-hand side, $[(3a + c) + (3a - c)] \cdot [(3a + c) - (3a - c)] = 0$, that is, $(6a) \cdot (2c) = 0$. Since $a \neq 0$, $c = 0$. Now the distance formula gives

$$\overline{CA} = \sqrt{(0 - a)^2 + (d - 0)^2} = \sqrt{[0 - (-a)]^2 + (d - 0)^2} = \overline{CB}$$

as required.

3.44 Find the intersection of the line L through $(1, 2)$ and $(5, 8)$ with the line M through $(2, 2)$ and $(4, 0)$.

▮ The slope of L is $(8-2)/(5-1) = \frac{3}{2}$. Its slope-intercept equation has the form $y = \frac{3}{2}x + b$. Since it passes through $(1, 2)$, $2 = \frac{3}{2}(1) + b$, and, therefore, $b = \frac{1}{2}$. So, L has equation $y = \frac{3}{2}x + \frac{1}{2}$. Similarly, we find that the equation of M is $y = -x + 4$. So, we must solve the equations $y = -x + 4$ and $y = \frac{3}{2}x + \frac{1}{2}$ simultaneously. Then, $-x + 4 = \frac{3}{2}x + \frac{1}{2}$, $-2x + 8 = 3x + 1$, $7 = 5x$, $x = \frac{7}{5}$. When $x = \frac{7}{5}$, $y = -x + 4 = -\frac{7}{5} + 4 = \frac{13}{5}$. Hence, the point of intersection is $(\frac{7}{5}, \frac{13}{5})$.

3.45 Find the distance from the point $(1, 2)$ to the line $3x - 4y = 10$.

▮ Remember that the distance from a point (x_1, y_1) to a line $Ax + By + C = 0$ is $|Ax_1 + By_1 + C|/\sqrt{A^2 + B^2}$. In our case, $A = 3$, $B = -4$, $C = 10$, and $\sqrt{A^2 + B^2} = \sqrt{25} = 5$. So, the distance is $|3(1) - 4(2) - 10|/5 = \frac{15}{5} = 3$.

3.46 Find equations of the lines of slope $-\frac{3}{4}$ that form with the coordinate axes a triangle of area 24 square units.

▮ The slope-intercept equations have the form $y = -\frac{3}{4}x + b$. When $y = 0$, $x = \frac{4}{3}b$. So, the x-intercept a is $\frac{4}{3}b$. Hence, the area of the triangle is $\frac{1}{2}ab = \frac{1}{2}(\frac{4}{3}b)b = \frac{2}{3}b^2 = 24$. So, $b^2 = 36$, $b = \pm 6$, and the desired equations are $y = -\frac{3}{4}x \pm 6$; that is, $3x + 4y = 24$ and $3x + 4y = -24$.

3.47 A point (x, y) moves so that its distance from the line $x = 5$ is twice as great as its distance from the line $y = 8$. Find an equation of the path of the point.

▮ The distance of (x, y) from $x = 5$ is $|x - 5|$, and its distance from $y = 8$ is $|y - 8|$. Hence, $|x - 5| = 2|y - 8|$. So, $x - 5 = \pm 2(y - 8)$. There are two cases: $x - 5 = 2(y - 8)$ and $x - 5 = -2(y - 8)$, yielding the lines $x - 2y = -11$ and $x + 2y = 21$. A single equation for the path of the point would be $(x - 2y + 11)(x + 2y - 21) = 0$.

3.48 Find the equations of the lines through $(4, -2)$ and at a perpendicular distance of 2 units from the origin.

▮ A point-slope equation of a line through $(4, -2)$ with slope m is $y + 2 = m(x - 4)$ or $mx - y - (4m + 2) = 0$. The distance of $(0, 0)$ from this line is $|4m + 2|/\sqrt{m^2 + 1}$. Hence, $|4m + 2|/\sqrt{m^2 + 1} = 2$. So, $(4m + 2)^2 = 4(m^2 + 1)$, or $(2m + 1)^2 = m^2 + 1$. Simplifying, $m(3m + 4) = 0$, and, therefore, $m = 0$ or $m = -\frac{4}{3}$. The required equations are $y = -2$ and $4x + 3y - 10 = 0$.

In Problems 3.49–3.51, find a point-slope equation of the line through the given points.

3.49 $(2, 5)$ and $(-1, 4)$.

▮ $m = (5 - 4)/[2 - (-1)] = \frac{1}{3}$. So, an equation is $(y - 5)/(x - 2) = \frac{1}{3}$ or $y - 5 = \frac{1}{3}(x - 2)$.

3.50 $(1, 4)$ and the origin.

▮ $m = (4 - 0)/(1 - 0) = 4$. So, an equation is $y/x = 4$ or $y = 4x$.

3.51 $(7, -1)$ and $(-1, 7)$.

▮ $m = (-1 - 7)/[7 - (-1)] = -8/8 = -1$. So, an equation is $(y + 1)/(x - 7) = -1$ or $y + 1 = -(x - 7)$.

In Problems 3.52–3.60, find the slope-intercept equation of the line satisfying the given conditions.

3.52 Through the points $(-2, 3)$ and $(4, 8)$.

▮ $m = (3 - 8)/(-2 - 4) = -5/-6 = \frac{5}{6}$. The equation has the form $y = \frac{5}{6}x + b$. Hence, $8 = \frac{5}{6}(4) + b$, $b = \frac{14}{3}$. Thus, the equation is $y = \frac{5}{6}x + \frac{14}{3}$.

3.53 Having slope 2 and y-intercept -1.

▮ $y = 2x - 1$.

3.54 Through $(1, 4)$ and parallel to the x-axis.

▮ Since the line is parallel to the x-axis, the line is horizontal. Since it passes through $(1, 4)$, the equation is $y = 4$.

3.55 Through $(1, -4)$ and rising 5 units for each unit increase in x.

▮ Its slope must be 5. The equation has the form $y = 5x + b$. So, $-4 = 5(1) + b$, $b = -9$. Thus, the equation is $y = 5x - 9$.

3.56 Through $(5, 1)$ and falling 3 units for each unit increase in x.

▮ Its slope must be -3. So, its equation has the form $y = -3x + b$. Then, $1 = -3(5) + b$, $b = 16$. Thus, the equation is $y = -3x + 16$.

3.57 Through the origin and parallel to the line with the equation $y = 1$.

▮ Since the line $y = 1$ is horizontal, the desired line must be horizontal. It passes through $(0, 0)$, and, therefore, its equation is $y = 0$.

3.58 Through the origin and perpendicular to the line L with the equation $2x - 6y = 5$.

▮ The equation of L is $y = \frac{1}{3}x - \frac{5}{6}$. So, the slope of L is $\frac{1}{3}$. Hence, our line has slope -3. Thus, its equation is $y = -3x$.

3.59 Through $(4, 3)$ and perpendicular to the line with the equation $x = 1$.

▮ The line $x = 1$ is vertical. So, our line is horizontal. Since it passes through $(4, 3)$, its equation is $y = 3$.

3.60 Through the origin and bisecting the angle between the positive x-axis and the positive y-axis.

▮ Its points are equidistant from the positive x- and y-axes. So, $(1, 1)$ is on the line, and its slope is 1. Thus, the equation is $y = x$.

In Problems 3.61–3.65, find the slopes and y-intercepts of the line given by the indicated equations, and find the coordinates of a point other than $(0, b)$ on the line.

3.61 $y = 5x + 4$.

▮ From the form of the equation, the slope $m = 5$ and the y-intercept $b = 4$. To find another point, set $x = 1$; then $y = 9$. So, $(1, 9)$ is on the line.

3.62 $7x - 4y = 8$.

▮ $y = \frac{7}{4}x - 2$. So, $m = \frac{7}{4}$ and $b = -2$. To find another point, set $x = 4$; then $y = 5$. Hence, $(4, 5)$ is on the line.

3.63 $y = 2 - 4x$.

▮ $m = -4$ and $b = 2$. To find another point, set $x = 1$; then $y = -2$. So, $(1, -2)$ lies on the line.

3.64 $y = 2$.

▮ $m = 0$ and $b = 2$. Another point is $(1, 2)$.

3.65 $\dfrac{y}{4} + \dfrac{x}{3} = 1$.

▮ $y = -\frac{4}{3}x + 4$. So, $m = -\frac{4}{3}$ and $b = 4$. To find another point on the line, set $x = 3$; then $y = 0$. So, $(3, 0)$ is on the line.

In Problems 3.66–3.70, determine whether the given lines are parallel, perpendicular, or neither.

3.66 $y = 5x - 2$ and $y = 5x + 3$.

▮ Since the lines both have slope 5, they are parallel.

3.67 $y = x + 3$ and $y = 2x + 3$.

▮ Since the slopes of the lines are 1 and 2, the lines are neither parallel nor perpendicular.

3.68 $4x - 2y = 7$ and $10x - 5y = 1$.

❚ The slopes are both 2. Hence, the lines are parallel.

3.69 $4x - 2y = 7$ and $2x + 4y = 1$.

❚ The slope of the first line is 2 and the slope of the second line is $-\frac{1}{2}$. Since the product of the slopes is -1, the lines are perpendicular.

3.70 $7x + 3y = 6$ and $3x + 7y = 14$.

❚ The slope of the first line is $-\frac{7}{3}$ and the slope of the second line is $-\frac{3}{7}$. Since the slopes are not equal and their product is not -1, the lines are neither parallel nor perpendicular.

3.71 Temperature is usually measured either in Fahrenheit or in Celsius degrees. The relation between Fahrenheit and Celsius temperatures is given by a linear equation. The freezing point of water is 0° Celsius or 32° Fahrenheit, and the boiling point of water is 100° Celsius or 212° Fahrenheit. Find an equation giving Fahrenheit temperature y in terms of Celsius temperature x.

❚ Since the equation is linear, we can write it as $y = mx + b$. From the information about the freezing point of water, we see that $b = 32$. From the information about the boiling point, we have $212 = 100m + 32$, $180 = 100m$, $m = \frac{9}{5}$. So, $y = \frac{9}{5}x + 32$.

Problems 3.72–3.74 concern a triangle with vertices $A(1, 2)$, $B(8, 0)$, and $C(5, 3)$.

3.72 Find an equation of the median from A to the midpoint of BC.

❚ The midpoint M of BC is $((8 + 5)/2, (0 + 3)/2) = (\frac{13}{2}, \frac{3}{2})$. So, the slope of AM is $(2 - \frac{3}{2})/(1 - \frac{13}{2}) = -\frac{1}{11}$. Hence, the equation has the form $y = -\frac{1}{11}x + b$. Since A is on the line, $2 = -\frac{1}{11} + b$, $b = \frac{23}{11}$. Thus, the equation is $y = -\frac{1}{11}x + \frac{23}{11}$, or $x + 11y = 23$.

3.73 Find an equation of the altitude from B to AC.

❚ The slope of AC is $(3 - 2)/(5 - 1) = \frac{1}{4}$. Hence, the slope of the altitude is the negative reciprocal -4. So, the desired equation has the form $y = -4x + b$. Since B is on the line, $0 = -32 + b$, $b = 32$. So, the equation is $y = -4x + 32$.

3.74 Find an equation of the perpendicular bisector of AB.

❚ The slope of AB is $(2 - 0)/(1 - 8) = -\frac{2}{7}$. Hence, the slope of the desired line is the negative reciprocal $\frac{7}{2}$. The line passes through the midpoint M of AB: $M = (\frac{9}{2}, 1)$. The equation has the form $y = \frac{7}{2}x + b$. Since M is on the line, $1 = \frac{7}{2} \cdot \frac{9}{2} + b$, $b = -\frac{59}{4}$. Thus, the equation is $y = \frac{7}{2}x - \frac{59}{4}$, or $14x - 4y = 59$.

3.75 Highboard Car Rental charges $30 per day and 15¢ per mile for a car, while Lowroad charges $33 per day and 12¢ per mile for the same kind of car. If you expect to drive x miles per day, for what values of x would it cost less to rent the car from Highroad?

❚ The daily cost from Highroad is $3000 + 15x$ cents, and the daily cost from Lowroad is $3300 + 12x$. Then $3000 + 15x < 3300 + 12x$, $3x < 300$, $x < 100$.

3.76 Using coordinates, prove that a parallelogram with perpendicular diagonals is a rhombus.

❚ Refer to Fig. 3-6. Let the parallelogram $ABCD$ have A at the origin and B on the positive x-axis, and DC in the upper half plane. Let the length AB be a, so that B is $(a, 0)$. Let D be (b, c), so that C is $(b + a, c)$. The

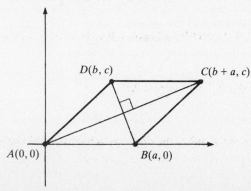

Fig. 3-6

slope of $AC = c/(b + a)$ and the slope of BD is $c/(b - a)$. Since AC and BD are perpendicular,

$$\frac{c}{b+a} \cdot \frac{c}{b-a} = -1 \qquad c^2 = a^2 - b^2$$

But, $AB = a$ and $AD = \sqrt{b^2 + c^2} = a = AB$. So, $ABCD$ is a rhombus.

3.77 Using coordinates, prove that a trapezoid with equal diagonals is isosceles.

▮ As shown in Fig. 3-7, let the trapezoid $ABCD$, with parallel bases $\overline{AB}$ and $\overline{CD}$, have A at the origin and B on the positive x-axis. Let D be (b, c) and C be (d, c), with $b < d$. By hypothesis, $\overline{AC} = \overline{BD}$, $\sqrt{c^2 + d^2} = \sqrt{(b-a)^2 + c^2}$, $d^2 = (b-a)^2$, $d = \pm(b-a)$. **Case 1.** $d = b - a$. Then $b = d + a > d$, contradicting $b < d$. **Case 2.** $d = a - b$. Then $b = a - d$, $b^2 = (d-a)^2$. Hence, $\overline{AD} = \sqrt{b^2 + c^2} = \sqrt{(d-a)^2 + c^2} = \overline{BC}$. So, the trapezoid is isosceles.

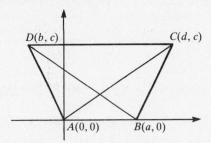

Fig. 3-7

3.78 Find the intersection of the lines $x - 2y = 2$ and $3x + 4y = 6$.

▮ We must solve $x - 2y = 2$ and $3x + 4y = 6$ simultaneously. Multiply the first equation by 2, obtaining $2x - 4y = 4$, and add this to the second equation. The result is $5x = 10$, $x = 2$. When $x = 2$, substitution in either equation yields $y = 0$. Hence, the intersection is the point $(2, 0)$.

3.79 Find the intersection of the lines $4x + 5y = 10$ and $5x + 4y = 8$.

▮ Multiply the first equation by 5 and the second equation by 4, obtaining $20x + 25y = 50$ and $20x + 16y = 32$. Subtracting the second equation from the first, we get $9y = 18$, $y = 2$. When $y = 2$, $x = 0$. So, the intersection is $(0, 2)$.

3.80 Find the intersection of the line $y = 8x - 6$ and the parabola $y = 2x^2$.

▮ Solve $y = 8x - 6$ and $y = 2x^2$ simultaneously. $2x^2 = 8x - 6$, $x^2 = 4x - 3$, $x^2 - 4x + 3 = 0$, $(x - 3)(x - 1) = 0$, $x = 3$ or $x = 1$. When $x = 3$, $y = 18$, and when $x = 1$, $y = 2$. Thus, the intersection consists of $(3, 18)$ and $(1, 2)$.

3.81 Find the intersection of the line $y = x - 3$ and the hyperbola $xy = 4$.

▮ We must solve $y = x - 3$ and $xy = 4$ simultaneously. Then $x(x - 3) = 4$, $x^2 - 3x - 4 = 0$, $(x - 4)(x + 1) = 0$, $x = 4$ or $x = -1$. When $x = 4$, $y = 1$, and when $x = -1$, $y = -4$. Hence, the intersection consists of the points $(4, 1)$ and $(-1, -4)$.

3.82 Let x represent the number of million pounds of chicken that farmers offer for sale per week, and let y represent the number of dollars per pound that consumers are willing to pay for chicken. Assume that the supply equation for chicken is $y = 0.02x + 0.25$, that is, $0.02x + 0.25$ is the price per pound at which farmers are willing to sell x million pounds. Assume also that the demand equation for chicken is $y = -0.025x + 2.5$, that is, $-0.025x + 2.5$ is the price per pound at which consumers are willing to buy x million pounds per week. Find the intersection of the graphs of the supply and demand equations.

▮ Set $0.02x + 0.25 = -0.025x + 2.5$, $0.045x = 2.25$, $x = 2.25/0.045 = 2250/45 = 50$ million pounds. Then $y = 1.25$ dollars is the price.

3.83 Find the coordinates of the point on the line $y = 2x + 1$ that is equidistant from $(0, 0)$ and $(5, -2)$.

▮ Setting the distances from (x, y) to $(0, 0)$ and $(5, -2)$ equal and squaring, we obtain $x^2 + y^2 = (x - 5)^2 + (y + 2)^2$, $x^2 + y^2 = x^2 - 10x + 25 + y^2 + 4y + 4$, $10x - 4y = 29$. Substituting $2x + 1$ for y in the last equation we obtain $10x - 4(2x + 1) = 29$, $2x = 33$, $x = \frac{33}{2}$. Then $y = 34$. So, the desired point is $\left(\frac{33}{2}, 34\right)$.

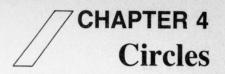

4.1 Write the standard equation for a circle with center at (a, b) and radius r.

▮ By the distance formula, a point (x, y) is on the circle if and only if $\sqrt{(x-a)^2 + (y-b)^2} = r$. Squaring both sides, we obtain the standard equation: $(x-a)^2 + (y-b)^2 = r^2$.

4.2 Write the standard equation for the circle with center $(3, 5)$ and radius 4.

▮ $(x-3)^2 + (y-5)^2 = 16$.

4.3 Write the standard equation for the circle with center $(4, -2)$ and radius 7.

▮ $(x-4)^2 + (y+2)^2 = 49$.

4.4 Write the standard equation for the circle with center at the origin and radius r.

▮ $x^2 + y^2 = r^2$.

4.5 Find the standard equation of the circle with center at $(1, -2)$ and passing through the point $(7, 4)$.

▮ The radius of the circle is the distance between $(1, -2)$ and $(7, 4)$: $\sqrt{(7-1)^2 + [4-(-2)]^2} = \sqrt{36+36} = \sqrt{72}$. Thus, the standard equation is: $(x-1)^2 + (y+2)^2 = 72$.

4.6 Identify the graph of the equation $x^2 + y^2 - 12x + 20y + 15 = 0$.

▮ Complete the square in x and in y: $(x-6)^2 + (y+10)^2 + 15 = 36 + 100$. [Here the -6 in $(x-6)$ is half of the coefficient, -12, of x in the original equation, and the $+10$ in $(y+10)$ is half of the coefficient 20, of y. The 36 and 100 on the right balance the squares of -6 and $+10$ that have in effect been added on the left.] Simplifying, we obtain $(x-6)^2 + (y+10)^2 = 121$, the standard equation of a circle with center at $(6, -10)$ and radius 11.

4.7 Identify the graph of the equation $x^2 + y^2 + 3x - 2y + 4 = 0$.

▮ Complete the square (as in Problem 4.6): $(x + \frac{3}{2})^2 + (y-1)^2 + 4 = \frac{9}{4} + 1$. Simplifying, we obtain $(x + \frac{3}{2})^2 + (y-1)^2 = -\frac{3}{4}$. But this equation has no solutions, since the left side is always nonnegative. In other words, the graph is the empty set.

4.8 Identify the graph of the equation $x^2 + y^2 + 2x - 2y + 2 = 0$.

▮ Complete the square: $(x+1)^2 + (y-1)^2 + 2 = 1 + 1$, which simplifies to $(x+1)^2 + (y-1)^2 = 0$. This is satisfied when and only when $x + 1 = 0$ and $y - 1 = 0$, that is, for the point $(-1, 1)$. Hence, the graph is a single point.

4.9 Show that any circle has an equation of the form $x^2 + y^2 + Dx + Ey + F = 0$.

▮ Consider the standard equation $(x-a)^2 + (y-b)^2 = r^2$. Squaring and simplifying, $x^2 + y^2 - 2ax - 2by + a^2 + b^2 - r^2 = 0$. Let $D = -2a$, $E = -2b$, and $F = a^2 + b^2 - r^2$.

4.10 Determine the graph of an equation $x^2 + y^2 + Dx + Ey + F = 0$.

▮ Complete the square: $\left(x + \dfrac{D}{2}\right)^2 + \left(y + \dfrac{E}{2}\right)^2 + F = \dfrac{D^2}{4} + \dfrac{E^2}{4}$. Simplifying: $\left(x + \dfrac{D}{2}\right)^2 + \left(y + \dfrac{E}{2}\right)^2 = \dfrac{D^2}{4} + \dfrac{E^2}{4} - F = \dfrac{D^2 + E^2 - 4F}{4}$. Now, let $d = D^2 + E^2 - 4F$. When $d > 0$, we obtain a circle with center at $\left(-\dfrac{D}{2}, -\dfrac{E}{2}\right)$ and radius $\sqrt{d}/2$. When $d = 0$, we obtain a single point $\left(-\dfrac{D}{2}, -\dfrac{E}{2}\right)$. When $d < 0$, graph contains no points at all.

4.11 Find the center and radius of the circle passing through the points $(3, 8)$, $(9, 6)$, and $(13, -2)$.

▮ By Problem 4.9, the circle has an equation $x^2 + y^2 + Dx + Ey + F = 0$. Substituting the values (x, y) at each of the three given points, we obtain three equations: $3D + 8E + F = -73$, $9D + 6E + F = -117$, $13D - 2E + F = -173$. First, we eliminate F (subtracting the second equation from the first, and subtracting the third equation from the first):

$$-6D + 2E = 44 \qquad \text{or, more simply} \qquad -3D + E = 22$$
$$-10D + 10E = 100 \qquad\qquad\qquad -D + E = 10$$

Subtracting the second equation from the first, $-2D = 12$, or $D = -6$. So, $E = 10 + D = 4$, and $F = -73 - 3D - 8E = -87$. Hence, we obtain $x^2 + y^2 - 6x + 4y - 87 = 0$. Complete the square: $(x - 3)^2 + (y + 2)^2 - 87 = 9 + 4$, or, $(x - 3)^2 + (y + 2)^2 = 100$. Hence, the center is $(3, -2)$ and the radius is $\sqrt{100} = 10$. **Answer**

4.12 Use a geometrical/coordinate method to find the standard equation of the circle passing through the points $A(0, 6)$, $B(12, 2)$, and $C(16, -2)$.

▮ Find the perpendicular bisector of $\overline{AB}$. The slope of $\overleftrightarrow{AB}$ is $-\frac{1}{3}$ and therefore the slope of the perpendicular bisector is 3. Since it passes through the midpoint $(6, 4)$ of $\overline{AB}$, a point-slope equation for it is $y - 4 = 3(x - 6)$, or, equivalently, $y = 3x - 14$. Now find the perpendicular bisector of $\overline{BC}$. A similar calculation yields $y = x - 14$. Since the perpendicular bisector of a chord of a circle passes through the center of the circle, the center of the circle will be the intersection point of $y = 3x - 14$ and $y = x - 14$. Setting $3x - 14 = x - 14$, we find $x = 0$. So, $y = x - 14 = -14$. Thus, the center is $(0, -14)$. The radius is the distance between the center $(0, -14)$ and any point on the circle, say $(0, 6)$: $\sqrt{(0 - 0)^2 + (-14 - 6)^2} = \sqrt{400} = 20$. Hence, the standard equation is $x^2 + (y + 14)^2 = 400$.

4.13 Find the graph of the equation $2x^2 + 2y^2 - x = 0$.

▮ First divide by 2: $x^2 + y^2 - \frac{1}{2}x = 0$, and then complete the square: $(x - \frac{1}{4})^2 + y^2 = \frac{1}{4}$. This is the standard equation of the circle with center $(\frac{1}{4}, 0)$ and radius $\frac{1}{2}$.

4.14 For what value(s) of k does the circle $(x - k)^2 + (y - 2k)^2 = 10$ pass through the point $(1, 1)$?

▮ $(1 - k)^2 + (1 - 2k)^2 = 10$. Squaring out and simplifying, we obtain $5k^2 - 6k - 8 = 0$. The left side factors into $(5k + 4)(k - 2)$. Hence, the solutions are $k = -4/5$ and $k = 2$.

4.15 Find the centers of the circles of radius 3 that are tangent to both the lines $x = 4$ and $y = 6$.

▮ Let (a, b) be a center. The conditions of tangency imply that $|a - 4| = 3$ and $|b - 6| = 3$ (see Fig. 4-1). Hence, $a = 1$ or $a = 7$, and $b = 3$ or $b = 9$. Thus, there are four circles.

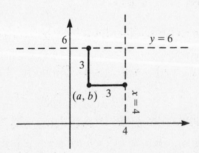

Fig. 4-1

4.16 Determine the value of k so that $x^2 + y^2 - 8x + 10y + k = 0$ is the equation of a circle of radius 7.

▮ Complete the square: $(x - 4)^2 + (y + 5)^2 + k = 16 + 25$. Thus, $(x - 4)^2 + (y + 5)^2 = 41 - k$. So, $\sqrt{41 - k} = 7$. Squaring, $41 - k = 49$, and, therefore, $k = -8$.

4.17 Find the standard equation of the circle which has as a diameter the segment joining the points $(5, -1)$ and $(-3, 7)$. The center is the midpoint $(1, 3)$ of the given segment. The radius is the distance between $(1, 3)$ and $(5, -1)$: $\sqrt{(5 - 1)^2 + (-1 - 3)^2} = \sqrt{16 + 16} = \sqrt{32}$. Hence, the equation is $(x - 1)^2 + (y - 3)^2 = 32$.

4.18 Find the standard equation of a circle with radius 13 that passes through the origin, and whose center has abscissa −12.

▮ Let the center be $(-12, b)$. The distance formula yields $13 = \sqrt{(-12 - 0)^2 + (b - 0)^2} = \sqrt{144 + b^2}$. So $144 + b^2 = 169$, $b^2 = 25$, and $b = \pm 5$. Hence, there are two circles, with equations $(x + 12)^2 + (y - 5)^2 = 169$ and $(x + 12)^2 + (y + 5)^2 = 169$.

4.19 Find the standard equation of the circle with center at $(1, 3)$ and tangent to the line $5x - 12y - 8 = 0$.

▮ The radius is the perpendicular distance from the center $(1, 3)$ to the line: $\dfrac{|5(1) - 12(3) - 8|}{13} = 3$. So, the standard equation is $(x - 1)^2 + (y - 3)^2 = 9$.

4.20 Find the standard equation of the circle passing through $(-2, 1)$ and tangent to the line $3x - 2y = 6$ at the point $(4, 3)$.

▮ Since the circle passes through $(-2, 1)$ and $(4, 3)$, its center (a, b) is on the perpendicular bisector of the segment connecting those points. The center must also be on the line perpendicular to $3x - 2y = 6$ at $(4, 3)$. The equation of the perpendicular bisector of the segment is found to be $3x + y = 5$. The equation of the line perpendicular to $3x - 2y = 6$ at $(4, 3)$ turns out to be $2x + 3y = 17$. Solving $3x + y = 5$ and $2x + 3y = 17$ simultaneously, we find $x = -\frac{2}{7}$ and $y = \frac{41}{7}$. Then the radius $r = \sqrt{(4 + \frac{2}{7})^2 + (3 - \frac{41}{7})^2} = \frac{10}{7}\sqrt{13}$. So, the required equation is $(x + \frac{2}{7})^2 + (y - \frac{41}{7})^2 = \frac{1300}{49}$.

4.21 Find a formula for the length l of the tangent from an exterior point $P(x_1, y_1)$ to the circle $(x - a)^2 + (y - b)^2 = r^2$. See Fig. 4-2.

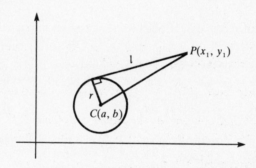

Fig. 4-2

▮ By the Pythagorean theorem, $l^2 = (\overline{PC})^2 - r^2$. By the distance formula, $(\overline{PC})^2 = (x_1 - a)^2 + (y_1 - b)^2$. Hence, $l = \sqrt{(x_1 - a)^2 + (y_1 - b)^2 - r^2}$. **Answer**

4.22 Find the standard equations of the circles passing through the points $A(1, 2)$ and $B(3, 4)$ and tangent to the line $3x + y = 3$.

▮ Let the center of the circle be $C(a, b)$. Since $\overline{CA} = \overline{CB}$,

$$(a - 1)^2 + (b - 2)^2 = (a - 3)^2 + (b - 4)^2 \qquad (1)$$

Since the radius is the perpendicular distance from C to the given line, $\dfrac{|3a + b - 3|}{\sqrt{10}}$, we have

$$(a - 1)^2 + (b - 2)^2 = \left(\frac{3a + b - 3}{\sqrt{10}}\right)^2 \qquad (2)$$

Expanding and simplifying (1) and (2), we have $a + b = 5$ and $a^2 + 9b^2 - 6ab - 2a - 34b + 41 = 0$, whose simultaneous solution yields $a = 4$, $b = 1$, and $a = \frac{3}{2}$, $b = \frac{7}{2}$. From $r = |3a + b - 3|/\sqrt{10}$, we get $r = (12 + 1 - 3)/\sqrt{10} = \sqrt{10}$ and $r = (\frac{9}{2} + \frac{7}{2} - 3)/\sqrt{10} = \sqrt{10}/2$. So, the standard equations are:

$$(x - 4)^2 + (y - 1)^2 = 10 \qquad \text{and} \qquad (x - \tfrac{3}{2})^2 + (y - \tfrac{7}{2})^2 = \tfrac{5}{2}.$$

4.23 Find the center and radius of the circle passing through $(2, 4)$ and $(-1, 2)$ and having its center on the line $x - 3y = 8$.

❚ Let (a, b) be the center. Then the distances from (a, b) to the given points must be equal, and, if we square those distances, we get $(a-2)^2 + (b-4)^2 = (a+1)^2 + (b-2)^2$, $-4a + 4 - 8b + 16 = 2a + 1 - 4b + 4$, $15 = 6a + 4b$. Since (a, b) also is on the line $x - 3y = 8$, we have $a - 3b = 8$. If we multiply this equation by -6 and add the result to $6a + 4b = 15$, we obtain $22b = -33$, $b = -\frac{3}{2}$. Then $a = \frac{7}{2}$, and the center is $(\frac{7}{2}, -\frac{3}{2})$. The radius is the distance between the center and $(-1, 2)$: $\sqrt{(\frac{7}{2}+1)^2 + (-\frac{3}{2}-2)^2} = \dfrac{\sqrt{130}}{2}$.

4.24 Find the points of intersection (if any) of the circles $x^2 + (y-4)^2 = 10$ and $(x-8)^2 + y^2 = 50$.

❚ The circles are $x^2 + y^2 - 8y + 6 = 0$ and $x^2 - 16x + y^2 + 14 = 0$. Subtract the second equation from the first: $16x - 8y - 8 = 0$, $2x - y - 1 = 0$, $y = 2x - 1$. Substitute this equation for y in the second equation: $(x-8)^2 + (2x-1)^2 = 50$, $5x^2 - 20x + 15 = 0$, $x^2 - 4x + 3 = 0$, $(x-3)(x-1) = 0$, $x = 3$ or $x = 1$. Hence, the points of intersection are $(3, 5)$ and $(1, 1)$.

4.25 Let $x^2 + y^2 + C_1 x + D_1 y + E_1 = 0$ be the equation of a circle $\mathscr{C}_1$, and $x^2 + y^2 + C_2 x + D_2 y + E_2 = 0$ the equation of a circle $\mathscr{C}_2$ that intersects $\mathscr{C}_1$ at two points. Show that, as k varies over all real numbers $\neq -1$, the equation $(x^2 + y^2 + C_1 x + D_1 y + E_1) + k(x^2 + y^2 + C_2 x + D_2 y + E_2) = 0$ yields all circles through the intersection points of $\mathscr{C}_1$ and $\mathscr{C}_2$ except $\mathscr{C}_2$ itself.

❚ Clearly, the indicated equation yields the equation of a circle that contains the intersection points. Conversely, given a circle $\mathscr{C} \neq \mathscr{C}_2$ that goes through those intersection points, take a point (x_0, y_0) of $\mathscr{C}$ that does not lie on $\mathscr{C}_2$ and substitute x_0 for x and y_0 for y in the indicated equation. By choice of (x_0, y_0) the coefficient of k is nonzero, so we can solve for k. If we then put this value of k in the indicated equation, we obtain an equation of a circle that is satisfied by (x_0, y_0) and by the intersection points of $\mathscr{C}_1$ and $\mathscr{C}_2$. Since three noncollinear points determine a circle, we have an equation for $\mathscr{C}$. [Again, it is the choice of (x_0, y_0) that makes the three points noncollinear; i.e., $k \neq -1$.]

4.26 Find an equation of the circle that contains the point $(3, 1)$ and passes through the points of intersection of the two circles $x^2 + y^2 - x - y - 2 = 0$ and $x^2 + y^2 + 4x - 4y - 8 = 0$.

❚ Using Problem 4.25, substitute $(3, 1)$ for (x, y) in the equation $(x^2 + y^2 - x - y - 2) + k(x^2 + y^2 + 4x - 4y - 8) = 0$. Then $4 + 10k = 0$, $k = -\frac{2}{5}$. So, the desired equation can be written as

$$5(x^2 + y^2 - x - y - 2) - 2(x^2 + y^2 + 4x - 4y - 8) = 0 \qquad \text{or} \qquad 3x^2 + 3y^2 - 13x + 3y + 6 = 0$$

4.27 Find the equation of the circle containing the point $(-2, 2)$ and passing through the points of intersection of the two circles $x^2 + y^2 + 3x - 2y - 4 = 0$ and $x^2 + y^2 - 2x - y - 6 = 0$.

❚ Using Problem 4.25, substitute $(-2, 2)$ for (x, y) in the equation $(x^2 + y^2 + 3x - 2y - 4) + k(x^2 + y^2 - 2x - y - 6) = 0$. Then $-6 + 4k = 0$, $k = \frac{3}{2}$. So, the desired equation is

$$2(x^2 + y^2 + 3x - 2y - 4) + 3(x^2 + y^2 - 2x - y - 6) = 0 \qquad \text{or} \qquad 5x^2 + 5y^2 - 7y - 26 = 0$$

4.28 Determine the locus of a point that moves so that the sum of the squares of its distances from the lines $5x + 12y - 4 = 0$ and $12x - 5y + 10 = 0$ is 5. [Note that the lines are perpendicular.]

❚ Let (x, y) be the point. The distances from the two lines are $\dfrac{|5x + 12y - 4|}{13}$ and $\dfrac{|12x - 5y + 10|}{13}$.

Hence, $\left(\dfrac{5x + 12y - 4}{13}\right)^2 + \left(\dfrac{12x - 5y + 10}{13}\right)^2 = 5$. Simplifying, we obtain $169x^2 + 169y^2 + 200x - 196y - 729 = 0$, the equation of a circle.

4.29 Find the locus of a point the sum of the squares of whose distances from $(2, 3)$ and $(-1, -2)$ is 34.

❚ Let (x, y) be the point. Then $(x-2)^2 + (y-3)^2 + (x+1)^2 + (y+2)^2 = 34$. Simplify: $x^2 + y^2 - x - y - 8 = 0$, the equation of a circle.

4.30 Find the locus of a point (x, y) the square of whose distance from $(-5, 2)$ is equal to its distance from the line $5x + 12y - 26 = 0$.

❚ $$(x+5)^2 + (y-2)^2 = \dfrac{|5x + 12y - 26|}{13} = \pm\dfrac{5x + 12y - 26}{13}$$

Simplifying, we obtain two equations $13x^2 + 13y^2 + 125x - 64y + 403 = 0$ and $13x^2 + 13y^2 + 135x - 40y + 351 = 0$, both equations of circles.

CHAPTER 5
Functions and their Graphs

In Problems 5.1–5.19, find the domain and range, and draw the graph, of the function determined by the given formula.

5.1 $h(x) = 4 - x^2$.

▌ The domain consists of all real numbers, since $4 - x^2$ is defined for all x. The range consists of all real numbers ≤ 4: solving the equation $y = 4 - x^2$ for x, we obtain $x = \pm\sqrt{4 - y}$, which is defined when and only when $y \leq 4$. The graph (Fig. 5-1) is a parabola with vertex at $(0, 4)$ and the y-axis as its axis of symmetry.

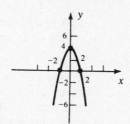

Fig. 5-1

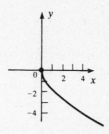

Fig. 5-2

5.2 $G(x) = -2\sqrt{x}$.

▌ The domain consists of all nonnegative real numbers. The range consists of all real numbers ≤ 0. The graph (Fig. 5-2) is the lower half of the parabola $4x = y^2$.

5.3 $H(x) = \sqrt{4 - x^2}$.

▌ The domain is the closed interval $[-2, 2]$, since $\sqrt{4 - x^2}$ is defined when and only when $x^2 \leq 4$. The graph (Fig. 5-3) is the upper half of the circle $x^2 + y^2 = 4$ with center at the origin and radius 2. The range is the closed interval $[0, 2]$.

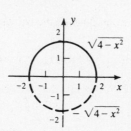

Fig. 5-3

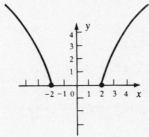

Fig. 5-4

5.4 $U(x) = \sqrt{x^2 - 4}$.

▌ The domain consists of all x such that $x \geq 2$ or $x \leq -2$, since we must have $x^2 \geq 4$. The graph (Fig. 5-4) is the part of the hyperbola $x^2 - y^2 = 4$ on or above the x-axis. The range consists of all nonnegative real numbers.

5.5 $V(x) = |x - 1|$.

▌ The domain is the set of all real numbers. The range is the set of all nonnegative real numbers. The graph (Fig. 5-5) is the graph of $y = |x|$ shifted one unit to the right.

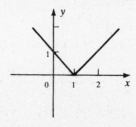

Fig. 5-5

5.6 $f(x) = [2x] =$ the greatest integer $\leq 2x$.

▮   The domain consists of all real numbers.   The range is the set of all integers.   The graph (Fig. 5-6) is the graph of a step function, with each step of length $\frac{1}{2}$ and height 1.

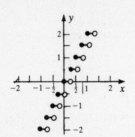

Fig. 5-6

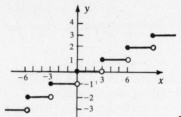

Fig. 5-7

5.7 $g(x) = [x/3]$ (see Problem 5.6).

▮   The domain is the set of all real numbers and the range is the set of all integers.   The graph (Fig. 5-7) is the graph of a step function with each step of length 3 and height 1.

5.8 $h(x) = \dfrac{1}{x}$.

▮   The domain is the set of all nonzero real numbers, and the range is the same set.   The graph (Fig. 5-8) is the hyperbola $xy = 1$.

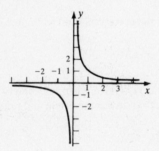

Fig. 5-8

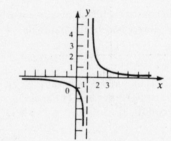

Fig. 5-9

5.9 $F(x) = \dfrac{1}{x-1}$.

▮   The domain is the set of all real numbers $\neq 1$.   The graph (Fig. 5-9) is Fig. 5-8 shifted one unit to the right. The range consists of all nonzero real numbers.

5.10 $H(x) = -\frac{1}{2}x^3$.

▮   The domain is the set of all real numbers, and the range is the same set.   See Fig. 5-10.

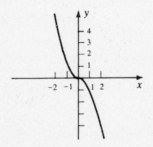

Fig. 5-10

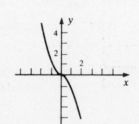

Fig. 5-11

5.11 $J(x) = -x\,|x|$.

▮   The domain and range are the set of all real numbers.   The graph (Fig. 5-11) is obtained by reflecting in the x-axis that part of the parabola $y = x^2$ that lies to the right of the y-axis.

5.12
$$U(x) = \begin{cases} 3 - x & \text{for } x \le 1 \\ 5x - 3 & \text{for } x > 1 \end{cases}$$

❚ The domain is the set of all real numbers. The graph (Fig. 5-12) consists of two half lines meeting at the point $(1, 2)$. The range is the set of all real numbers ≥ 2.

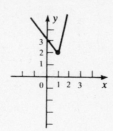

Fig. 5-12

Fig. 5-13

5.13 $f(1) = -1, \quad f(2) = 3, \quad f(4) = -1.$

❚ The domain is $\{1, 2, 4\}$. The range is $\{-1, 3\}$. The graph (Fig. 5-13) consists of three points.

5.14 $G(x) = \dfrac{x^2 - 4}{x + 2}$.

❚ The domain is the set of all real numbers except -2. Since $G(x) = \dfrac{(x + 2)(x - 2)}{x + 2} = x - 2$, the graph (Fig. 5-14) consists of all points on the line $y = x - 2$ except the point $(-2, -4)$. The range is the set of all real numbers except -4.

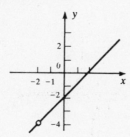

Fig. 5-14

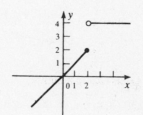

Fig. 5-15

5.15
$$H(x) = \begin{cases} x & \text{if } x \le 2 \\ 4 & \text{if } x > 2 \end{cases}$$

❚ The domain is the set of all real numbers. The graph (Fig. 5-15) is made up of the left half of the line $y = x$ for $x \le 2$ and the right half of the line $y = 4$ for $x > 2$. The range consists of all real numbers ≤ 2, plus the number 4.

5.16 $f(x) = \dfrac{|x|}{x}$.

❚ The domain is the set of all nonzero real numbers. The graph (Fig. 5-16) is the right half of the line $y = 1$ for $x > 0$, plus the left half of the line $y = -1$ for $x < 0$. The range is $\{1, -1\}$.

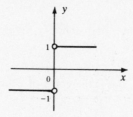

Fig. 5-16

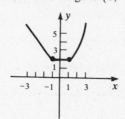

Fig. 5-17

5.17
$$g(x) = \begin{cases} 1 - x & \text{if } x \le -1 \\ 2 & \text{if } -1 < x < 1 \\ x^2 + 1 & \text{if } x \ge 1 \end{cases}$$

❚ The domain is the set of all real numbers. The graph (Fig. 5-17) is a continuous curve consisting of three pieces: the half of the line $y = 1 - x$ to the left of $x = -1$, the horizontal segment $y = 2$ between $x = -1$ and $x = 1$, and the part of the parabola $y = x^2 + 1$ to the right of $x = 1$. The range consists of all real numbers ≥ 2.

5.18

$$h(x) = \begin{cases} \dfrac{x^2 - 9}{x - 3} & \text{if } x \neq 3 \\ 6 & \text{if } x = 3 \end{cases}$$

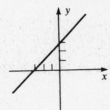

▮ The domain is the set of all real numbers. Since $h(x) = \dfrac{(x+3)(x-3)}{x-3} = x + 3$ for $x \neq 3$ and $h(x) = x + 3$ for $x = 3$, the graph (Fig. 5-18) is the straight line $y = x + 3$. The range is the set of all real numbers.

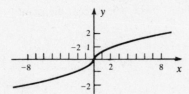

Fig. 5-18

5.19 $F(x) = \sqrt[3]{x}$.

▮ The domain is the set of all real numbers. The graph (Fig. 5-19) is the reflection in the line $y = x$ of the graph of $y = x^3$. [See Problem 5.100.] The range is the set of all real numbers.

Fig. 5-19

5.20 Is Fig. 5-20 the graph of a function?

▮ Since $(0,0)$ and $(0,2)$ are on the graph, this cannot be the graph of a function.

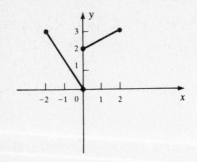

Fig. 5-20

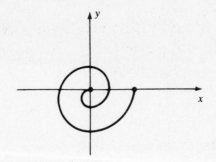

Fig. 5-21

5.21 Is Fig. 5-21 the graph of a function?

▮ Since some vertical lines cut the graph in more than one point, this cannot be the graph of a function.

5.22 Is Fig. 5-22 the graph of a function?

▮ Since each vertical line cuts the graph in at most one point, this is the graph of a function.

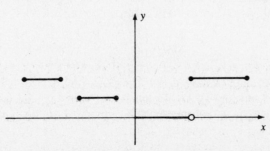

Fig. 5-22

5.23 Is Fig. 5-23 the graph of a function?

▮ Since each vertical line cuts the graph in at most one point, this is the graph of a function.

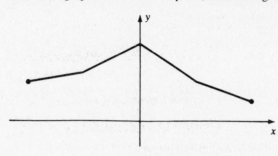

Fig. 5-23

5.24 Find a formula for the function $f(x)$ whose graph consists of all points (x, y) such that $x^3y - 2 = 0$, and specify the domain of $f(x)$.

▮ $f(x) = 2/x^3$. The domain is the set of all nonzero real numbers.

5.25 Find a formula for the function $f(x)$ whose graph consists of all points (x, y) such that $x = \dfrac{1+y}{1-y}$, and specify the domain of $f(x)$.

▮ $x(1 - y) = 1 + y$, $x - xy = 1 + y$, $y(x + 1) = x - 1$, $y = \dfrac{x-1}{x+1}$. So, $f(x) = \dfrac{x-1}{x+1}$ and the domain is the set of all real numbers different from -1.

5.26 Find a formula for the function $f(x)$ whose graph consists of all points (x, y) such that $x^2 - 2xy + y^2 = 0$, and specify the domain of $f(x)$.

▮ The given equation is equivalent to $(x - y)^2 = 0$, $x - y = 0$, $y = x$. Thus, $f(x) = x$, and the domain is the set of all real numbers.

In Problems 5.27–5.31, specify the domain and range of the given function.

5.27 $f(x) = \dfrac{1}{(x-2)(x-3)}$.

▮ The domain consists of all real numbers except 2 and 3. To determine the range, set $y = \dfrac{1}{(x-2)(x-3)} = \dfrac{1}{x^2 - 5x + 6}$ and solve for x. $x^2 - 5x + \left(6 - \dfrac{1}{y}\right) = 0$. This has a solution when and only when $b^2 - 4ac = (-5)^2 - 4(1)\left(6 - \dfrac{1}{y}\right) = 25 - 24 + \dfrac{4}{y} = 1 + \dfrac{4}{y} \geq 0$. This holds if and only if $4/y \geq -1$. This holds when $y > 0$, and, if $y < 0$, when $y \leq -4$. Hence the range is $(0, +\infty) \cup (-\infty, -4]$.

5.28 $g(x) = \dfrac{1}{\sqrt{1 - x^2}}$.

▮ x is in the domain if and only if $x^2 < 1$. Thus, the domain is $(-1, 1)$. To find the range, first note that $g(x) > 0$. Then set $y = 1/\sqrt{1 - x^2}$ and solve for x. $y^2 = 1/(1 - x^2)$, $x^2 = 1 - 1/y^2 \geq 0$, $1 \geq 1/y^2$, $y^2 \geq 1$, $y \geq 1$. Thus, the range is $[1, +\infty)$.

5.29
$$h(x) = \begin{cases} x + 1 & \text{if } -1 < x < 1 \\ 2 & \text{if } 1 \leq x \end{cases}$$

▮ The domain is $(-1, +\infty)$. The graph consists of the open segment from $(-1, 0)$ to $(1, 2)$, plus the half line of $y = 2$ with $x \geq 1$. Hence, the range is the half-open interval $(0, 2]$.

5.30
$$F(x) = \begin{cases} x - 1 & \text{if } 0 \leq x \leq 3 \\ x - 2 & \text{if } 3 < x < 4 \end{cases}$$

▮ The domain is $[0, 4)$. Inspection of the graph shows that the range is $[-1, 2]$.

5.31 $G(x) = |x| - x$.

▮ The domain is the set of all real numbers. To determine the range, note that $G(x) = 0$ if $x \geq 0$, and $G(x) = -2x$ if $x < 0$. Hence, the range consists of all nonnegative real numbers.

5.32   Let
$$f(x) = x - 4 \qquad g(x) = \begin{cases} \dfrac{x^2 - 16}{x + 4} & \text{if} \quad x \neq -4 \\ k & \text{if} \quad x = -4 \end{cases}$$
Determine k so that $f(x) = g(x)$ for all x.

▮ $g(x) = \dfrac{(x+4)(x-4)}{x+4} = x - 4 = f(x)$ if $x \neq -4$. Since $f(-4) = -8$, we must set $k = -8$.

5.33   Let $f(x) = \dfrac{x^2 - x}{x}$ and $g(x) = x - 1$. Why is it wrong to assert that f and g are the same function?

▮ f is not defined when $x = 0$, but g is defined when $x = 0$.

In Problems 5.34–5.37, define *one* function having set $\mathcal{D}$ as its domain and set $\mathcal{R}$ as its range.

5.34 $\mathcal{D} = (0, 1)$ and $\mathcal{R} = (0, 2)$.

▮ Let $f(x) = 2x$ for $0 < x < 1$.

5.35 $\mathcal{D} = [0, 1)$ and $\mathcal{R} = [-1, 4)$.

▮ Look for a linear function $f(x) = mx + b$, taking 0 into -1 and 1 into 4. Then $b = -1$, and $4 = m - 1$, $m = 5$. Hence, $f(x) = 5x - 1$.

5.36 $\mathcal{D} = [0, +\infty)$ and $\mathcal{R} = \{0, 1\}$.

▮ $f(0) = 0$, and $f(x) = 1$ for $x > 0$.

5.37 $\mathcal{D} = (-\infty, 1) \cup (1, 2)$ and $\mathcal{R} = (1, +\infty)$.

▮ Let $f(x) = \dfrac{1}{x + 1} + 1$ for $x < -1$, and $f(x) = x$ for $1 < x < 2$. See Fig. 5-24.

In Problems 5.38–5.47, determine whether the function is even, odd, or neither even nor odd.

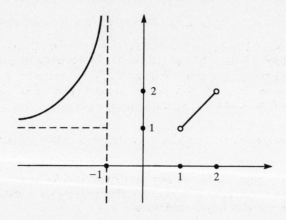

Fig. 5-24

5.38 $f(x) = 9 - x^2$.

▮ Since $f(-x) = 9 - (-x)^2 = 9 - x^2 = f(x)$, $f(x)$ is even.

5.39 $f(x) = \sqrt{x}$.

▮ For a function $f(x)$ to be considered even or odd, it must be defined at $-x$ whenever it is defined at x. Since $f(1)$ is defined but $f(-1)$ is not defined, $f(x)$ is neither even nor odd.

5.40 $f(x) = 4 - x^2$.

▮ This function is even, since $f(-x) = f(x)$.

5.41 $f(x) = |x|$.

▮ Since $|-x| = |x|$, this function is even.

5.42 $f(x) = [x]$.

 ❚ $[\frac{1}{2}] = 0$ and $[-\frac{1}{2}] = -1$. Hence, this function is neither even nor odd.

5.43 $f(x) = \dfrac{1}{x}$.

 ❚ $f(-x) = \dfrac{1}{-x} = -\dfrac{1}{x} = -f(x)$. Hence, this function is odd.

5.44 $f(x) = \sqrt[3]{x}$.

 ❚ $f(-x) = \sqrt[3]{-x} = -\sqrt[3]{x} = -f(x)$. So, $f(x)$ is odd.

5.45 $f(x) = |x - 1|$.

 ❚ $f(1) = 0$ and $f(-1) = |-1 - 1| = 2$. So, $f(x)$ is neither even nor odd.

5.46 The function $J(x)$ of Problem 5.11.

 ❚ $J(-x) = -(-x)\,|-x| = x\,|x| = -J(x)$, so this is an odd function.

5.47 $f(x) = 2x + 1$.

 ❚ $f(1) = 3$ and $f(-1) = -1$. So, $f(x)$ is neither even nor odd.

5.48 Show that a function $f(x)$ is even if and only if its graph is symmetric with respect to the y-axis.

 ❚ Assume that $f(x)$ is even. Let (x, y) be on the graph of f. We must show that $(-x, y)$ is also on the graph of f. Since (x, y) is on the graph of f, $f(x) = y$. Hence, since f is even, $f(-x) = f(x) = y$, and, thus, $(-x, y)$ is on the graph of f. Conversely, assume that the graph of f is symmetric with respect to the y-axis. Assume that $f(x)$ is defined and $f(x) = y$. Then (x, y) is on the graph of f. By assumption, $(-x, y)$ also is on the graph of f. Hence, $f(-x) = y$. Then, $f(-x) = f(x)$, and $f(x)$ is even.

5.49 Show that $f(x)$ is odd if and only if the graph of f is symmetric with respect to the origin.

 ❚ Assume that $f(x)$ is odd. Let (x, y) be on the graph of f. Then $f(x) = y$. Since $f(x)$ is odd, $f(-x) = -f(x) = -y$, and, therefore, $(-x, -y)$ is on the graph of f. But, (x, y) and $(-x, -y)$ are symmetric with respect to the origin. Conversely, assume the graph of f is symmetric with respect to the origin. Assume $f(x) = y$. Then, (x, y) is on the graph of f. Hence, by assumption, $(-x, -y)$ is on the graph of f. Thus, $f(-x) = -y = -f(x)$, and, therefore, $f(x)$ is odd.

5.50 Show that, if a graph is symmetric with respect to both the x-axis and the y-axis, then it is symmetric with respect to the origin.

 ❚ Assume (x, y) is on the graph. Since the graph is symmetric with respect to the x-axis, $(x, -y)$ is also on the graph, and, therefore, since the graph is symmetric with respect to the y-axis, $(-x, -y)$ is also on the graph. Thus, the graph is symmetric with respect to the origin.

5.51 Show that the converse of Problem 5.50 is false.

 ❚ The graph of the odd function $f(x) = x$ is symmetric with respect to the origin. However, $(1, 1)$ is on the graph but $(-1, 1)$ is not; therefore, the graph is not symmetric with respect to the y-axis. It is also not symmetric with respect to the x-axis.

5.52 If f is an odd function and $f(0)$ is defined, must $f(0) = 0$?

 ❚ Yes. $f(0) = f(-0) = -f(0)$. Hence, $f(0) = 0$.

5.53 If $f(x) = x^2 + kx + 1$ for all x and f is an even function, find k.

 ❚ $f(1) = 2 + k$ and $f(-1) = 2 - k$. By the evenness of f, $f(-1) = f(1)$. Hence, $2 + k = 2 - k$, $k = -k$, $k = 0$.

5.54 If $f(x) = x^3 - kx^2 + 2x$ for all x and if f is an odd function, find k.

▌ $f(1) = 3 - k$ and $f(-1) = -3 - k$. Since f is odd, $-3 - k = -(3 - k) = -3 + k$. Hence, $-k = k$, $k = 0$.

5.55 Show that any function $F(x)$ that is defined for all x may be expressed as the sum of an even function and an odd function: $F(x) = E(x) + O(x)$.

▌ Take $E(x) = \frac{1}{2}[F(x) + F(-x)]$ and $O(x) = \frac{1}{2}[F(x) - F(-x)]$.

5.56 Prove that the representation of $F(x)$ in Problem 5.55 is unique.

▌ If $F(x) = E(x) + O(x)$ and $F(x) = E^*(x) + O^*(x)$, then, by subtraction,

$$0 = e(x) + o(x) \tag{1}$$

where $e(x) \equiv E^*(x) - E(x)$ is even and $o(x) = O^*(x) - O(x)$ is odd. Replace x by $-x$ in (1) to obtain

$$0 = e(x) - o(x) \tag{2}$$

But (1) and (2) together imply $e(x) \equiv o(x) \equiv 0$; that is, $E^*(x) \equiv E(x)$ and $O^*(x) \equiv O(x)$.

In Problems 5.57–5.63, determine whether the given function is one-one.

5.57 $f(x) = mx + b$ for all x, where $m \neq 0$.

▌ Assume $f(u) = f(v)$. Then, $mu + b = mv + b$, $mu = mv$, $u = v$. Thus, f is one-one.

5.58 $f(x) = \sqrt{x}$ for all nonnegative x.

▌ Assume $f(u) = f(v)$. Then, $\sqrt{u} = \sqrt{v}$. Square both sides; $u = v$. Thus, f is one-one.

5.59 $f(x) = x^2$ for all x.

▌ $f(-1) = 1 = f(1)$. Hence, f is not one-one.

5.60 $f(x) = \dfrac{1}{x}$ for all nonzero x.

▌ Assume $f(u) = f(v)$. Then $\dfrac{1}{u} = \dfrac{1}{v}$. Hence, $u = v$. Thus, f is one-one.

5.61 $f(x) = |x|$ for all x.

▌ $f(-1) = 1 = f(1)$. Hence, f is not one-one.

5.62 $f(x) = [x]$ for all x.

▌ $f(0) = 0 = f(\frac{1}{2})$. Hence, f is not one-one.

5.63 $f(x) = x^3$ for all x.

▌ Assume $f(u) = f(v)$. Then $u^3 = v^3$. Taking cube roots (see Problem 5.84), we obtain $u = v$. Hence, f is one-one.

In Problems 5.64–5.68, evaluate the expression $\dfrac{f(x + h) - f(x)}{h}$ for the given function f.

5.64 $f(x) = x^2 - 2x$.

▌ $f(x + h) = (x + h)^2 - 2(x + h) = (x^2 + 2xh + h^2) - 2x - 2h$. So, $f(x + h) - f(x) = [(x^2 + 2xh + h^2) - 2x - 2h] - (x^2 - 2x) = 2xh + h^2 - 2h = h(2x + h - 2)$. Hence, $\dfrac{f(x + h) - f(x)}{h} = 2x + h - 2$.

5.65 $f(x) = x + 4$.

▌ $f(x + h) = x + h + 4$. So, $f(x + h) - f(x) = (x + h + 4) - (x + 4) = h$. Hence, $\dfrac{f(x + h) - f(x)}{h} = 1$.

5.66 $f(x) = x^3 + 1$.

▮ $f(x + h) = (x + h)^3 + 1 = x^3 + 3x^2h + 3xh^2 + h^3 + 1$. So, $f(x + h) - f(x) = (x^3 + 3x^2h + 3xh^2 + h^3 + 1) - (x^3 + 1) = 3x^2h + 3xh^2 + h^3 = h(3x^2 + 3xh + h^2)$. Hence, $\dfrac{f(x + h) - f(x)}{h} = 3x^2 + 3xh + h^2$.

5.67 $f(x) = \sqrt{x}$.

▮ $f(x + h) - f(x) = \sqrt{x + h} - \sqrt{x} = (\sqrt{x + h} - \sqrt{x})\dfrac{\sqrt{x + h} + \sqrt{x}}{\sqrt{x + h} + \sqrt{x}} = \dfrac{(x + h) - x}{\sqrt{x + h} + \sqrt{x}} = \dfrac{h}{\sqrt{x + h} + \sqrt{x}}$

So, $\dfrac{f(x + h) - f(x)}{h} = \dfrac{1}{\sqrt{x + h} + \sqrt{x}}$

5.68 $f(x) = \dfrac{1}{x}$.

▮
$$f(x + h) - f(x) = \frac{1}{x + h} - \frac{1}{x} = \frac{x - (x + h)}{x(x + h)} = \frac{-h}{x(x + h)}$$

Hence, $\dfrac{f(x + h) - f(x)}{h} = -\dfrac{1}{x(x + h)}$

In Problems 5.69–5.74, for each of the given one-one functions $f(x)$, find a formula for the inverse function $f^{-1}(y)$.

5.69 $f(x) = x$.

▮ Let $y = f(x) = x$. So, $x = y$. Thus, $f^{-1}(y) = y$.

5.70 $f(x) = 2x + 1$.

▮ Let $y = 2x + 1$ and solve for x. $x = \frac{1}{2}(y - 1)$. Thus, $f^{-1}(y) = \frac{1}{2}(y - 1)$.

5.71 $f(x) = x^3$.

▮ Let $y = x^3$. Then $x = \sqrt[3]{y}$. So, $f^{-1}(y) = \sqrt[3]{y}$.

5.72 $f(x) = \dfrac{1 + x}{1 - x}$.

▮ Let $y = \dfrac{1 + x}{1 - x}$. Then $y(1 - x) = 1 + x$, $y - yx = 1 + x$, $y - 1 = x(1 + y)$, $x = (y - 1)/(y + 1)$. Thus, $f^{-1}(y) = (y - 1)/(y + 1)$.

5.73 $f(x) = \sqrt{1 - x}$.

▮ Let $y = \sqrt{1 - x}$. Then $y^2 = 1 - x$, $x = 1 - y^2$. So, $f^{-1}(y) = 1 - y^2$ (for $y \geq 0$).

5.74 $f(x) = \dfrac{1}{x}$.

▮ Let $y = 1/x$. Then $x = 1/y$. So, $f^{-1}(y) = 1/y$.

5.75 Does a self-inverse function exist? Is there more than one?

▮ See Problems 5.69 and 5.74.

In Problems 5.76–5.82, find all real roots of the given polynomial.

5.76 $x^4 - 10x^2 + 9$.

▮ $x^4 - 10x^2 + 9 = (x^2 - 9)(x^2 - 1) = (x - 3)(x + 3)(x - 1)(x + 1)$. Hence, the roots are $3, -3, 1, -1$.

5.77 $x^3 + 2x^2 - 16x - 32$.

▮ Inspection of the divisors of the constant term 32 reveals that -2 is a root. Division by $x + 2$ yields the factorization $(x + 2)(x^2 - 16) = (x + 2)(x^2 + 4)(x^2 - 4) = (x + 2)(x^2 + 4)(x - 2)(x + 2)$. So, the roots are 2 and -2.

5.78 $x^4 - x^3 - 10x^2 + 4x + 24$.

▮ Testing the divisors of 24 yields the root 2. Division by $x - 2$ yields the factorization $(x - 2)(x^3 + x^2 - 8x - 12)$. It turns out that -2 is another root; division of $x^3 + x^2 - 8x - 12$ by $x + 2$ gives $(x - 2)(x + 2)(x^2 - x - 6) = (x - 2)(x + 2)(x - 3)(x + 2)$. So, the roots are 2, -2, and 3.

5.79 $x^3 - 2x^2 + x - 2$.

▮ $x^3 - 2x^2 + x - 2 = x^2(x - 2) + x - 2 = (x - 2)(x^2 + 1)$. Thus, the only real root is 2.

5.80 $x^3 + 9x^2 + 26x + 24$.

▮ Testing the divisors of 24, reveals the root -2. Dividing by $x + 2$ yields the factorization $(x + 2)(x^2 + 7x + 12) = (x + 2)(x + 3)(x + 4)$. Thus, the roots are -2, -3, and -4.

5.81 $x^3 - 5x - 2$.

▮ -2 is a root. Dividing by $x + 2$ yields the factorization $(x + 2)(x^2 - 2x - 1)$. The quadratic formula applied to $x^2 - 2x - 1$ gives the additional roots $1 \pm \sqrt{2}$.

5.82 $x^3 - 4x^2 - 2x + 8$.

▮ $x^3 - 4x^2 - 2x + 8 = x^2(x - 4) - 2(x - 4) = (x - 4)(x^2 - 2)$. Thus, the roots are 4 and $\pm\sqrt{2}$.

5.83 Establish the factorization

$$u^n - v^n = (u - v)(u^{n-1} + u^{n-2}v + u^{n-3}v^2 + \cdots + uv^{n-2} + v^{n-1})$$

for $n = 2, 3, \ldots$.

▮ Simply multiply out the right-hand side. The cross-product terms will cancel in pairs (u times the kth term of the second factor will cancel with $-v$ times the $(k - 1)$st term).

5.84 Prove algebraically that a real number x has a unique cube root.

▮ Suppose there were two cube roots, u and v, so that $u^3 = v^3 = x$, or $u^3 - v^3 = 0$. Then, by Problem 5.83,

$$0 = u^3 - v^3 = (u - v)(u^2 + uv + v^2) = (u - v)\left[\frac{(u + v)^2 + u^2 + v^2}{2}\right]$$

Unless both u and v are zero, the factor in brackets is positive (being a sum of squares); hence the other factor must vanish, giving $u = v$. If both u and v are zero, then again $u = v$.

5.85 If $f(x) = (x + 3)(x + k)$, and the remainder is 16 when $f(x)$ is divided by $x - 1$, find k.

▮ $f(x) = (x - 1)q(x) + 16$. Hence, $f(1) = 16$. But, $f(1) = (1 + 3)(1 + k) = 4(1 + k)$. So, $1 + k = 4$, $k = 3$.

5.86 If $f(x) = (x + 5)(x - k)$ and the remainder is 28 when $f(x)$ is divided by $x - 2$, find k.

▮ $f(x) = (x - 2)q(x) + 28$. Hence, $f(2) = 28$. But, $f(2) = (2 + 5)(2 - k) = 7(2 - k)$. So, $2 - k = 4$, $k = -2$.

5.87 If the zeros of a function $f(x)$ are 3 and -4, what are the zeros of the function $g(x) = f(x/3)$?

▮ $f(x/3) = 0$ if and only if $x/3 = 3$ or $x/3 = -4$, that is, if and only if $x = 9$ or $x = -12$.

5.88 Describe the function $f(x) = |x| + |x - 1|$ and draw its graph.

▮ **Case 1.** $x \geq 1$. Then $f(x) = x + x - 1 = 2x - 1$. **Case 2.** $0 \leq x < 1$. Then $f(x) = x - (x - 1) = 1$. **Case 3.** $x < 0$. Then $f(x) = -x - (x - 1) = -2x + 1$. So, the graph (Fig. 5-25) consists of a horizontal line segment and two half lines.

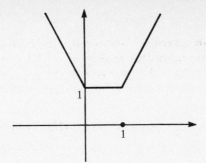

Fig. 5-25

5.89 Find the domain and range of $f(x) = \sqrt{5 - 4x - x^2}$.

▮ By completing the square, $x^2 + 4x - 5 = (x + 2)^2 - 9$. So, $5 - 4x - x^2 = 9 - (x + 2)^2$. For the function to be defined we must have $(x + 2)^2 \le 9$, $-3 \le x + 2 \le 3$, $-5 \le x \le 1$. Thus, the domain is $[-5, 1]$. For x in the domain, $9 \ge 9 - (x + 2)^2 \ge 0$, and, therefore, the range will be $[0, 3]$.

5.90 Show that the product of two even functions and the product of two odd functions are even functions.

▮ If f and g are even, then $f(-x) \cdot g(-x) = f(x) \cdot g(x)$. On the other hand, if f and g are odd, then $f(-x) \cdot g(-x) = [-f(x)] \cdot [-g(x)] = f(x) \cdot g(x)$.

5.91 Show that the product of an even function and an odd function is an odd function.

▮ Let f be even and g odd. Then $f(-x) \cdot g(-x) = f(x) \cdot [-g(x)] = -f(x) \cdot g(x)$.

5.92 Prove that if an odd function $f(x)$ is one-one, the inverse function $g(y)$ is also odd.

▮ Write $y = f(x)$; then $x = g(y)$ and, by oddness, $f(-x) = -y$, or $-x = g(-y)$. Thus, $g(-y) = -g(y)$, and g is odd.

5.93 What can be said about the inverse of an *even*, one-one function?

▮ Anything you wish, since no even function is one-one $[f(-x) = f(x)]$.

5.94 Find an equation of the new curve C^* when the graph of the curve C with the equation $x^2 - xy + y^2 = 1$ is reflected in the x-axis.

▮ (x, y) is on C^* if and only if $(x, -y)$ is on C, that is, if and only if $x^2 - x(-y) + (-y)^2 = 1$, which reduces to $x^2 + xy + y^2 = 1$.

5.95 Find the equation of the new curve C^* when the graph of the curve C with the equation $y^3 - xy^2 + x^3 = 8$ is reflected in the y-axis.

▮ (x, y) is on C^* if and only if $(-x, y)$ is on C, that is, if and only if $y^3 - (-x)y^2 + (-x)^3 = 8$, which reduces to $y^3 + xy^2 - x^3 = 8$.

5.96 Find the equation of the new curve C^* obtained when the graph of the curve C with the equation $x^2 - 12x + 3y = 1$ is reflected in the origin.

▮ (x, y) is on C^* if and only if $(-x, -y)$ is on C, that is, if and only if $(-x)^2 - 12(-x) + 3(-y) = 1$, which reduces to $x^2 + 12x - 3y = 1$.

5.97 Find the reflection of the line $y = mx + b$ in the y-axis.

▮ We replace x by $-x$, obtaining $y = -mx + b$. Thus, the y-intercept remains the same and the slope changes to its negative.

5.98 Find the reflection of the line $y = mx + b$ in the x-axis.

▮ We replace y by $-y$, obtaining $-y = mx + b$, that is, $y = -mx - b$. Thus, both the y-intercept and the slope change to their negatives.

5.99 Find the reflection of the line $y = mx + b$ in the origin.

▎ We replace x by $-x$ and y by $-y$, obtaining $-y = -mx + b$, that is, $y = mx - b$. Thus, the y-intercept changes to its negative and the slope remains unchanged.

5.100 Show geometrically that when the graph of a one-one function is reflected in the 45° line $y = x$, the result is the graph of the inverse function.

▎ It is evident from Fig. 5-26 that right triangles ORP and $OR'P'$ are congruent. Hence,

$$\overline{R'P'} = \overline{RP} = y \qquad \text{and} \qquad \overline{OR'} = \overline{OR} = x$$

Thus the locus of P' is the graph of x as a function of y; i.e., of $x = f^{-1}(y)$. Note that because $y = f(x)$ meets the horizontal-line test (f being one-one), $x = f^{-1}(y)$ meets the vertical-line test.

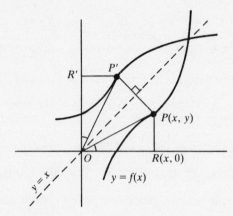

Fig. 5-26

5.101 Graph the function $f(x) = \sqrt{|x - 1| - 1}$.

▎ The complement of the domain is given by $|x - 1| < 1$, or $-1 < x - 1 < 1$, or $0 < x < 2$. Hence, the domain consists of all x such that $x \le 0$ or $x \ge 2$. **Case 1.** $x \ge 2$. Then $y = \sqrt{x - 2}$, $y^2 = x - 2$. So, we have the top half of a parabola with its vertex at $(2, 0)$ and the x-axis as axis of symmetry. **Case 2.** $x \le 0$. Then, $y = \sqrt{-x}$, $y^2 = -x$. So, we have the top half of a parabola with vertex at the origin and with as axis of symmetry. The graph is shown in Fig. 5-27.

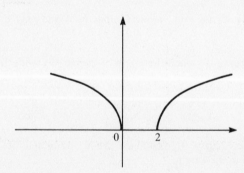

Fig. 5-27

6.1   Define $\lim\limits_{x \to a} f(x) = L$.

▌ Intuitively, this means that, as x gets closer and closer to a, $f(x)$ gets closer and closer to L. We can state this in more precise language as follows: For any $\varepsilon > 0$, there exists $\delta > 0$ such that, if $|x - a| < \delta$, then $|f(x) - L| < \varepsilon$. Here, we assume that, for any $\delta > 0$, there exists at least one x in the domain of $f(x)$ such that $|x - a| < \delta$.

6.2   Find $\lim\limits_{u \to 5} \dfrac{u^2 - 25}{u - 5}$.

▌ The numerator and denominator both approach 0. However, $u^2 - 25 = (u + 5)(u - 5)$. Hence, $\dfrac{u^2 - 25}{u - 5} = u + 5$. Thus, $\lim\limits_{u \to 5} \dfrac{u^2 - 25}{u - 5} = \lim\limits_{u \to 5} (u + 5) = 10$.

6.3   Find $\lim\limits_{x \to 1} \dfrac{x^3 - 1}{x - 1}$.

▌ Both the numerator and denominator approach 0. However, $x^3 - 1 = (x - 1)(x^2 + x + 1)$. Hence, $\lim\limits_{x \to 1} \dfrac{x^3 - 1}{x - 1} = \lim\limits_{x \to 1} (x^2 + x + 1) = 1 + 1 + 1 = 3$.

6.4   Find $\lim\limits_{u \to 0} \dfrac{5u^2 - 4}{u + 1}$.

▌ $\lim\limits_{u \to 0} 5u^2 - 4 = -4$ and $\lim\limits_{u \to 0} (u + 1) = 1$. Hence, by the quotient law for limits,

$$\lim_{u \to 0} \frac{5u^2 - 4}{u + 1} = \frac{-4}{1} = -4.$$

6.5   Find $\lim\limits_{x \to 2} [x]$. [As usual, $[x]$ is the greatest integer $\leq x$; see Fig. 6-1.]

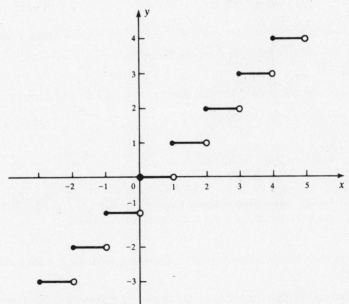

Fig. 6-1

▌ As x approaches 2 from the right (that is, with $x > 2$), $[x]$ remains equal to 2. However, as x approaches 2 from the left (that is, with $x < 2$), $[x]$ remains equal to 1. Hence, there is no unique number which is approached by $[x]$ as x approaches 2. Therefore, $\lim\limits_{x \to 2} [x]$ does not exist.

6.6   Find $\lim_{x \to 4} \dfrac{x^2 - x - 12}{x - 4}$.

▮ Both the numerator and denominator approach 0. However, $x^2 - x - 12 = (x + 3)(x - 4)$. Hence,
$\lim_{x \to 4} \dfrac{x^2 - x - 12}{x - 4} = \lim_{x \to 4} (x + 3) = 7$.

6.7   Find $\lim_{x \to 1} \dfrac{x^4 + 3x^3 - 13x^2 - 27x + 36}{x^2 + 3x - 4}$.

▮ Both the numerator and denominator approach 0. However, division of the numerator by the denominator
reveals that $x^4 + 3x^3 - 13x^2 - 27x + 36 = (x^2 + 3x - 4)(x^2 - 9)$. Hence, $\lim_{x \to 1} \dfrac{x^4 + 3x^3 - 13x^2 - 27x + 36}{x^2 + 3x - 4} =$
$\lim_{x \to 1} (x^2 - 9) = 1 - 9 = -8$.

6.8   Find $\lim_{x \to 2} \dfrac{x^3 - 5x^2 + 2x - 4}{x^2 - 3x + 3}$.

▮ In this case, neither the numerator nor the denominator approaches 0. In fact, $\lim_{x \to 2} (x^3 - 5x^2 + 2x - 4) =$
-12 and $\lim_{x \to 2} (x^2 - 3x + 3) = 1$. Hence, our limit is $\frac{-12}{1} = -12$.

6.9   Find $\lim_{x \to 0} \dfrac{\sqrt{x + 3} - \sqrt{3}}{x}$.

▮ Both numerator and denominator approach 0. "Rationalize" the numerator by multiplying both numerator
and denominator by $\sqrt{x + 3} + \sqrt{3}$.

$$\frac{\sqrt{x + 3} - \sqrt{3}}{x} \cdot \frac{\sqrt{x + 3} + \sqrt{3}}{\sqrt{x + 3} + \sqrt{3}} = \frac{(x + 3) - 3}{x(\sqrt{x + 3} + \sqrt{3})} = \frac{1}{\sqrt{x + 3} + \sqrt{3}}$$

So, we obtain $\lim_{x \to 0} \dfrac{1}{\sqrt{x + 3} + \sqrt{3}} = \dfrac{1}{\sqrt{3} + \sqrt{3}} = \dfrac{1}{2\sqrt{3}}$.

6.10   Find $\lim_{x \to 2} \left(\dfrac{1}{x - 2} - \dfrac{4}{x^2 - 4} \right)$.

▮ As $x \to 0$,   both terms $1/(x - 2)$ and $4/(x^2 - 4)$ "blow up" (that is, become infinitely large in magnitude).   Since $x^2 - 4 = (x + 2)(x - 2)$,   we can factor out $\dfrac{1}{x - 2}$ and simplify:

$$\frac{1}{x - 2} - \frac{4}{x^2 - 4} = \frac{1}{x - 2} \left(1 - \frac{4}{x + 2} \right) = \frac{1}{x - 2} \left(\frac{x + 2 - 4}{x + 2} \right) = \frac{1}{x - 2} \left(\frac{x - 2}{x + 2} \right) = \frac{1}{x + 2}$$

Hence, the limit reduces to $\lim_{x \to 2} [1/(x + 2)] = \frac{1}{4}$.

6.11   Give an ε-δ proof of the fact that $\lim_{x \to 4} (2x - 5) = 3$.

▮ Assume $\varepsilon > 0$. We wish to find $\delta > 0$ so that, if $|x - 4| < \delta$,   then $|(2x - 5) - 3| < \varepsilon$.   But,
$(2x - 5) - 3 = 2x - 8 = 2(x - 4)$. Thus, we must have $|2(x - 4)| < \varepsilon$,   or, equivalently, $|x - 4| < \varepsilon/2$.   So, it
suffices to choose $\delta = \varepsilon/2$ (or any positive number smaller than $\varepsilon/2$).

6.12   In an ε-δ proof of the fact that $\lim_{x \to 3} (2 + 5x) = 17$,   if we are given some ε, what is the largest value of δ that can
be used?

▮ δ must   be   chosen so that, if $|x - 3| < \delta$,   then $|(2 + 5x) - 17| < \varepsilon$.   But, $(2 + 5x) - 17 = 5x - 15 =$
$5(x - 3)$. So, we must have $|5(x - 3)| < \varepsilon$,   or, equivalently, $|x - 3| < \varepsilon/5$. Thus, the largest suitable value
of δ would be $\varepsilon/5$.

6.13   Give   an ε-δ proof   of   the   addition   property   of   limits: If $\lim_{x \to a} f(x) = L$ and $\lim_{x \to a} g(x) = K$,   then
$\lim_{x \to a} [f(x) + g(x)] = L + K$.

▮ Let $\varepsilon > 0$. Then $\varepsilon/2 > 0$. Since $\lim\limits_{x \to a} f(x) = L$, there exists $\delta_1 > 0$ such that, if $|x - a| < \delta_1$, then $|f(x) - L| < \varepsilon/2$. Also, since $\lim\limits_{x \to a} g(x) = K$, there exists $\delta_2 > 0$ such that, if $|x - a| < \delta_2$, then $|g(x) - K| < \varepsilon/2$. Let $\delta = \text{minimum} (\delta_1, \delta_2)$. Hence, if $|x - a| < \delta$, then $|x - a| < \delta_1$ and $|x - a| < \delta_2$, and, therefore, $|f(x) - L| < \varepsilon/2$ and $|g(x) - K| < \varepsilon/2$; so,

$$|[f(x) + g(x)] - (L + K)| = |[f(x) - L] + [g(x) - K]|$$
$$\leq |f(x) - L| + |g(x) - K| \quad \text{[triangle inequality]}$$
$$< \frac{\varepsilon}{2} + \frac{\varepsilon}{2} = \varepsilon$$

6.14 Find $\lim\limits_{x \to 3} \dfrac{1}{(x - 3)^2}$.

▮ As $x \to 3$, from either the right or the left, $(x - 3)^2$ remains positive and approaches 0. Hence, $1/(x - 3)^2$ becomes larger and larger without bound and is positive. Hence, $\lim\limits_{x \to 3} \dfrac{1}{(x - 3)^2} = +\infty$ (an improper limit).

6.15 Find $\lim\limits_{x \to 2} \dfrac{3}{x - 2}$.

▮ As $x \to 2$ from the right (that is, with $x > 2$), $x - 2$ approaches 0 and is positive; therefore $3/(x - 2)$ approaches $+\infty$. However, as $x \to 2$ from the left (that is, with $x < 2$), $x - 2$ approaches 0 and is negative; therefore, $3/(x - 2)$ approaches $-\infty$. Hence, nothing can be said about $\lim\limits_{x \to 2} \dfrac{3}{x - 2}$. $\left[\text{Some} \right.$ people prefer to write $\lim\limits_{x \to 2} \dfrac{3}{x - 2} = \infty$ to indicate that the magnitude $\left|\dfrac{3}{x - 2}\right|$ approaches $+\infty$. $\left.\right]$

6.16 Find $\lim\limits_{x \to 3} \dfrac{x + 2}{x - 3}$.

▮ The numerator approaches 5. The denominator approaches 0, but it is positive for $x > 3$ and negative for $x < 3$. Hence, the quotient approaches $+\infty$ as $x \to 3$ from the right and approaches $-\infty$ as $x \to 3$ from the left. Hence, there is no limit (neither an ordinary limit, nor $+\infty$, nor $-\infty$). However, as in Problem 6.15, we can write $\lim\limits_{x \to 3} \dfrac{x + 2}{x - 3} = \infty$.

6.17 Evaluate $\lim\limits_{x \to +\infty} (2x^{11} - 5x^6 + 3x^2 + 1)$.

▮ $2x^{11} - 5x^6 + 3x^2 + 1 = x^{11}\left(2 - \dfrac{5}{x^5} + \dfrac{3}{x^9} + \dfrac{1}{x^{11}}\right)$. But, $\dfrac{5}{x^5}$ and $\dfrac{3}{x^9}$ and $\dfrac{1}{x^{11}}$ all approach 0 as $x \to +\infty$. Thus, $2 - \dfrac{5}{x^5} + \dfrac{3}{x^9} + \dfrac{1}{x^{11}}$ approaches 2. At the same time x^{11} approaches $+\infty$. Hence, the limit is $+\infty$.

6.18 Evaluate $\lim\limits_{x \to -\infty} (2x^3 - 12x^2 + x - 7)$.

▮ $2x^3 - 12x^2 + x - 7 = x^3\left(2 - \dfrac{12}{x} + \dfrac{1}{x^2} - \dfrac{7}{x^3}\right)$. As $x \to -\infty$, $\dfrac{12}{x}$, $\dfrac{1}{x^2}$, and $\dfrac{7}{x^3}$ all approach 0. Hence, $2 - \dfrac{12}{x} + \dfrac{1}{x^2} - \dfrac{7}{x^3}$ approaches 2. But x^3 approaches $-\infty$. Therefore, the limit is $-\infty$. (Note that the limit will always be $-\infty$ when $x \to -\infty$ and the function is a polynomial of odd degree with positive leading coefficient.)

6.19 Evaluate $\lim\limits_{x \to -\infty} (3x^4 - x^2 + x - 7)$.

▮ $3x^4 - x^2 + x - 7 = x^4\left(3 - \dfrac{1}{x^2} + \dfrac{1}{x^3} - \dfrac{7}{x^4}\right)$. As $x \to -\infty$, $\dfrac{1}{x^2}$, $\dfrac{1}{x^3}$, and $\dfrac{7}{x^4}$ all approach 0. Hence, $3 - \dfrac{1}{x^2} + \dfrac{1}{x^3} - \dfrac{7}{x^4}$ approaches 3. At the same time, x^4 approaches $+\infty$. So, the limit is $+\infty$. (Note that the limit will always be $+\infty$ when $x \to -\infty$ and the function is a polynomial of even degree with positive leading coefficient.)

6.20 Find $\lim\limits_{x \to +\infty} \dfrac{2x + 5}{x^2 - 7x + 3}$.

▮ The numerator and denominator both approach ∞. Hence, we divide numerator and denominator by x^2, the highest power of x in the denominator. We obtain

$$\lim_{x \to +\infty} \frac{2/x + 5/x^2}{1 - 7/x + 3/x^2} = \frac{\lim\limits_{x \to +\infty} (2/x) + \lim\limits_{x \to +\infty} (5/x^2)}{\lim\limits_{x \to +\infty} 1 - \lim\limits_{x \to +\infty} (7/x) + \lim\limits_{x \to +\infty} (3/x^2)} = \frac{0 + 0}{1 - 0 + 0} = \frac{0}{1} = 0$$

6.21 Find $\lim\limits_{x \to +\infty} \dfrac{3x^3 - 4x + 2}{7x^3 + 5}$.

▌ Both numerator and denominator approach 0. So, we divide both of them by x^3, *the highest power of x in the denominator*.

$$\lim_{x \to +\infty} \frac{3 - 4/x^2 + 2/x^3}{7 + 5/x^3} = \frac{\lim\limits_{x \to +\infty} 3 - \lim\limits_{x \to +\infty} (4/x^2) + \lim\limits_{x \to +\infty} (2/x^3)}{\lim\limits_{x \to +\infty} 7 + \lim\limits_{x \to +\infty} (5/x^3)} = \frac{3 - 0 + 0}{7 + 0} = \tfrac{3}{7}$$

(For a generalization, see Problem 6.43.)

6.22 Evaluate $\lim\limits_{x \to +\infty} \dfrac{4x^5 - 1}{3x^3 + 7}$.

▌ Both numerator and denominator approach ∞. So, we divide both of them by x^3, *the highest power of x in the denominator*.

$$\lim_{x \to +\infty} \frac{4x^2 - 1/x^3}{3 + 7/x^3} = \frac{\lim\limits_{x \to +\infty} 4x^2 - \lim\limits_{x \to +\infty} (1/x^3)}{\lim\limits_{x \to +\infty} 3 + \lim\limits_{x \to +\infty} (7/x^3)} = \frac{\lim\limits_{x \to +\infty} 4x^2 - 0}{3 + 0} = \frac{\lim\limits_{x \to +\infty} 4x^2}{3} = +\infty$$

(For a generalization, see Problem 6.44.)

6.23 Find $\lim\limits_{x \to +\infty} \dfrac{4x - 1}{\sqrt{x^2 + 2}}$.

▌ When $f(x)$ is a polynomial of degree n, it is useful to think of the degree of $\sqrt{f(x)}$ as being $n/2$. Thus, in this problem, the denominator has degree 1, and, therefore, in line with the procedure that has worked before, we divide the numerator and denominator by x. Notice that, when $x > 0$ (as it is when $x \to +\infty$), $x = \sqrt{x^2}$. So, we obtain

$$\lim_{x \to +\infty} \frac{4 - (1/x)}{\sqrt{(x^2 + 2)/x^2}} = \lim_{x \to +\infty} \frac{4 - (1/x)}{\sqrt{1 + (2/x^2)}} = \frac{\lim\limits_{x \to +\infty} 4 - \lim\limits_{x \to +\infty} (1/x)}{\sqrt{\lim\limits_{x \to +\infty} 1 + \lim\limits_{x \to +\infty} (2/x^2)}} = \frac{4 - 0}{\sqrt{1 + 0}} = 4$$

6.24 Find $\lim\limits_{x \to +\infty} \dfrac{4x - 1}{\sqrt{x^2 + 2}}$.

▌ Exactly the same analysis applies as in Problem 6.23, except that, when $x \to -\infty$, $x < 0$, and, therefore, $x = -\sqrt{x^2}$. When we divide numerator and denominator by x and replace x by $-\sqrt{x^2}$ in the denominator, a minus sign is introduced. Thus, the answer is the negative, -4, of the answer to Problem 6.23.

6.25 Evaluate $\lim\limits_{x \to +\infty} \dfrac{7x - 4}{\sqrt{x^3 + 5}}$.

▌ We divide the numerator and denominator by $x^{3/2}$. Note that $x^{3/2} = \sqrt{x^3}$ when $x > 0$. So, we obtain

$$\lim_{x \to +\infty} \frac{7/\sqrt{x} - 4/x^{3/2}}{\sqrt{1 + 5/x^3}} = \frac{\lim\limits_{x \to +\infty} (7/\sqrt{x}) - \lim\limits_{x \to +\infty} (4/x^{3/2})}{\sqrt{1 + \lim\limits_{x \to +\infty} (5/x^3)}} = \frac{0 - 0}{\sqrt{1 + 0}} = 0$$

6.26 Evaluate $\lim\limits_{x \to -\infty} \dfrac{3x^3 + 2}{\sqrt{x^4 - 2}}$.

▌ We divide numerator and denominator by x^2. Note that $x^2 = \sqrt{x^4}$. We obtain

$$\lim_{x \to -\infty} \frac{3x + 2/x^2}{\sqrt{(x^4 - 2)/x^4}} = \lim_{x \to -\infty} \frac{3x + 2/x^2}{\sqrt{1 - 2/x^4}} = \frac{\lim\limits_{x \to -\infty} 3x + \lim\limits_{x \to -\infty} (2/x^2)}{\sqrt{1 - \lim\limits_{x \to -\infty} (2/x^4)}} = \frac{\lim\limits_{x \to -\infty} 3x + 0}{\sqrt{1 - 0}} = -\infty$$

6.27 Evaluate $\lim\limits_{x \to +\infty} \dfrac{\sqrt{x^2 + 5}}{3x^2 - 2}$.

▌ We divide numerator and denominator by x^2, obtaining

$$\lim_{x \to +\infty} \frac{\sqrt{(x^2 + 5)/x^4}}{3 - 2/x^2} = \lim_{x \to +\infty} \frac{\sqrt{(1/x^2) + (5/x^4)}}{3 - 2/x^2} = \frac{\sqrt{\lim\limits_{x \to +\infty} (1/x^2) + \lim\limits_{x \to +\infty} (5/x^4)}}{3 - \lim\limits_{x \to +\infty} (2/x^2)} = \frac{\sqrt{0 + 0}}{3 - 0} = 0$$

6.28 Find any vertical and horizontal asymptotes of the graph of the function $f(x) = (4x - 5)/(3x + 2)$.

▮ Remember that a vertical asymptote is a vertical line $x = c$ to which the graph gets closer and closer as x approaches c from the right or from the left. Hence, we obtain vertical asymptotes by setting the denominator $3x + 2 = 0$. Thus, the only vertical asymptote is the line $x = -\frac{2}{3}$. Recall that a horizontal asymptote is a line $y = d$ to which the graph gets closer and closer as $x \to +\infty$ or $x \to -\infty$. In this case, $\lim\limits_{x \to \pm\infty} \dfrac{4x - 5}{3x + 2} = \lim\limits_{x \to \pm\infty} \dfrac{4 - 5/x}{3 + 2/x} = \frac{4}{3}$. Thus, the line $y = \frac{4}{3}$ is a horizontal asymptote both on the right and the left.

6.29 Find the vertical and horizontal asymptotes of the graph of the function $f(x) = (2x + 3)/\sqrt{x^2 - 2x - 3}$.

▮ $x^2 - 2x - 3 = (x - 3)(x + 1)$. Hence, the denominator is 0 when $x = 3$ and when $x = -1$. So, those lines are the vertical asymptotes. (Observe that the numerator is not 0 when $x = 3$ and when $x = -1$.) To obtain horizontal asymptotes, we compute $\lim\limits_{x \to +\infty} \dfrac{2x + 3}{\sqrt{x^2 - 2x - 3}}$ and $\lim\limits_{x \to -\infty} \dfrac{2x + 3}{\sqrt{x^2 - 2x - 3}}$. In both cases, we divide numerator and denominator by x. The first limit becomes

$$\lim_{x \to +\infty} \frac{2 + 3/x}{\sqrt{(x^2 - 2x - 3)/x^2}} = \lim_{x \to +\infty} \frac{2 + 3/x}{\sqrt{1 - 2/x - 3/x^2}} = \frac{2 + 0}{\sqrt{1 - 0 - 0}} = 2$$

For the second limit, remember that $x = -\sqrt{x^2}$ when $x < 0$:

$$\lim_{x \to -\infty} \frac{2x + 3}{\sqrt{x^2 - 2x - 3}} = \lim_{x \to -\infty} \frac{2 + 3/x}{-\sqrt{(x^2 - 2x - 3)/x^2}} = \lim_{x \to -\infty} \frac{2 + 3/x}{-\sqrt{1 - 2/x - 3/x^2}} = \frac{2 + 0}{-\sqrt{1 - 0 - 0}} = -2$$

Hence, the horizontal asymptotes are $y = 2$ on the right and $y = -2$ on the left.

6.30 Find the vertical and horizontal asymptotes of the graph of the function $f(x) = (2x + 3)/\sqrt{x^2 - 2x + 3}$.

▮ Completing the square: $x^2 - 2x + 3 = (x - 1)^2 + 2$. Thus, the denominator is always positive, and therefore, there are no vertical asymptotes. The calculation of the horizontal asymptotes is essentially the same as that in Problem 6.29; $y = 2$ is a horizontal asymptote on the right and $y = -2$ a horizontal asymptote on the left.

6.31 Find the vertical and horizontal asymptotes of the graph of the function $f(x) = \sqrt{x + 1} - \sqrt{x}$.

▮ The function is defined only for $x \geq 0$. There are no values $x = c$ such that $f(x)$ approaches ∞ as $x \to c$. So, there are no vertical asymptotes. To find out whether there is a horizontal asymptote, we compute $\lim\limits_{x \to +\infty} \sqrt{x + 1} - \sqrt{x}$:

$$\lim_{x \to +\infty} (\sqrt{x + 1} - \sqrt{x}) \frac{\sqrt{x + 1} + \sqrt{x}}{\sqrt{x + 1} + \sqrt{x}} = \lim_{x \to +\infty} \frac{(x + 1) - x}{\sqrt{x + 1} + \sqrt{x}} = \lim_{x \to +\infty} \frac{1}{\sqrt{x + 1} + \sqrt{x}} = 0$$

Thus, $y = 0$ is a horizontal asymptote on the right.

6.32 Find the vertical and horizontal asymptotes of the graph of the function $f(x) = (x^2 - 5x + 6)/(x - 3)$.

▮ $x^2 - 5x + 6 = (x - 2)(x - 3)$. So, $(x^2 - 5x + 6)/(x - 3) = x - 2$. Thus, the graph is a straight line $y = x - 2$ [except for the point $(3, 1)$], and, therefore, there are neither vertical nor horizontal asymptotes.

6.33 Find the one-sided limits $\lim\limits_{x \to 2^+} f(x)$ and $\lim\limits_{x \to 2^-} f(x)$ if

$$f(x) = \begin{cases} 7x - 2 & \text{if } x \geq 2 \\ 3x + 5 & \text{if } x < 2 \end{cases}$$

▮ (See Fig. 6-2.) As x approaches 2 from the right, the value $f(x) = 7x - 2$ approaches $7(2) - 2 = 14 - 2 = 12$. Thus, $\lim\limits_{x \to 2^+} f(x) = 12$. As x approaches 2 from the left, the value $f(x) = 3x + 5$ approaches $3(2) + 5 = 6 + 5 = 11$. Thus, $\lim\limits_{x \to 2^-} f(x) = 11$.

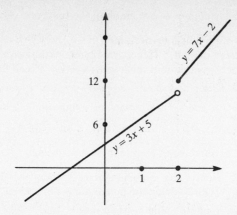

12

6

1   2

Fig. 6-2

$y = 7x - 2$

$y = 3x + 5$

6.34   Find $\lim\limits_{x \to 0^+} \dfrac{|x|}{x}$ and $\lim\limits_{x \to 0^-} \dfrac{|x|}{x}$.

▌ As x approaches 0 from the right, $x > 0$, and, therefore, $|x| = x$; hence, $|x|/x = 1$. Thus, the right-hand limit $\lim\limits_{x \to 0^+} (|x|/x)$ is 1. As x approaches 0 from the left, $x < 0$, and, therefore, $|x| = -x$; hence, $|x|/x = -1$. Thus, the left-hand limit $\lim\limits_{x \to 0^-} (|x|/x) = -1$.

6.35   Evaluate $\lim\limits_{x \to 4^+} \dfrac{3}{x - 4}$ and $\lim\limits_{x \to 4^-} \dfrac{3}{x - 4}$.

▌ As x approaches 4 from the right, $x - 4 > 0$, and, therefore, $3/(x - 4) > 0$; hence, since $3/(x - 4)$ is getting larger and larger in magnitude, $\lim\limits_{x \to 4^+} [3/(x - 4)] = +\infty$. As x approaches 4 from the left, $x - 4 < 0$, and, therefore, $3/(x - 4) < 0$; hence, since $3/(x - 4)$ is getting larger and larger in magnitude, $\lim\limits_{x \to 4^-} [3/(x - 4)] = -\infty$.

6.36   Evaluate $\lim\limits_{x \to 3^+} \dfrac{1}{x^2 - 7x + 12}$ and $\lim\limits_{x \to 3^-} \dfrac{1}{x^2 - 7x + 12}$.

▌ $x^2 - 7x + 12 = (x - 4)(x - 3)$. As x approaches 3 from the right, $x - 3 > 0$, and, therefore, $1/(x - 3) > 0$ and $1/(x - 3)$ is approaching $+\infty$; at the same, $1/(x - 4)$ is approaching -1 and is negative. So, as x approaches 3 from the right, $1/(x^2 - 7x + 12)$ is approaching $-\infty$. Thus, $\lim\limits_{x \to 3^+} \dfrac{1}{x^2 - 7x + 12} = -\infty$. As x approaches 3 from the left, the only difference from the case just analyzed is that $x - 3 < 0$, and, therefore, $1/(x - 3)$ approaches $-\infty$. Hence, $\lim\limits_{x \to 3^-} \dfrac{1}{x^2 - 7x + 12} = +\infty$.

6.37   Find $\lim\limits_{h \to 0} \dfrac{f(x + h) - f(x)}{h}$ when $f(x) = 4x^2 - x$.

▌ $f(x + h) = 4(x + h)^2 - (x + h) = 4(x^2 + 2xh + h^2) - x - h = 4x^2 + 8xh + 4h^2 - x - h$. Hence, $f(x + h) - f(x) = (4x^2 + 8xh + 4h^2 - x - h) - (4x^2 - x) = 8xh + 4h^2 - h$. So, $\dfrac{f(x + h) - f(x)}{h} = \dfrac{8xh + 4h^2 - h}{h} = 8x + 4h - 1$.

Hence, $\lim\limits_{h \to 0} \dfrac{f(x + h) - f(x)}{h} = \lim\limits_{h \to 0} (8x + 4h - 1) = 8x - 1$.   **Answer**

6.38   Find $\lim\limits_{h \to 0} \dfrac{f(x + h) - f(x)}{h}$ when $f(x) = \sqrt{2x}$.

▌ $f(x + h) - f(x) = \sqrt{2(x + h)} - \sqrt{2x} = (\sqrt{2(x + h)} - \sqrt{2x}) \dfrac{\sqrt{2(x + h)} + \sqrt{2x}}{\sqrt{2(x + h)} + \sqrt{2x}} = \dfrac{2(x + h) - 2x}{\sqrt{2(x + h)} + \sqrt{2x}}$

$= \dfrac{2h}{\sqrt{2(x + h)} + \sqrt{2x}}$

Hence $\lim\limits_{h \to 0} \dfrac{f(x + h) - f(x)}{h} = \lim\limits_{h \to 0} \dfrac{2}{\sqrt{2(x + h)} + \sqrt{2x}} = \dfrac{2}{2\sqrt{2x}} = \dfrac{1}{\sqrt{2x}}$

6.39   Find $\lim\limits_{x \to 2^+} f(x)$ and $\lim\limits_{x \to 2^-} f(x)$ for the function $f(x)$ whose graph is shown in Fig. 6-3.

▌ By inspection, $\lim\limits_{x \to 2^+} f(x) = 1$. [Notice that this is different from $f(2)$.] Also, $\lim\limits_{x \to 2^-} f(x) = 3$.

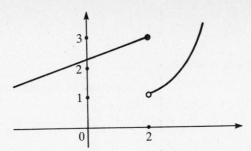

Fig. 6-3

6.40 Evaluate $\lim_{x \to +\infty} (\sqrt{x^2 + x} - x)$.

❚ As $x \to +\infty$, both $\sqrt{x^2 + x}$ and x approach $+\infty$. It is not obvious how their difference behaves. However, the limit equals

$$\lim_{x \to +\infty} (\sqrt{x^2 + x} - x) \frac{\sqrt{x^2 + x} + x}{\sqrt{x^2 + x} + x} = \lim_{x \to +\infty} \frac{(x^2 + x) - x^2}{\sqrt{x^2 + x} + x} = \lim_{x \to +\infty} \frac{x}{\sqrt{x^2 + x} + x}$$

Now we divide numerator and denominator by x (noting that $x = \sqrt{x^2}$):

$$\lim_{x \to +\infty} \frac{1}{\sqrt{(x^2 + x)/x^2}} + 1 = \lim_{x \to +\infty} \frac{1}{\sqrt{1 + (1/x)} + 1} = \frac{1}{\sqrt{1 + 0} + 1} = \frac{1}{2}$$

6.41 Evaluate $\lim_{x \to 2} \dfrac{\sqrt{x^2 + 5} - 3}{x^2 - 2x}$.

❚ Rationalize the numerator:

$$\frac{\sqrt{x^2 + 5} - 3}{x^2 - 2x} \cdot \frac{\sqrt{x^2 + 5} + 3}{\sqrt{x^2 + 5} + 3} = \frac{(x^2 + 5) - 9}{x(x - 2)[\sqrt{x^2 + 5} + 3]} = \frac{x^2 - 4}{x(x - 2)(\sqrt{x^2 + 5} + 3)} = \frac{(x + 2)(x - 2)}{x(x - 2)(\sqrt{x^2 + 5} + 3)}$$

$$= \frac{x + 2}{x(\sqrt{x^2 + 5} + 3)}$$

We obtain

$$\lim_{x \to 2} \frac{x + 2}{x(\sqrt{x^2 + 5} + 3)} = \frac{4}{2(\sqrt{9} + 3)} = \frac{1}{3} \quad \textbf{Answer}$$

6.42 Let $f(x) = a_n x^n + a_{n-1} x^{n-1} + \cdots + a_1 x + a_0$, with $a_n > 0$. Prove that $\lim_{x \to +\infty} f(x) = +\infty$.

❚ $f(x) = x^n \left(a_n + \dfrac{a_{n-1}}{x} + \cdots + \dfrac{a_1}{x^{n-1}} + \dfrac{a_0}{x^n} \right)$. Now, $\lim_{x \to +\infty} \dfrac{a_{n-1}}{x} = \cdots = \lim_{x \to +\infty} \dfrac{a_1}{x^{n-1}} = \lim_{x \to +\infty} \dfrac{a_0}{x^n} = 0$. So, the sum inside the parentheses approaches a_n. Since $\lim_{x \to +\infty} x^n = +\infty$, $\lim_{x \to +\infty} f(x) = +\infty$.

6.43 If $f(x) = \dfrac{a_n x^n + \cdots + a_1 x + a_0}{b_n x^n + \cdots + b_1 x + b_0}$, with $a_n \neq 0$ and $b_n \neq 0$, show that $\lim_{x \to \pm\infty} f(x) = \dfrac{a_n}{b_n}$.

❚ Dividing numerator and denominator by x^n,

$$f(x) = \frac{a_n + a_{x-1}/x + \cdots + a_1/x^{n-1} + a_0/x^n}{b_n + b_{n-1}/x + \cdots + b_1/x^{n-1} + b_0/x^n}$$

Since each a_{n-j}/x^j and b_{n-j}/x^j approaches 0, $\lim_{x \to \pm\infty} f(x) = a_n/b_n$.

6.44 If $f(x) = \dfrac{a_n x^n + \cdots + a_1 x + a_0}{b_k x^k + \cdots + b_1 x + b_0}$ with $a_n > 0$ and $b_k > 0$, prove that $\lim_{x \to +\infty} f(x) = +\infty$ if $n > k$.

❚ Factoring out x^n from the numerator and then dividing numerator and denominator by x^k, $f(x)$ becomes

$$x^{n-k} \left[\frac{a_n + (a_{n-1}/x) + \cdots + (a_1/x^{n-1}) + (a_0/x^n)}{b_k + (b_{k-1}/x) + \cdots + (b_1/x^{k-1}) + (b_0/x^k)} \right]$$

As $x \to +\infty$, all the quotients a_{n-j}/x^j and b_{k-j}/x^j approach 0, and, therefore, the quantity inside the parentheses approaches $a_n/b_k > 0$. Since x^{n-k} approaches $+\infty$, $\lim_{x \to +\infty} f(x) = +\infty$.

6.45 If $f(x) = \dfrac{a_n x^n + \cdots + a_1 x + a_0}{b_k x^k + \cdots + b_1 x + b_0}$ with $a_n > 0$ and $b_k > 0$, and $n < k$, then $\lim\limits_{k \to \pm\infty} f(x) = 0$.

▮ Dividing numerator and denominator by x^k, $f(x) = \dfrac{(a_n/x^{k-n}) + (a_{n-1}/x^{k-n+1}) + \cdots + (a_1/x^{k-1}) + (a_0/x^k)}{b_k + (b_{k-1}/x) + \cdots + (b_1/x^{k-1}) + (b_0/x^k)}$.

Since each of the quotients a_{n-j}/x^{k-n+j} and b_{k-j}/x^j approaches 0, the denominator approaches $b_k > 0$, and the numerator approaches 0. Hence, $\lim\limits_{x \to \pm\infty} f(x) = 0$.

6.46 Find $\lim\limits_{x \to \pm\infty} \dfrac{7x^3 + 2x^2}{4x^3 - x}$.

▮ By Problem 6.43, the limit is $\frac{7}{4}$.

6.47 Find $\lim\limits_{x \to +\infty} \dfrac{x^4 + 2x - 3}{x^2 + 100x}$.

▮ By Problem 6.44, the limit is $+\infty$.

6.48 Find $\lim\limits_{x \to \pm\infty} \dfrac{4x^3 + 20x^2}{x^4 - 1}$.

▮ By Problem 6.45, the limit is 0.

6.49 Find $\lim\limits_{x \to -\infty} \dfrac{x^4 + 2x - 3}{x^2 + 100x}$.

▮ (Compare with Problem 6.47.) Let $u = -x$. Then the given limit is equal to $\lim\limits_{u \to +\infty} \dfrac{u^4 - 2u - 3}{u^2 - 100u}$, which, by Problem 6.44, is $+\infty$.

6.50 Find $\lim \dfrac{3x - 4}{\sqrt[3]{x^2 - 1}}$.

▮ Divide the numerator and denominator by $x^{2/3}$, which is essentially the "highest power of x" in the denominator. (Pay attention only to the term of highest order.) We obtain $\dfrac{3x^{1/3} - 4/x^{2/3}}{\sqrt[3]{(x^2 - 1)/x^2}} = \dfrac{3x^{1/3} - 4/x^{2/3}}{\sqrt[3]{1 - 1/x^2}}$.

Since $1/x^2$ approaches 0, the denominator approaches 1. In the numerator, $4/x^{2/3}$ approaches 0. Since $x^{1/3}$ approaches $+\infty$, our limit is $+\infty$. (Note that the situation is essentially the same as in Problem 6.44).

6.51 Find $\lim\limits_{x \to +\infty} \dfrac{2x + 1}{\sqrt[3]{x^3 - 2}}$.

▮ Divide the numerator and denominator by x, which is essentially the highest power in the denominator (forgetting about -2 in $x^3 - 2$). We obtain $\dfrac{2 + 1/x}{\sqrt[3]{(x^3 - 2)/x^3}} = \dfrac{2 + 1/x}{\sqrt[3]{1 - 2/x^3}}$. Since $2/x^3$ and $1/x$ approach 0, our limit is 2. (This is essentially the same situation as in Problem 6.43.)

6.52 Find $\lim\limits_{x \to -\infty} \dfrac{3x + 2}{\sqrt{x^2 - 1}}$.

▮ We divide numerator and denominator by x, which is essentially the highest power of x in the denominator. Note that, for negative x (which we are dealing with when $x \to -\infty$), $x = -\sqrt{x^2}$. Hence, we obtain $\dfrac{3 + 2/x}{\sqrt{x^2 - 1}/(-\sqrt{x^2})} = \dfrac{3 + 2/x}{-\sqrt{1 - 1/x^2}}$. Since $2/x$ and $1/x^2$ both approach 0, our limit is -3.

7.1   Define: $f(x)$ is continuous at $x = a$.

▌ $f(a)$ is defined, $\lim\limits_{x \to a} f(x)$ exists, and $\lim\limits_{x \to a} f(x) = f(a)$.

7.2   Find the points of discontinuity (if any) of the function $f(x)$ whose graph is shown in Fig. 7-1.

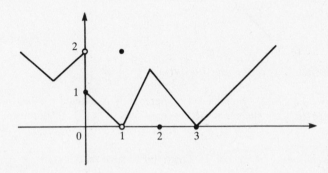

Fig. 7-1

▌ $x = 0$ is a point of discontinuity because $\lim\limits_{x \to 0} f(x)$ does not exist. $x = 1$ is a point of discontinuity because $\lim\limits_{x \to 1} f(x) \neq f(1)$ [since $\lim\limits_{x \to 1} f(x) = 0$ and $f(1) = 2$].

7.3   Determine the points of discontinuity (if any) of the function $f(x)$ such that $f(x) = x^2$ if $x \leq 0$ and $f(x) = x$ if $x > 0$.

▌ $f(x)$ is continuous everywhere.   In particular, $f(x)$ is continuous at $x = 0$ because $f(0) = (0)^2 = 0$ and $\lim\limits_{x \to 0} f(x) = 0$.

7.4   Determine the points of discontinuity (if any) of the function $f(x)$ such that $f(x) = 1$ if $x \geq 0$ and $f(x) = -1$ if $x < 0$.   (See Fig. 7-2.)

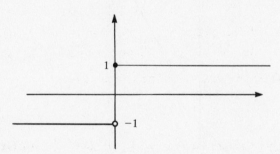

Fig. 7-2

▌ $f(x)$ is not continuous at $x = 0$ because $\lim\limits_{x \to 0} f(x)$ does not exist.

7.5   Determine the points of discontinuity (if any) of the function $f(x)$ such that $f(x) = \dfrac{x^2 - 4}{x + 2}$ if $x \neq -2$ and $f(x) = 0$ if $x = -2$.   (See Fig. 7-3.)

▌ Since $x^2 - 4 = (x - 2)(x + 2)$, $f(x) = x - 2$ if $x \neq -2$.   So, $f(x)$ is not continuous at $x = -2$ because $\lim\limits_{x \to -2} f(x) \neq f(-2)$ [since $f(-2) = 0$ but $\lim\limits_{x \to -2} f(x) = -4$].   [However, $x = -2$ is called a *removable discontinuity*, because, if we redefine $f(x)$ at $x = -2$ by setting $f(-2) = -4$,   then the new function is continuous at $x = -2$.   Compare Problem 7.2.]

43

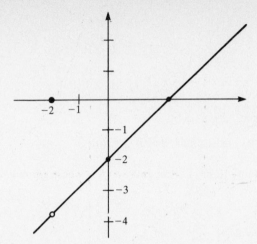

Fig. 7-3

7.6 Find the points of discontinuity of the function $f(x) = \dfrac{x^2 - 1}{x - 1}$.

▮ Since $x^2 - 1 = (x - 1)(x + 1)$, $f(x) = x + 1$ wherever it is defined. However, $f(x)$ is not defined when $x = 1$, since $(x^2 - 1)/(x - 1)$ does not make sense when $x = 1$. Therefore, $f(x)$ is not continuous at $x = 1$.

7.7 Find the points of discontinuity (if any) of the function $f(x)$ such that $f(x) = \dfrac{x^2 - 9}{x - 3}$ for $x \neq 3$ and $f(x) = 6$ for $x = 3$.

▮ Since $x^2 - 9 = (x - 3)(x + 3)$, $f(x) = x + 3$ for $x \neq 3$. However, $f(x) = x + 3$ also when $x = 3$, since $f(3) = 6 = 3 + 3$. Thus, $f(x) = x + 3$ for all x, and, therefore, $f(x)$ is continuous everywhere.

7.8 Find the points of discontinuity (if any) of the function $f(x)$ such that

$$f(x) = \begin{cases} x + 1 & \text{if } x \geq 2 \\ 2x - 1 & \text{if } 1 < x < 2 \\ x - 1 & \text{if } x \leq 1 \end{cases}$$

(See Fig. 7-4.)

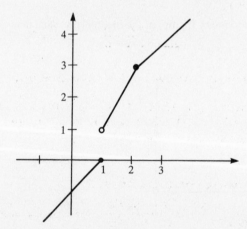

Fig. 7-4

▮ $f(x)$ is discontinuous at $x = 1$ because $\lim\limits_{x \to 1} f(x)$ does not exist. $f(x)$ is continuous at $x = 2$ because $f(2) = 2 + 1 = 3$ and $\lim\limits_{x \to 2} f(x) = 3$. Obviously $f(x)$ is continuous for all other x.

7.9 Find the points of discontinuity (if any) of $f(x) = \dfrac{3x + 3}{x^2 - 3x - 4}$, and write an equation for each vertical and horizontal asymptote of the graph of f.

▮ See Fig. 7-5. $f(x) = \dfrac{3(x + 1)}{(x + 1)(x - 4)} = \dfrac{3}{x - 4}$. $f(x)$ is discontinuous at $x = 4$ and $x = -1$ because it is not defined at those points. [However, $x = -1$ is a removable discontinuity. If we let $f(-1) = -\frac{3}{5}$, the new function is continuous at $x = -1$.] The only vertical asymptote is $x = 4$. Since $\lim\limits_{x \to \pm\infty} \dfrac{3x + 3}{x^2 - 3x - 4} = 0$, the x-axis, $y = 0$, is a horizontal asymptote to the right and to the left.

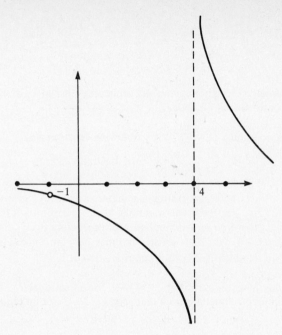

Fig. 7-5

7.10   If the function

$$f(x) = \begin{cases} \dfrac{x^2 - 16}{x - 4} & \text{if } x \neq 4 \\ C & \text{if } x = 4 \end{cases}$$

is continuous, what is the value of C?

▮ Since $x^2 - 16 = (x - 4)(x + 4)$, $f(x) = x + 4$ for $x \neq 4$. Hence, $\lim\limits_{x \to 4} f(x) = 8$. For continuity, we must have $\lim\limits_{x \to 4} f(x) = f(4)$, and, therefore, $8 = f(4) = C$.

7.11   Let $g(x)$ be the function such that

$$g(x) = \begin{cases} \dfrac{x^2 - b^2}{x - b} & \text{if } x \neq b \\ 0 & \text{if } x = b \end{cases}$$

(*a*) Does $g(b)$ exist?   (*b*) Does $\lim\limits_{x \to b} g(x)$ exist?   (*c*) Is $g(x)$ continuous at $x = b$?

▮ (*a*) $g(b) = 0$ by definition. (*b*) Since $x^2 - b^2 = (x - b)(x + b)$, $g(x) = x + b$ for $x \neq b$. Hence, $\lim\limits_{x \to b} g(x) = \lim\limits_{x \to b}(x + b) = 2b$. (*c*) For $g(x)$ to be continuous at $x = b$, we must have $\lim\limits_{x \to b} g(x) = g(b)$, that is, $2b = 0$. So, $g(x)$ is continuous at $x = b$ only when $b = 0$.

7.12   For what value of k is the following a continuous function?

$$f(x) = \begin{cases} \dfrac{\sqrt{7x + 2} - \sqrt{6x + 4}}{x - 2} & \text{if } x \geq -\tfrac{2}{7} \text{ and } x \neq 2 \\ k & \text{if } x = 2 \end{cases}$$

▮ Notice that $\sqrt{7x + 2}$ and $\sqrt{6x + 4}$ are defined when $x \geq -\tfrac{2}{7}$, since $7x + 2$ and $6x + 4$ are nonnegative for $x \geq -\tfrac{2}{7}$. Also,

$$\frac{\sqrt{7x + 2} - \sqrt{6x + 4}}{x - 2} = \frac{\sqrt{7x + 2} - \sqrt{6x + 4}}{x - 2} \cdot \frac{\sqrt{7x + 2} + \sqrt{6x + 4}}{\sqrt{7x + 2} + \sqrt{6x + 4}} = \frac{(7x + 2) - (6x + 4)}{(x - 2)(\sqrt{7x + 2} + \sqrt{6x + 4})}$$

$$= \frac{x - 2}{(x - 2)(\sqrt{7x + 2} + \sqrt{6x + 4})} = \frac{1}{\sqrt{7x + 2} + \sqrt{6x + 4}}$$

Hence, $\lim\limits_{x \to 2} f(x) = \lim\limits_{x \to 2} \dfrac{1}{\sqrt{7x + 2} + \sqrt{6x + 4}} = \dfrac{1}{\sqrt{16} + \sqrt{16}} = \tfrac{1}{8}$. Therefore, k must be $\tfrac{1}{8}$.

7.13 Determine the points of discontinuity (if any) of the following function $f(x)$.

$$f(x) = \begin{cases} 1 & \text{if } x \text{ is rational} \\ 0 & \text{if } x \text{ is irrational} \end{cases}$$

▮ Since there are both rational and irrational numbers arbitrarily close to a given number c, $\lim_{x \to c} f(x)$ does not exist. Hence, $f(x)$ is discontinuous at all points.

7.14 Determine the points of discontinuity (if any) of the following function $f(x)$.

$$f(x) = \begin{cases} x & \text{if } x \text{ is rational} \\ 0 & \text{if } x \text{ is irrational} \end{cases}$$

▮ Let c be any number. Since there are irrational numbers arbitrarily close to c, $f(x) = 0$ for values of x arbitrarily close to c. Hence, if $\lim_{x \to c} f(x)$ exists, it must be 0. Therefore, if $f(x)$ is to be continuous at c, we must have $f(c) = 0$. Since there are rational numbers arbitrarily close to c, $f(x) = x$ for some points that are arbitrarily close to c. Hence, if $\lim_{x \to c} f(x)$ exists, it must be $\lim_{x \to c} x = c$. So, if $f(x)$ is continuous at c, $f(c) = c = 0$. The only point at which $f(x)$ is continuous is $x = 0$.

7.15 (a) Let $f(x)$ be a continuous function such that $f(x) = 0$ for all rational x. Prove that $f(x) = 0$ for all x. (b) Let $f(x)$ and $g(x)$ be continuous functions such that $f(x) = g(x)$ for all rational x. Show that $f(x) = g(x)$ for all x.

▮ (a) Consider any real number c. Since $f(x)$ is continuous at c, $\lim_{x \to c} f(x) = f(c)$. But, since there are rational numbers arbitrarily close to c, $f(x) = 0$ for values of x arbitrarily close to c, and, therefore, $\lim_{x \to c} f(x) = 0$. Hence, $f(c) = 0$. (b) Let $h(x) = f(x) - g(x)$. Since $f(x)$ and $g(x)$ are continuous, so is $h(x)$. Since $f(x) = g(x)$ for all rational x, $h(x) = 0$ for all rational x, and, therefore, by part (a), $h(x) = 0$ for all x. Hence, $f(x) = g(x)$ for all x.

7.16 Let $f(x)$ be a continuous function such that $f(x + y) = f(x) + f(y)$ for all x and y. Prove that $f(x) = cx$ for some constant c.

▮ Let $f(1) = c$. (i) Let us show by induction that $f(n) = cn$ for all positive integers n. When $n = 1$, this is just the definition of c. Assume $f(n) = cn$ for some n. Then $f(n + 1) = f(n) + f(1) = cn + c = c(n + 1)$. (ii) $f(0) = f(0 + 0) = f(0) + f(0)$. So, $f(0) = 0 = c \cdot 0$. (iii) Consider any negative integer $-n$, where $n > 0$. Then, $0 = f(0) = f(n + (-n)) = f(n) + f(-n)$. So, $f(-n) = -f(n) = -cn = c(-n)$. (iv) Any rational number can be written in the form m/n, where m and n are integers and $n > 0$. Then,

$$cm = f(m) = f\left(\underbrace{\frac{m}{n} + \frac{m}{n} + \cdots + \frac{m}{n}}_{n \text{ times}}\right) = \underbrace{f\left(\frac{m}{n}\right) + f\left(\frac{m}{n}\right) + \cdots + f\left(\frac{m}{n}\right)}_{n \text{ times}} = n \cdot f\left(\frac{m}{n}\right)$$

Hence, $f\left(\dfrac{m}{n}\right) = c \cdot \dfrac{m}{n}$. (v) Let b be irrational. Then b is the limit of a sequence $r_1, r_2, \ldots$ of rational numbers. By (iv), $f(r_n) = c \cdot r_n$. By continuity, $f(b) = \lim_{x \to b} f(x) = \lim_{n \to \infty} c \cdot r_n = c \cdot \lim_{n \to \infty} r_n = c \cdot b$.

7.17 Find the discontinuities (if any) of the function $f(x)$ such that $f(x) = 0$ for $x = 0$ or x irrational, and $f(x) = \dfrac{1}{n}$ when x is a nonzero rational number $\dfrac{m}{n}$, $n > 0$, and $\dfrac{m}{n}$ is in lowest terms (that is, the integers m and n have no common integral divisor greater than 1).

▮ **Case 1.** c is rational. Assume $f(x)$ is continuous at c. Since there are irrational numbers arbitrarily close to c, $f(x) = 0$ for values of x arbitrarily close to c, and, therefore, by continuity, $f(c) = 0$. By definition of f, c cannot be a nonzero rational. So, $c = 0$. Now, $f(x)$ is in fact continuous at $x = 0$, since, as rational numbers m/n approach 0, their denominators approach $+\infty$, and, therefore, $f(m/n) = 1/n$ approaches 0, which is $f(0)$. **Case 2.** c is irrational. Then $f(c) = 0$. But, as rational numbers m/n approach c, their denominators n approach $+\infty$, and, therefore, the values $f(m/n) = 1/n$ approach $0 = f(c)$. Thus, any irrational number is a point of continuity, and the points of discontinuity are the nonzero rational numbers.

7.18 Define: (a) $f(x)$ is continuous on the left at $x = a$. (b) $f(x)$ is continuous on the right at $x = a$.

▮ (a) $f(a)$ is defined, $\lim_{x \to a^-} f(x)$ exists, and $\lim_{x \to a^-} f(x) = f(a)$. (b) $f(a)$ is defined, $\lim_{x \to a^+} f(x)$ exists, and $\lim_{x \to a^+} f(x) = f(a)$.

7.19 Consider the function $f(x)$ graphed in Fig. 7-6. At all points of discontinuity, determine whether $f(x)$ is continuous on the left and whether $f(x)$ is continuous on the right.

▌ At $x = 0$, $f(x)$ is not continuous on the left, since $\lim_{x \to 0^-} f(x) = 3 \neq 1 = f(0)$. At $x = 0$, $f(x)$ is continuous on the right, since $\lim_{x \to 0^+} f(x) = 1 = f(0)$. At $x = 2$, $f(x)$ is continuous neither on the left nor on the right, since $\lim_{x \to 2^-} f(x) = 2$, $\lim_{x \to 2^+} f(x) = 0$, but $f(2) = 3$. At $x = 3$, $f(x)$ is continuous on the left, since $\lim_{x \to 3^-} f(x) = 2 = f(3)$. At $x = 3$, $f(x)$ is not continuous on the right since $\lim_{x \to 3^+} f(x) = 0 \neq f(3)$.

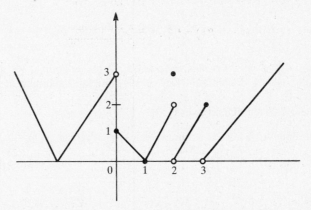

Fig. 7-6

7.20 Let $f(x)$ be a continuous function from the closed interval $[a, b]$ into itself. Show that $f(x)$ has a fixed point, that is, a point x such that $f(x) = x$.

▌ If $f(a) = a$ or $f(b) = b$, then we have a fixed point. So, we may assume that $a < f(a)$ and $f(b) < b$ (Fig. 7-7). Consider the continuous function $h(x) = f(x) - x$. Recall the Intermediate Value theorem: Any continuous function $h(x)$ on $[a, b]$ assumes somewhere in $[a, b]$ any value between $h(a)$ and $h(b)$. Now, $h(a) = f(a) - a > 0$, and $h(b) = f(b) - b < 0$. Since 0 lies between $h(a)$ and $h(b)$, there must be a point c in $[a, b]$ such that $h(c) = 0$. Hence, $f(c) - c = 0$, or $f(c) = c$.

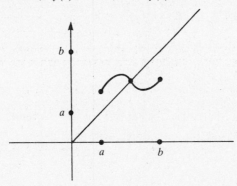

Fig. 7-7

7.21 Assume that $f(x)$ is continuous at $x = c$ and that $f(c) > 0$. Prove that there is an open interval around c on which $f(x)$ is positive.

▌ Let $\varepsilon = f(c)$. Since $\lim_{x \to c} f(x) = f(c)$, there exists $\delta > 0$ such that, if $|x - c| < \delta$, then $|f(x) - f(c)| < \varepsilon = f(c)$. So, $-f(c) < f(x) - f(c) < f(c)$. By the left-hand inequality, $f(x) > 0$. This holds for all x in the open interval $(c - \delta, c + \delta)$.

7.22 Show that the function $f(x) = 2x^3 - 4x^2 + 5x - 4$ has a zero between $x = 1$ and $x = 2$.

▌ $f(x)$ is continuous, $f(1) = -1$, and $f(2) = 6$. Since $f(1) < 0 < f(2)$, the Intermediate Value theorem implies that there must be a number c in the interval $(1, 2)$ such that $f(c) = 0$.

7.23 Verify the Intermediate Value theorem in the case of the function $f(x) = \sqrt{16 - x^2}$, the interval $[-4, 0]$, and the intermediate value $\sqrt{7}$.

▌ $f(-4) = 0$, $f(0) = 4$, f is continuous on $[-4, 0]$, and $0 < \sqrt{7} < 4$. We must find a number c in $[-4, 0]$ such that $f(c) = \sqrt{7}$. So, $\sqrt{16 - c^2} = \sqrt{7}$, $16 - c^2 = 7$, $c^2 = 9$, $c = \pm 3$. Hence, the desired value of c is -3.

7.24 Is $f(x) = [x]$ continuous over the interval $[1, 2]$?

▐ No. $f(2) = 2$, but $\lim_{x \to 2^-} f(x) = \lim_{x \to 2^-} 1 = 1$.

7.25 Is the function f such that $f(x) = \dfrac{1}{x}$ for $x > 0$ and $f(0) = 0$ continuous over $[0, 1]$?

▐ No. $f(0) = 0$, but $\lim_{x \to 0^+} f(x) = \lim_{x \to 0^+} \dfrac{1}{x} = +\infty$.

7.26 Consider the function f such that $f(x) = 2x$ if $0 \le x \le 1$ and $f(x) = x - 1$ if $x > 1$. Is f continuous over $[0, 1]$?

▐ Yes. When continuity *over an interval* is considered, at the endpoints we are concerned with only the one-sided limit. So, although f is discontinuous at $x = 1$, the left-hand limit at 1 is 2 and $f(1) = 2$.

7.27 Is the function of Problem 7.26 continuous over $[1, 2]$?

▐ No. The right-hand limit at $x = 1$ is $\lim_{x \to 1^+} (x - 1) = 0$, whereas $f(1) = 2$.

7.28 Let

$$f(x) = \begin{cases} 3x^2 - 1 & \text{if } x < 0 \\ cx + d & \text{if } 0 \le x \le 1 \\ \sqrt{x + 8} & \text{if } x > 1 \end{cases}$$

Determine c and d so that f is continuous everywhere (as indicated in Fig. 7-8).

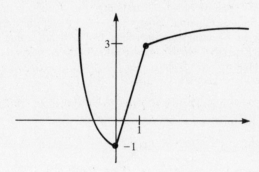

Fig. 7-8

▐ Since $\lim_{x \to 0^-} 3x^2 - 1 = -1$, the value of $cx + d$ at $x = 0$ must be -1, that is, $d = -1$. Since $\lim_{x \to 1^+} \sqrt{x + 8} = 3$, the value of $cx + d$ at $x = 1$ must be 3, that is, $3 = c(1) - 1$, $c = 4$.

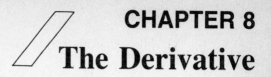
8.1 Using the Δ-definition, find the derivative $f'(x)$ of the function $f(x) = 2x - 7$.

▎$f(x + \Delta x) = 2(x + \Delta x) - 7 = 2x + 2\,\Delta x - 7$. So, $f(x + \Delta x) - f(x) = (2x + 2\,\Delta x - 7) - (2x - 7) = 2\,\Delta x$. Hence, $\dfrac{f(x + \Delta x) - f(x)}{\Delta x} = 2$. Thus, $f'(x) = \lim\limits_{\Delta x \to 0} \dfrac{f(x + \Delta x) - f(x)}{\Delta x} = \lim\limits_{\Delta x \to 0} 2 = 2$. **Answer**

8.2 Using the Δ-definition, show that the derivative of any linear function $f(x) = Ax + B$ is $f'(x) = A$.

▎$f(x + \Delta x) = A(x + \Delta x) + B = Ax + A\,\Delta x + B$. Then, $f(x + \Delta x) - f(x) = (Ax + A\,\Delta x + B) - (Ax + B) = A\,\Delta x$. Hence, $\dfrac{f(x + \Delta x) - f(x)}{\Delta x} = A$. Thus, $f'(x) = \lim\limits_{\Delta x \to 0} \dfrac{f(x + \Delta x) - f(x)}{\Delta x} = \lim\limits_{\Delta x \to 0} A = A$.

8.3 Using the Δ-definition, find the derivative $f'(x)$ of the function $f(x) = 2x^2 - 3x + 5$.

▎$f(x + \Delta x) = 2(x + \Delta x)^2 - 3(x + \Delta x) + 5 = 2[x^2 + 2x\,\Delta x + (\Delta x)^2] - 3x - 3\,\Delta x + 5 = 2x^2 + 4x\,\Delta x + 2(\Delta x)^2 - 3x - 3\,\Delta x + 5$. So, $f(x + \Delta x) - f(x) = [2x^2 + 4x\,\Delta x + 2(\Delta x)^2 - 3x - 3\,\Delta x + 5] - (2x^2 - 3x + 5) = 4x\,\Delta x + 2(\Delta x)^2 - 3\,\Delta x = (\Delta x)(4x + 2\,\Delta x - 3)$. Thus, $\dfrac{f(x + \Delta x) - f(x)}{\Delta x} = 4x + 2\,\Delta x - 3$. Hence, $f'(x) = \lim\limits_{\Delta x \to 0} \dfrac{f(x + \Delta x) - f(x)}{\Delta x} = \lim\limits_{\Delta x \to 0} (4x + 2\,\Delta x - 3) = 4x - 3$.

8.4 Using the Δ-definition, find the derivative $f'(x)$ of the function $f(x) = x^3$.

▎$f(x + \Delta x) = (x + \Delta x)^3 = x^3 + 3x^2\,\Delta x + 3x(\Delta x)^2 + (\Delta x)^3$. So, $f(x + \Delta x) - f(x) = [x^3 + 3x^2\,\Delta x + 3x(\Delta x)^2 + (\Delta x)^3] - x^3 = 3x^2\,\Delta x + 3x(\Delta x)^2 + (\Delta x)^3 = (\Delta x)(3x^2 + 3x\,\Delta x + (\Delta x)^2)$. So, $\dfrac{f(x + \Delta x) - f(x)}{\Delta x} = 3x^2 + 3x\,\Delta x + (\Delta x)^2$. Thus, $f'(x) = \lim\limits_{\Delta x \to 0} \dfrac{f(x + \Delta x) - f(x)}{\Delta x} = \lim\limits_{\Delta x \to 0} [3x^2 + 3x\,\Delta x + (\Delta x)^2] = 3x^2$.

8.5 State the formula for the derivative of an arbitrary polynomial function $f(x) = a_n x^n + a_{n-1} x^{n-1} + \cdots + a_2 x^2 + a_1 x + a_0$.

▎$f'(x) = na_n x^{n-1} + (n-1)a_{n-1} x^{n-2} + \cdots + 2a_2 x + a_1$.

8.6 Write the derivative of the function $f(x) = 7x^5 - 3x^4 + 6x^2 + 3x + 4$.

▎$f'(x) = 35x^4 - 12x^3 + 12x + 3$.

8.7 Given functions $f(x)$ and $g(x)$, state the formulas for the derivatives of the sum $f(x) + g(x)$, the product $f(x) \cdot g(x)$, and the quotient $f(x)/g(x)$.

▎
$$D_x[f(x) + g(x)] = D_x f(x) + D_x g(x)$$
$$D_x[f(x) \cdot g(x)] = f(x) \cdot D_x g(x) + g(x) \cdot D_x f(x)$$
$$D_x\left[\frac{f(x)}{g(x)}\right] = \frac{g(x) \cdot D_x f(x) - f(x) \cdot D_x g(x)}{[g(x)]^2}$$

[Notice the various ways of denoting a derivative: $f'(x)$, $D_x f(x)$, $\dfrac{dy}{dx}$, y'.]

8.8 Using the product rule, find the derivative of $f(x) = (5x^3 - 20x + 13)(4x^6 + 2x^5 - 7x^2 + 2x)$.

▎$F'(x) = (5x^3 - 20x + 13)(24x^5 + 10x^4 - 14x + 2) + (4x^6 + 2x^5 - 7x^2 + 2x)(15x^2 - 20)$. [In such cases, do not bother to carry out the tedious multiplications, unless a particular problem requires it.]

8.9 Using the formula from Problem 8-7, find the derivative of $G(x) = \dfrac{3x - 2}{x^2 + 7}$.

▎
$$G'(x) = \frac{(x^2 + 7)(3) - (3x - 2)(2x)}{(x^2 + 7)^2} = \frac{3x^2 + 21 - 6x^2 + 4x}{(x^2 + 7)^2} = \frac{-3x^2 + 4x + 21}{(x^2 + 7)^2}$$

8.10 Using the formula from Problem 8.7, find the derivative of $H(x) = \dfrac{3x - 2}{2x + 5}$.

▌ $$H'(x) = \frac{(2x + 5)(3) - (3x - 2)(2)}{(2x + 5)^2} = \frac{6x + 15 - 6x + 4}{(2x + 5)^2} = \frac{19}{(2x + 5)^2}$$

8.11 Using the Δ-definition, find the derivative of $f(x) = \dfrac{1}{x}$.

▌ $f(x + \Delta x) = \dfrac{1}{x + \Delta x}$. Hence

$$f(x + \Delta x) - f(x) = \frac{1}{x + \Delta x} - \frac{1}{x} = \frac{x - (x + \Delta x)}{x(x + \Delta x)} = \frac{-\Delta x}{x(x + \Delta x)}$$

So,

$$\frac{f(x + \Delta x) - f(x)}{\Delta x} = -\frac{1}{x(x + \Delta x)} \quad \text{and} \quad f'(x) = \lim_{\Delta x \to 0} \frac{f(x + \Delta x) - f(x)}{\Delta x} = \lim_{\Delta x \to 0} -\frac{1}{x(x + \Delta x)} = -\frac{1}{x^2}$$

8.12 Using formulas, find the derivatives of the following functions: (a) $-8x^5 + \sqrt{3}\, x^3 + 2\pi x^2 - 12$, (b) $2x^{51} + 3x^{12} - 14x^2 + \sqrt[3]{7}\, x + \sqrt{5}$.

▌ (a) $-40x^4 + 3\sqrt{3}\, x^2 + 4\pi x$. **Answer**
(b) $102x^{50} + 36x^{11} - 28x + \sqrt[3]{7}$. **Answer**

8.13 Find the slope-intercept equation of the tangent line to the graph of the function $f(x) = 4x^3 - 7x^2$ at the point corresponding to $x = 3$.

▌ When $x = 3$, $f(x) = 45$. So, the point is $(3, 45)$. Recall that the slope of the tangent line is the derivative $f'(x)$, evaluated for the given value of x. But, $f'(x) = 12x^2 - 14x$. Hence, $f'(3) = 12(9) - 14(3) = 66$. Thus, the slope-intercept equation of the tangent line has the form $y = 66x + b$. Since the point $(3, 45)$ is on the tangent line, $45 = 66(3) + b$, and, therefore, $b = -153$. Thus, the equation is $y = 66x - 153$. **Answer**

8.14 At what point(s) of the graph of $y = x^5 + 4x - 3$ does the tangent line to the graph also pass through the point $B(0, 1)$?

▌ The derivative is $y' = 5x^4 + 4$. Hence, the slope of the tangent line at a point $A(x_0, y_0)$ of the graph is $5x_0^4 + 4$. The line $\overleftrightarrow{AB}$ has slope $\dfrac{(x_0^5 + 4x_0 - 3) - 1}{x_0 - 0} = \dfrac{x_0^5 + 4x_0 - 4}{x_0}$. So, the line $\overleftrightarrow{AB}$ is the tangent line if and only if $(x_0^5 + 4x_0 - 4)/x_0 = 5x_0^4 + 4$. Solving, $x_0 = -1$. So, there is only one point $(-1, -8)$.

8.15 Specify all lines through the point $(1, 5)$ and tangent to the curve $y = 3x^3 + x + 4$.

▌ $y' = 9x^2 + 1$. Hence, the slope of the tangent line at a point (x_0, y_0) of the curve is $9x_0^2 + 1$. The slope of the line through (x_0, y_0) and $(1, 5)$ is $\dfrac{y_0 - 5}{x_0 - 1} = \dfrac{3x_0^3 + x_0 + 4 - 5}{x_0 - 1} = \dfrac{3x_0^3 + x_0 - 1}{x_0 - 1}$. So, the tangent line passes through $(1, 5)$ if and only if $\dfrac{3x_0^3 + x_0 - 1}{x_0 - 1} = 9x_0^2 + 1$, $3x_0^3 + x_0 - 1 = (x_0 - 1)(9x_0^2 + 1)$, $3x_0^3 + x_0 - 1 = 9x_0^3 - 9x_0^2 + x_0 - 1$, $9x_0^2 = 6x_0^3$, $6x_0^3 - 9x_0^2 = 0$, $3x_0^2(2x_0 - 3) = 0$. Hence, $x_0 = 0$ or $x_0 = \frac{3}{2}$, and the points on the curve are $(0, 4)$ and $(\frac{3}{2}, \frac{125}{8})$. The slopes at these points are, respectively, 1 and $\frac{85}{4}$. So, the tangent lines are $y - 4 = x$ and $y - \frac{125}{8} = \frac{85}{4}(x - \frac{3}{2})$, or, equivalently, $y = x + 4$ and $y = \frac{85}{4}x - \frac{65}{4}$.

8.16 Find the slope-intercept equation of the normal line to the graph of $y = x^3 - x^2$ at the point where $x = 1$.

▌ The normal line is the line perpendicular to the tangent line. Since $y' = 3x^2 - 2x$, the slope of the tangent line at $x = 1$ is $3(1)^2 - 2(1) = 1$. Hence, the slope of the normal line is the negative reciprocal of 1, namely -1. Thus, the required slope-intercept equation has the form $y = -x + b$. On the curve, when $x = 1$, $y = (1)^3 - (1)^2 = 0$. So, the point $(1, 0)$ is on the normal line, and, therefore, $0 = -1 + b$. Thus, $b = 1$, and the required equation is $y = -x + 1$.

8.17 Evaluate $\displaystyle\lim_{h \to 0} \frac{5(\frac{1}{3} + h)^4 - 5(\frac{1}{3})^4}{h}$.

❚ Recall the definition of the derivative. When $f(x) = 5x^4$, $f'(x) = \lim\limits_{\Delta x \to 0} \dfrac{5(x + \Delta x)^4 - 5x^4}{\Delta x}$. In particular, for $x = \frac{1}{3}$, $f'(\frac{1}{3}) = \lim\limits_{\Delta x \to 0} \dfrac{5(\frac{1}{3} + \Delta x)^4 - 5(\frac{1}{3})^4}{\Delta x}$. If we replace Δx by h in this limit, we obtain the limit to be evaluated, which is, therefore, equal to $f'(\frac{1}{3})$. But, $f'(x) = 20x^3$. So, the value of the limit is $20(\frac{1}{3})^3 = \frac{20}{27}$. **Answer**

8.18 If the line $4x - 9y = 0$ is tangent in the first quadrant to the graph of $y = \frac{1}{3}x^3 + c$, what is the value of c?

❚ $y' = x^2$. If we rewrite the equation $4x - 9y = 0$ as $y = \frac{4}{9}x$, we see that the slope of the line is $\frac{4}{9}$. Hence, the slope of the tangent line is $\frac{4}{9}$, which must equal the derivative x^2. So, $x = \pm\frac{2}{3}$. Since the point of tangency is in the first quadrant, $x = \frac{2}{3}$. The corresponding point on the line has y-coordinate $y = \frac{4}{9}x = \frac{4}{9}(\frac{2}{3}) = \frac{8}{27}$. Since this point of tangency is also on the curve $y = \frac{1}{3}x^3 + c$, we have $\frac{8}{27} = \frac{1}{3}(\frac{2}{3})^3 + c$. So, $c = \frac{16}{81}$.

8.19 For what nonnegative value(s) of b is the line $y = -\frac{1}{12}x + b$ normal to the graph of $y = x^3 + \frac{1}{3}$?

❚ $y' = 3x^2$. Since the slope of $y = -\frac{1}{12}x + b$ is $-\frac{1}{12}$, the slope of the tangent line at the point of intersection with the curve is the negative reciprocal of $-\frac{1}{12}$, namely 12. This slope is equal to the derivative $3x^2$. Hence, $x^2 = 4$, and $x = \pm 2$. The y-coordinate at the point of intersection is $y = x^3 + \frac{1}{3} = (\pm 2)^3 + \frac{1}{3} = \frac{25}{3}$ or $-\frac{23}{3}$. So, the possible points are $(2, \frac{25}{3})$ and $(-2, -\frac{23}{3})$. Substituting in $y = -\frac{1}{12}x + b$, we obtain $b = \frac{17}{2}$ and $b = -\frac{47}{6}$. Thus, $b = \frac{17}{2}$ is the only nonnegative value.

8.20 A certain point (x_0, y_0) is on the graph of $y = x^3 + x^2 - 9x - 9$, and the tangent line to the graph at (x_0, y_0) passes through the point $(4, -1)$. Find (x_0, y_0).

❚ $y' = 3x^2 + 2x - 9$ is the slope of the tangent line. This slope is also equal to $(y + 1)/(x - 4)$. Hence, $y + 1 = (3x^2 + 2x - 9)(x - 4)$. Multiplying out and simplifying, $y = 3x^3 - 10x^2 - 17x + 35$. But the equation $y = x^3 + x^2 - 9x - 9$ is also satisfied at the point of tangency. Hence, $3x^3 - 10x^2 - 17x + 35 = x^3 + x^2 - 9x - 9$. Simplifying, $2x^3 - 11x^2 - 8x + 44 = 0$. In searching for roots of this equation, we first try integral factors of 44. It turns out that $x = 2$ is a root. So, $x - 2$ is a factor of $2x^3 - 11x^2 - 8x + 44$. Dividing $2x^3 - 11x^2 - 8x + 44$ by $x - 2$, we obtain $2x^2 - 7x - 22$, which factors into $(2x - 11)(x + 2)$. Hence, the solutions are $x = 2$, $x = -2$, and $x = \frac{11}{2}$. The corresponding points are $(2, -15)$, $(-2, 5)$, and $(\frac{11}{2}, \frac{1105}{8})$. **Answer**

8.21 Let f be differentiable (that is, f' exists). Define a function f^* by the equation $f^*(x) = \lim\limits_{\Delta x \to 0} \dfrac{f(x + \Delta x) - f(x - \Delta x)}{\Delta x}$. Find the relationship between f^* and f'.

❚
$$f^*(x) = \lim_{\Delta x \to 0} \frac{f(x + \Delta x) - f(x) + f(x) - f(x - \Delta x)}{\Delta x} = \lim_{\Delta x \to 0} \frac{f(x + \Delta x)}{\Delta x} + \lim_{\Delta x \to 0} \frac{f(x) - f(x - \Delta x)}{\Delta x}$$
$$= f'(x) + \lim_{\Delta x \to 0} \frac{f(x) - f(x - \Delta x)}{\Delta x}$$

But $\lim\limits_{\Delta x \to 0} \dfrac{f(x) - f(x - \Delta x)}{\Delta x} = -\lim\limits_{\Delta x \to 0} \dfrac{f(x - \Delta x) - f(x)}{\Delta x} = -\lim\limits_{h \to 0} \dfrac{f(x + h) - f(x)}{-h}$ [where $h = -\Delta x$]

$$= \lim_{h \to 0} \frac{f(x + h) - f(x)}{h} = f'(x)$$

Thus, $f^*(x) = 2f'(x)$.

8.22 A function f, defined for all real numbers, is such that (i) $f(1) = 2$, (ii) $f(2) = 8$, and (iii) $f(u + v) - f(u) = kuv - 2v^2$ for all u and v, where k is some constant. Find $f'(x)$ for arbitrary x.

❚ Substituting $u = 1$ and $v = 1$ in (iii) and using (i) and (ii), we find that $k = 8$. Now, in (iii), let $u = x$ and $v = \Delta x$. Then $f(x + \Delta x) - f(x) = 8x \Delta x - 2(\Delta x)^2$. So, $[f(x + \Delta x) - f(x)]/\Delta x = 8x - 2 \Delta x$. Thus, $f'(x) = \lim\limits_{\Delta x \to 0} \dfrac{f(x + \Delta x) - f(x)}{\Delta x} = \lim\limits_{\Delta x \to 0} (8x - 2 \Delta x) = 8x$.

8.23 Find the points on the curve $y = \frac{1}{3}x^3 - x$ where the tangent line is parallel to the line $y = 3x$.

❚ $y' = x^2 - 1$ is the slope of the tangent line. To be parallel to the line $y = 3x$ having slope 3, it also must have slope 3. Hence, $x^2 - 1 = 3$, $x^2 = 4$, $x = \pm 2$. Thus, the points are $(2, \frac{2}{3})$ and $(-2, -\frac{2}{3})$.

8.24 Using the Δ-method, find the derivative of $f(x) = \sqrt{3x+2}$.

▮
$$f(x + \Delta x) - f(x) = \sqrt{3(x + \Delta x) + 2} - \sqrt{3x + 2} = \sqrt{3x + 3\Delta x + 2} - \sqrt{3x + 2}$$

$$= \frac{(\sqrt{3x + 3\,\Delta x + 2} - \sqrt{3x + 2})(\sqrt{3x + 3\,\Delta x + 2} + \sqrt{3x + 2})}{\sqrt{3x + 3\,\Delta x + 2} + \sqrt{3x + 2}}$$

$$= \frac{(3x + 3\,\Delta x + 2) - (3x + 2)}{\sqrt{3x + 3\,\Delta x + 2} + \sqrt{3x + 2}} = \frac{3\,\Delta x}{\sqrt{3x + 3\,\Delta x + 2} + \sqrt{3x + 2}}$$

So,

$$f'(x) = \lim_{\Delta x \to 0} \frac{f(x + \Delta x) - f(x)}{\Delta x} = \lim_{\Delta x \to 0} \frac{3}{\sqrt{3x + 3\,\Delta x + 2} + \sqrt{3x + 2}} = \frac{3}{2\sqrt{3x + 2}}$$

8.25 Show that a differentiable function $f(x)$ is continuous.

▮ $f(x + \Delta x) - f(x) = \{[f(x + \Delta x) - f(x)]/\Delta x\}\,\Delta x$. Thus,

$$\lim_{\Delta x \to 0} [f(x + \Delta x) - f(x)] = \lim_{\Delta x \to 0} \frac{f(x + \Delta x) - f(x)}{\Delta x} \cdot \lim_{\Delta x \to 0} \Delta x = f'(x) \cdot 0 = 0$$

Hence, $\quad 0 = \lim\limits_{\Delta x \to 0} [f(x + \Delta x) - f(x)] = \lim\limits_{\Delta x \to 0} f(x + \Delta x) - \lim\limits_{\Delta x \to 0} f(x) = \lim\limits_{\Delta x \to 0} f(x + \Delta x) - f(x)$

So, $\lim\limits_{\Delta x \to 0} f(x + \Delta x) = f(x)$, f is continuous at x.

8.26 Show that the converse of Problem 8.25 is false.

▮ Consider the function $f(x) = |x|$ at $x = 0$. Clearly, f is continuous everywhere. However, $\dfrac{f(0 + \Delta x) - f(0)}{\Delta x} = \dfrac{|\Delta x| - 0}{\Delta x} = \dfrac{|\Delta x|}{\Delta x}$. When $\Delta x > 0$, $|\Delta x|/\Delta x = 1$, and, when $\Delta x < 0$, $|\Delta x|/\Delta x = -1$.

Therefore, $\lim\limits_{\Delta x \to 0} \dfrac{|\Delta x|}{\Delta x} = f'(0)$ does not exist.

8.27 Find the derivative of $f(x) = x^{1/3}$.

▮ $f(x + \Delta x) - f(x) = (x + \Delta x)^{1/3} - x^{1/3}$

$$= \frac{[(x + \Delta x)^{1/3} - x^{1/3}][(x + \Delta x)^{2/3} + x^{1/3}(x + \Delta x)^{1/3} + x^{2/3}]}{(x + \Delta x)^{2/3} + x^{1/3}(x + \Delta x)^{1/3} + x^{2/3}} = \frac{x + \Delta x - x}{(x + \Delta x)^{2/3} + x^{1/3}(x + \Delta x)^{1/3} + x^{2/3}}$$

So,
$$\frac{f(x + \Delta x) - f(x)}{\Delta x} = \frac{1}{(x + \Delta x)^{2/3} + x^{1/3}(x + \Delta x)^{1/3} + x^{2/3}}$$

$$f'(x) = \lim_{\Delta x \to 0} \frac{1}{(x + \Delta x)^{2/3} + x^{1/3}(x + \Delta x)^{1/3} + x^{2/3}} = \frac{1}{3x^{2/3}} = \tfrac{1}{3} x^{-2/3}$$

8.28 Find the point(s) at which the tangent line to the parabola $y = ax^2 + bx + c$ is horizontal. (Notice that the solution to this problem locates the "nose" of the parabola.)

▮ $y' = 2ax + b$ is the slope of the tangent line. A line is horizontal if and only if its slope is 0. Therefore, we must solve $2ax + b = 0$. The solution is $x = -b/2a$. The corresponding value of y is $(4ac - b^2)/4a$.

8.29 Let $f(x)$ be a function with the property that $f(u + v) = f(u)f(v)$ for all u and v, and such that $f(0) = f'(0) = 1$. Show that $f'(x) = f(x)$ for all x.

▮ $f(x + \Delta x) - f(x) = f(x)f(\Delta x) - f(x) = f(x)[f(\Delta x) - 1]$. So,

$$f'(x) = \lim_{\Delta x \to 0} \frac{f(x + \Delta x) - f(x)}{\Delta x} = \lim_{\Delta x \to 0} \frac{f(x)[f(\Delta x) - 1]}{\Delta x} = f(x) \cdot \lim_{\Delta x \to 0} \frac{f(\Delta x) - 1}{\Delta x} = f(x) \cdot \lim_{\Delta x \to 0} \frac{f(0 + \Delta x) - f(0)}{\Delta x}$$

$$= f(x)f'(0) = f(x) \cdot 1 = f(x)$$

8.30 Determine whether the following function is differentiable at $x = 0$:

$$f(x) = \begin{cases} x^2 & \text{if } x \text{ is rational} \\ 0 & \text{if } x \text{ is irrational} \end{cases}$$

▌ $$f(0 + \Delta x) - f(0) = f(\Delta x) - 0^2 = f(\Delta x) = \begin{cases} (\Delta x)^2 & \text{if } \Delta x \text{ is rational} \\ 0 & \text{if } \Delta x \text{ is irrational} \end{cases}$$

So, $$\frac{f(0 + \Delta x) - f(0)}{\Delta x} = \begin{cases} \Delta x & \text{if } \Delta x \text{ is rational} \\ 0 & \text{if } \Delta x \text{ is irrational} \end{cases}$$

Hence, $f'(0) = \lim\limits_{\Delta x \to 0} \dfrac{f(0 + \Delta x) - f(0)}{\Delta x}$ exists (and equals 0).

8.31 Consider the function

$$f(x) = \begin{cases} x & \text{if } x \text{ is rational} \\ 0 & \text{if } x \text{ is irrational} \end{cases}$$

Determine whether f is differentiable at $x = 0$.

▌ $$f(0 + \Delta x) - f(0) = f(\Delta x) - 0 = f(\Delta x) = \begin{cases} \Delta x & \text{if } \Delta x \text{ is rational} \\ 0 & \text{if } x \text{ is irrational} \end{cases}$$

So, $$\frac{f(0 + \Delta x) - f(0)}{\Delta x} = \begin{cases} 1 & \text{if } \Delta x \text{ is rational} \\ 0 & \text{if } \Delta x \text{ is irrational} \end{cases}$$

Since there are both rational and irrational numbers arbitrarily close to 0, $f'(0) = \lim\limits_{\Delta x \to 0} \dfrac{f(0 + \Delta x) - f(0)}{\Delta x}$ does not exist.

8.32 Find the derivative of the function $f(x) = (2x - 3)^2$.

▌ $f(x) = 4x^2 - 12x + 9$. Hence, $f'(x) = 8x - 12$. [Notice that the same method would be difficult to carry out with a function like $(2x - 3)^{20}$.]

8.33 Where does the normal line to the curve $y = x - x^2$ at the point $(1, 0)$ intersect the curve a second time?

▌ $y' = 1 - 2x$. The tangent line at $(1, 0)$ has slope $1 - 2(1) = -1$. Hence, the normal line has slope 1, and a point-slope equation for it is $y = x - 1$. Solving $y = x - x^2$ and $y = x - 1$ simultaneously, $x - x^2 = x - 1$, $x^2 = 1$, $x = \pm 1$. Hence, the other point of intersection occurs when $x = -1$. Then $y = x - 1 = -1 - 1 = -2$. So, the other point is $(-1, -2)$.

8.34 Find the point(s) on the graph of $y = x^2$ at which the tangent line is parallel to the line $y = 6x - 1$.

▌ Since the slope of $y = 6x - 1$ is 6, the slope of the tangent line must be 6. Thus, the derivative $2x = 6$, $x = 3$. Hence, the desired point is $(3, 9)$.

8.35 Find the point(s) on the graph of $y = x^3$ at which the tangent line is perpendicular to the line $3x + 9y = 4$.

▌ The equation of the line can be rewritten as $y = -\frac{1}{3}x + \frac{4}{9}$, and so its slope is $-\frac{1}{3}$. Hence, the slope of the required tangent line must be the negative reciprocal of $-\frac{1}{3}$, namely, 3. So, the derivative $3x^2 = 3$, $x^2 = 1$, $x = \pm 1$. Thus, the solutions are $(1, 1)$ and $(-1, -1)$.

8.36 Find the slope-intercept equation of the normal line to the curve $y = x^3$ at the point at which $x = \frac{1}{3}$.

▌ The slope of the tangent line is the derivative $3x^2$, which, at $x = \frac{1}{3}$, is $\frac{1}{3}$. Hence, the slope of the normal line is the negative reciprocal of $\frac{1}{3}$, namely, -3. So, the required equation has the form $y = -3x + b$. On the curve, when $x = \frac{1}{3}$, $y = x^3 = \frac{1}{27}$. Thus, the point $(\frac{1}{3}, \frac{1}{27})$ lies on the line, and $\frac{1}{27} = -3(\frac{1}{3}) + b$, $b = \frac{28}{27}$. So, the required equation is $y = -3x + \frac{28}{27}$.

8.37 At what points does the normal line to the curve $y = x^2 - 3x + 5$ at the point $(3, 5)$ intersect the curve?

▌ The derivative $2x - 3$ has, at $x = 3$, the value 3. So, the slope of the normal line is $-\frac{1}{3}$, and its equation is $y = -\frac{1}{3}x + b$. Since $(3, 5)$ lies on the line, $5 = -1 + b$, or $b = 6$. Thus, the equation of the normal line is $y = -\frac{1}{3}x + 6$. To find the intersections of this line with the curve, we set $-\frac{1}{3}x + 6 = x^2 - 3x + 5$, $3x^2 - 8x - 3 = 0$, $(3x + 1)(x - 3) = 0$, $x = -\frac{1}{3}$ or $x = 3$. We already know about the point $(3, 5)$, the other intersection point is $(-\frac{1}{3}, \frac{55}{9})$.

8.38 Find the point(s) on the graph of $y = x^2$ at which the tangent line passes through $(2, -12)$.

    ❚ The slope of the tangent line is the derivative $2x$. Since (x, x^2) and $(2, -12)$ lie on the tangent line, its slope is $(x^2 + 12)/(x - 2)$. Hence, $(x^2 + 12)/(x - 2) = 2x$, $x^2 + 12 = 2x^2 - 4x$, $x^2 - 4x - 12 = 0$, $(x - 6)(x + 2) = 0$, $x = 6$ or $x = -2$. Thus, the two points are $(6, 36)$ and $(-2, 4)$.

8.39 Use the Δ-definition to calculate the derivative of $f(x) = x^4$.

    ❚ $f(x + \Delta x) = (x + \Delta x)^4 = x^4 + 4x^3 \Delta x + 6x^2 (\Delta x)^2 + 4x(\Delta x)^3 + (\Delta x)^4$. So, $\dfrac{f(x + \Delta x) - f(x)}{\Delta x} = 4x^3 + 6x^2 \Delta x + 4x(\Delta x)^2 + (\Delta x)^3$. Hence, $f'(x) = \lim\limits_{\Delta x \to 0} \dfrac{f(x + \Delta x) - f(x)}{x} = 4x^3$.

8.40 Find a formula for the derivative $D_x[f(x) g(x) h(x)]$.

    ❚ By the product rule, $D_x\{[f(x) g(x)]h(x)\} = f(x) g(x) h'(x) + D_x[f(x) g(x)]h(x) = f(x)g(x)h'(x) + [f(x)g'(x) + f'(x)g(x)]h(x) = f(x)g(x)h'(x) + f(x)g'(x)h(x) + f'(x) g(x) h(x)$.

8.41 Find $D_x[x(2x - 1)(x + 2)]$.

    ❚ By Problem 8.40, $x(2x - 1) + x \cdot 2 \cdot (x + 2) + (2x - 1)(x + 2) = x(2x - 1) + 2x(x + 2) + (2x - 1)(x + 2)$.

8.42 Let $f(x) = 3x^3 - 11x^2 - 15x + 63$. Find all points on the graph of f where the tangent line is horizontal.

    ❚ The slope of the tangent line is the derivative $f'(x) = 9x^2 - 22x - 15$. The tangent line is horizontal when and only when its slope is 0. Hence, we set $9x^2 - 22x - 15 = 0$, $(9x + 5)(x - 3) = 0$, $x = 3$ or $x = -\frac{5}{9}$. Thus, the desired points are $(3, 0)$ and $\left(-\dfrac{5}{9}, \dfrac{16\,384}{243}\right)$.

8.43 Determine the points at which the function $f(x) = |x - 3|$ is differentiable.

    ❚ The graph (Fig. 8-1), reveals a sharp point at $x = 3$, $y = 0$, where there is no unique tangent line. Thus the function is not differentiable at $x = 3$. (This can be verified in a more rigorous way by considering the Δ-definition.)

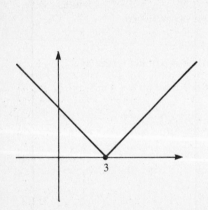

Fig. 8-1

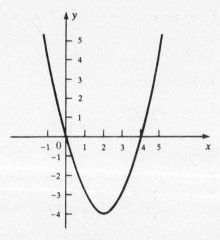

Fig. 8-2

8.44 Figure 8-2 shows the graph of the function $f(x) = x^2 - 4x$. Draw the graph of $y = |f(x)|$ and determine where y' does not exist.

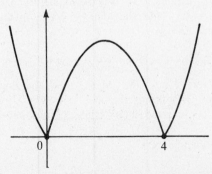

Fig. 8-3

I The graph of y (Fig. 8-3) is obtained when the part of Fig. 8-2 below the x-axis is reflected in the x-axis. We see that there is no unique tangent line (i.e., y' is not defined) at $x = 0$ and $x = 4$.

8.45 If f is differentiable and $f(x_0) = 0$, find $\lim\limits_{h \to 0} \dfrac{f(x_0 + h)}{7h}$.

I
$$\lim_{h \to 0} \frac{f(x_0 + h)}{7h} = \tfrac{1}{7} \lim_{h \to 0} \frac{f(x_0 + h) - f(x_0)}{h} = \tfrac{1}{7} f'(x_0)$$

8.46 If $f(x)$ is even and differentiable, prove that $f'(x)$ is odd.

I
$$f'(-x) = \lim_{\Delta x \to 0} \frac{f(-x + \Delta x) - f(-x)}{\Delta x}$$
$$= \lim_{\Delta x \to 0} \frac{f(x - \Delta x) - f(x)}{\Delta x}$$
$$= \lim_{h \to 0} \frac{f(x + h) - f(x)}{-h} \qquad [\text{setting } h = -\Delta x]$$
$$= -\lim_{h \to 0} \frac{f(x + h) - f(x)}{h} = -f'(x)$$

8.47 If $f(x)$ is odd and differentiable, prove that $f'(x)$ is even.

I
$$f'(-x) = \lim_{\Delta x \to 0} \frac{f(-x + \Delta x) - f(-x)}{\Delta x} = \lim_{\Delta x \to 0} \frac{-f(x - \Delta x) + f(x)}{\Delta x}$$
$$= \lim_{\Delta x \to 0} \frac{f(x - \Delta x) - f(x)}{-\Delta x} = \lim_{h \to 0} \frac{f(x + h) - f(x)}{h} \qquad [\text{setting } h = -\Delta x]$$
$$= f'(x)$$

In Problems 8.48–8.51, calculate the derivative of the given function, using the appropriate formula from Problem 8.7.

8.48 $(x^{100} + 2x^{50} - 3)(7x^8 + 20x + 5)$.

I By the product formula, the derivative is $(x^{100} + 2x^{50} - 3)(56x^7 + 20) + (100x^{99} + 100x^{49})(7x^8 + 20x + 5)$.

8.49 $\dfrac{x^2 - 3}{x + 4}$.

I By the quotient formula, the derivative is
$$\frac{(x + 4) \cdot D_x(x^2 - 3) - (x^2 - 3) \cdot D_x(x + 4)}{(x + 4)^2} = \frac{(x + 4)(2x) - (x^2 - 3)}{(x + 4)^2} = \frac{2x^2 + 8x - x^2 + 3}{(x + 4)^2} = \frac{x^2 + 8x + 3}{(x + 4)^2}$$

8.50 $\dfrac{x^5 - x + 2}{x^3 + 7}$.

I By the quotient rule, the derivative is
$$\frac{(x^3 + 7) \cdot D_x(x^5 - x + 2) - (x^5 - x + 2) \cdot D_x(x^3 + 7)}{(x^3 + 7)^2} = \frac{(x^3 + 7)(5x^4 - 1) - (x^5 - x + 2)(3x^2)}{(x^3 + 7)^2}$$
$$= \frac{5x^7 + 35x^4 - x^3 - 7 - 3x^7 + 3x^3 - 6x^2}{(x^3 + 7)^2}$$
$$= \frac{2x^7 + 35x^4 + 2x^3 - 6x^2 - 7}{(x^3 + 7)^2}$$

8.51 $\dfrac{3x^7 + x^5 - 2x^4 + x - 3}{x^4}$.

I Here, it is easier *not* to use the quotient rule. The given function is equal to $3x^3 + x - 2 + x^{-3} - 3x^{-4}$. Hence, its derivative is $9x^2 + 1 - 3x^{-4} + 12x^{-5} = \dfrac{9x^7 + x^5 - 3x + 12}{x^5}$.

CHAPTER 9
The Chain Rule

9.1 If $f(x) = x^2 + 2x - 5$ and $g(x) = x^3$, find formulas for the composite functions $f \circ g$ and $g \circ f$.

▌ $(f \circ g)(x) = f(g(x)) = f(x^3) = (x^3)^2 + 2(x^3) - 5 = x^6 + 2x^3 - 5$

$(g \circ f)(x) = g(f(x)) = g(x^2 + 2x - 5) = (x^2 + 2x - 5)^3$

9.2 Write the function $\sqrt{3x-5}$ as the composition of two functions.

▌ Let $g(x) = 3x - 5$ and let $f(x) = \sqrt{x}$. Then $(f \circ g)(x) = f(g(x)) = f(3x - 5) = \sqrt{3x - 5}$.

9.3 If $f(x) = 2x$ and $g(x) = 1/(x - 1)$, find all solutions of the equation $(f \circ g)(x) = (g \circ f)(x)$.

▌ $(f \circ g)(x) = f(g(x)) = f\left(\dfrac{1}{x-1}\right) = \dfrac{2}{x-1}$ and $(g \circ f)(x) = g(f(x)) = g(2x) = \dfrac{1}{2x-1}$

So, we must solve $\dfrac{2}{x-1} = \dfrac{1}{2x-1}$, $4x - 2 = x - 1$, $3x = 1$, $x = \tfrac{1}{3}$. **Answer**

9.4 Write the chain rule formula for the derivative of $f \circ g$

▌ $(f \circ g)'(x) = f'(g(x)) \cdot g'(x)$.

9.5 If $y = F(u)$ and $u = G(x)$, then we can write $y = F(G(x))$. Write the chain rule formula for dy/dx, where we think of y as a function of x.

▌ $\dfrac{dy}{dx} = \dfrac{dy}{du} \cdot \dfrac{du}{dx}$. Here, the first occurrence of y refers to y as a function of x, while the second occurrence of y (on the right side) refers to y as a function of u.

9.6 Find the derivative of $(x^3 - 2x^2 + 7x - 3)^4$.

▌ Use the chain rule. Think of the function as a composition $(f \circ g)(x)$, where $f(x) = x^4$ and $g(x) = x^3 - 2x^2 + 7x - 3$. Then $f'(x) = 4x^3$ and $g'(x) = 3x^2 - 4x + 7$. Hence, $\dfrac{d}{dx}(x^3 - 2x^2 + 7x - 3)^4 = 4(x^3 - 2x^2 + 7x - 3)^3 \cdot (3x^2 - 4x + 7)$.

9.7 Find the derivative of $\dfrac{1}{(3x^2 + 5)^4}$.

▌ We can write $\dfrac{1}{(3x^2 + 5)^4} = (3x^2 + 5)^{-4}$. Now we can use the chain rule. Remember that $\dfrac{d}{dx} x^r = rx^{r-1}$ for any real number r. In particular, $\dfrac{d}{dx} x^{-4} = -4x^{-5}$. By the chain rule,

$$\frac{d}{dx}(3x^2 + 5)^{-4} = -4(3x^2 + 5)^{-5} \cdot (6x) = -\frac{24x}{(3x^2 + 5)^5}$$

9.8 Find the derivative of $\sqrt{2x + 7}$.

▌ We can write $\sqrt{2x + 7} = (2x + 7)^{1/2}$. By the chain rule,

$$\frac{d}{dx}(2x + 7)^{1/2} = \frac{1}{2}(2x + 7)^{-1/2} \cdot (2) = (2x + 7)^{-1/2} = \frac{1}{(2x + 7)^{1/2}} = \frac{1}{\sqrt{2x + 7}}$$

$\left[\text{Here, we have used } \dfrac{d}{dv} v^{1/2} = \dfrac{1}{2} v^{-1/2} \text{ and } \dfrac{d}{dx}(2x + 7) = 2.\right]$

9.9 Find the derivative of $\left(\dfrac{x+2}{x-3}\right)^3$.

▮ Use the chain rule. $\dfrac{d}{dx}\left(\dfrac{x+2}{x-3}\right)^3 = 3\left(\dfrac{x+2}{x-3}\right)^2 \cdot \dfrac{d}{dx}\left(\dfrac{x+2}{x-3}\right)$. Here, we must calculate $\dfrac{d}{dx}\left(\dfrac{x+2}{x-3}\right)$ by the quotient rule: $\dfrac{d}{dx}\left(\dfrac{x+2}{x-3}\right) = \dfrac{(x-3)\cdot 1 - (x+2)\cdot 1}{(x-3)^2} = -\dfrac{5}{(x-3)^2}$. Hence, $\dfrac{d}{dx}\left(\dfrac{x+2}{x-3}\right)^3 = 3\left(\dfrac{x+2}{x-3}\right)^2 \cdot$ $\left[-\dfrac{5}{(x-3)^2}\right] = -15\dfrac{(x+2)^2}{(x-3)^3}$. **Answer**

9.10 Find the derivative of $(4x^2 - 3)^2(x+5)^3$.

▮ Think of this function as a product of $(4x^2 - 3)^2$ and $(x+5)^3$, and first apply the product rule: $\dfrac{d}{dx}[(4x^2 - 3)^2(x+5)^3] = (4x^2 - 3)^2 \cdot \dfrac{d}{dx}(x+5)^3 + (x+5)^3 \cdot \dfrac{d}{dx}(4x^2 - 3)^2$. By the chain rule, $\dfrac{d}{dx}(x+5)^3 = 3(x+5)^2 \cdot 1 = 3(x+5)^2$, and $\dfrac{d}{dx}(4x^2 - 3)^2 = 2(4x^2 - 3)\cdot(8x) = 16x(4x^2 - 3)$. Thus, $\dfrac{d}{dx}[(4x^2 - 3)^2(x+5)^3] = (4x^2 - 3)^2 \cdot 3(x+5)^2 + (x+5)^3 \cdot 16x(4x^2 - 3)$. We can factor out $(4x^2 - 3)$ and $(x+5)^2$ to obtain: $(4x^2 - 3)(x+5)^2[3(4x^2 - 3) + 16x(x+5)] = (4x^2 - 3)(x+5)^2(12x^2 - 9 + 16x^2 + 80x) = (4x^2 - 3)(x+5)^2(28x^2 + 80x - 9)$. **Answer**

9.11 Find the derivative of $\sqrt{\dfrac{4}{x}} - \sqrt{3x}$.

▮ The chain rule is unnecessary here. $\sqrt{\dfrac{4}{x}} = \dfrac{\sqrt{4}}{\sqrt{x}} = \dfrac{2}{\sqrt{x}} = 2x^{-1/2}$. So, $\dfrac{d}{dx}\sqrt{\dfrac{4}{x}} = 2(-\tfrac{1}{2}x^{-3/2}) = -x^{-3/2}$. Also, $\sqrt{3x} = \sqrt{3}\sqrt{x} = \sqrt{3}x^{1/2}$. So, $\dfrac{d}{dx}\sqrt{3x} = \sqrt{3}(\tfrac{1}{2}x^{-1/2}) = \dfrac{\sqrt{3}}{2}x^{-1/2}$. Thus, $\dfrac{d}{dx}\left(\sqrt{\dfrac{4}{x}} - \sqrt{3x}\right) = -x^{-3/2} - \dfrac{\sqrt{3}}{2}x^{-1/2} = x^{-1/2}\left(-x^{-1} - \dfrac{\sqrt{3}}{2}\right) = -\dfrac{1}{\sqrt{x}}\left(\dfrac{\sqrt{3}}{2} + \dfrac{1}{x}\right) = -\dfrac{1}{\sqrt{x}}\left(\dfrac{\sqrt{3}x + 2}{2x}\right) = -\dfrac{\sqrt{3}x + 2}{2x^{3/2}}$.

9.12 Find the derivative of $f(x) = \sqrt{4 - \sqrt{4+x}}$.

▮ $f(x) = [4 - (4+x)^{1/2}]^{1/2}$. By the chain rule, $f'(x) = \tfrac{1}{2}[4 - (4+x)^{1/2}]^{-1/2} \cdot \dfrac{d}{dx}[4 - (4+x)^{1/2}]$. But, $\dfrac{d}{dx}[4 - (4+x)^{1/2}] = -\tfrac{1}{2}(4+x)^{-1/2} \cdot \dfrac{d}{dx}(4+x) = -\tfrac{1}{2}(4+x)^{-1/2}$. Hence,

$$f'(x) = \tfrac{1}{2}[4 - (4+x)^{1/2}]^{-1/2} \cdot [-\tfrac{1}{2}(4+x)^{-1/2}] = \dfrac{-\tfrac{1}{4}}{\sqrt{4 - (\sqrt{4+x})}\sqrt{4+x}}$$

9.13 Find the slope-intercept equation of the tangent line to the graph of $y = \dfrac{\sqrt{x-1}}{x^2 + 1}$ at the point $(2, \tfrac{1}{5})$.

▮ By the quotient rule, $\qquad y' = \dfrac{(x^2 + 1)\dfrac{d}{dx}\sqrt{x-1} - \sqrt{x-1}(2x)}{(x^2 + 1)^2}$

By the chain rule, $\qquad \dfrac{d}{dx}\sqrt{x-1} = \dfrac{d}{dx}(x-1)^{1/2} = \tfrac{1}{2}(x-1)^{-1/2} = \dfrac{1}{2\sqrt{x-1}}$.

Thus, $\qquad y' = \dfrac{(x^2 + 1)\dfrac{1}{2\sqrt{x-1}} - 2x\sqrt{x-1}}{(x^2 + 1)^2}$

When $x = 2$, $\sqrt{x-1} = 1$, and, therefore, at the point $(2, \tfrac{1}{5})$, $y' = -\tfrac{3}{50}$. Hence, a point-slope equation of the tangent line is $y - \tfrac{1}{5} = -\tfrac{3}{50}(x - 2)$. Solving for y, we obtain the slope-intercept equation $y = -\tfrac{3}{50}x + \tfrac{8}{25}$.

9.14 Find the slope-intercept equation of the normal line to the curve $y = \sqrt{x^2 + 16}$ at the point $(3, 5)$.

▮ $y = (x^2 + 16)^{1/2}$. Hence, by the chain rule,

$$y' = \tfrac{1}{2}(x^2 + 16)^{-1/2} \cdot \dfrac{d}{dx}(x^2 + 16) = \tfrac{1}{2}(x^2 + 16)^{-1/2} \cdot (2x) = x(x^2 + 16)^{-1/2} = \dfrac{x}{\sqrt{x^2 + 16}}$$

At the point $(3, 5)$, $\sqrt{x^2 + 16} = 5$, and, therefore, $y' = \tfrac{3}{5}$. This is the slope of the tangent line. Hence, the slope of the normal line is $-\tfrac{5}{3}$, and a point-slope equation for it is $y - 5 = -\tfrac{5}{3}(x - 3)$. Solving for y, we obtain the slope-intercept equation $y = -\tfrac{5}{3}x + 10$.

9.15 If $y = x^3 - 2$ and $x = 3z^2 + 5$, then y can be considered a function of z. Express $\dfrac{dy}{dz}$ in terms of z.

▮ $\qquad\qquad \dfrac{dy}{dz} = \dfrac{dy}{dx} \cdot \dfrac{dx}{dz} = 3x^2 \cdot (6z) = 18zx^2 = 18z(3z^2 + 5)^2$

9.16 If $g(x) = x^{1/5}(x-1)^{3/5}$, find the domain of $g'(x)$.

▮ By the product and chain rules,

$$g'(x) = x^{1/5} \cdot \tfrac{3}{5}(x-1)^{-2/5} + (x-1)^{3/5} \cdot \tfrac{1}{5}x^{-4/5} = \frac{3x^{1/5}}{5(x-1)^{2/5}} + \frac{1}{5}\frac{(x-1)^{3/5}}{x^{4/5}}$$

Since a fraction is not defined when its denominator is 0, the domain of $g'(x)$ consists of all real numbers except 0 and 1.

9.17 Rework Problem 8.47 by means of the chain rule.

▮ $\dfrac{d}{dx} f(-x) = f'(-x) \cdot \dfrac{d}{dx}(-x) = f'(-x) \cdot (-1) = -f'(-x)$. But, since f is odd, $f(x) = -f(-x)$, and, therefore, $f'(x) = -\dfrac{d}{dx} f(-x) = -[-f'(-x)] = f'(-x)$.

9.18 Let F and G be differentiable functions such that $F(3) = 5$, $G(3) = 7$, $F'(3) = 13$, $G'(3) = 6$, $F'(7) = 2$, $G'(7) = 0$. If $H(x) = F(G(x))$, find $H'(3)$.

▮ By the chain rule, $H'(x) = F'(G(x)) \cdot G'(x)$. Hence, $H'(3) = F'(G(3)) \cdot G'(3) = F'(7) \cdot 6 = 2 \cdot 6 = 12$.

9.19 Let $F(x) = \sqrt{1+3x}$. Find the coordinates of the point(s) on the graph of F where the normal line is parallel to the line $4x + 3y = 1$.

▮ $\sqrt{1+3x} = (1+3x)^{1/2}$. Hence, by the chain rule, $F'(x) = \tfrac{1}{2}(1+3x)^{-1/2} \cdot (3) = 3/(2\sqrt{1+3x})$. This is the slope of the tangent line; hence, the slope of the normal line is $-\tfrac{2}{3}\sqrt{1+3x}$. The line $4x + 3y = 1$ has slope $-\tfrac{4}{3}$, and, therefore, the parallel normal line must also have slope $-\tfrac{4}{3}$. Thus, $-\tfrac{2}{3}\sqrt{1+3x} = -\tfrac{4}{3}$, $\sqrt{1+3x} = 2$, $1+3x = 4$, $x = 1$. So, $y = \sqrt{1+3x} = 2$. Thus, the point is $(1, 2)$. **Answer**

9.20 Find the derivative of $F(x) = \sqrt[3]{(1+x^2)^4}$.

▮ $F(x) = (1+x^2)^{4/3}$. By the chain rule,

$$F'(x) = \tfrac{4}{3}(1+x^2)^{1/3} \cdot \frac{d}{dx}(1+x^2) = \tfrac{4}{3}(1+x^2)^{1/3} \cdot (2x) = \tfrac{8}{3}x\sqrt[3]{1+x^2}$$

9.21 Given $y = \dfrac{u^2-1}{u^2+1}$ and $u = \sqrt[3]{x^2+2}$, find $\dfrac{dy}{dx}$.

▮ Using the quotient rule, we find that $\dfrac{dy}{du} = \dfrac{4u}{(u^2+1)^2}$. By the chain rule $\dfrac{du}{dx} = \dfrac{d}{dx}(x^2+2)^{1/3} = \tfrac{1}{3}(x^2+2)^{-2/3} \cdot (2x) = \tfrac{2}{3}x \dfrac{1}{(\sqrt[3]{x^2+2})^2} = \dfrac{2x}{3u^2}$. Again by the chain rule,

$$\frac{dy}{dx} = \frac{dy}{du} \cdot \frac{du}{dx} = \frac{4u}{(u^2+1)^2} \cdot \frac{2x}{3u^2} = \frac{8x}{3u(u^2+1)^2} \quad \textbf{Answer}$$

9.22 A point moves along the curve $y = x^3 - 3x + 5$ so that $x = \tfrac{1}{2}\sqrt{t} + 3$, where t is time. At what rate is y changing when $t = 4$?

▮ We are asked to find the value of dy/dt when $t = 4$. $dy/dx = 3x^2 - 3 = 3(x^2 - 1)$, and $dx/dt = 1/(4\sqrt{t})$. Hence, $\dfrac{dy}{dt} = \dfrac{dy}{dx} \cdot \dfrac{dx}{dt} = \dfrac{3(x^2-1)}{4\sqrt{t}}$. When $t = 4$, $x = 4$ and $\dfrac{dy}{dt} = 3(16-1)/(4 \cdot 2) = \tfrac{45}{8}$ units per unit of time. **Answer**

9.23 A particle moves in the plane according to the law $x = t^2 + 2t$, $y = 2t^3 - 6t$. Find the slope of the tangent line when $t = 0$.

▮ The slope of the tangent line is dy/dx. Since the first equation may be solved for t and this result substituted for t in the second equation, y is a function of x. $dy/dt = 6t^2 - 6$, $dx/dt = 2t + 2$, $dt/dx = 1/(2t+2)$ (see Problem 9.49). So by the chain rule, $\dfrac{dy}{dx} = \dfrac{dy}{dt} \cdot \dfrac{dt}{dx} = 6(t^2 - 1) \cdot \dfrac{1}{2(t+1)} = 3(t-1)$. When $t = 0$, $dy/dx = -3$. **Answer**

In Problems 9.24–9.28, find formulas for $(f \circ g)(x)$ and $(g \circ f)(x)$.

9.24 $f(x) = \dfrac{2}{x+1}$, $g(x) = 3x$.

$$(f \circ g)(x) = f(g(x)) = f(3x) = \frac{2}{3x+1}$$

$$(g \circ f)(x) = g(f(x)) = g\left(\frac{2}{x+1}\right) = 3\left(\frac{2}{x+1}\right) = \frac{6}{x+1}$$

9.25 $f(x) = 2x^3 - x^2 + 4$, $g(x) = 3$.

$$(f \circ g)(x) = f(g(x)) = f(3) = 49$$

$$(g \circ f)(x) = g(f(x)) = 3$$

9.26 $f(x) = x^3$, $g(x) = x^2$.

$$(f \circ g)(x) = f(g(x)) = f(x^2) = (x^2)^3 = x^6$$

$$(g \circ f)(x) = g(f(x)) = g(x^3) = (x^3)^2 = x^6$$

9.27 $f(x) = g(x) = \dfrac{1}{x}$.

$$(f \circ g)(x) = f\left(\frac{1}{x}\right) = \frac{1}{1/x} = x$$

9.28 $f(x) = x$, $g(x) = x^2 - 4$.

$$(f \circ g)(x) = f(g(x)) = f(x^2 - 4) = x^2 - 4$$

$$(g \circ f)(x) = g(f(x)) = g(x) = x^2 - 4$$

In Problems 9.29 and 9.30, find the set of solutions of $(f \circ g)(x) = (g \circ f)(x)$.

9.29 $f(x) = \dfrac{2}{x+1}$, $g(x) = 3x$.

$\blacksquare$ By Problem 9.24, we must solve $\dfrac{2}{3x+1} = \dfrac{6}{x+1}$, $2x + 2 = 18x + 6$, $-4 = 16x$, $x = -\frac{1}{4}$.

9.30 $f(x) = x^2$, $g(x) = \dfrac{1}{x^2 - 3}$.

$$(f \circ g)(x) = f(g(x)) = f\left(\frac{1}{x^2 - 3}\right) = \frac{1}{(x^2 - 3)^2}$$

$$(g \circ f)(x) = g(f(x)) = g(x^2) = \frac{1}{(x^2)^2 - 3} = \frac{1}{x^4 - 3}$$

So, we must solve $\dfrac{1}{(x^2 - 3)^2} = \dfrac{1}{x^4 - 3}$, $x^4 - 3 = x^4 - 6x^2 + 9$, $6x^2 = 12$, $x^2 = 2$, $x = \pm\sqrt{2}$.

In Problems 9.31–9.34, express the given function as a composition of two simpler functions.

9.31 $(x^3 - x^2 + 2)^7$.

$\blacksquare$ Let $g(x) = x^3 - x^2 + 2$ and $f(x) = x^7$.   Then $(f \circ g)(x) = f(g(x)) = f(x^3 - x^2 + 2) = (x^3 - x^2 + 2)^7$.

9.32 $(8 - x)^4$.

$\blacksquare$ Let $g(x) = 8 - x$ and $f(x) = x^4$.   Then $(f \circ g)(x) = f(g(x)) = f(8 - x) = (8 - x)^4$.

9.33 $\sqrt{1 + x^2}$.

$\blacksquare$ Let $g(x) = 1 + x^2$ and $f(x) = \sqrt{x}$.   Then $(f \circ g)(x) = f(g(x)) = f(1 + x^2) = \sqrt{1 + x^2}$.

9.34 $\dfrac{1}{x^2-4}$.

▌ Let $g(x)=x^2-4$ and $f(x)=1/x$. Then $(f\circ g)(x)=f(g(x))=f(x^2-4)=1/(x^2-4)$.

In Problems 9.35–9.44, use the chain rule (and possibly other rules) to find the derivative of the given function.

9.35 $(7+3x)^5$.

▌ $D_x[(7+3x)^5]=5(7+3x)^4\cdot D_x(7+3x)=5(7+3x)^4\cdot(3)=15(7+3x)^4$.

9.36 $(2x-3)^{-2}$.

▌ $D_x[(2x-3)^{-2}]=(-2)(2x-3)^{-3}\cdot D_x(2x-3)=-2(2x-3)^{-3}\cdot(2)=-4(2x-3)^{-3}$.

9.37 $(3x^2+5)^{-3}$.

▌ $D_x[(3x^2+5)^{-3}]=(-3)(3x^2+5)^{-4}\cdot D_x(3x^2+5)=-3(3x^2+5)^{-4}\cdot(6x)=-18x(3x^2+5)^{-4}$.

9.38 $\left(\dfrac{x^2-2}{2x^2+1}\right)^2$.

▌ $D_x\left[\left(\dfrac{x^2-2}{2x^2+1}\right)^2\right]=2\left(\dfrac{x^2-2}{2x^2+1}\right)D_x\left(\dfrac{x^2-2}{2x^2+1}\right)=2\left(\dfrac{x^2-2}{2x^2+1}\right)\left[\dfrac{(2x^2+1)\cdot D_x(x^2-2)-(x^2-2)\cdot D_x(2x^2+1)}{(2x^2+1)^2}\right]$

$=2\left(\dfrac{x^2-2}{2x^2+1}\right)\left[\dfrac{(2x^2+1)\cdot 2x-(x^2-2)\cdot 4x}{(2x^2+1)^2}\right]=\dfrac{2(x^2-2)}{(2x^2+1)^3}[2x(2x^2+1-2x^2+4)]$

$=\dfrac{4x(x^2-2)}{(2x^2+1)^3}\,(5)=\dfrac{20x(x^2-2)}{(2x^2+1)^3}$

9.39 $\dfrac{4}{3x^2-x+5}$.

▌ $D_x\left(\dfrac{4}{3x^2-x+5}\right)=4D_x[(3x^2-x+5)^{-1}]=4(-1)(3x^2-x+5)^{-2}D_x(3x^2-x+5)$

$=-4(3x^2-x+5)^{-2}(6x-1)=\dfrac{4(1-6x)}{(3x^2-x+5)^2}$.

9.40 $x^2(1-3x^3)^{1/3}$.

▌ $D_x[x^2(1-3x^3)^{1/3}]=x^2D_x(1-3x^3)^{1/3}+2x(1-3x^3)^{1/3}=x^2(\tfrac{1}{3})(1-3x^3)^{-2/3}\cdot D_x(1-3x^3)+2x(1-3x^3)^{1/3}$

$=\tfrac{1}{3}x^2(1-3x^3)^{-2/3}\cdot(-9x^2)+2x(1-3x^3)^{1/3}=-3x^4(1-3x^3)^{-2/3}+2x(1-3x^3)^{1/3}$

$=x(1-3x^3)^{-2/3}[-3x^3+2(1-3x^3)]=x(1-3x^3)^{-2/3}(2-9x^3)=\dfrac{x(2-9x^3)}{(1-3x^3)^{2/3}}$.

9.41 $\dfrac{x}{\sqrt{x^2+1}}$.

▌ $D_x\left(\dfrac{x}{\sqrt{x^2+1}}\right)=\dfrac{\sqrt{x^2+1}\cdot 1-x\cdot D_x(\sqrt{x^2+1})}{x^2+1}=\dfrac{\sqrt{x^2+1}-x\cdot D_x(x^2+1)^{1/2}}{x^2+1}$

$=\dfrac{\sqrt{x^2+1}-x\cdot\tfrac{1}{2}(x^2+1)^{-1/2}\cdot D_x(x^2+1)}{x^2+1}=\dfrac{\sqrt{x^2+1}-(x/2)(x^2+1)^{-1/2}(2x)}{x^2+1}$

$=\dfrac{(x^2+1)-x^2}{(x^2+1)^{3/2}}=\dfrac{1}{(x^2+1)^{3/2}}$

9.42 $(7x^3-4x^2+2)^{1/4}$.

▌ $D_x(7x^3-4x^2+2)^{1/4}=\tfrac{1}{4}(7x^3-4x^2+2)^{-3/4}\cdot D_x(7x^3-4x^2+2)=\tfrac{1}{4}(7x^3-4x^2+2)^{-3/4}(21x^2-8x)$

$=\dfrac{1}{4}\dfrac{x(21x-8)}{(7x^3-4x^2+2)^{3/4}}$.

9.43 $\dfrac{\sqrt{x+2}}{\sqrt{x-1}}$.

∎
$$\frac{\sqrt{x+2}}{\sqrt{x-1}} = \sqrt{\frac{x+2}{x-1}} = \left(\frac{x+2}{x-1}\right)^{1/2}$$

$$D_x\left(\frac{x+2}{x-1}\right)^{1/2} = \frac{1}{2}\left(\frac{x+2}{x-1}\right)^{-1/2}\cdot D_x\left(\frac{x+2}{x-1}\right) = \frac{1}{2}\left(\frac{x+2}{x-1}\right)^{-1/2}\cdot \frac{(x-1)-(x+2)}{(x-1)^2}$$

$$= \frac{1}{2}\left(\frac{x-1}{x+2}\right)^{1/2}\frac{-3}{(x-1)^2} = -\frac{3}{2}\frac{1}{\sqrt{x+2}(\sqrt{x-1})^3}$$

9.44 $\sqrt[3]{(4x^2+3)^2}$.

∎ $\sqrt[3]{(4x^2+3)^2} = (4x^2+3)^{2/3}$. $D_x(4x^2+3)^{2/3} = \frac{2}{3}(4x^2+3)^{-1/3}\cdot D_x(4x^2+3) = \dfrac{2}{3}\dfrac{1}{(4x^2+3)^{1/3}}\cdot 8x = \dfrac{16x}{3\sqrt[3]{4x^2+3}}$.

9.45   Assume that F and G are differentiable functions such that $F'(x) = -G(x)$ and $G'(x) = -F(x)$. Let $H(x) = [F(x)]^2 - [G(x)]^2$.   Find $H'(x)$.

∎ $H'(x) = 2F(x)\cdot D_xF(x) - 2G(x)\cdot D_xG(x) = 2F(x)[-G(x)] - 2G(x)[-F(x)] = -2F(x)G(x) + 2F(x)G(x) = 0$.

9.46   If f is continuous at a and g is continuous at $f(a)$, prove that $g\circ f$ is continuous at a.

∎   Assume $\varepsilon > 0$. Choose $\delta_1 > 0$ such that $|g(u) - g(f(a))| < \varepsilon$ whenever $|u - f(a)| < \delta_1$. Then choose $\delta > 0$ such that $|f(x) - f(a)| < \delta_1$ whenever $|x - a| < \delta$. Hence, if $|x - a| < \delta$, $|g(f(x)) - g(f(a))| < \varepsilon$.

9.47   Show that $D_x|x| = \dfrac{|x|}{x}$, for $x \neq 0$.

∎ When $x > 0$, $D_x|x| = D_x(x) = 1 = |x|/x$. When $x < 0$, $D_x|x| = D_x(-x) = -1 = -x/x = |x|/x$.

9.48   Find a formula for $D_x|x^2 + 2x|$ $(x \neq 0, -2)$.

∎ By the chain rule and Problem 9.47, $D_x|x^2 + 2x| = \dfrac{|x^2 + 2x|}{x^2 + 2x}\cdot D_x(x^2 + 2x) = \dfrac{|x^2 + 2x|}{x^2 + 2x}(2x + 2)$.

9.49   Give a justification of the rule

$$\frac{dv}{du} = \frac{1}{du/dv} \tag{1}$$

∎ Let $u = f(v)$ be a one-one, differentiable function. Then the inverse function $v = g(u) = g(f(v))$ is differentiable, and the chain rule gives

$$\frac{dv}{dv} = 1 = g'(u)\,f'(v) \tag{2}$$

Writing dv/du and du/dv for $g'(u)$ and $f'(v)$ in (2), we get (1).

CHAPTER 10

Trigonometric Functions and Their Derivatives

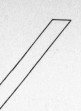

10.1 Define radian measure, that is, describe an angle of 1 radian.

▎ Consider a circle with a radius of one unit (Fig. 10-1). Let the center be C, and let CA and CB be two radii for which the intercepted arc $\overset{\frown}{AB}$ of the circle has length 1. Then the central angle $\angle ACB$ has a measure of one radian.

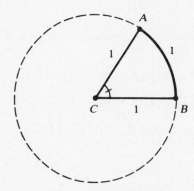

Fig. 10-1

10.2 Give the equations relating degree measure and radian measure of angles.

▎ 2π radians is the same as 360 degrees. Hence, 1 radian $= 180/\pi$ degrees, and 1 degree $= \pi/180$ radians. So, if an angle has a measure of D degrees and R radians, then $D = (180/\pi)R$ and $R = (\pi/180)D$.

10.3 Give the radian measure of angles of 30°, 45°, 60°, 90°, 120°, 135°, 180°, 270°, and 360°.

▎ We use the formula $R = (\pi/180)D.$ Hence 30° $= \pi/6$ radians, 45° $= \pi/4$ radians, 60° $= \pi/3$ radians, 90° $= \pi/2$ radians, 120° $= 2\pi/3$ radians, 135° $= 3\pi/4$ radians, 180° $= \pi$ radians, 270° $= 3\pi/2$ radians, 360° $= 2\pi$ radians.

10.4 Give the degree measure of angles of $3\pi/5$ radians and $5\pi/6$ radians.

▎ We use the formula $D = (180/\pi)R.$ Thus, $3\pi/5$ radians $= 108°$ and $5\pi/6$ radians $= 150°.$

10.5 In a circle of radius 10 inches, what arc length along the circumference is intercepted by a central angle of $\pi/5$ radians?

▎ The arc length s, the radius r, and the central angle θ (measured in radians) are related by the equation $s = r\theta$. In this case, $r = 10$ inches and $\theta = \pi/5$. Hence, $s = 2\pi$ inches.

10.6 If a bug moves a distance of 3π centimeters along a circular arc and if this arc subtends a central angle of 45°, what is the radius of the circle?

▎ $s = r\theta$. In this case, $s = 3\pi$ centimeters and $\theta = \pi/4$ (the radian measure equivalent of 45°). Thus, $3\pi = r \cdot \pi/4$. Hence, $r = 12$ centimeters.

10.7 Draw a picture of the rotation determining an angle of $-\pi/3$ radians.

▎ See Fig. 10-2. $\pi/3$ radians $= 60°$, and the minus sign indicates that a 60° rotation is to be taken in the clockwise direction. (Positive angles correspond to counterclockwise rotations.)

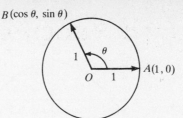

Fig. 10-2   **Fig. 10-3**

10.8   Give the definition of $\sin \theta$ and $\cos \theta$.

❚ Refer to Fig. 10-3. Place an arrow $\overrightarrow{OA}$ of unit length so that its initial point O is the origin of a coordinate system and its endpoint A is $(1, 0)$. Rotate $\overrightarrow{OA}$ about the point O through an angle with radian measure θ. Let $\overrightarrow{OB}$ be the final position of the arrow after the rotation. Then $\cos \theta$ is defined to be the x-coordinate of B, and $\sin \theta$ is defined to be the y-coordinate of B.

10.9   State the values of $\cos \theta$ and $\sin \theta$ for $\theta = 0, \pi/6, \pi/4, \pi/3, \pi/2, \pi, 3\pi/2, 2\pi, 9\pi/4$.

❚

θ	$\sin \theta$	$\cos \theta$
0	0	1
$\pi/6$	$1/2$	$\sqrt{3}/2$
$\pi/4$	$\sqrt{2}/2$	$\sqrt{2}/2$
$\pi/3$	$\sqrt{3}/2$	$1/2$
$\pi/2$	1	0
π	0	-1
$3\pi/2$	-1	0
2π	0	1

Notice that $9\pi/4 = 2\pi + \pi/4$, and the sine and cosine functions have a period of 2π, that is, $\sin(\theta + 2\pi) = \sin \theta$ and $\cos(\theta + 2\pi) = \cos \theta$. Hence, $\sin(9\pi/4) = \sin(\pi/4) = \sqrt{2}/2$ and $\cos(9\pi/4) = \cos(\pi/4) = \sqrt{2}/2$.

10.10   Evaluate: (*a*) $\cos(-\pi/6)$ (*b*) $\sin(-\pi/6)$ (*c*) $\cos(2\pi/3)$ (*d*) $\sin(2\pi/3)$

❚ (*a*) In general, $\cos(-\theta) = \cos \theta$. Hence, $\cos(-\pi/6) = \cos(\pi/6) = \sqrt{3}/2$. (*b*) In general, $\sin(-\theta) = -\sin \theta$. Hence, $\sin(-\pi/6) = -\sin(\pi/6) = -\frac{1}{2}$. (*c*) $2\pi/3 = \pi/2 + \pi/6$. We use the identity $\cos(\theta + \pi/2) = -\sin \theta$. Thus, $\cos(2\pi/3) = -\sin(\pi/6) = -\frac{1}{2}$. (*d*) We use the identity $\sin(\theta + \pi/2) = \cos \theta$. Thus, $\sin(2\pi/3) = \cos(\pi/6) = \sqrt{3}/2$.

10.11   Sketch the graph of the cosine and sine functions.

❚ We use the values calculated in Problem 10.9 to draw Fig. 10-4.

10.12   Sketch the graph of $y = \cos 3x$.

❚ Because $\cos 3(x + 2\pi/3) = \cos(3x + 2\pi) = \cos 3x$, the function is of period $p = 2\pi/3$. Hence, the length of each wave is $2\pi/3$. The number f of waves over an interval of length 2π is 3. (In general, this number f, called the *frequency* of the function, is given by the equation $f = 2\pi/p$.) Thus, the graph is as indicated in Fig. 10-5.

10.13   Sketch the graph of $y = 1.5 \sin 4x$.

❚ The period $p = \pi/2$. (In general, $p = 2\pi/b$, where b is the coefficient of x.) The coefficient 1.5 is the *amplitude*, the greatest height above the x-axis reached by points of the graph. Thus, the graph looks like Fig. 10-6.

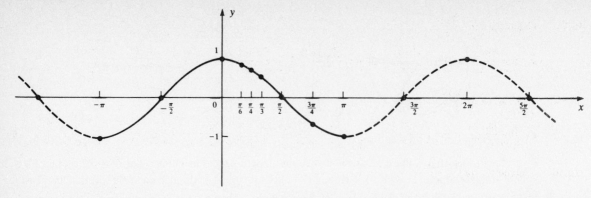

(a) $y = \cos x$

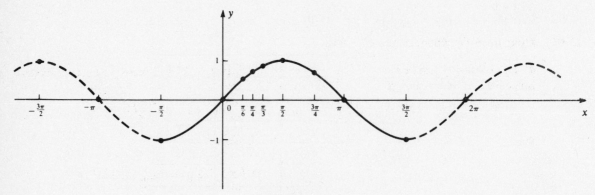

(b) $y = \sin x$

Fig. 10-4

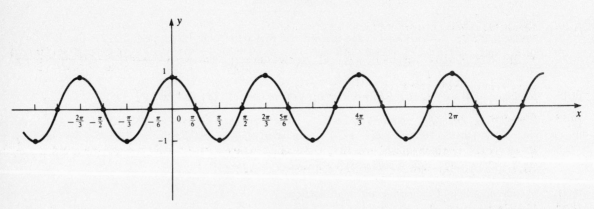

Fig. 10-5

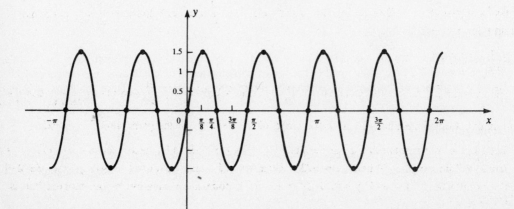

Fig. 10-6

10.14 Calculate $\lim\limits_{x\to 0} \dfrac{\sin 3x}{x}$.

▮ $\lim\limits_{x\to 0} \dfrac{\sin 3x}{x} = \lim\limits_{x\to 0} 3\,\dfrac{\sin 3x}{3x} = 3\cdot\lim\limits_{x\to 0} \dfrac{\sin 3x}{3x} = 3\cdot\lim\limits_{\theta\to 0} \dfrac{\sin\theta}{\theta} = 3\cdot 1 = 3$.

10.15 Calculate $\lim\limits_{x\to 0} \dfrac{\sin 2x}{\sin 3x}$.

▮ $\lim\limits_{x\to 0} \dfrac{\sin 2x}{\sin 3x} = \lim\limits_{x\to 0}\left(\dfrac{\sin 2x}{2x}\cdot\dfrac{3x}{\sin 3x}\cdot\dfrac{2}{3}\right) = \lim\limits_{x\to 0}\dfrac{\sin 2x}{2x}\cdot\lim\limits_{x\to 0}\dfrac{3x}{\sin 3x}\cdot\dfrac{2}{3} = \lim\limits_{\theta\to 0}\dfrac{\sin\theta}{\theta}\cdot\lim\limits_{\theta\to 0}\dfrac{\theta}{\sin\theta}\cdot\dfrac{2}{3} = 1\cdot 1\cdot\dfrac{2}{3} = \dfrac{2}{3}$.

10.16 Calculate $\lim\limits_{x\to 0} \dfrac{1-\cos x}{x}$.

▮ $\lim\limits_{x\to 0}\dfrac{1-\cos x}{x} = \lim\limits_{x\to 0}\dfrac{1-\cos x}{x}\cdot\dfrac{1+\cos x}{1+\cos x} = \lim\limits_{x\to 0}\dfrac{1-\cos^2 x}{x(1+\cos x)} = \lim\limits_{x\to 0}\dfrac{\sin^2 x}{x(1+\cos x)} = \lim\limits_{x\to 0}\dfrac{\sin x}{x}\cdot\lim\limits_{x\to 0}\dfrac{\sin x}{1+\cos x} = 1\cdot\dfrac{0}{2} = 0$.

10.17 Using the Δ-definition, calculate $\dfrac{d}{dx}(\sin x)$.

▮ By the identity $\sin(u+v) = \sin u\cos v + \cos u\sin v$, $\sin(x+\Delta x) = \sin x\cos(\Delta x) + \cos x\sin(\Delta x)$. Hence, $\sin(x+\Delta x) - \sin x = \sin x[\cos(\Delta x) - 1] + \cos x\sin(\Delta x)$, and

$$\dfrac{\sin(x+\Delta x) - \sin x}{\Delta x} = \sin x\cdot\dfrac{\cos(\Delta x) - 1}{\Delta x} + \cos x\cdot\dfrac{\sin(\Delta x)}{\Delta x}$$

Thus,
$$\dfrac{d}{dx}(\sin x) = \lim\limits_{\Delta x\to 0}\dfrac{\sin(x+\Delta x) - \sin x}{\Delta x}$$

$$= \sin x\cdot\lim\limits_{\Delta x\to 0}\dfrac{\cos(\Delta x) - 1}{\Delta x} + \cos x\cdot\lim\limits_{\Delta x\to 0}\dfrac{\sin(\Delta x)}{\Delta x}$$

$$= (\sin x)(0) + (\cos x)(1) = \cos x$$

Here, we have used $\lim\limits_{\theta\to 0}\dfrac{\sin\theta}{\theta} = 1$ and $\lim\limits_{\theta\to 0}\dfrac{\cos\theta - 1}{\theta} = 0$ (Problem 10.16).

10.18 Calculate $\dfrac{d}{dx}(\cos x)$ from the known derivative of $\sin x$.

▮ By the identity $\cos x = \sin\left(\dfrac{\pi}{2} - x\right)$, $\dfrac{d}{dx}(\cos x) = \dfrac{d}{dx}\left[\sin\left(\dfrac{\pi}{2} - x\right)\right] = \cos\left(\dfrac{\pi}{2} - x\right)\cdot\dfrac{d}{dx}\left(\dfrac{\pi}{2} - x\right)$ [chain rule] $= \sin x\cdot(-1)$ $\left[\sin x = \cos\left(\dfrac{\pi}{2} - x\right)\right] = -\sin x$.

10.19 Calculate $\dfrac{d}{dx}(\sin 3x)$.

▮ $\sin 3x$ is a composite function of $3x$ and the sine function. By the chain rule and the fact that $\dfrac{d}{dx}(\sin x) = \cos x$, $\dfrac{d}{dx}(\sin 3x) = \cos 3x\cdot\dfrac{d}{dx}(3x) = \cos 3x\cdot 3 = 3\cos 3x$.

10.20 Calculate $\dfrac{d}{dx}(\cos^2 x)$.

▮ $\cos^2 x = (\cos x)^2$. Hence, by the chain rule, $\dfrac{d}{dx}(\cos^2 x) = 2(\cos x)\cdot\dfrac{d}{dx}\cos x = 2\cos x\cdot(-\sin x) = -2\sin x\cos x = -\sin 2x$.

10.21 Find $\dfrac{d}{dx}(\sqrt{\sin x})$.

▮ $\sqrt{\sin x} = (\sin x)^{1/2}$. By the chain rule, $\dfrac{d}{dx}(\sqrt{\sin x}) = \tfrac{1}{2}(\sin x)^{-1/2}\cdot\dfrac{d}{dx}(\sin x) = \dfrac{1}{2\sqrt{\sin x}}\cdot\cos x = \dfrac{\cos x}{2\sqrt{\sin x}}$.

10.22 Find an equation of the tangent line to the graph of $y = \sin^2 x$ at the point where $x = \pi/3$.

▮ The slope of the tangent line is the derivative y'. By the chain rule, since $\sin^2 x = (\sin x)^2$, $y' = 2(\sin x)\cdot\dfrac{d}{dx}(\sin x) = 2\sin x\cos x$. When $x = \pi/3$, $\sin x = \sqrt{3}/2$ and $\cos x = \tfrac{1}{2}$. So $y' = 2\cdot\sqrt{3}/2\cdot\tfrac{1}{2} = \sqrt{3}/2$. At the point where $x = \pi/3$, $y = (\sqrt{3}/2)^2 = \tfrac{3}{4}$. So a point-slope equation of the tangent line is $y - \tfrac{3}{4} = (\sqrt{3}/2)(x - \pi/3)$.

10.23 Find an equation of the normal line to the curve $y = 1 + \cos x$ at the point $(\pi/3, \frac{3}{2})$.

▮ The slope of the tangent line is the derivative y'. But, $y' = -\sin x = -\sin(\pi/3) = -\sqrt{3}/2$. Hence, the slope of the normal line is the negative reciprocal $2/\sqrt{3}$. So a point-slope equation of the normal line is $y - \frac{3}{2} = (2/\sqrt{3})(x - \pi/3) = (2\sqrt{3}/3)x - 2\pi\sqrt{3}/9$.

10.24 Derive the formula $\dfrac{d}{dx}(\tan x) = \sec^2 x$.

▮ Remember that $\tan x = \sin x/\cos x$ and $\sec x = 1/\cos x$. By the quotient rule,

$$\frac{d}{dx}(\tan x) = \frac{d}{dx}\left(\frac{\sin x}{\cos x}\right) = \frac{(\cos x)\dfrac{d}{dx}(\sin x) - (\sin x)\dfrac{d}{dx}(\cos x)}{(\cos x)^2}$$

$$= \frac{(\cos x)(\cos x) - (\sin x)(-\sin x)}{(\cos x)^2} = \frac{\cos^2 x + \sin^2 x}{(\cos x)^2}$$

$$= \frac{1}{(\cos x)^2} = \sec^2 x$$

10.25 Find an equation of the tangent line to the curve $y = \tan^2 x$ at the point $(\pi/3, 3)$.

▮ Note that $\tan(\pi/3) = \sin(\pi/3)/\cos(\pi/3) = (\sqrt{3}/2)/\frac{1}{2} = \sqrt{3}$, and $\sec(\pi/3) = 1/\cos(\pi/3) = 1/\frac{1}{2} = 2$. By the chain rule, $y' = 2(\tan x) \cdot \dfrac{d}{dx}(\tan x) = 2(\tan x)(\sec^2 x)$. Thus, when $x = \pi/3$, $y' = 2\sqrt{3} \cdot 4 = 8\sqrt{3}$, so the slope of the tangent line is $8\sqrt{3}$. Hence, a point-slope equation of the tangent line is $y - 3 = 8\sqrt{3}(x - \pi/3)$.

10.26 Derive the formula $\dfrac{d}{dx}(\cot x) = -\csc^2 x$.

▮ By the identity $\cot x = \tan(\pi/2 - x)$ and the chain rule,

$$\frac{d}{dx}(\cot x) = \frac{d}{dx}\left[\tan\left(\frac{\pi}{2} - x\right)\right] = \sec^2\left(\frac{\pi}{2} - x\right) \cdot \frac{d}{dx}\left(\frac{\pi}{2} - x\right)$$

$$= \frac{1}{\cos^2(\pi/2 - x)} \cdot (-1) = -\frac{1}{\sin^2 x} = -\csc^2 x.$$

10.27 Show that $\dfrac{d}{dx}(\sec x) = \sec x \tan x$.

▮ By the chain rule,

$$\frac{d}{dx}(\sec x) = \frac{d}{dx}(\cos x)^{-1} = -(\cos x)^{-2} \cdot \frac{d}{dx}(\cos x) = -\frac{1}{\cos^2 x} \cdot (-\sin x)$$

$$= \frac{\sin x}{\cos^2 x} = \frac{1}{\cos x} \cdot \frac{\sin x}{\cos x} = \sec x \tan x$$

10.28 Find an equation of the normal line to the curve $y = 3\sec^2 x$ at the point $(\pi/6, 4)$.

▮ By the chain rule, $y' = 3[2\sec x \cdot \dfrac{d}{dx}(\sec x)] = 3(2\sec x \cdot \sec x \cdot \tan x) = 6\sec^2 x \tan x$. So the slope of the tangent line is $y' = 6(\frac{4}{3})(\sqrt{3}/3) = 8\sqrt{3}/3$. Hence, the slope of the normal line is the negative reciprocal $-\sqrt{3}/8$. Thus, a point-slope equation of the normal line is $y - 4 = -(\sqrt{3}/8)(x - \pi/6)$.

10.29 Find $D_x(\csc \sqrt{x})$.

▮ Recall that $D_x(\csc x) = -\csc x \cot x$. Hence, by the chain rule, $D_x(\csc \sqrt{x}) = -\csc(\sqrt{x})\cot(\sqrt{x}) \cdot D_x(\sqrt{x}) = -\csc(\sqrt{x})\cot(\sqrt{x})(1/2\sqrt{x})$.

10.30 Evaluate $\displaystyle\lim_{x \to 0} \frac{\tan x}{x}$.

▮ $\dfrac{\tan x}{x} = \dfrac{\sin x}{\cos x} \cdot \dfrac{1}{x} = \dfrac{\sin x}{x} \cdot \dfrac{1}{\cos x}$. Hence, $\displaystyle\lim_{x \to 0}\frac{\tan x}{x} = \lim_{x \to 0}\frac{\sin x}{x} \cdot \lim_{x \to 0}\frac{1}{\cos x} = 1 \cdot \frac{1}{1} = 1$.

10.31 Evaluate $\displaystyle\lim_{x \to 0} \frac{\tan^3 2x}{x^3}$.

▮ $\dfrac{\tan^3 2x}{x^3} = \dfrac{\sin^3 2x}{(2x)^3} \cdot \dfrac{1}{\cos^3 2x} \cdot 8.$

Hence, $\qquad \lim\limits_{x\to 0} \dfrac{\tan^3 2x}{x^3} = \lim\limits_{x\to 0} \dfrac{\sin^3 2x}{(2x)^3} \cdot \lim\limits_{x\to 0} \dfrac{1}{\cos^3 2x} \cdot 8 = \left(\lim\limits_{\theta\to 0} \dfrac{\sin \theta}{\theta}\right)^3 \cdot 1 \cdot 8 = (1)^3 \cdot 8 = 8.$

10.32 Show that the curve $y = x \sin x$ is tangent to the line $y = x$ whenever $x = (4n + 1)(\pi/2)$, where n is any integer.

▮ When $x = (4n + 1)(\pi/2) = \pi/2 + 2\pi n$, $\sin x = \sin \pi/2 = 1$, $\cos x = \cos \pi/2 = 0$, and $x \sin x = x$. Thus, at such points, the curve $y = x \sin x$ intersects $y = x$. For $y = x \sin x$, $y' = x \cdot D_x(\sin x) + \sin x \cdot D_x(x) = x \cos x + \sin x$. Thus, at the given points, $y' = x \cdot 0 + 1 = 1$. Hence, the slope of the tangent line to the curve $y = x \sin x$ at those points is 1. But the slope of the line $y = x$ is also 1, and, therefore, $y = x$ is the tangent line.

10.33 At what values of x does the graph of $y = \sec x$ have a horizontal tangent?

▮ A line is horizontal when and only when its slope is 0. The slope of the tangent line is $y' = D_x(\sec x) = \sec x \tan x$. Hence, we must solve $\sec x \tan x = 0$. Since $\sec x = 1/\cos x$, $\sec x$ is never 0. Hence, $\tan x = 0$. But, since $\tan x = \sin x/\cos x$, $\tan x = 0$ is equivalent to $\sin x = 0$. The latter occurs when and only when $x = n\pi$ for some integer n.

10.34 For what values of x are the tangent lines to the graphs of $y = \sin x$ and $y = \cos x$ perpendicular?

▮ The tangent line to the graph of $y = \sin x$ has slope $D_x(\sin x) = \cos x$, and the tangent line to the graph of $y = \cos x$ has slope $D_x(\cos x) = -\sin x$. Hence, the condition for perpendicularity is that $\cos x \cdot (-\sin x) = -1$, which is equivalent to $\cos x \sin x = 1$. Since $2 \cos x \sin x = \sin 2x$, this is equivalent to $\sin 2x = 2$, which is impossible, because $|\sin x| \le 1$ for all x. Hence, there are no values of x which satisfy the property.

10.35 Find the angle at which the curve $y = \frac{1}{3} \sin 3x$ crosses the x-axis.

▮ The curve crosses the x-axis when $y = \frac{1}{3} \sin 3x = 0$, which is equivalent to $\sin 3x = 0$, and thence to $3x = n\pi$, where n is an arbitrary integer. Thus, $x = n\pi/3$. The slope of the tangent line is $y' = \frac{1}{3} \cos 3x \cdot 3 = \cos 3x = \cos (n\pi) = \pm 1$. The lines with slope ± 1 make an angle of $\pm 45°$ with the x-axis.

In Problems 10.36 to 10.43, calculate the derivative of the given function.

10.36 $x \sin x$

▮ $D_x(x \sin x) = x \cdot D_x(\sin x) + D_x(x) \cdot \sin x = x \cos x + \sin x.$

10.37 $x^2 \cos 2x.$

▮ $D_x(x^2 \cos 2x) = x^2 \cdot D_x(\cos 2x) + 2x \cdot \cos 2x = x^2(-\sin 2x) \cdot D_x(2x) + 2x \cos 2x = -2x^2 \sin 2x + 2x \cos 2x.$

10.38 $\dfrac{\sin x}{x}.$

▮ $D_x\left(\dfrac{\sin x}{x}\right) = \dfrac{x \cos x - \sin x}{x^2}.$

10.39 $\sin^3 (5x + 4).$

▮ $D_x[\sin^3(5x + 4)] = 3 \sin^2 (5x + 4) \cdot D_x(5x + 4) = 3 \sin^2 (5x + 4) \cdot (5) = 15 \sin^2 (5x + 4).$

10.40 $2 \tan (x/2) - 5.$

▮ $D_x(2 \tan (x/2) - 5) = 2 \sec^2 (x/2) \cdot D_x(x/2) = \sec^2 (x/2).$

10.41 $\tan x - \sec x.$

▮ $D_x(\tan x - \sec x) = \sec^2 x - \sec x \tan x = (\sec x)(\sec x - \tan x).$

10.42 $\cot^2 x.$

┃ $D_x(\cot^2 x) = 2 \cot x \cdot D_x (\cot x) = 2 \cot x (-\csc^2 x) = -2 \cot x \csc^2 x.$

10.43 $\csc (3x - 5).$

┃ $D_x[\csc (3x - 5)] = [-\csc (3x - 5) \cot (3x - 5)] \cdot D_x(3x - 5) = -\csc (3x - 5) \cot (3x - 5) \cdot (3)$

$$= -3 \csc (3x - 5) \cot (3x - 5).$$

10.44 Evaluate $\displaystyle\lim_{h \to 0} \frac{\cos (\pi/3 + h) - \frac{1}{2}}{h}$

┃ Remember that $\frac{1}{2} = \cos (\pi/3)$ and use the definition of the derivative. $\displaystyle\lim_{h \to 0} \frac{\cos (\pi/3 + h) - \frac{1}{2}}{h} =$
$\displaystyle\lim_{h \to 0} \frac{\cos (\pi/3 + h) - \cos (\pi/3)}{h} = [D_x(\cos x)](\pi/3) = -\sin (\pi/3) = -\sqrt{3}/2.$

10.45 For what value of A does $3 \sin Ax$ have a period of 2?

┃ The period $p = 2\pi/A.$ Thus, $2 = 2\pi/A,\ \ 2A = 2\pi,\ \ A = \pi.$

10.46 Find the angle of intersection of the lines $\mathcal{L}_1$: $y = x - 3$ and $\mathcal{L}_2$: $y = -5x + 4.$

┃ The angle θ_1 that $\mathcal{L}_1$ makes with the x-axis has a tangent that is equal to the slope of the line. The angle θ_2 that $\mathcal{L}_2$ makes with the x-axis has a tangent equal to the slope of $\mathcal{L}_2.$ Thus $\tan \theta_1 = 1$ and $\tan \theta_2 = -5.$ The angle θ between $\mathcal{L}_1$ and $\mathcal{L}_2$ is $\theta_2 - \theta_1.$ So, $\tan \theta = \tan (\theta_2 - \theta_1) = \dfrac{\tan \theta_2 - \tan \theta_1}{1 + \tan \theta_2 \tan \theta_1} = \dfrac{-5 - 1}{1 + (-5)} = \dfrac{-6}{-4} = 1.5.$
Reference to a table of tangents reveals that $\theta \approx 56°.$

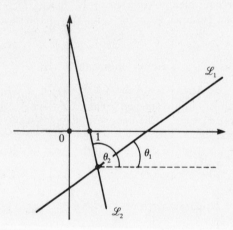

Fig. 10-7

10.47 Find the angle of intersection of the tangent lines to the curves $xy = 1$ and $y = x^3$ at the common point $(1, 1).$

┃ Let θ_1 be the angle between the horizontal and the tangent line to $y = x^3,$ and let θ_2 be the angle between the horizontal and the tangent line to $xy = 1.$ Now, $\tan \theta_1$ is the slope of the tangent line to $y = x^3,$ which is the derivative of x^3 evaluated at $(1, 1),$ that is, $3x^2$ evaluated at $x = 1$ or 3. So, $\tan \theta_1 = 3.$ Likewise, since the derivative of $1/x$ is $-(1/x^2),$ which, when evaluated at $x = 1,$ is $-1,$ we have $\tan \theta_2 = -1.$ Hence, $\tan (\theta_2 - \theta_1) = \dfrac{\tan \theta_2 - \tan \theta_1}{1 + \tan \theta_2 \tan \theta_1} = \dfrac{-1 - 3}{1 + (-3)} = 2.$ A table of tangents yields $\theta_2 - \theta_1 \approx 63°.$

10.48 Evaluate $\displaystyle\lim_{x \to +\infty} \frac{4x^3 - 1}{3x^2 + 2} \sin \left(\frac{1}{x}\right).$

┃ $\displaystyle\lim_{x \to +\infty} \frac{4x^3 - 1}{3x^2 + 2} \sin \left(\frac{1}{x}\right) = \lim_{x \to +\infty} \frac{4x^3 - 1}{3x^3 + 2x} \cdot \frac{\sin (1/x)}{1/x} = \lim_{x \to +\infty} \frac{4x^3 - 1}{3x^3 + 2x} \cdot \lim_{x \to +\infty} \frac{\sin (1/x)}{1/x}.$ But,
$\displaystyle\lim_{x \to +\infty} \frac{\sin (1/x)}{1/x} = \lim_{\theta \to 0^+} \frac{\sin \theta}{\theta} = 1,$ and $\displaystyle\lim_{x \to +\infty} \frac{4x^3 - 1}{3x^3 + 2x} = \lim_{x \to +\infty} \frac{4 - 1/x^3}{3 + 2/x^2} = \frac{4}{3}.$ Thus, the desired limit is $\frac{4}{3}.$

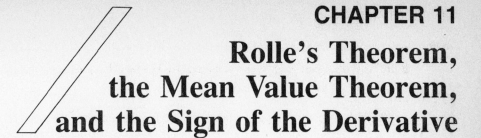

CHAPTER 11

Rolle's Theorem, the Mean Value Theorem, and the Sign of the Derivative

11.1 State Rolle's theorem.

▌ If f is continuous over a closed interval $[a, b]$ and differentiable on the open interval (a, b), and if $f(a) = f(b) = 0$, then there is at least one number c in (a, b) such that $f'(c) = 0$.

In Problems 11.2 to 11.9, determine whether the hypotheses of Rolle's theorem hold for the function f on the given interval, and, if they do, verify the conclusion of the theorem.

11.2 $f(x) = x^2 - 2x - 3$ on $[-1, 3]$.

▌ $f(x)$ is clearly differentiable everywhere, and $f(-1) = f(3) = 0$. Hence, Rolle's theorem applies. $f'(x) = 2x - 2$. Setting $f'(x) = 0$, we obtain $x = 1$. Thus, $f'(1) = 0$ and $-1 < 1 < 3$.

11.3 $f(x) = x^3 - x$ on $[0, 1]$.

▌ $f(x)$ is differentiable, with $f'(x) = 3x^2 - 1$. Also, $f(0) = f(1) = 0$. Thus, Rolle's theorem applies. Setting $f'(x) = 0$, $3x^2 = 1$; $x^2 = \frac{1}{3}$, $x = \pm\sqrt{3}/3$. The positive solution $x = \sqrt{3}/3$ lies between 0 and 1.

11.4 $f(x) = 9x^3 - 4x$ on $[-\frac{2}{3}, \frac{2}{3}]$.

▌ $f'(x) = 27x^2 - 4$ and $f(-\frac{2}{3}) = f(\frac{2}{3}) = 0$. Hence, Rolle's theorem is applicable. Setting $f'(x) = 0$, $27x^2 = 4$, $x^2 = \frac{4}{27}$, $x = \pm 2/3\sqrt{3} = \pm 2\sqrt{3}/9$. Both of these values lie in $[-\frac{2}{3}, \frac{2}{3}]$, since $2\sqrt{3}/9 < \frac{2}{3}$.

11.5 $f(x) = x^3 - 3x^2 + x + 1$ on $[1, 1 + \sqrt{2}]$.

▌ $f'(x) = 3x^2 - 6x + 1$ and $f(1) = f(1 + \sqrt{2}) = 0$. This means that Rolle's theorem applies. Setting $f'(x) = 0$ and using the quadratic formula, we obtain $x = 1 \pm \frac{1}{3}\sqrt{6}$ and observe that $1 < 1 + \frac{1}{3}\sqrt{6} < 1 + \sqrt{2}$.

11.6 $f(x) = \dfrac{x^2 - x - 6}{x - 1}$ on $[-2, 3]$.

▌ There is a discontinuity at $x = 1$, since $\lim\limits_{x \to 1} f(x)$ does not exist. Hence, Rolle's theorem does not apply.

11.7
$$f(x) = \begin{cases} \dfrac{x^3 - 2x^2 - 5x + 6}{x - 1} & \text{if } x \neq 1 \text{ and } x \text{ is in } [-2, 3] \\ -6 & \text{if } x = 1 \end{cases}$$

▌ Notice that $x^3 - 2x^2 - 5x + 6 = (x - 1)(x^2 - x - 6)$. Hence, $f(x) = x^2 - x - 6$ if $x \neq 1$ and x is in $[-2, 3]$. But $f(x) = -6 = x^2 - x - 6$ when $x = 1$. So $f(x) = x^2 - x - 6$ throughout the interval $[-2, 3]$. Also, note that $f(-2) = f(3) = 0$. Hence, Rolle's theorem applies. $f'(x) = 2x - 1$. Setting $f'(x) = 0$, we obtain $x = \frac{1}{2}$ which lies between -2 and 3.

11.8 $f(x) = x^{2/3} - 2x^{1/3}$ on $[0, 8]$.

▌ $f(x)$ is differentiable within $(0, 8)$, but not at 0. However, it is continuous at $x = 0$ and, therefore, throughout $[0, 8]$. Also, $f(0) = f(8) = 0$. Hence, Rolle's theorem applies. $f'(x) = 2/3\sqrt[3]{x} - 2/3(\sqrt[3]{x})^2$. Setting $f'(x) = 0$, we obtain $x = 1$, which is between 0 and 8.

11.9
$$f(x) = \begin{cases} x^2 & \text{if } 0 \leq x \leq 1 \\ 2 - x & \text{if } 1 < x \leq 2 \end{cases}$$

▌ $f(x)$ is not differentiable at $x = 1$. (To see this, note that, when $\Delta x < 0$, $[f(1 + \Delta x) - 1]/\Delta x = 2 + \Delta x \to 2$ as $\Delta x \to 0$. But, when $\Delta x > 0$, $[f(1 + \Delta x) - 1]/\Delta x = -1 \to -1$ as $\Delta x \to 0$.) Thus, Rolle's theorem does not apply.

11.10 State the mean value theorem.

▌ If $f(x)$ is continuous on the closed interval $[a, b]$ and differentiable on the open interval (a, b), then there is a number c in (a, b) such that $f'(c) = \dfrac{f(b) - f(a)}{b - a}$.

In Problems 11.11 to 11.16, determine whether the hypotheses of the mean value theorem hold for the function $f(x)$ on the given interval, and, if they do, find a value c satisfying the conclusion of the theorem.

11.11 $f(x) = 2x + 3$ on $[1, 4]$.

▌ $f'(x) = 2$. Hence, the mean value theorem applies. Note that $\dfrac{f(4) - f(1)}{4 - 1} = \dfrac{11 - 5}{4 - 1} = 2$. Thus, we can take c to be any point in $(1, 4)$.

11.12 $f(x) = 3x^2 - 5x + 1$ on $[2, 5]$.

▌ $f'(x) = 6x - 5$, and the mean value theorem applies. Setting $6c - 5 = \dfrac{f(5) - f(2)}{5 - 2} = \dfrac{51 - 3}{3} = 16$, we find $c = \frac{21}{6} = \frac{7}{2}$, which lies between 2 and 5.

11.13 $f(x) = x^{3/4}$ on $[0, 16]$.

▌ $f(x)$ is continuous for $x \geq 0$ and differentiable for $x > 0$. Thus, the mean value theorem is applicable. $f'(x) = \dfrac{3}{4\sqrt[4]{x}}$. Setting $\dfrac{3}{4\sqrt[4]{c}} = \dfrac{f(16) - f(0)}{16 - 0} = \dfrac{8 - 0}{16} = \dfrac{1}{2}$, we find $c = \dfrac{81}{16}$, which lies between 0 and 16.

11.14 $f(x) = \dfrac{x + 3}{x - 4}$ on $[1, 3]$.

▌ Since $\dfrac{1}{x - 4}$ is differentiable and nonzero on $[1, 3]$, $f(x)$ is differentiable on $[1, 3]$.

$$f'(x) = \frac{(x - 4) - (x + 3)}{(x - 4)^2} = -\frac{7}{(x - 4)^2}.$$

Setting $-\dfrac{7}{(c - 4)^2} = \dfrac{f(3) - f(1)}{3 - 1} = \dfrac{-6 + \frac{4}{3}}{2} = -\dfrac{14}{6} = -\dfrac{7}{3}$, we obtain $(c - 4)^2 = 3$, $c - 4 = \pm\sqrt{3}$, $c = 4 \pm \sqrt{3}$. The value $4 - \sqrt{3}$ lies between 1 and 3.

11.15 $f(x) = \sqrt{25 - x^2}$ on $[-3, 4]$.

▌ $f(x)$ is differentiable on $[-3, 4]$, since $25 - x^2 > 0$ on that interval. $f'(x) = -\dfrac{x}{\sqrt{25 - x^2}}$. Setting $-\dfrac{c}{\sqrt{25 - c^2}} = \dfrac{f(4) - f(-3)}{4 - (-3)} = \dfrac{3 - 4}{7} = -\dfrac{1}{7}$, we obtain $7c = \sqrt{25 - c^2}$, $49c^2 = 25 - c^2$, $c^2 = \dfrac{1}{2}$, $c = \pm\dfrac{\sqrt{2}}{2}$. Both of these values lie in $[-3, 4]$.

11.16 $f(x) = \dfrac{1}{x - 4}$ on $[0, 2]$.

▌ Since $x - 4$ is differentiable and nonzero on $[0, 2]$, so is $f(x)$. $f'(x) = -\dfrac{1}{(x - 4)^2}$. Setting $-\dfrac{1}{(c - 4)^2} = \dfrac{f(2) - f(0)}{2 - 0} = \dfrac{-\frac{1}{2} - (-\frac{1}{4})}{2} = -\dfrac{1}{8}$, we obtain $(c - 4)^2 = 8$, $c - 4 = \pm 2\sqrt{2}$, $c = 4 \pm 2\sqrt{2}$. The value $4 - 2\sqrt{2}$ lies between 0 and 2.

11.17 Prove that, if $f'(x) > 0$ for all x in the open interval (a, b), then $f(x)$ is an increasing function on (a, b).

▌ Assume $a < u < v < b$. Then the mean value theorem applies to $f(x)$ on the closed interval (u, v). So, for some c between u and v, $f'(c) = [f(v) - f(u)]/(v - u)$. Hence, $f(v) - f(u) = f'(c)(v - u)$. Since $u < v$, $v - u > 0$. By hypothesis, $f'(c) > 0$. Hence, $f(v) - f(u) > 0$, and $f(v) > f(u)$. Thus, $f(x)$ is increasing in (a, b).

In Problems 11.18 to 11.26, determine where the function f is increasing and where it is decreasing.

11.18 $f(x) = 3x + 1$.

▌ $f'(x) = 3$. Hence, $f(x)$ is increasing everywhere.

11.19 $f(x) = -2x + 1$.

▌ $f'(x) = -2 < 0$. Hence, $f(x)$ is always decreasing.

11.20 $f(x) = x^2 - 4x + 7$.

▌ $f'(x) = 2x - 4$. Since $2x - 4 > 0 \leftrightarrow x > 2$, $f(x)$ is increasing when $x > 2$. Similarly, since $2x - 4 < 0 \leftrightarrow x < 2$, $f(x)$ is decreasing when $x < 2$.

11.21 $f(x) = 1 - 4x - x^2$.

▌ $f'(x) = -4 - 2x$. Since $-4 - 2x > 0 \leftrightarrow x < -2$, $f(x)$ is increasing when $x < -2$. Similarly, $f(x)$ is decreasing when $x > -2$.

11.22 $f(x) = \sqrt{1 - x^2}$.

▌ $f(x)$ is defined only for $-1 \le x \le 1$. Now, $f'(x) = -x/\sqrt{1 - x^2}$. So, $f(x) > 0 \leftrightarrow x < 0$. Thus, $f(x)$ is increasing when $-1 < x < 0$. Similarly, $f(x)$ is decreasing when $0 < x < 1$.

11.23 $f(x) = \frac{1}{3}\sqrt{9 - x^2}$.

▌ $f(x)$ is defined only when $-3 \le x \le 3$. $f'(x) = \frac{1}{3}(-x)/\sqrt{9 - x^2}$. So, $f'(x) > 0 \leftrightarrow x < 0$. Thus, $f(x)$ is increasing when $-3 < x < 0$ and decreasing when $0 < x < 3$.

11.24 $f(x) = x^3 - 9x^2 + 15x - 3$.

▌ $f'(x) = 3x^2 - 18x + 15 = 3(x - 5)(x - 1)$. The key points are $x = 1$ and $x = 5$. $f'(x) > 0$ when $x > 5$, $f'(x) < 0$ for $1 < x < 5$, and $f'(x) > 0$ when $x < 1$. Thus, $f(x)$ is increasing when $x < 1$ or $x > 5$, and it is decreasing when $1 < x < 5$.

11.25 $f(x) = x + 1/x$.

▌ $f(x)$ is defined for $x \neq 0$. $f'(x) = 1 - (1/x^2)$. Hence, $f'(x) < 0 \leftrightarrow 1 < 1/x^2$, which is equivalent to $x^2 < 1$. Hence, $f(x)$ is decreasing when $-1 < x < 0$ or $0 < x < 1$, and it is increasing when $x > 1$ or $x < -1$.

11.26 $f(x) = x^3 - 12x + 20$.

▌ $f'(x) = 3x^2 - 12 = 3(x - 2)(x + 2)$. The key points are $x = 2$ and $x = -2$. For $x > 2$, $f'(x) > 0$; for $-2 < x < 2$, $f'(x) < 0$; for $x < -2$, $f'(x) > 0$. Hence, $f(x)$ is increasing when $x > 2$ or $x < -2$, and it is decreasing for $-2 < x < 2$.

11.27 Let $f(x)$ be a differentiable function such that $f'(x) \neq 0$ for all x in the open interval (a, b). Prove that there is at most one zero of $f(x)$ in (a, b).

▌ Assume that there exist two zeros u and v of $f(x)$ in (a, b) with $u < v$. Then Rolle's theorem applies to $f(x)$ in the closed interval $[u, v]$. Hence, there exists a number c in (u, v) such that $f'(c) = 0$. Since $a < c < b$, this contradicts the assumption that $f'(x) \neq 0$ for all x in (a, b).

11.28 Consider the polynomial $f(x) = 5x^3 - 2x^2 + 3x - 4$. Prove that $f(x)$ has a zero between 0 and 1 that is the only zero of $f(x)$.

▌ $f(0) = -4 < 0$, and $f(1) = 2 > 0$. Hence, by the intermediate value theorem, $f(x) = 0$ for some x between 0 and 1. $f'(x) = 15x^2 - 4x + 3$. By the quadratic formula, we see that $f'(x)$ has no real roots and is, therefore, always positive. Hence, $f(x)$ is an increasing function and, thus, can take on the value 0 at most once.

11.29 Let $f(x)$ and $g(x)$ be differentiable functions such that $f(a) \ge g(a)$ and $f'(x) > g'(x)$ for all x. Show that $f(x) > g(x)$ for all $x > a$.

▌ The function $h(x) = f(x) - g(x)$ is differentiable, $h(a) \ge 0$, and $h'(x) > 0$ for all x. By the latter condition, $h(x)$ is increasing, and, therefore, since $h(a) \ge 0$, $h(x) > 0$ for all $x > a$. Thus, $f(x) > g(x)$ for all $x > a$.

11.30 The mean value theorem ensures the existence of a certain point on the graph of $y = \sqrt[3]{x}$ between $(27, 3)$ and $(125, 5)$. Find the x-coordinate of the point.

▮ $f(x) = \sqrt[3]{x}$, $f'(x) = 1/3(\sqrt[3]{x})^2$. By the mean value theorem, there is a number c between 27 and 125 such that $1/3(\sqrt[3]{c})^2 = \dfrac{f(125) - f(27)}{125 - 27} = \dfrac{5 - 3}{98} = \dfrac{1}{49}$. So, $49 = 3(\sqrt[3]{c})^2$, $\sqrt[3]{c} = \dfrac{7}{\sqrt{3}}$, $c = \dfrac{343}{3\sqrt{3}} = \dfrac{343\sqrt{3}}{9}$.

11.31 Show that $g(x) = 8x^3 - 6x^2 - 2x + 1$ has a zero between 0 and 1.

▮ Notice that the intermediate value theorem does not help, since $g(0) = 1$ and $g(1) = 1$. Let $f(x) = 2x^4 - 2x^3 - x^2 + x$ and note that $f'(x) = g(x)$. Since $f(0) = f(1) = 0$, Rolle's theorem applies to $f(x)$ on the interval $[0, 1]$. Hence, there must exist c between 0 and 1 such that $f'(c) = 0$. Then $g(c) = 0$.

11.32 Show that $x^3 + 2x - 5 = 0$ has exactly one real root.

▮ Let $f(x) = x^3 + 2x - 5$. Since $f(0) = -5 < 0$ and $f(2) = 7 > 0$, the intermediate value theorem tells us that there is a root of $f(x) = 0$ between 0 and 2. Since $f'(x) = 3x^2 + 2 > 0$ for all x, $f(x)$ is an increasing function and, therefore, can assume the value 0 at most once. Hence, $f(x)$ assumes the value 0 exactly once.

11.33 Suppose that $f(x)$ is differentiable everywhere, that $f(2) = -3$, and that $1 < f'(x) < 2$ if $2 < x < 5$. Show that $0 < f(5) < 3$.

▮ By the mean value theorem, there exists a c between 2 and 5 such that $f'(c) = \dfrac{f(5) - f(2)}{5 - 2} = \dfrac{f(5) + 3}{3}$. So, $f(5) + 3 = 3f'(c)$. Since $2 < c < 5$, $1 < f'(c) < 2$, $3 < 3f'(c) < 6$, $3 < f(5) + 3 < 6$, and $0 < f(5) < 3$.

11.34 Use the mean value theorem to prove that $\tan x > x$ for $0 < x < \pi/2$.

▮ The mean value theorem applies to $\tan x$ on the interval $[0, x]$. Hence, there exists c between 0 and x such that $\sec^2 c = (\tan x - \tan 0)/(x - 0) = \tan x/x$. [Recall that $D_x(\tan x) = \sec^2 x$.] Since $0 < c < \pi/2$, $0 < \cos c < 1$, $\sec c > 1$, $\sec^2 c > 1$. Thus, $\tan x/x > 1$, and, therefore, $\tan x > x$.

11.35 If $f'(x) = 0$ throughout an interval $[a, b]$, prove that $f(x)$ is constant on that interval.

▮ Let $a < x \le b$. The mean value theorem applies to $f(x)$ on the interval $[a, x]$. Hence, there exists a c between a and x such that $f'(c) = \dfrac{f(x) - f(a)}{x - a}$. Since $f'(c) = 0$, $f(x) = f(a)$. Hence, $f(x)$ has the value $f(a)$ throughout the interval.

11.36 If $f'(x) = g'(x)$ for all x in an interval $[a, b]$, show that there is a constant K such that $f(x) = g(x) + K$ for all x in $[a, b]$.

▮ Let $h(x) = f(x) - g(x)$. Then $h'(x) = 0$ for all x in $[a, b]$. By Problem 11.35, there is a constant K such that $h(x) = K$ for all x in $[a, b]$. Hence, $f(x) = g(x) + K$ for all x in $[a, b]$.

11.37 Prove that $x^3 + px + q = 0$ has exactly one real root if $p > 0$.

▮ Let $f(x) = x^3 + px + q$. Then $f'(x) = 3x^2 + p > 0$. Hence, $f(x)$ is an increasing function. So $f(x)$ assumes the value 0 at most once. Now, $\lim\limits_{x \to +\infty} f(x) = +\infty$ and $\lim\limits_{x \to -\infty} f(x) = -\infty$. Hence, there are numbers u and v where $f(u) > 0$ and $f(v) < 0$. By the intermediate value theorem, $f(x)$ assumes the value 0 for some number between u and v. Thus, $f(x)$ has exactly one real root.

11.38 Prove the following generalized mean value theorem: If $f(x)$ and $g(x)$ are continuous on $[a, b]$, and if $f(x)$ and $g(x)$ are differentiable on (a, b) with $g'(x) \ne 0$, then there exists a c in (a, b) such that $\dfrac{f'(c)}{g'(c)} = \dfrac{f(b) - f(a)}{g(b) - g(a)}$.

▮ $g(a) \ne g(b)$. [Otherwise, if $g(a) = g(b) = K$, then Rolle's theorem applied to $g(x) - K$ would yield a number between a and b at which $g'(x) = 0$, contrary to our hypothesis.] Let $L = \dfrac{f(b) - f(a)}{g(b) - g(a)}$ and set $F(x) = f(x) - f(b) - L[g(x) - g(b)]$. It is easy to see that Rolle's theorem applies to $F(x)$. Therefore, there is a number c between a and b for which $F'(c) = 0$. Then, $f'(c) - Lg'(c) = 0$, and $\dfrac{f'(c)}{g'(c)} = L = \dfrac{f(b) - f(a)}{g(b) - g(a)}$.

11.39   Use the generalized mean value theorem to show that $\lim\limits_{x\to 0}\dfrac{\sin x}{x}=1$.

$\blacksquare$ Let $f(x)=\sin x$ and $g(x)=x$. Since $g'(x)=1$, the generalized mean value theorem applies to the interval $[0, x]$ when $x>0$. Hence, there is a number c such that $0<c<x$ for which $\dfrac{f'(c)}{g'(c)}=\dfrac{\cos c}{1}=\dfrac{\sin x-\sin 0}{x-0}=\dfrac{\sin x}{x}$. As $x\to 0^{+}$, $c\to 0$ and $\cos c\to 1$. Hence, $\lim\limits_{x\to 0^{+}}\dfrac{\sin x}{x}=1$. Since $\sin(-x)=-\sin x$, we also have $\lim\limits_{x\to 0^{-}}\dfrac{\sin x}{x}=1$. Hence, $\lim\limits_{x\to 0}\dfrac{\sin x}{x}=1$.

11.40   Show that $|\sin u-\sin v|\le|u-v|$.

$\blacksquare$ By the mean value theorem, there exists a c between u and v for which $\cos c=\dfrac{\sin u-\sin v}{u-v}$. Since $|\cos c|\le 1$, $|\sin u-\sin v|\le|u-v|$.

11.41   Apply the mean value theorem to the following functions on the interval $[-1, 8]$. (a) $f(x)=x^{4/3}$ (b) $g(x)=x^{2/3}$.

$\blacksquare$ (a) $f'(x)=\frac{4}{3}x^{1/3}$. By the mean value theorem, there is a number c between -1 and 8 such that $\frac{4}{3}c^{1/3}=\dfrac{f(8)-f(-1)}{8-(-1)}=\dfrac{16-1}{9}=\dfrac{5}{3}$. Hence, $\frac{4}{3}\sqrt[3]{c}=\frac{5}{3}$, $c=\frac{125}{64}$. (b) The mean value theorem is not applicable because $g'(x)$ does not exist at $x=0$.

11.42   Show that the equation $3\tan x+x^{3}=2$ has exactly one solution in the interval $[0, \pi/4]$.

$\blacksquare$ Let $f(x)=3\tan x+x^{3}$. Then $f'(x)=3\sec^{2}x+3x^{2}>0$, and, therefore, $f(x)$ is an increasing function. Thus, $f(x)$ assumes the value 2 at most once. But, $f(0)=0$ and $f(\pi/4)=3+(\pi/4)^{3}>2$. So, by the intermediate value theorem, $f(c)=2$ for some c between 0 and $\pi/4$. Hence, $f(x)=2$ for exactly one x in $[0, \pi/4]$.

11.43   Give an example of a function that is continuous on $[-1, 1]$ and for which the conclusion of the mean value theorem does not hold.

$\blacksquare$ Let $f(x)=|x|$. Then $\dfrac{f(1)-f(-1)}{1-(-1)}=\dfrac{1-1}{2}=0$. However, there is no number c in $(-1, 1)$ for which $f'(c)=0$, since $f'(x)=1$ for $x>0$ and $f'(x)=-1$ for $x<0$. Of course, $f'(0)$ does not exist, which is the reason that the mean value theorem does not apply.

11.44   Find a point on the graph of $y=x^{2}+x+3$, between $x=1$ and $x=2$, where the tangent line is parallel to the line connecting $(1, 5)$ and $(2, 9)$.

$\blacksquare$ This is essentially an application of the mean value theorem to $f(x)=x^{2}+x+3$ on the interval $[1, 2]$. The slope of the line connecting $(1, 5)$ and $(2, 9)$ is $\dfrac{f(2)-f(1)}{2-1}=\dfrac{9-5}{1}=4$. For that line to be parallel to the tangent line at a point $(c, f(c))$, the slope of the tangent line, $f'(c)$, must be equal to 4. But, $f'(x)=2x+1$. Hence, we must have $2c+1=4$, $c=\frac{3}{2}$. Hence, the point is $(\frac{3}{2}, \frac{27}{4})$.

11.45   For a function $f(x)=Ax^{2}+Bx+C$, with $A\neq 0$, on an interval $[a, b]$, find the number in (a, b) determined by the mean value theorem.

$\blacksquare$ $f'(x)=2Ax+B$.   On the other hand,

$$\frac{f(b)-f(a)}{b-a}=\frac{(Ab^{2}+Bb+C)-(Aa^{2}+Ba+C)}{b-a}=\frac{A(b^{2}-a^{2})+B(b-a)}{b-a}=A(b+a)+B$$

Hence, we must find c such that $2Ac+B=A(b+a)+B$. Then $c=\frac{1}{2}(b+a)$. Thus, the point is the midpoint of the interval.

11.46   If f is a differentiable function such that $\lim\limits_{x\to +\infty}f'(x)=0$, prove that $\lim\limits_{x\to +\infty}[f(x+1)-f(x)]=0$.

$\blacksquare$ By the mean value theorem, there exists a c with $x<c<x+1$ such that $f(x+1)-f(x)=f'(c)$. As $x\to +\infty$, $c\to +\infty$. Hence, $f'(c)$ approaches 0, since $\lim\limits_{x\to +\infty}f'(x)=0$. Therefore $\lim\limits_{x\to +\infty}[(f(x+1)-f(x))]=0$.

11.47 Prove that the zeros of sin x and cos x separate each other, that is, between any two zeros of sin x, there is a zero of cos x, and vice versa.

❚ Assume $\sin a = 0$ and $\sin b = 0$ with $a < b$. By Rolle's theorem, there exists a c with $a < c < b$ such that $\cos c = 0$, since $D_x(\sin x) = \cos x$. Similarly, if $\cos a = \cos b = 0$ with $a < b$, then there exists a c with $a < c < b$ such that $\sin c = 0$, since $D_x(\cos x) = -\sin x$.

11.48 An important function in calculus (the *exponential function*) may be defined by the conditions

$$f'(x) = f(x) \quad (-\infty < x < +\infty) \quad \text{and} \quad f(0) = 1 \tag{1}$$

Show that this function is (**a**) strictly positive, and (**b**) strictly increasing.

❚ (**a**) Let $h(x) = f(x)f(-x)$; then, using the product rule, the chain rule, and (*1*), $h'(x) = f'(x)f(-x) + f(x)f'(-x)(-1) = f(x)f(-x) - f(x)f(-x) = 0$. So by Problem 11.35, $h(x) = \text{const.} = h(0) = 1 \cdot 1 = 1$; that is, for all x,

$$f(x)f(-x) = 1 \tag{2}$$

By (*2*), $f(x)$ is never zero. Furthermore, the continuous (because it is differentiable) function $f(x)$ can never be negative; for $f(a) < 0$ and $f(0) = 1 > 0$ would imply an intermediate zero value, which we have just seen to be impossible. Hence $f(x)$ is strictly positive, (**b**) $f'(x) = f(x) > 0$, so (Problem 11.17) $f(x)$ is strictly increasing.

11.49 Give an example of a continuous function $f(x)$ on $[0, 1]$ for which the conclusion of Rolle's theorem fails.

❚ Let $f(x) = \frac{1}{2} - |x - \frac{1}{2}|$ (see Fig. 11-1). Then $f(0) = f(1) = 0$, but $f'(x)$ is not 0 for any x in $(0, 1)$. $f'(x) = \pm 1$ for all x in $(0, 1)$, except at $x = \frac{1}{2}$, where $f'(x)$ is not defined.

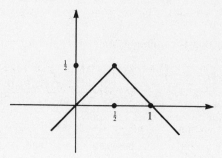

Fig. 11-1

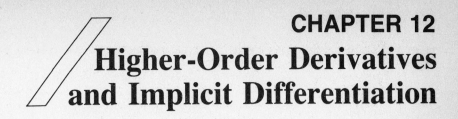

12.1 Find the second derivative y'' of the function $y = \sqrt{x^2 + 1}$ by direct computation.

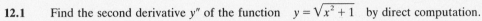 $y = (x^2 + 1)^{1/2}$. By the chain rule, $y' = \frac{1}{2}(x^2 + 1)^{-1/2} \cdot \frac{d}{dx}(x^2 + 1) = \frac{1}{2\sqrt{x^2 + 1}} \cdot (2x) = \frac{x}{\sqrt{x^2 + 1}}$. By the

quotient rule, $y'' = \dfrac{\sqrt{x^2 + 1}\,\dfrac{d}{dx}(x) - x\,\dfrac{d}{dx}(\sqrt{x^2 + 1})}{x^2 + 1} = \dfrac{\sqrt{x^2 + 1} - x \cdot \dfrac{x}{\sqrt{x^2 + 1}}}{x^2 + 1} = \dfrac{(x^2 + 1) - x^2}{(x^2 + 1)^{3/2}} = \dfrac{1}{(x^2 + 1)^{3/2}}$.

12.2 Use implicit differentiation to solve Problem 12.1.

$\blacksquare$ $y^2 = x^2 + 1$. Take the derivative of both sides with respect to x. By the chain rule, $\frac{d}{dx}(y^2) = 2y \cdot y'$. Thus, $2yy' = 2x$, and, therefore, $yy' = x$. Take the derivative with respect to x of both sides, using the product rule on the left: $y \cdot y'' + y' \cdot y' = 1$. So, $yy'' = 1 - (y')^2$. But, since $yy' = x$, $y' = x/y$. Hence, $yy'' = 1 - x^2/y^2 = (y^2 - x^2)/y^2 = 1/y^2 = 1/(x^2 + 1)$. Thus, $y'' = 1/y(x^2 + 1) = 1/(x^2 + 1)^{3/2}$.

12.3 Find all derivatives $y^{(n)}$ of the function $y = \pi x^3 - 7x$.

$\blacksquare$ $y' = 3\pi x^2 - 7$, $y'' = 6\pi x$, $y''' = 6\pi$, and $y^{(n)} = 0$ for $n \geq 4$.

12.4 Find all derivatives $y^{(n)}$ of the function $y = \sqrt{x + 5}$.

$\blacksquare$ $y = (x + 5)^{1/2}$, $y' = \frac{1}{2}(x + 5)^{-1/2}$, $y'' = -\frac{1}{4}(x + 5)^{-3/2}$, $y''' = \frac{3}{8}(x + 5)^{-5/2}$, and $y^{(4)} = -\frac{15}{16}(x + 5)^{-7/2}$. This is enough to detect the general pattern: $y^{(n)} = (-1)^{n+1} \dfrac{1 \cdot 3 \cdot 5 \cdot \cdots \cdot (2n - 3)}{2^n} (x + 5)^{-(2n-1)/2}$.

12.5 Find all derivatives $y^{(n)}$ of the function $y = 1/(3 + x)$.

$\blacksquare$
$$y = (3 + x)^{-1}$$
$$y' = -(3 + x)^{-2} = -\frac{1}{(3 + x)^2}$$
$$y'' = 2(3 + x)^{-3} = \frac{2}{(3 + x)^3}$$
$$y''' = -6(3 + x)^{-4} = -\frac{6}{(3 + x)^4}$$

The general pattern is $y^{(n)} = (-1)^n (n!)(3 + x)^{-(n+1)} = (-1)^n (n!)\dfrac{1}{(3 + x)^{n+1}}$

12.6 Find all derivatives $y^{(n)}$ of the function $y = (x + 1)/(x - 1)$.

$\blacksquare$
$$y = \frac{(x - 1) + 2}{x - 1} = 1 + \frac{2}{x - 1} = 1 + 2(x - 1)^{-1}$$
$$y' = -2(x - 1)^{-2}$$
$$y'' = 4(x - 1)^{-3}$$
$$y''' = -12(x - 1)^{-4}$$

The general pattern is $y^{(n)} = (-1)^n 2(n!)(x - 1)^{-(n+1)} = (-1)^n (n!)\dfrac{2}{(x - 1)^{n+1}}$

12.7 Find all derivatives $y^{(n)}$ of the function $y = \sqrt{2x - 1}$.

▮ Use implicit differentiation. $y^2 = 2x - 1$. Hence, $2yy' = 2$, $y' = y^{-1}$. So,

$$y'' = -y^{-2} \cdot y' = -y^{-2} \cdot y^{-1} = -y^{-3}$$
$$y''' = 3y^{-4} \cdot y' = 3y^{-4} \cdot y^{-1} = 3y^{-5}$$
$$y^{(4)} = -3 \cdot 5y^{-6} \cdot y' = -3 \cdot 5y^{-6} \cdot y^{-1} = -3 \cdot 5y^{-7}$$

So, the pattern that emerges is $y^{(n)} = (-1)^{n+1}[1 \cdot 3 \cdot 5 \cdot \cdots \cdot (2n - 3)]y^{-(2n-1)} = (-1)^{n+1}[1 \cdot 3 \cdot 5 \cdot \cdots \cdot (2n - 3)] \cdot \dfrac{1}{(\sqrt{2x - 1})^{2n-1}}$.

12.8 Find all derivatives $y^{(n)}$ of the function $y = \sin x$.

▮ $y' = \cos x$, $y'' = -\sin x$, $y''' = -\cos x$, $y^{(4)} = \sin x$, and then the pattern of these four functions keeps on repeating.

12.9 Find the smallest positive integer n such that $D_x^n(\cos x) = \cos x$.

▮ Let $y = \cos x$. $y' = -\sin x$, $y'' = -\cos x$, $y''' = \sin x$ and $y^{(4)} = \cos x$. Hence, $n = 4$.

12.10 Calculate $y^{(5)}$ for $y = \sin^2 x$.

▮ By the chain rule, $y' = 2 \sin x \cos x = \sin 2x$. Hence, $y'' = \cos 2x \cdot 2 = 2 \cos 2x$, $y''' = 2(-\sin 2x) \cdot 2 = -4 \sin 2x$, $y^{(4)} = -4 (\cos 2x) \cdot 2 = -8 \cos 2x$, $y^{(5)} = -8(-\sin 2x) \cdot 2 = 16 \sin 2x = 16(2 \sin x \cos x) = 32 \sin x \cos x$.

12.11 On the circle $x^2 + y^2 = a^2$, find y''.

▮ By implicit differentiation, $2x + 2yy' = 0$, $y' = -x/y$. By the quotient rule, $y'' = -\dfrac{y \cdot 1 - x \cdot y'}{y^2} = -\dfrac{y - x(-x/y)}{y^2} = -\dfrac{y^2 + x^2}{y^3} = -\dfrac{a^2}{y^3}$.

12.12 If $x^3 - y^3 = 1$, find y''.

▮ Use implicit differentiation. $3x^2 - 3y^2y' = 0$. So, $y' = x^2/y^2$. By the quotient rule,

$$y'' = \frac{y^2(2x) - x^2 \dfrac{d}{dx}(y^2)}{y^4} = \frac{2xy^2 - x^2(2yy')}{y^4} = \frac{2xy^2 - 2x^2y(x^2/y^2)}{y^4}$$

$$= \frac{2xy^3 - 2x^4}{y^5} = \frac{2x(y^3 - x^3)}{y^5} = \frac{2x(-1)}{y^5} = -\frac{2x}{y^5}.$$

12.13 If $xy + y^2 = 1$, find y' and y''.

▮ Use implicit differentiation. $xy' + y + 2yy' = 0$. Hence, $y'(x + 2y) = -y$, and $y' = -\dfrac{y}{x + 2y}$. By the quotient rule,

$$y'' = -\frac{(x + 2y)y' - y(1 + 2y')}{(x + 2y)^2} = -\frac{xy' - y}{(x + 2y)^2}$$

$$= -\frac{x[-y/(x + 2y)] - y}{(x + 2y)^2} = -\frac{-xy - y(x + 2y)}{(x + 2y)^3}$$

$$= \frac{2xy + 2y^2}{(x + 2y)^3} = \frac{2(xy + y^2)}{(x + 2y)^3} = \frac{2}{(x + 2y)^3}.$$

12.14 At the point $(1, 2)$ of the curve $x^2 - xy + y^2 = 3$, find an equation of the tangent line.

▮ Use implicit differentiation. $2x - (xy' + y) + 2yy' = 0$. Substitute 1 for x and 2 for y. $2 - (y' + 2) + 4y' = 0$. So, $y' = 0$. Hence, the tangent line has slope 0, and, since it passes through $(1, 2)$, its equation is $y = 2$.

12.15 If $x^2 + 2xy + 3y^2 = 2$, find y' and y'' when $y = 1$.

❚ Use implicit differentiation. (*) $2x + 2(xy' + y) + 6yy' = 0$. When $y = 1$, the original equation yields $x^2 + 2x + 3 = 2$, $x^2 + 2x + 1 = 0$, $(x + 1)^2 = 0$, $x + 1 = 0$, $x = -1$. Substitute -1 for x and 1 for y in (*), which results in $-2 + 2(-y' + 1) + 6y' = 0$; so, $y' = 0$ when $y = 1$. To find y'', first simplify (*) to $x + xy' + y + 3yy' = 0$, and then differentiate implicitly to get $1 + (xy'' + y') + y' + 3(yy'' + y'y') = 0$. In this equation, substitute -1 for x, 1 for y, and 0 for y', which results in $1 - y'' + 3y'' = 0$, $y'' = -\frac{1}{2}$.

12.16 Find the slope of the tangent line to the graph of $y = x + \cos xy$ at $(0, 1)$.

❚ Differentiate implicitly to get $y' = 1 - [\sin xy \cdot (xy' + y)]$. Replace x by 0 and y by 1. $y' = 1 - [\sin (0) \cdot 1] = 1 - 0 = 1$. Thus, the tangent line has slope 1.

12.17 If $\cos y = x$, find y'.

❚ Differentiate implicitly: $(-\sin y)y' = 1$. Hence, $y' = -1/(\sin y) = \pm 1/\sqrt{1 - x^2}$.

12.18 Find an equation of the tangent line to the curve $1 + 16x^2 y = \tan (x - 2y)$ at the point $(\pi/4, 0)$.

❚ Differentiating implicitly, $16(x^2 y' + 2xy) = [\sec^2(x - 2y)](1 - 2y')$. Substituting $\pi/4$ for x and 0 for y, $16(\pi^2/16)(y') = [\sec^2 (\pi/4)](1 - 2y')$. Since $\cos (\pi/4) = \sqrt{2}/2$, $\sec^2 (\pi/4) = 2$. Thus, $\pi^2 y' = 2(1 - 2y')$. Hence, $y' = 2/(\pi^2 + 4)$, which is the slope of the tangent line. A point-slope equation of the tangent line is $y = [2/(\pi^2 + 4)](x - \pi/4)$.

12.19 Evaluate y'' on the ellipse $b^2 x^2 + a^2 y^2 = a^2 b^2$.

❚ Use implicit differentiation to get $2b^2 x + 2a^2 yy' = 0$, $y' = -(b^2/a^2)(x/y)$. Now differentiate by the quotient rule.

$$y'' = -\frac{b^2}{a^2} \left(\frac{y - xy'}{y^2} \right) = -\frac{b^2}{a^2} \frac{y - x[(-b^2/a^2)(x/y)]}{y^2} = -\frac{b^2}{a^2} \frac{a^2 y^2 + b^2 x^2}{a^2 y^3} = -\frac{b^2}{a^2} \frac{a^2 b^2}{a^2 y^3} = -\frac{b^4}{a^2 y^3}.$$

12.20 Find y'' on the parabola $y^2 = 4px$.

❚ By implicit differentiation, $2yy' = 4p$, $yy' = 2p$. Differentiating again and using the product rule, $yy'' + y'y' = 0$. Multiply both sides by y^2. $y^3 y'' + y^2(y')^2 = 0$. However, since $yy' = 2p$, $y^2(y')^2 = 4p^2$. So, $y^3 y'' + 4p^2 = 0$, and, therefore, $y'' = -4p^2/y^3$.

12.21 Find a general formula for y'' on the curve $x^n + y^n = a^n$.

❚ By implicit differentiation, $nx^{n-1} + ny^{n-1}y' = 0$, that is, (*) $x^{n-1} + y^{n-1}y' = 0$. Hence, $y^{n-1}y' = -x^{n-1}$, and, therefore, squaring, (**) $y^{2n-2}(y')^2 = x^{2n-2}$. Now differentiate (*) and multiply by y^n. $(n - 1)x^{n-2}y^n + [y^{2n-1}y'' + (n - 1)^{2n-2}(y')^2] = 0$. Use (**) to replace $y^{2n-2}(y')^2$ by x^{2n-2}. $(n - 1)x^{n-2}y^n + [y^{2n-1}y'' + (n - 1)x^{2n-2}] = 0$. Hence, $y^{2n-1}y'' = -(n - 1)(x^{2n-2} + x^{n-2}y^n) = -(n - 1)x^{n-2}(x^n + y^n) = -(n - 1)x^{n-2}a^n$. Thus, $y'' = -(n - 1)a^n x^{n-2}/y^{2n-1}$. (Check this formula in the special case of Problem 12.11).

12.22 Find y'' on the curve $x^{1/2} + y^{1/2} = a^{1/2}$.

❚ Use the formula obtained in Problem 12.21 in the special case $n = \frac{1}{2}$. $y'' = -(-\frac{1}{2}a^{1/2}x^{-3/2})/y^0 = \frac{1}{2}a^{1/2}/x^{3/2}$.

12.23 Find the 10th and 11th derivatives of the function $f(x) = x^{10} - 14x^7 + 3x^5 + 2x^3 - x + 2$.

❚ By the 10th differentiation, the offspring of all the terms except x^{10} have been reduced to 0. The successive offspring of x^{10} are $10x^9, 9 \cdot 10x^8, 8 \cdot 9 \cdot 10x^7, \ldots, 1 \cdot 2 \cdot 3 \cdot \cdots \cdot 10$. Thus, the 10th derivative is 10!. The 11th derivative is 0.

12.24 For the curve $y^3 = x^2$, calculate y' (a) by implicit differentiation and (b) by first solving for y and then differentiating. Show that the two results agree.

❚ (a) $3y^2 y' = 2x$. Hence, $y' = 2x/3y^2$. (b) $y = x^{2/3}$. So, $y' = \frac{2}{3}x^{-1/3}$. Observe that, since $y^2 = x^{4/3}$, the two answers are the same.

12.25 Find a formula for the nth derivative of $y = 1/x(1 - x)$.

❚ Observe that $y = 1/x + 1/(1 - x)$. Now, the nth derivative of $1/x$ is easily seen to be $(-1)^n (n!) x^{-(n+1)}$ and that of $1/(x - 1)$ to be $(n!)(1 - x)^{-(n+1)}$. Hence, $y^{(n)} = (n!)[(-1)^n/x^{n+1} + 1/(1 - x)^{n+1})]$.

12.26 Consider the function $f(x)$ defined by

$$f(x) = \begin{cases} x^2 & \text{if } x \geq 0 \\ -x^2 & \text{if } x < 0 \end{cases}$$

Show that $f''(0)$ does not exist.

❚ $f'(x) = 2x$ if $x > 0$ and $f'(x) = -2x$ if $x < 0$. Direct computation by the Δ-definition, shows that $f'(0) = 0$. Hence, $f'(x) = 2|x|$ for all x. Since $|x|$ is not differentiable at $x = 0$, $f''(0)$ cannot exist.

12.27 Consider the circles C_1: $(x - a)^2 + y^2 = 8$ and C_2: $(x + a)^2 + y^2 = 8$. Determine the value of a so that C_1 and C_2 intersect at right angles.

❚ Solving the equations for C_1 and C_2 simultaneously, we find $(x - a)^2 = (x + a)^2$, and, therefore, $x = 0$. Hence, $y = \pm\sqrt{8 - a^2}$. On C_1, $2(x_1 - a) + 2y_1 y_1' = 0$, so at the intersection points, $y_1 y_1' = a$. (Here, the subscript indicates values on C_1.) On C_2, $2(x_2 + a) + 2y_2 y_2' = 0$, so at the intersection points, $y_2 y_2' = -a$. Hence, multiplying these equations at the intersection points, $y_1 y_1' y_2 y_2' = -a^2$. At these points, $y_1 = y_2 = y$; hence, $y^2 y_1' y_2' = -a^2$. Since C_1 and C_2 are supposed to be perpendicular at the intersection points, their tangent lines are perpendicular, and, therefore, the product $y_1' y_2'$ of the slopes of their tangent lines must be -1. Hence, $-y^2 = -a^2$, $y^2 = a^2$. But $y^2 = 8 - a^2$ at the intersection point. So, $a^2 = 8 - a^2$, $a^2 = 4$, $a = 2$.

12.28 Show that the curves C_1: $9y - 6x + y^4 + x^3 y = 0$ and C_2: $10y + 15x + x^2 - xy^3 = 0$ intersect at right angles at the origin.

❚ On C_1, $9y' - 6 + 4y^3 y' + x^3 y' + 3x^2 y = 0$. At the origin $(0, 0)$, $9y' - 6 = 0$, or $y' = \frac{2}{3}$. On C_2, $10y' + 15 + 2x - y^3 - 3xy^2 y' = 0$. At the origin, $10y' + 15 = 0$, or $y' = -\frac{3}{2}$. Since the values of y' on C_1 and C_2 at the origin are negative reciprocals of each other, the tangent lines of C_1 and C_2 are perpendicular at the origin.

In Problems 12.29 to 12.35, calculate the second derivative y''.

12.29 $y = \sqrt{x + 5}$.

❚ $y = (x + 5)^{1/2}$, $y' = \frac{1}{2}(x + 5)^{-1/2}$, $y'' = -\frac{1}{4}(x + 5)^{-3/2} = -1/[4(\sqrt{x + 5})^3]$.

12.30 $y = \sqrt[3]{x - 1}$.

❚ $y = (x - 1)^{1/3}$, $y' = \frac{1}{3}(x - 1)^{-2/3}$, $y'' = -\frac{2}{9}(x - 1)^{-5/3} = -2/[9(\sqrt[3]{x - 1})^5]$.

12.31 $y = \sqrt[4]{x^2 + 1}$.

❚
$$y = (x^2 + 1)^{1/4}$$
$$y' = \frac{1}{4}(x^2 + 1)^{-3/4} \cdot (2x) = \frac{1}{2}x(x^2 + 1)^{-3/4}$$
$$y'' = \frac{1}{2}\{x \cdot [-\frac{3}{4}(x^2 + 1)^{-7/4})](2x) + (x^2 + 1)^{-3/4}\}$$
$$= \frac{1}{4}[-3x^2(x^2 + 1)^{-7/4} + 2(x^2 + 1)^{-3/4}]$$
$$= \frac{1}{4}(x^2 + 1)^{-7/4}[-3x^2 + 2(x^2 + 1)]$$
$$= \frac{1}{4}(x^2 + 1)^{-7/4}(2 - x^2) = \frac{2 - x^2}{4(\sqrt[4]{x^2 + 1})^7}$$

12.32 $y = (x + 1)(x - 3)^3$.

❚ Here it is simplest to use the product rule $(uv)'' = u''v + 2u'v' + uv''$. Then $y'' = (0)(x - 3)^3 + 2(1)[3(x - 3)^2] + (x + 1)[6(x - 3)] = 12(x - 3)(x - 1)$.

12.33 $y = \dfrac{x}{(1-x)^2}$.

▮
$$y' = \frac{(1-x)^2 - x(2)(1-x)(-1)}{(1-x)^4} = \frac{(1-x)+2x}{(1-x)^3} = \frac{x+1}{(1-x)^3}$$

$$y'' = \frac{(1-x)^3 - (x+1)(3)(1-x)^2(-1)}{(1-x)^6} = \frac{(1-x)+3(x+1)}{(1-x)^4} = \frac{2(x+2)}{(1-x)^4}$$

12.34 $x^2 - y^2 = 1$.

▮ $2x - 2yy' = 0$, $\quad x - yy' = 0$, $\quad y' = \dfrac{x}{y}$. $\quad y'' = \dfrac{y - xy'}{y^2} = \dfrac{y - x(x/y)}{y^2} = \dfrac{y^2 - x^2}{y^3} = -\dfrac{1}{y^3}$.

12.35 $\sqrt{x} - \sqrt{y} = 1$.

▮ $x^{1/2} - y^{1/2} = 1$, $\quad \frac{1}{2}x^{-1/2} - \frac{1}{2}y^{-1/2}y' = 0$, $\quad x^{-1/2} - y^{-1/2}y' = 0$, $\quad x^{-1/2} = y^{-1/2}y'$, $\quad y' = x^{-1/2}/y^{-1/2} = y^{1/2}/x^{1/2}$.

$y'' = \dfrac{x^{1/2}(\frac{1}{2}y^{-1/2}y') - \frac{1}{2}y^{1/2}x^{-1/2}}{x} = \dfrac{1}{2}\dfrac{x^{1/2}(x^{-1/2}) - y^{1/2}x^{-1/2}}{x} = \dfrac{1}{2}x^{-1/2}\dfrac{x^{1/2} - y^{1/2}}{x} = \dfrac{1}{2x^{3/2}}$.

12.36 Find all derivatives of $y = 2x^2 + x - 1 + 1/x$.

▮ $y = 2x^2 + x - 1 + x^{-1}$, $\quad y' = 4x + 1 - x^{-2}$, $\quad y'' = 4 + 2x^{-3}$, $\quad y''' = -(3 \cdot 2)x^{-4}$, $\quad y^{(4)} = (4 \cdot 3 \cdot 2)x^{-5}$, and, for $n \geq 4$, $\quad y^{(n)} = (-1)^n(n!)x^{-(n+1)}$.

12.37 At the point $(1, 2)$ of the curve $x^2 - xy + y^2 = 3$, find the rate of change with respect to x of the slope of the tangent line to the curve.

▮ By implicit differentiation, (∗) $2x - (xy' + y) + 2yy' = 0$. Substitution of $(1, 2)$ for (x, y) yields $2 - (y' + 2) + 4y' = 0$, $\quad y' = 0$. Implicit differentiation of (∗) yields $2 - (xy'' + y' + y') + 2yy'' + 2(y')^2 = 0$, Substitution of $(1, 2)$ for (x, y), taking into account that $y' = 0$ at $(1, 2)$, yields $2 + (4 - 1)y'' = 0$, $\quad y'' = -\frac{2}{3}$. This is the rate of change of the slope y' of the tangent line.

In Problems 12.38 to 12.41, use implicit differentiation to find y'.

12.38 $\tan xy = y$.

▮ $(\sec^2 xy) \cdot (xy' + y) = y'$. Note that $\sec^2 xy = 1 + \tan^2 xy = 1 + y^2$. Hence, $(1 + y^2)(xy' + y) = y'$, $y'[x(1 + y^2) - 1] = -y(1 + y^2)$, $\quad y' = y(1 + y^2)/[1 - x(1 + y^2)]$.

12.39 $\sec^2 y + \cot^2 x = 3$.

▮ $(2 \sec y)(\sec y \tan y)y' + (2 \cot x)(-\csc^2 x) = 0$, $\quad y' = \cot x \csc^2 x / \sec^2 y \tan y$.

12.40 $\tan^2 (y + 1) = 3 \sin x$.

▮ $2 \tan(y + 1) \sec^2 (y + 1)y' = 3 \cos x$. Hence, $y' = \dfrac{3 \cos x}{2 \tan (y + 1)[\tan^2 (y + 1) + 1]}$. Since $\sec^2 (y + 1) = \tan^2 (y + 1) + 1 = 3 \sin x + 1$, the answer can also be written as $y' = \dfrac{3 \cos x}{2 \tan (y + 1)(3 \sin x + 1)}$.

12.41 $y = \tan^2 (x + y)$.

▮ Note that $\sec^2 (x + y) = \tan^2 (x + y) + 1 = y + 1$. $\quad y' = 2 \tan (x + y) \sec^2 (x + y)(1 + y') = 2 \tan (x + y) \times (y + 1)(1 + y')$. So, $y'[1 - 2 \tan (x + y)(y + 1)] = 2 \tan (x + y)(y + 1)$,

$$y' = \frac{2 \tan (x + y)(y + 1)}{1 - 2 \tan (x + y)(y + 1)} = \frac{2\sqrt{y}(y + 1)}{1 - 2\sqrt{y}(y + 1)}$$

12.42 Find the equations of the tangent lines to the ellipse $9x^2 + 16y^2 = 52$ that are parallel to the line $9x - 8y = 1$.

▮ By implicit differentiation, $18x + 32yy' = 0$, $\quad y' = -(9x/16y)$. The slope of $9x - 8y = 1$ is $\frac{9}{8}$. Hence, for the tangent line to be parallel to $9x - 8y = 1$, we must have $-(9x/16y) = \frac{9}{8}$, $\quad -x = 2y$. Substituting in the equation of the ellipse, we obtain $9(4y^2) + 16y^2 = 52$, $52y^2 = 52$, $y^2 = 1$, $y = \pm 1$. Since $x = -2y$, the points of tangency are $(-2, 1)$ and $(2, -1)$. Hence, the required equations are $y - 1 = \frac{9}{8}(x + 2)$ and $y + 1 = \frac{9}{8}(x - 2)$, or $9x - 8y = -26$ and $9x - 8y = 26$.

12.43 Show that the ellipse $4x^2 + 9y^2 = 45$ and the hyperbola $x^2 - 4y^2 = 5$ are orthogonal.

   ❚   To find the intersection points, multiply the equation of the hyperbola by 4 and subtract the result from the equation of the ellipse, obtaining $25y^2 = 25$, $y^2 = 1$, $y = \pm 1$, $x = \pm 3$. Differentiate both sides of the equation of the ellipse: $8x + 18yy' = 0$, $y' = -(4x/9y)$, which is the slope of the tangent line. Differentiate both sides of the equation of the hyperbola: $2x - 8yy' = 0$, $y' = x/4y$, which is the slope of the tangent line. Hence, the product of the slopes of the tangent lines is $-(4x/9y) \cdot (x/4y) = -(x^2/9y^2)$. Since $x^2 = 9$ and $y^2 = 1$ at the intersection points, the product of the slopes is -1, and, therefore, the tangent lines are perpendicular.

12.44 Find the slope of the tangent line to the curve $x^2 + 2xy - 3y^2 = 9$ at the point $(3, 2)$.

   ❚ $2x + 2(xy' + y) - 6yy' = 0$. Replace x by 3 and y by 2, obtaining $6 + 2(3y' + 2) - 12y' = 0$, $10 - 6y' = 0$, $y' = \frac{5}{3}$. Thus, the slope is $\frac{5}{3}$.

12.45 Show that the parabolas $y^2 = 4x + 4$ and $y^2 = 4 - 4x$ intersect at right angles.

   ❚   To find the intersection points, set $4x + 4 = 4 - 4x$. Then $x = 0$, $y^2 = 4$, $y = \pm 2$. For the first parabola, $2yy' = 4$, $y' = 2/y$. For the second parabola, $2yy' = -4$, $y' = -2/y$. Hence, the product of the slopes of the tangent lines is $(2/y)(-2/y) = -4/y^2$. At the points of intersection, $y^2 = 4$. Hence, the product of the slopes is -1, and, therefore, the tangent lines are perpendicular.

12.46 Show that the circles $x^2 + y^2 - 12x - 6y + 25 = 0$ and $x^2 + y^2 + 2x + y - 10 = 0$ are tangent to each other at the point $(2, 1)$.

   ❚   For the first circle, $2x + 2yy' - 12 - 6y' = 0$, and, therefore, at $(2, 1)$, $4 + 2y' - 12 - 6y' = 0$, $y' = -2$. For the second circle, $2x + 2yy' + 2 + y' = 0$, and, therefore, at $(2, 1)$, $4 + 2y' + 2 + y' = 0$, $y' = -2$. Since the tangent lines to the two circles at the point $(2, 1)$ have the same slope, they are identical, and, therefore, the circles are tangent at that point.

12.47 If the curve $\sin y = x^3 - x^5$ passes through the point $(1, 0)$, find y' and y'' at the point $(1, 0)$.

   ❚ $(\cos y)y' = 3x^2 - 5x^4$. At $(1, 0)$, $y' = 3 - 5 = -2$. Differentiating again, $(\cos y)y'' - (\sin y)y' = 6x - 20x^3$. So, at $(1, 0)$, $y'' = 6 - 20 = -14$.

12.48 If $x + y = xy$, show that $y'' = 2y^3/x^3$.

   ❚ $1 + y' = xy' + y$, $y'(1 - x) = y - 1$. Note that, from the original equation, $y - 1 = y/x$ and $x - 1 = x/y$. Hence, $y' = -y^2/x^2$. From the equation $y'(1 - x) = y - 1$ above, $y'(-1) + y''(1 - x) = y'$, $y''(1 - x) = 2y'$, $y''(-x/y) = 2(-y^2/x^2)$, $y'' = 2y^3/x^3$.

CHAPTER 13

Maxima and Minima

13.1 State the second derivative test for relative extrema.

▌ If $f'(c) = 0$ and $f''(c) < 0$, then $f(x)$ has a relative maximum at c. [See Fig. 13-1(a).] If $f'(c) = 0$ and $f''(c) > 0$, then $f(x)$ has a relative minimum at c. [See Fig. 13-1(b).] If $f'(c) = 0$ and $f''(c) = 0$, we cannot draw any conclusions at all.

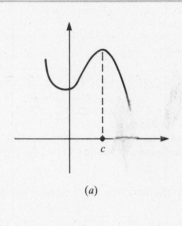

(a)

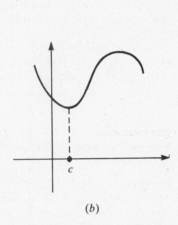

(b)

Fig. 13-1

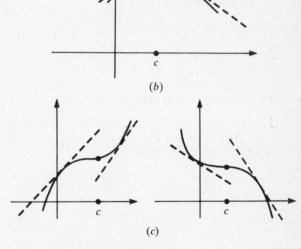

(a)

(b)

(c)

Fig. 13-2

13.2 State the first derivative test for relative extrema.

▌ Assume $f'(c) = 0$. If f' is negative to the left of c and positive to the right of c $\{-, +\}$, then f has a relative minimum at c. [See Fig. 13-2(a).] If f' is positive to the left of c and negative to the right of c $\{+, -\}$, then f has a relative maximum at c. [See Fig. 13-2(b).] If f' has the same sign to the left and to the right of c $\{+, +\}$ or $\{-, -\}$, then f has an inflection point at c. [See Fig. 13-2(c).]

13.3 Find the critical numbers of $f(x) = 5 - 2x + x^2$, and determine whether they yield relative maxima, relative minima, or inflection points.

▌ Recall that a critical number is a number c such that $f(c)$ is defined and either $f'(c) = 0$ or $f'(c)$ does not exist. Now, $f'(x) = -2 + 2x$. So, we set $-2 + 2x = 0$. Hence, the only critical number is $x = 1$. But $f''(x) = 2$. In particular, $f''(1) = 2 > 0$. Hence, by the second derivative test, $f(x)$ has a relative minimum at $x = 1$.

13.4 Find the critical numbers of $f(x) = x^2/(x-1)$ and determine whether they yield relative maxima, relative minima, or inflection points.

▮ $$f'(x) = \frac{(x-1)(2x) - x^2(1)}{(x-1)^2} = \frac{2x^2 - 2x - x^2}{(x-1)^2} = \frac{x^2 - 2x}{(x-1)^2} = \frac{x(x-2)}{(x-1)^2}$$

Hence, the critical numbers are $x = 0$ and $x = 2$. [$x = 1$ is not a critical number because $f(1)$ is not defined.] Now let us compute $f''(x)$.

$$f''(x) = \frac{(x-1)^2(2x-2) - (x^2 - 2x)[2(x-1)]}{(x-1)^4} = \frac{2(x-1)[(x-1)^2 - (x^2 - 2x)]}{(x-1)^4}$$

$$= \frac{2(x^2 - 2x + 1 - x^2 + 2x)}{(x-1)^3} = \frac{2}{(x-1)^3}$$

Hence, $f''(0) = -2 < 0$, and, therefore, by the second derivative test, $f(x)$ has a relative maximum at 0. Similarly, $f''(2) = 2 > 0$, and, therefore, $f(x)$ has a relative minimum at 2.

13.5 Find the critical numbers of $f(x) = x^3 - 5x^2 - 8x + 3$, and determine whether they yield relative maxima, relative minima, or inflection points.

▮ $f'(x) = 3x^2 - 10x - 8 = (3x + 2)(x - 4)$. Hence, the critical numbers are $x = 4$ and $x = -\frac{2}{3}$. Now, $f''(x) = 6x - 10$. So, $f''(4) = 14 > 0$, and, by the second derivative test, there is a relative minimum at $x = 4$. Similarly, $f''(-\frac{2}{3}) = -14$, and, therefore, there is a relative maximum at $x = -\frac{2}{3}$.

13.6 Find the critical numbers of $f(x) = x(x-1)^3$, and determine whether they yield relative maxima, relative minima, or inflection points.

▮ $f'(x) = x \cdot 3(x-1)^2 + (x-1)^3 = (x-1)^2(3x + x - 1) = (x-1)^2(4x-1)$. So, the critical numbers are $x = 1$ and $x = \frac{1}{4}$. Now, $f''(x) = (x-1)^2 \cdot 4 + 2(x-1)(4x-1) = 2(x-1)[2(x-1) + 4x - 1] = 2(x-1)(6x-3) = 6(x-1)(2x-1)$. Thus, $f''(\frac{1}{4}) = 6(-\frac{3}{4})(-\frac{1}{2}) = \frac{9}{4} > 0$, and, therefore, by the second derivative test, there is a relative minimum at $x = \frac{1}{4}$. On the other hand $f''(1)6 \cdot 0 \cdot 1 = 0$, and, therefore, the second derivative test is inapplicable. Let us use the first derivative test. $f'(x) = (x-1)^2(4x-1)$. For $x \neq 1$, $(x-1)^2$ is positive. Since $4x - 1$ has the value 3 when $x = 1$, $4x - 1 > 0$ just to the left and to the right of 1. Hence, $f'(x)$ is positive both on the left and on the right of $x = 1$, and this means that we have the case $\{+, +\}$. By the first derivative test, there is an inflection point at $x = 1$.

13.7 Find the critical numbers of $f(x) = \sin x - x$, and determine whether they yield relative maxima, relative minima, or inflection points.

▮ $f'(x) = \cos x - 1$. The critical numbers are the solutions of $\cos x = 1$, and these are the numbers $x = 2\pi n$ for any integer n. Now, $f''(x) = -\sin x$. So, $f''(2\pi n) = -\sin(2\pi n) = -0 = 0$, and, therefore, the second derivative test is inapplicable. Let us use the first derivative test. Immediately to the left and right of $x = 2\pi n$, $\cos x < 1$, and, therefore, $f'(x) = \cos x - 1 < 0$. Hence, the case $\{-, -\}$ holds, and there is an inflection point at $x = 2\pi n$.

[handwritten annotations: graph sketch labeled $y = \sin x + x$; $f'''(x) = -\cos x$ and $f'''(2\pi n) = -\cos(2\pi n)$ $= -1$. Thus ∃ inflection pt. at $x = 2\pi n$, and $f(x)$ is ↘, BY 201(15)TH.]

13.8 Find the critical numbers of $f(x) = (x-1)^{2/3}$ and determine whether they yield relative maxima, relative minima, or inflection points.

▮ $f'(x) = \frac{2}{3}(x-1)^{-1/3} = \frac{2}{3}[1/(x-1)^{1/3}]$. There are no values of x for which $f'(x) = 0$, but $x = 1$ is a critical number, since $f'(1)$ is not defined. Try the first derivative test [which is also applicable when $f'(c)$ is not defined]. To the left of $x = 1$, $(x-1)$ is negative, and, therefore, $f'(x)$ is negative. To the right of $x = 1$, $(x-1)$ is positive, and, therefore, $f(x)$ is positive. Thus, the case $\{-, +\}$ holds, and there is a relative minimum at $x = 1$.

13.9 Describe a procedure for finding the absolute maximum and absolute minimum values of a continuous function $f(x)$ on a closed interval $[a, b]$.

▮ Find all the critical numbers of $f(x)$ in $[a, b]$. List all these critical numbers, $c_1, c_2, \ldots$, and add the endpoints a and b to the list. Calculate $f(x)$ for each x in the list. The largest value thus obtained is the maximum value of $f(x)$ on $[a, b]$, and the minimal value thus obtained is the minimal value of $f(x)$ on $[a, b]$.

x	a	b	c_1	c_2	$\ldots$	c_n
$f(x)$	$f(a)$	$f(b)$	$f(c_1)$	$f(c_2)$	$\ldots$	$f(c_n)$

13.10 Find the absolute maximum and minimum of the function $f(x) = 4x^2 - 7x + 3$ on the interval $[-2, 3]$.

▮ $f'(x) = 8x - 7$. Solving $8x - 7 = 0$, we find the critical number $x = \frac{7}{8}$, which lies in the interval. So we list $\frac{7}{8}$ and the endpoints -2 and 3 in a table, and calculate the corresponding values $f(x)$. The absolute maximum 33 is assumed at $x = -2$. The absolute minimum $-\frac{1}{16}$ is assumed at $x = \frac{7}{8}$.

x	-2	3	$\frac{7}{8}$
$f(x)$	33	18	$-\frac{1}{16}$

13.11 Find the absolute maximum and minimum of $f(x) = 4x^3 - 8x^2 + 1$ on the closed interval $[-1, 1]$.

▮ $f'(x) = 12x^2 - 16x = 4x(3x - 4)$. So, the critical numbers are $x = 0$ and $x = \frac{4}{3}$. But $x = \frac{4}{3}$ does not lie in the interval. Hence, we list only 0 and the endpoints -1 and 1, and calculate the corresponding values of $f(x)$. So, the absolute maximum 1 is achieved at $x = 0$, and the absolute minimum -11 is achieved at $x = -1$.

x	-1	1	0
$f(x)$	-11	-3	1

13.12 Find the absolute maximum and minimum of $f(x) = x^4 - 2x^3 - x^2 - 4x + 3$ on the interval $[0, 4]$.

▮ $f'(x) = 4x^3 - 6x^2 - 2x - 4 = 2(2x^3 - 3x^2 - x - 2)$. We first search for roots of $2x^3 - 3x^2 - x - 2$ by trying integral factors of the constant term 2. It turns out that $x = 2$ is a root. Dividing $2x^3 - 3x^2 - x - 2$ by $x - 2$, we obtain the quotient $2x^2 + x + 1$. By the quadratic formula, the roots of the latter are $x = (-1 \pm \sqrt{-7})/4$, which are not real. Thus, the only critical number is $x = 2$. So, listing 2 and the endpoints 0 and 4, we calculate the corresponding values of $f(x)$. Thus, the absolute maximum 99 is attained at $x = 4$, and the absolute minimum -9 at $x = 2$.

x	0	4	2
$f(x)$	3	99	-9

13.13 Find the absolute maximum and minimum of $f(x) = x^3/(x + 2)$ on the interval $[-1, 1]$.

▮
$$f'(x) = \frac{(x + 2)(3x^2) - x^3}{(x + 2)^2} = \frac{3x^3 + 6x^2 - x^3}{(x + 2)^2} = \frac{2x^3 + 6x^2}{(x + 2)^2} = \frac{2x^2(x + 3)}{(x + 2)^2}$$

Thus, the critical numbers are $x = 0$ and $x = -3$. However, $x = -3$ is not in the given interval. So, we list 0 and the endpoints -1 and 1. The absolute maximum $\frac{1}{3}$ is assumed at $x = 1$, and the absolute minimum -1 is assumed at $x = -1$.

x	0	-1	1
$f(x)$	0	-1	$\frac{1}{3}$

13.14 For what value of k will $f(x) = x - kx^{-1}$ have a relative maximum at $x = -2$?

▮ $f'(x) = 1 + kx^{-2} = 1 + k/x^2$. We want -2 to be a critical number, that is, $1 + k/4 = 0$. Hence, $k = -4$. Thus, $f'(x) = 1 - 4/x^2$, and $f''(x) = 8/x^3$. Since $f''(-2) = -1$, there is a relative maximum at $x = -2$.

13.15 Find the absolute extrema of $f(x) = \sin x + x$ on $[0, 2\pi]$.

▮ $f'(x) = \cos x + 1$. For a critical number, $\cos x + 1 = 0$, or $\cos x = -1$. The only solution of this equation in $[0, 2\pi]$ is $x = \pi$. We list π and the two endpoints 0 and 2π, and compute the values of $f(x)$. Hence, the absolute maximum 2π is achieved at $x = 2\pi$, and the absolute minimum 0 at $x = 0$.

$$\begin{array}{c|ccc} x & \pi & 0 & 2\pi \\ \hline f(x) & \pi & 0 & 2\pi \end{array}$$

13.16 Find the absolute extrema of $f(x) = \sin x - \cos x$ on $[0, \pi]$.

▮ $f'(x) = \cos x + \sin x$. Setting this equal to 0, we have $\sin x = -\cos x$, or $\tan x = -1$. The only solution for this equation in $[0, \pi]$ is $3\pi/4$. Thus, the only critical number is $3\pi/4$. We list this and the endpoints 0 and π, and calculate the corresponding values of $f(x)$. Then, the absolute maximum $\sqrt{2}$ is attained at $x = 3\pi/4$, and the absolute minimum -1 is attained at $x = 0$.

$$\begin{array}{c|ccc} x & 3\pi/4 & 0 & \pi \\ \hline f(x) & \sqrt{2} & -1 & 1 \end{array}$$

13.17 (a) Find the absolute extrema of $f(x) = x - \sin x$ on $[0, \pi/2]$. (b) Show that $\sin x < x$ for all positive x.

▮ (a) $f'(x) = 1 - \cos x$. Setting $1 - \cos x = 0$, $\cos x = 1$, and the only solution in $[0, \pi/2]$ is $x = 0$. Thus, the only critical number is 0, which is one of the endpoints. Drawing up the usual table, we find that the absolute maximum $\pi/2 - 1$ is achieved at $x = \pi/2$ and the absolute minimum 0 is achieved at $x = 0$. (b) By Part (a), since the absolute minimum of $x - \sin x$ on $[0, \pi/2]$ is 0, which is achieved only at 0, then, for positive x in that interval, $x - \sin x > 0$, or $x > \sin x$. For $x > \pi/2$, $\sin x \le 1 < \pi/2 < x$.

$$\begin{array}{c|cc} x & 0 & \pi/2 \\ \hline f(x) & 0 & \pi/2 - 1 \end{array}$$

13.18 Find the points at which $f(x) = (x - 2)^4(x + 1)^3$ has relative extrema.

▮ $f'(x) = 3(x - 2)^4(x + 1)^2 + 4(x - 2)^3(x + 1)^3 = (x - 2)^3(x + 1)^2[3(x - 2) + 4(x + 1)] = (x - 2)^3(x + 1)^2(7x - 2)$. Hence, the critical numbers are $x = 2$, $x = -1$, $x = \frac{2}{7}$. We shall use the first derivative test. At $x = 2$, $(x + 1)^2(7x - 2)$ is positive, and, therefore, $(x + 1)^2(7x - 2)$ is positive immediately to the left and right of $x = 2$. For $x < 2$, $x - 2 < 0$, $(x - 2)^3 < 0$, and, therefore, $f'(x) < 0$. For $x > 2$, $(x - 2)^3 > 0$, and, therefore, $f'(x) > 0$. Thus, we have the case $\{-, +\}$; therefore, there is a relative minimum at $x = 2$. For $x = -1$, $x - 2 < 0$, $(x - 2)^3 < 0$, $7x - 2 < 0$, and, therefore, $(x - 2)^3(7x - 2) > 0$. Thus, immediately to the left and right of $x = -1$, $(x - 2)^3(7x - 2) > 0$. $(x + 1)^2 > 0$ on both sides of $x = -1$. Hence, $f'(x) > 0$ on both sides of $x = -1$. Thus, we have the case $\{+, +\}$, and there is an inflection point at $x = -1$. For $x = \frac{2}{7}$, $(x - 2)^3(x + 1)^2 < 0$. Hence, immediately to the left and right of $x = \frac{2}{7}$ $(x - 2)^3(x + 1)^2 < 0$. For $x < \frac{2}{7}$, $7x - 2 < 0$, and, therefore, $f'(x) > 0$ immediately to the left of $\frac{2}{7}$. For $x > \frac{2}{7}$, $7x - 2 > 0$, and, therefore, $f'(x) < 0$ immediately to the right of $\frac{2}{7}$. Thus, we have the case $\{+, -\}$, and, therefore, there is a relative maximum at $x = \frac{2}{7}$.

13.19 Show that $f(x) = (ax + b)/(cx + d)$ has no relative extrema [except in the trivial case when $f(x)$ is a constant].

▮ $f'(x) = \dfrac{(cx + d)(a) - (ax + b)(c)}{(cx + d)^2} = \dfrac{ad - bc}{(cx + d)^2}$. This is never 0, and, therefore, there are no critical numbers. [Note that, if $ad - bc = 0$, then $f(x)$ is a constant function. For, if $d \ne 0$, then $f(x) = \dfrac{adx + bd}{d(cx + d)} = \dfrac{bcx + bd}{d(cx + d)} = \dfrac{b}{d}$. If $d = 0$, then $c \ne 0$ and $b = 0$; then, $f(x) = \dfrac{a}{c}$.]

13.20 Test $f(x) = x^3 - 3px + q$ for relative extrema.

▮ $f'(x) = 3x^2 - 3p = 3(x^2 - p)$. Set $f'(x) = 0$. Then $x^2 = p$. If $p < 0$, there are no critical numbers and, therefore, no relative extrema. If $p \ge 0$, the critical numbers are $\pm\sqrt{p}$. Now, $f''(x) = 6x$. If $p > 0$, $f''(\sqrt{p}) = 6\sqrt{p} > 0$, and, therefore, there is a relative minimum at $x = \sqrt{p}$; while, $f''(-\sqrt{p}) = -6\sqrt{p} < 0$, and, therefore, there is a relative maximum at $x = -\sqrt{p}$. If $p = 0$, $f'(x) = 3x^2$, and, at the critical number $x = 0$, we have the case $\{+, +\}$ of the first derivative test; thus, if $p = 0$, there is only an inflection point at $x = 0$ and no extrema.

13.21 Show that $f(x) = (x - a_1)^2 + (x - a_2)^2 + \cdots + (x - a_n)^2$ has an absolute minimum when $x = (a_1 + a_2 + \cdots + a_n)/n$.

$f'(x) = 2(x - a_1) + 2(x - a_2) + \cdots + 2(x - a_n)$. Setting this equal to 0 and solving for x, $x = (a_1 + a_2 + \cdots + a_n)/n$. Now, $f''(x) = 2n > 0$. So, by the second derivative test, there is a relative minimum at $x = (a_1 + a_2 + \cdots + a_n)/n$. However, since this is the *only* relative extremum, the graph of the continuous function $f(x)$ must go up on both sides of $(a_1 + a_2 + \cdots + a_n)/n$ and must keep on going up (since, if it ever turned around and started going down, there would have to be another relative extremum).

13.22 Find the absolute maximum and minimum of $f(x) = -4x + 5$ on $[-2, 3]$.

$\blacksquare$ Since $f(x)$ is a decreasing linear function, the absolute maximum is attained at the left endpoint and the absolute minimum at the right endpoint. So, the absolute maximum is $-4(-2) + 5 = 13$, and the absolute minimum is $-4(3) + 5 = -7$.

13.23 Find the absolute maximum and minimum of $f(x) = 2x^2 - 7x - 10$ on $[-1, 3]$.

$\blacksquare$ $f'(x) = 4x - 7$. Setting $4x - 7 = 0$, we find the critical number $x = \frac{7}{4}$. We tabulate the values at the critical number and at the endpoints. Thus, the absolute minimum $-16\frac{1}{8}$ is attained at $x = \frac{7}{4}$ and the absolute maximum -1 at $x = -1$.

x	-1	$\frac{7}{4}$	3
$f(x)$	-1	$-16\frac{1}{8}$	-13

13.24 Find the absolute maximum and minimum of $f(x) = x^3 + 2x^2 + x - 1$ on $[-1, 1]$.

$\blacksquare$ $f'(x) = 3x^2 + 4x + 1 = (3x + 1)(x + 1)$. Setting $f'(x) = 0$, we obtain the critical numbers $x = -1$ and $x = -\frac{1}{3}$. Hence, we need only tabulate the values at -1, $-\frac{1}{3}$, and 1. From these values, we see that the absolute maximum is 3, attained at $x = 1$, and the absolute minimum is $-\frac{31}{27}$, attained at $x = -\frac{1}{3}$.

x	-1	$-\frac{1}{3}$	1
$f(x)$	-1	$-\frac{31}{27}$	3

13.25 Find the absolute maximum and minimum of $f(x) = (2x + 5)/(x^2 - 4)$ on $[-5, -3]$.

$\blacksquare$ $f'(x) = \dfrac{(x^2 - 4)(2) - (2x + 5)(2x)}{(x^2 - 4)^2} = \dfrac{-2(x^2 + 5x + 4)}{(x^2 - 4)^2} = \dfrac{-2(x + 1)(x + 4)}{(x^2 - 4)^2}$. Setting $f'(x) = 0$, we find the critical numbers $x = -1$ and $x = -4$, of which only -4 is in the given interval. Thus, we need only compute values of $f(x)$ for -4 and the endpoints. Since $-\frac{1}{4} < -\frac{5}{21} < -\frac{1}{5}$, the absolute maximum $-\frac{1}{5}$ is attained at $x = -3$, and the absolute minimum $-\frac{1}{4}$ at $x = -4$.

x	-3	-4	-5
$f(x)$	$-\frac{1}{5}$	$-\frac{1}{4}$	$-\frac{5}{21}$

13.26 Find the absolute maximum and minimum of $f(x) = x^2/16 + 1/x$ on $[1, 4]$.

$\blacksquare$ $f'(x) = x/8 - 1/x^2$. Setting $f'(x) = 0$, we have $x^3 = 8$, $x = 2$. We tabulate the values of $f(x)$ for the critical number $x = 2$ and the endpoints. Thus, the absolute maximum $\frac{5}{4}$ is attained at $x = 4$, and the absolute minimum $\frac{3}{4}$ at $x = 2$.

x	1	2	4
$f(x)$	$\frac{17}{16}$	$\frac{3}{4}$	$\frac{5}{4}$

13.27 Find the absolute maximum and minimum of the function f on $[0, 2]$, where

$$f(x) = \begin{cases} x^3 - \dfrac{x}{3} & \text{for } 0 \le x \le 1 \\ x^2 + x - \frac{4}{3} & \text{for } 1 < x \le 2. \end{cases}$$

$\blacksquare$ For $0 \le x < 1$, $f'(x) = 3x^2 - \frac{1}{3}$. Setting $f'(x) = 0$, we find $x^2 = \frac{1}{9}$, $x = \pm\frac{1}{3}$. Only $\frac{1}{3}$ is in the given interval. For $1 < x < 2$, $f'(x) = 2x + 1$. Setting $f'(x) = 0$, we obtain the critical number $x = -\frac{1}{2}$, which is not in the given interval. We also have to check the value of $f(x)$ at $x = 1$, where the derivative might not exist. We see that the absolute maximum $\frac{14}{3}$ is attained at $x = 2$, and the absolute minimum $-\frac{2}{27}$ at $x = \frac{1}{3}$.

x	0	$\frac{1}{3}$	1	2
$f(x)$	0	$-\frac{2}{27}$	$\frac{2}{3}$	$\frac{14}{3}$

13.28 Find the absolute maximum and minimum (if they exist) of $f(x) = (x^2 + 4)/(x - 2)$ on the interval $[0, 2)$.

▌ $f'(x) = \dfrac{(x - 2)(2x) - (x^2 + 4)}{(x - 2)^2} = \dfrac{x^2 - 4x - 4}{(x - 2)^2}$. By the quadratic formula, applied to $x^2 - 4x - 4 = 0$, we find the critical numbers $2 \pm 2\sqrt{2}$, neither of which is in the given interval. Since $f'(0) = -1$, $f'(x)$ remains negative in the entire interval, and, therefore, $f(x)$ is a decreasing function. Thus, its maximum is attained at the left endpoint 0, and this maximum value is -2. Since $f(x)$ approaches $-\infty$ as x approaches 2 from the left, there is no absolute minimum.

13.29 Find the absolute maximum and minimum (if they exist) of $f(x) = x/(x^2 + 1)^{3/2}$ on $[0, +\infty)$.

▌ Note that $f(0) = 0$ and $f(x)$ is positive for $x > 0$. Hence, 0 is the absolute minimum.

$$f'(x) = \frac{(x^2 + 1)^{3/2} - x(\frac{3}{2})(x^2 + 1)^{1/2} \cdot (2x)}{(x^2 + 1)^3} = \frac{(x^2 + 1)^{1/2}[(x^2 + 1) - 3x^2]}{(x^2 + 1)^3} = \frac{1 - 2x^2}{(x^2 + 1)^{5/2}}$$

Setting $f'(x) = 0$, we have $2x^2 = 1$, $x^2 = \frac{1}{2}$, $x = \pm\sqrt{2}/2$. So, the only critical number in $[0, +\infty)$ is $x = \sqrt{2}/2$. At that point, the first derivative test involves the case $\{+, -\}$, and, therefore, there is a relative maximum at $x = \sqrt{2}/2$, where $y = 2\sqrt{3}/9$. Since this is the only critical number in the given interval, the relative maximum is actually an absolute maximum.

13.30 Find the absolute maximum and minimum of $f(x) = \cos^2 x + \sin x$ on $[0, \pi]$.

▌ $f'(x) = 2 \cos x(-\sin x) + \cos x$. Setting $f'(x) = 0$, we have $\cos x(1 - 2 \sin x) = 0$, $\cos x = 0$ or $1 - 2 \sin x = 0$. In the given interval, $\cos x = 0$ at $x = \pi/2$. In the given interval, $1 - 2 \sin x = 0$ (that is, $\sin x = \frac{1}{2}$) only when $x = \pi/6$ or $x = 5\pi/6$. So, we must tabulate the values of $f(x)$ at these critical numbers and at the endpoints. We see that the absolute maximum is $\frac{5}{4}$, attained at $x = \pi/6$ and $x = 5\pi/6$. The absolute minimum is 1, attained at $x = 0$, $x = \pi/2$, and $x = \pi$.

x	0	$\pi/6$	$\pi/2$	$5\pi/6$	π
$f(x)$	1	$\frac{5}{4}$	1	$\frac{5}{4}$	1

13.31 Find the absolute maximum and minimum of $f(x) = 2 \sin x + \sin 2x$ on $[0, 2\pi]$.

▌ $f'(x) = 2 \cos x + 2 \cos 2x$. Setting $f'(x) = 0$, we obtain $\cos x + \cos 2x = 0$. Since $\cos 2x = 2 \cos^2 x - 1$, we have $2 \cos^2 x + \cos x - 1 = 0$, $(2 \cos x - 1)(\cos x + 1) = 0$, $\cos x = -1$ or $\cos x = \frac{1}{2}$. In the given interval, the solution of $\cos x = -1$ is $x = \pi$, and the solutions of $\cos x = \frac{1}{2}$ are $x = \pi/3$ and $x = 5\pi/3$. We tabulate the values of $f(x)$ for these critical numbers and the endpoints. So, the absolute maximum is $3\sqrt{3}/2$, attained at $x = \pi/3$, and the absolute minimum is $-3\sqrt{3}/2$, attained at $x = 5\pi/3$.

x	0	$\pi/3$	π	$5\pi/3$	2π
$f(x)$	0	$3\sqrt{3}/2$	0	$-3\sqrt{3}/2$	0

13.32 Find the absolute maximum and minimum of $f(x) = x/2 - \sin x$ on $[0, 2\pi]$.

▌ $f'(x) = \frac{1}{2} - \cos x$. Setting $f'(x) = 0$, we have $\cos x = \frac{1}{2}$. Hence, the critical numbers are $x = \pi/3$ and $x = 5\pi/3$. We tabulate the values of $f(x)$ for these numbers and the endpoints. Note that $\pi/6 < \sqrt{3}/2$ (since $\pi < 3\sqrt{3}$). Hence, $\pi/6 - \sqrt{3}/2 < 0 < \pi < 5\pi/6 + \sqrt{3}/2$. So, the absolute maximum is $5\pi/6 + \sqrt{3}/2$, attained at $x = 5\pi/3$, and the absolute minimum is $\pi/6 - \sqrt{3}/2$, attained at $x = \pi/3$.

13.33 Find the absolute maximum and minimum of $f(x) = 3 \sin x - 4 \cos x$ on $[0, 2\pi]$.

▌ $f'(x) = 3 \cos x + 4 \sin x$. Setting $f'(x) = 0$, we have $3 \cos x = -4 \sin x$, $\tan x = -0.75$. There are two critical numbers: x_0, between $\pi/2$ and π, and x_1, between $3\pi/2$ and 2π. We calculate the values of $f(x)$ for these numbers by using the 3-4-5 right triangle and noting that, $\sin x_0 = \frac{3}{5}$ and $\cos x_0 = -\frac{4}{5}$, and that $\sin x_1 = -\frac{3}{5}$ and $\cos x_1 = \frac{4}{5}$. So, the absolute maximum is 5, attained at $x = x_0$, and the absolute minimum -5 is attained at $x = x_1$. From a table of tangents, x_0 is approximately 143° and x_1 is approximately 323°.

x	0	x_0	x_1	2
$f(x)$	-4	5	-5	-4

13.34 Find the absolute maximum and minimum of $f(x) = |\cos x - \frac{1}{2}|$ on $[0, 2\pi]$.

▮ Clearly, the absolute minimum is 0, attained where $\cos x = \frac{1}{2}$, that is, at $x = \pi/3$ and $x = 5\pi/3$. So, we only have to determine the absolute maximum. Now, $f'(x) = \dfrac{|\cos x - \frac{1}{2}|}{\cos x - \frac{1}{2}}(-\sin x)$. We need only consider the critical numbers that are solutions of $\sin x = 0$ (since the solutions of $\cos x = \frac{1}{2}$ give the absolute minimum). Thus, the critical numbers are 0, π, and 2π. Tabulation of $f(x)$ for these numbers shows that the absolute maximum is $\frac{3}{2}$, achieved at $x = \pi$.

x	0	π	2π
$f(x)$	$\frac{1}{2}$	$\frac{3}{2}$	$\frac{1}{2}$

13.35 Find the absolute maximum and minimum (if they exist) of $f(x) = (x+2)/(x-1)$.

▮ Since $\lim\limits_{x \to 1^+} f(x) = +\infty$ and $\lim\limits_{x \to 1^-} f(x) = -\infty$, no absolute maximum or minimum exists.

13.36 Find the absolute maximum and minimum of $f(x) = \sqrt{|4x - 3|}$ on $[0, 1]$.

▮ Since $f(x)$ is not differentiable at $x = \frac{3}{4}$ (because $|4x - 3|$ is not differentiable at this point), $x = \frac{3}{4}$ is a critical number. $f'(x) = \dfrac{1}{2\sqrt{|4x-3|}} \cdot \dfrac{|4x-3|}{4x-3} \cdot (4)$. (Here, we have used the chain rule and Problem 9.47.) Thus, there are no other critical numbers. We need only compute $f(x)$ at $x = \frac{3}{4}$ and at the endpoints. We see that the absolute maximum is $\sqrt{3}$, attained at $x = 0$, and the absolute minimum is 0, attained at $x = \frac{3}{4}$.

x	0	$\frac{3}{4}$	1
$f(x)$	$\sqrt{3}$	0	1

CHAPTER 14
Related Rates

14.1 The top of a 25-foot ladder, leaning against a vertical wall is slipping down the wall at the rate of 1 foot per second. How fast is the bottom of the ladder slipping along the ground when the bottom of the ladder is 7 feet away from the base of the wall?

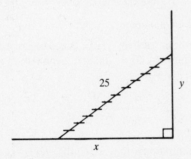

Fig. 14-1

▌ Let y be the distance of the top of the ladder from the ground, and let x be the distance of the bottom of the ladder from the base of the wall (Fig. 14-1). By the Pythagorean theorem, $x^2 + y^2 = (25)^2$. Differentiating with respect to time t, $2x \cdot D_t x + 2y \cdot D_t y = 0$; so, $x \cdot D_t x + y \cdot D_t y = 0$. The given information tells us that $D_t y = -1$ foot per second. (Since the ladder is sliding down the wall, y is decreasing, and, therefore, its derivative is negative.) When $x = 7$, substitution in $x^2 + y^2 = (25)^2$ yields $y^2 = 576$, $y = 24$. Substitution in $x \cdot D_t x + y \cdot D_t y = 0$ yields: $7 \cdot D_t x + 24 \cdot (-1) = 0$, $D_t x = \frac{24}{7}$ feet per second.

14.2 A cylindrical tank of radius 10 feet is being filled with wheat at the rate of 314 cubic feet per minute. How fast is the depth of the wheat increasing? (The volume of a cylinder is $\pi r^2 h$, where r is its radius and h is its height.)

▌ Let V be the volume of wheat at time t, and let h be the depth of the wheat in the tank. Then $V = \pi(10)^2 h$. So, $D_t V = 100\pi \cdot D_t h$. But we are given that $D_t V = 314$ cubic feet per minute. Hence, $314 = 100\pi \cdot D_t h$, $D_t h = 314/(100\pi)$. If we approximate π by 3.14, then $D_t h = 1$. Thus, the depth of the wheat is increasing at the rate of 1 cubic foot per minute.

14.3 A 5-foot girl is walking toward a 20-foot lamppost at the rate of 6 feet per second. How fast is the tip of her shadow (cast by the lamp) moving?

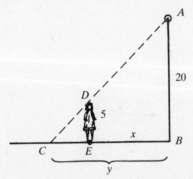

Fig. 14-2

▌ Let x be the distance of the girl from the base of the post, and let y be the distance of the tip of her shadow from the base of the post (Fig. 14-2). $\triangle ABC$ is similar to $\triangle DEC$. Hence, $AB/DE = y/(y-x)$, $\frac{20}{5} = y/(y-x)$, $4 = y/(y-x)$, $4y - 4x = y$, $3y = 4x$. Hence, $3 \cdot D_t y = 4 \cdot D_t x$. But, we are told that $D_t x = -6$ feet per second. (Since she is walking toward the base, x is decreasing, and $D_t x$ is negative.) So $3 \cdot D_t y = 4 \cdot (-6)$, $D_t y = -8$. Thus the tip of the shadow is moving at the rate of 8 feet per second toward the base of the post.

14.4 Under the same conditions as in Problem 14.3, how fast is the length of the girl's shadow changing?

▌ Use the same notation as in Problem 14.3. Let ℓ be the length of her shadow. Then $\ell = y - x$. Hence,

$D_t\ell = D_t y - D_t x = (-8) - (-6) = -2$. Thus, the length of the shadow is decreasing at the rate of 2 feet per second.

14.5 A rocket is shot vertically upward with an initial velocity of 400 feet per second. Its height s after t seconds is $s = 400t - 16t^2$. How fast is the distance changing from the rocket to an observer on the ground 1800 feet away from the launching site, when the rocket is still rising and is 2400 feet above the ground?

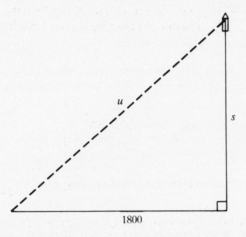

Fig. 14-3

▮ Let u be the distance from the rocket to the observer, as shown in Fig. 14-3. By the Pythagorean theorem, $u^2 = s^2 + (1800)^2$. Hence, $2u \cdot D_t u = 2s \cdot D_t s$, $u \cdot D_t u = s \cdot D_t s$. When $s = 2400$, $u^2 = (100)^2 \cdot (900)$, $u = 100 \cdot 30 = 3000$. Since $s = 400t - 16t^2$, when $s = 2400$, $2400 = 400t - 16t^2$, $t^2 - 25t + 150 = 0$, $(t - 10)(t - 15) = 0$. So, on the way up, the rocket is at 2400 feet when $t = 10$. But, $D_t s = 400 - 32t$. So, when $t = 10$, $D_t s = 400 - 32 \cdot 10 = 80$. Substituting in $u \cdot D_t u = s \cdot D_t s$, we obtain $3000 \cdot D_t u = 2400 \cdot 80$, $D_t u = 64$. So the distance from the rocket to the observer is increasing at the rate of 64 feet per second when $t = 10$.

14.6 A small funnel in the shape of a cone is being emptied of fluid at the rate of 12 cubic centimeters per second. The height of the funnel is 20 centimeters and the radius of the top is 4 centimeters. How fast is the fluid level dropping when the level stands 5 centimeters above the vertex of the cone? (Remember that the volume of a cone is $\frac{1}{3}\pi r^2 h$.)

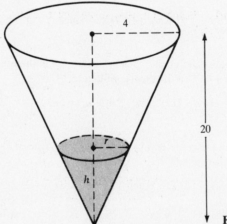

Fig. 14-4

▮ The fluid in the funnel forms a cone with radius r, height h, and volume V. By similar triangles, $r/4 = h/20$, $r = h/5$. So, $V = \frac{1}{3}\pi r^2 h = \frac{1}{3}\pi(h/5)^2 h = \frac{1}{75}\pi h^3$. Hence, $D_t V = \frac{1}{25}\pi h^2 \cdot D_t h$. We are given that $D_t V = -12$, since the fluid is leaving at the rate of 12 cubic centimeters per second. Hence, $-12 = \frac{1}{25}\pi h^2 \cdot D_t h$, $-300 = \pi h^2 \cdot D_t h$. When $h = 5$, $-300 = \pi \cdot 25 \cdot D_t h$, $D_t h = -12/\pi$, or approximately -3.82 centimeters per second. Hence, the fluid level is dropping at the rate of about 3.82 centimeters per second.

14.7 A balloon is being inflated by pumped air at the rate of 2 cubic inches per second. How fast is the diameter of the balloon increasing when the radius is $\frac{1}{2}$ inch?

▮ $V = \frac{4}{3}\pi r^3$. So, $D_t V = 4\pi r^2 \cdot D_t r$. We are told that $D_t V = 2$. So, $2 = 4\pi r^2 \cdot D_t r$. When $r = \frac{1}{2}$, $2 = 4\pi(\frac{1}{4}) \cdot D_t r$, $D_t r = 2/\pi$. Let d be the diameter. Then, $d = 2r$, $D_t d = 2 \cdot D_t r = 2 \cdot (2/\pi) = 4/\pi \approx 1.27$. So, the diameter is increasing at the rate of about 1.27 inches per second.

14.8 Oil from an uncapped well in the ocean is radiating outward in the form of a circular film on the surface of the water. If the radius of the circle is increasing at the rate of 2 meters per minute, how fast is the area of the oil film growing when the radius is 100 meters.

▮ The area $A = \pi r^2$. So, $D_t A = 2\pi r \cdot D_t r$. We are given that $D_t r = 2$. Hence, when $r = 100$, $D_t A = 2\pi \cdot 100 \cdot 2 = 400\pi$, which is about 1256 m^2/min.

14.9 The length of a rectangle of constant area 800 square millimeters is increasing at the rate of 4 millmeters per second. What is the width of the rectangle at the moment the width is decreasing at the rate of 0.5 millimeter per second?

▮ The area $800 = \ell w$. Differentiating, $0 = \ell \cdot D_t w + w \cdot D_t \ell$. We are given that $D_t \ell = 4$. So, $0 = \ell \cdot D_t w + 4w$. When $D_t w = -0.5$, $0 = -0.5\ell + 4w$, $4w = 0.5\ell$. But $\ell = 800/w$. So, $4w = 0.5(800/w) = 400/w$, $w^2 = 100$, $w = 10$ mm.

14.10 Under the same conditions as in Problem 14.9, how fast is the diagonal of the rectangle changing when the width is 20 mm?

▮ As in the solution of Problem 14.9. $0 = \ell \cdot D_t w + 4w$. Let u be the diagonal. Then $u^2 = w^2 + \ell^2$, $2u \cdot D_t u = 2w \cdot D_t w + 2\ell \cdot D_t \ell$, $u \cdot D_t u = w \cdot D_t w + \ell \cdot D_t \ell$. When $w = 20$, $\ell = 800/w = 40$. Substitute in $0 = \ell \cdot D_t w + 4w$: $0 = 40 \cdot D_t w + 80$, $D_t w = -2$. When $w = 20$, $u^2 = (20)^2 + (40)^2 = 2000$, $u = 20\sqrt{5}$. Substituting in $u \cdot D_t u = w \cdot D_t w + \ell \cdot D_t \ell$, $20\sqrt{5} \cdot D_t u = 20 \cdot (-2) + 40 \cdot 4 = 120$, $D_t u = 6\sqrt{5}/5 \approx 2.69$ mm/s.

14.11 A particle moves on the hyperbola $x^2 - 18y^2 = 9$ in such a way that its y-coordinate increases at a constant rate of 9 units per second. How fast is its x-coordinate changing when $x = 9$?

▮ $2x \cdot D_t x - 36y \cdot D_t y = 0$, $x \cdot D_t x = 18y \cdot D_t y$. We are given that $D_t y = 9$. Hence, $x \cdot D_t x = 18y \cdot 9 = 162y$. When $x = 9$, $(9)^2 - 18y^2 = 9$, $18y^2 = 72$, $y^2 = 4$, $y = \pm 2$. Substituting in $x \cdot D_t x = 162y$, $9 \cdot D_t x = \pm 324$, $D_t x = \pm 36$ units per second.

14.12 An object moves along the graph of $y = f(x)$. At a certain point, the slope of the curve is $\frac{1}{2}$ and the x-coordinate of the object is decreasing at the rate of 3 units per second. At that point, how fast is the y-coordinate of the object changing?

▮ $y = f(x)$. By the chain rule, $D_t y = f'(x) \cdot D_t x$. Since $f'(x)$ is the slope $\frac{1}{2}$ and $D_t x = -3$, $D_t y = \frac{1}{2} \cdot (-3) = -\frac{3}{2}$ units per second.

14.13 If the radius of a sphere is increasing at the constant rate of 3 millimeters per second, how fast is the volume changing when the surface area $4\pi r^2$ is 10 square millimeters?

▮ $V = \frac{4}{3}\pi r^3$. Hence, $D_t V = 4\pi r^2 \cdot D_t r$. We are given that $D_t r = 3$. So, $D_t V = 4\pi r^2 \cdot 3$. When $4\pi r^2 = 10$, $D_t V = 30$ mm^3/s.

14.14 What is the radius of an expanding circle at a moment when the rate of change of its area is numerically twice as large as the rate of change of its radius?

▮ $A = \pi r^2$. Hence, $D_t A = 2\pi r \cdot D_t r$. When $D_t A = 2 \cdot D_t r$, $2 \cdot D_t r = 2\pi r \cdot D_t r$, $1 = \pi r$, $r = 1/\pi$.

14.15 A particle moves along the curve $y = 2x^3 - 3x^2 + 4$. At a certain moment, when $x = 2$, the particle's x-coordinate is increasing at the rate of 0.5 unit per second. How fast is its y-coordinate changing at that moment?

▮ $D_t y = 6x^2 \cdot D_t x - 6x \cdot D_t x = 6x \cdot D_t x(x - 1)$. When $x = 2$, $D_t x = 0.5$. So, at that moment, $D_t y = 12(0.5)(1) = 6$ units per second.

14.16 A plane flying parallel to the ground at a height of 4 kilometers passes over a radar station R (Fig. 14-5). A short time later, the radar equipment reveals that the distance between the plane and the station is 5 kilometers and that the distance between the plane and the station is increasing at a rate of 300 kilometers per hour. At that moment, how fast is the plane moving horizontally?

▮ At time t, let x be the horizontal distance of the plane from the point directly over R, and let u be the distance between the plane and the station. Then $u^2 = x^2 + (4)^2$. So, $2u \cdot D_t u = 2x \cdot D_t x$, $u \cdot D_t u = x \cdot D_t x$. When $u = 5$, $(5)^2 = x^2 + (4)^2$, $x = 3$, and we are also told that $D_t u$ is 300. Substituting in $u \cdot D_t u = x \cdot D_t x$, $5 \cdot 300 = 3 \cdot D_t x$, $D_t x = 500$ kilometers per hour.

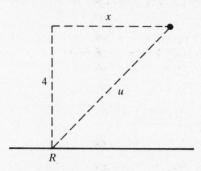

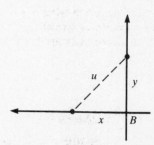

Fig. 14-5   **Fig. 14-6**

14.17 A boat passes a fixed buoy at 9 a.m. heading due west at 3 miles per hour. Another boat passes the same buoy at 10 a.m. heading due north at 5 miles per hour. How fast is the distance between the boats changing at 11:30 a.m.?

▮ Refer to Fig. 14-6. Let the time t be measured in hours after 9 a.m. Let x be the number of miles that the first boat is west of the buoy at time t, and let y be the number of miles that the second boat is north of the buoy at time t. Let u be the distance between the boats at time t. For any time $t \geq 1$, $u^2 = x^2 + y^2$. Then $2u \cdot D_t u = 2x \cdot D_t x + 2y \cdot D_t y$, $u \cdot D_t u = x \cdot D_t x + y \cdot D_t y$. We are given that $D_t x = 3$ and $D_t y = 5$. So, $u \cdot D_t u = 3x + 5y$. At 11:30 a.m. the first boat has travelled $2\frac{1}{2}$ hours at 3 miles per hour; so, $x = \frac{15}{2}$. Similarly, the second boat has travelled at 5 miles per hour for $1\frac{1}{2}$ hours since passing the buoy; so, $y = \frac{15}{2}$. Also, $u^2 = (\frac{15}{2})^2 + (\frac{15}{2})^2 = \frac{225}{2}$, $u = 15/\sqrt{2}$. Substituting in $u \cdot D_t u = 3x + 5y$, $(15/\sqrt{2}) \cdot D_t u = 3 \cdot \frac{15}{2} + 5 \cdot \frac{15}{2} = 60$, $D_t u = 4\sqrt{2} \approx 5.64$ miles per hour.

14.18 Water is pouring into an inverted cone at the rate of 3.14 cubic meters per minute. The height of the cone is 10 meters, and the radius of its base is 5 meters. How fast is the water level rising when the water stands 7.5 meters above the base?

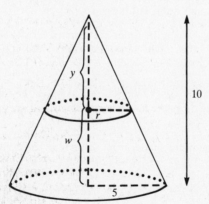

Fig. 14-7

▮ Let w be the level of the water above the base, and let r be the radius of the circle that forms the surface of the water. Let $y = 10 - w$. Then y is the height of the cone-shaped region above the water (see Fig. 14-7). So, the volume of that cone is $V_1 = \frac{1}{3}\pi r^2 y$. The total volume of the conical container is $V_2 = \frac{1}{3}\pi(5)^2 \cdot 10 = 250\pi/3$. Thus, the total volume of the water is $V = V_2 - V_1 = 250\pi/3 - \pi r^2 y/3$. By similar triangles, $10/5 = y/r$, $y = 2r$. So, $V = 250\pi/3 - \pi r^2(2r)/3 = 250\pi/3 - 2\pi r^3/3$. Hence, $D_t V = -2\pi r^2 \cdot D_t r$. We are given that $D_t V = 3.14$. So, $3.14 = -2\pi r^2 \cdot D_t r$. Thus, $D_t r = -3.14/2\pi r^2$. When the water stands 7.5 meters in the cone, $w = 7.5$, $y = 10 - 7.5 = 2.5$ $r = \frac{1}{2}y = 1.25$. So $D_t r = -3.14/2\pi(1.25)^2$. $D_t y = 2 \cdot D_t r = -3.14/\pi(1.25)^2$, $D_t w = -D_t y = 3.14/\pi(1.25)^2 \approx \frac{16}{25} = 0.64$ m/min.

14.19 A particle moves along the curve $y = x^2 + 2x$. At what point(s) on the curve are the x- and y-coordinates of the particle changing at the same rate?

▮ $D_t y = 2x \cdot D_t x + 2 \cdot D_t x = D_t x(2x + 2)$. When $D_t y = D_t x$, $2x + 2 = 1$, $2x = -1$, $x = -\frac{1}{2}$, $y = -\frac{3}{4}$.

14.20   A boat is being pulled into a dock by a rope that passes through a ring on the bow of the boat.   The dock is 8 feet higher than the bow ring.   How fast is the boat approaching the dock when the length of rope between the dock and the boat is 10 feet, if the rope is being pulled in at the rate of 3 feet per second?

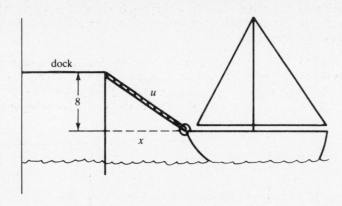

Fig. 14-8

❚ Let x be the horizontal distance from the bow ring to the dock, and let u be the length of the rope between the dock and the boat. Then, $u^2 = x^2 + (8)^2$. So $2u \cdot D_t u = 2x \cdot D_t x$, $u \cdot D_t u = x \cdot D_t x$. We are told that $D_t u = -3$. So $-3u = x \cdot D_t x$. When $u = 10$, $x^2 = 36$, $x = 6$. Hence, $-3 \cdot 10 = 6 \cdot D_t x$, $D_t x = -5$. So the boat is approaching the dock at the rate of 5 ft/s.

14.21   A girl is flying a kite, which is at a height of 120 feet.   The wind is carrying the kite horizontally away from the girl at a speed of 10 feet per second.   How fast must the kite string be let out when the string is 150 feet long?

❚ Let x be the horizontal distance of the kite from the point directly over the girl's head at 120 feet.   Let u be the length of the kite string from the girl to the kite. Then $u^2 = x^2 + (120)^2$. So, $2u \cdot D_t u = 2x \cdot D_t x$, $u \cdot D_t u = x \cdot D_t x$. We are told that $D_t x = 10$. Hence, $u \cdot D_t u = 10x$. When $u = 150$, $x^2 = 8100$, $x = 90$. So, $150 \cdot D_t u = 900$, $D_t u = 6$ ft/s.

14.22   A rectangular trough is 8 feet long, 2 feet across the top, and 4 feet deep.   If water flows in at a rate of 2 ft³/min, how fast is the surface rising when the water is 1 ft deep?

❚ Let x be the depth of the water.   Then the water is a rectangular slab of dimensions x, 2, and 8.   Hence, the volume $V = 16x$. So $D_t V = 16 \cdot D_t x$. We are told that $D_t V = 2$. So, $2 = 16 \cdot D_t x$. Hence, $D_t x = \frac{1}{8}$ ft/min.

14.23   A ladder 20 feet long leans against a house.   Find the rate at which the top of the ladder is moving downward if the foot of the ladder is 12 feet away from the house and sliding along the ground away from the house at the rate of 2 feet per second?

❚ Let x be the distance of the foot of the ladder from the base of the house, and let y be the distance of the top of the ladder from the ground.   Then $x^2 + y^2 = (20)^2$. So, $2x \cdot D_t x + 2y \cdot D_t y = 0$, $x \cdot D_t x + y \cdot D_t y = 0$. We are told that $x = 12$ and $D_t x = 2$. When $x = 12$, $y^2 = 256$, $y = 16$. Substituting in $x \cdot D_t x + y \cdot D_t y = 0$, $12 \cdot 2 + 16 \cdot D_t y = 0$, $D_t y = -\frac{3}{2}$. So the ladder is sliding down the wall at the rate of 1.5 ft/s.

14.24   In Problem 14.23, how fast is the angle α between the ladder and the ground changing at the given moment?

❚ $\tan \alpha = y/x$.   So, by the chain rule, $\sec^2 \alpha \cdot D_t \alpha = \dfrac{x \cdot D_t y - y \cdot D_t x}{x^2} = \dfrac{12 \cdot (-\frac{3}{2}) - 16 \cdot 2}{144} = -\dfrac{50}{144}$.   Also, $\tan \alpha = y/x = \frac{16}{12} = \frac{4}{3}$.   So, $\sec^2 \alpha = 1 + \tan^2 \alpha = 1 + \frac{16}{9} = \frac{25}{9}$.   Thus, $\frac{25}{9} \cdot D_t \alpha = -\frac{50}{144}$, $D_t \alpha = -\frac{1}{8}$.   Hence, the angle is decreasing at the rate of $\frac{1}{8}$ radian per second.

14.25   A train, starting at 11 a.m., travels east at 45 miles per hour, while another starting at noon from the same point travels south at 60 miles per hour.   How fast is the distance between them increasing at 3 p.m.?

❚ Let the time t be measured in hours, starting at 11 a.m.   Let x be the distance that the first train is east of the starting point, and let y be the distance that the second train is south of the starting point.   Let u be the distance between the trains.   Then $u^2 = x^2 + y^2$, $2u \cdot D_t u = 2x \cdot D_t x + 2y \cdot D_t y$, $u \cdot D_t u = x \cdot D_t x + y \cdot D_t y$. We are told that $D_t x = 45$ and $D_t y = 60$. So $u \cdot D_t u = 45x + 60y$. At 3 p.m., the first train has been travelling for 4 hours at 45 mi/h, and, therefore, $x = 180$;   the second train has been travelling for 3 hours at 60 mi/h, and, therefore, $y = 180$. Then, $u^2 = (180)^2 + (180)^2$, $u = 180\sqrt{2}$. Thus, $180\sqrt{2} \cdot D_t u = 45 \cdot 180 + 60 \cdot 180$, $D_t u = 105\sqrt{2}/2$ mi/h.

14.26 A light is at the top of a pole 80 feet high. A ball is dropped from the same height (80 ft) from a point 20 feet from the light. Assuming that the ball falls according to the law $s = 16t^2$, how fast is the shadow of the ball moving along the ground one second later?

▌ See Fig. 14-9. Let x be the distance of the shadow of the ball from the base of the lightpole. Let y be the height of the ball above the ground. By similar triangles, $y/80 = (x - 20)/x$. But, $y = 80 - 16t^2$. So, $1 - \frac{1}{5}t^2 = 1 - (20/x)$. Differentiating, $-\frac{2}{5}t = (20/x^2) \cdot D_t x$. When $t = 1$, $1 - \frac{1}{5}(1)^2 = 1 - 20/x$, $x = 100$. Substituting in $-\frac{2}{5}t = (20/x^2) \cdot D_t x$, $D_t x = -200$. Hence, the shadow is moving at 200 ft/s.

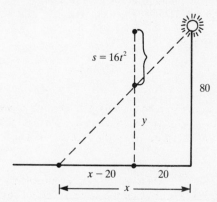

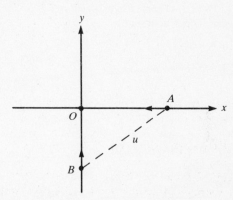

Fig. 14-9

Fig. 14-10

14.27 Ship A is 15 miles east of point O and moving west at 20 miles per hour. Ship B is 60 miles south of O and moving north at 15 miles per hour. Are they approaching or separating after 1 hour, and at what rate?

▌ Let the point O be the origin of a coordinate system, with A moving on the x-axis and B moving on the y-axis (Fig. 14-10). Since A begins at $x = 15$ and is moving to the left at 20 mi/h, its position is $x = 15 - 20t$. Likewise, the position of B is $y = -60 + 15t$. Let u be the distance between A and B. Then $u^2 = x^2 + y^2$, $2u \cdot D_t u = 2x \cdot D_t x + 2y \cdot D_t y$, $u \cdot D_t u = x \cdot D_t x + y \cdot D_t y$. Since $D_t x = -20$ and $D_t y = 15$, $u \cdot D_t u = -20x + 15y$. When $t = 1$, $x = 15 - 20 = -5$, $y = -60 + 15 = -45$, $u^2 = (-5)^2 + (-45)^2 = (25)(82)$, $u = 5\sqrt{82}$. Substituting in $u \cdot D_t u = -20x + 15y$, $5\sqrt{82} D_t u = -575$, $D_t u = -115/\sqrt{82} \approx -13$. Since the derivative of u is negative, the distance between the ships is getting smaller, at roughly 13 mi/h.

14.28 Under the same hypotheses as in Problem 14.27, when are the ships nearest each other?

▌ When the ships are nearest each other, their distance u assumes a relative minimum, and, therefore, $D_t u = 0$. Substituting in $u \cdot D_t u = -20x + 15y$, $0 = -20x + 15y$. But $x = 15 - 20t$ and $y = -60 + 15t$. So, $0 = -20(15 - 20t) + 15(-60 + 15t)$, $t = \frac{48}{25}$ hours, or approximately, 1 hour and 55 minutes.

14.29 Water, at the rate of 10 cubic feet per minute, is pouring into a leaky cistern whose shape is a cone 16 feet deep and

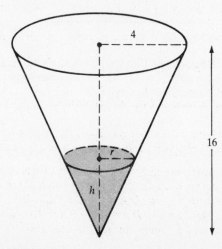

Fig. 14-11

8 feet in diameter at the top. At the time the water is 12 feet deep, the water level is observed to be rising 4 inches per minute. How fast is the water leaking out?

❚ Let h be the depth of the water, and let r be the radius of the water surface (Fig. 14-11). The water's volume $V = \frac{1}{3}\pi r^2 h$. By similar triangles, $r/4 = h/16$, $r = \frac{1}{4}h$, $V = \frac{1}{3}\pi(h/4)^2 h = \frac{1}{48}\pi h^3$. So $D_t V = \frac{1}{16}\pi h^2 \cdot D_t h$. We are told that when $h = 12$, $D_t h = \frac{1}{3}$. Hence, at that moment, $D_t V = \frac{1}{16}\pi(144)(\frac{1}{3}) = 3\pi$. Since the rate at which the water is pouring in is 10, the rate of leakage is $(10 - 3\pi)$ ft³/min.

14.30 An airplane is ascending at a speed of 400 kilometers per hour along a line making an angle of 60° with the ground. How fast is the altitude of the plane changing?

❚ Let h be the altitude of the plane, and let u be the distance of the plane from the ground along its flight path (Fig. 14-12). Then $h/u = \sin 60° = \sqrt{3}/2$, $2h = \sqrt{3}u$, $2 \cdot D_t h = \sqrt{3}D_t u = \sqrt{3} \cdot 400$. Hence, $D_t h = 200\sqrt{3}$ kilometers per hour.

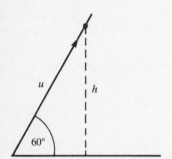

Fig. 14-12

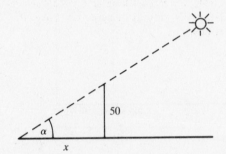

Fig. 14-13

14.31 How fast is the shadow cast on level ground by a pole 50 feet tall lengthening when the angle α of elevation of the sun is 45° and is decreasing by $\frac{1}{4}$ radian per hour? (See Fig. 14.13.)

❚ Let x be the length of the shadow. $\tan \alpha = 50/x$. By the chain rule, $\sec^2 \alpha \cdot D_t \alpha = (-50/x^2) \cdot D_t x$. When $\alpha = 45°$, $\tan \alpha = 1$, $\sec^2 \alpha = 1 + \tan^2 \alpha = 2$, $x = 50$. So, $2(-\frac{1}{4}) = -\frac{1}{50} \cdot D_t x$. Hence, $D_t x = 25$ ft/h.

14.32 A revolving beacon is situated 3600 feet off a straight shore. If the beacon turns at 4π radians per minute, how fast does its beam sweep along the shore at its nearest point A?

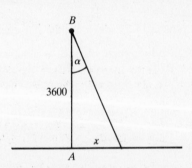

Fig. 14-14

❚ Let x be the distance from A to the point on the shore hit by the beacon, and let α be the angle between the line from the lighthouse B to A, and the beacon (Fig. 14-14). Then $\tan \alpha = x/3600$, so $\sec^2 \alpha \cdot D_t \alpha = \frac{1}{3600} \cdot D_t x$. We are told that $D_t \alpha = 4\pi$. When the beacon hits point A, $\alpha = 0$, $\sec \alpha = 1$, so $4\pi = \frac{1}{3600} D_t x$, $D_t x = 14,400\pi$ ft/min $= 240\pi$ ft/s.

14.33 Two sides of a triangle are 15 and 20 feet long, respectively. How fast is the third side increasing when the angle α between the given sides is 60° and is increasing at the rate of 2° per second?

❚ Let x be the third side. By the law of cosines, $x^2 = (15)^2 + (20)^2 - 2(15)(20) \cdot \cos \alpha$. Hence, $2x \cdot D_t x = 600 \sin \alpha \cdot D_t \alpha$. $x \cdot D_t x = 300 \sin \alpha \cdot D_t \alpha$. We are told that $D_t \alpha = 2 \cdot (\pi/180) = \pi/90$ rad/s. When $\alpha = 60°$, $\sin \alpha = \sqrt{3}/2$, $\cos \alpha = \frac{1}{2}$, $x^2 = 225 + 400 - 600 \cdot \frac{1}{2} = 325$, $x = 5\sqrt{13}$. Hence, $5\sqrt{13} \cdot D_t x = 300 \cdot (\sqrt{3}/2) \cdot (\pi/90)$, $D_t x = \pi/\sqrt{39}$ ft/s.

14.34 The area of an expanding rectangle is increasing at the rate of 48 square centimeters per second. The length of the rectangle is always equal to the square of its width (in centimeters). At what rate is the length increasing at the instant when the width is 2 cm?

▌ $A = \ell w$, and $\ell = w^2$. So, $A = w^3$. Hence, $D_t A = 3w^2 \cdot D_t w$. We are told that $D_t A = 48$. Hence, $48 = 3w^2 \cdot D_t w$, $16 = w^2 \cdot D_t w$. When $w = 2$, $16 = 4 \cdot D_t w$, $D_t w = 4$. Since $\ell = w^2$, $D_t \ell = 2w \cdot D_t w$. Hence, $D_t \ell = 2 \cdot 2 \cdot 4 = 16$ cm/s.

14.35 A spherical snowball is melting (symmetrically) at the rate of 4π centimeters per hour. How fast is the diameter changing when it is 20 centimeters?

▌ The volume $V = \frac{4}{3}\pi r^3$. So, $D_t V = 4\pi r^2 \cdot D_t r$. We are told that $D_t V = -4\pi$. Hence, $-4\pi = 4\pi r^2 \cdot D_t r$. Thus, $-1 = r^2 \cdot D_t r$. When the diameter is 20 centimeters, the radius $r = 10$. Hence, $-1 = 100 \cdot D_t r$, $D_t r = -0.01$. Since the diameter $d = 2r$, $D_t d = 2 \cdot D_t r = 2 \cdot (-0.01) = -0.02$. So, the diameter is decreasing at the rate of 0.02 centimeter per hour.

14.36 A trough is 10 feet long and has a cross section in the shape of an equilateral triangle 2 feet on each side (Fig. 14-15). If water is being pumped in at the rate of 20 ft³/min, how fast is the water level rising when the water is 1 ft deep?

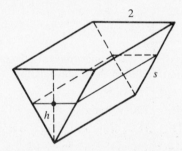

Fig. 14-15

▌ The water in the trough will have a cross section that is an equilateral triangle, say of height h and side s. In an equilateral triangle with side s, $s = 2h/\sqrt{3}$. Hence, the cross-sectional area of the water is $\frac{1}{2} \cdot (2h/\sqrt{3}) \cdot h = h^2/\sqrt{3}$. Therefore, the volume V of water is $10h^2/\sqrt{3}$. So, $D_t V = (20h/\sqrt{3}) \cdot D_t h$. We are told that $D_t V = 20$. So, $20 = (20h/\sqrt{3}) \cdot D_t h$, $\sqrt{3} = h \cdot D_t h$. When $h = 1$ ft, $D_t h = \sqrt{3}$ ft/min.

14.37 If a mothball evaporates at a rate proportional to its surface area $4\pi r^2$, show that its radius decreases at a constant rate.

▌ The volume $V = \frac{4}{3}\pi r^3$. So, $D_t V = 4\pi r^2 \cdot D_t r$. We are told that $D_t V = k \cdot 4\pi r^2$ for some constant k. Hence, $k = D_t r$.

14.38 Sand is being poured onto a conical pile at the constant rate of 50 cubic feet per minute. Frictional forces in the sand are such that the height of the pile is always equal to the radius of its base. How fast is the height of the pile increasing when the sand is 5 feet deep?

▌ The volume $V = \frac{1}{3}\pi r^2 h$. Since $h = r$, $V = \frac{1}{3}\pi h^3$. So, $D_t V = \pi h^2 \cdot D_t h$. We are told that $D_t V = 50$, so $50 = \pi h^2 \cdot D_t h$. When $h = 5$, $50 = \pi \cdot 25 \cdot D_t h$, $D_t h = 2/\pi$ ft/min.

14.39 At a certain moment, a sample of gas obeying Boyle's law, $pV = \text{constant}$, occupies a volume V of 1000 cubic inches at a pressure p of 10 pounds per square inch. If the gas is being compressed at the rate of 12 cubic inches per minute, find the rate at which the pressure is increasing at the instant when the volume is 600 cubic inches.

▌ Since $pV = \text{constant}$, $p \cdot D_t V + V \cdot D_t p = 0$. We are told that $D_t V = -12$, so $-12p + V \cdot D_t p = 0$. When $V = 1000$ and $p = 10$, $D_t p = 0.12$ pound per square inch.

14.40 A ladder 20 feet long is leaning against a wall 12 feet high with its top projecting over the wall (Fig. 14-16). Its bottom is being pulled away from the wall at the constant rate of 5 ft/min. How rapidly is the height of the top of the ladder decreasing when the top of the ladder reaches the top of the wall?

▌ Let y be the height of the top of the ladder, let x be the distance of the bottom of the ladder from the wall, and let u be the distance from the bottom of the ladder to the top of the wall. Now, $u^2 = x^2 + (12)^2$, $2u \cdot D_t u = 2x \cdot D_t x$, $u \cdot D_t u = x \cdot D_t x$. We are told that $D_t x = 5$. So, $u \cdot D_t u = 5x$. When the top of the ladder reaches the top of the wall, $u = 20$, $x^2 = (20)^2 - (12)^2 = 256$, $x = 16$. Hence, $20 \cdot D_t u = 5 \cdot 16$, $D_t u = 4$. By similar triangles, $y/12 = 20/u$, $y = 240/u$, $D_t y = -(240/u^2) \cdot D_t u = -\frac{240}{400} \cdot 4 = -2.4$ ft/min. Thus, the height of the ladder is decreasing at the rate of 2.4 feet per minute.

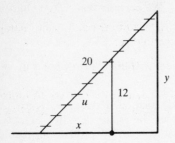

Fig. 14-16

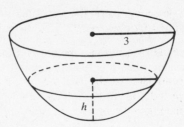

Fig. 14-17

14.41 Water is being poured into a hemispherical bowl of radius 3 inches at the rate of 1 cubic inch per second. How fast is the water level rising when the water is 1 inch deep? [The spherical segment of height h shown in Fig. 14-17 has volume $V = \pi h^2(r - h/3)$, where r is the radius of the sphere.]

▮ $V = \pi h^2(3 - h/3) = 3\pi h^2 - (\pi/3)h^3$. So, $D_t V = 6\pi h \cdot D_t h - \pi h^2 \cdot D_t h = \pi h D_t h(6 - h)$. We are told that $D_t V = 1$, so $1 = \pi h D_t h(6 - h)$. When $h = 1$, $D_t h = 1/5\pi$ in/s.

14.42 A metal ball of radius 90 centimeters is coated with a uniformly thick layer of ice, which is melting at the rate of 8π cubic centimeters per hour. Find the rate at which the thickness of the ice is decreasing when the ice is 10 centimeters thick?

▮ Let h be the thickness of the ice. The volume of the ice $V = \frac{4}{3}\pi(90 + h)^3 - \frac{4}{3}\pi(90)^3$. So, $D_t V = 4\pi(90 + h)^2 \cdot D_t h$. We are told that $D_t V = -8\pi$. Hence, $-2 = (90 + h)^2 \cdot D_t h$. When $h = 10$, $-2 = (100)^2 \cdot D_t h$, $D_t h = -0.0002$ cm/h.

14.43 A snowball is increasing in volume at the rate of 10 cm³/h. How fast is the surface area growing at the moment when the radius of the snowball is 5 cm?

▮ The surface area $A = 4\pi r^2$. So, $D_t A = 8\pi r \cdot D_t r$. Now, $V = \frac{4}{3}\pi r^3$, $D_t V = 4\pi r^2 \cdot D_t r$. We are told that $D_t V = 10$. So, $10 = 4\pi r^2 \cdot D_t r = \frac{1}{2}r \cdot 8\pi r \cdot D_t r = \frac{1}{2}r \cdot D_t A$. When $r = 5$, $10 = \frac{1}{2} \cdot 5 \cdot D_t A$, $D_t A = 4$ cm²/h.

14.44 If an object is moving on the curve $y = x^3$, at what point(s) is the y-coordinate of the object changing three times more rapidly than the x-coordinate?

▮ $D_t y = 3x^2 \cdot D_t x$. When $D_t y = 3 \cdot D_t x$, $x^2 = 1$, $x = \pm 1$. So, the points are $(1, 1)$ and $(-1, -1)$. (Other solutions occur when $D_t x = 0$, $D_t y = 0$. This happens within an interval of time when the object remains fixed at one point on the curve.)

14.45 If the diagonal of a cube is increasing at a rate of 3 cubic inches per minute, how fast is the side of the cube increasing?

▮ Let u be the length of the diagonal of a cube of side s. Then $u^2 = s^2 + s^2 + s^2 = 3s^2$, $u = s\sqrt{3}$, $D_t u = \sqrt{3}D_t s$. Thus, $3 = \sqrt{3}D_t s$, $D_t s = \sqrt{3}$ in/min.

14.46 The two equal sides of an isosceles triangle with fixed base b are decreasing at the rate of 3 inches per minute. How fast is the area decreasing when the two equal sides are equal to the base?

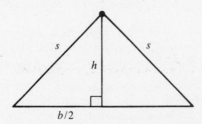

Fig. 14-18

❚ Let s be the length of the two equal sides, and let h be the height. Then, from Fig. 14-18, $h^2 = s^2 - b^2/4$, $2h \cdot D_t h = 2s \cdot D_t s$, $h \cdot D_t h = s \cdot D_t s$. When $s = b$, $h^2 = \frac{3}{4}b^2$, $h = (\sqrt{3}/2)b$, $(\sqrt{3}/2)b \cdot D_t h = b \cdot D_t s$, $(\sqrt{3}/2) \cdot D_t h = D_t s$. We are told that $D_t s = -3$. Hence, $D_t h = -2\sqrt{3}$. Now, $A = \frac{1}{2}bh$, $D_t A = \frac{1}{2}b \cdot D_t h = \frac{1}{2}b \cdot (-2\sqrt{3}) = -b\sqrt{3} \text{ in}^2/\text{min}$.

14.47 An object moves on the parabola $3y = x^2$. When $x = 3$, the x-coordinate of the object is increasing at the rate of 1 foot per minute. How fast is the y-coordinate increasing at that moment?

❚ $3 \cdot D_t y = 2x \cdot D_t x$. When $x = 3$, $D_t x = 1$. So $3 \cdot D_t y = 6$, $D_t y = 2 \text{ ft/min}$.

14.48 A solid is formed by a cylinder of radius r and altitude h, together with two hemispheres of radius r attached at each end (Fig. 14-19). If the volume V of the solid is constant but r is increasing at the rate of $1/(2\pi)$ meters per minute, how fast must h be changing when r and h are 10 meters?

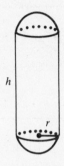

Fig. 14-19

❚ $V = \pi r^2 h + \frac{4}{3}\pi r^3$. $D_t V = \pi r^2 \cdot D_t h + 2\pi r h \cdot D_t r + 4\pi r^2 \cdot D_t r$. But, since V is constant, $D_t V = 0$. We are told that $D_t r = 1/(2\pi)$. Hence, $0 = \pi r^2 \cdot D_t h + rh + 2r^2$. When $r = h = 10$, $100\pi \cdot D_t h + 300 = 0$, $D_t h = -3/\pi$ meters per minute.

14.49. If $y = 7x - x^3$ and x increases at the rate of 4 units per second, how fast is the slope of the graph changing when $x = 3$?

❚ The slope $D_x y = 7 - 3x^2$. Hence, the rate of change of the slope is $D_t(D_x y) = -6x \cdot D_t x = -6x \cdot 4 = -24x$. When $x = 3$, $D_t(D_x y) = -72$ units per second.

14.50 A segment UV of length 5 meters moves so that its endpoints U and V stay on the x-axis and y-axis, respectively. V is moving away from the origin at the rate of 2 meters per minute. When V is 3 meters from the origin, how fast is U's position changing?

❚ Let x be the x-coordinate of U and let y be the y-coordinate of V. Then $y^2 + x^2 = 25$, $2y \cdot D_t y + 2x \cdot D_t x = 0$, $y \cdot D_t y + x \cdot D_t x = 0$. We are told that $D_t y = 2$. So, $2y + x \cdot D_t x = 0$. When $y = 3$, $x = 4$, $2 \cdot 3 + 4 \cdot D_t x = 0$, $D_t x = -\frac{3}{2}$ meters per minute.

14.51 A railroad track crosses a highway at an angle of 60°. A train is approaching the intersection at the rate of 40 mi/h, and a car is approaching the intersection from the same side as the train, at the rate of 50 mi/h. If, at a certain moment, the train and car are both 2 miles from the intersection, how fast is the distance between them changing?

❚ Refer to Fig. 14-20. Let x and y be the distances of the train and car, respectively, from the intersection, and

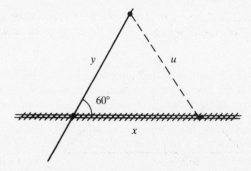

Fig. 14-20

let u be the distance between the train and the car. By the law of cosines, $u^2 = x^2 + y^2 - 2xy \cdot \cos 60° = x^2 + y^2 - xy$ (since $\cos 60° = \frac{1}{2}$). Hence, $2u \cdot D_t u = 2x \cdot D_t x + 2y \cdot D_t y - x \cdot D_t y - y \cdot D_t x$. We are told that $D_t x = -40$ and $D_t y = -50$, so $2u \cdot D_t u = -80x - 100y + 50x + 40y = -30x - 60y$. When $y = 2$ and $x = 2$, $u^2 = 4 + 4 - 4 = 4$, $u = 2$. Hence, $4 \cdot D_t u = -60 - 120 = -180$, $D_t u = -45$ mi/h.

14.52 A trough 20 feet long has a cross section in the shape of an equilateral trapezoid, with a base of 3 feet and whose sides make a 45° angle with the vertical. Water is flowing into it at the rate of 14 cubic feet per hour. How fast is the water level rising when the water is 2 feet deep?

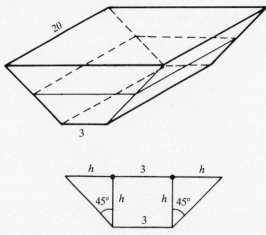

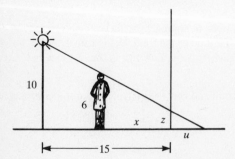

Fig. 14-21

▮ Let h be the depth of the water at time t. The cross-sectional area is $3h + h^2$ (see Fig. 14-21), and, therefore, the volume $V = 20(3h + h^2)$. So $D_t V = 20(3 \cdot D_t h + 2h \cdot D_t h) = 20 \cdot D_t h \cdot (3 + 2h)$. We are told that $D_t V = 14$, so $14 = 20 \cdot D_t h \cdot (3 + 2h)$. When $h = 2$, $14 = 20 \cdot D_t h \cdot 7$, $D_t h = 0.1$ ft/h.

14.53 A lamppost 10 feet tall stands on a walkway that is perpendicular to a wall. The distance from the post to the wall is 15 feet. A 6-foot man moves on the walkway toward the wall at the rate of 5 feet per second. When he is 5 feet from the wall, how fast is the shadow of his head moving up the wall?

▮ See Fig. 14-22. Let x be the distance from the man to the wall. Let u be the distance between the base of the wall and the intersection with the ground of the line from the lamp to the man's head. Let z be the height of the shadow of the man's head on the wall. By similar triangles, $6/(x + u) = 10/(15 + u) = z/u$. From the first equation, we obtain $u = \frac{5}{2}(9 - x)$. Hence, $x + u = \frac{3}{2}(15 - x)$. Since $z/u = 6/(x + u)$, $z = 6u/(x + u) = 10(9 - x)/(15 - x) = 10[1 - 6/(15 - x)]$. Thus, $D_t z = [10 \cdot 6/(15 - x)^2] \cdot (-D_t x)$. We are told that $D_t x = -5$. Hence, $D_t z = 300/(15 - x)^2$. When $x = 5$, $D_t z = 3$ ft/s. Thus, the shadow is moving up the wall at the rate of 3 feet per second.

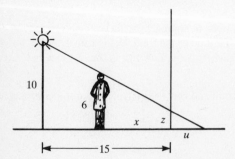

Fig. 14-22

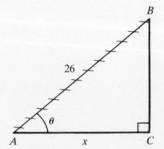

Fig. 14-23

14.54 In Fig. 14-23, a ladder 26 feet long is leaning against a vertical wall. If the bottom of the ladder, A, is slipping away from the base of the wall at the rate of 3 feet per second, how fast is the angle between the ladder and the ground changing when the bottom of the ladder is 10 feet from the base of the wall?

▮ Let x be the distance of A from the base of the wall at C. Then $D_t x = 3$. Since $\cos \theta = x/26$, $(-\sin \theta) \cdot D_t \theta = \frac{1}{26} D_t x = \frac{3}{26}$. When $x = 10$, $CB = \sqrt{(26)^2 - (10)^2} = \sqrt{576} = 24$, and $\sin \theta = \frac{24}{26}$. So, $-\frac{24}{26} D_t \theta = \frac{3}{26}$, $D_t \theta = -\frac{1}{8}$ radian per second.

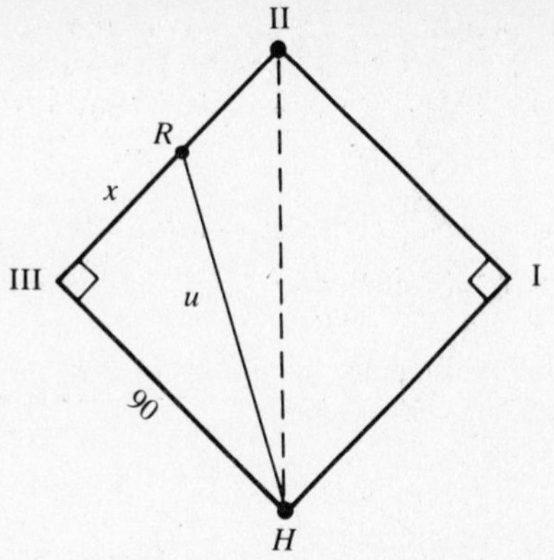

Fig. 14-24

14.55 In Fig. 14-24, a baseball field is a square of side 90 feet. If a runner on second base (II) starts running toward third base (III) at a rate of 20 ft/s. how fast is his distance from home plate (H) changing when he is 60 ft from II?

▌ Let x and u be the distances of the runner from III and H, respectively. Then $u^2 = x^2 + (90)^2$, $2u\dfrac{du}{dt} = 2x\dfrac{dx}{dt} = 2x(-20)$, $\dfrac{du}{dt} = -20\dfrac{x}{u}$. When $x = 90 - 60 = 30$, $u = \sqrt{(30)^2 + (90)^2} = 30\sqrt{10}$; therefore, $\dfrac{du}{dt} = -20\left(\dfrac{30}{30\sqrt{10}}\right) = -2\sqrt{10}$ ft/s.

14.56 An open pipe with length 3 meters and outer radius of 10 centimeters has an outer layer of ice that is melting at the rate of 2π cm^3/min. How fast is the thickness of ice decreasing when the ice is 2 centimeters thick?

▌ Let x be the thickness of the ice. Then the volume of the ice $V = 300[\pi(10 + x)^2 - 100\pi]$, So $D_t V = 300\pi[2(10 + x) \cdot D_t x]$. Since $D_t V = -2\pi$, we have $-2\pi = 600\pi(10 + x) \cdot D_t x$, $D_t x = -1/[300(10 + x)]$. When $x = 2$, $D_t x = -\frac{1}{3600}$ cm/min. So the thickness is decreasing at the rate of $\frac{1}{3600}$ centimeter per minute (4 millimeters per day).

CHAPTER 15
Curve Sketching (Graphs)

When sketching a graph, show all relative extrema, inflection points, and asymptotes; indicate concavity; and suggest the behavior at infinity.

In Problems 15.1 to 15.5, determine the intervals where the graphs of the following functions are concave upward and where they are concave downward. Find all inflection points.

15.1 $f(x) = x^2 - x + 12$.

▌ $f'(x) = 2x - 1$, $f''(x) = 2$. Since the second derivative is always positive, the graph is always concave upward and there are no inflection points.

15.2 $f(x) = x^3 + 15x^2 + 6x + 1$.

▌ $f'(x) = 3x^2 + 30x + 6$, $f''(x) = 6x + 30 = 6(x + 5)$. Thus, $f''(x) > 0$ when $x > -5$; hence, the graph is concave upward for $x > -5$. Since $f''(x) < 0$ for $x < -5$, the graph is concave downward for $x < -5$. Hence, there is an inflection point, where the concavity changes, at $(5, 531)$.

15.3 $f(x) = x^4 + 18x^3 + 120x^2 + x + 1$.

▌ $f'(x) = 4x^3 + 54x^2 + 240x + 1$, $f''(x) = 12x^2 + 108x + 240 = 12(x^2 + 9x + 20) = 12(x + 4)(x + 5)$. Thus, the important points are $x = -4$ and $x = -5$. For $x > -4$, $x + 4$ and $x + 5$ are positive, and, therefore, so is $f''(x)$. For $-5 < x < -4$, $x + 4$ is negative and $x + 5$ is positive; hence, $f''(x)$ is negative. For $x < -5$, both $x + 4$ and $x + 5$ are negative, and, therefore, $f''(x)$ is positive. Therefore, the graph is concave upward for $x > -4$ and for $x < -5$. The graph is concave downward for $-5 < x < -4$. Thus, the inflection points are $(-4, 1021)$ and $(-5, 1371)$.

15.4 $f(x) = x/(2x - 1)$.

▌ $f(x) = [(x - \frac{1}{2}) + \frac{1}{2}]/[2(x - \frac{1}{2})] = \frac{1}{2}\{1 + [1/(2x - 1)]\}$. Hence, $f'(x) = \frac{1}{2}[-1/(2x - 1)^2] \cdot 2 = -1/(2x - 1)^2$. Then $f''(x) = [1/(2x - 1)^3] \cdot 2 = 2/(2x - 1)^3$. For $x > \frac{1}{2}$, $2x - 1 > 0$, $f''(x) > 0$, and the graph is concave upward. For $x < \frac{1}{2}$, $2x - 1 < 0$, $f''(x) < 0$, and the graph is concave downward. There is no inflection point, since $f(x)$ is not defined when $x = \frac{1}{2}$.

15.5 $f(x) = 5x^4 - x^5$.

▌ $f'(x) = 20x^3 - 5x^4$, and $f''(x) = 60x^2 - 20x^3 = 20x^2(3 - x)$. So, for $0 < x < 3$ and for $x < 0$, $3 - x > 0$, $f''(x) > 0$, and the graph is concave upward. For $x > 3$, $3 - x < 0$, $f''(x) < 0$, and the graph is concave downward. There is an inflection point at $(3, 162)$. There is no inflection point at $x = 0$; the graph is concave upward for $x < 3$.

For Problems 15.6 to 15.10, find the critical numbers and determine whether they yield relative maxima, relative minima, inflection points, or none of these.

15.6 $f(x) = 8 - 3x + x^2$.

▌ $f'(x) = -3 + 2x$, $f''(x) = 2$. Setting $-3 + 2x = 0$, we find that $x = \frac{3}{2}$ is a critical number. Since $f''(x) = 2 > 0$, the second derivative test tells us that there is a relative minimum at $x = \frac{3}{2}$.

15.7 $f(x) = x^4 - 18x^2 + 9$.

▌ $f'(x) = 4x^3 - 36x = 4x(x^2 - 9) = 4x(x - 3)(x + 3)$. $f''(x) = 12x^2 - 36 = 12(x^2 - 3)$. The critical numbers are $0, 3, -3$. $f''(0) = -36 < 0$; hence, $x = 0$ yields a relative maximum. $f''(3) = 72 > 0$; hence, $x = 3$ yields a relative minimum. $f''(-3) = 72 > 0$; hence $x = -3$ yields a relative minimum. There are inflection points at $x = \pm\sqrt{3}$, $y = -36$.

100

15.8 $f(x) = x^3 - 5x^2 - 8x + 3$.

▮ $f'(x) = 3x^2 - 10x - 8 = (3x + 2)(x - 4)$. $f''(x) = 6x - 10$. The critical numbers are $x = -\frac{2}{3}$ and $x = 4$. $f''(-\frac{2}{3}) = -14 < 0$; hence, $x = -\frac{2}{3}$ yields a relative maximum. $f''(4) = 14 > 0$; hence, $x = 4$ yields a relative minimum. There is an inflection point at $x = \frac{5}{3}$.

15.9 $f(x) = x^2/(x - 1)$.

▮ $f(x) = \dfrac{(x^2 - 1) + 1}{x - 1} = \dfrac{(x - 1)(x + 1) + 1}{x - 1} = x + 1 + \dfrac{1}{x - 1}$. So, $f'(x) = 1 - 1/(x - 1)^2$, $f''(x) = 2/(x - 1)^3$. Thus, the critical numbers are the solutions of $1 = 1/(x - 1)^2$, $(x - 1)^2 = 1$, $x - 1 = \pm 1$, $x = 0$ or $x = 2$. $f''(0) = -2 < 0$; thus, $x = 0$ yields a relative maximum. $f''(2) = 2 > 0$; thus, $x = 2$ yields a relative minimum.

15.10 $f(x) = x^2/(x^2 + 1)$.

▮ $f(x) = 1 - 1/(x^2 + 1)$. So, $f'(x) = 2x/(x^2 + 1)^2$. The only critical number is $x = 0$. Use the first derivative test. To the right of 0, $f'(x) > 0$, and to the left of 0, $f'(x) < 0$. Thus, we have the case $\{-, +\}$, and, therefore, $x = 0$ yields a relative minimum. Using the quotient rule, $f''(x) = 2(1 - 3x^2)/(x^2 + 1)^3$. So, there are inflection points at $x = \pm 1/\sqrt{3}$, $y = \frac{1}{4}$.

In Problems 15.11 to 15.19, sketch the graph of the given function.

15.11 $f(x) = (x^2 - 1)^3$.

▮ $f'(x) = 3(x^2 - 1)^2 \cdot 2x = 6x(x^2 - 1)^2 = 6x(x - 1)^2(x + 1)^2$. There are three critical numbers $0, 1$, and -1. At $x = 0$, $f'(x) > 0$ to the right of 0, and $f'(x) < 0$ to the left of 0. Hence, we have the case $\{-, +\}$ of the first derivative test; thus $x = 0$ yields a relative minimum at $(0, -1)$. For both $x = 1$ and $x = -1$, $f'(x)$ has the same sign to the right and to the left of the critical number; therefore, there are inflection points at $(1, 0)$ and $(-1, 0)$. When $x \to \pm\infty$, $f(x) \to +\infty$.

It is obvious from what we have of the graph in Fig. 15-1 so far that there must be inflection points between $x = -1$ and $x = 0$, and between $x = 0$ and $x = 1$. To find them, we compute the second derivative:

$$f''(x) = 6(x - 1)^2(x + 1)^2 + 12x(x - 1)^2(x + 1) + 12x(x - 1)(x + 1)^2$$

$$= 6(x - 1)(x + 1)[(x - 1)(x + 1) + 2x(x - 1) + 2x(x + 1)]$$

$$= 6(x - 1)(x + 1)(5x^2 - 1)$$

Hence, the inflection points occur when $5x^2 - 1 = 0$, $x^2 = \frac{1}{5}$, $x = \pm 1/\sqrt{5} \approx 0.45$, $y = (\frac{1}{5} - 1)^3 = -\frac{64}{125} \approx -0.51$. The graph is in Fig. 15-1.

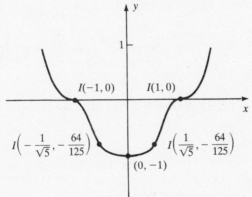

Fig. 15-1

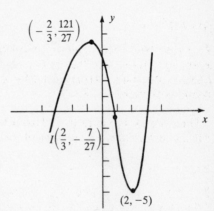

Fig. 15-2

15.12 $f(x) = x^3 - 2x^2 - 4x + 3$.

▮ $f'(x) = 3x^2 - 4x - 4 = (3x + 2)(x - 2)$. $f''(x) = 6x - 4 = 6(x - \frac{2}{3})$. The critical numbers are $x = -\frac{2}{3}$ and $x = 2$. $f''(-\frac{2}{3}) = -8 < 0$; hence, there is a relative maximum at $x = -\frac{2}{3}$, $y = \frac{121}{27} \approx 4.5$. $f''(2) = 8 > 0$; so there is a relative minimum at $x = 2$, $y = -5$. As $x \to +\infty$, $f(x) \to +\infty$. As $x \to -\infty$, $f(x) \to -\infty$. To find the inflection point(s), we set $f''(x) = 6x - 4 = 0$, obtaining $x = \frac{2}{3}$, $y = -\frac{7}{27} \approx -0.26$. The graph is shown in Fig. 15-2.

15.13 $f(x) = x(x-2)^2$.

❚ $f'(x) = x \cdot 2(x-2) + (x-2)^2 = (x-2)(3x-2)$, and $f''(x) = (x-2) \cdot 3 + (3x-2) = 6x - 8$. The critical numbers are $x = 2$ and $x = \frac{2}{3}$. $f''(2) = 4 > 0$; hence, there is a relative minimum at $x = 2$, $y = 0$. $f''(\frac{2}{3}) = -4 < 0$; hence, there is a relative maximum at $x = \frac{2}{3}$, $y = \frac{32}{27} \sim 1.2$. There is an inflection point where $f''(x) = 6x - 8 = 0$, $x = \frac{4}{3}$, $y = \frac{16}{27} \sim 0.6$. As $x \to +\infty$, $f(x) \to +\infty$. As $x \to -\infty$, $f(x) \to -\infty$. Notice that the graph intersects the x-axis at $x = 0$ and $x = 2$. The graph is shown in Fig. 15-3.

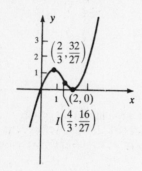

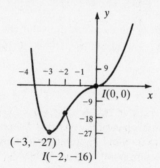

Fig. 15-3

Fig. 15-4

15.14 $f(x) = x^4 + 4x^3$.

❚ $f'(x) = 4x^3 + 12x^2 = 4x^2(x+3)$ and $f''(x) = 12x^2 + 24x = 12x(x+2)$. The critical numbers are $x = 0$ and $x = -3$. $f''(0) = 0$, so we have to use the first derivative test: $f'(x)$ is positive to the right and left of 0; hence, there is an inflection point at $x = 0$, $y = 0$. $f''(-3) = 36 > 0$; hence, there is a relative minimum at $x = -3$, $y = -27$. Solving $f''(x) = 0$, we see that there is another inflection point at $x = -2$, $y = -16$. Since $f(x) = x^3(x+4)$, the graph intersects the x-axis only at $x = 0$ and $x = -4$. As $x \to \pm\infty$, $f(x) \to +\infty$. The graph is shown in Fig. 15-4.

15.15 $f(x) = 3x^5 - 20x^3$.

❚ $f'(x) = 15x^4 - 60x^2 = 15x^2(x^2 - 4) = 15x^2(x-2)(x+2)$, and $f''(x) = 60x^3 - 120x = 60x(x^2 - 2) = 60x(x - \sqrt{2})(x + \sqrt{2})$. The critical numbers are $0, 2, -2$. $f''(0) = 0$. So, we must use the first derivative test for $x = 0$. $f'(x)$ is negative to the right and left of $x = 0$; hence, we have the $\{-, -\}$ case, and there is an inflection point at $x = 0$, $y = 0$. For $x = 2$, $f''(2) = 240 > 0$; thus, there is a relative minimum at $x = 2$, $y = -64$. Similarly, $f''(-2) = -240 < 0$, so there is a relative maximum at $x = -2$, $y = 64$. There are also inflection points at $x = \sqrt{2}$, $y = -28\sqrt{2} \approx -39.2$, and at $x = -\sqrt{2}$, $y = 28\sqrt{2} \approx 39.2$. As $x \to +\infty$, $f(x) \to +\infty$. As $x \to -\infty$, $f(x) \to -\infty$. See Fig. 15-5.

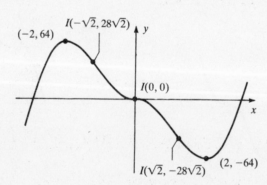

Fig. 15-5

15.16 $f(x) = \sqrt[3]{x-1} = (x-1)^{1/3}$.

❚ $f'(x) = \frac{1}{3}(x-1)^{-2/3}$, and $f''(x) = -\frac{2}{9}(x-1)^{-5/3}$. The only critical number is $x = 1$, where $f'(x)$ is not defined. $f(1) = 0$. $f'(x)$ is positive to the right and left of $x = 1$. $f''(x)$ is negative to the right of $x = 1$; so the graph is concave downward for $x > 1$. $f''(x)$ is positive to the left of $x = 1$; hence, the graph is concave upward for $x < 1$ and there is an inflection point at $x = 1$. As $x \to +\infty$, $f(x) \to +\infty$, and, as $x \to -\infty$, $f(x) \to -\infty$. See Fig. 15-6.

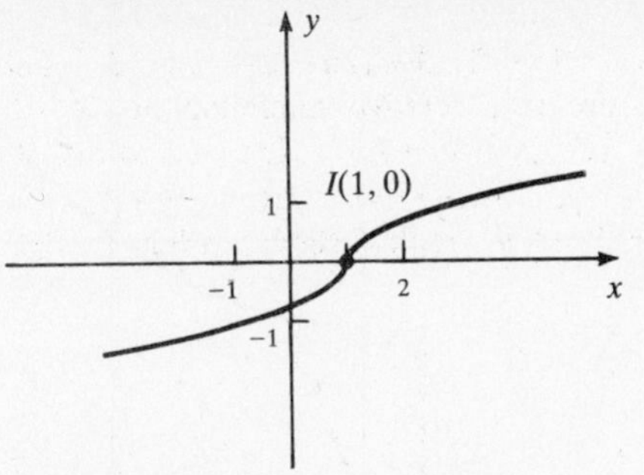

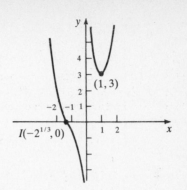

Fig. 15-6

Fig. 15-7

15.17 $f(x) = x^2 + 2/x$.

▌ $f'(x) = 2x - 2/x^2 = 2(x^3 - 1)/x^2 = 2(x - 1)(x^2 + x + 1)/x^2$, and $f''(x) = 2 + 4/x^3$. By the quadratic formula, $x^2 + x + 1$ has no real roots. Hence, the only critical number is $x = 1$. Since $f''(1) = 6 > 0$, there is a relative minimum at $x = 1$, $y = 3$. There is a vertical asymptote at $x = 0$. As $x \to 0^+$, $f(x) \to +\infty$. As $x \to 0^-$, $f(x) \to -\infty$. There is an inflection point where $2 + 4/x^3 = 0$, namely, at $x = -\sqrt[3]{2}$; the graph is concave downward for $-\sqrt[3]{2} < x < 0$ and concave upward for $x < -\sqrt[3]{2}$. As $x \to +\infty$, $f(x) \to +\infty$. As $x \to -\infty$, $f(x) \to +\infty$. See Fig. 15-7.

15.18 $f(x) = (x^2 - 3)/x^3$.

▌ Writing $f(x) = x^{-1} - 3x^{-3}$, we obtain $f'(x) = -1/x^2 + 9/x^4 = -(x - 3)(x + 3)/x^4$. Similarly, $f''(x) = -2(18 - x^2)/x^5$. So the critical numbers are $3, -3$. $f''(3) = -\frac{2}{27} < 0$. Thus, there is a relative maximum at $x = 3$, $y = \frac{2}{9}$. $f''(-3) = \frac{2}{27} > 0$. Thus, there is a relative minimum at $x = -3$, $y = -\frac{2}{9}$. There is a vertical asymptote at $x = 0$. As $x \to 0^+$, $f(x) \to -\infty$. As $x \to 0^-$, $f(x) \to +\infty$. As $x \to +\infty$, $f(x) \to 0$. As $x \to -\infty$, $f(x) \to 0$. Thus, the x-axis is a horizontal asymptote on the right and on the left. There are inflection points where $x^2 = 18$, that is, at $x = \pm 3\sqrt{2} \approx \pm 4.2$, $y = \pm \frac{5}{36}\sqrt{2} \approx \pm 0.2$. See Fig. 15-8.

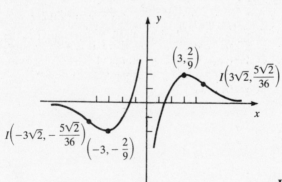

Fig. 15-8

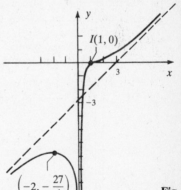

Fig. 15-9

15.19 $f(x) = (x - 1)^3/x^2$.

▌ $$f'(x) = \frac{x^2 \cdot 3(x - 1)^2 - (x - 1)^3 \cdot 2x}{x^4} = \frac{(x - 1)^2[3x - 2(x - 1)]}{x^3} = \frac{(x - 1)^2(x + 2)}{x^3}$$

Similarly, $f''(x) = \frac{6(x - 1)}{x^4}$. The critical numbers are 1 and -2. Since $f''(1) = 0$, we use the first derivative test for $x = 1$. Clearly, $f'(x)$ is positive on both sides of $x = 1$; hence, we have the case $\{+, +\}$, and there is an inflection point at $x = 1$. On the other hand, $f''(-2) = -\frac{9}{8} < 0$, and there is a relative maximum at $x = -2$, $y = -\frac{27}{4}$. There is a vertical asymptote at $x = 0$; notice that, as $x \to 0$, $f(x) \to -\infty$ from both sides. As $x \to +\infty$, $f(x) \to +\infty$. As $x \to -\infty$, $f(x) \to -\infty$. Note also that $f(x) - (x - 3) \to 0$ as $x \to \pm\infty$. Hence, the line $y = x - 3$ is an asymptote. See Fig. 15-9.

15.20 If, for all x, $f'(x) > 0$ and $f''(x) < 0$, which of the curves in Fig. 15-10 could be part of the graph of f?

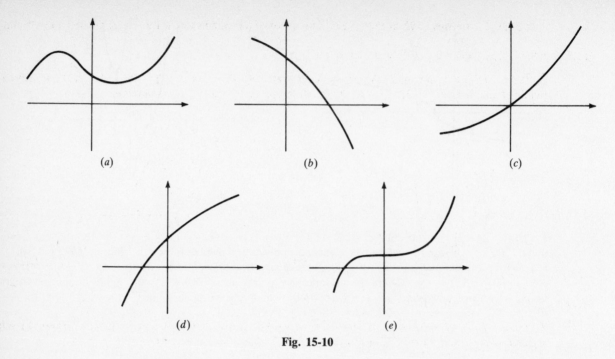

Fig. 15-10

▮ Since $f''(x) < 0$, the graph must be concave downward. This eliminates (a), (c), and (e). Since $f'(x) > 0$, the slope of the tangent line must be positive. This eliminates (b). The only possibility is (d).

15.21 At which of the five indicated points of the graph in Fig 15-11 do y' and y'' have the same sign?

▮ At A and B, the graph is concave downward, and, therefore, $y'' < 0$. The slope y' of the tangent line is > 0 at A and < 0 at B. Thus, B is one of the desired points. At C, there is an inflection point, and therefore, $y'' = 0$; however, the slope y' of the tangent line at C is not 0, and, therefore, C does not qualify. At D and E, the graph is concave upward, and, therefore, $y'' > 0$. At E, the slope y of the tangent line is > 0, and, therefore, E is one of the desired points. At D, the slope y' of the tangent line seems to be 0, and, therefore, D does not qualify. Thus, B and E are the only points that qualify.

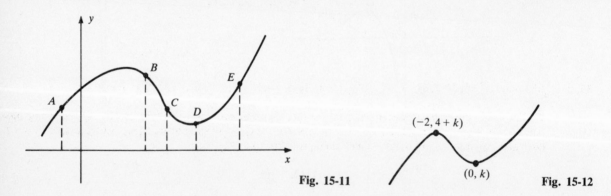

Fig. 15-11   **Fig. 15-12**

15.22 If $f(x) = x^3 + 3x^2 + k$ has three distinct real roots, what are the bounds on k?

▮ $f'(x) = 3x^2 + 6x = 3x(x + 2)$, and $f''(x) = 6x + 6 = 6(x + 1)$. The critical numbers are 0 and -2. Since $f''(0) = 6 > 0$, there is a relative minimum at $x = 0$, $y = k$. Since $f''(-2) = -6 < 0$, there is a relative maximum at $x = -2$, $y = 4 + k$. Since there are no critical numbers > 0 and there is a relative minimum at $x = 0$, the graph to the right of $x = 0$ keeps on going up toward $+\infty$. Since there are no critical numbers < -2 and there is a relative maximum at $x = -2$, the graph to the left of $x = -2$ keeps on going down toward $-\infty$. Hence, a partial sketch of the graph, with the axes missing, is shown in Fig. 15-12. Since there are three distinct real roots, the graph must intersect the x-axis at three distinct points. Thus, the x-axis must lie strictly between $y = k$ and $y = 4 + k$, that is, $k < 0 < 4 + k$. Hence, $-4 < k < 0$.

In each of Problems 15.23 to 15-29, sketch the graph of a continuous function f satisfying the given conditions.

15.23 $f(1) = -2$, $f'(1) = 0$, $f''(x) > 0$ for all x.

▌ Since $f''(x) > 0$ for all x, the graph is concave upward. Since $f'(1) = 0$ and $f''(1) > 0$, there is a relative minimum at $x = 1$. Hence, the graph in Fig. 15-13 satisfies the requirements.

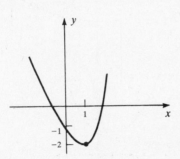

Fig. 15-13

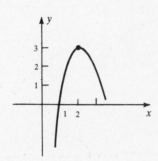

Fig. 15-14

15.24 $f(2) = 3$, $f'(2) = 0$, $f''(x) < 0$ for all x.

▌ Since $f''(x) < 0$, the graph is concave downward. Since $f'(2) = 0$ and $f''(2) < 0$, there is a relative maximum at $x = 2$, $y = 3$. The graph in Fig. 15-14 satisfies these requirements.

15.25 $f(1) = 1$, $f''(x) < 0$ for $x > 1$, $f''(x) > 0$ for $x < 1$, $\lim\limits_{x \to +\infty} f(x) = +\infty$, $\lim\limits_{x \to -\infty} f(x) = -\infty$.

▌ The graph is concave upward for $x < 1$ and concave downward for $x > 1$; therefore, it has an inflection point at $x = 1$, $y = 1$. A possible graph is shown in Fig. 15.15.

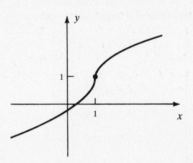

Fig. 15-15

15.26 $f(0) = 0$, $f''(x) < 0$ for $x > 0$, $f''(x) > 0$ for $x < 0$, $\lim\limits_{x \to +\infty} f(x) = 1$, $\lim\limits_{x \to -\infty} f(x) = -1$.

▌ The graph must be concave upward for $x < 0$ and concave downward for $x > 0$, so there is an inflection point at $x = 0$, $y = 0$. The line $y = 1$ is a horizontal asymptote to the right, and the line $y = -1$ is a horizontal asymptote to the left. Such a graph is shown in Fig. 15-16.

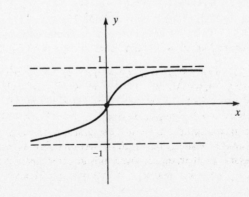

Fig. 15-16

15.27 $f(0) = 1$, $f''(x) < 0$ for $x \neq 0$, $\lim\limits_{x \to 0^+} f'(x) = +\infty$, $\lim\limits_{x \to 0^-} f'(x) = -\infty$.

▮ For $x \neq 0$, the graph is concave downward. As the graph approaches $(0, 1)$ from the right, the slope of the tangent line approaches $+\infty$. As the graph approaches $(0, 1)$ from the left, the slope of the tangent line approaches $-\infty$. Such a graph is shown in Fig. 15-17.

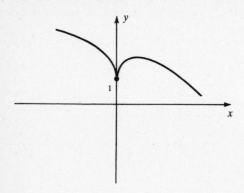

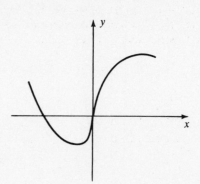

Fig. 15-17   **Fig. 15-18**

15.28 $f(0) = 0$, $f''(x) > 0$ for $x < 0$, $f''(x) < 0$ for $x > 0$, $\lim\limits_{x \to 0^-} f'(x) = +\infty$, $\lim\limits_{x \to 0^+} f'(x) = +\infty$.

▮ The graph is concave upward for $x < 0$ and concave downward for $x > 0$. As the graph approaches the origin from the left or from the right, the slope of the tangent line approaches $+\infty$. Such a graph is shown in Fig. 15-18.

15.29 $f(0) = 1$, $f''(x) < 0$ if $x \neq 0$, $\lim\limits_{x \to 0^+} f'(x) = 0$, $\lim\limits_{x \to 0^-} f'(x) = -\infty$.

▮ The graph is concave downward. As the graph approaches $(0, 1)$ from the right, it levels off. As the graph approaches $(0, 1)$ from the left, the slope of the tangent line approaches $-\infty$. Such a graph is shown in Fig. 15-19.

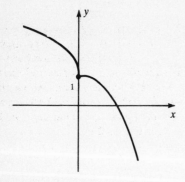

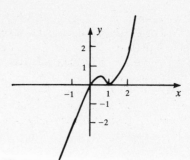

Fig. 15-19   **Fig. 15-20**

15.30 Sketch the graph of $f(x) = x|x - 1|$.

▮ First consider $x > 1$. Then $f(x) = x(x - 1) = x^2 - x = (x - \frac{1}{2})^2 - \frac{1}{4}$. The graph is part of a parabola with vertex at $(\frac{1}{2}, -\frac{1}{4})$ and passing through $(1, 0)$. Now consider $x < 1$. Then $f(x) = -x(x - 1) = -(x^2 - x) = -[(x - \frac{1}{2})^2 - \frac{1}{4}] = -(x - \frac{1}{2})^2 + \frac{1}{4}$. Thus, we have part of a parabola with vertex at $(\frac{1}{2}, \frac{1}{4})$. The graph is shown in Fig. 15-20.

15.31 Let $f(x) = x^4 + Ax^3 + Bx^2 + Cx + D$. Assume that the graph of $y = f(x)$ is symmetric with respect to the y-axis, has a relative maximum at $(0, 1)$, and has an absolute minimum at $(k, -3)$. Find A, B, C, and D, as well as the possible value(s) of k.

▮ It is given that $f(x)$ is an even function, so $A = C = 0$. Thus, $f(x) = x^4 + Bx^2 + D$. Since the graph passes through $(0, 1)$, $D = 1$. So, $f(x) = x^4 + Bx^2 + 1$. Then $f'(x) = 4x^3 + 2Bx = 2x(2x^2 + B)$. 0 and k are critical numbers, where $2k^2 + B = 0$. Since $(k, -3)$ is on the graph, $k^4 + Bk^2 + 1 = -3$. Replacing B by $-2k^2$, $k^4 - 2k^4 = -4$, $k^4 = 4$, $k^2 = 2$, $B = -4$. k can be $\pm\sqrt{2}$.

In Problems 15.32 to 15.54, sketch the graphs of the given functions.

15.32 $f(x) = \sin^2 x$.

▮ Since $\sin (x + \pi) = -\sin x$, $f(x)$ has a period of π. So, we need only show the graph for $-\pi/2 \le x \le \pi/2$. Now, $f'(x) = 2 \sin x \cdot \cos x = \sin 2x$. $f''(x) = 2 \cos 2x$. Within $[-\pi/2, \pi/2]$, we only have the critical numbers 0, $-\pi/2$, $\pi/2$. $f''(0) = 2 > 0$; hence, there is a relative minimum at $x = 0$, $y = 0$. $f''(\pi/2) = -2 < 0$; hence, there is a relative maximum at $x = \pi/2$, $y = 1$, and similarly at $x = -\pi/2$, $y = 1$. Inflection points occur where $f''(x) = 2 \cos 2x = 0$, $2x = \pm \pi/2$, $x = \pm \pi/4$, $y = \frac{1}{2}$. The graph is shown in Fig. 15-21.

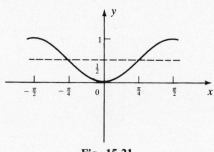

Fig. 15-21

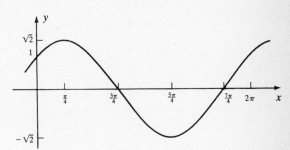

Fig. 15-22

15.33 $f(x) = \sin x + \cos x$.

▮ $f(x)$ has a period of 2π. Hence, we need only consider the interval $[0, 2\pi]$. $f'(x) = \cos x - \sin x$, and $f''(x) = -(\sin x + \cos x)$. The critical numbers occur where $\cos x = \sin x$ or $\tan x = 1$, $x = \pi/4$ or $x = 5\pi/4$. $f''(\pi/4) = -(\sqrt{2}/2 + \sqrt{2}/2) = -\sqrt{2} < 0$. So, there is a relative maximum at $x = \pi/4$, $y = \sqrt{2}$. $f''(5\pi/4) = -(-\sqrt{2}/2 - \sqrt{2}/2) = \sqrt{2} > 0$. Thus, there is a relative minimum at $x = 5\pi/4$, $y = -\sqrt{2}$. The inflection points occur where $f''(x) = -(\sin x + \cos x) = 0$, $\sin x = -\cos x$, $\tan x = -1$, $x = 3\pi/4$ or $x = 7\pi/4$, $y = 0$. See Fig. 15-22.

15.34 $f(x) = 3 \sin x - \sin^3 x$.

▮ $f'(x) = 3 \cos x - 3 \sin^2 x \cdot \cos x = 3 \cos x (1 - \sin^2 x) = 3 \cos x \cdot \cos^2 x = 3 \cos^3 x$. $f''(x) = 9 \cos^2 x \cdot (-\sin x) = -9 \cos^2 x \cdot \sin x$. Since $f(x)$ has a period of 2π, we need only look at, say, $(-\pi, \pi)$. The critical numbers are the solutions of $3 \cos^3 x = 0$, $\cos x = 0$, $x = -\pi/2$ or $x = \pi/2$. $f''(-\pi/2) = 0$, so we must use the first derivative test. $f'(x)$ is negative to the left of $-\pi/2$ and positive to the right of $-\pi/2$. Hence, we have the case $\{-, +\}$, and there is a relative minimum at $x = -\pi/2$, $y = -2$. Similarly, there is a relative maximum at $x = \pi/2$, $y = 2$. [Notice that $f(x)$ is an odd function; that is, $f(-x) = -f(x)$.] To find inflection points, set $f''(x) = -9 \cos^2 x \cdot \sin x = 0$. Aside from the critical numbers $\pm \pi/2$, this yields the solutions $-\pi$, 0, π of $\sin x = 0$. Thus, there are inflection points at $(-\pi, 0)$, $(0, 0)$, $(\pi, 0)$. See Fig. 15-23.

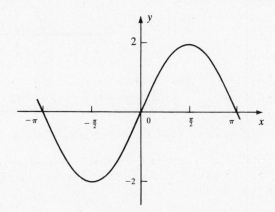

Fig. 15-23

15.35 $f(x) = \cos x - \cos^2 x$.

▮ $f'(x) = -\sin x - 2 (\cos x)(-\sin x) = (\sin x)(2 \cos x - 1)$, and $f''(x) = (\sin x)(-2 \sin x) + (2 \cos x - 1)(\cos x) =$

$2(\cos^2 x - \sin^2 x) - \cos x = 2(2\cos^2 x - 1) - \cos x = 4\cos^2 x - \cos x - 2$.   Since $f(x)$ has a period of 2π, we need only consider, say, $[-\pi, \pi]$.   Moreover, since $f(-x) = f(x)$,   we only have to consider $[0, \pi]$, and then reflect in the y-axis.   The critical numbers are the solutions in $[0, \pi]$ of $\sin x = 0$ or $2\cos x - 1 = 0$.   The equation $\sin x = 0$ has the solutions $0, \pi$.   The equation $2\cos x - 1 = 0$ is equivalent to $\cos x = \frac{1}{2}$,   having the solution $x = \pi/3$. $f''(0) = 1 > 0$,   so there is a relative minimum at $x = 0$, $y = 0$. $f''(\pi) = 3 > 0$,   so there is a relative minimum at $x = \pi$, $y = -2$. $f''(\pi/3) = -\frac{3}{2} < 0$,   so there is a relative maximum at $x = \pi/3$, $y = \frac{1}{4}$.   There are inflection points between 0 and $\pi/3$ and between $\pi/3$ and π; they can be approximated by using the quadratic formula to solve $f''(x) = 4\cos^2 x - \cos x - 2 = 0$ for $\cos x$, and then using a cosine table to approximate x.   See Fig. 15-24.

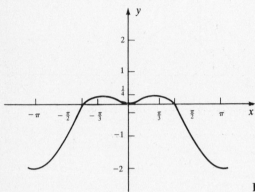

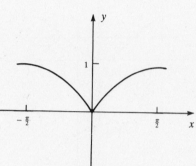

Fig. 15-24   **Fig. 15-25**

15.36 $f(x) = |\sin x|$.

    ▌   Since $f(x)$ is even, has a period of π, and coincides with $\sin x$ on $[0, \pi/2]$, we get the curve of Fig. 15-25.

15.37 $f(x) = \sin x + x$.

    ▌ $f'(x) = \cos x + 1$. $f''(x) = -\sin x$.   The critical numbers are the solutions of $\cos x = -1$, $x = (2n + 1)\pi$. The first derivative test yields the case $\{+, +\}$, and, therefore, we obtain only inflection points at $x = (2n + 1)\pi$, $y = (2n + 1)\pi$.   See Fig. 15-26.

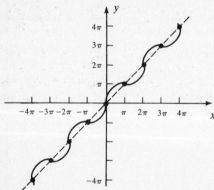

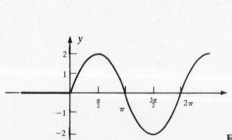

Fig. 15-26   **Fig. 15-27**

15.38 $f(x) = \sin x + \sin |x|$.

    ▌   **Case 1.** $x \geq 0$.   Then $f(x) = 2\sin x$.   **Case 2.** $x < 0$.   Then $f(x) = 0$ [since $\sin(-x) = -\sin x$].   See Fig. 15-27.

15.39 $f(x) = \dfrac{9 + x^2}{9 - x^2}$

    ▌

$$f(x) = \frac{9 + x^2}{(3 + x)(3 - x)}$$

$$f'(x) = \frac{(9 - x^2)2x - (9 + x^2)(-2x)}{(9 - x^2)^2} = \frac{2x(18)}{(9 - x^2)^2} = \frac{36x}{(9 - x^2)^2}$$

$$f''(x) = 36 \frac{(9-x^2)^2 - x[2(9-x^2)(-2x)]}{(9-x^2)^4} = 36 \frac{(9-x^2)+4x^2}{(9-x^2)^3} = 108 \frac{3+x^2}{(9-x^2)^3}$$

The only critical number is 0. $f''(0) = \frac{4}{9} > 0$, so there is a relative minimum at $x = 0$, $y = 1$. There are no inflection points. Vertical asymptotes occur at $x = -3$ and $x = 3$. As $x \to -3^-$, $f(x) \to -\infty$, since $9 + x^2$ and $3 - x$ are positive, while $3 + x$ is negative. Similarly, as $x \to -3^+$, $f(x) \to +\infty$. Note that $f(-x) = f(x)$, so the graph is symmetric with respect to the y-axis. In particular, at the vertical asymptote $x = 3$, as $x \to 3^-$, $f(x) \to +\infty$, and, as $x \to 3^+$, $f(x) \to -\infty$. As $x \to \pm\infty$, $f(x) = (9/x^2 + 1)/(9/x^2 - 1) \to -1$. Thus, the line $y = -1$ is a horizontal asymptote on the left and right. Observe that $f''(x) > 0$ for $-3 < x < 3$; hence, the graph is concave upward on that interval. In addition, $f''(x) < 0$ for $|x| > 3$; hence, the graph is concave downward for $x > 3$ and $x < -3$. See Fig. 15-28.

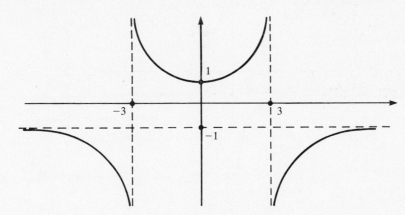

Fig. 15-28

15.40 $f(x) = \dfrac{x}{x^2 - 1}$.

▌ $$f(x) = \frac{x}{(x-1)(x+1)}$$

$$f'(x) = \frac{(x^2-1) - x(2x)}{(x^2-1)^2} = -\frac{1+x^2}{(x^2-1)^2}$$

$$f''(x) = -\frac{(x^2-1)^2(2x) - (1+x^2)[2(x^2-1)\cdot 2x]}{(x^2-1)^4} = -\frac{2x[(x^2-1) - (1+x^2)\cdot 2]}{(x^2-1)^3} = \frac{2x(3+x^2)}{(x^2-1)^3}$$

There are no critical numbers. There are vertical asymptotes at $x = 1$ and $x = -1$. As $x \to 1^+$, $f(x) \to +\infty$. As $x \to 1^-$, $f(x) \to -\infty$. As $x \to -1^+$, $f(x) \to +\infty$. As $x \to -1^-$, $f(x) \to -\infty$. As $x \to \pm\infty$, $f(x) = (1/x)/(1 - 1/x^2) \to 0$. Hence, the x-axis is a horizontal asymptote to the right and left. There is an inflection point at $x = 0$, $y = 0$. The concavity is upward for $x > 1$ and for $-1 < x < 0$, where $f''(x) > 0$; elsewhere, the concavity is downward. See Fig. 15-29.

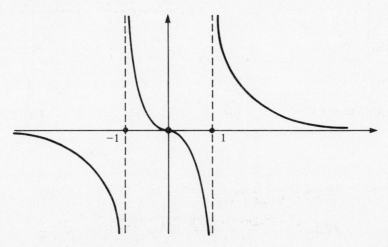

Fig. 15-29

15.41 $f(x) = x + 9/x$.

▮ $f'(x) = 1 - 9/x^2 = (x^2 - 9)/x^2 = (x - 3)(x + 3)/x^2$. $f''(x) = 18/x^3$. The critical numbers are ± 3. $f''(3) = \frac{2}{3} > 0$, so there is a relative minimum at $x = 3$, $y = 6$. $f''(-3) = -\frac{2}{3} < 0$, so there is relative maximum at $x = -3$, $y = -6$. There is a vertical asymptote at $x = 0$: as $x \to 0^+$, $f(x) \to +\infty$; as $x \to 0^-$, $f(x) \to -\infty$. As $x \to \pm\infty$, $f(x) - x = 9/x \to 0$; so the line $y = x$ is an asymptote. The concavity is upward for $x > 0$ and downward for $x < 0$. Note that $f(x) = -f(-x)$, so the graph is symmetric with respect to the origin. See Fig. 15-30.

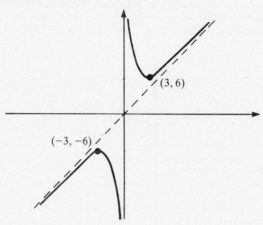

Fig. 15-30

15.42 $f(x) = \dfrac{2x^2}{x^2 + 3}$.

▮ $$f'(x) = \frac{(x^2 + 3)(4x) - 2x^2(2x)}{(x^2 + 3)^2} = \frac{12x}{(x^2 + 3)^2}$$

$$f''(x) = \frac{(x^2 + 3)^2(12) - 12x[2(x^2 + 3)(2x)]}{(x^2 + 3)^4} = 12 \cdot \frac{(x^2 + 3) - 4x^2}{(x^2 + 3)^3} = 12 \frac{3 - 3x^2}{(x^2 + 3)^3} = 36 \frac{1 - x^2}{(x^2 + 3)^3}$$

The only critical number is 0. $f''(0) = \frac{4}{3} > 0$, so there is a relative minimum at $x = 0$, $y = 0$. Note that $f(x) = f(-x)$; hence, the graph is symmetric with respect to the y-axis. As $x \to \pm\infty$, $f(x) \to 2$. Hence, the line $y = 2$ is a horizontal asymptote on the right and left. There are inflection points where $f''(x) = 0$, that is, at $x = \pm 1$, $y = \frac{1}{2}$. See Fig. 15-31.

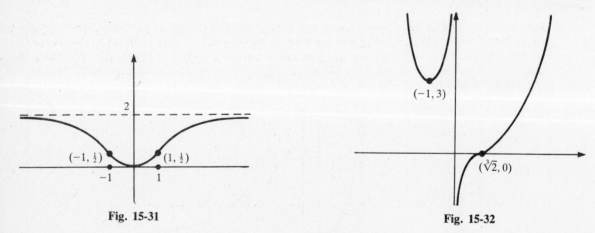

Fig. 15-31   Fig. 15-32

15.43 $f(x) = x^2 - 2/x$.

▮ $f(x) = (x^3 - 2)/x$. $f'(x) = 2x + 2/x^2 = 2(x^3 + 1)/x^2$. $f''(x) = 2 - 4/x^3 = 2(1 - 2/x^3) = 2(x^3 - 2)/x^3$. The only critical number is the solution -1 of $x^3 + 1 = 0$. $f''(-1) = 6 > 0$, so there is a relative minimum at $x = -1$, $y = 3$. There is a vertical asymptote at $x = 0$. As $x \to 0^+$, $f(x) \to -\infty$. As $x \to 0^-$, $f(x) \to +\infty$. As $x \to \pm\infty$, $f(x) \to +\infty$. There is an inflection point at $x = \sqrt[3]{2}$, $y = 0$; the graph is concave upward to the right of that point, as well as for $x < 0$. See Fig. 15-32.

15.44 $f(x) = \frac{3}{5}x^{5/3} - 3x^{2/3}$

▮ $f(x) = 3x^{2/3}(\frac{1}{5}x - 1)$. $f'(x) = x^{2/3} - 2x^{-1/3} = x^{-1/3}(x - 2) = (x - 2)/\sqrt[3]{x}$. $f''(x) = \frac{2}{3}x^{-1/3} + \frac{2}{3}x^{-4/3} = \frac{2}{3}x^{-4/3}(x + 1) = \frac{2}{3}(x + 1)/\sqrt[3]{x^4}$. There are critical numbers at 2 and 0. $f''(2) > 0$; hence, there is a relative minimum at $x = 2$, $y = 3\sqrt[3]{4}(-\frac{3}{5}) \approx -3.2$. Near $x = 0$, $f'(x)$ is positive to the left and negative to the right, so we have the case $\{+, -\}$ of the first derivative test and there is a relative maximum at $x = 0$, $y = 0$. As $x \to +\infty$, $f(x) \to +\infty$. As $x \to -\infty$, $f(x) \to -\infty$. Note that the graph cuts the x-axis at the solution $x = 5$ of $\frac{1}{5}x - 1 = 0$. There is an inflection point at $x = -1$, $y = -3.6$. The graph is concave downward for $x < -1$ and concave upward elsewhere. Observe that there is a cusp at the origin. See Fig. 15-33.

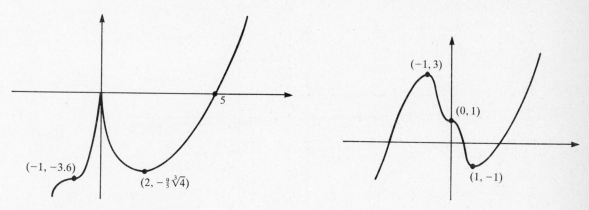

Fig. 15-33   Fig. 15-34

15.45 $f(x) = 3x^5 - 5x^3 + 1$.

▮ $f'(x) = 15x^4 - 15x^2 = 15x^2(x^2 - 1) = 15x^2(x - 1)(x + 1)$, and $f''(x) = 15(4x^3 - 2x) = 30x(2x^2 - 1)$. The critical numbers are 0, ± 1. $f''(1) = 30 > 0$, so there is a relative minimum at $x = 1$, $y = -1$. $f''(-1) = -30 < 0$, so there is a relative maximum at $x = -1$, $y = 3$. Near $x = 0$, $f'(x)$ is negative to the right and left of $x = 0$ (since $x^2 > 0$ and $x^2 - 1 < 0$). Thus, we have the case $\{-, -\}$ of the first derivative test, and therefore, there is an inflection point at $x = 0$, $y = 1$. As $x \to +\infty$, $f(x) \to +\infty$. As $x \to -\infty$, $f(x) \to -\infty$. There are also inflection points at the solutions of $2x^2 - 1 = 0$, $x = \pm\sqrt{2}/2 \approx \pm 0.7$. See Fig. 15-34.

15.46 $f(x) = x^4 - 2x^2 + 1$.

▮ Note that the function is even. $f(x) = (x^2 - 1)^2$, $f'(x) = 4x^3 - 4x = 4x(x^2 - 1) = 4x(x - 1)(x + 1)$, and $f''(x) = 4(3x^2 - 1)$. The critical numbers are 0, ± 1. $f''(0) = -4 < 0$, so there is a relative maximum at $x = 0$, $y = 1$. $f''(\pm 1) = 8 > 0$, so there are relative minima at $x = \pm 1$, $y = 0$. There are inflection points where $3x^2 - 1 = 0$, $x = \pm\sqrt{3}/3 \approx 0.6$, $y = \frac{4}{9} \approx 0.4$. As $x \to \pm\infty$, $f(x) \to +\infty$. See Fig. 15-35.

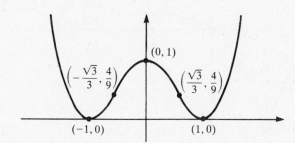

Fig. 15-35

15.47 $f(x) = \dfrac{x}{\sqrt{x^2 - 9}}$.

▮ The function is not defined when $x^2 < 9$, that is, for $-3 < x < 3$. Observe also that $f(-x) = -f(x)$, so the graph is symmetric with respect to the origin.

$$f(x) = \frac{x}{(x^2 - 9)^{1/2}}$$

$$f'(x) = \frac{(x^2-9)^{1/2} - x \cdot \frac{1}{2}(x^2-9)^{-1/2}(2x)}{x^2-9} = \frac{(x^2-9)-x^2}{(x^2-9)^{3/2}} = -\frac{9}{(x^2-9)^{3/2}}$$

$$f''(x) = \frac{27}{2}(x^2-9)^{-5/2} \cdot 2x = \frac{27x}{(x^2-9)^{5/2}}$$

There are no critical numbers. There are vertical asymptotes $x = 3$ and $x = -3$. As $x \to 3^+$, $f(x) \to +\infty$. As $x \to -3^-$, $f(x) \to -\infty$. As $x \to +\infty$, $f(x) = \dfrac{1}{\sqrt{1-9/x^2}} \to 1$. Hence, as $x \to -\infty$, $f(x) \to -1$. Thus, $y = 1$ is a horizontal asymptote on the right, and $y = -1$ is a horizontal asymptote on the left. See Fig. 15-36.

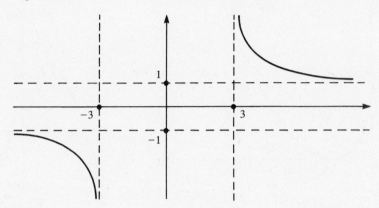

Fig. 15-36

15.48 $f(x) = \dfrac{x}{(x+1)^2}.$

$$f'(x) = \frac{(x+1)^2 - x \cdot 2(x+1)}{(x+1)^4} = \frac{(x+1)-2x}{(x+1)^3} = \frac{1-x}{(x+1)^3}$$

$$f''(x) = \frac{(x+1)^3(-1) - (1-x) \cdot 3(x+1)^2}{(x+1)^6} = -\frac{(x+1)+3(1-x)}{(x+1)^4} = \frac{2(x-2)}{(x+1)^4}$$

The only critical number is 1. $f''(1) = -\frac{1}{8} < 0$, so there is a relative maximum at $x = 1$, $y = \frac{1}{4}$. The line $x = -1$ is a vertical asymptote. As $x \to -1$, $f(x) \to -\infty$. As $x \to \pm\infty$, $f(x) = (1/x)/(1+1/x)^2 \to 0$. Thus, the x-axis is a horizontal asymptote on the right and left. There is an inflection point at $x = 2$, $y = \frac{2}{9}$. The curve is concave downward for $x < 2$. See Fig. 15-37.

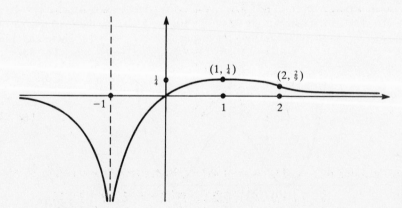

Fig. 15-37

15.49 $f(x) = x^4 - 6x^2 + 8x + 8.$

$f'(x) = 4x^3 - 12x + 8 = 4(x^3 - 3x + 2)$, and $f''(x) = 12x^2 - 12 = 12(x^2-1) = 12(x-1)(x+1)$. The critical numbers are the solutions of $x^3 - 3x + 2 = 0$. Inspecting the integral factors of the constant term 2, we see that 1 is a root. Dividing $x^3 - 3x + 2$ by $x - 1$, we obtain the quotient $x^2 + x - 2 = (x-1)(x+2)$. Hence, $f'(x) = 4(x-1)^2(x+2)$. Thus, the critical numbers are 1 and -2. $f''(-2) = 36 > 0$, so there is a relative minimum at $x = -2$, $y = -16$. At $x = 1$, use the first derivative test. To the right and left of $x = 1$,

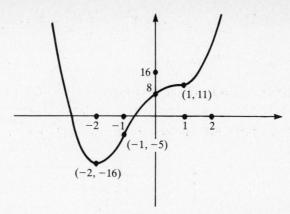

Fig. 15-38

$f'(x) > 0$, so we have the case $\{+, +\}$, and there is an inflection point at $x = 1$, $y = 11$. There is another inflection point at $x = -1$, $y = -5$. As $x \to \pm\infty$, $f(x) \to +\infty$. See Fig. 15-38.

15.50 $f(x) = \dfrac{2x^2}{(x-2)(x-6)}$.

▮ $$f'(x) = \frac{(x-2)(x-6)(4x) - 2x^2(2x-8)}{(x-2)^2(x-6)^2} = \frac{16x(3-x)}{(x-2)^2(x-6)^2}$$

$$f''(x) = \frac{(x-2)^2(x-6)^2(48-32x) - 16x(3-x)\cdot 2(x-2)(x-6)(2x-8)}{(x-2)^4(x-6)^4} = \frac{16(2x^3 - 9x^2 + 36)}{(x-2)^3(x-6)^3}$$

The critical numbers are 0 and 3. $f''(0) > 0$, so there is a relative minimum at $x = 0$, $y = 0$. $f''(3) < 0$; hence, there is a relative maximum at $x = 3$, $y = -6$. The lines $x = 2$ and $x = 6$ are vertical asymptotes. As $x \to 6^+$, $f(x) \to +\infty$. As $x \to 6^-$, $f(x) \to -\infty$. As $x \to 2^+$, $f(x) \to -\infty$. As $x \to 2^-$, $f(x) \to +\infty$. As $x \to \pm\infty$, $f(x) = 2/(1 - 2/x)(1 - 6/x) \to 2$. Hence, the line $y = 2$ is a horizontal asymptote on the right and left. There is a root of $2x^3 - 9x^2 + 36 = 0$ between $x = -1$ and $x = -2$ that yields an inflection point. See Fig. 15-39.

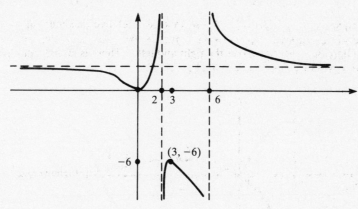

Fig. 15-39

15.51 $f(x) = \dfrac{\sin x}{2 - \cos x}$.

▮ Since $f(x)$ has a period of 2π and is an odd function, we can restrict attention to $[0, \pi]$.

$$f'(x) = \frac{(2 - \cos x)(\cos x) - (\sin x)(\sin x)}{(2 - \cos x)^2} = \frac{2\cos x - 1}{(2 - \cos x)^2}$$

$$f''(x) = \frac{(2 - \cos x)^2(-2\sin x) - (2\cos x - 1)\cdot 2(2 - \cos x)(\sin x)}{(2 - \cos x)^4}$$

$$= (-2\sin x)\frac{2 - \cos x + 2\cos x - 1}{(2 - \cos x)^3} = -\frac{2\sin x(1 + \cos x)}{(2 - \cos x)^3}$$

The critical numbers are the solutions in $[0, \pi]$ of $\cos x = \frac{1}{2}$, that is, $\pi/3$. $f''(\pi/3) = -2\sqrt{3}/3 < 0$; thus, there is a relative maximum at $x = \pi/3$, $y = \sqrt{3}/3$. The graph intersects the x-axis at $x = 0$ and π.

There is an inflection point where $f''(x) = 0$, that is, where $\sin x = 0$ or $\cos x = -1$, namely, $x = 0$, $x = \pi$. See Fig. 15-40.

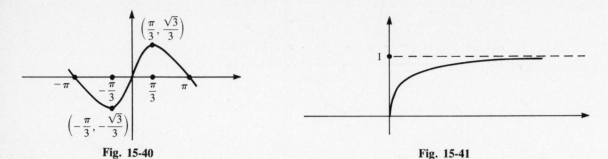

Fig. 15-40 **Fig. 15-41**

15.52 $f(x) = \dfrac{\sqrt{x}}{1 + \sqrt{x}}$.

❚ Note that $f(x)$ is defined only for $x \geq 0$.

$$f'(x) = \frac{(1 + \sqrt{x})(1/2\sqrt{x}) - \sqrt{x}(1/2\sqrt{x})}{(1 + \sqrt{x})^2} = \frac{1}{2\sqrt{x}(1 + \sqrt{x})^2}$$

$$f''(x) = \frac{1}{2}\left(\frac{1}{\sqrt{x}}\right)\left[\frac{-2}{(1 + \sqrt{x})^3}\right]\left(\frac{1}{2\sqrt{x}}\right) + \frac{1}{(1 + \sqrt{x})^2}\left(-\frac{1}{4}\right)\left(\frac{1}{x^{3/2}}\right) = \frac{-\sqrt{x}}{4x^2(1 + \sqrt{x})^3}(3\sqrt{x} + 1)$$

The only critical number is $x = 0$. Since $f(x)$ is always positive, $f(x)$ is an increasing function. In addition, as $x \to +\infty$, $f(x) = 1/(1/\sqrt{x} + 1) \to 1$. Thus, the line $y = 1$ is an asymptote on the right. Since $f''(x)$ is always negative, the graph is always concave downward. See Fig. 15-41.

15.53 $f(x) = \sin x + \sqrt{3} \cos x$.

❚ $f(x)$ has a period of 2π, so we consider only $[0, 2\pi]$. $f'(x) = \cos x - \sqrt{3} \sin x$, and $f''(x) = -\sin x - \sqrt{3}\cos x = -f(x)$. The critical numbers are the solutions of $\cos x - \sqrt{3}\sin x = 0$, $\tan x = 1/\sqrt{3}$, $x = \pi/6$ or $x = 7\pi/6$. $f''(\pi/6) = -2 < 0$, so there is a relative maximum at $x = \pi/6$, $y = 2$. $f''(7\pi/6) = 2 > 0$, so there is a relative minimum at $x = 7\pi/6$, $y = -2$. The graph cuts the x-axis at the solutions of $\sin x + \sqrt{3}\cos x = 0$, $\tan x = -\sqrt{3}$, $x = 2\pi/3$ or $x = 5\pi/3$, which also yield the inflection points [since $f''(x) = -f(x)$]. See Fig. 15-42.

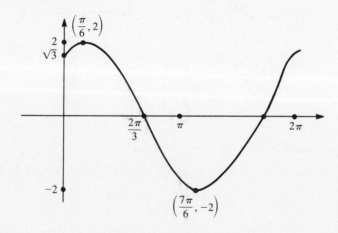

Fig. 15-42

15.54 $f(x) = x\sqrt{1 - x}$.

❚ $y = f(x)$ is defined for $x \leq 1$. When $0 < x < 1$, $y > 0$, and, when $x < 0$, $y < 0$, $y^2 = x^2(1 - x) = x^2 - x^3$. Hence, $2yy' = 2x - 3x^2 = x(2 - 3x)$. So, $x = \frac{2}{3}$ is a critical number. Differentiating again, $2(yy'' + y' \cdot y') = 2 - 6x$, $yy'' + (y')^2 = 1 - 3x$. When $x = \frac{2}{3}$, $y = 2\sqrt{3}/9$, $(2\sqrt{3}/9)y'' = 1 - 3(\frac{2}{3}) = -1$, $y'' < 0$; so there is a relative maximum at $x = \frac{2}{3}$. When $x \neq 0$, $4y^2y'^2 = x^2(2 - 3x)^2$, $4x^2(1 - x)y'^2 = x^2(2 - $

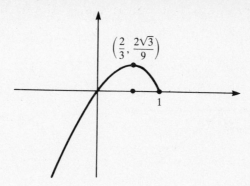

Fig. 15-43

$3x)^2$, $4(1-x)y'^2 = (2-3x)^2$. So, as $x \to 0$, $y'^2 \to 1$. Since $2yy' = x(2-3x)$, as $x \to 0^+$, $y' \to 1$, and, as $x \to 0^-$, $y' \to 1$. As $x \to -\infty$, $y \to -\infty$. Let us look for inflection points. Assume $y'' = 0$. Then, $y'^2 = 1 - 3x$, $4y^2(1-3x) = x^2(2-3x)^2$, $4x^2(1-x)(1-3x) = x^2(2-3x)^2$, $4(1-x)(1-3x) = (2-3x)^2$, $4 - 16x + 12x^2 = 9x^2 - 12x + 4$, $-16 + 12x = 9x - 12$, $3x = 4$, $x = \frac{4}{3}$. Hence, there are no inflection points. See Fig. 15-43.

15.55 Sketch the graph of a function $f(x)$ having the following properties: $f(0) = 0$, $f(x)$ is continuous except at $x = 2$, $\lim_{x \to 2} f(x) = +\infty$, $\lim_{x \to -\infty} f(x) = 0$, $\lim_{x \to +\infty} f(x) = 3$, $f(x)$ is differentiable except at $x = 2$ and $x = -1$, $f'(x) > 0$ if $-1 < x < 2$, $f'(x) < 0$ if $x < -1$ or $x > 2$, $f''(x) < 0$ if $x < 0$ and $x \neq -1$, $f''(x) > 0$ if $x > 0$ and $x \neq 2$.

∥ See Fig. 15-44.

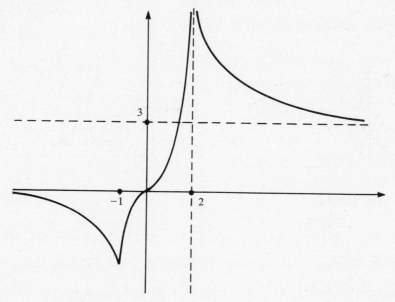

Fig. 15.44

15.56 Sketch the graph of a function $f(x)$ such that: $f(0) = 0$, $f(2) = f(-2) = 1$, $f(0) = 0$, $f'(x) > 0$ for $x > 0$, $f'(x) < 0$ for $x < 0$, $f''(x) > 0$ for $|x| < 2$, $f''(x) < 0$ for $|x| > 2$, $\lim_{x \to +\infty} f(x) = 2$, $\lim_{x \to -\infty} f(x) = 2$.

∥ See Fig. 15-45.

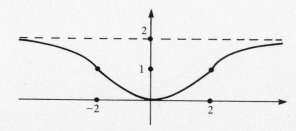

Fig. 15-45

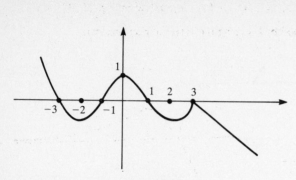

Fig. 15-46

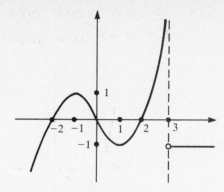

Fig. 15-47

15.57 Sketch the graph of $f'(x)$, if Fig. 15-46 gives the graph of $f(x)$.

▮ A possible graph for $f'(x)$ is shown in Fig. 15-47. At $x = -2$, $f'(x) = 0$. To the left of $x = -2$, $f'(x)$ is negative and approaches $-\infty$ as $x \to -\infty$. From $x = -2$ to $x = -1$, $f'(x)$ is positive and increasing. At $x = -1$, the graph of f has an inflection point, where $f'(x)$ reaches a relative maximum. From $x = -1$ to $x = 0$, $f'(x)$ is positive and decreasing to 0. From $x = 0$ to $x = 1$, $f'(x)$ is negative and decreasing. At $x = 1$, where the graph of f has an inflection point, the graph of f' has a relative minimum. From $x = 1$ to $x = 2$, $f'(x)$ is negative and increasing, reaching 0 at $x = 2$. From $x = 2$ to $x = 3$, $f'(x)$ is positive and increasing toward $+\infty$. At $x = 3$, f is not differentiable, so f' is not defined. The graph of f' has $x = 3$ as a vertical asymptote. For $x > 3$, $f'(x)$ has a constant negative value (approximately -1).

Problems 15.58 to 15.61 refer to the function $f(x)$ graphed in Fig. 15-48.

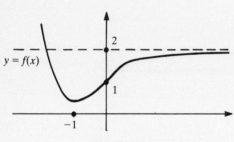

Fig. 15-48

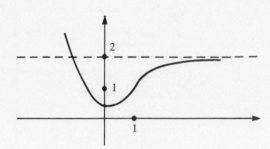

Fig. 15-49

15.58 Which of the functions $f(x) - 1$, $f(x - 1)$, $f(-x)$, or $f'(x)$ is graphed in Fig. 15-49?

▮ This is the graph of $f(x - 1)$. It is obtained by shifting the graph of $f(x)$ one unit to the right.

15.59 Which of the functions $f(x) - 1$, $f(x - 1)$, $f(-x)$, or $f'(x)$ is graphed in Fig. 15-50?

▮ This is the graph of $f(-x)$. It is obtained by reflecting the graph of $f(x)$ in the y-axis.

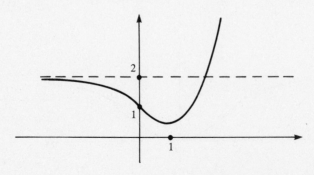

Fig. 15-50

15.60 Which of the functions $f(x) - 1$, $f(x - 1)$, $f(-x)$, or $f'(x)$ is graphed in Fig. 15-51?

▮ This is the graph of $f(x) - 1$. It is obtained by lowering the graph of $f(x)$ one unit.

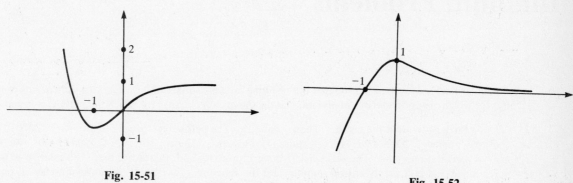

Fig. 15-51

Fig. 15-52

15.61 Which of the functions $f(x) - 1$, $f(x - 1)$, $f(-x)$, or $f'(x)$ is graphed in Fig. 15-52?

▮ This is the graph of $f'(x)$. At $x = -1$, $f'(x) = 0$. To the left of $x = -1$, $f'(x)$ is negative and decreases to $-\infty$ as $x \to -\infty$. Where the graph of f has an inflection point at $x = 0$, $f'(x)$ reaches a relative (actually, absolute) maximum. For $x > 0$, $f'(x)$ is positive and decreasing toward 0. Thus, the positive x-axis is a horizontal asymptote of the graph of $f'(x)$.

CHAPTER 16
Applied Maximum and Minimum Problems

16.1 A rectangular field is to be fenced in so that the resulting area is c square units. Find the dimensions of that field (if any) for which the perimeter is a minimum, and the dimensions (if any) for which the perimeter is a maximum.

▮ Let ℓ be the length and w the width. Then $\ell w = c$. The perimeter $p = 2\ell + 2w = 2\ell + 2c/\ell$. ℓ can be any positive number. $D_\ell p = 2 - 2c/\ell^2$, and $D_\ell^2 p = 4c/\ell^3$. Hence, solving $2 - 2c/\ell^2 = 0$, we see that $\ell = \sqrt{c}$ is a critical number. The second derivative is positive, and, therefore, there is a relative minimum at $\ell = \sqrt{c}$. Since that is the only critical number and the function $2\ell + 2c/\ell$ is continuous for all positive ℓ, there is an absolute minimum at $\ell = \sqrt{c}$. (If p achieved a still smaller value at some other point ℓ_0, there would have to be a relative maximum at some point between $\sqrt{c}$ and ℓ_0, yielding another critical number.) When $\ell = \sqrt{c}$, $w = \sqrt{c}$. Thus, for a fixed area, the square is the rectangle with the smallest perimeter. Notice that the perimeter does not achieve a maximum, since $p = 2\ell + 2c/\ell \rightarrow +\infty$ as $\ell \rightarrow +\infty$.

16.2 Find the point(s) on the parabola $2x = y^2$ closest to the point $(1, 0)$.

▮ Refer to Fig. 16-1. Let u be the distance between $(1, 0)$ and a point (x, y) on the parabola. Then $u = \sqrt{(x-1)^2 + y^2}$. To minimize u it suffices to minimize $u^2 = (x-1)^2 + y^2$. Now, $u^2 = (x-1)^2 + 2x$. Since (x, y) is a point on $2x = y^2$, x can be any nonnegative number. Now, $D_x(u^2) = 2(x-1) + 2 = 2x > 0$ for $x > 0$. Hence, u^2 is an increasing function, and, therefore, its minimum value is attained at $x = 0$, $y = 0$.

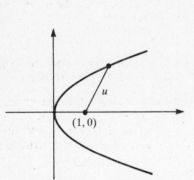

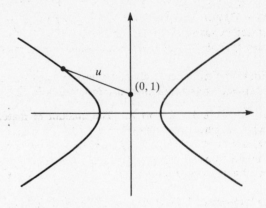

Fig. 16-1

Fig. 16-2

16.3 Find the point(s) on the hyperbola $x^2 - y^2 = 2$ closest to the point $(0, 1)$.

▮ Refer to Fig. 16-2. Let u be the distance between $(0, 1)$ and a point (x, y) on the hyperbola. Then $u = \sqrt{x^2 + (y-1)^2}$. To minimize u, it suffices to minimize $u^2 = x^2 + (y-1)^2 = 2 + y^2 + (y-1)^2$. Since $x^2 = y^2 + 2$, y can be any real number. $D_y(u^2) = 2y + 2(y-1) = 4y - 2$. Also, $D_y^2(u^2) = 4$. The only critical number is $\frac{1}{2}$, and, since the second derivative is positive, there is a relative minimum at $y = \frac{1}{2}$, $x = \pm\frac{3}{2}$. Since there is only one critical number, this point yields the absolute minimum.

16.4 A closed box with a square base is to contain 252 cubic feet. The bottom costs $5 per square foot, the top costs $2 per square foot, and the sides cost $3 per square foot. Find the dimensions that will minimize the cost.

▮ Let s be the side of the square base and let h be the height. Then $s^2 h = 252$. The cost of the bottom is $5s^2$, the cost of the top is $2s^2$, and the cost of each of the four sides is $3sh$. Hence, the total cost $C = 5s^2 + 2s^2 + 4(3sh) = 7s^2 + 12sh = 7s^2 + 12s(252/s^2) = 7s^2 + 3024/s$. s can be any positive number. Now, $D_s C = 14s - 3024/s^2$, and $D_s^2 C = 14 + 6048/s^3$. Solving $14s - 3024/s^2 = 0$, $14s^3 = 3024$, $s^3 = 216$, $s = 6$. Thus, $s = 6$ is the only critical number. Since the second derivative is always positive for $s > 0$, there is a relative minimum at $s = 6$. Since $s = 6$ is the *only* critical number, it yields an absolute minimum. When $s = 6$, $h = 7$.

16.5 A printed page must contain 60 cm² of printed material. There are to be margins of 5 cm on either side and margins of 3 cm on the top and bottom (Fig. 16-3). How long should the printed lines be in order to minimize the amount of paper used?

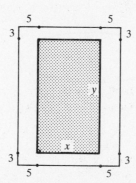

Fig. 16-3

▮ Let x be the length of the line and let y be the height of the printed material. Then $xy = 60$. The amount of paper $A = (x + 10)(y + 6) = (x + 10)(60/x + 6) = 6(10 + x + 100/x + 10) = 6(20 + x + 100/x)$. x can be any positive number. Then $D_x A = 6(1 - 100/x^2)$ and $D_x^2 A = 1200/x^3$. Solving $1 - 100/x^2 = 0$, we see that the only critical number is 10. Since the second derivative is positive, there is a relative minimum at $x = 10$, and, since this is the *only* critical number, there is an absolute minimum at $x = 10$.

16.6 A farmer wishes to fence in a rectangular field of 10,000 ft². The north-south fences will cost \$1.50 per foot, while the east-west fences will cost \$6.00 per foot. Find the dimensions of the field that will minimize the cost.

▮ Let x be the east-west dimension, and let y be the north-south dimension. Then $xy = 10,000$. The cost $C = 6(2x) + 1.5(2y) = 12x + 3y = 12x + 3(10,000/x) = 12x + 30,000/x$. x can be any positive number. $D_x C = 12 - 30,000/x^2$. $D_x^2 C = 60,000/x^3$. Solving $12 - 30,000/x^2 = 0$, $2500 = x^2$, $x = 50$. Thus, 50 is the only critical number. Since the second derivative is positive, there is a relative minimum at $x = 50$. Since this is the only critical number, this is an absolute minimum. When $x = 50$, $y = 200$.

16.7 Find the dimensions of the closed cylindrical can that will have a capacity of k units of volume and will use the minimum amount of material. Find the ratio of the height h to the radius r of the top and bottom.

▮ The volume $k = \pi r^2 h$. The amount of material $M = 2\pi r^2 + 2\pi rh$. (This is the area of the top and bottom, plus the lateral area.) So $M = 2\pi r^2 + 2\pi r(k/\pi r^2) = 2\pi r^2 + 2k/r$. Then $D_r M = 4\pi r - 2k/r^2$, $D_r^2 M = 4\pi + 4k/r^3$. Solving $4\pi r - 2k/r^2 = 0$, we find that the only critical number is $r = \sqrt[3]{k/2\pi}$. Since the second derivative is positive, this yields a relative minimum, which, by the uniqueness of the critical number, is an absolute minimum. Note that $k = \pi r^2 h = \pi r^3 (h/r) = \pi(k/2\pi)(h/r)$. Hence, $h/r = 2$.

16.8 In Problem 16.7, find the ratio h/r that will minimize the amount of material used if the bottom and top of the can have to be cut from square pieces of metal and the rest of these squares are wasted. Also find the resulting ratio of height to radius.

▮ $k = \pi r^2 h$. Now $M = 8r^2 + 2\pi rh = 8r^2 + 2\pi r(k/\pi r^2) = 8r^2 + 2k/r$. $D_r M = 16r - 2k/r^2$. $D_r^2 M = 16 + 4k/r^3$. Solving for the critical number, $r^3 = k/8$, $r = \sqrt[3]{k}/2$. As before, this yields an absolute minimum. Again, $k = \pi r^2 h = \pi r^3 (h/r) = \pi(k/8)(h/r)$. So, $h/r = 8/\pi$.

16.9 A thin-walled cone-shaped cup (Fig. 16-4) is to hold 36π in³ of water when full. What dimensions will minimize the amount of material needed for the cup?

▮ Let r be the radius and h be the height. Then the volume $36\pi = \frac{1}{3}\pi r^2 h$. The lateral surface area $A = \pi rs$, where s is the slant height of the cone. $s^2 = r^2 + h^2$ and $h = 108/r^2$. Hence, $A = \pi r\sqrt{r^2 + (108/r^2)^2} = (\pi/r)\sqrt{r^6 + (108)^2}$. Then,

$$D_r A = \frac{\pi[r \cdot \frac{1}{2} \cdot 6r^5/\sqrt{r^6 + (108)^2} - \sqrt{r^6 + (108)^2}]}{r^2}$$

$$= \frac{\pi\{3r^6 - [r^6 + (108)^2]\}}{r^2\sqrt{r^6 + (108)^2}} = \frac{\pi[2r^6 - (108)^2]}{r^2\sqrt{r^6 + (108)^2}}$$

Solving $2r^6 - (108)^2 = 0$ for the critical number, $r = 3\sqrt{2}$. The first derivative test yields the case $\{-, +\}$, showing that $r = 3\sqrt{2}$ gives a relative minimum, which, by the uniqueness of the critical number, must be an absolute minimum. When $r = 3\sqrt{2}$, $h = 6$.

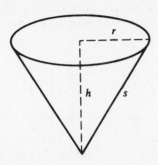

Fig. 16-4

16.10 A rectangular bin, open at the top, is required to contain 128 cubic meters. If the bottom is to be a square, at a cost of $2 per square meter, while the sides cost $0.50 per square meter, what dimensions will minimize the cost?

▮ Let s be the side of the bottom square and let h be the height. Then $128 = s^2 h$. The cost (in dollars) $C = 2s^2 + \frac{1}{2}(4sh) = 2s^2 + 2s(128/s^2) = 2s^2 + 256/s$, so $D_s C = 4s - 256/s^2$, $D_s^2 C = 4 + 512/s^3$. Solving $4s - 256/s^2 = 0$, $s^3 = 64$, $s = 4$. Since the second derivative is positive, the critical number $s = 4$ yields a relative minimum, which, by the uniqueness of the critical number, is an absolute minimum. When $s = 4$, $h = 8$.

16.11 The selling price P of an item is $100 - 0.02x$ dollars, where x is the number of items produced per day. If the cost C of producing and selling x items is $40x + 15,000$ dollars per day, how many items should be produced and sold every day in order to maximize the profit?

▮ The total income per day is $x(100 - 0.02x)$. Hence the profit $G = x(100 - 0.02x) - (40x + 15,000) = 60x - 0.02x^2 - 15,000$, and $D_x G = 60 - 0.04x$ and $D_x^2 G = -0.04$. Hence, the unique critical number is the solution of $60 - 0.04x = 0$, $x = 1500$. Since the second derivative is negative, this yields a relative maximum, which, by the uniqueness of the critical number, is an absolute maximum.

16.12 Find the point(s) on the graph of $3x^2 + 10xy + 3y^2 = 9$ closest to the origin.

▮ It suffices to minimize $u = x^2 + y^2$, the square of the distance from the origin. By implicit differentiation, $D_x u = 2x + 2yD_x y$ and $6x + 10(xD_x y + y) + 6yD_x y = 0$. From the second equation, $D_x y = -(3x + 5y)/(5x + 3y)$, and, then, substituting in the first equation, $D_x u = 2x + 2y[-(3x + 5y)/(5x + 3y)]$. Setting $D_x u = 0$, $x(5x + 3y) - y(3x + 5y) = 0$, $5(x^2 - y^2) = 0$, $x^2 = y^2$, $x = \pm y$. Substituting in the equation of the graph, $6x^2 \pm 10x^2 = 9$. Hence, we have the + sign, and $16x^2 = 9$, $x = \pm\frac{3}{4}$ and $y = \pm\frac{3}{4}$. Thus, the two points closest to the origin are $(\frac{3}{4}, \frac{3}{4})$ and $(-\frac{3}{4}, -\frac{3}{4})$.

16.13 A man at a point P on the shore of a circular lake of radius 1 mile wants to reach the point Q on the shore diametrically opposite P (Fig. 16-5). He can row 1.5 miles per hour and walk 3 miles per hour. At what angle θ $(0 \le \theta \le \pi/2)$ to the diameter PQ should he row in order to minimize the time required to reach Q?

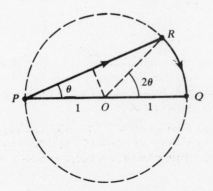

Fig. 16-5

❚ Let R be the point where the boat lands, and let O be the center of the circle. Since $\triangle OPR$ is isosceles, $PR = 2\cos\theta$. The arc length $\overset{\frown}{RQ} = 2\theta$. Hence, the time $T = PR/1.5 + \overset{\frown}{RQ}/3 = \frac{4}{3}\cos\theta + \frac{2}{3}\theta$. So, $D_\theta T = -\frac{4}{3}\sin\theta + \frac{2}{3}$. Setting $D_\theta T = 0$, we find $\sin\theta = \frac{1}{2}$, $\theta = \pi/6$. Since T is a continuous function on the closed interval $[0, \pi/2]$, we can use the tabular method. List the critical number $\pi/6$ and the endpoints 0 and $\pi/2$, and compute the corresponding values of T. The smallest of these values is the absolute minimum. Clearly, $\pi/3 < \frac{4}{3}$, and it is easy to check that $\pi/3 < (6\sqrt{3} + \pi)/9$. (Assume the contrary and obtain the false consequence that $\pi > 3\sqrt{3}$.) Thus, the absolute minimum is attained when $\theta = \pi/2$. That means that the man walks all the way.

θ	$\pi/6$	0	$\pi/2$
T	$(6\sqrt{3} + \pi)/9$	$\frac{4}{3}$	$\pi/3$

16.14 Find the answer to Problem 16.13 when, instead of rowing, the man can paddle a canoe at 4 miles per hour.

❚ Using the same notation as in Problem 16.13, we find $T = \frac{1}{2}\cos\theta + \frac{2}{3}\theta$, $D_\theta T = -\frac{1}{2}\sin\theta + \frac{2}{3}$. Setting $D_\theta T = 0$, We obtain $\sin\theta = \frac{4}{3}$, which is impossible. Hence, we use the tabular method for just the endpoints 0 and $\pi/2$. Then, since $\frac{1}{2} < \pi/3$, the absolute minimum is $\frac{1}{2}$, attained when $\theta = 0$. Hence, in this case, the man paddles all the way.

θ	0	$\pi/2$
T	$\frac{1}{2}$	$\pi/3$

16.15 A wire 16 feet long has to be formed into a rectangle. What dimensions should the rectangle have to maximize the area?

❚ Let x and y be the dimensions. Then $16 = 2x + 2y$, $8 = x + y$. Thus, $0 \le x \le 8$. The area $A = xy = x(8 - x) = 8x - x^2$, so $D_x A = 8 - 2x$, $D_x^2 A = -2$. Hence, the only critical number is $x = 4$. We can use the tabular method. Then the maximum value 16 is attained when $x = 4$. When $x = 4$, $y = 4$. Thus, the rectangle is a square.

x	4	0	8
A	16	0	0

16.16 Find the height h and radius r of a cylinder of greatest volume that can be cut within a sphere of radius b.

❚ The axis of the cylinder must lie on a diameter of the sphere. From Fig. 16-6, $b^2 = r^2 + (h/2)^2$, so the volume of the cylinder $V = \pi r^2 h = \pi(b^2 - h^2/4)h = \pi(b^2 h - h^3/4)$. Then $D_h V = \pi(b^2 - 3h^2/4)$ and $D_h^2 V = -(3\pi/2)h$, so the critical number is $h = 2b/\sqrt{3}$. Since the second derivative is negative, there is a relative maximum at $h = 2b/\sqrt{3}$, which, by virtue of the uniqueness of the critical number, is an absolute maximum. When $h = 2b/\sqrt{3}$, $r = b\sqrt{\frac{2}{3}}$.

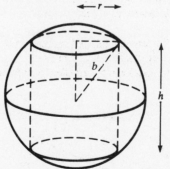

Fig. 16-6

16.17 Among all pairs of nonnegative numbers that add up to 5, find the pair that maximizes the product of the square of the first number and the cube of the second number.

❚ Let x and y be the numbers. Then $x + y = 5$. We wish to maximize $P = x^2 y^3 = (5 - y)^2 y^3$. Clearly, $0 \le y \le 5$. $D_y P = (5 - y)^2(3y^2) + y^3[2(5 - y)(-1)] = (5 - y)y^2[3(5 - y) - 2y] = (5 - y)y^2(15 - 5y)$. Hence, the critical numbers are 0, 3, and 5. By the tabular method, the absolute maximum for P is 108, corresponding to $y = 3$. When $y = 3$, $x = 2$.

y	0	3	5
P	0	108	0

16.18 A solid steel cylinder is to be produced so that the sum of its height h and diameter $2r$ is to be at most 3 units. Find the dimensions that will maximize its volume.

❚ We may assume that $h + 2r = 3$. So, $0 \le r \le \frac{3}{2}$. Then $V = \pi r^2 h = \pi r^2 (3 - 2r) = 3\pi r^2 - 2\pi r^3$, so $D_r V = 6\pi r - 6\pi r^2$. Hence, the critical numbers are 0 and 1. We use the tabular method. The maximum value π is attained when $r = 1$. When $r = 1$, $h = 1$.

r	0	1	2
V	0	π	0

16.19 Among all right triangles with fixed perimeter p, find the one with maximum area.

❚ Let the triangle $\triangle ABC$ have a right angle at C, and let the two sides have lengths x and y (Fig. 16-7). Then the hypotenuse $AB = p - x - y$. Therefore, $(p - x - y)^2 = x^2 + y^2$, so $2(p - x - y)(-1 - D_x y) = 2x + 2y D_x y$, $D_x y[(p - x - y) + y] = (-p + x - y) + x$, $D_x y(p - x) = y - p$, $D_x y = (y - p)/(p - x)$. Now, the area $A = \frac{1}{2}xy$, $D_x A = \frac{1}{2}(xD_x y + y) = \frac{1}{2}[x(y - p)/(p - x) + y] = \frac{1}{2}\{[x(y - p) + y(p - x)]/(p - x)\} = \frac{1}{2}[p(y - x)/(p - x)]$. Then, when $D_x A = 0$, $y = x$, and the triangle is isosceles. Then, $(p - 2x)^2 = 2x^2$, $p - 2x = x\sqrt{2}$, $x = p/(2 + \sqrt{2})$. Thus, the only critical number is $x = p/(2 + \sqrt{2})$. Since $0 \le x \le p$, we can use the tabular method. When $x = p/(2 + \sqrt{2})$, $y = p/(2 + \sqrt{2})$, and $A = \frac{1}{2}[p(2 + \sqrt{2})]^2$. When $x = 0$, $A = 0$. When $x = p/2$, $y = 0$ and $A = 0$. Hence, the maximum is attained when $x = y = p/(2 + \sqrt{2})$.

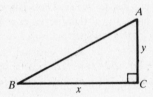

Fig. 16-7

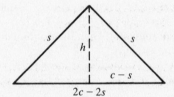

Fig. 16-8

16.20 Of all isosceles triangles with a fixed perimeter, which one has the maximum area?

❚ Let s be the length of the equal sides, and let h be the altitude to the base. Let the fixed perimeter be $2c$. Then the base is $2c - 2s$, and (Fig. 16-8) $h^2 = s^2 - (c - s)^2 = 2cs - c^2$. Hence, $2hD_s h = 2c$, $hD_s h = c$. The area $A = \frac{1}{2}h(2c - 2s) = h(c - s)$. So, $D_s A = (c - s)D_s h - h = (c - s)c/h - h = \dfrac{c(c - s) - h^2}{h} = \dfrac{c^2 - cs - 2cs + c^2}{h} = \dfrac{2c^2 - 3cs}{h}$. Solving $2c^2 - 3cs = 0$, we find the critical number $s = \frac{2}{3}c$. Now, $D_s^2 A = \dfrac{hc(-3) - c(2c - 3s)D_s h}{h^2}$, which eventually evaluates to $-(c^2/h^3)(3s - c)$. When $s = \frac{2}{3}c$, the second derivative becomes $-(c^3/h^3) < 0$. Hence, $s = \frac{2}{3}c$ yields a relative maximum, which by virtue of the uniqueness of the critical number, must be an absolute maximum. When $s = \frac{2}{3}c$, the base $2c - 2s = \frac{2}{3}c$. Hence, the triangle that maximizes the area is equilateral. [Can you see from the ellipse of Fig. 16-9 that of *all* triangles with a fixed perimeter, the equilateral has the greatest area?]

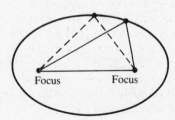

Focus   Focus

Fig. 16-9

16.21 A rectangular yard is to be built which encloses 400 ft^2. Two opposite sides are to be made from fencing which costs \$1 per foot, while the other two opposite sides are to be made from fencing which costs \$2 per foot. Find the least possible cost.

▌ Let x be the length of the each side costing $1 per foot, and let y be the length of each side costing $2 per foot. Then $400 = xy$. The cost $C = 4y + 2x = 1600/x + 2x$. Hence, $D_x C = -1600/x^2 + 2$ and $D_x^2 C = 1600/x^3$. Solving $-1600/x^2 + 2 = 0$, $x^2 = 800$, $x = 20\sqrt{2}$. Since the second derivative is positive, this yields a relative minimum, which, by virtue of the uniqueness of the critical number, must be an absolute minimum. Then the least cost is $1600/20\sqrt{2} + 40\sqrt{2} = 40\sqrt{2} + 40\sqrt{2} = 80\sqrt{2}$, which is approximately $112.

16.22 A closed, right cylindrical container is to have a volume of 5000 in³. The material for the top and bottom of the container will cost $2.50 per in², while the material for the rest of the container will cost $4 per in². How should you choose the height h and the radius r in order to minimize the cost?

▌ $5000 = \pi r^2 h$. The lateral surface area is $2\pi rh$. Hence, the cost $C = 2(2.50)\pi r^2 + 4(2\pi rh) = 5\pi r^2 + 8\pi rh = 5\pi r^2 + 8\pi r(5000/\pi r^2) = 5\pi r^2 + 40,000/r$. Hence, $D_r C = 10\pi r - 40,000/r^2$ and $D_r^2 C = 10\pi + (80,000/r^3)$. Solving $10\pi r - 40,000/r^2 = 0$, we find the unique critical number $r = 10\sqrt[3]{4/\pi}$. Since the second derivative is positive, this yields a relative minimum, which, by virtue of the uniqueness of the critical number, is an absolute minimum. The height $h = 5000/\pi r^2 = 25/\sqrt[3]{2\pi}$.

16.23 The sum of the squares of two nonnegative numbers is to be 4. How should they be chosen so that the product of their cubes is a maximum?

▌ Let x and y be the numbers. Then $x^2 + y^2 = 4$, so $0 \le x \le 2$. Also, $2x + 2y D_x y = 0$, $D_x y = -x/y$. The product of their cubes $P = x^3 y^3$, so $D_x P = x^3(3y^2 D_x y) + 3x^2 y^3 = x^3[3y^2(-x/y)] + 3x^2 y^3 = -3x^4 y + 3x^2 y^3 = 3x^2 y(-x^2 + y^2)$. Hence, when $D_x P = 0$, either $x = 0$ or $y = 0$ (and $x = 2$), or $y = x$ (and then, by $x^2 + y^2 = 4$, $x = \sqrt{2}$). Thus, we have three critical numbers $x = 0$, $x = 2$, and $x = \sqrt{2}$. Using the tabular method, with the endpoints 0 and 2, we find that the maximum value of P is achieved when $x = \sqrt{2}$, $y = \sqrt{2}$.

x	0	$\sqrt{2}$	2
P	0	8	0

16.24 Two nonnegative numbers are such that the first plus the square of the second is 10. Find the numbers if their sum is as large as possible.

▌ Let x be the first and y the second number. Then $x + y^2 = 10$. Their sum $S = x + y = 10 - y^2 + y$. Hence, $D_y S = -2y + 1$, $D_y^2 S = -2$, so the critical number is $y = \frac{1}{2}$. Since the second derivative is negative, this yields a relative maximum, which, by virtue of the uniqueness of the critical number, is an absolute maximum. $x = 10 - (\frac{1}{2})^2 = \frac{39}{4}$.

16.25 Find two nonnegative numbers x and y whose sum is 300 and for which $x^2 y$ is a maximum.

▌ $x + y = 300$. The product $P = x^2 y = x^2(300 - x) = 300x^2 - x^3$. $D_x P = 600x - 3x^2$, so the critical numbers are $x = 0$ and $x = 200$. Clearly, $0 \le x \le 300$, so using the tabular method, we find that the absolute maximum is attained when $x = 200$, $y = 100$.

x	0	200	300
P	0	4×10^6	0

16.26 A publisher decides to print the pages of a large book with $\frac{1}{2}$-inch margins on the top, bottom, and one side, and a 1-inch margin on the other side (to allow for the binding). The area of the entire page is to be 96 square inches. Find the dimensions of the page that will maximize the printed area of the page.

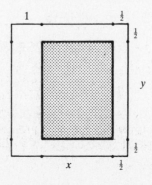

Fig. 16-10

▌ Let x be the width and y be the height of the page. Then $96 = xy$, and (Fig. 16-10) the printed area $A = (x - \frac{3}{2})(y - 1)$. Hence, $0 = xD_xy + y$, $D_xy = -y/x$. Now $D_xA = (x - \frac{3}{2})D_xy + y - 1 = (x - \frac{3}{2})(-y/x) + y - 1 = 3y/2x - 1$. Therefore, $D_x^2A = \frac{3}{2}(xD_xy - y)/x^2 = \frac{3}{2}(-2y/x^2) < 0$. Setting $D_xA = 0$, we obtain $y = \frac{2}{3}x$, $96 = x(\frac{2}{3}x)$, $x^2 = 144$, $x = 12$, $y = 8$. Since the second derivative is negative, the unique critical number $x = 12$ yields an absolute maximum for A.

16.27 A paint manufacturer produces anywhere from 10 to 30 cubic meters of paint per day. The profit for the day (in hundreds of dollars) is given by $P = (x - 15)^3/1000 - 3(x - 15)/10 + 300$, where x is the volume produced and sold. What value of x maximizes the profit?

▌ $D_xP = \frac{3}{1000}(x - 15)^2 - \frac{3}{10}$. Setting $D_xP = 0$, $(x - 15)^2 = 100$, $x - 15 = \pm 10$, $x = 25$ or $x = 5$. Since $x = 5$ is not within the permissible range, the only critical number is $x = 25$. Using the tabular method, we find that the maximum profit is achieved when $x = 10$.

x	10	25	30
P	301.375	298	298.875

16.28 A printed page is to have a total area of 80 in^2 and margins of 1 inch at the top and on each side and of 1.5 inches at the bottom. What should the dimensions of the page be so that the printed area will be a maximum?

▌ Let x be the width and y be the height of the page. Then $80 = xy$, $0 = xD_xy + y$, $D_xy = -y/x$. The area of the printed page $A = (x - 2)(y - 2.5)$, so $D_xA = (x - 2)D_xy + y - 2.5 = (x - 2)(-y/x) + y - 2.5 = -y + 2y/x + y - 2.5 = 2y/x - 2.5$. Also, $D_x^2A = 2(xD_xy - y)/x^2 = 2(-y - y)/x^2 = -4y/x^2 < 0$. Solving $D_xA = 0$, we find $y = 1.25x$, $80 = 1.25x^2$, $64 = x^2$, $x = 8$, $y = 10$. Since the second derivative is negative, this unique critical number yields an absolute maximum for A.

16.29 One side of an open field is bounded by a straight river. Determine how to put a fence around the other sides of a rectangular plot in order to enclose as great an area as possible with 2000 feet of fence.

▌ Let x be the length of the side parallel to the river, and let y be the length of each of the other sides. Then $2y + x = 2000$. The area $A = xy = y(2000 - 2y) = 2000y - 2y^2$, $D_yA = 2000 - 4y$, and $D_y^2A = -4$. Solving $D_yA = 0$, we find the critical number $y = 500$. Since the second derivative is negative, this unique critical number yields an absolute maximum. $x = 2000 - 2(500) = 1000$.

16.30 A box will be built with a square base and an open top. Material for the base costs $8 per square foot, while material for the sides costs $2 per square foot. Find the dimensions of the box of maximum volume that can be built for $2400.

▌ Let s be the side of the base and h be the height. Then $V = s^2h$. We are told that $2400 = 8s^2 + 2(4hs)$, so $300 = s^2 + hs$, $h = 300/s - s$. Hence, $V = s^2(300/s - s) = 300s - s^3$. Then $D_sV = 300 - 3s^2$, $D_s^2V = -6s$. Solving $D_sV = 0$, we find the critical number $s = 10$. Since the second derivative is negative, the unique critical number yields an absolute maximum. $h = \frac{300}{10} - 10 = 20$.

16.31 Find the maximum area of any rectangle which may be inscribed in a circle of radius 1.

▌ Let the center of the circle be the origin. We may assume that the sides of the rectangle are parallel to the coordinate axes. Let $2x$ be the length of the horizontal sides and $2y$ be the length of the vertical sides. Then $x^2 + y^2 = 1$, so $2x + 2yD_xy = 0$, $D_xy = -x/y$. The area $A = (2x)(2y) = 4xy$. So, $D_xA = 4(xD_xy + y) = 4\left(-\frac{x^2}{y} + y\right) = \frac{4}{y}(-x^2 + y^2)$. Also, $D_x^2A = 4\left(-\frac{2xy - x^2D_xy}{y^2} + D_xy\right) = 4\left(-\frac{2xy + x^3/y}{y^2} - \frac{x}{y}\right) = -4\left(\frac{2xy^2 + x^3 - xy^2}{y^3}\right) = -4\left(\frac{x^3 + xy^2}{y^3}\right) = -\frac{4x}{y^3}(x^2 + y^2) = -\frac{4x}{y^3}$. Solving $D_xA = 0$, we find $y = x$, $2x^2 = 1$, $x = 1/\sqrt{2}$, $y = 1/\sqrt{2}$. Since the second derivative is negative, the unique critical number yields an absolute maximum. The maximum area is $(2 \cdot 1/\sqrt{2})(2 \cdot 1/\sqrt{2}) = 2$.

16.32 A factory producing a certain type of electronic component has fixed costs of $250 per day and variable costs of $90x$, where x is the number of components produced per day. The demand function for these components is $p(x) = 250 - x$, and the feasible production levels satisfy $0 \le x \le 90$. Find the level of production for maximum profit.

▌ The daily income is $x(250 - x)$, since $250 - x$ is the price at which x units are sold. The profit $G = x(250 - x) - (250 + 90x) = 250x - x^2 - 250 - 90x = 160x - x^2 - 250$. Hence, $D_x G = 160 - 2x$, $D_x^2 G = -2$. Solving $D_x G = 0$, we find the critical number $x = 80$. Since the second derivative is negative, the unique critical number yields an absolute maximum. Notice that this maximum, taken over a continuous variable x, is assumed for the integral value $x = 80$. So it certainly has to remain the maximum when x is restricted to integral values (whole numbers of electronic components).

16.33 A gasoline station selling x gallons of fuel per month has fixed cost of \$2500 and variable costs of $0.90x$. The demand function is $1.50 - 0.00002x$ and the station's capacity allows no more than 20,000 gallons to be sold per month. Find the maximum profit.

▌ The price that x gallons can be sold at is the value of the demand function. Hence, the total income is $x(1.50 - 0.00002x)$, and the profit $G = x(1.50 - 0.00002x) - 2500 - 0.90x = 0.60x - 0.00002x^2 - 2500$. Hence, $D_x G = 0.60 - 0.00004x$, and $D_x^2 G = -0.00004$. Solving $D_x G = 0$, we find $0.60 = 0.00004x$, $60,000 = 4x$, $x = 15,000$. Since the second derivative is negative, the unique critical number $x = 15,000$ yields the maximum profit \$2000.

16.34 Maximize the volume of a box, open at the top, which has a square base and which is composed of 600 square inches of material.

▌ Let s be the side of the base and h be the height. Then $V = s^2 h$. We are told that $600 = s^2 + 4hs$. Hence, $h = (600 - s^2)/4s$. So $V = s^2[(600 - s^2)/4s] = (s/4)(600 - s^2) = 150s - \frac{1}{4}s^3$. Then $D_s V = 150 - \frac{3}{4}s^2$, $D_s^2 V = -\frac{3}{2}s$. Solving $D_s V = 0$, we find $200 = s^2$, $10\sqrt{2} = s$. Since the second derivative is negative, this unique critical number yields an absolute maximum. When $s = 10\sqrt{2}$, $h = 5\sqrt{2}$.

16.35 A rectangular garden is to be completely fenced in, with one side of the garden adjoining a neighbor's yard. The neighbor has agreed to pay for half of the section of the fence that separates the plots. If the garden is to contain 432 ft², find the dimensions that minimize the cost of the fence to the garden's owner.

▌ Let y be the length of the side adjoining the neighbor, and let x be the other dimension. Then $432 = xy$, $0 = x D_x y + y$, $D_x y = -y/x$. The cost $C = 2x + y + \frac{1}{2}y = 2x + \frac{3}{2}y$. Then, $D_x C = 2 + \frac{3}{2}(D_x y) = 2 + \frac{3}{2}(-y/x)$ and $D_x^2 C = -\frac{3}{2}(x D_x y - y)/x^2 = -\frac{3}{2}(-y - y)/x^2 = 3y/x^2$. Setting $D_x C = 0$, We obtain $2 = 3y/2x$, $4x = 3y$, $y = \frac{4}{3}x$, $432 = x(4x/3)$, $324 = x^2$, $x = 18$, $y = 24$. Since the second derivative is positive, the unique critical number $x = 18$ yields the absolute minimum cost.

16.36 A rectangular box with open top is to be formed from a rectangular piece of cardboard which is 3 inches × 8 inches. What size square should be cut from each corner to form the box with maximum volume? (The cardboard is folded along the dotted lines to form the box.)

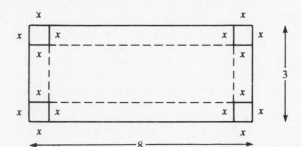

Fig. 16-11

▌ Let x be the side of the square that is cut out. The length will be $8 - 2x$, the width $3 - 2x$, and the height x. Hence, the volume $V = x(3 - 2x)(8 - 2x)$; so $D_x V = (1)(3 - 2x)(8 - 2x) + x(-2)(8 - 2x) + x(3 - 2x)(-2) = 4(3x - 2)(x - 3)$, and $D_x^2 V = 24x - 44$. Setting $D_x V = 0$, we find $x = \frac{2}{3}$ or $x = 3$. Since the width 3 of the cardboard is greater than $2x$, we must have $x < \frac{3}{2}$. Hence, the value $x = 3$ is impossible. Thus, we have a unique critical number $x = \frac{2}{3}$, and, for that value, the second derivative turns out to be negative. Hence, that critical number determines an absolute maximum for the volume.

16.37 Refer to Fig. 16-12. At 9 a.m., ship B was 65 miles due east of another ship, A. Ship B was then sailing due west at 10 miles per hour, and A was sailing due south at 15 miles per hour. If they continue their respective courses, when will they be nearest one another?

▮ Let the time t be measured in hours from 9 a.m. Choose a coordinate system with B moving along the x-axis and A moving along the y-axis. Then the x-coordinate of B is $65 - 10t$, and the y-coordinate of B is $-15t$. Let u be the distance between the ships. Then $u^2 = (15t)^2 + (65 - 10t)^2$. It suffices to minimize u^2. $D_t(u^2) = 2(15t)(15) + 2(65 - 10t)(-10) = 650t - 1300$, and $D_t^2(u^2) = 650$. Setting $D_t(u^2) = 0$, we obtain $t = 2$, Since the second derivative is positive, the unique critical number yields an absolute minimum. Hence, the ships will be closest at 11 a.m.

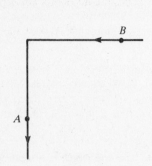

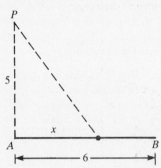

Fig. 16-12   **Fig. 16-13**

16.38 A woman in a rowboat at P, 5 miles from the nearest point A on a straight shore, wishes to reach a point B, 6 miles from A along the shore (Fig. 16-13). If she wishes to reach B in the shortest time, where should she land if she can row 2 mi/h and walk 4 mi/h?

▮ Let x be the distance between A and the landing point. Then the distance rowed is $\sqrt{25 + x^2}$ and the distance walked is $6 - x$. Hence the total time $t = (\sqrt{25 + x^2})/2 + (6 - x)/4$. Then $D_x t = \dfrac{1}{2} \dfrac{x}{\sqrt{25 + x^2}} - \dfrac{1}{4}$. $D_x^2 t = \dfrac{1}{2} \dfrac{\sqrt{25 + x^2} - x^2/\sqrt{25 + x^2}}{25 + x^2} = \dfrac{1}{2} \dfrac{25}{(25 + x^2)^{3/2}}$. Setting $D_x t = 0$, we obtain $2x = \sqrt{25 + x^2}$, $4x^2 = 25 + x^2$, $3x^2 = 25$, $x = 5\sqrt{3}/3 \approx 2.89$. Since the second derivative is positive, the unique critical number yields the absolute minimum time.

16.39 A wall 8 feet high is 3.375 feet from a house. Find the shortest ladder that will reach from the ground to the house when leaning over the wall.

▮ Let x be the distance from the foot of the ladder to the wall. Let y be the height above the ground of the point where the ladder touches the house. Let L be the length of the ladder. Then $L^2 = (x + 3.375)^2 + y^2$. It suffices to minimize L^2. By similar triangles, $y/8 = (x + 3.375)/x$. Then $D_x y = -27/x^2$. Now, $D_x(L^2) = 2(x + 3.375) + 2yD_x y = 2(x + 3.375) + 2(8/x)(x + 3.375)(-27/x^2) = 2(x + 3.375)(1 - 216/x^3)$. Solving $D_x(L^2) = 0$, we find the unique positive critical number $x = 6$. Calculation of $D_x^2(L^2)$ yields $2 + (2/x^4)[(27)^2 + yx] > 0$. Hence, the unique positive critical number yields the minimum length. When $x = 6$, $y = \frac{25}{2}$, $L = \frac{125}{8} = 15.625$ ft.

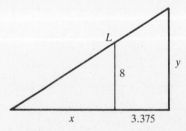

Fig. 16-14

16.40 A company offers the following schedule of charges: $30 per thousand for orders of 50,000 or less, with the charge per thousand decreased by 37.5 cents for each thousand above 50,000. Find the order that will maximize the company's income.

▮ Let x be the number of orders in thousands. Then the price per thousand is 30 for $x \le 50$ and $30 - \frac{3}{8}(x - 50)$ for $x > 50$. Hence, for $x \le 50$, the income $I = 30x$, and, for $x > 50$, $I = x[30 - \frac{3}{8}(x - 50)] = \frac{390}{8}x - \frac{3}{8}x^2$. So, for $x \le 50$, the maximum income is 1500 thousand. For $x > 50$, $D_x I = \frac{390}{8} - \frac{3}{4}x$ and $D_x^2 I = -\frac{3}{4}$. Solving $D_x I = 0$, $x = 65$. Since the second derivative is negative, $x = 65$ yields the maximum income for $x > 50$. That maximum is 3084.375 thousand. Hence, the maximum income is achieved when 65,000 orders are received.

16.41 A rectangle is inscribed in the ellipse $x^2/400 + y^2/225 = 1$ with its sides parallel to the axes of the ellipse (Fig. 16-15). Find the dimensions of the rectangle of maximum perimeter which can be so inscribed.

▌ $x/200 + (2y/225)D_x y = 0$, $D_x y = -(9x/16y)$. The perimeter $P = 4x + 4y$, so $D_x P = 4 + 4D_x y = 4(1 - 9x/16y) = 4(16y - 9x)/16y$ and $D_x^2 P = -\frac{9}{4}[y - x(-9x/16y)]/y^2 = -\frac{9}{4}(16y^2 + 9x^2)/16y^3 < 0$. Solving $D_x P = 0$, $16y = 9x$ and, then, substituting in the equation of the ellipse, we find $x^2 = 256$, $x = 16$, $y = 9$. Since the second derivative is negative, this unique critical number yields the maximum perimeter.

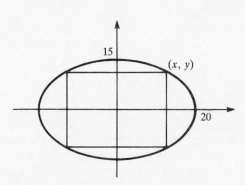

Fig. 16-15

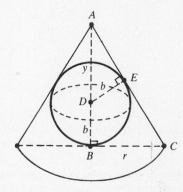

Fig. 16-16

16.42 Find the dimensions of the right circular cone of minimum volume which can be circumscribed about a sphere of radius b.

▌ See Fig. 16-16. Let r be the radius of the base of the cone, and let $y + b$ be the height of the cone. From the similar triangles ABC and AED, $\dfrac{r}{b} = -\dfrac{y + b}{\sqrt{y^2 - b^2}}$. Then $r^2 = \dfrac{b^2(y + b)^2}{y^2 - b^2} = \dfrac{b^2(y + b)}{y - b}$. The volume of the cone $V = \dfrac{(\pi r^2)(y + b)}{3} = \dfrac{\pi b^2(y + b)^2}{3(y - b)}$. Hence, $D_y V = \dfrac{\pi b^2(y + b)(y - 3b)}{3(y - b)^2}$. The only critical number $>b$ is $y = 3b$, and, by the first derivative test, this yields a relative minimum, which, by the uniqueness of the critical number, must be an absolute minimum.

16.43 Find the dimensions of the right circular cylinder of maximum volume that can be inscribed in a right circular cone of radius R and height H (Fig. 16-17).

▌ Let r and h be the radius and height of the cylinder. By similar triangles, $r/(H - h) = R/H$, $r = (R/H)(H - h)$. The volume of the cylinder $V = \pi r^2 h = \pi(R^2/H^2)(H - h)^2 h$. Then $D_h V = (\pi R^2/H^2)(H - h)(H - 3h)$, so the only critical number for $h < H$ is $h = H/3$. By the first derivative test, this yields a relative maximum, which, by the uniqueness of the critical number, is an absolute maximum. The radius $r = \frac{2}{3}R$.

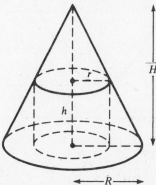

Fig. 16-17

16.44 A rectangular yard must be enclosed by a fence and then divided into two yards by a fence parallel to one of the sides. If the area A is given, find the ratio of the sides that will minimize the total length of the fencing.

▌ Let y be the length of the side with the parallel inside fence, and let x be the length of the other side. Then $A = xy$. The length of fencing is $F = 3y + 2x = 3(A/x) + 2x$. So, $D_x F = -3A/x^2 + 2$, and $D_x^2 F = 6A/x^3$. Solving $D_x F = 0$, we obtain $x^2 = \frac{3}{2}A$, $x = \sqrt{\frac{3}{2}A}$. Since the second derivative is positive, this unique critical number yields the absolute minimum for F. When $x = \sqrt{\frac{3}{2}A}$, $y = \sqrt{\frac{2}{3}A}$, and $y/x = \frac{2}{3}$.

16.45 Two vertices of a rectangle are on the positive x-axis. The other two vertices are on the lines $y = 4x$ and $y = -5x + 6$ (Fig. 16-18). What is the maximum possible area of the rectangle?

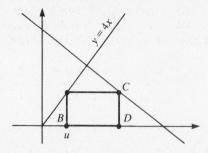

Fig. 16-18

▌ Let u be the x-coordinate of the leftmost vertex B of the rectangle on the x-axis. Then the y-coordinate of the other two vertices is $4u$. The x-coordinate of the vertex C opposite B is obtained by solving the equation $y = -5x + 6$ for x when $y = 4u$. This yields $x = (6 - 4u)/5$. Hence, this is the x-coordinate of the other vertex D on the x-axis. Thus, the base of the rectangle is equal to $(6 - 4u)/5 - u = (6 - 9u)/5$. Therefore, the area of the rectangle $A = 4u(6 - 9u)/5 = \frac{24}{5}u - \frac{36}{5}u^2$. Then $D_u A = \frac{24}{5} - \frac{72}{5}u$ and $D_u^2 A = -\frac{72}{5}$. Solving $D_u A = 0$, we find that the only positive critical number is $u = \frac{1}{3}$. Since the second derivative is negative, this yields the maximum area. When $u = \frac{1}{3}$, $A = \frac{4}{5}$.

16.46 A window formed by a rectangle surmounted by a semicircle is to have a fixed perimeter P. Find the dimensions that will admit the most light.

▌ Let $2y$ be the length of the side on which the semicircle rests, and let x be the length of the other side. Then $P = 2x + 2y + \pi y$. Hence, $0 = 2D_y x + 2 + \pi$, $D_y x = -(2 + \pi)/2$. To admit the most light, we must maximize the area $A = 2xy + \pi y^2/2$. $D_y A = 2(x + D_y x \cdot y) + \pi y = 2x - 2y$, and $D_y^2 A = 2D_y x - 2 = -\pi - 4 < 0$. Solving $D_y A = 0$, we obtain $x = y$, $P = (4 + \pi)x$, $x = P/(4 + \pi)$. Since the second derivative is negative, this unique critical number yields the maximum area, so the side on which the semicircle rests is twice the other side.

16.47 Find the y-coordinate of the point on the parabola $x^2 = 2py$ that is closest to the point $(0, b)$ on the axis of the parabola (Fig. 16-19).

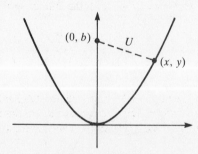

Fig. 16-19

▌ It suffices to find the point (x, y) that minimizes the square of the distance between (x, y) and $(0, b)$. $U = x^2 + (y - b)^2$, and $D_x U = 2x + 2(y - b) \cdot D_x y$. But $2x = 2p D_x y$, $D_x y = x/p$. So $D_x U = (2x/p)(p + y - b)$. Also, $D_x^2 U = (2/p)(x^2/p + p + y - b)$. Setting $D_x U = 0$, we obtain $x = 0$ or $y = b - p$. **Case 1.** $b \le p$. Then $b - p \le 0$, and, therefore, the only possible critical number is $x = 0$. By the first derivative test, we see that $x = 0$, $y = 0$ yields the absolute minimum for U. **Case 2.** $b > p$. When $x = 0$, $U = b^2$. When $y = b - p$, $U = p(2b - p) < b^2$. When $y > b - p$, $D_x U > 0$ (for positive x) and, therefore, the value of U is greater than its value when $y = b - p$. Thus, the minimum value occurs when $y = b - p$.

16.48 A wire of length L is cut into two pieces, one is formed into a square and the other into a circle. How should the wire be divided to maximize or minimize the sum of the areas of the pieces?

▮ Let the part used to form the circle be of length x. Then the radius of the circle is $x/2\pi$ and its area is $\pi(x/2\pi)^2 = x^2/4\pi$. The part used to form the square is $L - x$, the side of the square is $(L - x)/4$, and its area is $[(L - x)/4]^2$. So the total area $A = x^2/4\pi + [(L - x)/4]^2$. Then $D_xA = x/2\pi - \frac{1}{8}(L - x)$. Solving $D_xA = 0$, we obtain the critical value $x = \pi L/(4 + \pi)$. Notice that $0 \le x \le L$. So, to obtain the minimum and maximum values for A, we need only calculate the values of A at the critical number and at the endpoints 0 and L. Clearly, $L^2/4(4 + \pi) < L^2/16 < L^2/4\pi$. Hence, the maximum area is attained when $x = L$, that is, when all the wire is used for the circle. The minimum area is obtained when $x = \pi L/(4 + \pi)$.

x	0	$\pi L/(4 + \pi)$	L
A	$L^2/16$	$L^2/4(4 + \pi)$	$L^2/4\pi$

16.49 Find the positive number x that exceeds its square by the largest amount.

▮ We must maximize $f(x) = x - x^2$ for positive x. Then $f'(x) = 1 - 2x$ and $f''(x) = -2$. Hence, the only critical number is $x = \frac{1}{2}$. Since the second derivative is negative, this unique critical number yields an absolute maximum.

16.50 An east-west and a north-south road intersect at a point O. A diagonal road is to be constructed from a point E east of O to a point N north of O passing through a town C that is a miles east and b miles north of O. Find the distances of E and N from O if the area of $\triangle NOE$ is to be as small as possible.

▮ Let x be the x-coordinate of E. Let y be the y-coordinate of N. By similar triangles, $y/b = x/(x - a)$, $y = bx/(x - a)$. Hence, the area A of $\triangle NOE$ is given by $A = \frac{1}{2}x \cdot bx/(x - a) = (b/2)x^2/(x - a)$. Using the quotient rule, $D_xA = (b/2)(x^2 - 2ax)/(x - a)^2$, and $D_x^2A = a^2h/(x - a)^3$. Solving $D_xA = 0$, we obtain the critical number $x = 2a$. The second derivative is positive, since $x > a$ is obviously necessary. Hence, $x = 2a$ yields the minimum area A. When $x = 2a$ $y = 2b$.

16.51 A wire of length L is to be cut into two pieces, one to form a square and the other to form an equilateral triangle. How should the wire be divided to maximize or to minimize the sum of the areas of the square and triangle?

▮ Let x be the part used for the triangle. Then the side of the triangle is $x/3$ and its height is $x\sqrt{3}/6$. Hence, the area of the triangle is $\frac{1}{2}(x/3)(x\sqrt{3}/6) = \sqrt{3}x^2/36$. The side of the square is $(L - x)/4$, and its area is $[(L - x)/4]^2$. Hence, the total area $A = \sqrt{3}x^2/36 + [(L - x)/4]^2$. Then $D_xA = x\sqrt{3}/18 - (L - x)/8$. Setting $D_xA = 0$, we obtain the critical number $x = 9L/(9 + 4\sqrt{3})$. Since $0 \le x \le L$, to find the maximum and minimum values of A we need only compute the values of A at the critical number and the endpoints. It is clear that, since $16 < 12\sqrt{3} < 16 + 12\sqrt{3}$, the maximum area corresponds to $x = 0$, where everything goes into the square, and the minimum value corresponds to the critical number.

x	0	$9L/(9 + 4\sqrt{3})$	L
A	$L^2/16$	$L^2/(16 + 12\sqrt{3})$	$L^2/12\sqrt{3}$

16.52 Two towns A and B are, respectively, a miles and b miles from a railroad line (Fig. 16-20). The points C and D on the line nearest to A and B, respectively, are at a distance of c miles from each other. A station S is to be located on the line so that the sum of the distances from A and B to S is minimal. Find the position of S.

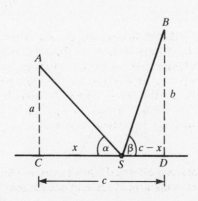

Fig. 16-20

Let x be the distance of S from C. Then the sum of the distances from A and B to S is given by the function $f(x) = \sqrt{a^2 + x^2} + \sqrt{b^2 + (c - x)^2}$. Hence, $f'(x) = x/\sqrt{a^2 + x^2} - (c - x)/\sqrt{b^2 + (c - x)^2}$. Setting $f'(x) = 0$, we eventually obtain the equation (*) $a/x = b/(c - x)$, $x = ac/(a + b)$. To see that this yields the absolute minimum, computation of $f''(x)$ yields (after extensive simplifications) $a^2/(a^2 + x^2)^{3/2} + b^2/[b^2 + (c - x)^2]^{3/2}$, which is positive. [Notice that the equation (*) also tells us that the angles α and β are equal. If we reinterpret this problem in terms of a light ray from A being reflected off a mirror to B, we have found that the angle of incidence α is equal to the angle of reflection β.]

16.53 A telephone company has to run a line from a point A on one side of a river to another point B that is on the other side, 5 miles down from the point opposite A (Fig. 16-21). The river is uniformly 12 miles wide. The company can run the line along the shoreline to a point C and then run the line under the river to B. The cost of laying the line along the shore is $1000 per mile, and the cost of laying it under water is twice as great. Where should the point C be located to minimize the cost?

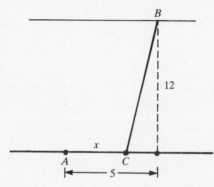

Fig. 16-21

▮ Let x be the distance from A to C. Then the cost of running the line is $f(x) = 1000x + 2000\sqrt{144 + (5 - x)^2}$.
$f'(x) = 1000 - \dfrac{2000(5 - x)}{\sqrt{144 + (5 - x)^2}}$, and $f''(x) = 2000\dfrac{144 + 2(5 - x)^2}{[144 + (5 - x)^2]^{3/2}} > 0$. Setting $f'(x) = 0$ and solving for x, $48 = (x - 5)^2$, $x = 5 \pm 4\sqrt{3}$. Since x cannot be negative or greater than 5, neither critical number is feasible. So, the minimum occurs at an endpoint. Since $f(0) = 26,000$ and $f(5) = 29,000$, the minimum occurs at $x = 0$.

16.54 Let m and n be given positive integers. If x and y are positive numbers such that $x + y$ is a constant S, find the values of x and y that maximize $P = x^m y^n$.

▮ $P = x^m(S - x)^n$. $D_x P = mx^{m-1}(S - x)^n - nx^m(S - x)^{n-1}$. Setting $D_x P = 0$, we obtain $x = mS/(m + n)$. The first derivative test shows that this yields a relative maximum, and, therefore, by the uniqueness of the critical number, an absolute maximum. When $x = mS/(m + n)$, $y = nS/(m + n)$.

16.55 Show that of all triangles with given base and given area, the one with the least perimeter is isosceles. (Compare with Problem 16.20.)

▮ Let the base of length $2c$ lie on the x-axis with the origin as its midpoint, and let the other vertex (x, y) lie in the upper half-plane (Fig. 16-22). By symmetry we may assume $x \geq 0$. To minimize the perimeter, we must minimize $AC + BC$, which is given by the function $f(x) = \sqrt{(c + x)^2 + h^2} + \sqrt{(c - x)^2 + h^2}$, where h is the altitude. Then $f'(x) = \dfrac{c + x}{\sqrt{(c + x)^2 + h^2}} - \dfrac{c - x}{\sqrt{(c - x)^2 + h^2}}$. Setting $f'(x) = 0$, we obtain $c + x = \pm(c - x)$. The minus sign leads to the contradiction $c = 0$. Therefore, $c + x = c - x$, $x = 0$. Thus, the third vertex lies on the y-axis and the triangle is isosceles. That the unique critical number $x = 0$ yields an absolute minimum follows from a computation of the second derivative, which turns out to be positive.

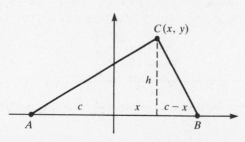

Fig. 16-22

16.56 Let two corridors of widths a and b intersect at a right angle. Find the minimum length of all segments DE that touch the outer walls of the corridors and are tangent to the corner C.

▌ Let θ be the angle between DE and the vertical (see Fig. 16-23). Then $0 < \theta < \pi/2$. Let L be the length of DE. $L = b \sec \theta + a \csc \theta$, and $D_\theta L = b \sec \theta \tan \theta - a \csc \theta \cot \theta$. Setting $D_\theta L = 0$, $b \sec \theta \tan \theta = a \csc \theta \cot \theta$, $b \sin \theta / \cos^2 \theta = a \cos \theta / \sin^2 \theta$, $b \sin^3 \theta = a \cos^3 \theta$, $\tan^3 \theta = a/b$, $\tan \theta = \sqrt[3]{a/b} = \sqrt[3]{a}/\sqrt[3]{b}$. Consider the hypotenuse u of a right triangle with legs $\sqrt[3]{a}$ and $\sqrt[3]{b}$. Then $u^2 = a^{2/3} + b^{2/3}$, $u = (a^{2/3} + b^{2/3})^{1/2}$. $\sec \theta = (a^{2/3} + b^{2/3})^{1/2}/b^{1/3}$, $\csc \theta = (a^{2/3} + b^{2/3})^{1/2}/a^{1/3}$. So, $L = (a^{2/3} + b^{2/3})^{1/2}(b^{2/3} + a^{2/3}) = (a^{2/3} + b^{2/3})^{3/2}$. Observe that $D_\theta L = (b \cos \theta / \sin^2 \theta)(\tan^3 \theta - a/b)$. Hence, the first derivative test yields the case $\{-, +\}$, which, by virtue of the uniqueness of the critical number, shows that $L = (a^{2/3} + b^{2/3})^{3/2}$ is the absolute minimum. Notice that this value of L is the minimal length of all poles that cannot turn the corner from one corridor into the other.

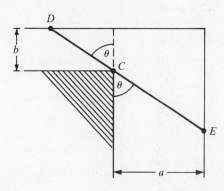

Fig. 16-23

16.57 A rectangular yard is to be laid out and fenced in, and then divided into 10 enclosures by fences parallel to one side of the yard. If a fixed length K of fencing is available, what dimensions will maximize the area?

▌ Let x be the length of the sides of the enclosure fences, and let y be the other side. Then $K = 11x + 2y$. The area $A = xy = x(K - 11x)/2 = (K/2)x - \frac{11}{2}x^2$. Hence, $D_x A = K/2 - 11x$, and $D_x^2 A = -11$. Setting $D_x A = 0$, we obtain the critical number $x = K/22$. Since the second derivative is negative, we have a relative maximum, and, since the critical number is unique, the relative maximum is an absolute maximum. When $x = K/22$, $y = K/4$.

16.58 Two runners A and B start at the origin and run along the positive x-axis, with B running 3 times as fast as A. An observer, standing one unit above the origin, keeps A and B in view. What is the maximum angle of sight θ between the observer's view of A and B? (See Fig. 16-24.)

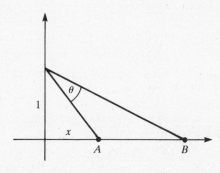

Fig. 16-24

▌ Let x be the distance of A from the origin. Then B is $3x$ units from the origin. Let θ_1 be the angle between the y-axis and the line of sight of A, and let θ_2 be the corresponding angle for B. Then $\theta = \theta_2 - \theta_1$. Note that $\tan \theta_1 = x$ and $\tan \theta_2 = 3x$. So $\tan \theta = \dfrac{\tan \theta_2 - \tan \theta_1}{1 + \tan \theta_1 \tan \theta_2} = \dfrac{3x - x}{1 + x(3x)} = \dfrac{2x}{1 + 3x^2}$. Since θ is between 0 and $\pi/2$, maximizing θ is equivalent to maximizing $\tan \theta$. Now, $D_x(\tan \theta) = \dfrac{(1 + 3x^2)(2) - 2x(6x)}{(1 + 3x^2)^2} = \dfrac{2(1 + 3x^2 - 6x^2)}{(1 + 3x^2)^2} = \dfrac{2(1 - 3x^2)}{(1 + 3x^2)^2}$. Setting $D_x(\tan \theta) = 0$, we obtain $1 = 3x^2$, $x = 1/\sqrt{3}$. The first derivative test shows that we have a relative maximum, which, by uniqueness, must be the absolute maximum. When $\tan \theta = 1/\sqrt{3}$, $\theta = 30°$.

16.59 A painting of height 3 feet hangs on the wall of a museum, with the bottom of the painting 6 feet above the floor. If the eyes of an observer are 5 feet above the floor, how far from the base of the wall should the observer stand to maximize his angle of vision θ? See Fig. 16-25.

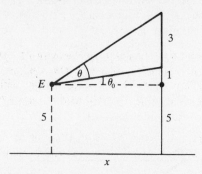

Fig. 16-25

▌ Let x be the distance from the observer to the base of the wall, and let θ_0 be the angle between the line of sight of the bottom of the painting and the horizontal. Then $\tan(\theta + \theta_0) = 4/x$ and $\tan \theta_0 = 1/x$. Hence,
$$\tan \theta = \frac{\tan(\theta + \theta_0) - \tan \theta_0}{1 + \tan(\theta + \theta_0)\tan \theta_0} = \frac{4/x - 1/x}{1 + 4/x^2} = \frac{3x}{x^2 + 4}.$$ Since maximizing θ is equivalent to maximizing $\tan \theta$, it suffices to do the latter. Now, $D_x(\tan \theta) = 3\left[\frac{(x^2 + 4) - x(2x)}{(x^2 + 4)^2}\right] = \frac{3(4 - x^2)}{(x^2 + 4)^2}$. Hence, the unique positive critical number is $x = 2$. The first derivative test shows this to be a relative maximum, and, by the uniqueness of the positive critical number, this is an absolute maximum.

16.60 A large window consists of a rectangle with an equilateral triangle resting on its top (Fig. 16-26). If the perimeter P of the window is fixed at 33 feet, find the dimensions of the rectangle that will maximize the area of the window.

▌ Let s be the side of the rectangle on which the triangle rests, and let y be the other side. Then $33 = 2y + 3s$. The height of the triangle is $(\sqrt{3}/2)s$. So the area $A = sy + \frac{1}{2}s(\sqrt{3}/2)s = s(33 - 3s)/2 + (\sqrt{3}/4)s^2 = \frac{33}{2}s + [(\sqrt{3} - 6)/4]s^2$. $D_s A = \frac{33}{2} + [(\sqrt{3} - 6)/2]s$. Setting $D_s A = 0$, we find the critical number $s = 33/(6 - \sqrt{3}) = 6 + \sqrt{3}$. The first derivative test shows that this yields a relative maximum, which, by virtue of the uniqueness of the critical number, must be an absolute maximum. When $s = 6 + \sqrt{3}$, $y = \frac{3}{2}(5 - \sqrt{3})$.

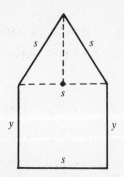

Fig. 16-26

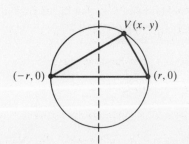

Fig. 16-27

16.61 Consider triangles with one side on a diameter of a circle of radius r and with the third vertex V on the circle (Fig. 16-27). What location of V maximizes the perimeter of the triangle?

▌ Let the origin be the center of the circle, with the diameter along the x-axis, and let (x, y), the third vertex, lie in the upper half-plane. Then the perimeter $P = 2r + \sqrt{(x + r)^2 + y^2} + \sqrt{(x - r)^2 + y^2}$, with $-r < x < r$.
$D_x P = \frac{(x + r) + y D_x y}{\sqrt{(x + r)^2 + y^2}} + \frac{(x - r) + y D_x y}{\sqrt{(x - r)^2 + y^2}}$. But, from the equation of the circle $x^2 + y^2 = r^2$, we find by implicit differentiation that $D_x y = -x/y$. Hence, $D_x P$ becomes $\frac{x + r - x}{\sqrt{(x + r)^2 + y^2}} + \frac{x - r - x}{\sqrt{(x - r)^2 + y^2}} = r\left(\frac{\sqrt{(x - r)^2 + y^2} - \sqrt{(x + r)^2 + y^2}}{\sqrt{(x + r)^2 + y^2}\sqrt{(x - r)^2 + y^2}}\right)$. Then we set $D_x P = 0$, and solving for x, find that the critical number is $x = 0$. The corresponding value of P is $2r(1 + \sqrt{2})$. At the endpoints $x = -r$ and $x = r$, the value of P is $4r$. Since $4r < 2r(1 + \sqrt{2})$, the maximum perimeter is attained when $x = 0$ and $y = r$, that is, V is on the diameter perpendicular to the base of the triangle.

17.1 The equation of <u>free fall of an object</u> (under the influence of gravity alone) is $\boxed{s = s_0 + v_0 t - 16t^2,}$ where s_0 is the initial position and v_0 is the initial velocity at <u>time</u> $t = 0$. (We assume that the s-axis is directed upward away from the earth, along the vertical line on which the object moves, with $s = 0$ at the earth's surface. s is measured in <u>feet</u> and t in <u>seconds</u>.) Show that, if an object is released from rest at any given height, it will have dropped $16t^2$ feet after t seconds.

▌ To say that the object is <u>released from rest</u> means that the initial velocity $v_0 = 0$, so its position after t seconds is $s_0 - 16t^2$. The difference between that position and its initial position s_0 is $16t^2$.

17.2 How many seconds does it take the object released from rest to fall 64 feet?

▌ By Problem 17.1, $64 = 16t^2$. Hence, $t^2 = 4$, and, since t is positive, $t = 2$. $\quad (\; s_0 - 64 = s_0 - 16t^2 \;)$

17.3 A rock is dropped down a well that is 256 feet deep. When will it hit the bottom of the well?

▌ If t is the time until it hits the bottom, $256 = 16t^2$, so $t^2 = 16$, $t = 4$.

17.4 Assuming that one story of a building is 10 feet, with what <u>speed</u>, in miles per hour, does an object dropped from the top of a 40-story building hit the ground?

▌ Let t be the time until the object hits the ground. Since the building is 400 feet tall, $400 = 16t^2$, $t^2 = 25$, $t = 5$. The velocity $v = D_t s$. Since $s = s_0 - 16t^2$, $v = -32t$. When $t = 5$, $v = -160$. Thus, the speed $|v|$ is 160 ft/s. To change to mi/h, we calculate as follows:

$$x\,\text{ft/s} = 60x\,\text{ft/min} = 3600x\,\text{ft/h} = \frac{3600x}{5280}\,\text{mi/h}$$

$$= \frac{15x}{22}\,\text{mi/h} \approx 0.68x\,\text{mi/h}$$

In particular, when $x = 160$ ft/s, the speed is about 108.8 mi/h.

17.5 A rocket is shot straight up into the air with an initial velocity of 128 ft/s. How far has it traveled in 1 second?

▌ The height $s = s_0 + v_0 t - 16t^2$. Since $s_0 = 0$ and $v_0 = 128$, $s = 128t - 16t^2$. When $t = 1$, $s = 112$ ft.

17.6 In Problem 17.5, when does the rocket reach its maximum height?

▌ At the maximum value of s, $v = D_t s = 0$, but $v = 128 - 32t$. Setting $v = 0$, we obtain $t = 4$ seconds.

17.7 In Problem 17.5, when does the rocket strike the ground again and what is its <u>velocity</u> when it hits the ground?

▌ Setting $s = 0$, $128t - 16t^2 = 0$, $16t(8 - t) = 0$, $t = 0$ or $t = 8$. So the rocket strikes the ground again after 8 seconds. When $t = 8$, the velocity $v = 128 - 32t = 128 - 256 = -128$ ft/s. The velocity is negative (because the rocket is moving downward) and of the same magnitude as the initial velocity (see Problem 17.28).

17.8 A rock is thrown straight down from a height of 480 feet with an initial velocity of 16 ft/s. How long does it take to hit the ground and with what speed does it hit the ground?

▌ The height $s = s_0 + v_0 t - 16t^2$. In this case, $s_0 = 480$ and $v_0 = -16$. Thus, $s = 480 - 16t - 16t^2 = 16(30 - t - t^2) = 16(6 + t)(5 - t)$. Setting $s = 0$, we obtain $t = -6$ or $t = 5$. Hence, the rock hits the ground after 5 seconds. The velocity $v = D_t s = -16 - 32t$. When $t = 5$, $v = -16 - 160 = -176$, so the rock hits the ground with a speed of 176 ft/s. (The minus sign in the velocity indicates that the rock is moving downward.)

17.9 Under the same conditions as in Problem 17.8, how long does it take before the rock is moving at a <u>speed</u> of 112 ft/s?

▮ From Problem 17.8, we know that $v = D_t s = -16 - 32t$. Since the rock is moving downward, a speed of 112 ft/s corresponds to a velocity v of -112. Setting $-112 = -16 - 32t$, $32t = 96$, $t = 3$ seconds.

17.10 Under the same conditions as in Problem 17.8, when has the rock traveled a distance of 60 feet?

▮ Since the rock starts at a height of 480 feet, it has traveled 60 feet when it reaches a height of 420 feet. Since $s = 480 - 16t - 16t^2$, we set $420 = 480 - 16t - 16t^2$, obtaining $4t^2 + 4t - 15 = 0$, $(2t + 5)(2t - 3) = 0$, $t = -2.5$ or $t = 1.5$. Hence, the rock traveled 1.5 seconds.

17.11 An automobile moves along a straight highway, with its position s given by $s = 12t^3 - 18t^2 + 9t - 1.5$ (s in feet, t in seconds). When is the car moving to the right, when to the left, and where and when does it change direction?

▮ Since s increases as we move right, the car moves right when $v = D_t s > 0$, and moves left when $v = D_t s < 0$. $v = 36t^2 - 36t + 9 = 9(4t^2 - 4t + 1) = 9(2t - 1)^2$. Since $v > 0$ (except at $t = 0.5$, where $v = 0$), the car always moves to the right and never changes direction. (It slows down to an instantaneous velocity of 0 at $t = 0.5$ second, but then immediately speeds up again.)

17.12 Refer to Problem 17.11. What distance has the car traveled in the one second from $t = 0$ to $t = 1$?

▮ From the solution to Problem 17.11, we know that the car is always moving right. Hence, the distance traveled from $t = 0$ to $t = 1$ is obtained by taking the difference between its position at time $t = 1$ and its position at time $t = 0$: $s(1) - s(0) = 1.5 - (-1.5) = 3$ ft.

17.13 The position of a moving object on a line is given by the formula $s = (t - 1)^3(t - 5)$. When is the object moving to the right, when is it moving left, when does it change direction, and when is it at rest? What is the farthest to the left of the origin that it moves?

▮ $v = D_t s = (t - 1)^3 + 3(t - 1)^2(t - 5) = (t - 1)^2[t - 1 + 3(t - 5)] = 4(t - 1)^2(t - 4)$. Thus, $v > 0$ when $t > 4$, and $v < 0$ when $t < 4$ (except at $t = 1$, when $v = 0$). Hence, the object is moving left when $t < 4$, and it is moving right when $t > 4$. Thus, it changes direction when $t = 4$. It is never at rest. (To be at rest means that s is constant for an interval of time, or, equivalently, that $v = 0$ for an entire interval of time.) The object reaches its farthest position to the left when it changes direction at $t = 4$. When $t = 4$, $s = -27$.

17.14 A particle moves on a straight line so that its position s (in miles) at time t (in hours) is given by $s = (4t - 1)(t - 1)^2$. When is the particle moving to the right, when to the left, and when does it change direction? When the particle is moving to the left, what is the maximum speed that it achieves?

▮ $v = D_t s = 4(t - 1)^2 + 2(t - 1)(4t - 1) = 2(t - 1)[2(t - 1) + 4t - 1] = 2(t - 1)(6t - 3) = 6(t - 1)(2t - 1)$. Thus, the key values are $t = 1$ and $t = 0.5$. When $t > 1$, $v > 0$; when $0.5 < t < 1$, $v < 0$; when $t < 0.5$, $v > 0$. Thus, the particle is moving right when $t < 0.5$ and when $t > 1$. It is moving left when $0.5 < t < 1$. So it changes direction when $t = 0.5$ and when $t = 1$. To find out what the particle's maximum speed is when it is moving left, note that the speed is $|v|$. Hence, when v is negative, as it is when the particle is moving left, the maximum speed is attained when the velocity reaches its absolute *minimum*. Now, $D_t v = 6[2(t - 1) + 2t - 1] = 6(4t - 3)$, and $D_t^2 v = 24 > 0$. Hence, by the second derivative test, v reaches an absolute minimum when $t = 0.75$ hour. When $t = 0.75$, $v = -0.75$ mi/h. So the desired maximum speed is 0.75 mi/h.

17.15 Under the assumptions of Problem 17.14, what is the total distance traveled by the particle from $t = 0$ to $t = 1$?

▮ The problem cannot be solved by simply finding the difference between the particle's positions at $t = 1$ and $t = 0$, because it is moving in different directions during that period. We must add the distance d_r traveled while it is moving right (from $t = 0$ to $t = 0.5$) to the distance d_ℓ traveled while it is moving left (from $t = 0.5$ to $t = 1$). Now, $d_r = s(0.5) - s(0) = 0.25 - (-1) = 1.25$. Similarly, $d_\ell = s(0.5) - s(1) = 0.25 - 0 = 0.25$. Thus, the total distance is 1.5 miles.

17.16 A particle moves along the x-axis according to the equation $x = 10t - 2t^2$. What is the total distance covered by the particle between $t = 0$ and $t = 3$?

❚ The velocity $v = D_t x = 10 - 4t$. Thus, $v > 0$ when $t < 2.5$, and $v < 0$ when $t > 2.5$. Hence, the particle is moving right for $t < 2.5$ and it is moving left for $t > 2.5$. The distance d_r that it covers while it is moving right from $t = 0$ to $t = 2.5$ is $x(2.5) - x(0) = 12.5 - 0 = 12.5$. The distance d_ℓ that it covers while it is moving left from $t = 2.5$ to $t = 3$ is $x(2.5) - x(3) = 12.5 - 12 = 0.5$. Hence, the total distance is $d_r + d_\ell = 12.5 + 0.5 = 13$.

17.17 A rocket was shot straight up from the ground. What must its initial velocity have been if it returned to earth in 20 seconds?

❚ Its height $s = s_0 + v_0 t - 16t^2$. In this case, $s_0 = 0$ and v_0 is unknown, so $s = v_0 t - 16t^2$. We are told that $s = 0$ when $t = 20$. Hence, $0 = v_0(20) - 16(20)^2$, $v_0 = 320$ ft/s.

17.18 Two particles move along the x-axis. Their positions $f(t)$ and $g(t)$ are given by $f(t) = 6t - t^2$ and $g(t) = t^2 - 4t$. (a) When do they have the same position? (b) When do they have the same velocity? (c) When they have the same position, are they moving in the same direction?

❚ (a) Set $6t - t^2 = t^2 - 4t$. Then $t^2 - 5t = 0$, $t(t - 5) = 0$, $t = 0$ or $t = 5$. (b) The velocities are $f'(t) = 6 - 2t$ and $g'(t) = 2t - 4$. Setting $6 - 2t = 2t - 4$, we have $t = 2.5$. (c) When they meet at $t = 0$, $f'(t) = f'(0) = 6$ and $g'(t) = g'(0) = -4$. Since $f'(0)$ and $g'(0)$ have opposite signs, they are moving in opposite directions when $t = 0$. When they meet at $t = 5$, $f'(t) = f'(5) = -4$ and $g'(t) = g'(5) = 6$. Hence, when $t = 5$, they are moving in opposite directions.

17.19 A particle moves along the x-axis according to the equation $x = \frac{1}{3}t^3 - \frac{1}{2}\cos 2t + 3.5$. Find the distance traveled between $t = 0$ and $t = \pi/2$.

❚ The velocity $v = t^2 + \sin 2t$. For $0 < t < \pi/2$, $\sin 2t > 0$, and, therefore, $v > 0$. Hence, the particle moves right between $t = 0$ and $t = \pi/2$. So the distance traveled is $x(\pi/2) - x(0) = (\pi^3/24 + 4) - 3 = \pi^3/24 + 1$.

17.20 A ball is thrown vertically into the air so that its height s after t seconds is given by $s = -\frac{1}{27}t^2 + 4\sqrt{t}$. Find its maximum height.

❚ $D_t s = -\frac{2}{27}t + 2/\sqrt{t}$, $D_t^2 s = -\frac{2}{27} - 1/(\sqrt{t})^3$. Setting $D_t s = 0$, we obtain $t/27 = 1/\sqrt{t}$, $t^{3/2} = 27$, $t = 9$. Since the second derivative is negative, we must have a relative maximum at $t = 9$, and, since that is the unique critical number, it must be an absolute maximum. At $t = 9$, $s = 9$.

17.21 A particle moving along a straight line is accelerating at the constant rate of 3 m/s^2. Find the initial velocity v_0 if the displacement during the first two seconds is 10 m.

❚ The acceleration $a = D_t v = 3$. Hence, $v = 3t + C$. When $t = 0$, $v = v_0$. So $C = v_0$. Thus, $v = 3t + v_0$, but $v = D_t s$. So $s = \frac{3}{2}t^2 + v_0 t + K$. When $t = 0$, $s = s_0$, the initial position, so $K = s_0$. Thus, $s = \frac{3}{2}t^2 + v_0 t + s_0$. The displacement during the first two seconds is $s(2) - s(0) = (6 + 2v_0 + s_0) - s_0 = 6 + 2v_0$. Hence, $6 + 2v_0 = 10$, $v_0 = 2$ m/s.

17.22 A particle moving on a line is at position $s = t^3 - 6t^2 + 9t - 4$ at time t. At which time(s) t, if any, does it change direction?

❚ $v = D_t s = 3t^2 - 12t + 9 = 3(t^2 - 4t + 3) = 3(t - 1)(t - 3)$. Since the velocity changes sign at $t = 1$ and $t = 3$, the particle changes direction at those times.

17.23 A ball is thrown vertically upward. Its height s (in feet) after t seconds is given by $s = 40t - 16t^2$. Find (a) when the ball hits the ground, (b) the instantaneous velocity at $t = 1$, (c) the maximum height.

❚ $v = D_t s = 40 - 32t$, $D_t^2 s = -32$. (a) To find out when the ball hits the ground, we set $s = 40t - 16t^2 = 0$. Then $t = 0$ or $t = 2.5$. So the ball hits the ground after 2.5 seconds. (b) When $t = 1$, $v = 8$ ft/s. (c) Set $v = 40 - 32t = 0$. Then $t = 1.25$. Since the second derivative is negative, this unique critical number yields an absolute maximum. When $t = 1.25$, $s = 25$ ft.

17.24 A diver jumps off a springboard 10 feet above water with an initial upward velocity of 12 ft/s. Find (a) her maximum height, (b) when she will hit the water, (c) her velocity when she hits the water.

▮ Since she is moving only under the influence of gravity, her height $s = s_0 + v_0 t - 16t^2$. In this case, $s_0 = 10$ and $v_0 = 12$. So $s = 10 + 12t - 16t^2$, $v = D_t s = 12 - 32t$, $D_t^2 s = -32$. (a) Setting $v = 0$, we obtain $t = 0.375$. Since the second derivative is negative, this unique critical number yields an absolute maximum. When $t = 0.375$, $s = 16.75$ ft. (b) To find when she hits the water, set $s = 10 + 12t - 16t^2 = 0$. So $(5 - 4t)(1 + 2t) = 0$, and, therefore, she hits the water at $t = 1.25$ seconds. (c) At $t = 1.25$, $v = -28$ ft/s.

17.25 A ball is thrown vertically upward. Its height s (in feet) after t seconds is given by $s = 48t - 16t^2$. For which values of t will the height exceed 32 feet?

▮ We must have $48t - 16t^2 > 32$, $3t - t^2 > 2$, $t^2 - 3t + 2 < 0$, $(t - 2)(t - 1) < 0$. The latter inequality holds precisely when $1 < t < 2$.

17.26 The distance a locomotive is from a fixed point on a straight track at time t is given by $s = 3t^4 - 44t^3 + 144t^2$. When was it in reverse?

▮ $v = D_t s = 12t^3 - 132t^2 + 288t = 12t(t^2 - 11t + 24) = 12t(t - 3)(t - 8)$. The locomotive goes backwards when $v < 0$. Clearly, $v > 0$ when $t > 8$; $v < 0$ when $3 < t < 8$; $v > 0$ when $0 < t < 3$; $v < 0$ when $t < 0$. Thus, it was in reverse when $3 < t < 8$ (and, if we allow negative time, when $t < 0$).

17.27 An object is thrown straight up from the ground with an initial velocity v_0 ft/s. Show that the time taken on the upward flight is equal to the time taken on the way down.

▮ $s = s_0 + v_0 t - 16t^2$. In this case, $s_0 = 0$. So $s = v_0 t - 16t^2$, $v = D_t s = v_0 - 32t$, $a = D_t v = D_t^2 s = -32$. So the unique critical number is $t = v_0/32$, and, since the second derivative is negative, this yields the maximum height. Thus, the time of the upward flight is $v_0/32$. The object hits the ground again when $s = v_0 t - 16t^2 = 0$, $v_0 = 16t$, $t = v_0/16$. Hence, the total time of the flight was $v_0/16$, and half of that time, $v_0/32$, was used up in the upward flight. Hence, the time taken on the way down was also $v_0/32$.

17.28 Under the conditions of Problem 17.27, show that the object hits the ground with the same speed at which it was initially thrown.

▮ By Problem 17.27, the object hits the ground after $v_0/16$ seconds. At that time, $v = v_0 - 32t = v_0 - 32(v_0/16) = -v_0$. Thus, the velocity when it hits the ground is the negative of the initial velocity, and, therefore, the speeds are the same.

17.29 With what velocity must an object be thrown straight up from the ground in order for it to hit the ground t_0 seconds later?

▮ From Problem 17.27, we know that the object hits the ground $v_0/16$ seconds after it was thrown. Hence, $t_0 = v_0/16$, $v_0 = 16t_0$.

17.30 With what velocity must an object be thrown straight up from the ground in order to reach a maximum height of h feet?

▮ From Problem 17.27, we know that the object reaches its maximum height after $v_0/32$ seconds. When $t = v_0/32$, $s = v_0 t - 16t^2 = v_0^2/64$. Hence, $h = v_0^2/64$, $v_0 = 8\sqrt{h}$.

17.31 A woman standing on a bridge throws a stone straight up. Exactly 5 seconds later the stone passes the woman on the way down, and 1 second after that it hits the water below. Find the initial velocity of the stone and the height of the bridge above the water.

▮ The height of the stone $s = s_0 + v_0 t - 16t^2$, where s_0 is the height of the bridge above the water. When $t = 5$, $s = s_0$. So $s_0 = s_0 + v_0(5) - 16(5)^2$, $5v_0 = 400$, $v_0 = 80$ ft/s. Hence, $s = s_0 + 80t - 16t^2$. When $t = 6$, $s = 0$. So $0 = s_0 + 80(6) - 16(6)^2$, $s_0 = 96$ ft.

17.32 A stone is dropped from the roof of a building 256 ft high. Two seconds later a second stone is thrown downward from the roof of the same building with an initial velocity of v_0 ft/s. If both stones hit the ground at the same time, what is v_0?

▮ For the first stone, $s = 256 - 16t^2$. It hits the ground when $0 = s = 256 - 16t^2$, $t^2 = 16$, $t = 4$ seconds. Since the second stone was thrown 2 seconds later than the first and hit the ground at the same time as the first, the second stone's flight took 2 seconds. So, for the second stone, $0 = 256 + v_0(2) - 16(2)^2$, $v_0 = -192$ ft/s.

17.33 An object is dropped from a height 25 ft above the ground. At the same time another object is thrown straight down from a height 50 ft above the ground. Both objects hit the ground at the same time. Find the initial velocity of the second object.

▐ For the first object, $s = 25 - 16t^2$. When $s = 0$, $t = \frac{5}{4}$. For the second object, $s = 50 + v_0 t - 16t^2$. Since the second object also hits the ground after $\frac{5}{4}$ seconds, $0 = 50 + v_0(\frac{5}{4}) - 16(\frac{5}{4})^2$, $v_0 = 20$ ft/s.

17.34 If the position s of an object moving on a straight line is given by $s = \sqrt{t+2}$, show that its velocity is positive, and its acceleration is negative and proportional to the cube of the velocity.

▐ The velocity $v = D_t s = 1/(2\sqrt{t+2}) > 0$, and the acceleration $a = D_t v = -1/(4(\sqrt{t+2})^3) = -2v^3 < 0$.

17.35 An object moves along the x-axis so that its x-coordinate obeys the law $x = 3t^3 + 8t + 1$. Find the time(s) when its velocity and acceleration are equal.

▐ $v = D_t x = 9t^2 + 8$. $a = D_t v = 18t$. Setting $v = a$, $9t^2 + 8 = 18t$, $9t^2 - 18t + 8 = 0$, $(3t - 2)(3t - 4) = 0$, $t = \frac{2}{3}$ or $t = \frac{4}{3}$.

CHAPTER 18
Approximation by Differentials

18.1 State the approximation principle for a differentiable function $f(x)$.

❚ Let x be a number in the domain of f, let Δx be a small change in the value of x, and let $\Delta y = f(x + \Delta x) - f(x)$ be the corresponding change in the value of the function. Then the approximation principle asserts that $\Delta y \approx f'(x) \cdot \Delta x$, that is, Δy is very close to $f'(x) \cdot \Delta x$ for small values of Δx.

In Problems 18.2 to 18.8, estimate the value of the given quantity.

18.2 $\sqrt{51}$.

❚ Let $f(x) = \sqrt{x}$, let $x = 49$, and let $\Delta x = 2$. Then $x + \Delta x = 51$, $\Delta y = f(x + \Delta x) - f(x) = \sqrt{51} - \sqrt{49} = \sqrt{51} - 7$. Note that $f'(x) = \dfrac{1}{2\sqrt{x}} = \dfrac{1}{2 \cdot 7} = \dfrac{1}{14}$. The approximation principle tells us that $\Delta y \approx f'(x) \cdot \Delta x$, $\sqrt{51} - 7 \approx \frac{1}{14} \cdot 2$, $\sqrt{51} \approx 7 + \frac{1}{7} \approx 7.14$. (Checking a table of square roots shows that this is actually correct to two decimal places.)

18.3 $\sqrt{78}$.

❚ Let $f(x) = \sqrt{x}$, $x = 81$, $\Delta x = -3$. Then $x + \Delta x = 78$, $\Delta y = \sqrt{78} - \sqrt{81} = \sqrt{78} - 9$, $f'(x) = \dfrac{1}{2\sqrt{x}} = \dfrac{1}{2 \cdot 9} = \dfrac{1}{18}$. So, by the approximation principle, $\sqrt{78} - 9 \approx \frac{1}{18} \cdot (-3) = -\frac{1}{6}$. Hence, $\sqrt{78} \approx 9 - \frac{1}{6} \approx 9 - 0.17 = 8.83$. (Comparison with a square root table shows that this is correct to two decimal places.)

18.4 $\sqrt[3]{123}$.

❚ Let $f(x) = \sqrt[3]{x}$, $x = 125$, $\Delta x = -2$. Then $x + \Delta x = 123$, $\Delta y = \sqrt[3]{123} - \sqrt[3]{125} = \sqrt[3]{123} - 5$. $f'(x) = \dfrac{1}{3(\sqrt[3]{x})^2} = \dfrac{1}{3 \cdot (5)^2} = \dfrac{1}{75}$. So, by the approximation principle, $\sqrt[3]{123} - 5 \approx \frac{1}{75} \cdot (-2) = -\frac{2}{75}$. $\sqrt[3]{123} \approx 5 - \frac{2}{75} \approx 5 - 0.03 = 4.97$. (This is actually correct to two decimal places.)

18.5 $(8.35)^{2/3}$.

❚ Let $f(x) = x^{2/3}$, $x = 8$, $\Delta x = 0.35$. Then $x + \Delta x = 8.35$, $\Delta y = (8.35)^{2/3} - 8^{2/3} = (8.35)^{2/3} - 4$. Also, $f'(x) = \dfrac{2}{3\sqrt[3]{x}} = \dfrac{2}{3 \cdot 2} = \dfrac{1}{3}$. So, by the approximation principle, $(8.35)^{2/3} - 4 \approx \frac{1}{3} \cdot (0.35)$, $(8.35)^{2/3} \approx 4 + 0.35/3 \approx 4 + 0.117 = 4.117$. (The actual answer is 4.116 to three decimal places.)

18.6 $(33)^{-1/5}$.

❚ Let $f(x) = x^{-1/5}$, $x = 32$, $\Delta x = 1$. Then $x + \Delta x = 33$, $\Delta y = (33)^{-1/5} - (32)^{-1/5} = (33)^{-1/5} - 1/\sqrt[5]{32} = (33)^{-1/5} - \frac{1}{2} = (33)^{-1/5} - 0.5$. Also, $f'(x) = -\dfrac{1}{5(\sqrt[5]{x})^6} = -\dfrac{1}{5(2)^6} = -\dfrac{1}{320}$. So, by the approximation principle, $(33)^{-1/5} - 0.5 \approx -\frac{1}{320}$, $(33)^{-1/5} \approx 0.5 - \frac{1}{320} \approx 0.5 - 0.003 = 0.497$. (This is correct to three decimal places.)

18.7 $\sqrt[4]{\frac{17}{81}}$.

❚ Let $f(x) = \sqrt[4]{x}$, $x = \frac{16}{81}$, $\Delta x = \frac{1}{81}$. Then $x + \Delta x = \frac{17}{81}$, $\Delta y = \sqrt[4]{\frac{17}{81}} - \sqrt[4]{\frac{16}{81}} = \sqrt[4]{\frac{17}{81}} - \frac{2}{3}$. Also, $f'(x) = \dfrac{1}{4(\sqrt[4]{x})^3} = \dfrac{1}{4(\frac{2}{3})^3} = \dfrac{27}{32}$. So, by the approximation principle, $\sqrt[4]{\frac{17}{81}} - \frac{2}{3} \approx \frac{27}{32} \cdot \frac{1}{81} = \frac{1}{96}$, $\sqrt[4]{\frac{17}{81}} \approx \frac{2}{3} + \frac{1}{96} = \frac{65}{96} \approx 0.677$. (This is correct to three decimal places.)

18.8 $\sqrt[3]{0.065}$.

▮ Let $f(x) = \sqrt[3]{x}$, $x = 0.064$, $\Delta x = 0.001$. Then $x + \Delta x = 0.065$, $\Delta y = \sqrt[3]{0.065} - \sqrt[3]{0.064} = \sqrt[3]{0.065} - 0.4$. Also, $f'(x) = \dfrac{1}{3(\sqrt[3]{x})^2} = \dfrac{1}{3(\sqrt[3]{0.064})^2} = \dfrac{1}{3(0.4)^2} = \dfrac{1}{3(0.16)} = \dfrac{1}{0.48}$. So, by the approximation principle, $\sqrt[3]{0.065} - 0.4 \approx \dfrac{1}{0.48} \cdot (0.001) = \dfrac{1}{480} \approx 0.002$. Hence, $\sqrt[3]{0.065} \approx 0.4 + 0.002 = 0.402$. (This is correct to three decimal places.)

18.9   Measurement of the side of a cubical container yields the result 8.14 cm, with a possible error of at most 0.005 cm. Give an estimate of the possible error in the value $V = (8.14)^3 = 539.35314 \text{ cm}^3$ for the volume of the container.

▮ Let x be 8.14 and let $x + \Delta x$ be the actual length of the side, with $|\Delta x| \le 0.005$. Let $f(x) = x^3$. Then $|\Delta y| = (x + \Delta x)^3 - x^3$ is the error in the measurement of the volume. Now, $f'(x) = 3x^2 = 3(8.14)^2 \approx 3(66.26) = 198.78$. By the approximation principle, $|\Delta y| \approx 198.78 |\Delta x| \le 198.78 \cdot 0.005 \approx 0.994 \text{ cm}^3$.

18.10   It is desired to give a spherical tank of diameter 20 feet (240 inches) a coat of point 0.1 inch thick. Estimate how many gallons of paint will be required, if 1 gallon is about 231 cubic inches.

▮ The radius $r = 120$ in. and $\Delta r = 0.1$. $V = \frac{4}{3}\pi r^3$. So $D_r V = 4\pi r^2 = 4\pi(120)^2$. So, by the approximation principle, the extra volume $\Delta V \approx D_r V \cdot \Delta r = 4\pi(120)^2(0.01) = 4\pi(12)^2 = 576\pi$. So the number of gallons required is $\frac{576}{231}\pi \approx 7.83$.

18.11   A solid steel cylinder has a radius of 2.5 cm and a height of 10 cm. A tight-fitting sleeve is to be made that will extend the radius to 2.6 cm. Find the amount of steel needed for the sleeve.

▮ The volume $V = \pi r^2 h = 10\pi r^2$. Let $r = 2.5$ and $\Delta r = 0.1$. $\Delta V = 10\pi(2.6)^2 - 62.5\pi$. $D_r V = 20\pi r = 20\pi(2.5) = 50\pi$. So, by the approximation principle. $\Delta V \approx 50\pi(0.1) = 5\pi$. (An exact calculation yields $\Delta V = 5.1\pi$.)

18.12   If the side of a cube is measured with an error of at most 3 percent, estimate the percentage error in the volume of the cube.

▮ $V = s^3$. By the approximation principle, $|\Delta V| \approx 3s^2 \cdot |\Delta s|$. So $\dfrac{|\Delta V|}{V} \approx \dfrac{3s^2|\Delta s|}{s^3} = \dfrac{3|\Delta s|}{s} \le 3(0.03) = 0.09$. So, 9 percent is an approximate bound on the percentage error in the volume.

18.13   Assume, contrary to fact, that the earth is a perfect sphere, with a radius of 4000 miles. The volume of ice at the north and south poles is estimated to be about 8,000,000 cubic miles. If this ice were melted and if the resulting water were distributed uniformly over the globe, approximately what would be the depth of the added water at any point on the earth?

▮ $V = \frac{4}{3}\pi r^3$, $D_r V = 4\pi r^2$. By the approximation rule, $\Delta V \approx 4\pi r^2 \cdot \Delta r$. Since $\Delta V = 8,000,000$ and $r = 4000$, we have $8,000,000 \approx 4\pi(4000)^2 \cdot \Delta r$, $\Delta r \approx 1/(8\pi) \approx 0.0398$ mile ≈ 210 feet.

18.14   Let $y = x^{3/2}$. When $x = 4$ and $dx = 2$, find the value of dy.

▮ $dy = D_x y \cdot dx = (\frac{3}{2}x^{1/2})(2) = (\frac{3}{2}\sqrt{4})(2) = 6$. [Here we appeal to the *definition* of dy; there is no approximation involved.]

18.15   Let $y = 2x\sqrt{1 + x^2}$. When $x = 0$ and $dx = 3$, find dy.

▮ $D_x y = 2\left(\sqrt{1 + x^2} + \dfrac{1}{2}\dfrac{x \cdot 2x}{\sqrt{1 + x^2}}\right) = \dfrac{2(1 + 2x^2)}{\sqrt{1 + x^2}}$. Then, $dy = D_x y \cdot dx = \dfrac{2(1 + 2x^2)}{\sqrt{1 + x^2}} \cdot dx = 2 \cdot 3 = 6$.

18.16   Establish the very useful approximation formula $(1 + u)^r \approx 1 + ru$, where r is any rational exponent and $|u|$ is small compared to 1.

▮ Let $f(x) = x^r$, $x = 1$, and $\Delta x = u$. $f'(x) = rx^{r-1}$. Note that $\Delta y = f(1 + u) - f(1) = (1 + u)^r - 1$. By the approximation principle, $(1 + u)^r - 1 \approx (rx^{r-1}) \cdot u = ru$. Thus, $(1 + u)^r \approx 1 + ru$.

18.17   Approximate $\sqrt[3]{63}$.

▌ Let $f(x) = \sqrt[3]{x}$, $x = 64$, $\Delta x = -1$. Then $f'(x) = \dfrac{1}{3(\sqrt[3]{x})^2}$ and $\Delta y = \sqrt[3]{63} - \sqrt[3]{64} = \sqrt[3]{63} - 4$. By the approximation principle, $\sqrt[3]{63} - 4 \approx \dfrac{1}{3(\sqrt[3]{x})^2} \cdot (-1) = \dfrac{1}{3(\sqrt[3]{64})^2} \cdot (-1) = -\dfrac{1}{48}$. So $\sqrt[3]{63} \approx 4 - \frac{1}{48} \approx 3.98$. (This is correct to two decimal places.)

18.18   Approximate $\sqrt[4]{80}$.

▌ Let $f(x) = x^{1/4}$, $x = 81$, $\Delta x = -1$. Then $f'(x) = \dfrac{1}{4(\sqrt[4]{x})^3}$, and $\Delta y = \sqrt[4]{80} - \sqrt[4]{81} = \sqrt[4]{80} - 3$. By the approximation principle, $\sqrt[4]{80} - 3 \approx \dfrac{1}{4(\sqrt[4]{x})^3} \cdot (-1) = -\dfrac{1}{108}$. So $\sqrt[4]{80} \approx 3 - \frac{1}{108} \approx 3 - 0.0093 = 2.9907$. (This is correct to four decimal places.)

18.19   Estimate $\sqrt[3]{7.8}$.

▌ Let $f(x) = \sqrt[3]{x}$, $x = 8$, $\Delta x = -0.2$. Then $f'(x) = \dfrac{1}{3(\sqrt[3]{x})^2}$, and $\Delta y = \sqrt[3]{7.8} - \sqrt[3]{8} = \sqrt[3]{7.8} - 2$. By the approximation principle, $\sqrt[3]{7.8} - 2 \approx \dfrac{1}{3(\sqrt[3]{x})^2} \cdot (-0.2) = \dfrac{1}{12} \cdot (-0.2) = -\dfrac{1}{60}$. So $\sqrt[3]{7.8} \approx 2 - \frac{1}{60} \approx 2 - 0.0167 = 1.9833$. (The correct answer to four decimals places is 1.9832.)

18.20   Estimate $\sqrt{25.4}$.

▌ Let $f(x) = \sqrt{x}$, $x = 25$, $\Delta x = 0.4$. Then $f'(x) = 1/2\sqrt{x}$ and $\Delta y = \sqrt{25.4} - \sqrt{25} = \sqrt{25.4} - 5$. By the approximation principle, $\sqrt{25.4} - 5 \approx (1/2\sqrt{x})(0.4) = \frac{1}{10}(0.4) = 0.04$. So $\sqrt{25.4} \approx 5.04$. (This is correct to two decimal places.)

18.21   Approximate $\sqrt[3]{26.46}$.

▌ Let $f(x) = \sqrt[3]{x}$, $x = 27$, $\Delta x = -0.54$. Then $f'(x) = \dfrac{1}{3(\sqrt[3]{x})^2}$ and $\Delta y = \sqrt[3]{26.46} - \sqrt[3]{27} = \sqrt[3]{26.46} - 3$. By the approximation principle, $\sqrt[3]{26.46} - 3 \approx \dfrac{1}{3(\sqrt[3]{x})^2} \cdot (-0.54) = \dfrac{1}{27} \cdot (-0.54) = -0.02$. So $\sqrt[3]{26.46} \approx 3 - 0.02 = 2.98$. (This is correct to two decimal places.)

18.22   Estimate $(26.5)^{2/3}$.

▌ Let $f(x) = x^{2/3}$, $x = 27$, $x = -0.5$. Then $f'(x) = 2/3\sqrt[3]{x}$ and $\Delta y = (26.5)^{2/3} - (27)^{2/3} = (26.5)^{2/3} - 9$. By the approximation principle, $(26.5)^{2/3} - 9 \approx (2/3\sqrt[3]{x})(-0.5) = \frac{2}{9} \cdot (-0.5) = -\frac{1}{9} \approx -0.1111$. So $(26.5)^{2/3} \approx 9 - 0.1111 = 8.8889$. (The correct answer to four decimal places is 8.8883.)

18.23   Estimate $\sin 60°1'$.

▌ $1' = \frac{1}{60}$ of a degree $= \frac{1}{60}(\pi/180)$ radians $= \pi/10{,}800$ radians. Let $f(x) = \sin x$, $x = \pi/3$, $\Delta x = \pi/10{,}800$. Then $f'(x) = \cos x$, and $\Delta y = \sin 60°1' - \sin 60° = \sin 60°1' - \sqrt{3}/2$. By the approximation principle, $\sin 60°1' - \sqrt{3}/2 \approx \cos x \cdot (\pi/10{,}800) = \frac{1}{2} \cdot (\pi/10{,}800) \approx 0.00015$. So $\sin 60°1' \approx \sqrt{3}/2 + 0.00015 \approx 0.86603 + 0.00015 = 0.86618$. (The correct answer to five decimal places is 0.86617.)

18.24   Find the approximate change in the area of a square of side s caused by increasing the side by 1 percent.

▌ $A = s^2$. Then $\Delta s = 0.01s$. $D_s A = 2s$. By the approximation principle, $\Delta A \approx 2s \cdot (0.01s) = 0.02s^2$.

18.25   Approximate $\cos 59°$.

▌ Let $f(x) = \cos x$, $x = \pi/3$, $\Delta x = -\pi/180$. Then $f'(x) = -\sin x$ and $\Delta y = \cos 59° - \cos 60°$. By the approximation principle, $\cos 59° - \frac{1}{2} \approx -\sin x \cdot (-\pi/180) = (\sqrt{3}/2) \cdot (\pi/180) \approx 0.0151$. So $\cos 59° \approx 0.5151$. (The actual answer is 0.5150 to four decimal places.)

18.26   Estimate $\tan 44°$.

▮ Let $f(x) = \tan x$, $x = 45°$, $\Delta x = -1°$. Then $f'(x) = \sec^2 x$, and $\Delta y = \tan 44° - \tan 45° = \tan 44° - 1$. By the approximation principle, $\tan 44° - 1 \approx \sec^2 x \cdot (-\pi/180) = 2 \cdot (-\pi/180) = -\pi/90 \approx -0.0349$. So, $\tan 44° \approx 1 - 0.0349 = 0.9651$. (The correct answer to four decimal places is 0.9657.)

18.27 A coat of paint of thickness 0.01 inch is applied to the faces of a cube whose edge is 10 inches, thereby producing a slightly larger cube. Estimate the number of cubic inches of paint used.

▮ The volume $V = s^3$, where s is the side. $s = 10$ and $\Delta s = 0.02$. Then $D_s V = 3s^2$. By the approximation principle, $\Delta V \approx 3s^2 \cdot (0.02) = 300 \cdot (0.02) = 6$ in^3.

18.28 Estimate $\cos^6 (\pi/4 + 0.01)$.

▮ Let $f(x) = \cos^6 x$, $x = \pi/4$, $\Delta x = 0.01$. Then $f'(x) = 6(\cos^5 x)(-\sin x)$, and $\Delta y = \cos^6(\pi/4 + 0.01) - \cos^6 (\pi/4) = \cos^6 (\pi/4 + 0.01) - \frac{1}{8}$. By the approximation principle, $\cos^6 (\pi/4 + 0.01) - \frac{1}{8} \approx 6(\cos^5 x)(-\sin x) \cdot (0.01) = -\frac{3}{4}(0.01) = -0.0075$. So $\cos^6(\pi/4 + 0.01) \approx \frac{1}{8} - 0.0075 = 0.1175$.

18.29 Estimate $(730)^{1/2} + (730)^{2/3} - 8$.

▮ Let $f(x) = x^{1/2} + x^{2/3} - 8$. We must find $f(730)$. Note that $f(729) = \sqrt{729} + (\sqrt[3]{729})^2 - 8 = 27 + 81 - 8 = 100$. $\Delta x = 1$. $f'(x) = \dfrac{1}{2\sqrt{x}} + \dfrac{2}{3\sqrt[3]{x}} = \dfrac{1}{2 \cdot 27} + \dfrac{2}{3 \cdot 9} = \dfrac{5}{54}$. $\Delta y = f(730) - f(729) = f(730) - 100$. By the approximation principle, $f(730) - 100 \approx \frac{5}{54} \cdot 1 \approx 0.09$. So $f(730) \approx 100.09$.

18.30 Estimate $\tan 2°$.

▮ Let $f(x) = \tan x$, $x = 0$, $\Delta x = 2° = \pi/90$ radians. Then $f'(x) = \sec^2 x = \sec^2 0 = 1$, $\Delta y = \tan 2° - \tan 0° = \tan 2°$. By the approximation theorem, $\tan 2° \approx 1 \cdot (\pi/90) = 0.0349$.

18.31 For $f(x) = x^2$, compare Δy with its approximation by means of the approximation principle.

▮ $\Delta y = f(x + \Delta x) - f(x) = (x + \Delta x)^2 - x^2 = 2x \, \Delta x + (\Delta x)^2$. By the approximation principle, $\Delta y \approx f'(x) \cdot \Delta x = 2x \, \Delta x$. Thus, the error is $(\Delta x)^2$, which, for small values of Δx, will be very small.

18.32 Estimate $\sin 28°$.

▮ Let $f(x) = \sin x$, $x = 30°$, $\Delta x = -2° = -\pi/90$ radians. Then $f'(x) = \cos x = \sqrt{3}/2 \approx 0.8660$, and $\Delta y = \sin 28° - \sin 30° = \sin 28° - \frac{1}{2}$. By the approximation principle, $\sin 28° - \frac{1}{2} \approx 0.8660(-\pi/90) \approx -0.0302$. So $\sin 28° \approx \frac{1}{2} - 0.0302 = 0.4698$. (The correct figure is 0.4695 to four decimal places.)

18.33 Estimate $1/\sqrt{15}$.

▮ Let $f(x) = 1/\sqrt{x}$, $x = 16$, $\Delta x = -1$. Then $f'(x) = -1/2(\sqrt{x})^3 = -1/128$, and $\Delta y = 1/\sqrt{15} - 1/\sqrt{16} = 1/\sqrt{15} - 1/4$. By the approximation principle, $1/\sqrt{15} - 1/4 \approx (-1/128)(-1) = 1/128$. So $1/\sqrt{15} \approx \frac{1}{4} + \frac{1}{128} \approx 0.2578$. (The correct answer to four decimal places is 0.2582.)

18.34 A cube is to be built so that it holds 125 cm^3. How precisely should the edge be made so that the volume will be correct to within 3 cm^3?

▮ $V = s^3$. Let $s = 5$. $D_s V = 3s^2 = 75$. By the approximation principle, $\Delta V \approx 75 \cdot \Delta s$. We desire $|\Delta V| \le 3$, that is, $75 \cdot |\Delta s| \le 3$, $|\Delta s| \le \frac{1}{25} = 0.04$.

18.35 Show that the relative error in the nth power of a number is about n times the relative error in the number.

▮ Let $f(x) = x^n$. Then $f'(x) = nx^{n-1}$. By the approximation principle, $\Delta y \approx nx^{n-1} \Delta x$. Hence, $\left| \dfrac{\Delta y}{y} \right| \approx \left| \dfrac{nx^{n-1} \Delta x}{x^n} \right| = n \left| \dfrac{\Delta x}{x} \right|$.

CHAPTER 19
Antiderivatives (Indefinite Integrals)

19.1 Evaluate $\int (g(x))^r g'(x) \, dx.$

    ■ By the chain rule, $D_x((g(x))^{r+1}) = (r+1)(g(x))^r \cdot g'(x).$ Hence, $\int (g(x))^r g'(x) \, dx = \dfrac{1}{r+1} (g(x))^{r+1} + C,$ where C is an arbitrary constant.

19.2 Evaluate $\int x^r \, dx$ for $r \neq -1.$

    ■ $\int x^r \, dx = \dfrac{1}{r+1} x^{r+1} + C,$ since $D_x(x^{r+1}) = (r+1)x^r.$

19.3 Find $\int (2x^3 - 5x^2 + 3x + 1) \, dx.$

    ■ $\int (2x^3 - 5x^2 + 3x + 1) \, dx = 2\left(\dfrac{x^4}{4}\right) - 5\left(\dfrac{x^3}{3}\right) + 3\left(\dfrac{x^2}{2}\right) + x + C = \frac{1}{2}x^4 - \frac{5}{3}x^3 + \frac{3}{2}x^2 + x + C.$

19.4 Find $\int \left(5 - \dfrac{1}{\sqrt{x}}\right) dx.$

    ■ $\int \left(5 - \dfrac{1}{\sqrt{x}}\right) dx = \int (5 - x^{-1/2}) \, dx = 5x - 2x^{1/2} + C = 5x - 2\sqrt{x} + C.$

19.5 Find $\int 2\sqrt[4]{x} \, dx.$

    ■ $\int 2\sqrt[4]{x} \, dx = 2 \int x^{1/4} \, dx = 2 \cdot \frac{4}{5}x^{5/4} + C = \frac{8}{5}(\sqrt[4]{x})^5 + C.$

19.6 Evaluate $\int 5\sqrt[3]{x^2} \, dx.$

    ■ $\int 5\sqrt[3]{x^2} \, dx = 5 \int x^{2/3} \, dx = 5 \cdot \frac{3}{5} \cdot x^{5/3} + C = 3x^{5/3} + C.$

19.7 Find $\int \dfrac{3}{x^4} \, dx.$

    ■ $\int \dfrac{3}{x^4} \, dx = 3 \int x^{-4} \, dx = 3\left(\dfrac{1}{-3}\right)x^{-3} + C = -x^{-3} + C = -\dfrac{1}{x^3} + C.$

19.8 Find $\int (x^2 - 1)\sqrt{x} \, dx.$

    ■ $\int (x^2 - 1)\sqrt{x} \, dx = \int (x^2 - 1)x^{1/2} \, dx = \int (x^{5/2} - x^{1/2}) \, dx = \frac{2}{7}x^{7/2} - \frac{2}{3}x^{3/2} + C = \frac{2}{7}(\sqrt{x})^7 - \frac{2}{3}(\sqrt{x})^3 + C$
           $= 2(\sqrt{x})^3(\frac{1}{7}x^2 - \frac{1}{3}) + C.$

19.9 Find $\int \left(\dfrac{1}{x^3} - \dfrac{1}{x^5}\right) dx.$

    ■ $\int \left(\dfrac{1}{x^3} - \dfrac{1}{x^5}\right) dx = \int (x^{-3} - x^{-5}) \, dx = -\frac{1}{2}x^{-2} - (-\frac{1}{4})x^{-4} + C = -\dfrac{1}{2x^2} + \dfrac{1}{4x^4} + C = \dfrac{1}{4x^4}(1 - 2x^2) + C.$

19.10 Find $\int \dfrac{3x^2 - 2x + 1}{\sqrt{x}} \, dx.$

    ■ $\int \dfrac{3x^2 - 2x + 1}{\sqrt{x}} \, dx = \int (3x^{3/2} - 2x^{1/2} + x^{-1/2}) \, dx$

                 $= 3(\frac{2}{5})x^{5/2} - 2(\frac{2}{3})x^{3/2} + 2x^{1/2} + C = \frac{6}{5}x^{5/2} - \frac{4}{3}x^{3/2} + 2x^{1/2} + C$

                 $= \frac{6}{5}(\sqrt{x})^5 - \frac{4}{3}(\sqrt{x})^3 + 2\sqrt{x} + C = \frac{2}{15}\sqrt{x}(9x^2 - 10x + 15) + C.$

19.11 Evaluate $\int (3 \sin x + 5 \cos x)\, dx.$

▮ $\int (3 \sin x + 5 \cos x)\, dx = 3(-\cos x) + 5 \sin x + C = -3 \cos x + 5 \sin x + C.$

19.12 Find $\int (7 \sec^2 x - \sec x \tan x)\, dx.$

▮ $\int (7 \sec^2 x - \sec x \tan x)\, dx = 7 \tan x - \sec x + C.$

19.13 Evaluate $\int (\csc^2 x + 3x^2)\, dx.$

▮ $\int (\csc^2 x + 3x^2)\, dx = -\cot x + x^3 + C.$

19.14 Find $\int x\sqrt{3x}\, dx.$

▮ $\int x\sqrt{3x}\, dx = \int x(\sqrt{3}\sqrt{x})\, dx = \sqrt{3} \int x^{3/2}\, dx = \sqrt{3}(\frac{2}{5})x^{5/2} + C = \dfrac{2\sqrt{3}}{5}\, x^{5/2} + C.$

19.15 Find $\int \dfrac{1}{\sec x}\, dx.$

▮ $\int \dfrac{1}{\sec x}\, dx = \int \cos x\, dx = \sin x + C.$

19.16 Find $\int \tan^2 x\, dx.$

▮ $\int \tan^2 x\, dx = \int (\sec^2 x - 1)\, dx = \tan x - x + C.$

19.17 Evaluate $\int \sqrt{7x + 4}\, dx.$

▮ Use substitution. Let $u = 7x + 4$. Then $du = 7\, dx$, and $\int \sqrt{7x + 4}\, dx = \int \sqrt{u}\,\frac{1}{7}\, du = \frac{1}{7} \int u^{1/2}\, du = \frac{1}{7}(\frac{2}{3})u^{3/2} + C = \frac{2}{21}(7x + 4)^{3/2} + C = \frac{2}{21}(\sqrt{7x + 4})^3 + C.$

19.18 Find $\int \dfrac{1}{\sqrt{x - 1}}\, dx.$

▮ Let $u = x - 1$. Then $du = dx$, and $\int \dfrac{1}{\sqrt{x - 1}}\, dx = \int \dfrac{1}{\sqrt{u}}\, du = \int u^{-1/2}\, du = 2u^{1/2} + C = 2\sqrt{u} + C = 2\sqrt{x - 1} + C.$

19.19 Find $\int (3x - 5)^{12}\, dx.$

▮ Let $u = 3x - 5$, $du = 3\, dx$. Then $\int (3x - 5)^{12}\, dx = \int u^{12}(\frac{1}{3})\, du = \frac{1}{3} \int u^{12}\, du = (\frac{1}{3})(\frac{1}{13})u^{13} + C = \frac{1}{39}(3x - 5)^{13} + C.$

19.20 Evaluate $\int \sin (3x - 1)\, dx.$

▮ Let $u = 3x - 1$, $du = 3\, dx$. Then $\int \sin (3x - 1)\, dx = \int \sin u(\frac{1}{3})\, du = \frac{1}{3} \int \sin u\, du = \frac{1}{3}(-\cos u) + C = -\frac{1}{3} \cos (3x - 1) + C.$

19.21 Find $\int \sec^2 (x/2)\, dx.$

▮ Let $u = x/2$, $du = \frac{1}{2}\, dx$. Then $\int \sec^2 (x/2)\, dx = \int \sec^2 u \cdot 2\, du = 2 \int \sec^2 u\, du = 2 \tan u + C = 2 \tan (x/2) + C.$

19.22 Find $\int \dfrac{\cos \sqrt{x}}{\sqrt{x}}\, dx.$

▮ Let $u = \sqrt{x}$, $du = \frac{1}{2}(1/\sqrt{x})\, dx$. Then $\int \dfrac{\cos \sqrt{x}}{\sqrt{x}}\, dx = \int \cos u \cdot 2\, du = 2 \int \cos u\, du = 2 \sin u + C = 2 \sin (\sqrt{x}) + C.$

19.23 Evaluate $\int (4 - 2t^2)^7 t\, dt.$

▮ Let $u = 4 - 2t^2$, $du = -4t\, dt$. Then $\int (4 - 2t^2)^7 t\, dt = \int u^7(-\frac{1}{4})\, du = -\frac{1}{4} \int u^7\, du = -\frac{1}{4}(\frac{1}{8})u^8 + C = -\frac{1}{32}(4 - 2t^2)^8 + C.$

19.24 Solve Problem 19.23 by using Problem 19.1. *(p 143)*

▌ Notice that $D_t(4-2t^2) = -4t$. Then $\int (4-2t^2)^7 t\, dt = -\frac{1}{4}\int (4-2t^2)^7(-4t)\, dt = -\frac{1}{4}(\frac{1}{8})(4-2t^2)^8 + C = -\frac{1}{32}(4-2t^2)^8 + C$.

19.25 Find $\int x^2\sqrt[3]{x^3+5}\, dx$.

▌ Let $u = x^3+5$, $du = 3x^2\, dx$. Then $\int x^2\sqrt[3]{x^3+5}\, dx = \int \sqrt[3]{u}\cdot\frac{1}{3}\, du = \frac{1}{3}\int u^{1/3}\, d = \frac{1}{3}(\frac{3}{4})u^{4/3} + C = \frac{1}{4}(x^3+5)^{4/3} + C = \frac{1}{4}(\sqrt[3]{x^3+5})^4 + C$.

19.26 Evaluate $\int \dfrac{x}{\sqrt{x+1}}\, dx$.

[handwritten annotation:] $\int x(x+1)^{-1/2}dx$, By Parts: $u=x$, $dv=(x+1)^{-1/2}dx$ | $du=dx$, $v=2(x+1)^{1/2}$ | $uv - \int v\,du = $ | $2x(x+1)^{1/2} - \frac{4}{3}(x+1)^{3/2}$
$= (x+1)^{1/2}[2x - \frac{4}{3}(x+1)]$
$= (x+1)^{1/2}[\frac{2}{3}x - \frac{4}{3}] = (\frac{2}{3})(x+1)^{1/2}(x-2)+C$

▌ Let $u = x+1$, $du = dx$. Then $\int \dfrac{x}{\sqrt{x+1}}\, dx = \int \dfrac{u-1}{\sqrt{u}}\, du = \int (u^{1/2} - u^{-1/2})\, du = \frac{2}{3}u^{3/2} - 2u^{1/2} + C = \frac{2}{3}(\sqrt{u})^3 - 2\sqrt{u} + C = \frac{2}{3}(\sqrt{x+1})^3 - 2\sqrt{x+1} + C = \frac{2}{3}\sqrt{x+1}(x-2) + C$.

19.27 Find $\int \sqrt[3]{x^2-2x+1}\, dx$.

▌ Note that $x^2-2x+1 = (x-1)^2$. So, let $u = x-1$, $du = dx$. Then $\int \sqrt[3]{x^2-2x+1}\, dx = \int \sqrt[3]{u^2}\, du = \int u^{2/3}\, du = \frac{3}{5}u^{5/3} + C = \frac{3}{5}(\sqrt[3]{u})^5 + C = \frac{3}{5}(\sqrt[3]{x-1})^5 + C$.

19.28 Find $\int (x^4+1)^{1/3}x^7\, dx$.

[margin note: CS: SUPPL to ≥2H §10.4 (2)]

▌ Let $u = x^4+1$, $du = 4x^3\, dx$. Note that $x^7\, dx = x^4\cdot x^3\, dx = (u-1)\cdot\frac{1}{4}\, du$. Hence, $\int (x^4+1)^{1/3}x^7\, dx = \int u^{1/3}(u-1)\cdot\frac{1}{4}\, du = \frac{1}{4}\int (u^{4/3} - u^{1/3})\, du = \frac{1}{4}(\frac{3}{7}u^{7/3} - \frac{3}{4}u^{4/3}) + C = \frac{3}{4}u^{4/3}(\frac{1}{7}u - \frac{1}{4}) = \frac{3}{4}(x^4+1)^{4/3}[\frac{1}{7}(x^4+1) - \frac{1}{4}] + C = \frac{3}{112}(x^4+1)^{4/3}(4x^4-3) + C$.

19.29 Evaluate $\int \dfrac{x}{\sqrt{1+5x^2}}\, dx$.

▌ Let $u = 1+5x^2$, $du = 10x\, dx$. Then $\int \dfrac{x}{\sqrt{1+5x^2}}\, dx = \int \dfrac{1}{\sqrt{u}}\cdot\frac{1}{10}\, du = \frac{1}{10}\int u^{-1/2}\, du = \frac{1}{10}\cdot 2u^{1/2} + C = \frac{1}{5}\sqrt{1+5x^2} + C$.

19.30 Find $\int x\sqrt{ax+b}\, dx$, when $a \neq 0$.

▌ Let $u = ax+b$, $du = a\, dx$. Note that $x = (u-b)/a$. Then

$$\int x\sqrt{ax+b}\, dx = \int \frac{u-b}{a}\sqrt{u}\cdot\frac{1}{a}\, du = \frac{1}{a^2}\int (u^{3/2} - bu^{1/2})\, du = \frac{1}{a^2}\left(\frac{2}{5}u^{5/2} - \frac{2b}{3}u^{3/2}\right) + C$$

$$= \frac{2}{a^2}u^{3/2}\left(\frac{1}{5}u - \frac{b}{3}\right) + C = \frac{2}{a^2}(\sqrt{ax+b})^3\left[\frac{1}{5}(ax+b) - \frac{b}{3}\right] + C$$

$$= \frac{2}{a^2}(\sqrt{ax+b})^3\left(\frac{a}{5}x - \frac{2}{15}b\right) + C = \frac{2}{15a^2}(\sqrt{ax+b})^3(3ax-2b) + C$$

[handwritten: 2   +C]

19.31 Evaluate $\int \dfrac{\cos 3x}{\sin^2 3x}\, dx$.

▌ Let $u = \sin 3x$. By the chain rule, $du = 3\cos 3x\, dx$. So,

$$\int \frac{\cos 3x}{\sin^2 3x}\, dx = \int \frac{1}{u^2}\cdot\frac{1}{3}\, du = \frac{1}{3}\int u^{-2}\, du = \frac{1}{3}(-u^{-1}) + C = -\frac{1}{3u} + C = -\frac{1}{3\sin 3x} + C = -\frac{1}{3}\csc 3x + C.$$

19.32 Find $\int \sqrt{1-x}\, x^2\, dx$.

▌ Let $u = 1-x$, $du = -dx$. Note that $x = 1-u$. Then $\int \sqrt{1-x}\, x^2\, dx = \int \sqrt{u}(1-u)^2\cdot(-1)\, du = -\int \sqrt{u}(1-2u+u^2)\, du = -\int (u^{1/2} - 2u^{3/2} + u^{5/2})\, du = -[\frac{2}{3}u^{3/2} - 2(\frac{2}{5})u^{5/2} + \frac{2}{7}u^{7/2}] + C = -2u^{3/2}(\frac{1}{3} - \frac{2}{5}u + \frac{1}{7}u^2) + C = -2(\sqrt{1-x})^3[\frac{1}{3} - \frac{2}{5}(1-x) + \frac{1}{7}(1-x)^2] + C = -\frac{2}{105}(\sqrt{1-x})^3(8+12x+15x^2) + C$.

19.33 Find an equation of the curve passing through the point $(0, 1)$ and having slope $12x + 1$ at any point (x, y).

▮ $D_x y = 12x + 1$. Hence, $y = 6x^2 + x + C$. Since $(0, 1)$ lies on the curve, $1 = C$. Thus, $y = 6x^2 + x + 1$ is the equation of the curve. (*NOTE: Do not use point-slope form of a line, where slope is constant. This is a curve where slope is variable.*)

19.34 A particle moves along the x-axis with acceleration $a = 2t - 3$ ft/s². At time $t = 0$ it is at the origin and moving with a speed of 4 ft/s in the positive direction. Find formulas for its velocity v and position s, and determine where it changes direction and where it is moving to the left.

▮ $D_t v = a = 2t - 3$. So, $v = \int (2t - 3)\, dt = t^2 - 3t + C$. Since $v = 4$ when $t = 0$, $C = 4$. Thus, $v = t^2 - 3t + 4$. Since $v = D_t s$, $s = \int (t^2 - 3t + 4)\, dt = \frac{1}{3}t^3 - \frac{3}{2}t^2 + 4t + C_1$. Since $s = 0$ when $t = 0$, $C_1 = 0$. Thus, $s = \frac{1}{3}t^3 - \frac{3}{2}t^2 + 4t$. Changes of direction occur where s reaches a relative maximum or minimum. To look for critical numbers for s, we set $v = 0$. The quadratic formula shows that $v = 0$ has no real roots. [Alternatively, note that $t^2 - 3t + 4 = (t - \frac{3}{2})^2 + \frac{7}{4} > 0$.] Hence, the particle never changes direction. Since it is moving to the right at $t = 0$, it always moves to the right.

19.35 Rework Problem 19.34 when the acceleration $a = t^2 - \frac{13}{3}$ ft/s².

▮ $v = \int (t^2 - \frac{13}{3})\, dt = \frac{1}{3}t^3 - \frac{13}{3}t + C$. Since $v = 4$ when $t = 0$, $C = 4$. Thus, $v = \frac{1}{3}t^3 - \frac{13}{3}t + 4$. Then $s = \int v\, dt = \frac{1}{3} \cdot \frac{1}{4}t^4 - \frac{13}{3} \cdot \frac{1}{2}t^2 + 4t + C_1 = \frac{1}{12}t^4 - \frac{13}{6}t^2 + 4t + C_1$. Since $s = 0$ when $t = 0$, $C_1 = 0$. Thus, $s = \frac{1}{12}t^4 - \frac{13}{6}t^2 + 4t$. To find critical numbers for s, we set $v = 0$, obtaining $t^3 - 13t + 12 = 0$. Clearly, $t = 1$ is a root. Dividing $t^3 - 13t + 12$ by $t - 1$, we obtain $t^2 + t - 12 = (t + 4)(t - 3)$. Thus, $t = 3$ and $t = -4$ also are critical numbers. When $t = 1$, $D_t^2 s = a = -\frac{10}{3} < 0$. Hence, s has a relative maximum at $t = 1$. Similarly, $a = \frac{14}{3} > 0$ when $t = 3$, and, therefore, s has a relative minimum at $t = 3$; $a = \frac{35}{3} > 0$ when $t = -4$, and, therefore, s has a relative minimum at $t = -4$. Hence, the particle changes direction at $t = -4$, $t = 1$, and $t = 3$. It moves left for $t < -4$, where it reaches a minimum; it moves right from $t = -4$ to $t = 1$, where it reaches a maximum; it moves left from $t = 1$ to $t = 3$, where it reaches a minimum; then it moves right for $t > 3$.

19.36 A motorist applies the brakes on a car moving at 45 miles per hour on a straight road, and the brakes cause a constant deceleration of 22 ft/s². In how many seconds will the car stop, and how many feet will the car have traveled after the time the brakes were applied?

▮ Let $t = 0$ be the time the brakes were applied, let the positive s direction be the direction that the car was traveling, and let the origin $s = 0$ be the point at which the brakes were applied. Then the acceleration $a = -22$. So, $v = \int a\, dt = -22t + C$. The velocity at $t = 0$ was 45 mi/h, which is the same as $\frac{45 \cdot 5280}{3600} = 66$ ft/s. Hence, $C = 66$. Thus, $v = -22t + 66$. Then $s = \int v\, dt = -11t^2 + 66t + C_1$. Since $s = 0$ when $t = 0$, $C_1 = 0$ and $s = -11t^2 + 66t$. The car stops when $v = 0$, that is, when $t = 3$. At $t = 3$, $s = 99$. So, the car stops in 3 seconds and travels 99 feet during that time.

19.37 A particle moving on a straight line has acceleration $a = 5 - 3t$, and its velocity is 7 at time $t = 2$. If $s(t)$ is the distance from the origin, find $s(2) - s(1)$.

▮ $v = \int a\, dt = 5t - \frac{3}{2}t^2 + C$. Since the velocity is 7 when $t = 2$, $7 = 10 - 6 + C$, $C = 3$. So, $v = 5t - \frac{3}{2}t^2 + 3$. Then $s = \frac{5}{2}t^2 - \frac{1}{2}t^3 + 3t + C_1$. Hence, $s(2) = 12 + C_1$, $s(1) = 5 + C_1$, and $s(2) - s(1) = 7$.

19.38 Find an equation of the curve passing through the point $(3, 2)$ and having slope $2x^2 - 5$ at any point (x, y).

▮ $D_x y = 2x^2 - 5$. Hence, $y = \frac{2}{3}x^3 - 5x + C$. Since $(3, 2)$ is on the curve, $2 = \frac{2}{3}(3)^3 - 5(3) + C$, $C = -1$. Thus, $y = \frac{2}{3}x^3 - 5x - 1$ is an equation of the curve.

19.39 A particle is moving along a line with acceleration $a = \sin 2t + t^2$ ft/s². At time $t = 0$, its velocity is 3 ft/s. What is the distance between the particle's location at time $t = 0$ and its location at time $t = \pi/2$, and what is the particle's speed at time $t = \pi/2$?

▮ $v = \int a\, dt = -\frac{1}{2}\cos 2t + \frac{1}{3}t^3 + C$. Since $v = 3$ when $t = 0$, $C = 3$. Thus, $v = -\frac{1}{2}\cos 2t + \frac{1}{3}t^3 + 3$. Hence, at time $t = \pi/2$, $v = -\frac{1}{2}\cos \pi + \frac{1}{3}(\pi/2)^3 + 3 = \frac{7}{2} + \pi^3/24$. Its position $s = \int v\, dt = -\frac{1}{4}\sin 2t + \frac{1}{3} \cdot \frac{1}{4}t^4 + 3t + C_1 = -\frac{1}{4}\sin 2t + \frac{1}{12}t^4 + 3t + C_1$. Then $s(0) = C_1$ and $s(\pi/2) = -\frac{1}{4}\sin \pi + \frac{1}{12}(\pi/2)^4 + 3(\pi/2) + C_1$. Hence, $s(\pi/2) - s(0) = (\pi^4 + 288\pi)/192$.

19.40 Find $\int \sin^4 x \cos x \, dx$.

▎ Let $u = \sin x$, $du = \cos x \, dx$. Then $\int \sin^4 x \cos x \, dx = \int u^4 \, du = \frac{1}{5}u^5 + C = \frac{1}{5} \sin^5 x + C$.

19.41 Suppose that a particle moves along the x-axis and its velocity at time t is given by $v = t^2 - t - 2$ for $1 \le t \le 4$. Find the total distance traveled in the period from $t = 1$ to $t = 4$.

▎ $v = (t - 2)(t + 1)$. Hence, $v = 0$ when $t = 2$ or $t = -1$. Since $a = D_t v = 2t - 1$ is equal to 3 when $t = 2$, the position s of the particle is a relative minimum when $t = 2$. So, the particle moves to the left from $t = 1$ to $t = 2$, and to the right from $t = 2$ to $t = 4$. Now, $s = \int v \, dt = \frac{1}{3}t^3 - \frac{1}{2}t^2 - 2t + C$. By direct computation, $s(1) = -\frac{13}{6} + C$, $s(2) = -\frac{10}{3} + C$, and $s(4) = \frac{16}{3} + C$. Hence, the distance traveled from $t = 1$ to $t = 2$ is $|s(1) - s(2)| = \frac{7}{6}$ and the distance traveled from $t = 2$ to $t = 4$ is $|s(2) - s(4)| = \frac{26}{3}$. Thus, the total distance traveled is $\frac{59}{6}$.

19.42 Evaluate $\int \frac{x^3 - 1}{x^3} \, dx$.

▎
$$\int \frac{x^3 - 1}{x^3} \, dx = \int (1 - x^{-3}) \, dx = x - (-\tfrac{1}{2})x^{-2} + C = x + \frac{1}{2x^2} + C$$

19.43 Find $\int \csc^6 x \cot x \, dx$.

▎ Let $u = \csc x$. Then $du = -\csc x \cot x \, dx$. Hence, $\int \csc^6 x \cot x \, dx = -\int u^5 \, du = -\frac{1}{6}u^6 + C = -\frac{1}{6} \csc^6 x + C$.

19.44 A particle moving along a straight line is accelerating at the rate of 3 ft/s². Find the initial velocity if the distance traveled during the first 2 seconds is 10 feet.

▎ $v = \int 3 \, dt = 3t + C$. C is the initial velocity v_0 when $t = 0$. Thus, $v = 3t + v_0$. The position $s = \int v \, dt = \frac{3}{2}t^2 + v_0 t + C_1$. Hence, $s(2) = \frac{3}{2} \cdot 4 + 2v_0 + C_1 = 6 + 2v_0 + C_1$. On the other hand, $s(0) = C_1$. So, $10 = s(2) - s(0) = 6 + 2v_0$, $v_0 = 2$ ft/s.

19.45 Evaluate $\int \sin 3x \cos 3x \, dx$.

▎ Let $u = \sin 3x$, $du = 3 \cos 3x \, dx$. Then $\int \sin 3x \cos 3x \, dx = \frac{1}{3} \int u \, du = \frac{1}{3} \cdot \frac{1}{2}u^2 + C = \frac{1}{6} \sin^2 3x + C$.

19.46 Find $\int (x^4 - 4x^3)^3 (x^3 - 3x^2) \, dx$.

▎ Note that, if we let $g(x) = x^4 - 4x^3$, then $g'(x) = 4x^3 - 12x^2 = 4(x^3 - 3x^2)$. Thus, our integral has the form $\int (g(x))^3 \cdot \frac{1}{4}g'(x) \, dx = \frac{1}{4} \int (g(x))^3 g'(x) \, dx = \frac{1}{4} \cdot \frac{1}{4} \cdot (g(x))^4 + C = \frac{1}{16}(x^4 - 4x^3)^4 + C$, using the result of Problem 19.1.

19.47 Evaluate $\int \sec^2 4x \tan 4x \, dx$.

▎ Let $u = \tan 4x$. Then $du = \sec^2 4x \cdot 4 \, dx$. So, $\int \sec^2 4x \tan 4x \, dx = \frac{1}{4} \int u \, du = \frac{1}{4} \cdot \frac{1}{2}u^2 + C = \frac{1}{8} \tan^2 4x + C$.

19.48 Evaluate $\int \cos x \sin x \sqrt{1 + \sin^2 x} \, dx$.

▎ Let $u = 1 + \sin^2 x$. Then $du = 2 \sin x \cos x \, dx$. So, $\int \cos x \sin x \sqrt{1 + \sin^2 x} \, dx = \frac{1}{2} \int \sqrt{u} \, du = \frac{1}{2} \cdot \frac{2}{3}u^{3/2} + C = \frac{1}{3}(1 + \sin^2 x)^{3/2} + C$.

19.49 Find $\int \left(\frac{x^2 + 1}{x} \right)^2 \, dx$.

▎ $\int \left(\frac{x^2 + 1}{x} \right)^2 \, dx = \int \frac{x^4 + 2x^2 + 1}{x^2} \, dx = \int (x^2 + 2 + x^{-2}) \, dx = \frac{1}{3}x^3 + 2x - x^{-1} + C = \frac{1}{3}x^3 + 2x - \frac{1}{x} + C$.

19.50 Find $\int (t + 1)(t - 1) \, dt$.

▎ $\int (t + 1)(t - 1) \, dt = \int (t^2 - 1) \, dt = \frac{1}{3}t^3 - t + C$.

19.51 Find $\int (\sqrt{x} + 1)^2 \, dx$.

▎ $\int (\sqrt{x} + 1)^2 \, dx = \int (x + 2\sqrt{x} + 1) \, dx = \int (x + 2x^{1/2} + 1) \, dx = \frac{1}{2}x^2 + 2(\frac{2}{3})x^{3/2} + x + C = \frac{1}{2}x^2 + \frac{4}{3}x^{3/2} + x + C$.

19.52 Evaluate $\int \left(\sqrt[3]{t} + \dfrac{1}{\sqrt[3]{t}} \right) dt$.

▮ $\int \left(\sqrt[3]{t} + \dfrac{1}{\sqrt[3]{t}} \right) dt = \int (t^{1/3} + t^{-1/3}) \, dt = \frac{3}{4} t^{4/3} + \frac{3}{2} t^{2/3} + C$.

19.53 Evaluate $\int \sec^2 x \sqrt{\tan^3 x} \, dx$.

▮ Let $u = \tan x$, $du = \sec^2 x \, dx$. Then $\int \sec^2 x \sqrt{\tan^3 x} \, dx = \int u^{3/2} \, du = \frac{2}{5} u^{5/2} + C = \frac{2}{5} (\sqrt{\tan^5 x}) + C$.

19.54 Find $\int \dfrac{x^3}{\sqrt{x^2 + 25}} \, dx$.

▮ Let $u = x^2 + 25$, $du = 2x \, dx$. Note that $x^2 = u - 25$. Then $\int \dfrac{x^3}{\sqrt{x^2 + 25}} \, dx = \int \dfrac{u - 25}{\sqrt{u}} \cdot \frac{1}{2} \, du = \frac{1}{2} \int (u^{1/2} - 25u^{-1/2}) \, du = \frac{1}{2} (\frac{2}{3} u^{3/2} - 50u^{1/2}) + C = \frac{1}{3} u^{1/2} (u - 75) + C = \frac{1}{3} \sqrt{x^2 + 25} (x^2 - 50) + C$.

19.55 Find $\int \tan^2 \theta \sec^4 \theta \, d\theta$.

▮ Let $u = \tan \theta$, $du = \sec^2 \theta \, d\theta$. Note that $\sec^2 \theta = 1 + u^2$. Then $\int \tan^2 \theta \sec^4 \theta \, d\theta = \int u^2 (1 + u^2) du = \int (u^2 + u^4) \, du = \frac{1}{3} u^3 + \frac{1}{5} u^5 + C = \frac{1}{3} \tan^3 \theta + \frac{1}{5} \tan^5 \theta + C$.

19.56 Find $\int \cos^3 5x \sin^2 5x \, dx$.

▮ Let $u = \sin 5x$, $du = 5 \cos 5x \, dx$. Note that $\cos^2 5x = 1 - u^2$. So, $\int \cos^3 5x \sin^2 5x \, dx = \int (1 - u^2) u^2 \cdot \frac{1}{5} du = \frac{1}{5} \int (u^2 - u^4) \, du = \frac{1}{5} (\frac{1}{3} u^3 - \frac{1}{5} u^5) + C = \frac{1}{15} \sin^3 5x - \frac{1}{25} \sin^5 5x + C$.

19.57 A particle moves on a straight line with velocity $v = (4 - 2t)^3$ at time t. Find the distance traveled from $t = 0$ to $t = 3$.

▮ The position $s = \int v \, dt = \int (4 - 2t)^3 \, dt$. Let $u = 4 - 2t$, $du = -2 \, dt$. Then $s = \int (4 - 2t)^3 \, dt = \int u^3 (-\frac{1}{2}) \, du = -\frac{1}{2} \int u^3 \, du = -\frac{1}{2} \cdot \frac{1}{4} u^4 + C = -\frac{1}{8} (4 - 2t)^4 + C$. Notice that $v = 0$ when $t = 2$; at that time, the particle changes direction. Now $s(0) = -32 + C$, $s(2) = C$, $s(3) = -2 + C$. Hence, the distance traveled is $|s(2) - s(0)| + |s(3) - s(2)| = 32 + 2 = 34$.

19.58 Find $\int (2x^3 + x)(x^4 + x^2 + 1)^{49} \, dx$.

▮ Let $u = x^4 + x^2 + 1$, $du = (4x^3 + 2x) \, dx = 2(2x^3 + x) \, dx$. Then $\int (2x^3 + x)(x^4 + x^2 + 1)^{49} \, dx = \int u^{49} \cdot \frac{1}{2} \, du = (\frac{1}{2})(\frac{1}{50}) u^{50} + C = \frac{1}{100} (x^4 + x^2 + 1)^{50} + C$.

19.59 A space ship is moving in a straight line at 36,000 miles per second. Suddenly it accelerates at $a = 18t$ mi/s^2. Assume that the speed of light is 180,000 mi/s and that relativistic effects do not influence the velocity of the space ship. How long does it take the ship to reach the speed of light and how far does it travel during that time?

▮ $v = \int a \, dt = \int 18t \, dt = 9t^2 + C$. Let $t = 0$ be the time at which the ship begins to accelerate. Then $v = 36{,}000$ when $t = 0$, and, therefore, $C = 36{,}000$. Setting $9t^2 + 36{,}000 = 180{,}000$, we find that $t^2 = 16{,}000$, $t = 40\sqrt{10}$. The position $s = \int v \, dt = 3t^3 + 36{,}000t + s_0$, where s_0 is the position at time $t = 0$. The position at time $t = 40\sqrt{10}$ is $3t(t^2 + 12{,}000) + s_0 = 120\sqrt{10}(28{,}000) + s_0$. Hence, the distance traveled is $3{,}360{,}000 \sqrt{10}$ miles $\approx 10{,}625{,}253$ miles.

19.60 Evaluate $\int \sec^5 x \tan x \, dx$.

▮ Let $u = \sec x$, $du = \sec x \tan x \, dx$. Then $\int \sec^5 x \tan x \, dx = \int u^4 \, du = \frac{1}{5} u^5 + C = \frac{1}{5} \sec^5 x + C$.

19.61 Find $\int (3x^2 + 2)^{1/4} x \, dx$.

▮ Let $u = 3x^2 + 2$, $du = 6x \, dx$. Then $\int (3x^2 + 2)^{1/4} x \, dx = \frac{1}{6} \int u^{1/4} \, du = \frac{1}{6} (\frac{4}{5}) u^{5/4} + C = \frac{2}{15} (3x^2 + 2)^{5/4} + C$.

19.62 Find $\int (2 - x^3)^2 x \, dx$.

▮ $\int (2 - x^3)^2 x \, dx = \int (4 - 4x^3 + x^6) x \, dx = \int (4x - 4x^4 + x^7) \, dx = 2x^2 - \frac{4}{5} x^5 + \frac{1}{8} x^8 + C$.

19.63 Find $\int (2 - x^3)^2 x^2 \, dx$.

∎ Let $u = 2 - x^3$, $du = -3x^2 \, dx$. Then $\int (2 - x^3)^2 x^2 \, dx = -\frac{1}{3} \int u^2 \, du = -\frac{1}{3} \cdot \frac{1}{3} u^3 + C = -\frac{1}{9}(2 - x^3)^3 + C$.

19.64 Evaluate $\int \dfrac{4t \, dt}{\sqrt[5]{t^2 + 3}}$.

∎ Let $u = t^2 + 3$, $du = 2t \, dt$. Then $\int \dfrac{4t \, dt}{\sqrt[5]{t^2 + 3}} = 2 \int u^{-1/5} \, du = 2(\frac{5}{4}) u^{4/5} + C = \frac{5}{2}(t^2 + 3)^{4/5} + C$.

19.65 Find $\int \dfrac{(x - 2)(x + 3)}{\sqrt{x}} \, dx$.

∎ $\int \dfrac{(x - 2)(x + 3)}{\sqrt{x}} \, dx = \int \dfrac{x^2 + x - 6}{\sqrt{x}} \, dx = \int (x^{3/2} + x^{1/2} - 6x^{-1/2}) \, dx = \frac{2}{5}x^{5/2} + \frac{2}{3}x^{3/2} - 12x^{1/2} + C$

$= \frac{2}{15}x^{1/2}(3x^2 + 5x - 90) + C$.

19.66 Find the equation of the curve passing through $(1, 5)$ and whose tangent line at (x, y) has slope $4x$.

∎ $y' = 4x$. Hence, $y = \int 4x \, dx = 2x^2 + C$. Substituting $(1, 5)$ for (x, y), we have $5 = 2(1)^2 + C$, $C = 3$. Hence, $y = 2x^2 + 3$.

19.67 Find the equation of the curve passing through $(9, 18)$ and whose tangent line at (x, y) has slope $\sqrt{x}$.

∎ $y' = \sqrt{x}$. So, $y = \int \sqrt{x} \, dx = \frac{2}{3}x^{3/2} + C$. Substituting $(9, 18)$ for (x, y), we obtain $18 = \frac{2}{3}(\sqrt{9})^3 + C$, $C = 0$. Hence, $y = \frac{2}{3}x^{3/2}$.

19.68 Find the equation of the curve passing through $(4, 2)$ and whose tangent line at (x, y) has slope x/y.

∎ $dy/dx = x/y$. $\int y \, dy = \int x \, dx$, $\frac{1}{2}y^2 = \frac{1}{2}x^2 + C$, $y^2 = x^2 + C_1$. Substituting $(4, 2)$ for (x, y), we have $4 = 16 + C_1$, $C_1 = -12$, $y^2 = x^2 - 12$, or $x^2 - y^2 = 12$ (a hyperbola).

19.69 Find the equation of the curve passing through $(3, 2)$ and whose tangent line at (x, y) has slope x^2/y^3.

∎ $dy/dx = x^2/y^3$. $\int y^3 \, dy = \int x^2 \, dx$, $\frac{1}{4}y^4 = \frac{1}{3}x^3 + C$. Substituting $(3, 2)$ for (x, y), we obtain $4 = 9 + C$, $C = -5$, $\frac{1}{4}y^4 = \frac{1}{3}x^3 - 5$, $y^4 = \frac{4}{3}x^3 - 20$.

19.70 Find the equation of a curve such that y'' is always 2 and, at the point $(2, 6)$, the slope of the tangent line is 10.

∎ $y' = \int y'' \, dx = \int 2 \, dx = 2x + C$. When $x = 2$, $y' = 10$; so, $10 = 4 + C$, $C = 6$, $y' = 2x + 6$. Hence, $y = \int y' \, dx = x^2 + 6x + C_1$. When $x = 2$, $y = 6$; thus, $6 = 16 + C_1$, $C_1 = -10$, $y = x^2 + 6x - 10$.

19.71 Find the equation of a curve such that $y'' = 6x - 8$ and, at the point $(1, 0)$, $y' = 4$.

∎ $y' = \int y'' \, dx = \int (6x - 8) \, dx = 3x^2 - 8x + C$. When $x = 1$, $y' = 4$; so, $4 = -5 + C$, $C = 9$. Thus, $y' = 3x^2 - 8x + 9$. Then $y = \int y' \, dx = x^3 - 4x^2 + 9x + C_1$. When $x = 1$, $y = 0$. Hence, $0 = 6 + C_1$, $C_1 = -6$. So, $y = x^3 - 4x^2 + 9x - 6$.

19.72 A car is slowing down at the rate of $0.8 \, \text{ft/s}^2$. How far will the car move before it stops if its speed was initially $15 \, \text{mi/h}$?

∎ $a = -0.8$. Hence, $v = \int a \, dt = -0.8t + C$. When $t = 0$, $v = 22 \, \text{ft/s}$ [$15 \, \text{mi/h} = (15 \cdot 5280)/3600 \, \text{ft/s} = 22 \, \text{ft/s}$.] So, $v = -0.8t + 22$. The car stops when $v = 0$, that is, when $t = 27.5$. The position $s = \int v \, dt = -0.4t^2 + 22t + s_0$, where s_0 is the initial position. Hence, the distance traveled in 27.5 seconds is $-0.4(27.5)^2 + 22(27.5) = 302.5 \, \text{ft}$.

19.73 A block of ice slides down a 60-meter chute with an acceleration of $4 \, \text{m/s}^2$. What was the initial velocity of the block if it reaches the bottom in 5 s?

∎ $v = \int a \, dt = \int 4 \, dt = 4t + v_0$, where v_0 is the initial velocity. Then the position $s = \int v \, dt = 2t^2 + v_0 t$. [We let $s = 0$ at the top of the chute.] Then $60 = 2(25) + 5v_0$, $v_0 = 2 \, \text{m/s}$.

19.74 If a particle starts from rest, what constant acceleration is required to move the particle 50 mm in 5 s along a straight line?

▌ $v = \int a\, dt = at + v_0$. Since the particle starts from rest, $v_0 = 0$, $v = at$. Then $s = \int v\, dt = \frac{1}{2}at^2$. [We set $s = 0$ at $t = 0$.] So, $50 = \frac{1}{2}a(25)$, $a = 4\,\text{m/s}^2$.

19.75 What constant deceleration is needed to slow a particle from a velocity of 45 ft/s to a dead stop in 15 ft?

▌ $v = \int a\, dt = at + v_0$. Since $v_0 = 45$, $v = at + 45$. When the particle stops, $v = 0$, that is, $t = -45/a$. Now, $s = \int v\, dt = \frac{1}{2}at^2 + 45t$. [We let $s = 0$ at $t = 0$.] Then $15 = \frac{1}{2}a\left(-\dfrac{45}{a}\right)^2 + 45\left(-\dfrac{45}{a}\right) = \dfrac{(45)^2}{2a} - \dfrac{(45)^2}{a} = -\dfrac{(45)^2}{2a}$. Hence, $a = -\frac{135}{2} = -67.5\,\text{ft/s}^2$.

19.76 A ball is rolled in a straight line over a level lawn, with an initial velocity of 10 ft/s. If, because of friction, the velocity decreases at the rate of 4 ft/s², how far will the ball roll?

▌ The deceleration $a = -4$. $v = \int a\, dt = -4t + v_0$. When $t = 0$, $v = 10$; hence, $v_0 = 10$, $v = -4t + 10$. Then the position $s = \int v\, dt = -2t^2 + 10t$ [We assume $s = 0$ when $t = 0$.] The ball stops when $v = 0$, that is, when $t = 2.5$. Hence, the distance rolled is $-2(2.5)^2 + 10(2.5) = 12.5\,\text{ft}$.

19.77 Find the equation of the family of curves whose tangent line at any point (x, y) has slope equal to $-3x^2$.

▌ $y' = -3x^2$. $y = \int y'\, dx = -x^3 + C$. Thus, the family is a family of cubic curves $y = -x^3 + C$.

19.78 Find $\displaystyle\int \left(\dfrac{\sec x}{1 + \tan x}\right)^2 dx$.

▌ Let $u = 1 + \tan x$, $du = \sec^2 x\, dx$. Then

$$\int \left(\dfrac{\sec x}{1 + \tan x}\right)^2 dx = \int u^{-2}\, du = -u^{-1} + C = -\dfrac{1}{1 + \tan x} + C$$

19.79 Evaluate $\int x^2 \csc^2 x^3\, dx$.

▌ Let $u = \cot x^3$, $du = (-\csc^2 x^3)(3x^2\, dx)$. Then $\int x^2 \csc^2 x^3\, dx = -\frac{1}{3}\int 1\, du = -\frac{1}{3}u + C = -\frac{1}{3}\cot x^3 + C$.

19.80 Find $\int (\tan \theta + \cot \theta)^2\, d\theta$.

▌ $\int (\tan \theta + \cot \theta)^2\, d\theta = \int (\tan^2 \theta + 2 + \cot^2 \theta)\, d\theta = \int (\sec^2 \theta - 1 + 2 + \csc^2 \theta - 1)\, d\theta = \int (\sec^2 \theta + \csc^2 \theta)\, d\theta$
$$= \tan \theta - \cot \theta + C.$$

19.81 Evaluate $\displaystyle\int \dfrac{dx}{1 + \cos x}$.

▌ $\dfrac{1}{1 + \cos x} = \dfrac{1}{1 + \cos x} \cdot \dfrac{1 - \cos x}{1 - \cos x} = \dfrac{1 - \cos x}{1 - \cos^2 x} = \dfrac{1 - \cos x}{\sin^2 x} = \csc^2 x - \cot x \csc x$

Hence, $\displaystyle\int \dfrac{dx}{1 + \cos x} = \int (\csc^2 x - \cot x \csc x)\, dx = -\cot x + \csc x + C$

19.82 Evaluate $\displaystyle\int \dfrac{dx}{1 + \sin x}$.

▌ $\dfrac{1}{1 + \sin x} = \dfrac{1}{1 + \sin x} \cdot \dfrac{1 - \sin x}{1 - \sin x} = \dfrac{1 - \sin x}{1 - \sin^2 x} = \dfrac{1 - \sin x}{\cos^2 x} = \sec^2 x - \tan x \sec x$. Hence, $\displaystyle\int \dfrac{dx}{1 + \sin x} = \int (\sec^2 x - \tan x \sec x)\, dx = \tan x - \sec x + C$ [Or: substitute $x = \pi/2 - v$ and use Problem 19.81.]

19.83 Find the family of curves for which the slope of the tangent line at (x, y) is $(1 + x)/(1 - y)$.

▌ $dy/dx = (1 + x)/(1 - y)$. Then $\int (1 - y)\, dy = \int (1 + x)\, dx$, $y - \frac{1}{2}y^2 = x + \frac{1}{2}x^2 + C$, $x^2 + y^2 + 2x - 2y + C = 0$, $(x + 1)^2 + (y - 1)^2 = C_1$. With $C_1 > 0$, this is the family of circles with center at $(-1, 1)$.

19.84 Find the family of curves for which the slope of the tangent line at (x, y) is $(1 + x)/(1 + y)$.

▮ $dy/dx = (1 + x)/(1 + y)$. Then $\int (1 + y)\, dy = \int (1 + x)\, dx$, $y + \frac{1}{2}y^2 = x + \frac{1}{2}x^2 + C$, $x^2 - y^2 + 2x - 2y + C = 0$, $(x + 1)^2 - (y + 1)^2 + C_1 = 0$. This consists of two families of hyperbolas with center at $(-1, -1)$.

19.85 Find the family of curves for which the slope of the tangent line at (x, y) is $x\sqrt{y}$.

▮ $dy/dx = x\sqrt{y}$. Then $\int y^{-1/2}\, dy = \int x\, dx$, $2y^{1/2} = \frac{1}{2}x^2 + C$, $y^{1/2} = \frac{1}{4}x^2 + C_1$.

19.86 If $y'' = 24/x^3$ at all points of a curve, and, at the point $(1, 0)$, the tangent line is $12x + y = 12$, find the equation of the curve.

▮ $y'' = 24x^{-3}$. $y' = \int y''\, dx = -12x^{-2} + C$. Since the slope of $12x + y = 12$ is -12, when $x = 1$, $y' = -12$. Thus, $C = 0$, $y' = -12x^{-2}$. Then $y = \int y'\, dx = 12x^{-1} + C_1$. Since the curve passes through $(1, 0)$, $C_1 = -12$. So, the equation of the curve is $y = 12\left(\dfrac{1}{x} - 1\right)$, or $xy = 12(1 - x)$.

19.87 A rocket is shot from the top of a tower at an angle of 45° above the horizontal (Fig. 19-1). It hits the ground in 5 seconds at a horizontal distance from the foot of the tower equal to three times the height of the tower. Find the height of the tower.

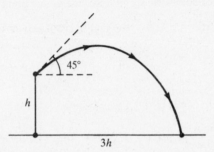

Fig. 19-1

▮ Let h be the height, in feet, of the tower. The height of the rocket $s = h + v_0 t - 16t^2$, where v_0 is the vertical component of the initial velocity. When $t = 5$, $s = 0$. So, $0 = h + 5v_0 - 400$, $v_0 = (400 - h)/5$. Since the angle of projection is 45°, the horizontal component of the velocity has the initial value v_0, and this value is maintained. Hence, the horizontal distance covered in 5 seconds is $5v_0$. Thus, $5[(400 - h)/5] = 3h$, $h = 100$ ft.

In Problems 19.88–19.94, find the general solution of the indicated differential equation.

19.88 $\dfrac{dy}{dx} = 6x^2 + 4x - 5$.

▮ $y = \int (6x^2 + 4x - 5)\, dx = 2x^3 + 2x^2 - 5x + C$.

19.89 $\dfrac{dy}{dx} = (3x + 1)^3$.

▮ $y = \int (3x + 1)^3\, dx$. Let $u = 3x + 1$, $du = 3\, dx$. Then $y = \frac{1}{3} \int u^3\, du = \frac{1}{3} \cdot \frac{1}{4} \cdot u^4 + C = \frac{1}{12}(3x + 1)^4 + C$.

19.90 $\dfrac{dy}{dx} = 24x^3 + 18x^2 - 8x + 3$.

▮ $y = \int (24x^3 + 18x^2 - 8x + 3)\, dx = 6x^4 + 6x^3 - 4x^2 + 3x + C$.

19.91 $\dfrac{dy}{dx} = 2\sqrt{y}$.

▮ $\int \dfrac{dy}{\sqrt{y}} = \int 2\, dx$, $\int y^{-1/2}\, dy = 2x + C$, $2y^{1/2} = 2x + C$, $x = \sqrt{y} + C_1$.

19.92 $\dfrac{dy}{dx} = \dfrac{x + \sqrt{x}}{y - \sqrt{y}}$.

▌ $\int (y - \sqrt{y})\, dy = \int (x + \sqrt{x})\, dx$, $\quad \frac{1}{2}y^2 - \frac{2}{3}y^{3/2} = \frac{1}{2}x^2 + \frac{2}{3}x^{3/2} + C$, $\quad$ or $\quad 3(y^2 - x^2) - 4(y^{3/2} + x^{3/2}) = C_1$.

19.93 $\dfrac{dy}{dx} = \left(\dfrac{y}{x}\right)^{1/3}$.

▌ $\int y^{-1/3}\, dy = \int x^{-1/3}\, dx$, $\quad \frac{3}{2}y^{2/3} = \frac{3}{2}x^{2/3} + C$, $\quad y^{2/3} - x^{2/3} = C_1$.

19.94 $\dfrac{dy}{dx} = \dfrac{1}{x^2} + x$.

▌ $y = \int \left(\dfrac{1}{x^2} + x\right) dx = \int (x^{-2} + x)\, dx = -x^{-1} + \frac{1}{2}x^2 + C = \frac{1}{2}x^2 - \dfrac{1}{x} + C$.

19.95 Find the escape velocity for an object shot vertically upward from the surface of a sphere of radius R and mass M. [Assume the inverse square law for gravitational attraction $F = -G(m_1 m_2/s^2)$, where G is a positive constant (dependent on the units used for force, mass, and distance), and m_1 and m_2 are two masses at a distance s. Assume also Newton's law $F = ma$.]

▌ Let m be the mass of the object. $ma = -G(mM/s^2)$, where s is the distance of the object from the center of the sphere. Hence, $a = -GM/s^2$. Now, $a = \dfrac{dv}{dt} = \dfrac{dv}{ds} \cdot \dfrac{ds}{dt} = \dfrac{dv}{ds} v$. Thus, $v \dfrac{dv}{ds} = -GM/s^2$, $\int v\, dv = -\int (GM/s^2)\, ds$, $\quad \frac{1}{2}v^2 = GM/s + C$. When $s = R$, $v = v_0$, the initial velocity. Hence, $C = \frac{1}{2}v_0^2 - GM/R$. Thus, $\frac{1}{2}v^2 = GM/s + \frac{1}{2}v_0^2 - GM/R$. In order for the object never to return to the surface of the sphere, v must never be 0. Since GM/s approaches 0 as $s \to +\infty$, we must have $\frac{1}{2}v_0^2 - GM/R \geq 0$, or $v_0 \geq \sqrt{2GM/R}$. Thus, $\sqrt{2GM/R}$ is the escape velocity.

19.96 Find the escape velocity for an object shot vertically upward from the surface of the Earth. (Let $-g$ be the acceleration due to the Earth's gravity at the surface of the Earth; $g \approx 32\ \text{ft/s}^2$.)

▌ By Problem 19.95, $-g = a = -GM/R^2$, $g = GM/R^2$, $GM = gR^2$. Hence, $2GM/R = 2gR$. Therefore, the escape velocity is $\sqrt{2gR}$. Now, $g \approx 32\ \text{ft/s}^2 = \frac{32}{5280}\ \text{mi/s}^2$. If we approximate the radius of the Earth by 4000 miles, then the escape velocity is about $\sqrt{2 \cdot \frac{32}{5280} \cdot 4000} \approx 7\ \text{mi/s}$.

19.97 The equation $xy = c$ represents the family of all equilateral hyperbolas with center at the origin. Find the equation of the family of curves that intersect the curves of the given family at right angles.

▌ For the given equation, $xy' + y = 0$, $y' = -y/x$. Hence, for the orthogonal family, $dy/dx = x/y$, $\int y\, dy = \int x\, dx$, $\frac{1}{2}y^2 = \frac{1}{2}x^2 + C$, $y^2 - x^2 = C_1$. This is a family of hyperbolas with axes of symmetry on the x- or y-axis.

19.98 Compare the values of $\int 2 \cos x \sin x\, dx$ obtained by the substitutions (a) $u = \sin x$ and (b) $u = \cos x$, and reconcile the results.

▌ (a) Let $u = \sin x$, $du = \cos x\, dx$. Then $\int 2 \sin x \cos x\, dx = 2 \int u\, du = 2 \cdot \frac{1}{2}u^2 + C = \sin^2 x + C$. (b) Let $u = \cos x$, $du = -\sin x\, dx$. Then $\int 2 \sin x \cos x\, dx = -2 \int u\, du = -2 \cdot \frac{1}{2}u^2 + C = -\cos^2 x + C$. There is no contradiction between the results of (a) and (b). $\sin^2 x$ and $-\cos^2 x$ differ by a constant, since $\sin^2 x + \cos^2 x = 1$. So, it is not surprising that they have the same derivative.

19.99 Compute $\int \cos^2 x\, dx$.

▌ Remember the trigonometric identity $\cos^2 x = (1 + \cos 2x)/2$. Hence, $\int \cos^2 x\, dx = \frac{1}{2} \int (1 + \cos 2x)\, dx = \frac{1}{2}(x + \frac{1}{2} \sin 2x) + C = \frac{1}{2}(x + \sin x \cdot \cos x) + C$. For the last equation, we used the trigonometric identity $\sin 2x = 2 \sin x \cos x$.

19.100 Compute $\int \sin^2 x\, dx$.

▌ $\int \sin^2 x\, dx = \int (1 - \cos^2 x)\, dx = x - \frac{1}{2}(x + \sin x \cos x) + C$ [by Problem 19.99] $= \frac{1}{2}(x - \sin x \cos x) + C$.

CHAPTER 20
The Definite Integral and the Fundamental Theorem of Calculus

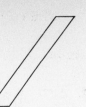

20.1 Evaluate $\int_2^5 4\,dx$ by the direct definition of the integral.

   ∎ Let $2 = x_0 < x_1 < \cdots < x_{n-1} < x_n = 5$ be any partition of $[2, 5]$, and let $\Delta_i x = x_i - x_{i-1}$. Then an approximating sum for $\int_2^5 4\,dx$ is

$$\sum_{i=1}^{n} f(x_i^*)\,\Delta_i x = \sum_{i=1}^{n} 4\,\Delta_i x = 4 \sum_{i=1}^{n} \Delta_i x = 4[(x_1 - x_0) + (x_2 - x_1) + \cdots + (x_n - x_{n-1})]$$

$$= 4(x_n - x_0) = 4(5 - 2) = 4 \cdot 3 = 12.$$

Hence, the integral, which is approximated arbitrarily closely by the approximating sums, must be 12.

20.2 Calculate $\int_0^1 5x^2\,dx$ by the original definition of the integral.

   ∎ Divide $[0, 1]$ into n equal subintervals, each of length $\Delta_i x = 1/n$. In the ith subinterval, choose x_i^* to be the right endpoint i/n. Then the approximating sum is

$$\sum_{i=1}^{n} f(x_i^*)\,\Delta_i x = \sum_{i=1}^{n} 5\left(\frac{i}{n}\right)^2 \frac{1}{n} = \frac{5}{n^3}\sum_{i=1}^{n} i^2 = \frac{5}{n^3}\frac{n(n+1)(2n+1)}{6} = \frac{5}{6}\left(\frac{n+1}{n}\right)\left(\frac{2n+1}{n}\right) = \frac{5}{6}\left(1 + \frac{1}{n}\right)\left(2 + \frac{1}{n}\right)$$

As we make the subdivision finer by letting $n \to +\infty$, the approximating sum approaches $\frac{5}{6} \cdot 1 \cdot 2 = \frac{5}{3}$, which is the value of the integral.

20.3 Prove the formula $\displaystyle\sum_{i=1}^{n} i^2 = \frac{n(n+1)(2n+1)}{6}$ that was used in the solution of Problem 20.2.

   ∎ Use induction with respect to n. For $n = 1$, the sum consists of one term $(1)^2 = 1$. The right side is $1 \cdot 2 \cdot 3/6 = 1$. Now assume that the formula holds for a given positive integer n. We must prove it for $n + 1$. Adding $(n + 1)^2$ to both sides of the formula $\displaystyle\sum_{i=1}^{n} i^2 = \frac{n(n+1)(2n+1)}{6}$, we have

$$\sum_{i=1}^{n+1} i^2 = \frac{n(n+1)(2n+1)}{6} + (n+1)^2 = (n+1)\left[\frac{n(2n+1)}{6} + (n+1)\right] = (n+1)\left[\frac{n(2n+1) + 6(n+1)}{6}\right]$$

$$= (n+1)\left(\frac{2n^2 + n + 6n + 6}{6}\right) = (n+1)\left(\frac{2n^2 + 7n + 6}{6}\right) = (n+1)\left[\frac{(n+2)(2n+3)}{6}\right]$$

$$= \frac{(n+1)(n+2)(2n+3)}{6}$$

which is the case of the formula for $n + 1$.

20.4 Prove the formula $1 + 2 + \cdots + n = \dfrac{n(n+1)}{2}$.

   ∎ Let $S = 1 + 2 + \cdots + (n - 1) + n$. Then we also can write $S = n + (n - 1) + \cdots + 2 + 1$. If we add these two equations column by column, we see that $2S$ is equal to the number $n + 1$ added to itself n times. Thus, $2S = n(n + 1)$, $S = n(n + 1)/2$.

20.5 Show that $\int_0^b x\,dx = \dfrac{b^2}{2}$ by the original definition of the integral.

   ∎ Divide the interval $[0, b]$ into n equal subintervals of length b/n, by the points $0 = x_0 < b/n < 2b/n < \cdots < nb/n = x_n = b$. In the ith subinterval choose x_i^* to be the right-hand endpoint ib/n. Then an approximating sum is $\displaystyle\sum_{i=1}^{n} \frac{ib}{n} \cdot \frac{b}{n} = \frac{b^2}{n^2}\sum_{i=1}^{n} i = \frac{b^2}{n^2}\frac{n(n+1)}{2} = \frac{b^2}{2}\left(1 + \frac{1}{n}\right)$. As $n \to +\infty$, the approximating sum approaches $b^2/2$, which is, therefore, the value of the integral.

20.6 Evaluate $\sum\limits_{i=1}^{n} (3i - 1)$.

$\blacksquare$ $\sum\limits_{i=1}^{n} (3i - 1) = \sum\limits_{i=1}^{n} 3i - \sum\limits_{i=1}^{n} 1 = 3 \cdot \sum\limits_{i=1}^{n} i - n = 3 \dfrac{n(n + 1)}{2} - n = \dfrac{n}{2} [3(n + 1) - 2] = \dfrac{n(3n + 1)}{2}$.

20.7 Evaluate $\sum\limits_{i=1}^{n} (3i^2 + 4)$.

$\blacksquare$ $\sum\limits_{i=1}^{n} (3i^2 + 4) = \sum\limits_{i=1}^{n} 3i^2 + \sum\limits_{i=1}^{n} 4 = 3 \cdot \sum\limits_{i=1}^{n} i^2 + 4n = 3 \dfrac{n(n + 1)(2n + 1)}{6} + 4n$

$= \dfrac{n}{2} [(n + 1)(2n + 1) + 8] = \dfrac{n(2n^2 + 3n + 9)}{2}$.

20.8 For the function f graphed in Fig. 20-1, express $\int_0^5 f(x)\, dx$ in terms of the areas A_1, A_2, and A_3.

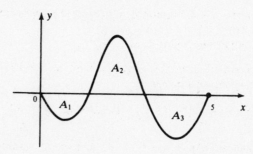

Fig. 20-1

$\blacksquare$ The integral is equal to the sum of the areas above the x-axis and under the graph, minus the sum of the areas under the x-axis and above the graph. Hence, $\int_0^5 f(x)\, dx = A_2 - A_1 - A_3$.

In Problems 20.9–20.14, use the fundamental theorem of calculus to compute the given definite integral.

20.9 $\int_{-1}^{3} (3x^2 - 2x + 1)\, dx$.

$\blacksquare$ $\int (3x^2 - 2x + 1)\, dx = x^3 - x^2 + x$. (We omit the arbitrary constant in all such cases.) So $\int_{-1}^{3} (3x^2 - 2x + 1)\, dx = (x^3 - x^2 + x) \rbrack_{-1}^{3} = (3^3 - 3^2 + 3) - [(-1)^3 - (-1)^2 + (-1)] = 21 - (-3) = 24$.

20.10 $\int_0^{\pi/4} \cos x\, dx$.

$\blacksquare$ $\int \cos x\, dx = \sin x$. Hence, $\int_0^{\pi/4} \cos x\, dx = \sin x \rbrack_0^{\pi/4} = \sin \dfrac{\pi}{4} - \sin 0 = \dfrac{\sqrt{2}}{2} - 0 = \dfrac{\sqrt{2}}{2}$.

20.11 $\int_0^{\pi/3} \sec^2 x\, dx$.

$\blacksquare$ $\int \sec^2 x\, dx = \tan x$. Hence, $\int_0^{\pi/3} \sec^2 x\, dx = \tan x \rbrack_0^{\pi/3} = \tan \dfrac{\pi}{3} - \tan 0 = \sqrt{3} - 0 = \sqrt{3}$.

20.12 $\int_1^{16} x^{3/2}\, dx$.

$\blacksquare$ $\int x^{3/2}\, dx = \frac{2}{5} x^{5/2}$. Hence, $\int_1^{16} x^{3/2}\, dx = \frac{2}{5} x^{5/2} \rbrack_1^{16} = \frac{2}{5} [(16)^{5/2} - (1)^{5/2}] = \frac{2}{5} [(\sqrt{16})^5 - (\sqrt{1})^5] = \frac{2}{5} (1024 - 1) = \frac{2}{5} (1023) = \frac{2046}{5}$.

20.13 $\int_4^5 \left(\dfrac{2}{\sqrt{x}} - x \right) dx$.

$\blacksquare$ $\int \left(\dfrac{2}{\sqrt{x}} - x \right) dx = \int (2x^{-1/2} - x)\, dx = 4x^{1/2} - \frac{1}{2} x^2$. Hence, $\int_4^5 \left(\dfrac{2}{\sqrt{x}} - x \right) dx = (4x^{1/2} - \frac{1}{2} x^2) \rbrack_4^5 = [4\sqrt{5} - \frac{1}{2} (5)^2] - [4\sqrt{4} - \frac{1}{2} (4)^2] = (4\sqrt{5} - \frac{25}{2}) - (8 - 8) = 4\sqrt{5} - \frac{25}{2}$.

• **20.14** $\int_0^1 \sqrt{x^2 - 6x + 9}\, dx$.

▮ $\sqrt{x^2 - 6x + 9} = \sqrt{(x-3)^2} = |x - 3| = 3 - x$ when $0 \le x \le 1$. So, the integral is $\int_0^1 (3 - x)\, dx = (3x - \frac{1}{2}x^2)\,]_0^1 = (3 - \frac{1}{2}) - (0 - 0) = \frac{5}{2}$.

In Problems 20.15–20.20, calculate the area A under the graph of the function $f(x)$, above the x-axis, and between the two indicated values a and b. In each case, one must check that $f(x) \ge 0$ for $a \le x \le b$.

• **20.15** $f(x) = \sin x$, $a = \pi/6$, $b = \pi/3$.

▮ $A = \int_{\pi/6}^{\pi/3} \sin x\, dx = (-\cos x)\,]_{\pi/6}^{\pi/3} = \left(-\cos \frac{\pi}{3}\right) - \left(-\cos \frac{\pi}{6}\right) = \left(-\frac{1}{2}\right) - \left(-\frac{\sqrt{3}}{2}\right) = \frac{\sqrt{3} - 1}{2}$.

• **20.16** $f(x) = x^2 + 4x$, $a = 0$, $b = 3$.

▮ $A = \int_0^3 (x^2 + 4x)\, dx = (\frac{1}{3}x^3 + 2x^2)\,]_0^3 = [\frac{1}{3}(3)^3 + 2(3)^2] = 9 + 18 = 27$.

• **20.17** $f(x) = 1/\sqrt[3]{x}$, $a = 1$, $b = 8$.

▮ $A = \int_1^8 \frac{1}{\sqrt[3]{x}}\, dx = \frac{3}{2}x^{2/3}\,]_1^8 = \frac{3}{2}(8^{2/3} - 1^{2/3}) = \frac{3}{2}(4 - 1) = \frac{9}{2}$.

• **20.18** $f(x) = \sqrt{4x + 1}$, $a = 0$, $b = 2$.

▮ $A = \int_0^2 \sqrt{4x + 1}\, dx$. To find $\int \sqrt{4x + 1}\, dx$, let $u = 4x + 1$, $du = 4\, dx$. $\int \sqrt{4x + 1}\, dx = \frac{1}{4}\int u^{1/2}\, du = \frac{1}{4} \cdot \frac{2}{3}u^{3/2} = \frac{1}{6}(\sqrt{4x + 1})^3$. So, $A = \frac{1}{6}(\sqrt{4x + 1})^3\,]_0^2 = \frac{1}{6}[(\sqrt{9})^3 - (\sqrt{1})^3] = \frac{1}{6}(27 - 1) = \frac{13}{3}$.

• **20.19** $f(x) = x^2 - 3x$, $a = 3$, $b = 5$.

▮ $A = \int_3^5 (x^2 - 3x)\, dx = (\frac{1}{3}x^3 - \frac{3}{2}x^2)\,]_3^5 = [\frac{1}{3}(5)^3 - \frac{3}{2}(5)^2] - [\frac{1}{3}(3)^3 - \frac{3}{2}(3)^2] = \frac{25}{6} + \frac{9}{2} = \frac{26}{3}$.

• **20.20** $f(x) = \sin^2 x \cos x$, $a = 0$, $b = \pi/2$.

▮ $A = \int_0^{\pi/2} \sin^2 x \cos x\, dx = \frac{1}{3}\sin^3 x\,]_0^{\pi/2} = \frac{1}{3}[\sin^3 (\pi/2) - \sin^3 0] = \frac{1}{3}$.

In Problems 20.21–20.31, compute the definite integrals.

• **20.21** $\int_0^{\pi/2} \cos x \sin x\, dx$.

▮ $\int_0^{\pi/2} \cos x \sin x\, dx = \frac{1}{2}\sin^2 x\,]_0^{\pi/2} = \frac{1}{2}[\sin^2 (\pi/2) - \sin^2 0] = \frac{1}{2}$ (using Problem 19.1 to find the antiderivative).

• **20.22** $\int_0^{\pi/4} \tan x \sec^2 x\, dx$.

▮ $\int_0^{\pi/4} \tan x \sec^2 x\, dx = \frac{1}{2}\tan^2 x\,]_0^{\pi/4} = \frac{1}{2}[\tan^2 (\pi/4) - \tan^2 0] = \frac{1}{2}(1 - 0) = \frac{1}{2}$.

• **20.23** $\int_{-1}^1 \sqrt{3x^2 - 2x + 3}\,(3x - 1)\, dx$.

▮ To find $\int \sqrt{3x^2 - 2x + 3}\,(3x - 1)\, dx$, let $u = 3x^2 - 2x + 3$, $du = (6x - 2)\, dx = 2(3x - 1)\, dx$. So, $\int \sqrt{3x^2 - 2x + 3}\,(3x - 1)\, dx = \frac{1}{2}\int \sqrt{u}\, du = \frac{1}{2} \cdot \frac{2}{3}u^{3/2} = \frac{1}{3}(3x^2 - 2x + 3)^{3/2}$. Hence, $\int_{-1}^1 \sqrt{3x^2 - 2x + 3}\,(3x - 1)\, dx = \frac{1}{3}(3x^2 - 2x + 3)^{3/2}\,]_{-1}^1 = \frac{1}{3}[(3 - 2 + 3)^{3/2} - (3 + 2 + 3)^{3/2}] = \frac{1}{3}(8 - 16\sqrt{2}) = \frac{8}{3}(1 - 2\sqrt{2})$.

• **20.24** $\int_0^{\pi/2} \sqrt{\sin x + 1} \cos x\, dx$.

▮ $\int \sqrt{\sin x + 1} \cos x\, dx = \int (\sin x + 1)^{1/2} \cos x\, dx = \frac{2}{3}(\sin x + 1)^{3/2}$ [by Problem 19.1]. Hence, $\int_0^{\pi/2} \sqrt{\sin x + 1} \cos x\, dx = \frac{2}{3}(\sin x + 1)^{3/2}\,]_0^{\pi/2} = \frac{2}{3}\{[\sin(\pi/2) + 1]^{3/2} - (\sin 0 + 1)^{3/2}\} = \frac{2}{3}(2\sqrt{2} - 1)$.

20.25 $\int_{-1}^2 \sqrt{x + 2}\,x^2\, dx$.

▮ Let $u = x + 2$, $x = u - 2$, $du = dx$. When $x = -1$, $u = 1$, and, when $x = 2$, $u = 4$. Then, by change of variables, $\int_{-1}^2 \sqrt{x + 2}\,x^2\, dx = \int_1^4 \sqrt{u}(u - 2)^2\, du = \int_1^4 \sqrt{u}(u^2 - 4u + 4)\, du = \int_1^4 (u^{5/2} - 4u^{3/2} + 4u^{1/2})\, du = [\frac{2}{7}u^{7/2} - 4(\frac{2}{5})u^{5/2} + 4(\frac{2}{3})u^{3/2}]_1^4 = 2u^{3/2}(\frac{1}{7}u^2 - \frac{4}{5}u + \frac{4}{3})]_1^4 = 2[8(\frac{16}{7} - \frac{16}{5} + \frac{4}{3}) - (\frac{1}{7} - \frac{4}{5} + \frac{4}{3})] = 2(\frac{127}{7} - \frac{124}{5} + \frac{28}{3}) = \frac{562}{105}$.

20.26 $\int_2^5 \sqrt{x^3 - 4}\, x^5 \, dx$.

▌ Let $u = x^3 - 4$, $x^3 = u + 4$, $du = 3x^2\, dx$. When $x = 2$, $u = 4$, and, when $x = 5$, $u = 121$. Then $\int_2^5 \sqrt{x^3-4}\, x^5\, dx = \frac13 \int_4^{121} \sqrt{u}(u+4)\, du = \frac13 \int_4^{121} (u^{3/2} + 4u^{1/2})\, du = \frac13 (\frac25 u^{5/2} + \frac83 u^{3/2})\,]_4^{121} = \frac{2}{45} u^{3/2}(3u + 20)\,]_4^{121}$
$= \frac{2}{45}[(11)^3(383) - 8(32)] = \dfrac{113{,}226}{5}$.

20.27 $\int_3^{15} \sqrt[3]{x^2 - 9}\, x^3\, dx$.

▌ Let $u = x^2 - 9$, $x^2 = u + 9$, $du = 2x\, dx$. Then $\int_3^{15} \sqrt[3]{x^2-9}\, x^3\, dx = \frac12 \int_0^{216} u^{1/3}(u+9)\, du = \frac12 \int_0^{216} (u^{4/3} + 9u^{1/3})\, du = \frac12 (\frac37 u^{7/3} + 9 \cdot \frac34 u^{4/3})\,]_0^{216} = \frac{3}{56} u^{4/3}(4u + 63)\,]_0^{216} = \frac{3}{56}\{(6)^4[4(216) + 63] - 0\} = \dfrac{349{,}522}{7}$.

20.28 $\displaystyle\int_0^1 \frac{x}{(2x^2 + 1)^3}\, dx$.

▌ Let $u = 2x^2 + 1$, $du = 4x\, dx$. Then $\displaystyle\int_0^1 \frac{x}{(2x^2+1)^3}\, dx = \frac14 \int_1^3 u^{-3}\, du = -\frac18 u^{-2}\,]_1^3 = -\frac18(\frac19 - 1) = \frac19$.

20.29 $\displaystyle\int_0^8 \frac{x}{(x+1)^{3/2}}\, dx$.

▌ Let $u = x + 1$, $x = u - 1$, $du = dx$. Then $\displaystyle\int_0^8 \frac{x}{(x+1)^{3/2}}\, dx = \int_1^9 \frac{u-1}{u^{3/2}}\, du = \int_1^9 (u^{-1/2} - u^{-3/2})\, du = (2u^{1/2} + 2u^{-1/2})\,]_1^9 = 2[(3 + \frac13) - (1 + 1)] = \frac83$.

20.30 $\int_{-1}^2 |x - 1|\, dx$.

▌ $\int_{-1}^2 |x-1|\, dx = \int_{-1}^1 |x-1|\, dx + \int_1^2 |x-1|\, dx = \int_{-1}^1 (1 - x)\, dx + \int_1^2 (x - 1)\, dx = (x - \frac12 x^2)\,]_{-1}^1 + (\frac12 x^2 - x)\,]_1^2$
$= [(1 - \frac12) - (-1 - \frac12)] + [(2 - 2) - (\frac12 - 1)] = 2 + \frac12 = \frac52$.

20.31 $\displaystyle\int_1^2 \frac{x^7 - 2x + 1}{4x^3}\, dx$.

▌ $\displaystyle\int_1^2 \frac{x^7 - 2x + 1}{4x^3}\, dx = \frac14 \int_1^2 (x^4 - 2x^{-2} + x^{-3})\, dx = \frac14(\frac15 x^5 + 2x^{-1} - \frac12 x^{-2})\,]_1^2 = \frac14[(\frac{32}{5} + 1 - \frac18) - (\frac15 + 2 - \frac12)]$
$= \frac14(\frac{31}{5} - 1 + \frac38) = \frac{223}{160}$.

20.32 Find the average value of $f(x) = \sqrt[3]{x}$ on $[0, 1]$.

▌ By definition, the average value of a function $f(x)$ on an interval $[a, b]$ is $\dfrac{1}{b-a} \int_a^b f(x)\, dx$. Hence, we must compute $\int_0^1 \sqrt[3]{x}\, dx = \int_0^1 x^{1/3}\, dx = \frac34 x^{4/3}\,]_0^1 = \frac34(1 - 0) = \frac34$.

20.33 Compute the average value of $f(x) = \sec^2 x$ on $[0, \pi/4]$.

▌ The average value is $\dfrac{1}{\pi/4} \int_0^{\pi/4} \sec^2 x\, dx = \dfrac{4}{\pi} \tan x\,]_0^{\pi/4} = \dfrac{4}{\pi}(1 - 0) = \dfrac{4}{\pi}$.

20.34 State the Mean-Value Theorem for integrals.

▌ If a function f is continuous on $[a, b]$, it assumes its average value in $[a, b]$; that is, $\dfrac{1}{b-a} \int_a^b f(x)\, dx = f(c)$ for some c in $[a, b]$.

20.35 Verify the Mean-Value Theorem for integrals for the function $f(x) = x + 2$ on $[1, 2]$.

▌ $\dfrac{1}{b-a} \int_2^b f(x)\, dx = \int_1^2 (x + 2)\, dx = (\frac12 x^2 + 2x)\,]_1^2 = (2 + 4) - (\frac12 + 2) = \frac72$. But, $\frac72 = x + 2$ when $x = \frac32$, $1 < \frac32 < 2$.

20.36 Verify the Mean-Value Theorem for integrals for the function $f(x) = x^3$ on $[0, 1]$.

▌ $\dfrac{1}{b-a} \int_a^b f(x)\, dx = \int_0^1 x^3\, dx = \frac14 x^4\,]_0^1 = \frac14(1 - 0) = \frac14$. But $\frac14 = x^3$ when $x = 1/\sqrt[3]{4}$, and $0 < 1/\sqrt[3]{4} < 1$.

20.37   Verify the Mean-Value Theorem for integrals for the function $f(x) = x^2 + 5$ on $[0, 3]$.

▮ $\frac{1}{b-a} \int_a^b f(x)\, dx = \frac{1}{3} \int_0^3 (x^2 + 5)\, dx = \frac{1}{3} (\frac{1}{3}x^3 + 5x)]_0^3 = \frac{1}{3}(9 + 15) = 8.$ But, $8 = x^2 + 5$ when $x = \sqrt{3},$
and $0 < \sqrt{3} < 3.$

20.38   Evaluate $\int_{1/2}^3 \sqrt{2x+1}\, x^2\, dx.$

▮ Let $u = 2x + 1,$ $x = (u-1)/2,$ $du = 2\, dx.$ Then

$$\int_{1/2}^3 \sqrt{2x+1}\, x^2\, dx = \frac{1}{2} \int_2^7 \sqrt{u} \left(\frac{u-1}{2}\right)^2 du = \frac{1}{8} \int_2^7 \sqrt{u}(u^2 - 2u + 1)\, du = \frac{1}{8} \int_2^7 (u^{5/2} - 2u^{3/2} + u^{1/2})\, du$$

$$= \frac{1}{8} [\frac{2}{7}u^{7/2} - 2(\frac{2}{5})u^{5/2} + \frac{2}{3}u^{3/2}] \Big]_2^7 = \frac{1}{4} u^{3/2}(\frac{1}{7}u^2 - \frac{2}{5}u + \frac{1}{3}) \Big]_2^7$$

$$= \frac{1}{4} [7\sqrt{7}(7 - \frac{14}{5} + \frac{1}{3}) - 2\sqrt{2}(\frac{4}{7} - \frac{4}{5} + \frac{1}{3})] = \frac{1}{210}(1666\sqrt{7} - 11\sqrt{2}).$$

20.39   Evaluate $\int_0^{\pi/2} \sin^5 x \cos x\, dx.$

▮ Let $u = \sin x,$ $du = \cos x\, dx.$ Then $\int_0^{\pi/2} \sin^5 x \cos x\, dx = \int_0^1 u^5\, du = \frac{1}{6}u^6]_0^1 = \frac{1}{6}(1 - 0) = \frac{1}{6}.$

20.40   Using only geometric reasoning, calculate the average value of $f(x) = \sqrt{2x - x^2}$ on $[0, 2]$.

▮ We must find $\frac{1}{2} \int_0^2 \sqrt{2x - x^2}\, dx.$ Setting $y = \sqrt{2x - x^2},$ $y^2 = -(x^2 - 2x) = -[(x-1)^2 - 1] = -(x - 1)^2 + 1.$ Thus, $y^2 = -(x-1)^2 + 1,$ $(x-1)^2 + y^2 = 1.$ This is the equation of the circle with center at $(1, 0)$ and radius 1. Hence, the graph of $y = f(x)$ between $x = 0$ and $x = 2$ is the upper half of that circle. So, the integral $\int_0^2 \sqrt{2x - x^2}\, dx$ is the area $\pi/2$ of that semicircle, and, therefore, the average value is $\pi/4$.

20.41   If, in a period of time T, an object moves along the x-axis from x_1 to x_2, find a formula for its average velocity.

▮ Let the initial and final time be t_1 and t_2, with $T = t_2 - t_1.$ The average velocity is

$$\frac{1}{t_2 - t_1} \int_{t_1}^{t_2} v\, dt = \frac{1}{T} x \Big]_{t_1}^{t_2} = \frac{1}{T} [x(t_2) - x(t_1)] = \frac{1}{T}(x_2 - x_1)$$

Thus, as usual, the average velocity is the distance (more precisely, the displacement) divided by the time.

20.42   Prove that, if f is continuous on $[a, b]$, $D_x[\int_a^x f(t)\, dt] = f(x).$

▮ Let $g(x) = \int_a^x f(t)\, dt.$ Then $g(x + \Delta x) - g(x) = \int_a^{x+\Delta x} f(t)\, dt - \int_a^x f(t)\, dt = \int_x^{x+\Delta x} f(t)\, dt.$ By the Mean-Value Theorem for integrals, the last integral is $\Delta x \cdot f(x^*)$ for some x^* between x and $x + \Delta x$. Hence,

$$\frac{g(x + \Delta x) - g(x)}{\Delta x} = f(x^*) \qquad \text{and} \qquad D_x\left[\int_a^x f(t)\, dt\right] = \lim_{\Delta x \to 0} \frac{g(x + \Delta x) - g(x)}{\Delta x} = \lim_{\Delta x \to 0} f(x^*)$$

But as $\Delta x \to 0,$ $x^* \to x,$ and, by the continuity of f, $\lim_{\Delta x \to 0} f(x^*) = f(x).$

20.43   Find $D_x\left[\int_x^b f(t)\, dt\right].$

▮ $D_x[\int_x^b f(t)\, dt] = D_x[-\int_b^x f(t)\, dt] = -D_x[\int_b^x f(t)\, dt] = -f(x)$ [by Problem 20.42].

20.44   Find $D_x[\int_{g(x)}^{h(x)} f(t)\, dt].$

▮ $D_x[\int_{g(x)}^{h(x)} f(t)\, dt] = D_x[\int_0^{h(x)} f(t)\, dt + \int_{g(x)}^0 f(t)\, dt] = D_x[\int_0^{h(x)} f(t)\, dt] + D_x[\int_{g(x)}^0 f(t)\, dt]$

$$= f(h(x)) \cdot h'(x) - f(g(x)) \cdot g'(x), \quad \text{by Problems 20.42 and 20.43 and the Chain Rule.}$$

20.45   Calculate $D_x(\int_2^x \sqrt{5 + 7t^2}\, dt).$

▮ By Problem 20.42, the derivative is $\sqrt{5 + 7x^2}.$

20.46   Calculate $D_x(\int_x^1 \sin^3 t\, dt).$

▮ By Problem 20.43, the derivative is $-\sin^3 x.$

20.47 Calculate $D_x(\int_{-x}^{x} \sqrt[3]{t^4+1}\, dt)$.

▌ By Problem 20.44, the derivative is $\sqrt[3]{x^4+1} - \sqrt[3]{(-x)^4+1}\cdot(-1) = 2\sqrt[3]{x^4+1}$.

20.48 If f is an odd function, show that $\int_{-a}^{a} f(x)\, dx = 0$.

▌ Let $u = -x$, $du = -dx$. Then $\int_{-a}^{0} f(x)\, dx = -\int_{a}^{0} f(-u)\, du = \int_{a}^{0} f(u)\, du = -\int_{0}^{a} f(u)\, du = -\int_{0}^{a} f(x)\, dx$. Hence, $\int_{-a}^{a} f(x)\, dx = \int_{-a}^{0} f(x)\, dx + \int_{0}^{a} f(x)\, dx = 0$.

20.49 Evaluate $\int_{-3}^{3} x^2 \sin x\, dx$.

▌ $x^2 \sin x$ is an odd function, since $\sin(-x) = -\sin x$. So, by Problem 20.48, the integral is 0.

20.50 If f is an even function, show that $\int_{-a}^{a} f(x)\, dx = 2\int_{0}^{a} f(x)\, dx$.

▌ Let $u = -x$, $du = -dx$. Then $\int_{-a}^{0} f(x)\, dx = -\int_{a}^{0} f(-u)\, du = -\int_{a}^{0} f(u)\, du = \int_{0}^{a} f(u)\, du = \int_{0}^{a} f(x)\, dx$. Hence, $\int_{-a}^{a} f(x)\, dx = \int_{-a}^{0} f(x)\, dx + \int_{0}^{a} f(x)\, dx = 2\int_{0}^{a} f(x)\, dx$.

20.51 Find $D_x(\int_{1}^{3x^2} \sqrt{t^5+1}\, dt)$.

▌ By Problem 20.44, the derivative is $\sqrt{(3x^2)^5+1}\cdot D_x(3x^2) = \sqrt{243x^{10}+1}\cdot(6x) = 6x\sqrt{243x^{10}+1}$.

20.52 Solve $\int_{1}^{b} x^{n-1}\, dx = 2/n$ for b.

▌ $\dfrac{2}{n} = \dfrac{1}{n}x^n \Big]_{1}^{b} = \dfrac{1}{n}(b^n - 1)$. Hence, $2 = b^n - 1$, $b^n = 3$, $b = \sqrt[n]{3}$.

20.53 If $\int_{3}^{5} f(x-k)\, dx = 1$, compute $\int_{3-k}^{5-k} f(x)\, dx$.

▌ Let $x = u - k$, $dx = du$. Then $\int_{3-k}^{5-k} f(x)\, dx = \int_{3}^{5} f(u-k)\, du = \int_{3}^{5} f(x-k)\, dx = 1$.

20.54 If
$$f(x) = \begin{cases} \sin x & \text{for } x < 0 \\ 3x^2 & \text{for } x \ge 0 \end{cases}$$
find $\int_{-\pi/2}^{1} f(x)\, dx$.

▌ $\int_{-\pi/2}^{1} f(x)\, dx = \int_{-\pi/2}^{0} \sin x\, dx + \int_{0}^{1} 3x^2\, dx = -\cos x\,]_{-\pi/2}^{0} + x^3\,]_{0}^{1} = \left[-\cos\left(\dfrac{\pi}{2}\right) - (-\cos 0)\right] + (1-0) = (-0+1)+1 = 2$.

20.55 Given that $2x^2 - 8 = \int_{a}^{x} f(t)\, dt$, find a formula for $f(x)$ and evaluate a.

▌ First, set $x = a$ to obtain $2a^2 - 8 = 0$, $a^2 = 4$, $a = \pm 2$. Then, differentiating, we find $4x = f(x)$.

20.56 Given $H(x) = \int_{1}^{x} \dfrac{1}{1+t^2}\, dt$, find $H(1)$ and $H'(1)$, and show that $H(4) - H(2) < \frac{2}{5}$.

▌ $H(1) = \int_{1}^{1} \dfrac{1}{1+t^2}\, dt = 0$. $H'(x) = \dfrac{1}{1+x^2}$; so, $H'(1) = \frac{1}{2}$. By the Mean-Value Theorem, $H(4) - H(2) = (4-2)\cdot H'(c)$ for some c in $(2,4)$. Now, $H'(c) = \dfrac{1}{1+c^2} < \dfrac{1}{1+2^2} = \dfrac{1}{5}$. Hence, $H(4) - H(2) < \frac{2}{5}$.

20.57 If the average value of $f(x) = x^3 + bx - 2$ on $[0,2]$ is 4, find b.

▌ $4 = \dfrac{1}{2-0}\int_{0}^{2}(x^3+bx-2)\, dx = \frac{1}{2}(\frac{1}{4}x^4 + \frac{1}{2}bx^2 - 2x)\,]_{0}^{2} = \frac{1}{2}[(4+2b-4)-0] = b$. Thus, $b = 4$.

20.58 Find $\lim\limits_{h\to 0}\left(\dfrac{1}{h}\int_{2}^{2+h}\sqrt{t^2+2}\, dt\right)$.

▌ Let $g(x) = \int_{0}^{x}\sqrt{t^2+2}\, dt$. Then

$$\dfrac{1}{h}\int_{2}^{2+h}\sqrt{t^2+2}\, dt = \dfrac{g(2+h)-g(2)}{h}$$

Therefore, the desired limit is

$$\lim_{h\to 0}\dfrac{g(2+h)-g(2)}{h} = g'(2) = \sqrt{2^2+2} = \sqrt{6}$$

20.59 If g is continuous, which of the following integrals are equal?

$$(a) \quad \int_a^b g(x)\, dx \qquad (b) \quad \int_{a+1}^{b+1} g(x-1)\, dx \qquad (c) \quad \int_0^{b-a} g(x+a)\, dx$$

▮ Let $u = x - 1$, $du = dx$. Then $\int_{a+1}^{b+1} g(x-1)\, dx = \int_a^b g(u)\, du = \int_a^b g(x)\, dx$. Let $v = x + a$, $dv = dx$. Then $\int_0^{b-a} g(x+a)\, dx = \int_a^b g(v)\, dv = \int_a^b g(x)\, dx$. Thus, all three integrals are equal to each other.

20.60 The region above the x-axis and under the curve $y = \sin x$, between $x = 0$ and $x = \pi$, is divided into two parts by the line $x = c$. If the area of the left part is one-third the area of the right part, find c.

▮ $\int_0^c \sin x\, dx = \frac{1}{3} \int_c^\pi \sin x\, dx$, $-\cos x\,]_0^c = \frac{1}{3}(-\cos x)\,]_c^\pi$, $-(\cos c - \cos 0) = -\frac{1}{3}(\cos \pi - \cos c)$, $\cos c - 1 = \frac{1}{3}(-1 - \cos c)$, $3 \cos c - 3 = -1 - \cos c$, $4 \cos c = 2$, $\cos c = \frac{1}{2}$, $c = \pi/3$.

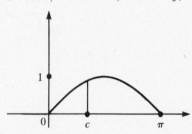

0   c   π   **Fig. 20-2**

20.61 Find the value(s) of k for which $\int_0^2 x^k\, dx = \int_0^2 (2 - x)^k\, dx$.

▮ Let $u = 2 - x$, $du = -dx$. Hence, $\int_0^2 (2-x)^k\, dx = -\int_2^0 u^k\, du = \int_0^2 u^k\, du = \int_0^2 x^k\, dx$. Thus, the equation holds for all k.

20.62 The velocity v of an object moving on the x-axis is $\cos 3t$, and the object is at the origin at $t = 0$. Find the average value of the position x over the interval $0 \le t \le \pi/3$.

▮ $x = \int v\, dt = \int \cos 3t\, dt = \frac{1}{3} \sin 3t + C$. But $x = 0$ when $t = 0$. Hence, $C = 0$ and $x = \frac{1}{3} \sin 3t$. The average value of x on $[0, \pi/3]$ is

$$\frac{1}{\pi/3} \int_0^{\pi/3} \frac{1}{3} \sin 3t\, dt = \frac{1}{\pi} \left(-\frac{1}{3} \cos 3t \right) \Big]_0^{\pi/3} = -\frac{1}{3\pi} (\cos \pi - \cos 0) = -\frac{1}{3\pi} (-1 - 1) = \frac{2}{3\pi}$$

20.63 Evaluate $\displaystyle \lim_{n \to +\infty} \frac{1}{n} \left(\sin \frac{\pi}{n} + \sin \frac{2\pi}{n} + \cdots + \sin \frac{n\pi}{n} \right)$.

▮ Partition the interval $[0, \pi]$ into n equal parts. Then the corresponding partial sum for $\int_0^\pi \sin x\, dx$, in which we choose the right endpoint in each subinterval, is

$$\sum_{j=1}^n \left(\sin \frac{j\pi}{n} \right)\left(\frac{\pi}{n} \right) = \frac{\pi}{n} \sum_{j=1}^n \sin \frac{j\pi}{n} = \frac{\pi}{n} \left(\sin \frac{\pi}{n} + \sin \frac{2\pi}{n} + \cdots + \sin \frac{n\pi}{n} \right)$$

This approximating sum approaches $\int_0^\pi \sin x\, dx = -\cos x\,]_0^\pi = -(\cos \pi - \cos 0) = -(-1 - 1) = 2$. Thus, π times the desired limit is 2. Hence, the required limit is $2/\pi$.

20.64 An object moves on a straight line with velocity $v = 3t - 1$, where v is measured in meters per second. How far does the object move in the period $0 \le t \le 2$ seconds?

▮ The distance traveled is $\int_{t_1}^{t_2} |v|\, dt$, in this case, $\int_0^2 |3t - 1|\, dt$. Since $3t - 1 < 0$ for $t < \frac{1}{3}$, we divide the integral into two parts: $\int_0^{1/3} |3t - 1|\, dt + \int_{1/3}^2 |3t - 1|\, dt = \int_0^{1/3} (1 - 3t)\, dt + \int_{1/3}^2 (3t - 1)\, dt = (t - \frac{3}{2}t^2)\,]_0^{1/3} + (\frac{3}{2}t^2 - t)\,]_{1/3}^2 = (\frac{1}{3} - \frac{1}{8}) + [(6 - 2) - (\frac{1}{6} - \frac{1}{3})] = \frac{13}{3}$.

20.65 Prove the formula $1^3 + 2^3 + \cdots + n^3 = \left[\dfrac{n(n+1)}{2} \right]^2$.

▮ For $n = 1$, both sides are 1. Assume the formula true for a given n, and add $(n+1)^3$ to both sides:

$$1^3 + 2^3 + \cdots + n^3 + (n+1)^3 = \left[\frac{n(n+1)}{2} \right]^2 + (n+1)^3 = (n+1)^2 \left[\frac{n^2}{4} + (n+1) \right] = \frac{(n+1)^2}{4} (n^2 + 4n + 4)$$

$$= \frac{(n+1)^2}{4} (n+2)^2 = \left[\frac{(n+1)(n+2)}{2} \right]^2$$

which is the case of the formula for $n + 1$. Hence, the formula has been proved by induction.

20.66 State the trapezoidal rule for approximation of integrals.

▎ Let $f(x) \geq 0$ be integrable on $[a, b]$. Divide $[a, b]$ into n equal parts of length $\Delta x = (b - a)/n$, by means of points $x_1, x_2, \ldots, x_{n-1}$. Then

$$\int_a^b f(x)\, dx \approx \frac{\Delta x}{2} \left[f(a) + f(b) + 2 \sum_{i=1}^{n-1} f(x_i) \right]$$

20.67 Use the trapezoidal rule with $n = 10$ to approximate $\int_0^1 x^3\, dx$.

▎

$$\int_0^1 x^3\, dx \approx \frac{1}{20} \left[0^3 + 1^3 + 2 \sum_{i=1}^{9} \left(\frac{i}{10} \right)^3 \right] = \frac{1}{20} \left(1 + 2 \sum_{i=1}^{9} \frac{i^3}{1000} \right)$$

$$= \frac{1}{20} \left(1 + \frac{1}{500} \sum_{i=1}^{9} i^3 \right) = \frac{1}{20} \left[1 + \frac{1}{500} \left(\frac{9 \cdot 10}{2} \right)^2 \right] \qquad \text{[by Problem 20.65]}$$

$$= \tfrac{1}{20} (1 + \tfrac{81}{20}) = 0.05(5.05) = 0.2525$$

The actual value is $\frac{1}{4} x^4 \big]_0^1 = \frac{1}{4}$, by the fundamental theorem of calculus.

20.68 State Simpson's rule for approximation of integrals.

▎ Let $f(x)$ be integrable and nonnegative on $[a, b]$. Divide $[a, b]$ into $n = 2k$ equal subintervals of length $\Delta x = (b - a)/n$. Then $\int_a^b f(x)\, dx \approx \frac{\Delta x}{3} [f(a) + 4f(a + \Delta x) + 2f(a + 2\Delta x) + 4f(a + 3\Delta x) + 2f(a + 4\Delta x) + \cdots + 2f(a + (2k - 2) \Delta x) + 4f(a + (2k - 1) \Delta x) + f(b)]$.

20.69 Apply Simpson's rule with $n = 4$ to approximate $\int_0^1 x^4\, dx$.

▎ We obtain, with $\Delta x = \frac{1}{4}$, $\frac{1}{12}[0^4 + 4(\frac{1}{4})^4 + 2(\frac{1}{2})^4 + 4(\frac{3}{4})^4 + 1^4] = \frac{1}{12}(\frac{1}{64} + \frac{1}{8} + \frac{81}{64} + 1) = \frac{154}{768} \approx 0.2005$, which is close to the exact value $\int_0^1 x^4\, dx = \frac{1}{5} x^5 \big]_0^1 = \frac{1}{5} = 0.2$.

20.70 Use the trapezoidal rule with $n = 10$ to approximate $\int_0^1 x^2\, dx$.

▎

$$\int_0^1 x^2\, dx \approx \frac{1}{20} \left(0^2 + 1^2 + 2 \sum_{i=1}^{9} \frac{i^2}{100} \right) = \frac{1}{20} \left(1 + \frac{1}{50} \sum_{i=1}^{9} i^2 \right)$$

$$= \frac{1}{20} \left(1 + \frac{1}{50} \cdot \frac{9 \cdot 10 \cdot 19}{6} \right) \qquad \text{[by Problem 20.66]}$$

$$= \tfrac{1}{20} \left(\tfrac{201}{30} \right) = 0.335$$

which is close to the exact value $\frac{1}{3}$.

20.71 Use geometric reasoning to calculate $\int_{-a}^{a} \sqrt{a^2 - x^2}\, dx$.

▎ The graph of $y = \sqrt{a^2 - x^2}$ is the upper half of the circle $x^2 + y^2 = a^2$ with center at the origin and radius a. $\int_{-a}^{a} \sqrt{a^2 - x^2}$ is, therefore, the area of the semicircle, that is, $\pi a^2 / 2$.

20.72 Find the area inside the ellipse $x^2/a^2 + y^2/b^2 = 1$.

▎ The area is twice the area above the x-axis and under the ellipse, which is given by

$$\int_{-a}^{a} y\, dx = \int_{-a}^{a} \frac{b}{a} \sqrt{a^2 - x^2}\, dx = \frac{b}{a} \int_{-a}^{a} \sqrt{a^2 - x^2}\, dx = \frac{b}{a} \frac{\pi a^2}{2} \qquad \text{[by Problem 20.71]}$$

$$= \frac{\pi a b}{2}$$

Hence, the total area inside the ellipse is $\pi a b$.

20.73 Find $D_x^2(\int_{1988}^{x} \cos 2t\, dt)$.

▎ $D_x(\int_{1988}^{x} \cos 2t\, dt) = \cos 2x$. Hence, $D_x^2(\int_{1988}^{x} \cos 2t\, dt) = D_x(\cos 2x) = (-\sin 2x)(2) = -2 \sin 2x$.

20.74 Compute $D_x^2\left(\int_{x^3}^{1789} \dfrac{1}{t}\, dt\right)$.

▮ $D_x\left(\int_{x^3}^{1789} \dfrac{1}{t}\, dt\right) = -\dfrac{1}{x^3}\cdot 3x^2 = -3/x = -3x^{-1}$. Hence, $D_x^2\left(\int_{x^3}^{1789} \dfrac{1}{\pi}\, dt\right) = 3x^{-2} = 3/x^2$.

20.75 Draw a region whose area is given by $\int_1^3 (2x+1)\, dx$, and find the area by geometric reasoning.

▮ See Fig. 20-3. The region is the area under the line $y = 2x+1$ between $x=1$ and $x=3$, above the x-axis. The region consists of a 2×3 rectangle, of area 6, and a right triangle of base 2 and height 4, with area $\frac{1}{2}\cdot 2\cdot 4 = 4$. Hence, the total area is 10, which is equal to $\int_1^3 (2x+1)\, dx$.

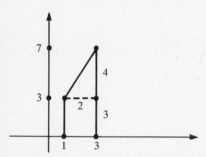

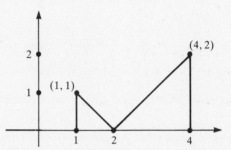

Fig. 20-3   Fig. 20-4

20.76 Draw a region whose area is given by $\int_1^4 |x-2|\, dx$, and find the area by geometric reasoning.

▮ See Fig. 20-4. The region consists of two triangles with bases on the x-axis, one under the line segment from $(1,1)$ to $(2,0)$, and the other under the line segment from $(2,0)$ to $(4,2)$. The first triangle has base and height equal to 1, and therefore, area $\frac{1}{2}$. The second triangle has base and height equal to 2, and, therefore, area 2. Thus, the total area is $\frac{5}{2}$, which is $\int_1^4 |x-2|\, dx$.

20.77 Find a region whose area is given by $\int_{-1}^2 [\sqrt{9-(x+1)^2}+2]\, dx$, and compute the area by geometric reasoning.

▮ If we let $y = \sqrt{9-(x+1)^2}+2$, then $(x+1)^2+(y-2)^2 = 9$, which is a circle with center $(-1,2)$ and radius 3. A suitable region is that above the x-axis and under the quarter arc of the above circle running from $(-1,5)$ to $(2,2)$—see Fig. 20-5. The region consists of a 3×2 rectangle of area 6, surmounted by a quarter circle of area $\frac{1}{4}(9\pi)$. Hence, the total area is $6+9\pi/4$.

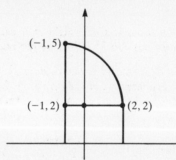

Fig. 20-5

20.78 Find the distance traveled by an object moving along a line with velocity $v = (2-t)/\sqrt{t}$ from $t=4$ to $t=9$.

▮ The distance is $s = \int_4^9 |v|\, dt$. Between $t=4$ and $t=9$, $v = (2-t)/\sqrt{t}$ is negative. Hence, $s = \int_4^9 \dfrac{t-2}{\sqrt{t}}\, dt = \int_4^9 (t^{1/2}-2t^{-1/2})\, dt = (\frac{2}{3}t^{3/2}-4t^{1/2})\big]_4^9 = (18-12)-(\frac{16}{3}-8) = \frac{26}{3}$.

20.79 Find the distance traveled by an object moving along a line with velocity $v = \sin \pi t$ from $t=\frac{1}{2}$ to $t=2$.

The distance is $s = \int_{1/2}^2 |v|\, dt$. Observe that $\sin \pi t$ is positive for $\frac{1}{2} < t < 1$ and negative for $1 < t < 2$. Hence,

$$s = \int_{1/2}^1 \sin \pi t\, dt + \int_1^2 (-\sin \pi t)\, dt = -\frac{1}{\pi}\cos \pi t \Big]_{1/2}^1 + \frac{1}{\pi}\cos \pi t\Big]_1^2$$

$$= -\frac{1}{\pi}(-1-0) + \frac{1}{\pi}[1-(-1)] = \frac{1}{\pi} + \frac{2}{\pi} = \frac{3}{\pi}$$

20.80 Find a function f such that $\int_0^x f(t)\, dt = \sin x - x^2$.

▌ Differentiating both sides of the given equation, we see that $f(x) = \cos x - 2x$. In fact, $\int_0^x (\cos t - 2t)\, dt = (\sin t - t^2)\,]_0^x = (\sin x - x^2) - (0 - 0) = \sin x - x^2$.

20.81 Find a function f such that $\int_0^x f(t)\, dt = \cos x - x^2$.

▌ Differentiating both sides of the given equation, we obtain $f(x) = -\sin x - 2x$. Checking back, $\int_0^x (-\sin t - 2t)\, dt = (\cos t - t^2)\,]_0^x = (\cos x - x^2) - (1 - 0) \neq \cos x - x^2$. Thus, there is no solution. In order for there to be a solution of $\int_0^x f(t)\, dt = g(x)$, we must have $g(0) = 0$.

20.82 Find $\displaystyle\int_0^{\pi/4} \frac{dx}{1 + \cot^2 x}$.

▌ $\displaystyle\frac{1}{1 + \cot^2 x} = \frac{1}{\csc^2 x} = \sin^2 x = \frac{1}{2}(1 - \cos 2x)$. Hence, $\displaystyle\int_0^{\pi/4} \frac{dx}{1 + \cot^2 x} = \frac{1}{2} \int_0^{\pi/4} (1 - \cos 2x)\, dx = \frac{1}{2}(x - \frac{1}{2} \sin 2x)\,]_0^{\pi/4} = \frac{1}{2}\left(\frac{\pi}{4} - \frac{1}{2}\right) = \frac{1}{8}(\pi - 2)$.

20.83 Evaluate $\int_0^{\pi/12} \cos 5x \cdot \cos x\, dx$.

▌ By the addition formula,

$$\cos(n-1)x = \cos(nx - x) = \cos nx \cos x + \sin nx \sin x$$

$$\cos(n+1)x = \cos(nx + x) = \cos nx \cos x - \sin nx \sin x$$

whence $\cos(n-1)x + \cos(n+1)x = 2 \cos nx \cos x$. In particular, $\cos 5x \cdot \cos x = \frac{1}{2}(\cos 6x + \cos 4x)$. Hence, $\int_0^{\pi/12} \cos 5x \cdot \cos x\, dx = \frac{1}{2} \int_0^{\pi/12} (\cos 6x + \cos 4x)\, dx = \frac{1}{2}(\frac{1}{6} \sin 6x + \frac{1}{4} \sin 4x)\,]_0^{\pi/12} = \frac{1}{2}(\frac{1}{6} + \frac{1}{4} \cdot \sqrt{3}/2) = \frac{1}{48}(4 + 3\sqrt{3})$.

20.84 If an object moves along a line with velocity $v = \sin t - \cos t$ from time $t = 0$ to $t = \pi/2$, find the distance traveled.

▌ The distance is $s = \int_0^{\pi/2} |v|\, dt$. Now, for $0 < t < \pi/2$, $\sin t - \cos t \geq 0$ when and only when $\sin t \geq \cos t$, that is, if and only if $\tan t \geq 1$, which is equivalent to $\pi/4 \leq t < \pi/2$. Hence, $s = \int_0^{\pi/4} (\cos t - \sin t)\, dt + \int_{\pi/4}^{\pi/2} (\sin t - \cos t)\, dt = (\sin t + \cos t)\,]_0^{\pi/4} + (-\cos t - \sin t)\,]_{\pi/4}^{\pi/2} = \left[\left(\frac{\sqrt{2}}{2} + \frac{\sqrt{2}}{2}\right) - (0 + 1)\right] + \left[(-0 - 1) - \left(-\frac{\sqrt{2}}{2} - \frac{\sqrt{2}}{2}\right)\right] = 2(\sqrt{2} - 1)$.

20.85 Let $y = f(x)$ be a function whose graph consists of straight lines connecting the points $P_1(0, 3)$, $P_2(3, -3)$, $P_3(4, 3)$, and $P_4(5, 3)$. Sketch the graph and find $\int_0^5 f(x)\, dx$ by geometry.

▌ See Fig. 20-6. The point B where $P_1 P_2$ intersects the x-axis is $(\frac{3}{2}, 0)$. $\int_0^5 f(x)\, dx = A_1 - A_2 + A_3$. The area A_1 of $\triangle OBP_1$ is $\frac{1}{2} \cdot \frac{3}{2} \cdot 3 = \frac{9}{4}$. The area A_2 of $\triangle BP_2 C$ is $\frac{1}{2} \cdot 2 \cdot 3 = 3$. [Note that $C = (\frac{7}{2}, 0)$.] The area A_3 of trapezoid $CP_3 P_4 D$ is $\frac{1}{2} \cdot 3 \cdot (\frac{3}{2} + 1) = \frac{15}{4}$. Hence, $\int_0^5 f(x)\, dx = \frac{9}{4} - 3 + \frac{15}{4} = 3$.

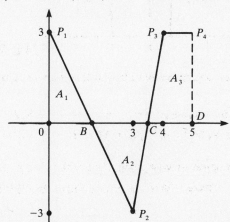

Fig. 20-6

20.86 Find by geometric reasoning $\int_0^5 |f(x)|\, dx$, where f is the function of Problem 20.85.

▌ The graph of $|f(x)|$ is obtained from that of $f(x)$ by reflecting $BP_2 C$ in the x-axis. Hence, $\int_0^5 |f(x)|\, dx = A_1 + A_2 + A_3 = \frac{9}{4} + 3 + \frac{15}{4} = 9$.

20.87 If $g(x) = \int_0^x f(t)\, dt$, where f is the function of Problem 20.85, find $g'(4)$.

    ▮ $g'(x) = D_x(\int_0^x f(t)\, dt) = f(x)$. Hence, $g'(4) = f(4) = 3$.

Problems 20.88–20.90 refer to the function $t(x)$ whose graph is shown in Fig. 20-7.

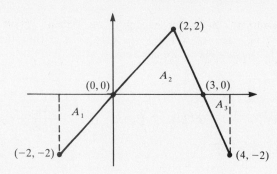

Fig. 20-7

20.88 Find $\int_0^3 f(x)\, dx$.

    ▮ $\int_0^3 f(x)\, dx$ is the area $A_2 = \frac{1}{2} \cdot 3 \cdot 2 = 3$.

20.89 Find $\int_{-2}^0 f(x)\, dx$.

    ▮ $\int_{-2}^0 f(x)\, dx = -A_1 = -\frac{1}{2} \cdot 2 \cdot 2 = -2$.

20.90 Find $\int_{-2}^4 f(x)\, dx$.

    ▮ $\int_{-2}^4 f(x)\, dx = -A_1 + A_2 - A_3$. Note that $A_3 = \frac{1}{2} \cdot 1 \cdot 2 = 1$. So, $-A_1 + A_2 - A_3 = -2 + 3 - 1 = 0$.

20.91 Verify that the average value of a linear function $f(x) = ax + b$ over an interval $[x_1, x_2]$ is equal to the value of the function at the midpoint of the interval.

    ▮ By definition, the indicated average value

$$\bar{f} = \frac{1}{x_2 - x_1} \int_{x_1}^{x_2} (ax + b)\, dx = \frac{1}{x_2 - x_1} \left(a\frac{x^2}{2} + bx \right) \Big]_{x_1}^{x_2}$$

$$= \frac{1}{x_2 - x_1} \left(a\frac{x_2^2}{2} + bx_2 - a\frac{x_1^2}{2} - bx_1 \right)$$

$$= \frac{1}{x_2 - x_1} \left[\frac{a}{2}(x_2 - x_1)(x_2 + x_1) + b(x_2 - x_1) \right]$$

$$= a\left(\frac{x_2 + x_1}{2} \right) + b = f\left(\frac{x_2 + x_1}{2} \right)$$

20.92 Find the volume of the solid generated when the region between the semicircle $y = 1 - \sqrt{1 - x^2}$ and the line $y = 1$ is rotated around the x-axis (see Fig. 20-8).

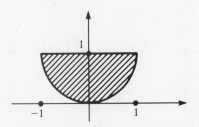

Fig. 20-8

    ▮ By the disk formula, $V = \pi \int_{-1}^1 [1^2 - (1 - \sqrt{1 - x^2})^2]\, dx = \pi \int_{-1}^1 [1 - (1 - 2\sqrt{1 - x^2} + 1 - x^2)]\, dx = \pi \int_{-1}^1 (2\sqrt{1 - x^2} + x^2 - 1)\, dx = 2\pi \int_{-1}^1 \sqrt{1 - x^2}\, dx + \pi(\frac{1}{3}x^3 - x)]_{-1}^1 = 2\pi(\frac{1}{2}\pi) + \pi[(\frac{1}{3} - 1) - (-\frac{1}{3} + 1)] = \pi^2 - \frac{4}{3}\pi$. (The integral $\int_{-1}^1 \sqrt{1 - x^2}\, dx = \frac{1}{2}\pi$ by Problem 20.71.)

CHAPTER 21

Area and Arc Length

21.1 Sketch and find the area of the region to the left of the parabola $x = 2y^2$, to the right of the y-axis, and between $y = 1$ and $y = 3$.

▌ See Fig. 21-1. The base of the region is the y-axis. The area is given by the integral $\int_1^3 2y^2\, dy = \frac{2}{3} y^3\,]_1^3 = \frac{2}{3}(27 - 1) = \frac{52}{3}$.

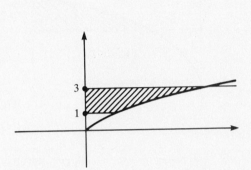

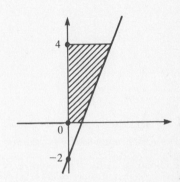

Fig. 21-1   Fig. 21-2

21.2 Sketch and find the area of the region above the line $y = 3x - 2$, in the first quadrant, and below the line $y = 4$.

▌ See Fig. 21-2. The region has a base on the y-axis. We must solve $y = 3x - 2$ for x: $x = \frac{1}{3}(y + 2)$. Then the area is $\int_0^4 \frac{1}{3}(y + 2)\, dy = \frac{1}{3}(\frac{1}{2} y^2 + 2y)\,]_0^4 = \frac{1}{3}(8 + 8) = \frac{16}{3}$.

21.3 Sketch and find the area of the region between the curve $y = x^3$ and the lines $y = -x$ and $y = 1$.

▌ See Fig. 21-3. The lower boundary of the region is $y = -x$ and the upper boundary is $y = x^3$. Hence, the area is given by the integral $\int_0^1 [y^{1/3} - (-y)]\, dy = (\frac{3}{4} y^{4/3} + \frac{1}{2} y^2)\,]_0^1 = \frac{3}{4} + \frac{1}{2} = \frac{5}{4}$.

In Problems 21.4–21.16, sketch the indicated region and find its area.

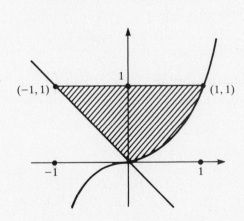

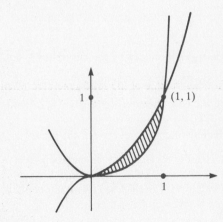

Fig. 21-3   Fig. 21-4

21.4 The bounded region between the curves $y = x^2$ and $y = x^3$.

▌ See Fig. 21-4. The curves intersect at $(0, 0)$ and $(1, 1)$. Between $x = 0$ and $x = 1$, $y = x^2$ lies above $y = x^3$. The area of the region between them is $\int_0^1 (x^2 - x^3)\, dx = (\frac{1}{3} x^3 - \frac{1}{4} x^4)\,]_0^1 = \frac{1}{3} - \frac{1}{4} = \frac{1}{12}$.

21.5   The bounded region between the parabola $y = 4x^2$ and the line $y = 6x - 2$.

▌ See Fig. 21-5.   First we find the points of intersection: $4x^2 = 6x - 2$, $2x^2 - 3x + 1 = 0$, $(2x - 1)(x - 1) = 0$, $x = \frac{1}{2}$ or $x = 1$.   So, the points of intersection are $(\frac{1}{2}, 1)$ and $(1, 4)$.   Hence, the area is $\int_{1/2}^{1} [(6x - 2) - 4x^2]\, dx = (3x^2 - 2x - \frac{4}{3}x^3) \big]_{1/2}^{1} = (3 - 2 - \frac{4}{3}) - (\frac{3}{4} - 1 - \frac{1}{6}) = \frac{1}{12}$.

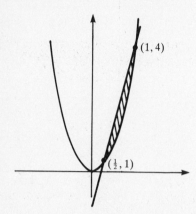

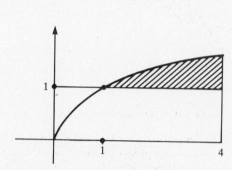

Fig. 21-5

Fig. 21-6

21.6   The region bounded by the curves $y = \sqrt{x}$, $y = 1$,   and $x = 4$.

▌ See Fig. 21-6.   The region is bounded above by $y = \sqrt{x}$ and below by $y = 1$.   Hence, the area is given by $\int_{1}^{4} (\sqrt{x} - 1)\, dx = (\frac{2}{3}x^{3/2} - x) \big]_{1}^{4} = (\frac{16}{3} - 4) - (\frac{2}{3} - 1) = \frac{14}{3} - 3 = \frac{5}{3}$.

21.7   The region under the curve $\sqrt{x} + \sqrt{y} = 1$ and in the first quadrant.

▌ See Fig. 21-7.   The region has its base on the x-axis.   The area is given by $\int_{0}^{1} (1 - \sqrt{x})^2\, dx = \int_{0}^{1} (1 - 2x^{1/2} + x)\, dx = (x - \frac{4}{3}x^{3/2} + \frac{1}{2}x^2) \big]_{0}^{1} = 1 - \frac{4}{3} + \frac{1}{2} = \frac{1}{6}$.

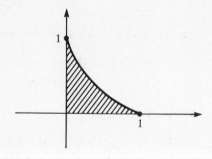

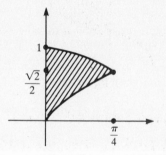

Fig. 21-7

Fig. 21-8

21.8   The region bounded by the curves $y = \sin x$, $y = \cos x$, $x = 0$,   and $x = \pi/4$.

▌ See Fig. 21-8.   The upper boundary is $y = \cos x$,   the lower boundary is $y = \sin x$,   and the left side is the y-axis.   The area is given by $\int_{0}^{\pi/4} (\cos x - \sin x)\, dx = (\sin x + \cos x) \big]_{0}^{\pi/4} = \left(\dfrac{\sqrt{2}}{2} + \dfrac{\sqrt{2}}{2}\right) - (0 + 1) = \sqrt{2} - 1$.

21.9   The bounded region between the parabola $x = -y^2$ and the line $y = x + 6$.

▌ See Fig. 21-9.   First we find the points of intersection: $y = -y^2 + 6$, $y^2 + y - 6 = 0$, $(y - 2)(y + 3) = 0$, $y = 2$ or $y = -3$.   Thus, the points of intersection are $(-4, 2)$ and $(-9, -3)$.   It is more convenient to integrate with respect to y, with the parabola as the upper boundary and the line as the lower boundary.   The area is given by the integral $\int_{-3}^{2} [-y^2 - (y - 6)]\, dy = (-\frac{1}{3}y^3 - \frac{1}{2}y^2 + 6y) \big]_{-3}^{2} = (-\frac{8}{3} - 2 + 12) - (9 - \frac{9}{2} - 18) = 19 + \frac{11}{6} = \frac{125}{6}$.

21.10   The bounded region between the parabola $y = x^2 - x - 6$ and the line $y = -4$.

▌ See Fig. 21-10.   First we find the points of intersection: $-4 = x^2 - x - 6$, $x^2 - x - 2 = 0$, $(x - 2)(x + 1) = 0$, $x = 2$ or $x = -1$.   Thus, the intersection points are $(2, -4)$ and $(-1, -4)$.   The upper boundary of the region is $y = -4$,   and the lower boundary is the parabola.   The area is given by $\int_{-1}^{2} [-4 - (x^2 - x - 6)]\, dx = \int_{-1}^{2} (2 - x^2 + x)\, dx = (2x - \frac{1}{3}x^3 + \frac{1}{2}x^2) \big]_{-1}^{2} = (4 - \frac{8}{3} + 2) - (-2 + \frac{1}{3} + \frac{1}{2}) = \frac{9}{2}$.

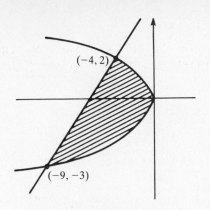

Fig. 21-9

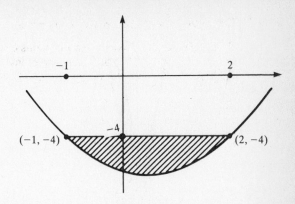

Fig. 21-10

21.11   The bounded region between the curve $y = \sqrt{x}$ and $y = x^3$.

❚   See Fig. 21-11. First we find the points of intersection: $\sqrt{x} = x^3$, $x = x^6$, $x(x^5 - 1) = 0$, $x = 0$ or $x = 1$. Thus, the points of intersection are $(0, 0)$ and $(1, 1)$. The upper curve is $y = \sqrt{x}$ and the lower curve is $y = x^3$. Hence, the area is $\int_0^1 (\sqrt{x} - x^3)\, dx = (\frac{2}{3}x^{3/2} - \frac{1}{4}x^4)\big]_0^1 = \frac{2}{3} - \frac{1}{4} = \frac{5}{12}$.

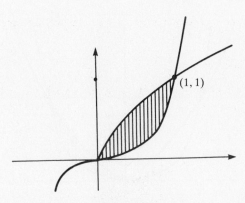

Fig. 21-11

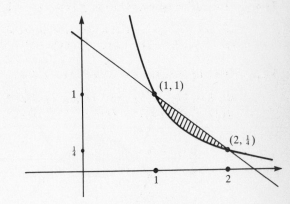

Fig. 21-12

21.12   The bounded region in the first quadrant between the curves $4y + 3x = 7$ and $y = x^{-2}$.

❚   See Fig. 21-12. First we find the points of intersection: $\frac{4}{x^2} + 3x = 7$, $4 + 3x^3 = 7x^2$, $3x^3 - 7x^2 + 4 = 0$. $x = 1$ is an obvious root. Dividing $3x^3 - 7x^2 + 4$ by $x - 1$, we obtain $3x^2 - 4x - 4 = (3x + 2)(x - 2)$. Hence, the other roots are $x = 2$ and $x = -\frac{2}{3}$. So, the intersection points in the first quadrant are $(1, 1)$ and $(2, \frac{1}{4})$. The upper boundary is the line and the lower boundary is $y = x^{-2}$. The area is given by $\int_1^2 [\frac{1}{4}(7 - 3x) - x^{-2}]\, dx = (\frac{1}{4}(7x - \frac{3}{2}x^2) + x^{-1})\big]_1^2 = [\frac{1}{4}(14 - 6) + \frac{1}{2}] - [\frac{1}{4}(7 - \frac{3}{2}) + 1] = \frac{1}{8}$.

21.13   The region bounded by the parabolas $y = x^2$ and $y = -x^2 + 6x$.

❚   See Fig. 21-13. First let us find the intersection points: $x^2 = -x^2 + 6x$, $x^2 = 3x$, $x^2 - 3x = 0$, $x(x - 3) = 0$, $x = 0$ or $x = 3$. Hence, the points of intersection are $(0, 0)$ and $(3, 9)$. The second parabola $y = -x^2 + 6x = -(x^2 - 6x) = -[(x - 3)^2 - 9] = -(x - 3)^2 + 9$ has its vertex at $(3, 9)$, $x = 3$ is its axis of symmetry, and it opens downward. That parabola is the upper boundary of our region, and $y = x^2$ is the lower boundary. The area is given by $\int_0^3 [(-x^2 + 6x) - x^2]\, dx = \int_0^3 (6x - 2x^2)\, dx = (3x^2 - \frac{2}{3}x^3)\big]_0^3 = 27 - 18 = 9$.

21.14   The region bounded by the parabola $x = y^2 + 2$ and the line $y = x - 8$.

❚   See Fig. 21-14. Let us find the points of intersection: $y + 8 = y^2 + 2$, $y^2 - y - 6 = 0$, $(y - 3)(y + 2) = 0$, $y = 3$ or $y = -2$. So, the points of intersection are $(11, 3)$ and $(6, -2)$. It is more convenient to integrate with respect to y. The area is $\int_{-2}^3 [(y + 8) - (y^2 + 2)]\, dy = \int_{-2}^3 (y + 6 - y^2)\, dy = (\frac{1}{2}y^2 + 6y - \frac{1}{3}y^3)\big]_{-2}^3 = (\frac{9}{2} + 18 - 9) - (2 - 12 + \frac{8}{3}) = \frac{125}{6}$.

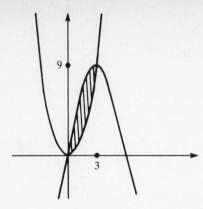

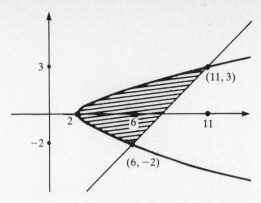

Fig. 21-13

Fig. 21-14

21.15 The region bounded by the parabolas $y = x^2 - x$ and $y = x - x^2$.

▮ See Fig. 21-15. Let us find the points of intersection: $x^2 - x = x - x^2$, $2x^2 - 2x = 0$, $x(x-1) = 0$, $x = 0$ or $x = 1$. Thus, the intersection points are $(0, 0)$ and $(1, 0)$. The parabola $y = x^2 - x = (x - \frac{1}{2})^2 - \frac{1}{4}$ has its vertex at $(\frac{1}{2}, -\frac{1}{4})$ and opens upward, while $y = x - x^2$ has its vertex at $(\frac{1}{2}, \frac{1}{4})$ and opens downward. The latter parabola is the upper boundary, and the first parabola is the lower boundary. Hence, the area is $\int_0^1 [(x - x^2) - (x^2 - x)] \, dx = 2 \int_0^1 (x - x^2) \, dx = 2(\frac{1}{2}x^2 - \frac{1}{3}x^3) \big]_0^1 = 2(\frac{1}{2} - \frac{1}{3}) = \frac{1}{3}$.

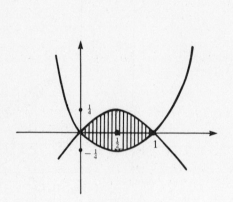

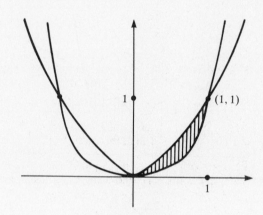

Fig. 21-15

Fig. 21-16

21.16 The region in the first quadrant bounded by the curves $y = x^2$ and $y = x^4$.

▮ See Fig. 21-16. Let us find the points of intersection: $x^4 = x^2$, $x^4 - x^2 = 0$, $x^2(x^2 - 1) = 0$, $x = 0$ or $x = \pm 1$. So, the intersection points in the first quadrant are $(0, 0)$ and $(1, 1)$. $y = x^2$ is the upper curve. Hence, the area is $\int_0^1 (x^2 - x^4) \, dx = (\frac{1}{3}x^3 - \frac{1}{5}x^5) \big]_0^1 = \frac{1}{3} - \frac{1}{5} = \frac{2}{15}$.

> In Problems 21.17–21.21, find the arc length of the given curve.

21.17 $y = \dfrac{x^4}{8} + \dfrac{1}{4x^2}$ from $x = 1$ to $x = 2$.

▮ Recall that the arc length formula is $L = \int_a^b \sqrt{1 + (y')^2} \, dx$. In this case,

$$y' = \frac{1}{2}x^3 - \frac{1}{2}\frac{1}{x^3} = \frac{1}{2}\left(\frac{x^6 - 1}{x^3}\right) \qquad (y')^2 = \frac{(x^6 - 1)^2}{4x^6} = \frac{x^{12} - 2x^6 + 1}{4x^6}$$

Hence,

$$1 + (y')^2 = \frac{4x^6 + x^{12} - 2x^6 + 1}{4x^6} = \frac{x^{12} + 2x^6 + 1}{4x^6} = \left(\frac{x^6 + 1}{2x^3}\right)^2$$

Thus, $\quad L = \displaystyle\int_1^2 \frac{x^6 + 1}{2x^3} \, dx = \frac{1}{2}\int_1^2 (x^3 + x^{-3}) \, dx = \frac{1}{2}\left(\frac{x^4}{4} - \frac{1}{2}x^{-2}\right)\Big]_1^2 = \frac{1}{2}[(4 - \frac{1}{8}) - (\frac{1}{4} - \frac{1}{2})] = \frac{33}{16}$

21.18 $y = 3x - 2$ from $x = 0$ to $x = 1.$

▮ $y' = 3.$ So, $L = \int_0^1 \sqrt{1 + (3)^2}\, dx = \sqrt{10} \int_0^1 1\, dx = \sqrt{10}\, x \big]_0^1 = \sqrt{10}\,(1 - 0) = \boxed{\sqrt{10}.}$

21.19 $y = x^{2/3}$ from $x = 1$ to $x = 8.$

▮ $y' = \frac{2}{3} x^{-1/3}.$ $(y')^2 = \frac{4}{9} x^{-2/3}.$ Hence, $1 + (y')^2 = 1 + (4/9 x^{2/3}) = (9 x^{2/3} + 4)/9 x^{2/3}.$ Thus, $L = \frac{1}{3} \int_1^8 \frac{\sqrt{9 x^{2/3} + 4}}{x^{1/3}}\, dx.$ Let $u = 9 x^{2/3} + 4,$ $du = 6 x^{-1/3}\, dx.$ Then, $L = \frac{1}{3} \int_{13}^{40} \sqrt{u} \cdot \frac{1}{6}\, du = \frac{1}{18} \cdot \frac{2}{3} u^{3/2} \big]_{13}^{40} = \frac{1}{27}[(40)^{3/2} - (13)^{3/2}] = \frac{1}{27}(80\sqrt{10} - 13\sqrt{13}).$

21.20 $x^{2/3} + y^{2/3} = 4$ from $x = 1$ to $x = 8.$

▮ By implicit differentiation, $\frac{2}{3} x^{-1/3} + \frac{2}{3} y^{-1/3} y' = 0,$ $y' = -(x^{-1/3}/y^{-1/3}) = -(y^{1/3}/x^{1/3}).$ So, $(y')^2 = y^{2/3}/x^{2/3}.$ Hence, $1 + (y')^2 = 1 + y^{2/3}/x^{2/3} = (x^{2/3} + y^{2/3})/x^{2/3} = 4/x^{2/3}.$ Therefore, $L = \int_1^8 \frac{2}{x^{1/3}}\, dx = 2 \int_1^8 x^{-1/3}\, dx = 2 \cdot \frac{3}{2} x^{2/3} \big]_1^8 = 3(4 - 1) = 9.$

21.21 $y = \dfrac{x^5}{15} + \dfrac{1}{4x^3}$ from $x = 1$ to $x = 2.$

▮ $y' = \frac{1}{3} x^4 - \frac{3}{4} x^{-4} = \frac{1}{3} x^4 - 3/4 x^4 = (4 x^8 - 9)/12 x^4.$ So,

$$(y')^2 = \frac{16 x^{16} - 72 x^8 + 81}{144 x^8} \quad \text{and} \quad 1 + (y')^2 = \frac{144 x^8 + 16 x^{16} - 72 x^8 + 81}{144 x^8} = \frac{16 x^{16} + 72 x^8 + 81}{144 x^8} = \left(\frac{4 x^8 + 9}{12 x^4}\right)^2$$

Hence, $L = \dfrac{1}{12} \int_1^2 \dfrac{4 x^8 + 9}{x^4}\, dx = \frac{1}{12} \int_1^2 (4 x^4 + 9 x^{-4})\, dx = \frac{1}{12}(\frac{4}{5} x^5 - 3 x^{-3}) \big]_1^2 = \frac{1}{12}[(\frac{128}{5} - \frac{3}{8}) - (\frac{4}{5} - 3)] = \frac{1}{12}(\frac{124}{5} + \frac{21}{8}) = \frac{97}{480}.$

21.22 Let $\mathcal{R}$ consist of all points in the plane that are above the x-axis and below the curve whose equation is $y = -x^2 + 2x + 8.$ Find the area of $\mathcal{R}.$

▮ $y = -(x^2 - 2x - 8) = -[(x - 1)^2 - 9] = -(x - 1)^2 + 9.$ The parabola's vertex is $(1, 9)$ and it opens downward (see Fig. 21-17). To find where it cuts the x-axis, let $-x^2 + 2x + 8 = 0,$ $x^2 - 2x - 8 = 0,$ $(x - 4)(x + 2) = 0,$ $x = 4$ or $x = -2.$ Hence, the area of $\mathcal{R}$ is $\int_{-2}^4 (-x^2 + 2x + 8)\, dx = (-\frac{1}{3} x^3 + x^2 + 8x) \big]_{-2}^4 = (-\frac{64}{3} + 16 + 32) - (\frac{8}{3} + 4 - 16) = 36.$

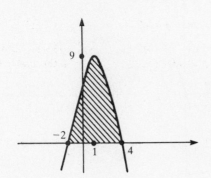

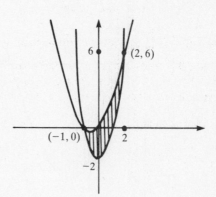

Fig. 21-17   **Fig. 21-18**

21.23 Find the area bounded by the curves $y = 2x^2 - 2$ and $y = x^2 + x.$

▮ See Fig. 21-18. $y = 2x^2 - 2$ is a parabola with vertex at $(0, -2).$ On the other hand, $y = x^2 + x = (x + \frac{1}{2})^2 - \frac{1}{4}$ is a parabola with vertex $(-\frac{1}{2}, -\frac{1}{4}).$ To find the points of intersection, set $2x^2 - 2 = x^2 + x,$ $x^2 - x - 2 = 0,$ $(x - 2)(x + 1) = 0,$ $x = 2$ or $x = -1.$ Thus, the points are $(2, 6)$ and $(-1, 0).$ Hence, the area is $\int_{-1}^2 [(x^2 + x) - (2x^2 - 2)]\, dx = \int_{-1}^2 (2 + x - x^2)\, dx = (2x + \frac{1}{2} x^2 - \frac{1}{3} x^3) \big]_{-1}^2 = (4 + 2 - \frac{8}{3}) - (-2 + \frac{1}{2} + \frac{1}{3}) = -\frac{9}{2}.$

21.24 Find the area of the region between the x-axis and $y = (x-1)^3$ from $x = 0$ to $x = 2$.

❚ As shown in Fig. 21-19, the region consists of two pieces, one below the x-axis from $x = 0$ to $x = 1$, and the other above the x-axis from $x = 1$ to $x = 2$. Hence, the total area is $\int_0^1 -(x-1)^3 \, dx + \int_1^2 (x-1)^3 \, dx = -\frac{1}{4}(x-1)^4 \,]_0^1 + \frac{1}{4}(x-1)^4 \,]_1^2 = [-\frac{1}{4}(0-1)] + [\frac{1}{4}(1-0)] = \frac{1}{2}$.

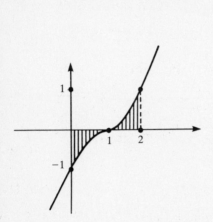

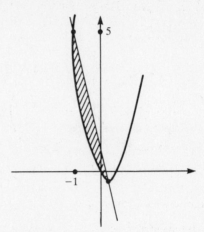

Fig. 21-19   Fig. 21-20

21.25 Find the area bounded by the curves $y = 3x^2 - 2x$ and $y = 1 - 4x$.

❚ See Fig. 21-20. $y = 3x^2 - 2x = 3(x^2 - \frac{2}{3}x) = 3[(x - \frac{1}{3})^2 - \frac{1}{9}] = 3(x - \frac{1}{3})^2 - \frac{1}{3}$. Thus, that curve is a parabola with vertex $(\frac{1}{3}, -\frac{1}{3})$. To find the intersection, let $3x^2 - 2x = 1 - 4x$, $3x^2 + 2x - 1 = 0$, $(3x - 1)(x + 1) = 0$, $x = \frac{1}{3}$ or $x = -1$. Hence, the intersection points are $(\frac{1}{3}, -\frac{1}{3})$ and $(-1, 5)$. Thus, the area is $\int_{-1}^{1/3} [(1 - 4x) - (3x^2 - 2x)] \, dx = \int_{-1}^{1/3} (1 - 2x - 3x^2) \, dx = (x - x^2 - x^3) \,]_{-1}^{1/3} = (\frac{1}{3} - \frac{1}{9} - \frac{1}{27}) - (-1 - 1 + 1) = \frac{32}{27}$.

21.26 Find the area of the region bounded by the curves $y = x^2 - 4x$ and $x + y = 0$.

❚ See Fig. 21-21. The parabola $y = x^2 - 4x = (x - 2)^2 - 4$ has vertex $(2, -4)$. Let us find the intersection of the curves: $x^2 - 4x = -x$, $x^2 - 3x = 0$, $x(x - 3) = 0$, $x = 0$ or $x = 3$. So, the points of intersection are $(0, 0)$ and $(3, -3)$. Hence, the area is $\int_0^3 [-x - (x^2 - 4x)] \, dx = \int_0^3 (3x - x^2) \, dx = (\frac{3}{2}x^2 - \frac{1}{3}x^3) \,]_0^3 = \frac{27}{2} - 9 = \frac{9}{2}$.

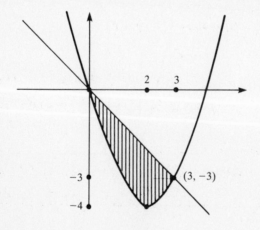

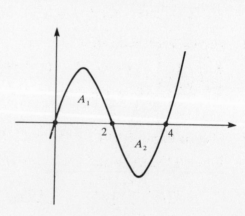

Fig. 21-21   Fig. 21-22

21.27 Find the area of the bounded region between the curve $y = x^3 - 6x^2 + 8x$ and the x-axis.

❚ $y = x^3 - 6x^2 + 8x = x(x^2 - 6x + 8) = x(x - 2)(x - 4)$. So, the curve cuts the x-axis at $x = 0$, $x = 2$, and $x = 4$. Since $\lim_{x \to +\infty} f(x) = +\infty$ and $\lim_{x \to -\infty} f(x) = -\infty$, the graph can be roughly sketched as in Fig. 21-22. Hence the required area is $A_1 + A_2 = \int_0^2 (x^3 - 6x^2 + 8x) \, dx + \int_2^4 -(x^3 - 6x^2 + 8x) \, dx = (\frac{1}{4}x^4 - 2x^3 + 4x^2) \,]_0^2 - (\frac{1}{4}x^4 - 2x^3 + 4x^2) \,]_2^4 = (4 - 16 + 16) - [(64 - 128 + 64) - (4 - 16 + 16)] = 4 - (-4) = 8$.

21.28 Find the area enclosed by the curve $y^2 = x^2 - x^4$.

▮ Since $y^2 = x^2(1 - x)(1 + x)$, the curve intersects the x-axis at $x = 0$, $x = 1$, and $x = -1$. Since the graph is symmetric with respect to the coordinate axes, it is as indicated in Fig. 21-23. The total area is four times the area in the first quadrant, which is $\int_0^1 x\sqrt{1 - x^2}\, dx = -\frac{1}{2}\int_0^1 (1 - x^2)^{1/2} D_x(1 - x^2)\, dx = -\frac{1}{2} \cdot \frac{2}{3}(1 - x^2)^{3/2}\,]_0^1 = -\frac{1}{3}(0 - 1) = \frac{1}{3}$. Hence, the total area is $\frac{4}{3}$.

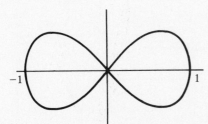

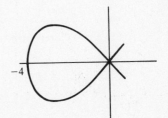

Fig. 21-23　　　　　　　　　　　　　**Fig. 21-24**

21.29 Find the area of the loop of the curve $y^2 = x^4(4 + x)$ between $x = -4$ and $x = 0$. (See Fig. 21-24.)

▮ By symmetry with respect to the x-axis, the required area is $2\int_{-4}^0 y\, dx = 2\int_{-4}^0 x^2\sqrt{4 + x}\, dx$. Let $u = \sqrt{4 + x}$, $u^2 = 4 + x$, $x = u^2 - 4$, $dx = 2u\, du$. Hence, we have $2\int_0^2 (u^2 - 4)^2 u \cdot 2u\, du = 4\int_0^2 (u^6 - 8u^4 + 16u^2)\, du = 4(\frac{1}{7}u^7 - \frac{8}{5}u^5 + \frac{16}{3}u^3)\,]_0^2 = 4(\frac{128}{7} - \frac{256}{5} + \frac{128}{3}) = \frac{4096}{105}$.

21.30 Find the length of the arc of the curve $x = 3y^{3/2} - 1$ from $y = 0$ to $y = 4$.

▮ The arc length $L = \int_0^4 \sqrt{1 + (dx/dy)^2}\, dy$, $dx/dy = \frac{9}{2}y^{1/2}$, $1 + (dx/dy)^2 = 1 + 81y/4 = (4 + 81y)/4$. So, $L = \frac{1}{2}\int_0^4 \sqrt{4 + 81y}\, dy$. Let $u = 4 + 81y$, $du = 81\, dy$. Then $L = \frac{1}{2}\int_4^{328} u^{1/2} \cdot \frac{1}{81}\, du = \frac{1}{162}(\frac{2}{3}u^{3/2})\,]_4^{328} = \frac{1}{243}(328 \cdot 2\sqrt{82} - 8) = \frac{8}{243}(82\sqrt{82} - 1)$.

21.31 Find the length of the arc of $24xy = x^4 + 48$ from $x = 2$ to $x = 4$.

▮ $y = (x^4 + 48)/24x = \frac{1}{24}(x^3 + 48x^{-1})$. Then $y' = \frac{1}{24}(3x^2 - 48/x^2) = (3x^4 - 48)/24x^2 = (x^4 - 16)/8x^2$. Hence,

$$1 + (y')^2 = 1 + \frac{x^8 - 32x^4 + 256}{64x^4} = \frac{64x^4 + x^8 - 32x^4 + 256}{64x^4} = \frac{x^8 + 32x^4 + 256}{64x^4} = \left(\frac{x^4 + 16}{8x^2}\right)^2.$$

So, $L = \frac{1}{8}\int_2^4 (x^2 + 16x^{-2})\, dx = \frac{1}{8}(\frac{1}{3}x^3 - 16x^{-1})\,]_2^4 = \frac{1}{8}[(\frac{64}{3} - 4) - (\frac{8}{3} - 8)] = \frac{17}{6}$.

21.32 Find the length of the arc of $y^3 = 8x^2$ from $x = 1$ to $x = 8$.

▮ $y = 2x^{2/3}$, $y' = \frac{4}{3}x^{-1/3}$, $(y')^2 = 16/9x^{2/3}$. So, $1 + (y')^2 = (9x^{2/3} + 16)/9x^{2/3}$. Hence, $L = \frac{1}{3}\int_1^8 \frac{\sqrt{9x^{2/3} + 16}}{x^{1/3}}\, dx = \frac{1}{18}\int_1^8 (9x^{2/3} + 16)^{1/2} D_x(9x^{2/3} + 16)\, dx = \frac{1}{18} \cdot \frac{2}{3}(9x^{2/3} + 16)^{3/2}\,]_1^8 = \frac{1}{27}(104\sqrt{13} - 125)$.

21.33 Find the length of the arc of $6xy = x^4 + 3$ from $x = 1$ to $x = 2$.

▮ $y = (x^4 + 3)/6x = \frac{1}{6}(x^3 + 3x^{-1})$. Then $y' = \frac{1}{6}(3x^2 - 3x^{-2}) = \frac{1}{2}[x^2 - (1/x^2)] = (x^4 - 1)/2x^2$. So,

$$(y')^2 = \frac{x^8 - 2x^4 + 1}{4x^4} \qquad \text{and} \qquad 1 + (y')^2 = \frac{4x^4 + x^8 - 2x^4 + 1}{4x^4} = \frac{x^8 + 2x^4 + 1}{4x^4} = \left(\frac{x^4 + 1}{2x^2}\right)^2.$$

Then $L = \frac{1}{2}\int_1^2 \frac{x^4 + 1}{x^2}\, dx = \frac{1}{2}\int_1^2 (x^2 + x^{-2})\, dx = \frac{1}{2}(\frac{1}{3}x^3 - x^{-1})\,]_1^2 = \frac{1}{2}[(\frac{8}{3} - \frac{1}{2}) - (\frac{1}{3} - 1)] = \frac{17}{12}$.

21.34 Find the length of the arc of $27y^2 = 4(x - 2)^3$ from $(2, 0)$ to $(11, 6\sqrt{3})$.

▮ $y = \frac{2}{3\sqrt{3}}(x - 2)^{3/2}$, $y' = \frac{1}{\sqrt{3}}(x - 2)^{1/2}$, $(y')^2 = \frac{1}{3}(x - 2)$, $1 + (y')^2 = (x + 1)/3$. So, $L = \frac{1}{\sqrt{3}}\int_2^{11} (x + 1)^{1/2}\, dx = \frac{1}{\sqrt{3}} \cdot \frac{2}{3}(x + 1)^{3/2}\,]_2^{11} = \frac{2}{3\sqrt{3}}(12^{3/2} - 3^{3/2}) = 2(8 - 1) = 14$.

21.35 Find the area bounded by the curve $y = 1 - x^{-2}$ and the lines $y = 1$, $x = 1$, and $x = 4$. (See Fig. 21-25.)

▮ The upper boundary is the line $y = 1$. So, the area is $\int_1^4 [1 - (1 - x^{-2})]\, dx = \int_1^4 x^{-2}\, dx = -x^{-1}\,]_1^4 = -(\frac{1}{4} - 1) = \frac{3}{4}$.

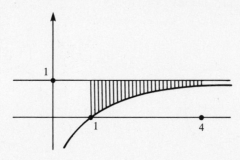

Fig. 21-25

21.36 Find the area in the first quadrant lying under the arc from the y-axis to the first point where the curve $x + y + y^2 = 2$ cuts the positive x-axis.

▮ The curve hits the y-axis when $x = 0$, that is, $y^2 + y - 2 = 0$, $(y + 2)(y - 1) = 0$, $y = -2$ or $y = 1$. It hits the x-axis when $y = 0$, that is, when $x = 2$. Hence, the arc extends in the first quadrant from $(0, 1)$ to $(2, 0)$, and the required area is $\int_0^1 (2 - y - y^2)\, dy = (2y - \frac{1}{2}y^2 - \frac{1}{3}y^3)\,]_0^1 = 2 - \frac{1}{2} - \frac{1}{3} = \frac{7}{6}$.

21.37 Find the area under the arch of $y = \sin x$ between $x = 0$ and $x = \pi$.

▮ The area is $\int_0^\pi \sin x\, dx = -\cos x\,]_0^\pi = -(-1 - 1) = 2$.

21.38 Find the area of the bounded region between $y = x^2$ and $y = 2x$ (see Fig. 21-26).

▮ Setting $x^2 = 2x$, we find $x = 0$ or $x = 2$. Hence, the curves intersect at $(0, 0)$ and $(2, 4)$. For $0 \le x \le 2$, $x^2 \le 2$, and, therefore, $y = 2x$ is the upper curve. The area is $\int_0^2 (2x - x^2)\, dx = (x^2 - \frac{1}{3}x^3)\,]_0^2 = 4 - \frac{8}{3} = \frac{4}{3}$.

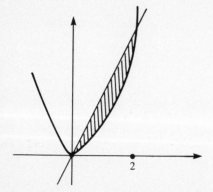

Fig. 21-26

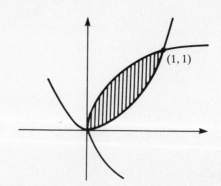

Fig. 21-27

21.39 Find the area of the region bounded by the parabolas $y = x^2$ and $x = y^2$.

▮ See Fig. 21-27. Solving simultaneously, $x = y^2 = x^4$, $x = 0$ or $x = 1$. Hence, the curves intersect at $(0, 0)$ and $(1, 1)$. Since $\sqrt{x} \ge x^2$ for $0 \le x \le 1$, $x = y^2$ is the upper curve. The area is $\int_0^1 (x^{1/2} - x^2)\, dx = (\frac{2}{3}x^{3/2} - \frac{1}{3}x^3)\,]_0^1 = \frac{2}{3} - \frac{1}{3} = \frac{1}{3}$.

21.40 Find the area of the bounded region between the curves $y = 2$ and $y = 4x^3 + 3x^2 + 2$.

▮ For $y = 4x^3 + 3x^2 + 2$, $y' = 12x^2 + 6x = 6x(2x + 1)$, and $y'' = 24x + 6$. Hence, the critical number $x = 0$ yields a relative minimum, and the critical number $-\frac{1}{2}$ yields a relative maximum. To find intersection points, $4x^3 + 3x^2 = 0$, $x = 0$ or $x = -\frac{3}{4}$. Thus, the region is as indicated in Fig. 21-28, and the area is $\int_{-3/4}^0 [(4x^3 + 3x^2 + 2) - 2]\, dx = \int_{-3/4}^0 (4x^3 + 3x^2)\, dx = (x^4 + x^3)\,]_{-3/4}^0 = -(\frac{81}{256} - \frac{27}{64}) = \frac{27}{256}$.

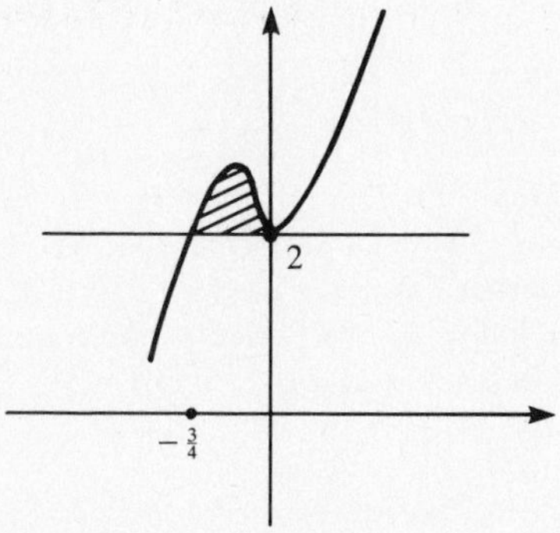

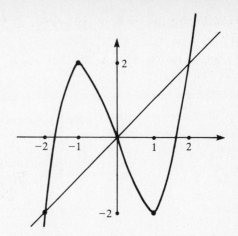

Fig. 21-28

Fig. 21-29

21.41 Find the area of the region bounded by $y = x^3 - 3x$ and $y = x$.

▮ For $y = x^3 - 3x$, $y' = 3x^2 - 3 = 3(x - 1)(x + 1)$, and $y'' = 6x$. Hence, the critical number $x = 1$ yields a relative minimum, and the critical number $x = -1$ yields a relative maximum. Setting $x^3 - 3x = x$, $x^3 = 4x$, $x = 0$ or $x = \pm 2$. Hence, the intersection points are $(0, 0)$, $(2, 2)$, and $(-2, -2)$. Thus, the region consists of two equal pieces, as shown in Fig. 21-29. The piece between $x = -2$ and $x = 0$ has area $\int_{-2}^{0} [(x^3 - 3x) - x]\, dx = \int_{-2}^{0} (x^3 - 4x)\, dx = (\frac{1}{4}x^4 - 2x^2)\,]_{-2}^{0} = -(4 - 8) = 4$. So the total area is 8.

21.42 The area bounded by $y = x^2$ and $y = 4$ (two even functions) is divided into two equal parts by a line $y = c$. Determine c.

▮ See Fig. 21-30. The upper part is, by symmetry, $2\int_{c}^{4} y^{1/2}\, dy = 2 \cdot \frac{2}{3} y^{3/2}\,]_{c}^{4} = \frac{4}{3}(8 - c^{3/2})$. The lower part is $2\int_{0}^{c} y^{1/2}\, dy = 2 \cdot \frac{2}{3} y^{3/2}\,]_{0}^{c} = \frac{4}{3}c^{3/2}$. Hence, $8 - c^{3/2} = c^{3/2}$, $8 = 2c^{3/2}$, $4 = c^{3/2}$, $c = 4^{2/3} = \sqrt[3]{16}$.

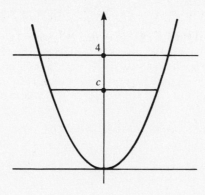

Fig. 21-30

21.43 Let $p > 1$. What happens to the area above the x-axis bounded by $y = x^{-p}$, $x = 1$, and $x = b$, as $b \to +\infty$?

▮ The area is

$$\int_{1}^{b} x^{-p}\, dx = \frac{1}{1-p} x^{-p+1}\,\Big]_{1}^{b} = \frac{1}{1-p}(b^{-p+1} - 1) = \frac{1}{1-p}\left(\frac{1}{b^{p-1}} - 1\right)$$

As $b \to +\infty$, the limit is $1/(p-1)$.

21.44 Find the arc length of $y = \frac{1}{3}\sqrt{x}(3 - x)$ for $0 \le x \le 3$.

▮ $y = x^{1/2} - \frac{1}{3}x^{3/2}$, $y' = \frac{1}{2}x^{-1/2} - \frac{1}{2}x^{1/2} = \frac{1}{2}\left(\frac{1}{\sqrt{x}} - \sqrt{x}\right) = \frac{1}{2}\left(\frac{1-x}{\sqrt{x}}\right)$. So,

$$(y')^2 = \frac{1}{4}\frac{1 - 2x + x^2}{x} \qquad \text{and} \qquad 1 + (y')^2 = \frac{4x + 1 - 2x + x^2}{4x} = \frac{x^2 + 2x + 1}{4x} = \left(\frac{x+1}{2\sqrt{x}}\right)^2$$

Hence, $L = \frac{1}{2}\int_{0}^{3} \frac{x+1}{\sqrt{x}}\, dx = \frac{1}{2}\int_{0}^{3} (x^{1/2} + x^{-1/2})\, dx = \frac{1}{2}(\frac{2}{3}x^{3/2} + 2x^{1/2})\,]_{0}^{3} = x^{1/2}(\frac{1}{3}x + 1)\,]_{0}^{3} = 2\sqrt{3}$.

21.45 Find the arc length of $y = \frac{2}{3}(1+x^2)^{3/2}$ for $0 \le x \le 3$.

▮ $y' = (1+x^2)^{1/2} \cdot 2x$, and $(y')^2 = 4x^2(1+x^2)$. So, $1+(y')^2 = 1+4x^2+4x^4 = (1+2x^2)^2$. Hence, $L = \int_0^3 (1+2x^2)\,dx = (x + \frac{2}{3}x^3) \big]_0^3 = 3 + 18 = 21$.

21.46 Find the area bounded by $y = x^3$ and its tangent line at $x = 1$ (Fig. 21-31).

▮ At $x = 1$, $y' = 3x^2 = 3$. Hence, the tangent line is $(y-1)/(x-1) = 3$, or $y = 3x - 2$. To find out where this line intersects $y = x^3$, we solve $x^3 = 3x - 2$, or $x^3 - 3x + 2 = 0$. One root is $x = 1$ (the point of tangency). Dividing $x^3 - 3x + 2$ by $x = 1$, we obtain $x^2 + x - 2 = (x+2)(x-1)$. Hence, $x = -2$ is a root, and, therefore, the intersection points are $(1, 1)$ and $(-2, -8)$. Hence, the required area is $\int_{-2}^1 [x^3 - (3x-2)]\,dx = \int_{-2}^1 (x^3 - 3x + 2)\,dx = (\frac{1}{4}x^4 - \frac{3}{2}x^2 + 2x) \big]_{-2}^1 = (\frac{1}{4} - \frac{3}{2} + 2) - (4 - 6 - 4) = \frac{27}{4}$.

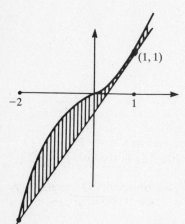

Fig. 21-31

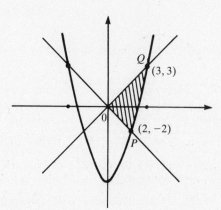

Fig. 21-32

21.47 Find the area of the region above the curve $y = x^2 - 6$, below $y = x$, and above $y = -x$ (Fig. 21-32).

▮ We must find the area of region OPQ. To find P, solve $y = -x$ and $y = x^2 - 6$: $x^2 - 6 = -x$, $x^2 + x - 6 = 0$, $(x+3)(x-2) = 0$, $x = -3$ or $x = 2$. Hence, P is $(2, -2)$. To find Q, solve $y = x$ and $y = x^2 - 6$: $x^2 - 6 = x$, $x^2 - x - 6 = 0$, $(x-3)(x+2) = 0$, $x = 3$ or $x = -2$. Hence, Q is $(3, 3)$. The area of the region is $\int_0^2 [x - (-x)]\,dx + \int_2^3 [x - (x^2-6)]\,dx = \int_0^2 2x\,dx + \int_2^3 (x - x^2 + 6)\,dx = x^2 \big]_0^2 + (\frac{1}{2}x^2 - \frac{1}{3}x^3 + 6x) \big]_2^3 = 4 + (\frac{9}{2} - 9 + 18) - (2 - \frac{8}{3} + 12) = \frac{37}{6}$. Notice that the region had to be broken into two pieces before we could integrate.

CHAPTER 22
Volume

22.1 Derive the formula $V = \frac{4}{3}\pi r^3$ for the volume of a sphere of radius r.

▮ Consider the upper semicircle $y = \sqrt{r^2 - x^2}$ (Fig. 22-1). If we rotate it about the x-axis, the sphere of radius r results. By the disk formula, $V = \pi \int_{-r}^{r} y^2\, dx = \pi \int_{-r}^{r} (r^2 - x^2)\, dx = \pi(r^2 x - \frac{1}{3}x^3)\,]_{-r}^{r} = \pi[(r^3 - \frac{1}{3}r^3) - (-r^3 + \frac{1}{3}r^3)] = \frac{4}{3}\pi r^3$.

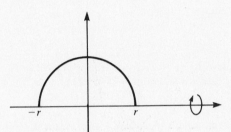
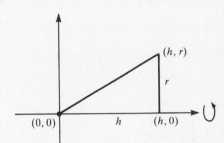

Fig. 22-1 **Fig. 22-2**

22.2 Derive the formula $V = \frac{1}{3}\pi r^2 h$ for the volume of a right circular cone of height h and radius of base r.

▮ Refer to Fig. 22-2. Consider the right triangle with vertices $(0, 0)$, $(h, 0)$, and (h, r). If this is rotated about the x-axis, a right circular cone of height h and radius of base r results. Note that the hypotenuse of the triangle lies on the line $y = (r/h)x$. Then, by the disk formula,

$$V = \pi \int_0^h y^2\, dx = \pi \int_0^h \frac{r^2}{h^2} x^2\, dx = \frac{\pi r^2}{h^2} \cdot \frac{1}{3} x^3 \Big]_0^h = \frac{\pi r^2}{h^2} \cdot \frac{1}{3} h^3 = \frac{1}{3}\pi r^2 h$$

In Problems 22.3–22.19, find the volume generated by revolving the given region about the given axis.

22.3 The region above the curve $y = x^3$, under the line $y = 1$, and between $x = 0$ and $x = 1$; about the x-axis.

▮ See Fig. 22-3. The upper curve is $y = 1$, and the lower curve is $y = x^3$. We use the circular ring formula: $V = \pi \int_0^1 [1^2 - (x^3)^2]\, dx = \pi(x - \frac{1}{7}x^7)\,]_0^1 = \pi(1 - \frac{1}{7}) = \frac{6}{7}\pi$.

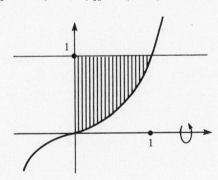

Fig. 22-3

22.4 The region of Problem 22.3, about the y-axis.

▮ We integrate along the y-axis from 0 to 1. The upper curve is $x = y^{1/3}$, the lower curve is the y-axis, and we use the disk formula: $V = \pi \int_0^1 x^2\, dy = \pi \int_0^1 y^{2/3}\, dy = \pi \cdot \frac{3}{5} y^{5/3}\,]_0^1 = \frac{3\pi}{5}$.

22.5 The region below the line $y = 2x$, above the x-axis, and between $x = 0$ and $x = 1$; about the x-axis. (See Fig. 22-4.)

▮ We use the disk formula: $V = \pi \int_0^1 y^2\, dx = \pi \int_0^1 4x^2\, dx = \pi \cdot \frac{4}{3}x^3\,]_0^1 = \frac{4}{3}\pi$.

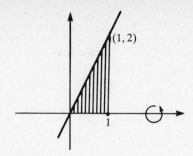

Fig. 22-4

22.6 The region of Problem 22.5, about the y-axis.

▮ We use the cylindrical shell formula: $V = 2\pi \int_0^1 xy\, dx = 2\pi \int_0^1 x \cdot 2x\, dx = 4\pi \int_0^1 x^2\, dx = 4\pi \cdot \frac{1}{3}x^3 \,]_0^1 = \frac{4}{3}\pi$.

22.7 The region of Problem 21.39, about the x-axis.

▮ The curves intersect at $(0,0)$ and $(1,1)$. The upper curve is $y = \sqrt{x}$, and the lower curve is $y = x^2$. We use the circular ring formula: $V = \pi \int_0^1 [(\sqrt{x})^2 - (x^2)^2]\, dx = \pi \int_0^1 (x - x^4)\, dx = \pi(\frac{1}{2}x^2 - \frac{1}{5}x^5) \,]_0^1 = \pi(\frac{1}{2} - \frac{1}{5}) = \frac{3}{10}\pi$.

22.8 The region inside the circle $x^2 + y^2 = r^2$ with $0 \le x \le a < r$; about the y-axis. (This gives the volume cut from a sphere of radius r by a pipe of radius a whose axis is a diameter of the sphere.)

▮ We shall consider only the region above the x-axis (Fig. 22-5), and then, by symmetry, double the result. We use the cylindrical shell formula: $V = 2\pi \int_0^a xy\, dx = 2\pi \int_0^a x\sqrt{r^2 - x^2}\, dx = 2\pi \cdot (-\frac{1}{2}) \int_0^a (r^2 - x^2)^{1/2}(-2x)\, dx = -\pi \cdot \frac{2}{3}(r^2 - x^2)^{3/2} \,]_0^a = -\frac{2\pi}{3}[(r^2 - a^2)^{3/2} - (r^2)^{3/2}] = \frac{2\pi}{3}[r^3 - (r^2 - a^2)^{3/2}]$. We multiply by 2 to obtain the answer $\frac{4\pi}{3}[r^3 - (r^2 - a^2)^{3/2}]$.

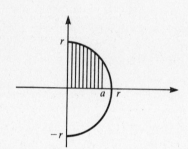

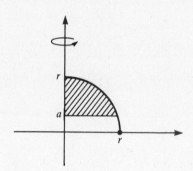

Fig. 22-5   Fig. 22-6

22.9 The region below the quarter-circle $x^2 + y^2 = r^2$ ($x \ge 0$, $y \ge 0$) and above the line $y = a$, where $0 < a < r$; about the y-axis. (This gives the volume of a polar cap of a sphere.)

▮ See Fig. 22-6. We use the disk formula along the y-axis: $V = \pi \int_a^r x^2\, dy = \pi \int_a^r (r^2 - y^2)\, dy = \pi(r^2 y - \frac{1}{3}y^3) \,]_a^r = \pi[(r^3 - \frac{1}{3}r^3) - (r^2 a - \frac{1}{3}a^3)] = \pi(\frac{2}{3}r^3 - ar^2 + \frac{1}{3}a^3)$.

22.10 The region bounded by $y = 1 + x^2$ and $y = 5$; about the x-axis. (See Fig. 22-7.)

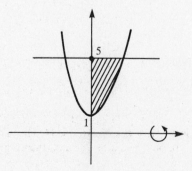

Fig. 22-7

▍ We use the circular ring formula for the region in the first quadrant, and then double the result: $V = \pi \int_0^2 [(5)^2 - (1 + x^2)^2]\, dx = \pi \int_0^2 [25 - (1 + 2x^2 + x^4)]\, dx = \pi \int_0^2 (24 - 2x^2 - x^4)\, dx = \pi(24x - \frac{2}{3}x^3 - \frac{1}{5}x^5)\,]_0^2$ $= \pi(48 - \frac{16}{3} - \frac{32}{5}) = 544\pi/15$. Doubling this, we obtain $1088\pi/15$.

22.11 Solve Problem 22.10 by means of the cylindrical shell formula.

▍ We compute the volume in the first quadrant and then double it. $V = 2\pi \int_1^5 xy\, dy = 2\pi \int_1^5 \sqrt{y-1}\, y\, dy$. Let $u = y - 1$, $y = u + 1$, $du = dy$. Then $V = 2\pi \int_0^4 \sqrt{u}(u + 1)\, du = 2\pi \int_0^4 (u^{3/2} + u^{1/2})\, du = 2\pi(\frac{2}{5}u^{5/2} + \frac{2}{3}u^{3/2})\,]_0^4 = 2\pi(\frac{64}{5} + \frac{16}{3}) = 544\pi/15$, which, doubled, yields the same answer as in Problem 22.10.

22.12 The region inside the circle $x^2 + (y - b)^2 = a^2$ $(0 < a < b)$; about the x-axis. (This yields the volume of a doughnut.)

▍ Refer to Fig. 22-8; we deal with the region in the first quadrant, and then double it. Use the cylindrical shell formula: $V = 2\pi \int_{b-a}^{b+a} xy\, dy = 2\pi \int_{b-a}^{b+a} \sqrt{a^2 - (y - b)^2}\, y\, dy$. Let $u = y - b$, $y = u + b$, $du = dy$. Then $V = 2\pi \int_{-a}^a \sqrt{a^2 - u^2}\,(u + b)\, du = 2\pi (\int_{-a}^a \sqrt{a^2 - u^2}\, u\, du + b \int_{-a}^a \sqrt{a^2 - u^2}\, du)$. The first integral is 0, since the integrand is an odd function (see Problem 20.48). The second integral is the area of a semicircle of radius a (see Problem 20.71) and is therefore equal to $\frac{1}{2}\pi a^2$. Hence, $V = 2\pi b \cdot \frac{1}{2}\pi a^2 = \pi^2 ba^2$, which, doubled, yields the answer $2\pi^2 ba^2$.

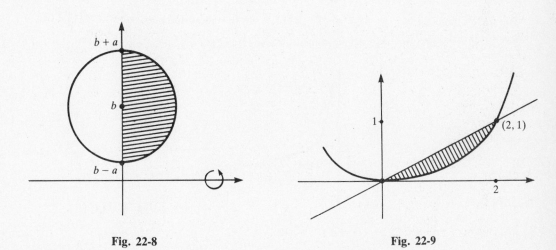

Fig. 22-8   Fig. 22-9

22.13 The region bounded by $x^2 = 4y$ and $y = \frac{1}{2}x$; about the y-axis.

▍ See Fig. 22-9. The curves $x^2 = 4y$ and $y = \frac{1}{2}x$ intersect at $(0, 0)$ and $(2, 1)$. We use the circular ring formula: $V = \pi \int_0^1 [4y - (2y)^2]\, dy = \pi \int_0^1 (4y - 4y^2)\, dy = \pi(2y^2 - \frac{4}{3}y^3)\,]_0^1 = \pi(2 - \frac{4}{3}) = 2\pi/3$.

22.14 Solve Problem 22.13 by means of the cylindrical shell formula.

▍
$$V = 2\pi \int_0^2 x(\tfrac{1}{2}x - \tfrac{1}{4}x^2)\, dx = \frac{\pi}{2} \int_0^2 (2x^2 - x^3)\, dx = \frac{\pi}{2}(\tfrac{2}{3}x^3 - \tfrac{1}{4}x^4)\,]_0^2 = \frac{\pi}{2}(\tfrac{16}{3} - 4) = \frac{2\pi}{3}$$

22.15 The region of Problem 22.13; about the x-axis.

▍ Use the circular ring formula: $V = \pi \int_0^2 [(\frac{1}{2}x)^2 - (\frac{1}{4}x^2)^2]\, dx = \pi \int_0^2 (\frac{1}{4}x^2 - \frac{1}{16}x^4)\, dx = \pi(\frac{1}{12}x^3 - \frac{1}{80}x^5)\,]_0^2 = \pi(\frac{2}{3} - \frac{2}{5}) = 4\pi/15$.

22.16 The region bounded by $y = 4/x$ and $y = (x - 3)^2$; about the x-axis. (See Fig. 22-10.)

▍ Solving $y = 4/x$ and $y = (x - 3)^2$, we get $4 = x^3 - 6x^2 + 9x$, $x^3 - 6x^2 + 9x - 4 = 0$. $x = 1$ is a root, and, dividing $x^3 - 6x^2 + 9x - 4$ by $x - 1$, we obtain $x^2 - 5x + 4 = (x - 1)(x - 4)$, with the additional root $x = 4$. Hence, the intersection points are $(1, 4)$ and $(4, 1)$. The slopes of the tangent lines at $(1, 4)$ (that is, the derivatives at $x = 1$) are both equal to -4, and, therefore, the curves are tangent at $(1, 4)$. The hyperbola $xy = 4$ is the upper curve. The circular ring formula yields $V = \pi \int_1^4 \{(4/x)^2 - [(x - 3)^2]^2\}\, dx = \pi \int_1^4 [16x^{-2} - (x - 3)^4]\, dx = \pi(-16x^{-1} - \frac{1}{5}(x - 3)^5)\,]_1^4 = \pi[(-4 - \frac{1}{5}) - (-16 + \frac{32}{5})] = 27\pi/5$.

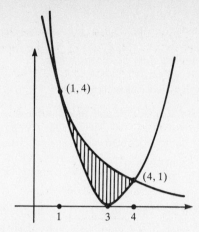

(1, 4)

(4, 1)

1   3   4

Fig. 22-10

22.17   The region of Problem 22.16; about the y-axis.

▐ Use the difference of cylindrical shells: $V = 2\pi \int_1^4 x\left[\dfrac{4}{x} - (x-3)^2\right] dx = 2\pi \int_1^4 (4 - x^3 + 6x^2 - 9x)\, dx = 2\pi(4x - \frac{1}{4}x^4 + 2x^3 - \frac{9}{2}x^2)\,]_1^4 = 2\pi[(16 - 64 + 128 - 72) - (4 - \frac{1}{4} + 2 - \frac{9}{2})] = 2\pi(\frac{27}{4}) = 27\pi/2$.

22.18   The region bounded by $xy = 1$, $x = 1$, $x = 3$, $y = 0$; about the x-axis.

▐ See Fig. 22-11. By the disk formula, $V = \pi \int_1^3 y^2\, dx = \pi \int_1^3 x^{-2}\, dx = \pi(-x^{-1})\,]_1^3 = -\pi(\frac{1}{3} - 1) = 2\pi/3$.

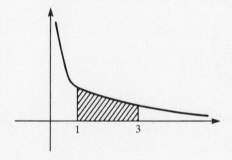

1   3

Fig. 22-11

22.19   The region of Problem 22.18; about the y-axis.

▐ Use the cylindrical shell formula: $V = 2\pi \int_1^3 xy\, dx = 2\pi \int_1^3 1\, dx = 2\pi x\,]_1^3 = 2\pi(3 - 1) = 4\pi$.

In Problems 22.20–22.23, use the cross-section formula to find the volume of the given solid.

22.20   The solid has a base which is a circle of radius r. Each cross section perpendicular to a fixed diameter of the circle is an isosceles triangle with altitude equal to one-half of its base.

▐ Let the center of the circular base be the origin, and the fixed diameter the x-axis (Fig. 22-12). The circle has the equation $x^2 + y^2 = r^2$. Then the base of the triangle is $2\sqrt{r^2 - x^2}$, the altitude is $\sqrt{r^2 - x^2}$, and the area A of the triangle is $\frac{1}{2}(2\sqrt{r^2 - x^2})(\sqrt{r^2 - x^2}) = r^2 - x^2$. Hence, by the cross-section formula, $V = \int_{-r}^r A\, dx = \int_{-r}^r (r^2 - x^2)\, dx = (r^2 x - \frac{1}{3}x^3)\,]_{-r}^r = (r^3 - \frac{1}{3}r^3) - (-r^3 + \frac{1}{3}r^3) = \frac{4}{3}r^3$.

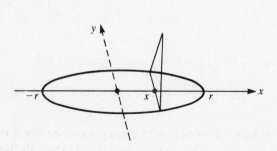

Fig. 22-12

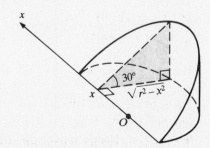

Fig. 22-13

22.21 The solid is a wedge, cut from a perfectly round tree of radius r by two planes, one perpendicular to the axis of the tree and the other intersecting the first plane at an angle of 30° along a diameter. (See Fig. 22-13.)

▮ Let the x-axis be the intersection of the two planes, with the origin on the tree's axis. Then a typical cross section is a right triangle with base $\sqrt{r^2-x^2}$ and height $h=\sqrt{r^2-x^2}\tan 30°=\frac{1}{\sqrt{3}}\sqrt{r^2-x^2}$. So, the area A is $\frac{1}{2}\sqrt{r^2-x^2}\cdot\frac{1}{\sqrt{3}}\sqrt{r^2-x^2}=\frac{1}{2\sqrt{3}}(r^2-x^2)$. By symmetry, we can compute the volume for $x\ge 0$ and then double the result. The cross-section formula yields the volume

$$V=2\int_0^r \frac{1}{2\sqrt{3}}(r^2-x^2)\,dx=\frac{1}{\sqrt{3}}(r^2x-\tfrac{1}{3}x^3)\Big]_0^r=\frac{1}{\sqrt{3}}(r^3-\tfrac{1}{3}r^3)=\frac{1}{\sqrt{3}}\cdot\frac{2}{3}r^3=\frac{2\sqrt{3}}{9}r^3$$

22.22 A square pyramid with a height of h units and a base of side r units.

▮ Locate the x-axis perpendicular to the base, with the origin at the center of the base (Fig. 22-14). By similar right triangles, $\dfrac{d}{e}=\dfrac{h-x}{h}$ and $\dfrac{A(x)}{r^2}=\left(\dfrac{d}{e}\right)^2$. So, $A(x)=r^2\left(\dfrac{h-x}{h}\right)^2$, and, by the cross-section formula,

$$V=\int_0^h A(x)\,dx=\frac{r^2}{h^2}\int_0^h (h-x)^2\,dx=\frac{r^2}{h^2}(-\tfrac{1}{3})(h-x)^3\Big]_0^h=-\frac{r^2}{3h^2}(0-h^3)=\tfrac{1}{3}r^2h$$

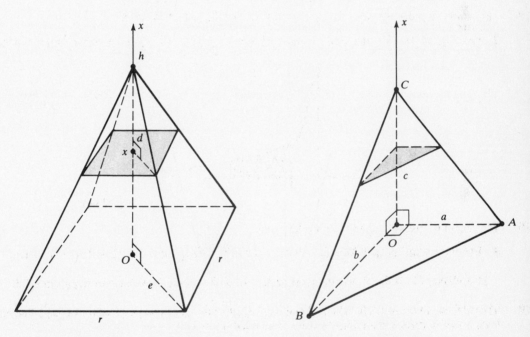

Fig. 22-14   **Fig. 22-15**

22.23 The tetrahedron formed by three mutually perpendicular edges of lengths a, b, c.

▮ Let the origin be the intersection of the edges, and let the x-axis lie along the edge of length c (Fig. 22-15). A typical cross section is a right triangle with legs of lengths d and e, parallel respectively to the edges of lengths a and b. By similar triangles, $\dfrac{d}{a}=\dfrac{c-x}{c}$ and $\dfrac{e}{b}=\dfrac{c-x}{c}$. So, the area $A(x)=\frac{1}{2}\left[\dfrac{a(c-x)}{c}\right]\left[\dfrac{b(c-x)}{c}\right]=\dfrac{ab}{2c^2}(c-x)^2$. By the cross-section formula

$$V=\frac{ab}{2c^2}\int_0^c (c-x)^2\,dx=\frac{ab}{2c^2}\left(-\frac{1}{3}\right)(c-x)^3\Big]_0^c=-\frac{ab}{6c^2}(0-c^3)=\frac{abc}{6}$$

22.24 Let $\mathcal{R}$ be the region between $y=x^3$, $x=1$, and $y=0$. Find the volume of the solid obtained by rotating region $\mathcal{R}$ about $y=-1$.

▮ The same volume is obtained by rotating about the x-axis ($y=0$) the region obtained by raising $\mathcal{R}$ one unit, that is, the region bounded by $y=x^3+1$, $y=1$, and $x=1$ (see Fig. 22-16). By the circular ring formula, this is $V=\pi\int_0^1 [(x^3+1)^2-1^2]\,dx=\pi\int_0^1 (x^6+2x^3)\,dx=\pi(\tfrac{1}{7}x^7+\tfrac{1}{2}x^4)\Big]_0^1=\pi(\tfrac{1}{7}+\tfrac{1}{2})=9\pi/14$.

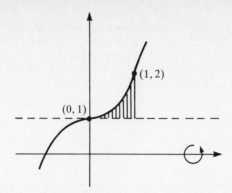

Fig. 22-16

22.25 Let $\mathcal{R}$ be the region in the first quadrant between the curves $y = x^2$ and $y = 2x$. Find the volume of the solid obtained by rotating $\mathcal{R}$ about the x-axis.

❙ By simultaneously solving $y = x^2$ and $y = 2x$, we see that the curves intersect at $(0, 0)$ and $(2, 4)$. The line $y = 2x$ is the upper curve. So, the circular ring formula yields $V = \pi \int_0^2 [(2x)^2 - (x^2)^2]\, dx = \pi \int_0^2 (4x^2 - x^4)\, dx = \pi(\frac{4}{3}x^3 - \frac{1}{5}x^5)\,]_0^2 = \pi(\frac{32}{3} - \frac{32}{5}) = 64\pi/15$.

22.26 Same as Problem 22.25, but the rotation is around the y-axis.

❙ Here let us use the difference of cylindrical shells: $V = 2\pi \int_0^2 x(2x - x^2)\, dx = 2\pi \int_0^2 (2x^2 - x^3)\, dx = 2\pi(\frac{2}{3}x^3 - \frac{1}{4}x^4)\,]_0^2 = 2\pi(\frac{16}{3} - 4) = 8\pi/3$.

22.27 Let $\mathcal{R}$ be the region above $y = (x - 1)^2$ and below $y = x + 1$ (see Fig. 22-17). Find the volume of the solid obtained by rotating $\mathcal{R}$ about the x-axis.

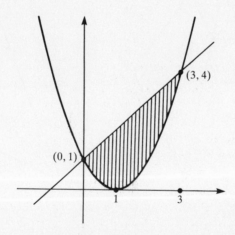

Fig. 22-17

❙ By setting $x + 1 = (x - 1)^2$ and solving, we obtain the intersection points $(0, 1)$ and $(3, 4)$. The circular ring formula yields $V = \pi \int_0^3 \{(x + 1)^2 - [(x - 1)^2]^2\}\, dx = \pi \int_0^3 [(x + 1)^2 - (x - 1)^4]\, dx = \pi(\frac{1}{3}(x + 1)^3 - \frac{1}{5}(x - 1)^5)\,]_0^3 = \pi[(\frac{64}{3} - \frac{32}{5}) - (\frac{1}{3} + \frac{1}{5})] = 216\pi/15$.

22.28 Same as Problem 22.27, but the rotation is around the line $y = -1$.

❙ Raise the region one unit and rotate around the x-axis. The bounding curves are now $y = x + 2$ and $y = (x - 1)^2 + 1$. By the circular ring formula, $V = \pi \int_0^3 \{(x + 2)^2 - [(x - 1)^2 + 1]^2\}\, dx = \pi \int_0^3 \{(x + 2)^2 - [(x - 1)^2 + 2(x - 1) + 1]\}\, dx = \pi \int_0^3 [(x + 2)^2 - (x - 1)^2 - 2x + 1]\, dx = \pi(\frac{1}{3}(x + 2)^3 - \frac{1}{3}(x - 1)^3 - x^2 + x)\,]_0^3 = \pi[(\frac{125}{3} - \frac{8}{3} - 9 + 3) - (\frac{8}{3} + \frac{1}{3})] = 30\pi$.

22.29 Find the volume of the solid generated when the region bounded by $y^2 = 4x$ and $y = 2x - 4$ is revolved about the y-axis.

❚ Solving $y = 2x - 4$ and $y^2 = 4x$ simultaneously, we obtain $y^2 - 2y - 8 = 0$, $y = 4$ or $y = -2$. Thus, the curves intersect at $(4, 4)$ and $(1, -2)$, as shown in Fig. 22-18. We integrate along the y-axis, using the circular ring formula: $V = \pi \int_{-2}^{4} \left[\left(\frac{y+4}{2} \right)^2 - \left(\frac{y^2}{4} \right)^2 \right] dy = \pi \int_{-2}^{4} \left[\frac{1}{4}(y+4)^2 - \frac{1}{16} y^4 \right] dy = \pi \left(\frac{1}{12}(y+4)^3 - \frac{1}{80} y^5 \right) \big]_{-2}^{4} = \pi \left[\left(\frac{128}{3} - \frac{64}{5} \right) - \left(\frac{2}{3} + \frac{2}{5} \right) \right] = 144\pi/5$.

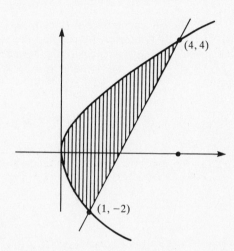

Fig. 22-18

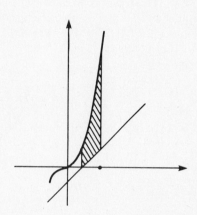

Fig. 22-19

22.30 Let $\mathcal{R}$ be the region bounded by $y = x^3$, $x = 1$, $x = 2$, and $y = x - 1$. Find the volume of the solid generated when $\mathcal{R}$ is revolved about the x-axis.

❚ $y = x^3$ lies above $y = x - 1$ for $1 \le x \le 2$ (see Fig. 22-19). So, we can use the circular ring formula: $V = \pi \int_{1}^{2} [(x^3)^2 - (x-1)^2] \, dx = \pi \int_{1}^{2} [x^6 - (x-1)^2] \, dx = \pi \left(\frac{1}{7} x^7 - \frac{1}{3}(x-1)^3 \right) \big]_{1}^{2} = \pi \left[\left(\frac{128}{7} - \frac{1}{3} \right) - \left(\frac{1}{7} - 0 \right) \right] = 374\pi/21$.

22.31 Same as Problem 22.30, but revolving about the y-axis.

❚ We use the difference of cylindrical shells: $V = 2\pi \int_{1}^{2} x[x^3 - (x-1)] \, dx = 2\pi \int_{1}^{2} (x^4 - x^2 + x) \, dx = 2\pi \left(\frac{1}{5} x^5 - \frac{1}{3} x^3 + \frac{1}{2} x^2 \right) \big]_{1}^{2} = 2\pi \left[\left(\frac{32}{5} - \frac{8}{3} + 2 \right) - \left(\frac{1}{5} - \frac{1}{3} + \frac{1}{2} \right) \right] = 161\pi/15$.

22.32 Let $\mathcal{R}$ be the region bounded by the curves $y = x^2 - 4x + 6$ and $y = x + 2$. Find the volume of the solid generated when $\mathcal{R}$ is rotated about the x-axis.

❚ Solving $y = x^2 - 4x + 6$ and $y = x + 2$, we obtain $x^2 - 5x + 4 = 0$, $x = 4$ or $x = 1$. So, the curves meet at $(4, 6)$ and $(1, 3)$. Note that $y = x^2 - 4x + 6 = (x - 2)^2 + 2$. Hence, the latter curve is a parabola with vertex $(2, 2)$ (see Fig. 22-20). We use the circular ring formula: $V = \pi \int_{1}^{4} \{ (x+2)^2 - [(x-2)^2 + 2]^2 \} \, dx = \pi \int_{1}^{4} [(x+2)^2 - (x-2)^4 - 4(x-2)^2 - 4] \, dx = \pi \left(\frac{1}{3}(x+2)^3 - \frac{1}{5}(x-2)^5 - \frac{4}{3}(x-2)^3 - 4x \right) \big]_{1}^{4} = \pi \left[\left(72 - \frac{32}{5} - \frac{32}{3} - 16 \right) - \left(9 + \frac{1}{5} + \frac{4}{3} - 4 \right) \right] = 162\pi/5$.

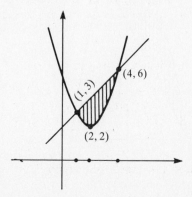

Fig. 22-20

22.33 Same as Problem 22.32, but with the rotation around the y-axis.

▌ We use the difference of cylindrical shells: $V = 2\pi \int_1^4 x[(x+2) - (x^2 - 4x + 6)] \, dx = 2\pi \int_1^4 x(5x - x^2 - 4) \, dx = 2\pi \int_1^4 (5x^2 - x^3 - 4x) \, dx = 2\pi(\frac{5}{3}x^3 - \frac{1}{4}x^4 - 2x^2)\,]_1^4 = 2\pi\{[\frac{5}{3}(64) - 64 - 32] - (\frac{5}{3} - \frac{1}{4} - 2)]\} = 45\pi/2$.

22.34 Let $\mathcal{R}$ be the region bounded by $y = 12 - x^3$ and $y = 12 - 4x$. Find the volume generated by revolving $\mathcal{R}$ about the x-axis.

▌ Solving $12 - x^3 = 12 - 4x$, we get $x = 0$ and $x = \pm 2$. So, the curves intersect at $(0, 12)$, $(2, 4)$, and $(-2, 20)$. The region consists of two pieces as shown in Fig. 22-21. By symmetry, we can find the volume generated by the part in the first quadrant and double the result. Using the circular ring formula, we have $V = 2\pi \int_0^2 [(12 - x^3)^2 - (12 - 4x)^2] \, dx = 2\pi \int_0^2 [(144 - 24x^3 + x^6) - (144 - 96x + 16x^2)] \, dx = 2\pi \int_0^2 (x^6 - 24x^3 - 16x^2 + 96x) \, dx = 2\pi(\frac{1}{7}x^7 - 6x^4 - \frac{16}{3}x^3 + 48x^2)\,]_0^2 = 2\pi(\frac{128}{7} - 96 - \frac{128}{3} + 192) = 3008\pi/21$.

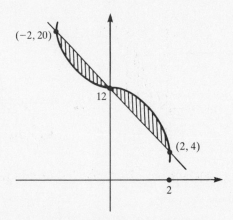

Fig. 22-21

22.35 Same as Problem 22.34, but the region is revolved about the y-axis.

▌ By symmetry, we need only double the volume generated by the piece in the first quadrant. We use the difference of cylindrical shells: $V = 2 \cdot 2\pi \int_0^2 x[(12 - x^3) - (12 - 4x)] \, dx = 4\pi \int_0^2 x(4x - x^3) \, dx = 4\pi \int_0^2 (4x^2 - x^4) \, dx = 4\pi(\frac{4}{3}x^3 - \frac{1}{5}x^5)\,]_0^2 = 4\pi(\frac{32}{3} - \frac{32}{5}) = 256\pi/15$.

22.36 Let $\mathcal{R}$ be the region bounded by $y = 9 - x^2$ and $y = 2x + 6$ (Fig. 22-22). Find, using the circular ring formula, the volume generated when $\mathcal{R}$ is revolved about the x-axis.

▌ Solving $9 - x^2 = 2x + 6$, we get $x^2 + 2x - 3 = 0$, $(x + 3)(x - 1) = 0$, $x = -3$ or $x = 1$. Thus, the curves meet at $(1, 8)$ and $(-3, 0)$. Then $V = \pi \int_{-3}^1 [(9 - x^2)^2 - (2x + 6)^2] \, dx = \pi \int_{-3}^1 (81 - 18x^2 + x^4 - 4x^2 - 24x - 36) \, dx = \pi \int_{-3}^1 (45 - 24x - 22x^2 + x^4) \, dx = \pi(45x - 12x^2 - \frac{22}{3}x^3 + \frac{1}{5}x^5)\,]_{-3}^1 = \pi[(45 - 12 - \frac{22}{3} + \frac{1}{5}) - (-135 - 108 + 198 - \frac{243}{5})] = 1792\pi/15$.

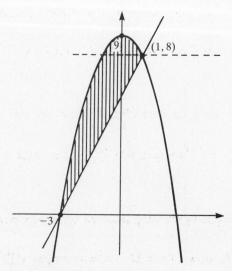

Fig. 22-22

22.37 Solve Problem 22.36 by the cylindrical shell formula.

❚ We must integrate along the y-axis, and it is necessary to break the region into two pieces by the line $y = 8$. Then we obtain $V = 2\pi \int_0^8 y\left[\dfrac{y-6}{2} - (-\sqrt{9-y})\right] dy + 2\pi \int_8^9 y[\sqrt{9-y} - (-\sqrt{9-y})] dy = \pi \int_0^8 (\frac{1}{2}y^2 - 3y + y\sqrt{9-y}) dy + 2\pi \int_8^9 2y\sqrt{9-y}\, dy$. We have to find $\int y\sqrt{9-y}\, dy$. Let $u = 9 - y$, $y = 9 - u$, $du = -dy$. Then $\int y\sqrt{9-y}\, dy = -\int (9-u)\sqrt{u}\, du = \int (u^{3/2} - 9u^{1/2})\, du = \frac{2}{5}u^{5/2} - 6u^{3/2} = \frac{2}{5}u^{3/2}(u-15) = -\frac{2}{5}(9-y)^{3/2}(y+6)$. Hence, $V = 2\pi\{(\frac{1}{6}y^3 - \frac{3}{2}y^2 - \frac{2}{5}(9-y)^{3/2}(y+6))\,]_0^8 - \frac{4}{5}(9-y)^{3/2}(y+6)\,]_8^9\} = 2\pi\{(\frac{256}{3} - 96 - \frac{28}{5}) - \frac{4}{5}(0-14)\} = 2\pi(\frac{296}{5} + \frac{256}{3} - 96 + \frac{56}{5}) = 1792\pi/15$.

22.38 Let $\mathcal{R}$ be the region bounded by $y = x^3 + x$, $y = 0$, and $x = 1$ (Fig. 22-23). Find the volume of the solid obtained by rotating $\mathcal{R}$ about the y-axis.

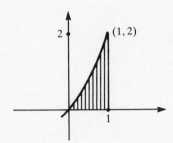

Fig. 22-23

❚ We use the cylindrical shell formula: $V = 2\pi \int_0^1 x(x^3 + x)\, dx = 2\pi \int_0^1 (x^4 + x^2)\, dx = 2\pi(\frac{1}{5}x^5 + \frac{1}{3}x^3)\,]_0^1 = 2\pi(\frac{1}{5} + \frac{1}{3}) = 16\pi/15$.

22.39 Same as Problem 22.38, but rotating about the x-axis.

❚ We use the disk formula: $V = \pi \int_0^1 (x^3 + x)^2\, dx = \pi \int_0^1 (x^6 + 2x^4 + x^2)\, dx = \pi(\frac{1}{7}x^7 + \frac{2}{5}x^5 + \frac{1}{3}x^3)\,]_0^1 = \pi(\frac{1}{7} + \frac{2}{5} + \frac{1}{3}) = 92\pi/105$.

22.40 Same as Problem 22.38, but rotating about the line $x = -2$.

❚ The same volume can be obtained by moving the region two units to the right and rotating about the y-axis. The new curve is $y = (x-2)^3 + x - 2$, and the interval of integration is $2 \le x \le 3$. By the cylindrical shell formula, $V = 2\pi \int_2^3 x[(x-2)^3 + x - 2]\, dx = 2\pi \int_2^3 x(x^3 - 6x^2 + 12x - 8 + x - 2)\, dx = 2\pi \int_2^3 (x^4 - 6x^3 + 13x^2 - 10x)\, dx = 2\pi(\frac{1}{5}x^5 - \frac{3}{2}x^4 + \frac{13}{3}x^3 - 5x^2)\,]_2^3 = 2\pi[(\frac{243}{5} - \frac{243}{2} + 117 - 45) - (\frac{32}{5} - 24 + \frac{104}{3} - 20)] = 61\pi/15$.

22.41 Find the volume of the solid generated when one arch of the curve $y = \sin x$, from $x = 0$ to $x = \pi$, is rotated about the x-axis.

❚ The disk formula yields $V = \pi \int_0^\pi \sin^2 x\, dx = \pi \int_0^\pi \frac{1}{2}(1 - \cos 2x)\, dx = \dfrac{\pi}{2}(x - \frac{1}{2}\sin 2x)\,]_0^\pi = \dfrac{\pi}{2}[(\pi - 0) - (0 - 0)] = \pi^2/2$.

22.42 Find the volume of the solid of rotation generated when the curve $y = \tan x$, from $x = 0$ to $x = \pi/4$, is rotated about the x-axis.

❚ The disk formula gives $V = \pi \int_0^{\pi/4} \tan^2 x\, dx = \pi \int_0^{\pi/4} (\sec^2 x - 1)\, dx = \pi(\tan x - x)\,]_0^{\pi/4} = \pi\left[\left(1 - \dfrac{\pi}{4}\right) - (0 - 0)\right] = \pi\left(1 - \dfrac{\pi}{4}\right)$.

22.43 Find the volume of the solid generated by revolving about the x-axis the region bounded by $y = \sec x$, $y = 0$, $x = 0$, and $x = \pi/4$.

❚ See Fig. 22-24. By the disk formula, $V = \pi \int_0^{\pi/4} \sec^2 x\, dx = \pi(\tan x)\,]_0^{\pi/4} = \pi(1 - 0) = \pi$.

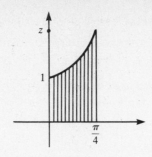

Fig. 22-24

22.44 Calculate the volume of the solid paraboloid generated when the region in the first quadrant under the parabola $x = y^2$, between $x = 0$ and $x = b$, is rotated about the x-axis.

▌ Use the disk formula: $V = \pi \int_0^b y^2 \, dx = \pi \int_0^b x \, dx = \pi(\frac{1}{2}x^2)]_0^b = \frac{1}{2}\pi b^2$.

22.45 Same as Problem 22.44, but the rotation is about the y-axis.

▌ Use the cylindrical shells formula: $V = \pi \int_0^b xy \, dx = 2\pi \int_0^b x\sqrt{x} \, dx = 2\pi \int_0^b x^{3/2} \, dx = 2\pi(\frac{2}{5}x^{5/2})]_0^b = \frac{4}{5}\pi b^{5/2}$.

22.46 Find the volume of the solid generated when the region in the first quadrant under the hyperbola $xy = 1$, between $x = 1$ and $x = b > 1$, is rotated about the x-axis.

▌ The disk formula applies: $V = \pi \int_1^b \left(\frac{1}{x}\right)^2 dx = \pi \int_1^b x^{-2} \, dx = \pi\left(-\frac{1}{x}\right)\Big]_1^b = \pi\left[-\frac{1}{b} - (-1)\right] = \pi\left(1 - \frac{1}{b}\right)$. Note that this approaches π as $b \to +\infty$.

22.47 Same as Problem 22.46, but with the rotation around the y-axis.

▌ Use the cylindrical shell formula: $V = 2\pi \int_1^b xy \, dx = 2\pi \int_1^b x(1/x) \, dx = 2\pi \int_1^b 1 \, dx = 2\pi x\,]_1^b = 2\pi(b - 1)$.

22.48 Find the volume of the ellipsoid obtained when the ellipse $\dfrac{x^2}{a^2} + \dfrac{y^2}{b^2} = 1$ is rotated (a) about the x-axis, (b) about the y-axis.

▌ (a) We double the value obtained from the disk formula applied to the part of the region in the first quadrant (see Fig. 22-25):

$$V = 2 \cdot \pi \int_0^a y^2 \, dx = 2\pi \int_0^a b^2\left(1 - \frac{x^2}{a^2}\right) dx = 2\pi b^2\left(x - \frac{1}{3a^2} x^3\right)\Big]_0^a = 2\pi b^2\left(a - \frac{a}{3}\right) = \frac{4}{3}\pi ab^2$$

(b) Interchange a and b in (a): $V = \frac{4}{3}\pi ba^2$.

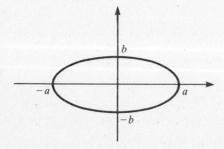

Fig. 22-25

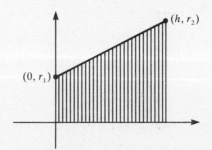

Fig. 22-26

22.49 Find the volume of the solid obtained by rotating about the x-axis the region in the first quadrant under the line segment from $(0, r_1)$ to (h, r_2), where $0 < r_1 < r_2$ and $0 < h$. (See Fig. 22-26. Note that this is the volume of a frustum of a cone with height h and radii r_1 and r_2 of the bases.)

▌ The equation of the line is $y = \dfrac{1}{h}[(r_2 - r_1)x + r_1 h]$. By the disk formula,

$$V = \pi \int_0^h y^2 \, dx = \frac{\pi}{h^2} \int_0^h [(r_2 - r_1)x + r_1 h]^2 \, dx = \frac{\pi}{h^2} \int_0^h [(r_2 - r_1)^2 x^2 + 2r_1 h(r_2 - r_1)x + r_1^2 h^2] \, dx$$

$$= \frac{\pi}{h^2}\left(\frac{1}{3}(r_2 - r_1)^2 x^3 + r_1 h(r_2 - r_1)x^2 + r_1^2 h^2 x\right)\Big]_0^h = \frac{\pi}{h^2}\left[\frac{1}{3}(r_2 - r_1)^2 h^3 + r_1(r_2 - r_1)h^3 + r_1^2 h^3\right]$$

$$= \pi h[\frac{1}{3}(r_2 - r_1)^2 + r_1(r_2 - r_1) + r_1^2] = \frac{1}{3}\pi h(r_2^2 + r_1 r_2 + r_1^2)$$

22.50 A solid has a circular base of radius r. Find the volume of the solid if every planar section perpendicular to a fixed diameter is a semicircle.

▌ Let the x-axis be the fixed diameter, with the center of the circle as the origin. Then the radius of the semicircle at abscissa x is $\sqrt{r^2 - x^2}$ (see Fig. 22-27), and, therefore, its area $A(x)$ is $\frac{1}{2}\pi(r^2 - x^2)$. The cross-section formula yields $V = \int_{-r}^{r} A(x)\, dx = \frac{1}{2}\pi \int_{-r}^{r} (r^2 - x^2)\, dx = \frac{1}{2}\pi(r^2 x - \frac{1}{3}x^3)\,]_{-r}^{r} = \frac{1}{2}\pi[(r^3 - \frac{1}{3}r^3) - (-r^3 + \frac{1}{3}r^3)] = \frac{2}{3}\pi r^3$.

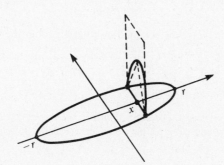

Fig. 22-27

22.51 Same as Problem 22.50, but the planar cross section is a square.

▌ The area $A(x)$ of the dashed square in Fig. 22-27 is $4(r^2 - x^2)$. Hence the volume will be $8/\pi$ times that found in Problem 22.50, or $16r^3/3$.

22.52 Same as Problem 22.50, except that the planar cross section is an isosceles right triangle with its hypotenuse on the base.

▌ The area $A(x)$ of the dashed triangle in Fig. 22-27 (which is inscribed in the semicircle) is $\frac{1}{2}(2\sqrt{r^2 - x^2})(\sqrt{r^2 - x^2}) = r^2 - x^2$. Hence the volume will be one-fourth that of Problem 22.51, or $4r^3/3$.

22.53 Find the volume of a solid whose base is the region in the first quadrant bounded by the line $4x + 5y = 20$ and the coordinate axes, if every planar section perpendicular to the x-axis is a semicircle (Fig. 22-28).

▌ The radius of the semicircle is $\frac{2}{5}(5 - x)$, and its area $A(x)$ is, therefore, $\frac{4}{25}\pi(5 - x)^2$. By the cross-section formula, $V = \frac{4}{25}\pi \int_0^5 (5 - x)^2\, dx = \frac{4}{25}\pi(-\frac{1}{3})(5 - x)^3\,]_0^5 = -\frac{4\pi}{75}(5 - x)^3\,]_0^5 = -\frac{4\pi}{75}(0 - 125) = 20\pi/3$.

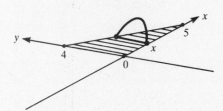

Fig. 22-28

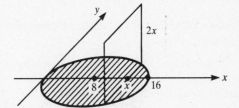

Fig. 22-29

22.54 The base of a solid is the circle $x^2 + y^2 = 16x$, and every planar section perpendicular to the x-axis is a rectangle whose height is twice the distance of the plane of the section from the origin. Find the volume of the solid.

▌ Refer to Fig. 22-29. By completing the square, we see that the equation of the circle is $(x - 8)^2 + y^2 = 64$. So, the center is $(8, 0)$ and the radius is 8. The height of each rectangle is $2x$, and its base is $2y = 2\sqrt{64 - (x - 8)^2}$. So, the area $A(x)$ is $4x\sqrt{64 - (x - 8)^2}$. By the cross-section formula, $V = 4\int_0^{16} x[64 - (x - 8)^2]^{1/2}\, dx$. Let $u = x - 8$, $x = u + 8$, $du = dx$. Then $V = 4\int_{-8}^{8} (u + 8)(64 - u^2)^{1/2}\, du = 4\int_{-8}^{8} u(64 - u^2)^{1/2}\, du + 32\int_{-8}^{8} (64 - u^2)^{1/2}\, du$. The integral in the first summand is 0, since its integrand is an odd function. The integrand in the second summand is the area, 32π, of a semicircle of radius 8 (by Problem 20.71). Hence, $V = 32(32\pi) = 1024\pi$.

22.55 The section of a certain solid cut by any plane perpendicular to the x-axis is a square with the ends of a diagonal lying on the parabolas $y^2 = 9x$ and $x^2 = 9y$. Find its volume.

▌ The parabolas intersect at $(0, 0)$ and $(9, 9)$. The diagonal d of the square is $3x^{1/2} - \frac{1}{9}x^2 = \frac{1}{9}(27x^{1/2} - x^2)$. Then the area of the square is $A(x) = \frac{1}{2}d^2 = \frac{1}{162}[(27)^2 x - 54x^{5/2} + x^4]$, and the cross-section formula yields the volume $V = \frac{1}{162}\int_0^9 [(27)^2 x - 54x^{5/2} + x^4]\, dx = \frac{1}{162}((27)^2 \cdot \frac{1}{2}x^2 - \frac{108}{7}x^{7/2} + \frac{1}{5}x^5)\,]_0^9 = \frac{1}{2}(81)^2$.

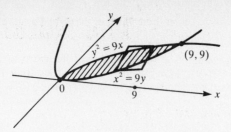

Fig. 22-30

22.56 Find the volume of the solid of revolution obtained by rotating about the x-axis the region in the first quadrant bounded by the curve $x^{2/3} + y^{2/3} = a^{2/3}$ and the coordinate axes.

▮ By the disk formula, $V = \pi \int_0^a y^2 \, dx = \pi \int_0^a (a^{2/3} - x^{2/3})^3 \, dx = \pi \int_0^a (a^2 - 3a^{4/3}x^{2/3} + 3a^{2/3}x^{4/3} - x^2) \, dx = \pi(a^2 x - 3a^{4/3} \cdot \frac{3}{5}x^{5/3} + 3a^{2/3} \cdot \frac{3}{7}x^{7/3} - \frac{1}{3}x^3) \big]_0^a = \pi(a^3 - \frac{9}{5}a^3 + \frac{9}{7}a^3 - \frac{1}{3}a^3) = 16\pi a^3/105.$

22.57 The base of a certain solid is an equilateral triangle of side b, with one vertex at the origin and an altitude along the positive x-axis. Each plane perpendicular to the x-axis intersects the solid in a square with one side in the base of the solid. Find the volume.

▮ See Fig. 22-31. The altitude $h = b \cos 30° = \frac{1}{2}b\sqrt{3}$. For each x, $y = x \tan 30° = (1/\sqrt{3})x$. Hence, the side of the square is $2y = (2/\sqrt{3})x$ and its area is $A = \frac{4}{3}x^2$. Hence, the cross-section formula yields

$$V = \frac{4}{3} \int_0^{\sqrt{3}b/2} x^2 \, dx = \frac{4}{9}x^3 \bigg]_0^{\sqrt{3}b/2} = \frac{\sqrt{3}b^3}{6}$$

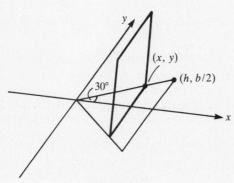

Fig. 22-31

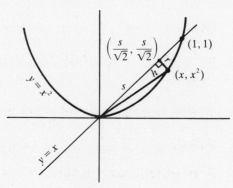

Fig. 22-32

22.58 What volume is obtained when the area bounded by the line $y = x$ and the parabola $y = x^2$ is rotated about the bounding line?

▮ Refer to Fig. 22-32. The required volume is given by the disk formula as $V = \pi \int_0^{\sqrt{2}} r^2 \, ds$; so our strategy will be to find r^2 and s as functions of x, and then to change the integration variable from s to x. Now, by the Pythagorean theorem,

$$r^2 + s^2 = h^2 = x^2 + x^4 \tag{1}$$

and, by the distance formula,

$$r^2 = \left(x - \frac{s}{\sqrt{2}}\right)^2 + \left(x^2 - \frac{s}{\sqrt{2}}\right)^2 = x^2 + x^4 + s^2 - \sqrt{2}\, s(x + x^2) \tag{2}$$

Eliminating r^2 between (1) and (2), we obtain

$$0 = s\left[s - \frac{1}{\sqrt{2}}(x + x^2)\right] \qquad \text{so} \qquad s = \frac{1}{\sqrt{2}}(x + x^2)$$

and then, from (1), $r^2 = \frac{1}{2}x^2 - x^3 + \frac{1}{2}x^4$. Carrying out the change of variable, we have $ds = \frac{1}{\sqrt{2}}(1 + 2x) \, dx$, with x ranging from 0 to 1. Hence

$$V = \pi \int_0^1 (\tfrac{1}{2}x^2 - x^3 + \tfrac{1}{2}x^4) \frac{1}{\sqrt{2}}(1 + 2x) \, dx = \frac{\pi}{2\sqrt{2}} \int_0^1 (x^2 - 2x^3 + x^4 + 2x^3 - 4x^4 + 2x^5) \, dx$$

$$= \frac{\pi}{2\sqrt{2}} \left[\tfrac{1}{3} - 2(\tfrac{1}{4}) + \tfrac{1}{5} + 2(\tfrac{1}{4}) - 4(\tfrac{1}{5}) + 2(\tfrac{1}{6})\right] = \frac{\pi}{30\sqrt{2}}$$

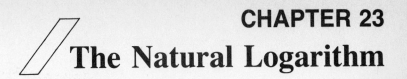

CHAPTER 23
The Natural Logarithm

23.1 State the definition of $\ln x$, and show that $D_x(\ln x) = \dfrac{1}{x}$.

> ▌ $\ln x = \displaystyle\int_1^x \frac{1}{t}\,dt$ for $x > 0$. Hence (Problem 20.42), $D_x(\ln x) = D_x\left(\displaystyle\int_1^x \frac{1}{t}\,dt\right) = \dfrac{1}{x}$.

23.2 Show that $\displaystyle\int \frac{1}{x}\,dx = \ln|x| + C$ for $x \neq 0$.

> ▌ **Case 1.** $x > 0$. Then $D_x(\ln|x| + C) = D_x(\ln x) = \dfrac{1}{x}$. **Case 2.** $x < 0$. Then $D_x(\ln|x| + C) = D_x[\ln(-x)] = \dfrac{1}{-x} \cdot D_x(-x) = -\dfrac{1}{x} \cdot (-1) = \dfrac{1}{x}$.

> In Problems 23.3–23.9, find the derivative of the given function.

23.3 $\ln(4x - 1)$.

> ▌ By the chain rule, $D_x[\ln(4x-1)] = \dfrac{1}{4x-1} \cdot D_x(4x-1) = \dfrac{4}{4x-1}$.

23.4 $(\ln x)^3$.

> ▌ By the chain rule, $D_x[(\ln x)^3] = 3(\ln x)^2 \cdot D_x(\ln x) = 3(\ln x)^2 \cdot \dfrac{1}{x} = \dfrac{3}{x}(\ln x)^2$.

23.5 $\sqrt{\ln x}$.

> ▌ By the chain rule, $D_x(\sqrt{\ln x}) = D_x[(\ln x)^{1/2}] = \tfrac{1}{2}(\ln x)^{-1/2} \cdot D_x(\ln x) = \tfrac{1}{2}(\ln x)^{-1/2} \cdot \dfrac{1}{x} = \dfrac{1}{2x\sqrt{\ln x}}$.

23.6 $\ln(\ln x)$.

> ▌ By the chain rule, $D_x[\ln(\ln x)] = \dfrac{1}{\ln x} \cdot D_x(\ln x) = \dfrac{1}{\ln x} \cdot \dfrac{1}{x} = \dfrac{1}{x \ln x}$.

23.7 $x^2 \ln x$.

> ▌ By the product rule, $D_x(x^2 \ln x) = x^2 \cdot D_x(\ln x) + \ln x \cdot D_x(x^2) = x^2 \cdot \dfrac{1}{x} + \ln x \cdot (2x) = x + 2x \ln x = x(1 + 2\ln x)$.

23.8 $\ln\left(\dfrac{x-1}{x+1}\right)$.

> ▌ By the chain rule,
> $$D_x\left[\ln\left(\frac{x-1}{x+1}\right)\right] = \frac{1}{(x-1)/(x+1)} \cdot D_x\left(\frac{x-1}{x+1}\right) = \frac{x+1}{x-1} \cdot \frac{(x+1)-(x-1)}{(x+1)^2} = \frac{1}{x-1} \cdot \frac{2}{x+1} = \frac{2}{(x-1)(x+1)}$$

23.9 $\ln|5x - 2|$.

> ▌ By the chain rule, $D_x(\ln|5x-2|) = \dfrac{1}{5x-2} \cdot D_x(5x-2) = \dfrac{5}{5x-2}$.

> In Problems 23.10–23.19, find the indicated antiderivative.

23.10 $\displaystyle\int \frac{1}{3x}\,dx$.

> ▌ $\displaystyle\int \frac{1}{3x}\,dx = \frac{1}{3}\int \frac{1}{x}\,dx = \tfrac{1}{3}\ln|x| + C$.

23.11 $\int \dfrac{1}{7x-2}\,dx.$

▮ Let $u = 7x - 2$, $du = 7\,dx$. Then $\int \dfrac{1}{7x-2}\,dx = \dfrac{1}{7}\int \dfrac{1}{u}\,du = \frac{1}{7}\ln|u| + C = \frac{1}{7}\ln|7x - 2| + C.$

23.12 $\int \dfrac{x^3}{x^4-1}\,dx.$

▮ Let $u = x^4 - 1$, $du = 4x^3\,dx$. Then $\int \dfrac{x^3}{x^4-1}\,dx = \dfrac{1}{4}\int \dfrac{1}{u}\,du = \frac{1}{4}\ln|u| + C = \frac{1}{4}\ln|x^4 - 1| + C.$

23.13 $\int \cot x\,dx.$

▮ Let $u = \sin x$, $du = \cos x\,dx$. Then $\int \cot x\,dx = \int \dfrac{\cos x}{\sin x}\,dx = \int \dfrac{1}{u}\,du = \ln|u| + C = \ln|\sin x| + C.$

23.14 $\int \dfrac{1}{x \ln x}\,dx.$

▮ Let $u = \ln x$, $du = \dfrac{1}{x}\,dx$. Then $\int \dfrac{1}{x \ln x}\,dx = \int \dfrac{1}{u}\,du = \ln|u| + C = \ln(\ln|x|) + C.$ [Compare Problem 23.6.]

23.15 $\int \dfrac{\cos 2x}{1-\sin 2x}\,dx.$

▮ Let $u = 1 - \sin 2x$, $du = -2\cos 2x\,dx$. Then

$$\int \dfrac{\cos 2x}{1-\sin 2x}\,dx = -\dfrac{1}{2}\int \dfrac{1}{u}\,du = -\tfrac{1}{2}\ln|u| + C = -\tfrac{1}{2}\ln|1 - \sin 2x| + C$$

23.16 $\int \dfrac{3x^5 + 2x^2 - 3}{x^3}\,dx.$

▮ $\int \dfrac{3x^5 + 2x^2 - 3}{x^3}\,dx = 3\int x^2\,dx + 2\int \dfrac{1}{x}\,dx - 3\int x^{-3}\,dx = x^3 + 2\ln|x| + \tfrac{3}{2}x^{-2} + C.$

23.17 $\int \dfrac{\sec^2 x}{\tan x}\,dx.$

▮ Let $u = \tan x$, $du = \sec^2 x\,dx$. Then

$$\int \dfrac{\sec^2 x}{\tan x}\,dx = \int \dfrac{1}{u}\,du = \ln|u| + C = \ln|\tan x| + C$$

23.18 $\int \dfrac{1}{\sqrt{x}(1-\sqrt{x})}\,dx.$

▮ Let $u = 1 - \sqrt{x}$, $du = -(1/2\sqrt{x})\,dx$. Then

$$\int \dfrac{1}{\sqrt{x}(1-\sqrt{x})}\,dx = -2\int \dfrac{1}{u}\,du = -2\ln|u| + C = -2\ln|1 - \sqrt{x}| + C$$

23.19 $\int \dfrac{\ln x}{x}\,dx.$

▮ $\int \dfrac{\ln x}{x}\,dx = \int (\ln x)\left(\dfrac{1}{x}\right)dx = \tfrac{1}{2}(\ln x)^2 + C.$

23.20 Show that $\int \dfrac{g'(x)}{g(x)}\,dx = \ln|g(x)| + C.$

▮ Let $u = g(x)$, $du = g'(x)\,dx$. Then

$$\int \dfrac{g'(x)}{g(x)}\,dx = \int \dfrac{1}{u}\,du = \ln|u| + C = \ln|g(x)| + C$$

23.21 Find $\int \dfrac{x}{3x^2+1}\,dx$.

▮ Use Problem 23.20: $\displaystyle\int \dfrac{x}{3x^2+1}\,dx = \dfrac{1}{6}\int \dfrac{6x}{3x^2+1}\,dx = \tfrac{1}{6}\ln|3x^2+1| + C$.

23.22 Find $\int \tan x\,dx$.

▮ Use Problem 23.20: $\displaystyle\int \tan x\,dx = \int \dfrac{\sin x}{\cos x}\,dx = -\int \dfrac{-\sin x}{\cos x}\,dx = -\ln|\cos x| + C$. Since $-\ln|\cos x| = \ln(|\cos x|^{-1}) = \ln(|\sec x|)$, the answer can be written as $\ln|\sec x| + C$.

In Problems 23.23–23.26, use logarithmic differentiation to find y'.

23.23 $y = x^3\sqrt{4-x^2}$.

▮ $\ln y = \ln x^3 + \ln(4-x^2)^{1/2} = 3\ln x + \tfrac{1}{2}\ln(4-x^2)$. By implicit differentiation,

$$\frac{1}{y}\,y' = 3\cdot\frac{1}{x} + \frac{1}{2}\cdot\frac{1}{4-x^2}\cdot(-2x) = \frac{3}{x} - \frac{x}{4-x^2} = \frac{12-3x^2-x^2}{x(4-x^2)} = \frac{12-4x^2}{x(4-x^2)} = \frac{4(3-x^2)}{x(4-x^2)}$$

Hence, $\qquad y' = \dfrac{4(3-x^2)}{x(4-x^2)}\cdot y = \dfrac{4(3-x^2)}{x(4-x^2)}\cdot x^3(4-x^2)^{1/2} = \dfrac{4x^2(3-x^2)}{\sqrt{4-x^2}}$

23.24 $y = \dfrac{(x-2)^4\sqrt[3]{x+5}}{\sqrt{x^2+4}}$.

▮ $\ln y = \ln(x-2)^4 + \ln(x+5)^{1/3} - \ln(x^2+4)^{1/2} = 4\ln(x-2) + \tfrac{1}{3}\ln(x+5) - \tfrac{1}{2}\ln(x^2+4)$. By implicit differentiation,

$$\frac{1}{y}\,y' = \frac{4}{x-2} + \frac{1}{3(x+5)} - \frac{1}{2}\frac{1}{(x^2+4)}\cdot 2x = \frac{4}{x-2} + \frac{1}{3(x+5)} - \frac{x}{x^2+4}$$

So, $\qquad y' = \left[\dfrac{4}{x-2} + \dfrac{1}{3(x+5)} - \dfrac{x}{x^2+4}\right]y$

23.25 $y = \dfrac{\sqrt{x^2-1}\cdot\sin x}{(2x+3)^4}$.

▮ $\ln y = \ln(x^2-1)^{1/2} + \ln(\sin x) - \ln(2x+3)^4 = \tfrac{1}{2}\ln(x^2-1) + \ln(\sin x) - 4\ln(2x+3)$. Hence,

$$\frac{1}{y}\,y' = \frac{1}{2}\frac{1}{x^2-1}\cdot(2x) + \frac{1}{\sin x}\cdot\cos x - \frac{4}{2x+3}\cdot 2 = \frac{x}{x^2-1} + \cot x - \frac{8}{2x+3}$$

Hence, $y' = \left(\dfrac{x}{x^2-1} + \cot x - \dfrac{8}{2x+3}\right)y$.

23.26 $y = \sqrt[3]{\dfrac{x+2}{x-1}}$.

▮ $\ln y = \dfrac{1}{3}\ln\left(\dfrac{x+2}{x-1}\right) = \tfrac{1}{3}[\ln(x+2) - \ln(x-1)]$. Hence,

$$\frac{1}{y}\,y' = \frac{1}{3}\left(\frac{1}{x+2} - \frac{1}{x-1}\right) = \frac{1}{3}\left[\frac{-3}{(x+2)(x-1)}\right] = -\frac{1}{(x+2)(x-1)}$$

So, $\qquad y' = -\dfrac{y}{(x+2)(x-1)} = -\dfrac{1}{(x+2)(x-1)}\dfrac{(x+2)^{1/3}}{(x-1)^{1/3}} = -\dfrac{1}{(x+2)^{2/3}(x-1)^{4/3}}$

In Problems 23.27–23.34, express the given number in terms of $\ln 2$ and $\ln 5$.

23.27 $\ln 10$.

▮ $\ln 10 = \ln(2\cdot 5) = \ln 2 + \ln 5$.

23.28 $\ln\tfrac{1}{2}$.

▮ $\ln\tfrac{1}{2} = \ln 2^{-1} = -\ln 2$.

23.29 $\ln \frac{1}{5}$.

▮ $\ln \frac{1}{5} = \ln 5^{-1} = -\ln 5$.

23.30 $\ln 25$.

▮ $\ln 25 = \ln 5^2 = 2 \ln 5$.

23.31 $\ln \sqrt{2}$.

▮ $\ln \sqrt{2} = \ln 2^{1/2} = \frac{1}{2} \ln 2$.

23.32 $\ln \sqrt[3]{5}$.

▮ $\ln \sqrt[3]{5} = \ln 5^{1/3} = \frac{1}{3} \ln 5$.

23.33 $\ln \frac{1}{20}$.

▮ $\ln \frac{1}{20} = \ln (20)^{-1} = -\ln 20 = -\ln(4 \cdot 5) = -(\ln 4 + \ln 5) = -(\ln 2^2 + \ln 5) = -(2 \ln 2 + \ln 5)$.

23.34 $\ln 2^{12}$.

▮ $\ln 2^{12} = 12 \ln 2$.

23.35 Find an equation of the tangent line to the curve $y = \ln x$ at the point $(1, 0)$.

▮ The slope of the tangent line is $y' = 1/x = 1$. Hence, a point-slope equation of the tangent line is $y = x - 1$.

23.36 Find the area of the region bounded by the curves $y = x^2$, $y = 1/x$, and $x = \frac{1}{2}$ (see Fig. 23-1).

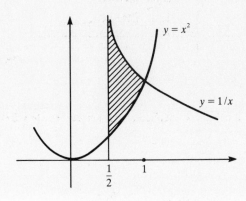

Fig. 23-1

▮ The curves intersect at $(1, 1)$. The region is bounded above by $y = 1/x$. Hence, the area is given by
$$\int_{1/2}^1 \left(\frac{1}{x} - x^2 \right) dx = (\ln x - \tfrac{1}{3}x^3) \,]_{1/2}^1 = (\ln 1 - \tfrac{1}{3}) - (\ln \tfrac{1}{2} - \tfrac{1}{24}) = (-\tfrac{1}{3}) - (-\ln 2 - \tfrac{1}{24}) = \ln 2 - \tfrac{7}{24}.$$

23.37 Find the average value of $1/x$ on $[1, 4]$.

▮ The average value $= \dfrac{1}{4-1} \int_1^4 \dfrac{1}{x} \, dx = \frac{1}{3} \ln x \,]_1^4 = \frac{1}{3}(\ln 4 - \ln 1) = \frac{1}{3}(2 \ln 2 - 0) = \frac{2}{3} \ln 2$.

23.38 Find the volume of the solid obtained by revolving about the x-axis the region in the first quadrant under $y = x^{-1/2}$ between $x = \frac{1}{4}$ and $x = 1$.

▮ We use the disk formula: $V = \pi \int_{1/4}^1 y^2 \, dx = \pi \int_{1/4}^1 \dfrac{1}{x} \, dx = \pi \ln x \,]_{1/4}^1 = \pi(\ln 1 - \ln \tfrac{1}{4}) = \pi(0 + \ln 4) = \pi(2 \ln 2) = 2\pi \ln 2$.

23.39 Sketch the graph of $y = \ln(x + 1)$.

▮ See Fig. 23-2. The graph is that of $y = \ln x$, moved one unit to the left.

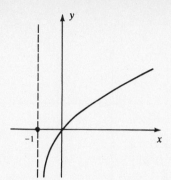

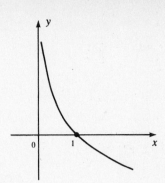

Fig. 23-2   Fig. 23-3

23.40 Sketch the graph of $y = \ln{(1/x)}$.

▮ See Fig. 23-3. Since $\ln{(1/x)} = -\ln x$, the graph is that of $y = \ln x$, reflected in the x-axis.

23.41 Show that $1 - \dfrac{1}{x} \le \ln x \le x - 1$.

▮ **Case 1.** $x \ge 1$. By looking at areas in Fig. 23-4, we see that $1 - \dfrac{1}{x} = \dfrac{1}{x}(x-1) \le \ln x \le x - 1$. **Case 2.** $0 < x < 1$. Then $1/x > 1$. So, by Case 1, $1 - x \le \ln{(1/x)} \le 1/x - 1$. Thus, $1 - x \le -\ln x \le 1/x - 1$, and multiplying by -1, we obtain $x - 1 \ge \ln x \ge 1 - 1/x$.

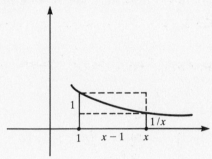

Fig. 23-4

23.42 Show that $\dfrac{\ln x}{x} < \dfrac{2}{\sqrt{x}}$.

▮ By Problem 23-41, $\ln x \le x - 1 < x$. Substituting $\sqrt{x}$ for x, $\ln \sqrt{x} < \sqrt{x}$, $\ln x^{1/2} < \sqrt{x}$, $\frac{1}{2} \ln x < \sqrt{x}$, $\dfrac{\ln x}{x} < \dfrac{2\sqrt{x}}{x} = \dfrac{2}{\sqrt{x}}$.

23.43 Prove that $\lim\limits_{x \to +\infty} \dfrac{\ln x}{x} = 0$.

▮ $\lim\limits_{x \to +\infty} \dfrac{2}{\sqrt{x}} = 0$. Hence, by Problem 23.42, $\lim\limits_{x \to +\infty} \dfrac{\ln x}{x} = 0$.

23.44 Prove that $\lim\limits_{x \to 0^+} x \ln x = 0$.

▮ Let $y = 1/x$. As $x \to 0^+$, $y \to +\infty$. By Problem 23.43, $\lim\limits_{y \to +\infty} \dfrac{\ln y}{y} = 0$. But, $\dfrac{\ln y}{y} = \dfrac{\ln{(1/x)}}{1/x} = x(-\ln x) = -x \ln x$. Hence, $\lim\limits_{x \to 0^+} x \ln x = 0$.

23.45 Prove that $\lim\limits_{x \to +\infty} (x - \ln x) = \infty$.

▮ $x - \ln x = x\left(1 - \dfrac{\ln x}{x}\right)$. By Problem 23.43, $\lim\limits_{x \to +\infty} \dfrac{\ln x}{x} = 0$. Hence, $\lim\limits_{x \to +\infty} (x - \ln x) = +\infty$.

23.46 Sketch the graph of $y = x - \ln x$.

▮ See Fig. 23-5. $y' = 1 - 1/x$. Setting $y' = 0$, we find that $x = 1$ is the only critical number. $y'' = 1/x^2$. Hence, by the second derivative test, there is a relative minimum at $(1, 1)$. To the right of $(1, 1)$, the curve increases without bound, since $\lim\limits_{x \to +\infty} (x - \ln x) = +\infty$, by Problem 23.45. As $x \to 0^+$, $x - \ln x \to +\infty$, since $\ln x \to -\infty$.

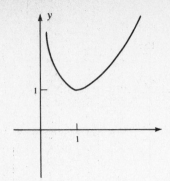

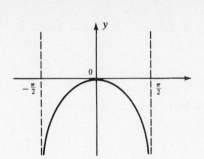

Fig. 23-5   Fig. 23-6

23.47   Graph $y = \ln(\cos x)$.

▌ See Fig. 23-6. Since $\cos x$ has period 2π, we need only consider $[-\pi, \pi]$. The function is defined only when $\cos x > 0$, that is, in $\left(-\dfrac{\pi}{2}, \dfrac{\pi}{2}\right)$. $y' = \dfrac{1}{\cos x} \cdot (-\sin x) = -\tan x$, and $y'' = -\sec^2 x$. Setting $y' = 0$, we see that the only critical number is $x = 0$. By the second derivative test, there is a relative maximum at $(0, 0)$. As $x \to \pm\pi/2$, $\cos x \to 0$ and $y \to -\infty$.

23.48   An object moves along the x-axis with acceleration $a = t - 1 + 6/t$. Find the maximum velocity v for $1 \le t \le 9$, if the velocity at $t = 1$ is 1.5.

▌ $v = \int a\, dt = \int \left(t - 1 + \dfrac{6}{t}\right) dt = \tfrac{1}{2}t^2 - t + 6\ln t + 2$, where the constant of integration is chosen to make $v(1) = 1.5$. Since the acceleration is positive over $[1, 9]$, $v(t)$ is increasing on that interval, with maximum value $v(9) = \tfrac{81}{2} - 9 + 6\ln 9 + 2 = \tfrac{67}{2} + 12\ln 3$.

23.49   Find y' when $y^2 = \ln(x^2 + y^2)$.

▌ By implicit differentiation, $2yy' = \dfrac{1}{x^2 + y^2}(2x + 2yy')$, $yy'(x^2 + y^2) = x + yy'$, $yy'(x^2 + y^2 - 1) = x$, $y' = \dfrac{x}{y(x^2 + y^2 - 1)}$.

23.50   Find y' if $\ln xy + 2x - y = 1$.

▌ By implicit differentiation,

$$\dfrac{1}{xy}(xy' + y) + 2 - y' = 0 \qquad xy' + y + 2xy - xyy' = 0 \qquad xy'(1 - y) = -y - 2xy \qquad y' = \dfrac{y(1 + 2x)}{x(y - 1)}$$

23.51   If $\ln(x + y^2) = y^3$, find y'.

▌ By implicit differentiation, $\dfrac{1}{x + y^2}(1 + 2yy') = 3y^2 y'$, $1 + 2yy' = 3y^2 y'x + 3y^4 y'$, $y'(2y - 3y^2x - 3y^4) = -1$, $y' = \dfrac{1}{y(3xy + 3y^3 - 2)}$.

23.52   Evaluate $\displaystyle\lim_{h \to 0}\left(\dfrac{1}{h}\ln\dfrac{3 + h}{3}\right)$.

▌
$$\dfrac{1}{h}\ln\dfrac{3 + h}{3} = \dfrac{1}{h}\left(\ln\dfrac{x + h}{3} - \ln\dfrac{x}{3}\right) \qquad \text{for} \quad x = 3$$

Hence, $\displaystyle\lim_{h \to 0}\left(\dfrac{1}{h}\ln\dfrac{3 + h}{3}\right) = \lim_{h \to 0}\dfrac{1}{h}\left(\ln\dfrac{x + h}{3} - \ln\dfrac{x}{3}\right) = D_x\left(\ln\dfrac{x}{3}\right) = D_x(\ln x - \ln 3) = D_x(\ln x) = \dfrac{1}{x} = \dfrac{1}{3}$.

23.53   Derive the formula $\int \csc x\, dx = \ln|\csc x - \cot x| + C$.

▌ $\csc x \cdot \dfrac{\csc x - \cot x}{\csc x - \cot x} = \dfrac{\csc^2 x - \csc x \cot x}{\csc x - \cot x}$.   Let $u = \csc x - \cot x$, $du = (-\csc x \cot x + \csc^2 x)\, dx$. Then $\int \csc x\, dx = \int \dfrac{1}{u}\, du = \ln|u| + C = \ln|\csc x - \cot x| + C$.

23.54   Find the length of the curve $y = \frac{1}{2}x^2 - \frac{1}{4}\ln x$ between $x = 1$ and $x = 8$.

$$y' = x - \frac{1}{4x} = \frac{4x^2 - 1}{4x} \qquad (y')^2 = \frac{16x^4 - 8x^2 + 1}{16x^2}$$

$$1 + (y')^2 = \frac{16x^2 + 16x^4 - 8x^2 + 1}{16x^2} = \frac{16x^4 + 8x^2 + 1}{16x^2} = \left(\frac{4x^2 + 1}{4x}\right)^2$$

So, the arc length is $L = \int_1^8 \frac{4x^2 + 1}{4x}\, dx = \int_1^8 \left(x + \frac{1}{4x}\right) dx = (\frac{1}{2}x^2 + \frac{1}{4}\ln x)\,]_1^8 = (32 + \frac{3}{4}\ln 2) - (\frac{1}{2} + 0) = \frac{63}{2} + \frac{3}{4}\ln 2$.

23.55   Graph $y = x^2 - 18\ln x$.

See Fig. 23-7. $y' = 2x - 18/x$, $y'' = 2 + 18/x^2$.   Setting $y' = 0$,   we find that $x^2 = 9$, $x = 3$, $y = 9 - 18\ln 3 \approx -10$.   Hence, by the second derivative test, there is a minimum at $x = 3$.   As $x \to +\infty$, $y = x^2\left(1 - \frac{18\ln x}{x^2}\right) \to +\infty$.   As $x \to 0^+$, $\ln x \to -\infty$ and $y \to +\infty$.

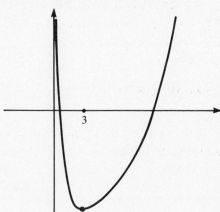

Fig. 23-7

23.56   Show that the area under $y = 1/x$ from $x = a$ to $x = b$ is the same as the area under that curve from $x = ka$ to $x = kb$ for any $k > 0$.

$\int_a^b \frac{1}{x}\, dx = \ln x\,]_a^b = \ln b - \ln a = \ln(b/a)$.   On the other hand, $\int_{ka}^{kb} \frac{1}{x}\, dx = \ln x\,]_{ka}^{kb} = \ln kb - \ln ka = \ln(kb/ka) = \ln(b/a)$.

23.57   Use the Trapezoidal Rule, with $n = 10$,   to approximate $\ln 2 = \int_1^2 \frac{1}{x}\, dx$.

$\Delta x = \frac{1}{10}$.   For $\int_1^2 \frac{1}{x}\, dx$,   The Trapezoidal Rule yields $\frac{1}{20}[1 + \frac{1}{2} + 2(\frac{10}{11} + \frac{10}{12} + \cdots + \frac{10}{19})] = \frac{3}{40} + (\frac{1}{11} + \frac{1}{12} + \cdots + \frac{1}{19}) \approx 0.0750 + (0.0909 + 0.0833 + \cdots + 0.0526) = 0.0750 + 0.6187 = 0.6937$.   (The actual value, correct to four decimal places, is 0.6931.)

23.58   If $y = \dfrac{x(1 - x^2)^2}{(1 + x^2)^{1/2}}$,   find y'.

Use logarithmic differentiation. $\ln y = \ln x + 2\ln(1 - x^2) - \frac{1}{2}\ln(1 + x^2)$.   Hence,

$$\frac{1}{y}\, y' = \frac{1}{x} + 2 \cdot \frac{1}{1 - x^2} \cdot (-2x) - \frac{1}{2}\frac{1}{1 + x^2} \cdot 2x = \frac{1}{x} - \frac{4x}{1 - x^2} - \frac{x}{1 + x^2}$$

So,

$$y' = \frac{x(1 - x^2)^2}{(1 + x^2)^{1/2}}\left(\frac{1}{x} - \frac{4x}{1 - x^2} - \frac{x}{1 + x^2}\right) = \frac{(1 - x^2)^2}{(1 + x^2)^{1/2}} - \frac{4x^2(1 - x^2)}{(1 + x^2)^{1/2}} - \frac{x^2(1 - x^2)^2}{(1 + x^2)^{3/2}} = \frac{(1 - 5x^2 - 4x^4)(1 - x^2)}{(1 + x^2)^{3/2}}$$

23.59   Find the volume of the solid generated when the region under $y = 1/x^2$,   above the x-axis, between $x = 1$ and $x = 2$,   is rotated around the y-axis.

By the cylindrical shell formula, $V = 2\pi \int_1^2 xy\, dx = 2\pi \int_1^2 \frac{1}{x}\, dx = 2\pi \ln x\,\Big]_1^2 = 2\pi(\ln 2 - 0) = 2\pi \ln 2$.

23.60 Prove that $|\ln b - \ln a| < |b - a|$ for any distinct a, b in $[1, +\infty)$.

❚ **Case 1.** $a < b$. By the Mean-Value Theorem, $\dfrac{\ln b - \ln a}{b - a} = \dfrac{1}{c}$ for some c in (a, b). Since $c > a \geq 1$, $\dfrac{1}{c} < 1$. Hence, $\left|\dfrac{\ln b - \ln a}{b - a}\right| < 1$, and, therefore, $|\ln b - \ln a| < |b - a|$. **Case 2.** $b < a$. By Case 1, $|\ln a - \ln b| < |a - b|$. But $|\ln a - \ln b| = |\ln b - \ln a|$ and $|a - b| = |b - a|$.

23.61 Evaluate $\displaystyle\int_1^2 \frac{x - 1}{x^2 - 2x + 2}\, dx$.

❚ $\displaystyle\int_1^2 \frac{x - 1}{x^2 - 2x + 2}\, dx = \frac{1}{2}\int_1^2 \frac{2x - 2}{x^2 - 2x + 2}\, dx = \frac{1}{2}\ln|x^2 - 2x + 2|\,]_1^2 = \frac{1}{2}(\ln 2 - \ln 1) = \frac{1}{2}\ln 2.$

23.62 Find $\displaystyle\int_1^2 \frac{x}{3x^2 - 2}\, dx$.

❚ $\displaystyle\int_1^2 \frac{x}{3x^2 - 2}\, dx = \frac{1}{6}\int_1^2 \frac{6x}{3x^2 - 2}\, dx = \frac{1}{6}\ln|3x^2 - 2|\,]_1^2 = \frac{1}{6}(\ln 10 - \ln 1) = \frac{1}{6}\ln 10.$

23.63 Evaluate $\displaystyle\int_{\pi/6}^{\pi/2} \frac{\cos x}{2 - \sin x}\, dx$.

❚ $\displaystyle\int_{\pi/6}^{\pi/2} \frac{\cos x}{2 - \sin x}\, dx = -\int_{\pi/6}^{\pi/2} \frac{-\cos x}{2 - \sin x}\, dx = -\ln|2 - \sin x|\,]_{\pi/6}^{\pi/2} = -(\ln 1 - \ln \frac{3}{2}) = \ln \frac{3}{2} = \ln 3 - \ln 2.$

23.64 Prove the basic property of logarithms: $\ln uv = \ln u + \ln v$.

❚ In $\ln v = \displaystyle\int_1^v \frac{1}{t}\, dt$, make the change of variable $w = ut$ (u fixed). Then $dw = u\, dt$ and the limits of integration $t = 1$ and $t = v$ go over into $w = u$ and $w = uv$, respectively. Hence,

$$\ln v = \int_u^{uv} \frac{u}{w}\frac{1}{u}\, dw = \int_u^{uv} \frac{1}{w}\, dw = \int_u^{uv} \frac{1}{t}\, dt$$

So, $$\ln u + \ln v = \int_1^u \frac{1}{t}\, dt + \int_u^{uv} \frac{1}{t}\, dt = \int_1^{uv} \frac{1}{t}\, dt = \ln uv$$

23.65 If a is a positive constant, find the length of the curve $y = \dfrac{x^2}{2a} - \dfrac{a}{4}\ln x$ between $x = 1$ and $x = 2$.

❚ $$y' = \frac{x}{a} - \frac{a}{4x} = \frac{4x^2 - a^2}{4ax} \qquad (y')^2 = \frac{16x^4 - 8a^2x^2 + a^4}{16a^2x^2}$$

$$1 + (y')^2 = \frac{16a^2x^2 + 16x^4 - 8a^2x^2 + a^4}{16a^2x^2} = \left(\frac{4x^2 + a^2}{4ax}\right)^2$$

So, the length

$$L = \int_1^2 \frac{4x^2 + a^2}{4ax}\, dx = \int_1^2 \left(\frac{x}{a} + \frac{a}{4x}\right) dx = \left(\frac{1}{2a}x^2 + \frac{a}{4}\ln|x|\right)\Bigg]_1^2$$

$$= \left[\left(\frac{2}{a} + \frac{a\ln 2}{4}\right) - \frac{1}{2a}\right] = \frac{3}{2a} + \frac{a\ln 2}{4} = \frac{6 + a^2\ln 2}{4a}$$

23.66 If a and b are positive, find the arc length of $\dfrac{y}{b} = \left(\dfrac{x}{a}\right)^2 - \dfrac{1}{8}\left(\dfrac{a^2}{b^2}\right)\ln\left(\dfrac{x}{a}\right)$ from $x = a$ to $x = 3a$.

❚ $\dfrac{y'}{b} = \dfrac{2}{a}\left(\dfrac{x}{a}\right) - \dfrac{1}{8}\left(\dfrac{a^2}{b^2}\right)\cdot\left(\dfrac{1}{x/a}\right)\cdot\dfrac{1}{a} = \dfrac{2x}{a^2} - \dfrac{a^2}{8b^2}\cdot\dfrac{1}{x} = \dfrac{16b^2x^2 - a^4}{8a^2b^2x}$ $\qquad (y')^2 = \dfrac{256b^4x^4 - 32b^2a^4x^2 + a^8}{64a^4b^2x^2}$

Hence, $$1 + (y')^2 = \frac{64a^4b^2x^2 + 256b^4x^4 - 32b^2a^4x^2 + a^8}{64a^4b^2x^2} = \left(\frac{16b^2x^2 + a^4}{8a^2bx}\right)^2$$

So, the arc length

$$L = \int_a^{3a} \left(\frac{2b}{a^2}x + \frac{a^2}{8b}\cdot\frac{1}{x}\right) dx = \frac{b}{a^2}x^2 + \frac{a^2}{8b}\ln|x|\Bigg]_a^{3a} = \left(9b + \frac{a^2}{8b}\ln 3a\right) - \left(b + \frac{a^2}{8b}\ln a\right)$$

$$= 8b + \frac{a^2}{8b}\ln 3 + \frac{a^2}{8b}\ln a - \frac{a^2}{8b}\ln a = 8b + \frac{a^2}{8b}\ln 3$$

23.67   If $y = (1 - 3x^2)^3 (\cos 2x)^4$, find y'.

▮ Use logarithmic differentiation. $\ln y = 3 \ln (1 - 3x^2) + 4 \ln (\cos 2x)$.   Hence,

$$\frac{1}{y}\, y' = \frac{3}{1 - 3x^2} \cdot (-6x) + \frac{4}{\cos 2x} \cdot (-\sin 2x) \cdot (2) = -\frac{18x}{1 - 3x^2} - 8 \tan 2x$$

So, $\qquad\qquad y' = -(1 - 3x^2)^3 (\cos 2x)^4 \left(\frac{18x}{1 - 3x^2} + 8 \tan 2x \right)$

23.68   Evaluate $\int \frac{x}{(ax + b)^2}\, dx$.

▮ $\int \frac{1}{(ax + b)^2}\, dx = \frac{1}{a} \int \frac{ax}{(ax + b)^2}\, dx = \frac{1}{a} \int \frac{(ax + b) - b}{(ax + b)^2}\, dx = \frac{1}{a} \left[\frac{1}{a} \int \frac{a}{ax + b}\, dx - \frac{b}{a} \int \frac{a}{(ax + b)^2}\, dx \right]$

$\qquad = \frac{1}{a} \left[\frac{1}{a} \ln |ax + b| + \frac{b}{a} (ax + b)^{-1} \right] + C = \frac{1}{a^2} \left(\ln |ax + b| + \frac{b}{ax + b} \right) + C$

23.69   Evaluate $\int \frac{dx}{\sin ax \cos ax}$.

▮ $\int \frac{dx}{\sin ax \cos ax} = \frac{1}{a} \int \frac{a \sec^2 ax}{\tan ax}\, dx = \frac{1}{a} \int \frac{D_x(\tan ax)}{\tan ax}\, dx = \frac{1}{a} \ln |\tan ax| + C$

23.70   Find the area under $y = 1/(1 - x^2)$ between $x = 0$ and $x = \frac{1}{2}$.

▮ $\frac{1}{1 - x^2} = \frac{1}{2} \left(\frac{1}{1 - x} + \frac{1}{1 + x} \right)$.   Hence, the area is

$$\frac{1}{2} \int_0^{1/2} \left(\frac{1}{1 - x} + \frac{1}{1 + x} \right) dx = \tfrac{1}{2}(-\ln |1 - x| + \ln |1 + x|) \,]_0^{1/2} = \frac{1}{2} \ln \left| \frac{1 + x}{1 - x} \right| \,\Big]_0^{1/2} = \tfrac{1}{2}(\ln 3 - \ln 1) = \tfrac{1}{2} \ln 3$$

23.71   Evaluate $\lim\limits_{x \to 0} \frac{\ln (1 + x)}{x}$.

▮ $\lim\limits_{x \to 0} \frac{\ln (1 + x)}{x} = \lim\limits_{h \to 0} \frac{\ln (1 + h) - \ln 1}{h} = D_x(\ln x) \Big|_{x=1} = \frac{1}{x} \Big|_{x=1} = 1.$

23.72   Evaluate $\lim\limits_{x \to 3} \frac{\ln (1 + 2x) - \ln 7}{x - 3}$.

▮ Let $h = 2x - 6$.   Then the given limit is $2 \lim\limits_{h \to 0} \frac{\ln (7 + h) - \ln 7}{h} = 2 \cdot D_x(\ln x) \Big|_{x=7} = 2 \cdot \frac{1}{x} \Big|_{x=7} = \tfrac{2}{7}.$

23.73   Estimate $\ln 2 = \int_1^2 \frac{1}{x}\, dx$ by Simpson's rule, with $n = 4$.

▮ $\Delta x = \frac{1}{4}$.   The points of subdivision are $\frac{5}{4}$, $\frac{6}{4}$, $\frac{7}{4}$.   Then Simpson's rule yields $\frac{1}{12}(1 + 4 \cdot \frac{4}{5} + 2 \cdot \frac{4}{6} + 4 \cdot \frac{4}{7} + \frac{1}{2}) = \frac{1}{12}(1 + \frac{16}{5} + \frac{4}{3} + \frac{16}{7} + \frac{1}{2}) = \frac{1747}{2520} \approx 0.6933.$ (Compare this with the result of Problem 23.57.)

23.74   Solve the equation $3 \ln x = \ln 3x$ for x.

▮ $3 \ln x = \ln 3x = \ln 3 + \ln x$, $2 \ln x = \ln 3$, $\ln x^2 = \ln 3$, $x^2 = 3$, $x = \sqrt{3}$.

23.75   Solve the equation $5 \ln x + 2x = 4 + \ln x^5$ for x.

▮ $5 \ln x + 2x = 4 + 5 \ln x$, $2x = 4$, $x = 2$.

23.76   Sketch the graph of $y = 3x + 1 - 5 \ln (1 + x^2)$.

▮ See Fig. 23-8. $\qquad\qquad y' = 3 - \frac{10x}{1 + x^2}$ $\qquad y'' = \frac{10(x^2 - 1)}{(1 + x^2)^2}$

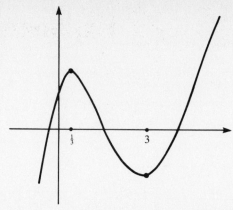

Fig. 23-8

Setting $y' = 0$, we find $3x^2 - 10x + 3 = 0$, $(3x - 1)(x - 3) = 0$, $x = \frac{1}{3}$ or $x = 3$. By the Second-Derivative Test, we have a relative minimum at $x = 3$, and a relative maximum at $x = \frac{1}{3}$. As $x \to \pm\infty$,

$$y = x\left[3 + \frac{1}{x} - \frac{5 \ln(1 + x^2)}{x}\right] \to \pm\infty. \quad \left[\text{Note that } \ln(1 + x^2) < \ln x^3 = 3 \ln x, \text{ and } \frac{\ln x}{x} \to 0.\right]$$

23.77 Prove that $\ln x < \sqrt{x}$ for $x > 0$.

▌ By the Second Derivative Test, the function $y = \sqrt{x} - \ln x$ has an absolute minimum value of $2 - 2 \ln 2$ when $x = 4$. Therefore, $\sqrt{x} - \ln x \geq 2 - 2 \ln 2 > 0$ for all $x > 0$. Hence, $\sqrt{x} > \ln x$ for $x > 0$.

23.78 Evaluate $\lim\limits_{n \to +\infty} \sum\limits_{j=1}^{n} \dfrac{1}{2n + j}$.

▌
$$\sum_{j=1}^{n} \frac{1/n}{2 + j/n} = \frac{1}{n} \sum_{j=1}^{n} \frac{1}{2 + j/n}$$

The latter is an approximating sum for $\displaystyle\int_0^1 \frac{dx}{2 + x}$. Hence, the limit is $\displaystyle\int_0^1 \frac{dx}{2 + x} = \ln|2 + x| \,]_0^1 = \ln 3 - \ln 2$.

23.79 Find $\displaystyle\int \frac{x^4 + 3x^3 - 2x - 3}{x^2 + 3x}\, dx$.

▌ Divide numerator by denominator, obtaining $x^2 - \dfrac{2x + 3}{x^2 + 3x}$. Hence, the integral reduces to $\int x^2\, dx - \int \dfrac{2x + 3}{x^2 + 3x}\, dx = \frac{1}{3}x^3 - \ln|x^2 + 3x| + C$.

23.80 Find $\displaystyle\int \frac{dx}{(x + 2)\sqrt{x + 11}}$.

▌ Let $u = \sqrt{x + 11}$, $u^2 = x + 11$, $2u\, du = dx$. Then the integral becomes

$$\int \frac{2u\, du}{(u^2 - 9)u} = \int \frac{2\, du}{u^2 - 9} = \frac{1}{3} \int \left(\frac{1}{u - 3} - \frac{1}{u + 3}\right) du = \frac{1}{3}(\ln|u - 3| - \ln|u + 3|) + C$$

$$= \frac{1}{3} \ln \frac{|u - 3|}{|u + 3|} + C = \frac{1}{3} \ln \left|\frac{\sqrt{x + 11} - 3}{\sqrt{x + 11} + 3}\right| + C$$

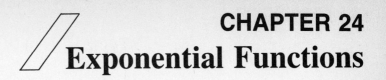

CHAPTER 24
Exponential Functions

24.1 Evaluate $e^{-\ln x}$.

▮ By Problem 23.81, $e^{-\ln x} = e^{\ln(1/x)} = 1/x$.

24.2 Evaluate $\ln e^{-x}$.

▮ $\ln e^{-x} = -x$ by virtue of the identity $\ln e^{u} = u$.

24.3 Find $(e^2)^{\ln x}$.

▮ $(e^2)^{\ln x} = (e^{\ln x})^2 = x^2$. Here, we have used the laws $(e^u)^v = e^{uv}$ and $e^{\ln u} = u$.

24.4 Evaluate $(3e)^{\ln x}$.

▮ $(3e)^{\ln x} = (e^{\ln 3e})^{\ln x} = (e^{\ln 3 + 1})^{\ln x} = (e^{\ln x})^{\ln 3 + 1} = x^{\ln 3 + 1}$.

24.5 Evaluate $e^{1 - \ln x}$.

▮ $e^{1 - \ln x} = e^1 e^{-\ln x} = e/e^{\ln x} = e/x$.

24.6 Find $\ln(e^x/x)$.

▮ $\ln(e^x/x) = \ln e^x - \ln x = x - \ln x$. We have used the identities $\ln(u/v) = \ln u - \ln v$ and $\ln e^u = u$.

In Problems 24.7–24.16, find the derivative of the given function.

24.7 e^{-x}.

▮ By the chain rule, $D_x(e^{-x}) = e^{-x} \cdot D_x(-x) = e^{-x} \cdot (-1) = -e^{-x}$. Here, we have used the fact that $D_u(e^u) = e^u$.

24.8 $e^{1/x}$.

▮ By the chain rule, $D_x(e^{1/x}) = e^{1/x} \cdot D_x(1/x) = e^{1/x} \cdot (-1/x^2) = -e^{1/x}/x^2$.

24.9 $e^{\cos x}$.

▮ By the chain rule, $D_x(e^{\cos x}) = e^{\cos x} \cdot D_x(\cos x) = e^{\cos x} \cdot (-\sin x) = -e^{\cos x} \sin x$.

24.10 $\tan e^x$.

▮ By the chain rule, $D_x(\tan e^x) = \sec^2 e^x \cdot D_x(e^x) = \sec^2 e^x \cdot e^x = e^x \sec^2 e^x$.

24.11 e^x/x.

▮ By the quotient rule, $D_x\left(\dfrac{e^x}{x}\right) = \dfrac{xe^x - e^x}{x^2} = e^x\left(\dfrac{x-1}{x^2}\right)$.

24.12 $e^x \ln x$.

▮ By the product rule, $D_x(e^x \ln x) = e^x \cdot D_x(\ln x) + \ln x \cdot D_x(e^x) = e^x \cdot \dfrac{1}{x} + \ln x \cdot e^x = e^x\left(\dfrac{1}{x} + \ln x\right)$.

24.13 x^π.

▮ $D_x(x^\pi) = D_x(e^{\pi \ln x}) = e^{\pi \ln x} \cdot D_x(\pi \ln x) = e^{\pi \ln x} \cdot \pi \cdot \dfrac{1}{x} = \pi \dfrac{x^\pi}{x} = \pi x^{\pi - 1}$. [In like manner, $D_x(x^r) = rx^{r-1}$ for any real number r.]

24.14 π^x.

   ▮ $D_x(\pi^x) = D_x(e^{x \ln \pi}) = e^{x \ln \pi} \cdot D_x(x \ln \pi) = e^{x \ln \pi} \cdot \ln \pi = \ln \pi \cdot \pi^x$.

24.15 $\ln e^{2x}$.

   ▮ $D_x(\ln e^{2x}) = D_x(2x) = 2$.

24.16 $e^x - e^{-x}$.

   ▮ $D_x(e^x - e^{-x}) = D_x(e^x) - D_x(e^{-x}) = e^x - (-e^{-x}) = e^x + e^{-x}$.   Here, $D_x(e^{-x}) = -e^{-x}$ is taken from Problem 24.7.

   In Problems 24.17–24.29, evaluate the given antiderivative.

24.17 $\int e^{3x} \, dx$.

   ▮ Let $u = 3x$, $du = 3 \, dx$.   Then $\int e^{3x} \, dx = \frac{1}{3} \int e^u \, du = \frac{1}{3} e^u + C = \frac{1}{3} e^{3x} + C$.

24.18 $\int e^{-x} \, dx$.

   ▮ Let $u = -x$, $du = -dx$.   Then $\int e^{-x} \, dx = -\int e^u \, du = -e^u + C = -e^{-x} + C$.

24.19 $\int e^x \sqrt{e^x - 2} \, dx$.

   ▮ Let $u = e^x - 2$, $du = e^x \, dx$.   Then $\int e^x \sqrt{e^x - 2} \, dx = \int u^{1/2} \, du = \frac{2}{3} u^{3/2} + C = \frac{2}{3} (\sqrt{e^x - 2})^3 + C$.

24.20 $\int e^{\cos x} \sin x \, dx$.

   ▮ $\int e^{\cos x} \sin x \, dx = -e^{\cos x} + C$,   by Problem 24.9.

24.21 $\int a^x \, dx$,   for $a \neq 1$.

   ▮ $a^x = e^{x \ln a}$.   So, let $u = (\ln a)x$, $du = (\ln a) \, dx$.   Then

$$\int a^x \, dx = \int e^{x \ln a} \, dx = \frac{1}{\ln a} \int e^u \, du = \frac{1}{\ln a} \cdot e^u + C = \frac{1}{\ln a} e^{x \ln a} + C = \frac{1}{\ln a} a^x + C$$

24.22 $\int 3^{2x} \, dx$.

   ▮ Choose $a = 3^2 = 9$ in Problem 24.21: $\int 3^{2x} \, dx = \frac{1}{2 \ln 3} 3^{2x} + C$.

24.23 $\int e^{ax} \, dx$.

   ▮ Let $u = ax$, $du = a \, dx$.   Then $\int e^{ax} \, dx = \frac{1}{a} \int e^u \, du = \frac{1}{a} e^u + C = \frac{1}{a} e^{ax} + C$.

24.24 $\int \sqrt{e^x} \, dx$.

   ▮ $\int \sqrt{e^x} \, dx = \int (e^x)^{1/2} \, dx = \int e^{(1/2)x} \, dx = 2e^{(1/2)x} + C$,   by Problem 24.23.

24.25 $\int x^\pi \, dx$.

   ▮ $\int x^\pi \, dx = \frac{1}{\pi + 1} x^{\pi + 1} + C$.   This is a special case of the general law $\int x^r \, dx = \frac{1}{r + 1} x^{r + 1} + C$ for any constant $r \neq -1$.

24.26 $\int e^x e^{2x} \, dx$.

   ▮ $\int e^x e^{2x} \, dx = \int e^{x + 2x} \, dx = \int e^{3x} \, dx = \frac{1}{3} e^{3x} + C$.

24.27 $\int \dfrac{e^{2x}}{e^x + 1} \, dx$.

   ▮ Let $u = e^x + 1$, $du = e^x \, dx$.   Then

$$\int \frac{e^{2x}}{e^x + 1}\, dx = \int \frac{e^x}{e^x + 1}\, e^x\, dx = \int \frac{u - 1}{u}\, du = \int \left(1 - \frac{1}{u}\right) du = u - \ln u + C = e^x + 1 - \ln\,(e^x + 1) + C$$
$$= e^x - \ln\,(e^x + 1) + C_1$$

Here, we designated $1 + C$ as the new arbitrary constant C_1.

24.28 $\int x^2\, 2^{x^3}\, dx.$

▌ Let $u = 2^{x^3},\quad du = \ln 2 \cdot 2^{x^3} \cdot 3x^2\, dx.$ [Here, we have used the fact that $D_x(a^x) = (\ln a)a^x.$] Then
$\int x^2 2^{x^3}\, dx = \dfrac{1}{3 \ln 2}\int 1\, du = \dfrac{1}{3 \ln 2}\, u + C = \dfrac{1}{3 \ln 2}\, 2^{x^3} + C.$

24.29 $\int x^3 e^{-x^4}\, dx.$

▌ Let $u = -x^4,\quad du = -4x^3\, dx.$ Then $\int x^3\, e^{-x^4}\, dx = -\frac{1}{4}\int e^u\, du = -\frac{1}{4} e^u + C = -\frac{1}{4} e^{-x^4} + C.$

In Problems 24.30–24.39, find y'.

24.30 $e^y = y + \ln x.$

▌ By implicit differentiation, $e^y y' = y' + \dfrac{1}{x},\quad y'(e^y - 1) = \dfrac{1}{x},\quad y' = \dfrac{1}{x(e^y - 1)} = \dfrac{1}{x(y - 1 + \ln x)}.$

24.31 $\tan e^{y-x} = x^2.$

▌ By implicit differentiation, $\sec^2 e^{y-x} \cdot e^{y-x} \cdot (y' - 1) = 2x.$ Note that $\sec^2 e^{y-x} = 1 + \tan^2 e^{y-x} = 1 + x^4.$
Thus, $(1 + x^4) \cdot e^{y-x} \cdot (y' - 1) = 2x,\quad y' = 1 + \dfrac{2x}{(1 + x^4)e^{y-x}}.$

24.32 $e^{1/y} + e^y = 2x.$

▌ By implicit differentiation,

$$e^{1/y} \cdot \left(-\frac{1}{y^2}\, y'\right) + e^y y' = 2 \qquad y'\left(e^y - \frac{1}{y^2}\, e^{1/y}\right) = 2 \qquad y'\left(\frac{y^2 e^y - e^{1/y}}{y^2}\right) = 2 \qquad y' = \frac{2y^2}{y^2 e^y - e^{1/y}}$$

24.33 $x^2 + e^{xy} + y^2 = 1.$

▌ $2x + e^{xy}(xy' + y) + 2yy' = 0,\quad y'(xe^{xy} + 2y) = -2x - ye^{xy},\quad y' = -\dfrac{2x + ye^{xy}}{2y + xe^{xy}}.$

24.34 $\sin x = e^y.$

▌ $\cos x = e^y y',\quad y' = \dfrac{\cos x}{e^y} = \dfrac{\cos x}{\sin x} = \cot x.$

24.35 $y = 3^{\sin x}.$

▌ Use logarithmic differentiation. $\ln y = \sin x \cdot \ln 3.$ [Here, we use the law $\ln\,(a^b) = b \ln a$]. Hence,
$(1/y)y' = (\ln 3)(\cos x),\quad y' = (\ln 3)(\cos x)(3^{\sin x}).$

24.36 $y = (\sqrt{2})^{e^x}.$

▌ $\ln y = e^x \cdot \ln \sqrt{2} = \frac{1}{2}(\ln 2)e^x.$ So, $(1/y)y' = \frac{1}{2}(\ln 2)e^x,\quad y' = \frac{1}{2}(\ln 2)e^x(\sqrt{2})^{e^x}.$

24.37 $y = x^{\ln x}.$

▌ $\ln y = \ln x \cdot \ln x = (\ln x)^2.$ So, $\dfrac{1}{y}\, y' = 2(\ln x) \cdot \dfrac{1}{x},\quad y' = \dfrac{2 \ln x}{x}\, x^{\ln x} = 2\,(\ln x)x^{\ln x - 1}.$

24.38 $y = (\ln x)^{\ln x}.$

▌ $\ln y = \ln x \cdot \ln\,(\ln x).$ So,

$$\frac{1}{y}\, y' = \ln x \cdot \frac{1}{\ln x} \cdot \frac{1}{x} + \ln\,(\ln x) \cdot \frac{1}{x} = \frac{1}{x}\,[1 + \ln\,(\ln x)] \qquad y' = \frac{1 + \ln\,(\ln x)}{x} \cdot (\ln x)^{\ln x}$$

24.39 $y^2 = (x + 1)(x + 2)$.

❚ $2 \ln y = \ln (x + 1) + \ln (x + 2)$, $\dfrac{2}{y} y' = \dfrac{1}{x + 1} + \dfrac{1}{x + 2} = \dfrac{2x + 3}{(x + 1)(x + 2)}$, $y' = \dfrac{1}{2} \dfrac{2x + 3}{(x + 1)(x + 2)} \cdot y$

$= \dfrac{1}{2} \dfrac{2x + 3}{\sqrt{(x + 1)(x + 2)}}$.

24.40   Solve $e^{3x} = 2$ for x.

❚ $\ln 2 = \ln (e^{3x}) = 3x$, $x = \frac{1}{3} \ln 2$.

24.41   Solve $\ln x^3 = -1$ for x.

❚ $-1 = 3 \ln x$, $\ln x = -\frac{1}{3}$, $e^{\ln x} = e^{-1/3}$, $x = e^{-1/3}$.

24.42   Solve $e^x - 2e^{-x} = 1$ for x.

❚ Multiply by e^x: $e^{2x} - 2 = e^x$, $e^{2x} - e^x - 2 = 0$, $(e^x - 2)(e^x + 1) = 0$.   Since $e^x > 0$, $e^x + 1 \neq 0$.   Hence, $e^x - 2 = 0$, $e^x = 2$, $x = \ln 2$.

24.43   Solve $\ln (\ln x) = 1$ for x.

❚ $e \doteq e^{\ln (\ln x)} = \ln x$,   since $e^{\ln u} = u$.   Hence, $e^e = e^{\ln x} = x$.

24.44   Solve $\ln (x - 1) = 0$ for x.

❚ $x - 1 = 1$,   since $\ln u = 0$ has the unique solution 1.   Hence, $x = 2$.

24.45   Let $\mathcal{R}$ be the region under the curve $y = e^x$,   above the x-axis, and between $x = 0$ and $x = 1$.   Find the area of $\mathcal{R}$.

❚ The area $A = \int_0^1 e^x \, dx = e^x \,]_0^1 = e^1 - e^0 = e - 1$.

24.46   Find the volume of the solid generated by rotating the region of Problem 24.45 around the x-axis.

❚ By the disk formula, $V = \pi \int_0^1 (e^x)^2 \, dx = \pi \int_0^1 e^{2x} \, dx = \pi \cdot \frac{1}{2} e^{2x} \,]_0^1 = \dfrac{\pi}{2} (e^2 - e^0) = \dfrac{\pi}{2} (e^2 - 1)$.

24.47   Let $\mathcal{R}$ be the region bounded by the curve $y = e^{x/2}$,   the y-axis, and the line $y = e$ (see Fig. 24-1).   Find the area of $\mathcal{R}$.

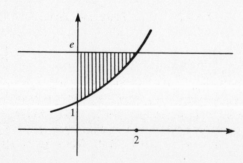

Fig. 24-1

❚ Setting $e = e^{x/2}$,   we find $x/2 = 1$, $x = 2$.   Hence, $y = e^{x/2}$ meets $y = e$ at the point $(2, e)$.   The area $A = \int_0^2 (e - e^{x/2}) \, dx = (ex - 2e^{x/2}) \,]_0^2 = (2e - 2e) - (0 - 2e^0) = 2$.

24.48   Find the volume when the region $\mathcal{R}$ of Problem 24.47 is rotated about the x-axis.

❚ By the circular ring formula, $V = \pi \int_0^2 [e^2 - (e^{x/2})^2] \, dx = \pi \int_0^2 (e^2 - e^x) \, dx = \pi (e^2 x - e^x) \,]_0^2 = \pi[(2e^2 - e^2) - (0 - e^0)] = \pi(e^2 + 1)$.

24.49   Let $\mathcal{R}$ be the region bounded by $y = e^{x^2}$,   the x-axis, the y-axis, and the line $x = 1$.   Find the volume of the solid generated when $\mathcal{R}$ is rotated about the y-axis.

❚ By the cylindrical shell formula $V = 2\pi \int_0^1 xy \, dx = 2\pi \int_0^1 xe^{x^2} \, dx$.   Let $u = x^2$, $du = 2x \, dx$.   Then $V = 2\pi \int_0^1 \frac{1}{2} e^u \, du = \pi e^u \,]_0^1 = \pi(e^1 - e^0) = \pi(e - 1)$.

24.50 Find the absolute extrema of $y = e^{\sin x}$ on $[-\pi, \pi]$.

▮ Since e^u is an increasing function of u, the maximum and minimum values of y correspond to the maximum and minimum values of the exponent $\sin x$, that is, 1 and -1. Hence, the absolute maximum is e (when $x = \pi/2$) and the absolute minimum is $e^{-1} = 1/e$ (when $x = -\pi/2$).

24.51 If $y = e^{nx}$, where n is a positive integer, find the nth derivative $y^{(n)}$.

▮ $y' = ne^{nx}, \quad y'' = n^2 e^{nx}, \ldots, \quad y^{(n)} = n^n e^{nx}$.

24.52 If $y = 2e^{\sin x}$, find y' and y''.

▮ $y' = 2e^{\sin x} \cdot \cos x, \quad y'' = 2[e^{\sin x}(-\sin x) + \cos x \cdot e^{\sin x} \cdot \cos x] = 2e^{\sin x}(\cos^2 x - \sin x)$.

24.53 Assume that the quantities x and y vary with time and are related by the equation $y = 2e^{\sin x}$. If y increases at a constant rate of 4 units per second, how fast is x changing when $x = \pi$?

▮ From Problem 24.52, $dy/dx = 2e^{\sin x} \cos x$. Hence,

$$\frac{dy}{dt} = \frac{dy}{dx}\frac{dx}{dt} = 2e^{\sin x} \cos x \cdot \frac{dx}{dt} \qquad 4 = 2e^{\sin x} \cos x \cdot \frac{dx}{dt} \qquad 2 = e^{\sin x} \cos x \cdot \frac{dx}{dt}$$

When $x = \pi$, $2 = e^0(-1) \cdot \dfrac{dx}{dt}$, $\dfrac{dx}{dt} = -2$.

24.54 The acceleration of an object moving on the x-axis is $9e^{3t}$. Find a formula for the velocity v, if the velocity at time $t = 0$ is 4 units per second.

▮ $v = \int a\, dt = \int 9e^{3t}\, dt = 3e^{3t} + C$. Hence, $4 = 3e^0 + C$, $4 = 3 + C$, $C = 1$. Hence, $v = 3e^{3t} + 1$.

24.55 How far does the object of Problem 24.54 move as its velocity increases from 4 to 10 units per second?

▮ From Problem 24.54, $v = 3e^{3t} + 1$. When $v = 4$, $t = 0$. When $v = 10$, $e^{3t} = 3$, $3t = \ln 3$, $t = \frac{1}{3}\ln 3$. The required distance is therefore

$$s = \int_0^{(\ln 3)/3} v\, dt = \int_0^{(\ln 3)/3} (3e^{3t} + 1)\, dt = e^{3t} + t\,]_0^{(\ln 3)/3} = (e^{\ln 3} + \tfrac{1}{3}\ln 3) - (1 + 0) = 3 + \tfrac{1}{3}\ln 3 - 1 = 2 + \tfrac{1}{3}\ln 3$$

24.56 Find an equation of the tangent line to the curve $y = 2e^x$ at the point $(0, 2)$.

▮ The slope is the derivative $y' = 2e^x = 2e^0 = 2$. Hence, an equation of the tangent line is $y - 2 = 2(x - 0)$, or $y = 2x + 2$.

24.57 Graph $y = e^{-x^2}$.

▮ $y' = e^{-x^2} \cdot (-2x) = -2xe^{-x^2}$. Hence, $x = 0$ is the only critical number. $y'' = -2[e^{-x^2} + x \cdot (-2xe^{-x^2})] = -2e^{-x^2}(1 - 2x^2)$. By the second-derivative test, there is a relative (and, therefore, absolute) maximum at $(0, 1)$. As $x \to \pm\infty$, $e^{x^2} \to +\infty$, and, therefore, $y \to 0$. Thus, the x-axis is a horizontal asymptote on the right and left. The graph is symmetric with respect to the y-axis, since e^{-x^2} is an even function. There are inflection points where $y'' = 0$, that is, at $x = \pm\sqrt{2}/2$. Thus the graph has the bell-shaped appearance indicated in Fig. 24-2.

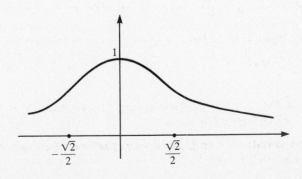

Fig. 24-2

24.58 Graph $y = x \ln x$.

▌ See Fig. 24-3. The function is defined only for $x > 0$. $y' = x \cdot \frac{1}{x} + \ln x = 1 + \ln x$. $y'' = 1/x$. Setting $y' = 0$, $\ln x = -1$, $x = e^{\ln x} = e^{-1} = 1/e$. This is the only critical number, and, by the second-derivative test, there is a relative (and, therefore, an absolute) minimum at $(1/e, -1/e)$. As $x \to +\infty$, $y \to +\infty$. As $x \to 0^+$, $y \to 0$, by Problem 23.44.

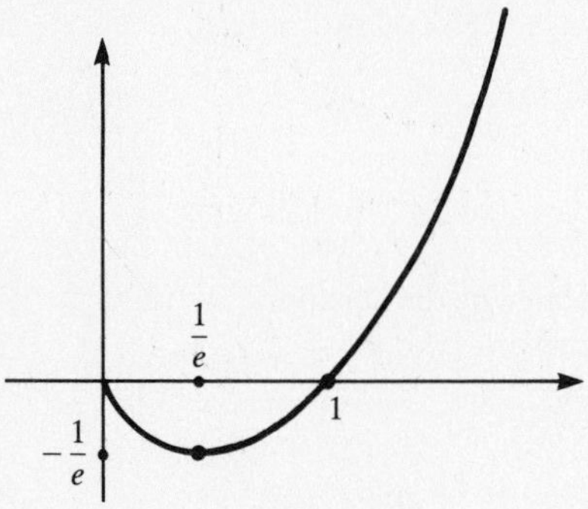

Fig. 24-3

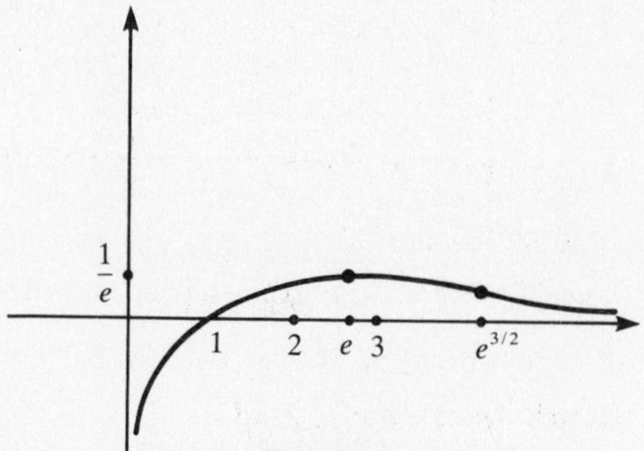

Fig. 24-4

24.59 Graph $y = \frac{\ln x}{x}$.

▌ $y' = \frac{1 - \ln x}{x^2}$ and $y'' = \frac{-x - 2x(1 - \ln x)}{x^4} = \frac{(2 \ln x - 3)}{x^3}$. The only critical number occurs when $\ln x = 1$, $x = e$. By the second-derivative test, there is a relative (and, therefore, an absolute) maximum at $(e, 1/e)$. As $x \to +\infty$, $y \to 0$, by Problem 23.43. As $x \to 0^+$, $y \to -\infty$. Hence, the positive x-axis is a horizontal asymptote and the negative y-axis is a vertical asymptote. There is an inflection point where $2 \ln x - 3 = 0$, that is, $\ln x = \frac{3}{2}$, $x = e^{3/2}$. See Fig. 24-4.

24.60 Sketch the graph of $y = e^{-x}$.

▌ The graph, Fig. 24-5, is obtained by reflecting the graph of $y = e^x$ in the y-axis.

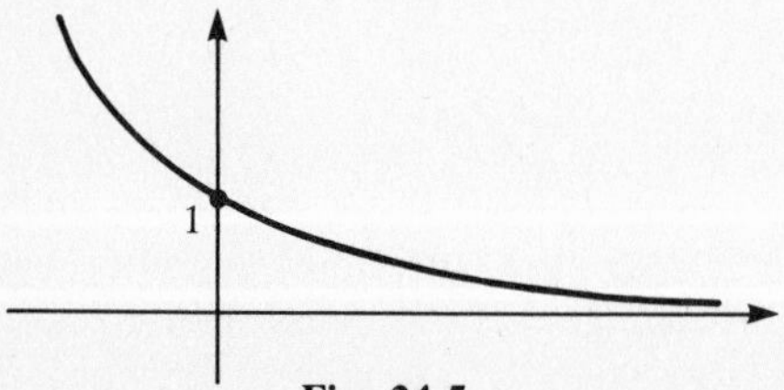

Fig. 24-5

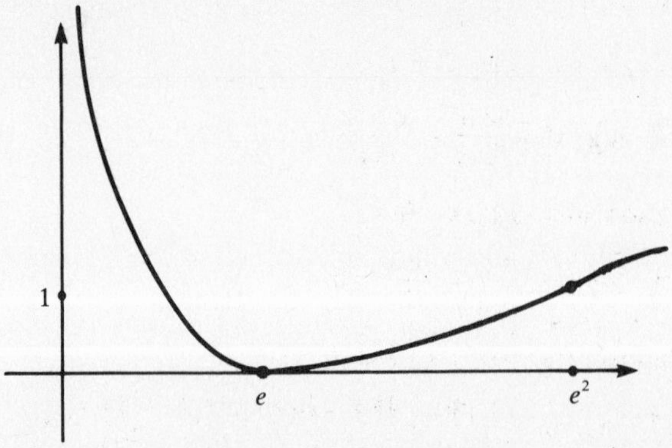

Fig. 24-6

24.61 Graph $y = (1 - \ln x)^2$.

▌ $y' = 2(1 - \ln x)\left(-\frac{1}{x}\right) = -\frac{2(1 - \ln x)}{x}$. Then $y'' = (-2)\frac{-1 - (1 - \ln x)}{x^2} = \frac{2(2 - \ln x)}{x^2}$. When $y' = 0$, $\ln x = 1$, $x = e$. This is the only critical number. By the second-derivative test, there is a relative (and, therefore, an absolute) minimum at $(e, 0)$. As $x \to +\infty$, $y \to +\infty$, and, as $x \to 0^+$, $y \to +\infty$. There is an inflection point when $2 - \ln x = 0$, $\ln x = 2$, $x = e^2$. The graph is shown in Fig. 24-6.

24.62 Graph $y = \frac{1}{x} + \ln x$.

▌ See Fig. 24-7. $y' = -\frac{1}{x^2} + \frac{1}{x} = \frac{1}{x^2}(x - 1)$. $y'' = \frac{2}{x^3} - \frac{1}{x^2} = \frac{1}{x^3}(2 - x)$. Hence, by the second-derivative test, the unique critical number $x = 1$ yields a relative (and, therefore, an absolute) minimum at $(1, 1)$. As $x \to +\infty$, $y \to +\infty$. As $x \to 0^+$, $y = (1 + x \ln x)/x \to +\infty$, since $x \ln x \to 0$ by Problem 23.44. There is an inflection point at $x = 2$, $y = \frac{1}{2} + \ln 2$.

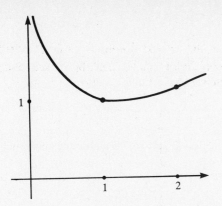

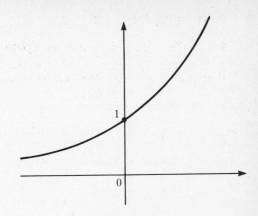

Fig. 24-7   Fig. 24-8

24.63   Sketch the graph of $y = 2^x$.

   ❚ Since $2^x = e^{x \ln 2}$,   the graph has the same general shape as that of $y = e^x$ (Fig. 24-8), a little lower for $x > 0$ and a little higher for $x < 0$ (because $2 < e$).

    Problems 24.64-24.71 refer to the function $\log_a x \equiv \dfrac{\ln x}{\ln a}$,   the so-called *logarithm of x to the base a*. (Assume $a > 0$ and $a \neq 1$.)

24.64   Show that $D_x(\log_a x) = \dfrac{1}{x \ln a}$.

   ❚ $D_x(\log_a x) = D_x\left(\dfrac{\ln x}{\ln a}\right) = \dfrac{1}{\ln a} D_x(\ln x) = \dfrac{1}{\ln a} \cdot \dfrac{1}{x} = \dfrac{1}{x \ln a}$.

24.65   Show that $a^{\log_a x} = x$.

   ❚ $a^{\log_a x} = a^{\ln x / \ln a} = e^{(\ln a)(\ln x / \ln a)} = e^{\ln x} = x$.

24.66   Show that $\log_a a^x = x$.

   ❚ $\log_a a^x = \dfrac{\ln a^x}{\ln a} = \dfrac{x \ln a}{\ln a} = x$.

24.67   Show that $\log_e x = \ln x$.

   ❚ $\log_e x = \dfrac{\ln x}{\ln e} = \dfrac{\ln x}{1} = \ln x$.

24.68   Show that $\log_a uv = \log_a u + \log_a v$.

   ❚ $\log_a uv = \dfrac{\ln uv}{\ln a} = \dfrac{\ln u + \ln v}{\ln a} = \dfrac{\ln u}{\ln a} + \dfrac{\ln v}{\ln a} = \log_a u + \log_a v$.

24.69   Show that $\log_a \dfrac{u}{v} = \log_a u - \log_a v$.

   ❚ By Problem 24.68, $\log_a \dfrac{u}{v} + \log_a v = \log_a\left(\dfrac{u}{v} \cdot v\right) = \log_a u$.   Subtract $\log_a v$ from both sides.

24.70   Show that $\log_a u^r = r \log_a u$.

   ❚ $\log_a u^r = \dfrac{\ln u^r}{\ln a} = \dfrac{r \ln u}{\ln a} = r \cdot \dfrac{\ln u}{\ln a} = r \log_a u$.

24.71   Prove that $\ln x = \dfrac{\log_a x}{\log_a e}$.

   ❚ $\dfrac{\log_a x}{\log_a e} = \dfrac{\ln x / \ln a}{\ln e / \ln a} = \dfrac{\ln x}{\ln a} \cdot \dfrac{\ln a}{1} = \ln x$.

24.72 Prove that the only solutions of the differential equation $f'(x) = f(x)$ are the functions Ce^x, where C is a constant.

▮ We know that one nonvanishing solution is e^x, so make the substitution $f(x) = e^x g(x)$: $e^x g' + e^x g = e^x g$, $e^x g' = 0$, $g' = 0$, $g = C$.

24.73 Find the absolute extrema of $f(x) = \dfrac{(\ln x)^2}{x}$ on $[1, e]$.

▮ $f' = \dfrac{x \cdot 2 \ln x \cdot (1/x) - (\ln x)^2}{x^2} = \dfrac{\ln x(2 - \ln x)}{x^2} > 0$ on $(1, e]$. Hence, the absolute minimum is $f(1) = 0$ and the absolute maximum is $f(e) = 1/e$.

24.74 Prove $e^x = \lim\limits_{u \to +\infty} \left(1 + \dfrac{x}{u}\right)^u$.

▮ Let $y = \left(1 + \dfrac{x}{u}\right)^u$. Then $\ln y = u \ln \left(1 + \dfrac{x}{u}\right) = u[\ln (u + x) - \ln u] = u \cdot x \cdot \dfrac{1}{u^*}$ where u^* is between u and $u + x$. (In the last step, we used the mean-value theorem.) Either $u < u^* < u + x$ or $u + x < u^* < u$. Then, either $1 < u^*/u < 1 + x/u$ or $1 + x/u < u^*/u < 1$. In either case, $u^*/u \to 1$ as $u \to +\infty$. Hence, $\ln y \to x$ as $u \to +\infty$. Therefore, $y = e^{\ln y} \to e^x$ as $u \to +\infty$.

24.75 Prove that, for any positive n, $\lim\limits_{x \to +\infty} \dfrac{x^n}{e^x} = 0$.

▮ $\dfrac{x^n}{e^x} = \dfrac{e^{n \ln x}}{e^x} = \dfrac{1}{e^{x - n \ln x}} = \dfrac{1}{e^{x[1 - n(\ln x/x)]}}$. But, $\ln x/x \to 0$ as $x \to +\infty$. Hence, since $e^x \to +\infty$, $x^n/e^x \to 0$.

24.76 Find $\displaystyle\int_e^{e^2} \dfrac{1}{x \ln x} \, dx$.

▮ $\displaystyle\int_e^{e^2} \dfrac{1}{x \ln x} \, dx = \ln |\ln x| \,]_e^{e^2} = \ln (\ln e^2) - \ln (\ln e) = \ln 2 - \ln 1 = \ln 2$.

24.77 Find the derivative of $y = x^{\sec x}$.

▮ $\ln y = \sec x \cdot \ln x$. Hence, $\dfrac{1}{y} y' = \sec x \cdot \dfrac{1}{x} + \ln x \cdot \sec x \tan x = \dfrac{\sec x}{x} (1 + x \ln x \tan x)$. So,

$$y' = x^{\sec x} \cdot \dfrac{\sec x}{x} (1 + x \ln x \tan x) = x^{\sec x - 1} \sec x(1 + x \ln x \tan x)$$

24.78 Evaluate $\lim\limits_{x \to +\infty} \left(\dfrac{x}{x + 1}\right)^{x+1}$.

▮ $\lim\limits_{x \to +\infty} \left(\dfrac{x}{x + 1}\right)^{x+1} = \lim\limits_{x + 1 \to +\infty} \left(1 - \dfrac{1}{x + 1}\right)^{x+1} = e^{-1}$, by Problem 24.74.

24.79 Evaluate $\lim\limits_{x \to 0^+} (\sin x)^{\tan x}$.

▮ Let $y = (\sin x)^{\tan x}$. Then $\ln y = \tan x \ln (\sin x) = \dfrac{\sin x \cdot \ln (\sin x)}{\cos x}$. By Problem 23.44, $u \ln u \to 0$ as $u \to 0^+$. Since $\sin x \to 0^+$ as $x \to 0^+$, it follows that $\sin x \cdot \ln (\sin x) \to 0$ as $x \to 0^+$. Since $\cos x \to 1$ as $x \to 0$, $\ln y \to 0$ as $x \to 0^+$. Therefore, $y = e^{\ln y} \to e^0 = 1$ as $x \to 0^+$.

24.80 Evaluate $\lim\limits_{x \to 0^+} x^{\sin x}$.

▮ Let $y = x^{\sin x}$. $\ln y = \sin x \cdot \ln x = \dfrac{\sin x}{x} \cdot (x \ln x)$. Since $\dfrac{\sin x}{x} \to 1$ and $x \ln x \to 0$ as $x \to 0^+$, $\ln y \to 0$ as $x \to 0^+$. Hence, $y = e^{\ln y} \to e^0 = 1$.

24.81 Evaluate $\lim\limits_{x \to 0} (\sin x)^{\cos x}$.

▮ Since $\cos x \to 1$ and $\sin x \to 0$ as $x \to 0$, $(\sin x)^{\cos x} \to 0^1 = 0$.

24.82 Evaluate $e^{3 \ln 2}$.

▮ $e^{3 \ln 2} = (e^{\ln 2})^3 = 2^3 = 8$.

24.83 Show that $e = \lim\limits_{u \to +\infty} \left(1 + \dfrac{1}{u}\right)^u$.

▌ Set $x = 1$ in the formula of Problem 24.74.

24.84 Graph $y = x^2 e^x$.

▌ See Fig. 24-9. $y' = x^2 e^x + 2xe^x = xe^x(x + 2)$. $y'' = xe^x + (x + 2)(xe^x + e^x) = e^x(x^2 + 4x + 2)$. The critical numbers are $x = 0$ and $x = -2$. The second-derivative test shows that there is a relative minimum at $(0, 0)$ and a relative maximum at $(-2, 4e^{-2})$. As $x \to +\infty$, $y \to +\infty$. As $x \to -\infty$, $y \to 0$ (by Problem 24.75). There are inflection points where $x^2 + 4x + 2 = 0$, that is, at $x = -2 \pm \sqrt{2}$.

24.85 Graph $y = x^2 e^{-x}$.

▌ The graph is obtained by reflecting Fig. 24-9 in the y-axis, since $y = x^2 e^{-x}$ is obtained from $y = x^2 e^x$ by replacing x by $-x$.

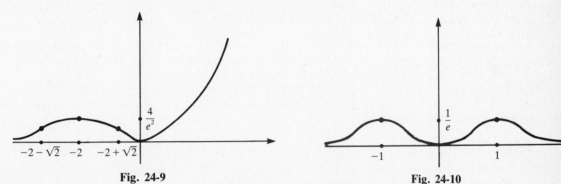

Fig. 24-9   **Fig. 24-10**

24.86 Graph $y = x^2 e^{-x^2}$.

▌ $y' = -2x^3 e^{-x^2} + 2xe^{-x^2} = 2xe^{-x^2}(1 - x^2)$. Then $y'' = 2xe^{-x^2}(-2x) + (1 - x^2)(-4x^2 e^{-x^2} + 2e^{-x^2}) = 2e^{-x^2}(2x^4 - 5x^2 + 1)$. The critical numbers are $x = 0$, and $x = \pm 1$. By the second-derivative test, there is a relative minimum at $(0, 0)$ and relative maxima at $(\pm 1, e^{-1})$. There are inflection points at $x = \pm\sqrt{5 + \sqrt{17}}/2$ and $x = \pm\sqrt{5 - \sqrt{17}}/2$. The graph is symmetric with respect to the y-axis. See Fig. 24-10.

24.87 Find the maximum area of a rectangle in the first quadrant, with base on the x-axis, one vertex at the origin and the opposite vertex on the curve $y = e^{-x^2}$ (see Fig. 24-11).

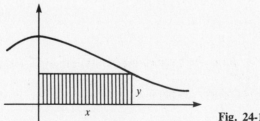

Fig. 24-11

▌ Let x be the length of the base. Then the area $A = xy = xe^{-x^2}$, $D_x A = -2x^2 e^{-x^2} + e^{-x^2} = e^{-x^2}(1 - 2x^2)$, $D_x^2 A = e^{-x^2}(-4x) + (1 - 2x^2)e^{-x^2}(-2x) = -2xe^{-x^2}(3 - 2x^2)$. Setting $D_x A = 0$, we see that the only positive critical number is $x = 1/\sqrt{2}$, and the second-derivative test shows that this is a relative (and, therefore, an absolute) maximum. Then the maximum area is $xe^{-x^2} = (1/\sqrt{2})e^{-1/2} = 1/\sqrt{2e}$.

24.88 Find $D_x(x^x)$.

▌ Let $y = x^x$. Then $\ln y = x \ln x$, $\dfrac{1}{y} y' = 1 + \ln x$, $y' = x^x(1 + \ln x)$.

24.89 Prove that $D_x^n(x^{n-1} \ln x) = \dfrac{(n-1)!}{x}$.

▮ We use mathematical induction. For $n=1$, $D_x(x^0 \ln x) = D_x(\ln x) = 1/x$, and $0!/x = 1/x$. Now assume the formula true for n: $D_x^n(x^{n-1} \ln x) = (n-1)!/x$, and we must prove it true for $n+1$: $D_x^{n+1}(x^n \ln x) = n!/x$. In fact,

$$D_x^{n+1}(x^n \ln x) = D_x^n[D_x(x \cdot x^{n-1} \ln x)]$$

$$= D_x^n[x \cdot D_x(x^{n-1} \ln x) + x^{n-1} \ln x] = D_x^n\{x[x^{n-2} + (n-1)x^{n-2} \ln x]\} + D_x^n(x^{n-1} \ln x)$$

$$= D_x^n\{x^{n-1}[1 + (n-1) \ln x]\} + \frac{(n-1)!}{x} = D_x^n[x^{n-1} \ln x(n-1)] + D_x^n(x^{n-1}) + \frac{(n-1)!}{x}$$

$$= (n-1)D_x^n(x^{n-1} \ln x) + 0 + \frac{(n-1)!}{x} = (n-1)\frac{(n-1)!}{x} + \frac{(n-1)!}{x} = \frac{n \cdot (n-1)!}{x} = \frac{n!}{x}$$

This completes the induction.

24.90 Prove that $D_x^n(xe^x) = (x+n)e^x$.

▮ Use mathematical induction. For $n=1$, $D_x(xe^x) = xe^x + e^x = (x+1)e^x$. Now, assume the formula true for n: $D_x^n(xe^x) = (x+n)e^x$, and we must prove it true for $n+1$: $D_x^{n+1}(xe^x) = (x+n+1)e^x$. In fact, $D_x^{n+1}(xe^x) = D_x[D_x^n(xe^x)] = D_x[(x+n)e^x] = (x+n)e^x + e^x = (x+n+1)e^x$.

24.91 Graph $y = e^{x/a} + e^{-x/a}$ $(a > 0)$.

▮ See Fig. 24-12. $y' = \dfrac{1}{a}e^{x/a} - \dfrac{1}{a}e^{-x/a} = \dfrac{1}{a}(e^{x/a} - e^{-x/a})$. $y'' = \dfrac{1}{a}\left(\dfrac{1}{a}e^{x/a} + \dfrac{1}{a}e^{-x/a}\right) = \dfrac{1}{a^2}(e^{x/a} + e^{-x/a}) > 0$. Setting $y' = 0$, we have $e^{x/a} = e^{-x/a}$, $(e^{x/a})^2 = 1$, $e^{x/a} = 1$, $x/a = 0$, $x = 0$. The second-derivative test shows that there is a minimum at $(0, 2)$. The graph is symmetric with respect to the y-axis. As $x \to \pm\infty$, $y \to +\infty$.

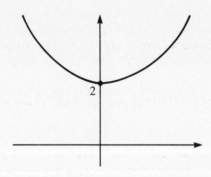

Fig. 24-12

24.92 Find the area under $y = e^{x/a} + e^{-x/a}$ (Problem 24.91), above the x-axis, and between $x = -a$ and $x = a$.

▮ $A = \int_{-a}^{a} (e^{x/a} + e^{-x/a}) \, dx = ae^{x/a} - ae^{-x/a} \Big]_{-a}^{a} = a(e^{x/a} - e^{-x/a}) \Big]_{-a}^{a} = a[(e - e^{-1}) - (e^{-1} - e)] = 2a\left(e - \dfrac{1}{e}\right)$.

24.93 Find the arc length of the curve $y = \dfrac{a}{2}(e^{x/a} + e^{-x/a})$ from $x = 0$ to $x = b$.

▮ $\qquad y' = \dfrac{a}{2}\left(\dfrac{1}{a}e^{x/a} - \dfrac{1}{a}e^{-x/a}\right) = \tfrac{1}{2}(e^{x/a} - e^{-x/a}) \qquad (y')^2 = \tfrac{1}{4}(e^{2x/a} - 2 + e^{-2x/a})$

$$1 + (y')^2 = \frac{4 + (e^{2x/a} - 2 + e^{-2x/a})}{4} = \frac{e^{2x/a} + 2 + e^{-2x/a}}{4} = [\tfrac{1}{2}(e^{x/a} + e^{-x/a})]^2$$

Hence, the arc length is $L = \int_0^b \tfrac{1}{2}(e^{x/a} + e^{-x/a}) \, dx = \tfrac{1}{2}(ae^{x/a} - ae^{-x/a})\big]_0^b = \dfrac{a}{2}(e^{b/a} - e^{-b/a})$.

24.94 Sketch the graph of $y = \dfrac{xe^{-x}}{x+1}$.

▌ See Fig. 24-13. $y' = \dfrac{(x+1)(-xe^{-x} + e^{-x}) - xe^{-x}}{(x+1)^2} = \dfrac{e^{-x}(1 - x - x^2)}{(x+1)^2}$. The critical numbers are $x = (-1 \pm \sqrt{5})/2$. The first-derivative test shows that $x = (-1 + \sqrt{5})/2$ yields a relative maximum and $x = (-1 - \sqrt{5})/2$ a relative minimum. There is a vertical asymptote at $x = -1$. $y > 0$ for $x < -1$ and $y \to +\infty$ as $x \to -1^-$, and $y \to -\infty$ as $x \to -1^+$. $y \to 0$ as $x \to +\infty$, since $x/(x+1) \to 1$ and $e^{-x} \to 0$. $y \to +\infty$ as $x \to -\infty$, since $e^{-x} \to +\infty$.

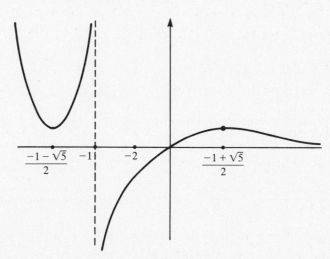

Fig. 24-13

Problems 24.95–24.110 concern the hyperbolic sine and cosine functions $\sinh x = \frac{1}{2}(e^x - e^{-x})$ and $\cosh x = \frac{1}{2}(e^x + e^{-x})$.

24.95 Find $D_x(\sinh x)$ and $D_x(\cosh x)$.

▌ $D_x(\sinh x) = D_x[\frac{1}{2}(e^x - e^{-x})] = \frac{1}{2}(e^x + e^{-x}) = \cosh x$. $D_x(\cosh x) = D_x[\frac{1}{2}(e^x + e^{-x})] = \frac{1}{2}(e^x - e^{-x}) = \sinh x$.

24.96 Find $D_x^2(\sinh x)$ and $D_x^2(\cosh x)$.

▌ By Problem 24.95, $D_x^2(\sinh x) = D_x(\cosh x) = \sinh x$ and $D_x^2(\cosh x) = D_x(\sinh x) = \cosh x$.

24.97 Graph $y = \sinh x$.

▌ See Fig. 24-14. Since $D_x(\sinh x) = \cosh x > 0$, $\sinh x$ is an increasing function. It is clearly an odd function, so $\sinh 0 = 0$. Since $D_x(\sinh x) = \sinh x$, the graph has an inflection point at $(0, 0)$, where the slope of the tangent line is $\cosh 0 = 1$.

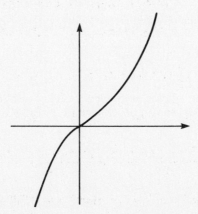

Fig. 24-14

CHAPTER 24

24.98   Show that $\cosh^2 x - \sinh^2 x = 1$.

 ▮ This follows by direct computation from the definitions.

24.99   Let $\tanh x \equiv \sinh x / \cosh x$ and $\operatorname{sech} x \equiv 1 / \cosh x$.   Find the derivative of $\tanh x$.

 ▮ $D_x(\tanh x) = D_x\!\left(\dfrac{\sinh x}{\cosh x}\right) = \dfrac{\cosh x \cdot \cosh x - \sinh x \cdot \sinh x}{\cosh^2 x} = \dfrac{\cosh^2 x - \sinh^2 x}{\cosh^2 x} = \dfrac{1}{\cosh^2 x} = \operatorname{sech}^2 x.$

24.100   Find $D_x(\operatorname{sech} x)$.

 ▮ $D_x(\operatorname{sech} x) = D_x\!\left(\dfrac{1}{\cosh x}\right) = -\dfrac{1}{\cosh^2 x} \cdot \sinh x = -\dfrac{\sinh x}{\cosh x} \dfrac{1}{\cosh x} = -\tanh x \cdot \operatorname{sech} x.$

24.101   Show that $1 - \tanh^2 x = \operatorname{sech}^2 x$.

 ▮ By Problem 24.98, $\cosh^2 x - \sinh^2 x = 1$.   Dividing both sides by $\cosh^2 x$, we get $1 - \tanh^2 x = \operatorname{sech}^2 x$.

In Problems 24.102–24.108, determine whether the given analogues of certain trigonometric identities also hold for hyperbolic functions.

24.102 $\sinh 2x \overset{?}{=} 2 \sinh x \cosh x$.

 ▮ $2 \sinh x \cosh x = 2 \dfrac{e^x - e^{-x}}{2} \dfrac{e^x + e^{-x}}{2} = \tfrac{1}{2}(e^{2x} - e^{-2x}) = \sinh 2x$.   The identity holds.

24.103 $\sinh (x + y) \overset{?}{=} \sinh x \cosh y + \cosh x \sinh y$.

 ▮ $\sinh x \cosh y + \cosh x \sinh y = \tfrac{1}{2}(e^x - e^x) \cdot \tfrac{1}{2}(e^y + e^{-y}) + \tfrac{1}{2}(e^x + e^{-x}) \cdot \tfrac{1}{2}(e^y - e^{-y})$

$= \tfrac{1}{4}(e^{x+y} - e^{-(x+y)} - e^{y-x} + e^{x-y}) + \tfrac{1}{4}(e^{x+y} - e^{-(x+y)} + e^{y-x} - e^{x-y})$

$= \tfrac{1}{2}(e^{x+y} - e^{-(x+y)}) = \sinh (x + y).$

The identity holds.

24.104 $\cosh (x + y) \overset{?}{=} \cosh x \cosh y - \sinh x \sinh y$.

 ▮ Think of y as fixed and take the derivatives of both sides of the identity in Problem 24.103.   Then $\cosh (x + y) = \cosh x \cosh y + \sinh x \sinh y$.   This is the correct identity, not the given one.

24.105 $\cosh (-x) \overset{?}{=} \cosh x$ and $\sinh(-x) \overset{?}{=} -\sinh (x)$

 ▮ $\cosh (-x) = \tfrac{1}{2}(e^{-x} + e^{-(-x)}) = \tfrac{1}{2}(e^{-x} + e^x) = \cosh x.$ $\sinh (-x) = \tfrac{1}{2}(e^{-x} - e^{-(-x)}) = \tfrac{1}{2}(e^{-x} - e^x) = -\tfrac{1}{2}(e^x + e^{-x}) = -\sinh x.$ Thus, both identities hold.

24.106 $\cosh 2x \overset{?}{=} \cosh^2 x - \sinh^2 x$.

 ▮ By the identity found in the solution of Problem 24.104, $\cosh 2x = \cosh^2 x + \sinh^2 x$.   This is the correct identity, not the given one.

24.107 $\cosh 2x \overset{?}{=} 2 \cosh^2 x - 1$.

 ▮ By the identity established in the solution of Problem 24.106, $\cosh 2x = \cosh^2 x + \sinh^2 x$.   By the identity $\cosh^2 x - \sinh^2 x = 1$, $\sinh^2 x = \cosh^2 x - 1$,   and, therefore, $\cosh 2x = \cosh^2 x + \cosh^2 x - 1 = 2 \cosh^2 x - 1$.   Thus, the identity is correct.

24.108 $\cosh 2x \overset{?}{=} 1 - 2 \sinh^2 x$.

 ▮ By the identity $\cosh^2 x - \sinh^2 x = 1$, $\cosh^2 x = \sinh^2 x + 1$,   and, substituting in the identity $\cosh 2x = \cosh^2 x + \sinh^2 x$,   we get $\cosh 2x = 1 + 2 \sinh^2 x$.   This is the correct identity, not the given one.

24.109 Find $\int \dfrac{\sinh x}{1 + \cosh x}\, dx$.

▮ Let $u = 1 + \cosh x$, $du = \sinh x\, dx$. Then $\int \dfrac{\sinh x}{1 + \cosh x}\, dx = \int \dfrac{1}{u}\, du = \ln |u| + C = \ln (1 + \cosh x) + C$.

24.110 Find $\int x \tanh x^2\, dx$.

▮ Let $u = x^2$, $du = 2x\, dx$. Then $\int x \tanh x^2\, dx = \frac{1}{2} \int \tanh u\, du = \dfrac{1}{2} \int \dfrac{\sinh u}{\cosh u}\, du = \frac{1}{2} \ln |\cosh u| + C = \frac{1}{2} \ln (\cosh x^2) + C$.

24.111 Find $\int \dfrac{1}{\sinh x}\, dx$.

▮ Let $u = \tanh (x/2)$; by Problems 24.99 and 24.101, $du = [\text{sech}^2 (x/2)](\frac{1}{2}\, dx) = (1 - u^2)(\frac{1}{2}\, dx)$. Moreover, by Problems 24.102 and 24.101,

$$\sinh x = 2\left(\sinh \frac{x}{2}\right)\left(\cosh \frac{x}{2}\right) = 2\left(\tanh \frac{x}{2}\right)\left(\cosh^2 \frac{x}{2}\right) = \frac{2 \tanh (x/2)}{1 - \tanh^2 (x/2)} = \frac{2u}{1 - u^2}$$

Thus, $\qquad \int \dfrac{1}{\sinh x}\, dx = \int \dfrac{1 - u^2}{2u} \cdot \dfrac{2}{1 - u^2}\, du = \int \dfrac{1}{u}\, du = \ln |u| + C = \ln \left|\tanh \dfrac{x}{2}\right| + C$

CHAPTER 25
L'Hôpital's Rule

25.1 State L'Hôpital's rule.

▌ First, let us state the *zero-over-zero case*. Under certain simple conditions, if $\lim\limits_{x \to b} f(x) = \lim\limits_{x \to b} g(x) = 0$ and $\lim\limits_{x \to b} \dfrac{f'(x)}{g'(x)} = L$, then $\lim\limits_{x \to b} \dfrac{f(x)}{g(x)} = L$. Here, $x \to b$ can be replaced by $x \to b^+$, $x \to b^-$, $x \to +\infty$, or $x \to -\infty$. The conditions are that f and g are differentiable in an open interval around b and that g' is not zero in that interval, except possibly at b. (In the case of one-sided limits, the interval can have b as an endpoint. In the case of $x \to \pm\infty$, the conditions on f and g hold for sufficiently large, or sufficiently small, values of x.) The second case is the *infinity-over-infinity case*. If $\lim\limits_{x \to b} f(x) = \lim\limits_{x \to b} g(x) = \pm\infty$ and $\lim\limits_{x \to b} \dfrac{f'(x)}{g'(x)} = L$, then $\lim\limits_{x \to b} \dfrac{f(x)}{g(x)} = L$. Here, again $x \to b$ can be replaced by $x \to b^+$, $x \to b^-$, $x \to +\infty$, or $x \to -\infty$. The conditions on f and g are the same as in the first case, except that we need not assume that g' is not zero. [That fact follows automatically from the assumption that $\lim\limits_{x \to b} g(x) = \pm\infty$.]

In Problems 25.2–25.53, evaluate the given limit.

25.2 $\lim\limits_{x \to 0} \dfrac{\sin x}{x}$.

▌ $\lim\limits_{x \to 0} \dfrac{\sin x}{x} = \lim\limits_{x \to 0} \dfrac{D_x(\sin x)}{D_x(x)} = \lim\limits_{x \to 0} \dfrac{\cos x}{1} = \dfrac{1}{1} = 1$.

25.3 $\lim\limits_{x \to 0} \dfrac{1 - \cos x}{x}$.

▌ $\lim\limits_{x \to 0} \dfrac{1 - \cos x}{x} = \lim\limits_{x \to 0} \dfrac{\sin x}{1} = \dfrac{0}{1} = 0$.

25.4 $\lim\limits_{x \to +\infty} \dfrac{5x^3 - 4x + 3}{2x^2 - 1}$.

▌ $\lim\limits_{x \to +\infty} \dfrac{5x^3 - 4x + 3}{2x^2 - 1} = \lim\limits_{x \to +\infty} \dfrac{15x^2 - 4}{4x} = \lim\limits_{x \to +\infty} \dfrac{30x}{4} = +\infty$. Here, we have applied L'Hôpital's rule twice in succession. In subsequent problems, successive use of L'Hôpital's rule will be made without explicit mention.

25.5 $\lim\limits_{x \to +\infty} \dfrac{\ln(1 + e^x)}{1 + x}$.

▌ $\lim\limits_{x \to +\infty} \dfrac{\ln(1 + e^x)}{1 + x} = \lim\limits_{x \to +\infty} \dfrac{e^x/(1 + e^x)}{1} = \lim\limits_{x \to +\infty} \dfrac{e^x}{1 + e^x} = \lim\limits_{x \to \infty} \dfrac{e^x}{e^x} = \lim\limits_{x \to \infty} 1 = 1$.

25.6 $\lim\limits_{x \to 0} \left(\dfrac{1}{x} - \dfrac{1}{\sin x} \right)$.

▌ Here we have the difference of two functions that both approach ∞. However, $\dfrac{1}{x} - \dfrac{1}{\sin x} = \dfrac{\sin x - x}{x \sin x}$, to which L'Hôpital's rule is applicable. $\lim\limits_{x \to 0} \dfrac{\sin x - x}{x \sin x} = \lim\limits_{x \to 0} \dfrac{\cos x - 1}{x \cos x + \sin x} = \lim\limits_{x \to 0} \dfrac{-\sin x}{-x \sin x + 2 \cos x} = \dfrac{0}{0 + 2} = 0$.

25.7 $\lim\limits_{x \to 0} \dfrac{1 - e^x}{x}$.

▌ $\lim\limits_{x \to 0} \dfrac{1 - e^x}{x} = \lim\limits_{x \to 0} \dfrac{-e^x}{1} = \dfrac{-1}{1} = -1$.

25.8 $\displaystyle\lim_{x\to+\infty}\frac{x^2}{(\ln x)^3}$.

▉ $\displaystyle\lim_{x\to+\infty}\frac{x^2}{(\ln x)^3}=\lim_{x\to+\infty}\frac{2x}{3(\ln x)^2\cdot(1/x)}=\lim_{x\to+\infty}\frac{2x^2}{3(\ln x)^2}=\lim_{x\to+\infty}\frac{4x}{6\ln x(1/x)}=\lim_{x\to+\infty}\frac{4x^2}{6\ln x}=\lim_{x\to+\infty}\frac{8x}{6(1/x)}$

$\displaystyle=\lim_{x\to+\infty}\tfrac{4}{3}x^2=+\infty.$

25.9 $\displaystyle\lim_{x\to1}\frac{x^3-x^2+x-1}{x+\ln x-1}$.

▉ $\displaystyle\lim_{x\to1}\frac{x^3-x^2+x-1}{x+\ln x-1}=\lim_{x\to1}\frac{3x^2-2x+1}{1+(1/x)}=\frac{2}{2}=1.$

25.10 $\displaystyle\lim_{x\to0}\left[\frac{1}{\ln(x+1)}-\frac{1}{x}\right]$.

▉ $\displaystyle\frac{1}{\ln(x+1)}-\frac{1}{x}=\frac{x-\ln(x+1)}{x\ln(x+1)}$,   to which L'Hôpital's rule applies (zero-over-zero case).

$\displaystyle\lim_{x\to0}\frac{x-\ln(x+1)}{x\ln(x+1)}=\lim_{x\to0}\frac{1-1/(x+1)}{x[1/(x+1)]+\ln(x+1)}=\lim_{x\to0}\frac{x}{x+(x+1)\ln(x+1)}=\lim_{x\to0}\frac{1}{1+1+\ln(x+1)}=\frac{1}{2}$

25.11 $\displaystyle\lim_{x\to0}\frac{\tan x}{x}$.

▉ $\displaystyle\lim_{x\to0}\frac{\tan x}{x}=\lim_{x\to0}\frac{\sec^2 x}{1}=\frac{1}{1}=1.$

25.12 $\displaystyle\lim_{x\to0^+}x\ln x$.

▉ $\displaystyle x\ln x=\frac{\ln x}{1/x}$,   to which L'Hôpital's rule applies. $\displaystyle\lim_{x\to0^+}\frac{\ln x}{1/x}=\lim_{x\to0^+}\frac{1/x}{-1/x^2}=-\lim_{x\to0^+}x=0.$ (This result was obtained in a different way in Problem 23.44.)

25.13 $\displaystyle\lim_{x\to0^+}x^x$.

▉ Let $y=x^x$. Then $\ln y=x\ln x$. By Problem 25.12, $\displaystyle\lim_{x\to0^+}\ln y=0$. Hence, $\displaystyle\lim_{x\to0^+}y=\lim_{x\to0^+}e^{\ln y}=e^0=1.$

25.14 $\displaystyle\lim_{x\to+\infty}\frac{\ln x}{x}$.

▉ $\displaystyle\lim_{x\to+\infty}\frac{\ln x}{x}=\lim_{x\to+\infty}\frac{1/x}{1}=0$ (as in Problem 23.43).

25.15 $\displaystyle\lim_{x\to+\infty}x^{1/x}$.

▉ Let $y=x^{1/x}$. Then $\ln y=\dfrac{1}{x}\ln x$. By Problem 25.14, $\displaystyle\lim_{x\to+\infty}\ln y=0$. Hence, $\displaystyle\lim_{x\to+\infty}y=\lim_{x\to+\infty}e^{\ln y}=e^0=1.$

25.16 $\displaystyle\lim_{x\to0}\frac{3^x-2^x}{x}$.

▉ $\displaystyle\lim_{x\to0}\frac{3^x-2^x}{x}=\lim_{x\to0}\frac{\ln3\cdot3^x-\ln2\cdot2^x}{1}=\ln3-\ln2.$

25.17 $\displaystyle\lim_{x\to0}\frac{e^x}{x^2}$.

▉ $\displaystyle\lim_{x\to0}\frac{e^x}{x^2}=+\infty$,   since $\displaystyle\lim_{x\to0}e^x=1$, $\displaystyle\lim_{x\to0}x^2=0$,   and $\dfrac{e^x}{x^2}>0$.   Note that L'Hôpital's rule did not apply.

25.18 $\lim\limits_{x\to\pi/2} \dfrac{1-\sin x}{x-\pi/2}$.

▐ $\lim\limits_{x\to\pi/2} \dfrac{1-\sin x}{x-\pi/2} = \lim\limits_{x\to\pi/2} \dfrac{-\cos x}{1} = \dfrac{0}{1} = 0$.

25.19 $\lim\limits_{x\to 0^+} x^{\sin x}$.

▐ Let $y = x^{\sin x}$. Then $\ln y = \sin x \cdot \ln x = \dfrac{\ln x}{\csc x}$, to which L'Hôpital's rule applies. $\lim\limits_{x\to 0^+} \dfrac{\ln x}{\csc x} =$
$\lim\limits_{x\to 0^+} \dfrac{1/x}{-\csc x \cot x} = -\lim\limits_{x\to 0^+} \dfrac{\sin x}{x} \cdot \dfrac{1}{\cos x} \cdot \sin x = -1 \cdot 1 \cdot 0 = 0$. Thus, $\lim\limits_{x\to 0^+} \ln y = 0$. Hence, $\lim\limits_{x\to 0^+} y =$
$\lim\limits_{x\to 0^+} e^{\ln y} = e^0 = 1$.

25.20 $\lim\limits_{x\to 1} \dfrac{x^3-1}{x+1}$.

▐ $\lim\limits_{x\to 1} \dfrac{x^3-1}{x+1} = \dfrac{0}{2} = 0$. Note that L'Hôpital's rule did not apply. Use of the rule in this case would have led to an incorrect answer.

25.21 $\lim\limits_{x\to a^+} \dfrac{1}{x-a} \displaystyle\int_a^x f(t)\,dt$.

▐ $\lim\limits_{x\to a^+} \dfrac{\int_a^x f(t)\,dt}{x-a} = \lim\limits_{x\to a^+} \dfrac{f(x)}{1} = f(a)$, with an obvious interpretation in terms of average values.

25.22 $\lim\limits_{x\to 0^+} \dfrac{\tan x}{x^2}$.

▐ $\lim\limits_{x\to 0^+} \dfrac{\tan x}{x^2} = \lim\limits_{x\to 0^+} \dfrac{\sec^2 x}{2x} = +\infty$, since $\lim\limits_{x\to 0^+} \sec^2 x = 1$.

25.23 $\lim\limits_{x\to 0^+} \dfrac{\sin x}{\sqrt{x}}$.

▐ $\lim\limits_{x\to 0^+} \dfrac{\sin x}{\sqrt{x}} = \lim\limits_{x\to 0^+} \dfrac{\cos x}{1/2\sqrt{x}} = \lim\limits_{x\to 0^+} 2\sqrt{x}\cos x = 0$.

25.24 $\lim\limits_{x\to 1} \dfrac{\ln x}{\tan \pi x}$.

▐ $\lim\limits_{x\to 1} \dfrac{\ln x}{\tan \pi x} = \lim\limits_{x\to 1} \dfrac{1/x}{\pi\sec^2 \pi x} = \dfrac{1}{\pi(-1)^2} = \dfrac{1}{\pi}$.

25.25 $\lim\limits_{x\to 0} \dfrac{\sin 3x}{\sin 7x}$.

▐ $\lim\limits_{x\to 0} \dfrac{\sin 3x}{\sin 7x} = \lim\limits_{x\to 0} \dfrac{3\cos 3x}{7\cos 7x} = \dfrac{3}{7}$.

25.26 $\lim\limits_{x\to 0} \dfrac{e^{3x}-1}{\tan x}$.

▐ $\lim\limits_{x\to 0} \dfrac{e^{3x}-1}{\tan x} = \lim\limits_{x\to 0} \dfrac{3e^{3x}}{\sec^2 x} = \dfrac{3\cdot 1}{1} = 3$.

25.27 $\lim\limits_{x\to 0} \dfrac{1-\cos^2 2x}{x^2}$.

▐ $\lim\limits_{x\to 0} \dfrac{1-\cos^2 2x}{x^2} = \lim\limits_{x\to 0} \dfrac{-2\cos 2x\cdot(-\sin 2x)\cdot 2}{2x} = \lim\limits_{x\to 0} \dfrac{2\cos 2x\sin 2x}{x} = \lim\limits_{x\to 0} \dfrac{\sin 4x}{x} = \lim\limits_{x\to 0} \dfrac{4\cos 4x}{1} = 4$.
(Here, we used the identity $2\cos u\sin u = \sin 2u$.)

25.28 $\displaystyle\lim_{x\to 0}\frac{\tan x - \sin x}{x^3}.$

▮ $\displaystyle\lim_{x\to 0}\frac{\tan x - \sin x}{x^3} = \lim_{x\to 0}\frac{\sec^2 x - \cos x}{3x^2} = \lim_{x\to 0}\frac{2\sec^2 x\tan x + \sin x}{6x} = \lim_{x\to 0}\frac{2\sec^4 x + 4\tan^2 x\sec^2 x + \cos x}{6} = $
$\frac{3}{6} = \frac{1}{2}.$

25.29 $\displaystyle\lim_{x\to 0}\frac{x\tan x}{1-\cos x}.$

▮ $\displaystyle\lim_{x\to 0}\frac{x\tan x}{1-\cos x} = \lim_{x\to 0}\frac{x\sec^2 x + \tan x}{\sin x}.$ But $\displaystyle\frac{x\sec^2 x + \tan x}{\sin x} = \frac{x}{\sin x}\sec^2 x + \sec x.$ Hence, the limit becomes $\displaystyle\lim_{x\to 0}\left(\frac{x}{\sin x}\sec^2 x + \sec x\right) = 1 + 1 = 2,$ since $\displaystyle\lim_{x\to 0}\frac{x}{\sin x} = 1.$

25.30 $\displaystyle\lim_{x\to 0}\frac{\tan 3x}{x\sqrt{3}}.$

▮ $\displaystyle\lim_{x\to 0}\frac{\tan 3x}{x\sqrt{3}} = \lim_{x\to 0}\frac{3\sec^2 3x}{\sqrt{3}} = \frac{3}{\sqrt{3}} = \sqrt{3}.$

25.31 $\displaystyle\lim_{x\to \pi/2}\frac{\cos x}{2x - \pi}.$

▮ $\displaystyle\lim_{x\to \pi/2}\frac{\cos x}{2x - \pi} = \lim_{x\to \pi/2}\frac{-\sin x}{2} = -\frac{1}{2}.$

25.32 $\displaystyle\lim_{x\to 0^+} x^2\ln x.$

▮ $\displaystyle\lim_{x\to 0^+} x^2\ln x = \lim_{x\to 0^+} x\cdot\lim_{x\to 0^+} x\ln x = 0\cdot 0 = 0,$ by Problem 25.12.

25.33 $\displaystyle\lim_{x\to +\infty}\frac{2^x}{x^{10}}.$

▮ $\displaystyle\lim_{x\to +\infty}\frac{2^x}{x^{10}} = \lim_{x\to +\infty}\frac{\ln 2\cdot 2^x}{10x^9} = \lim_{x\to +\infty}\frac{(\ln 2)^2\cdot 2^x}{10\cdot 9x^8} = \lim_{x\to +\infty}\frac{(\ln 2)^3 2^x}{10\cdot 9\cdot 8x^7} = \cdots = \lim_{x\to +\infty}\frac{(\ln 2)^9 2^x}{10!x}$
$= \lim_{x\to +\infty}\frac{(\ln 2)^{10} 2^x}{10!} = +\infty$

25.34 $\displaystyle\lim_{x\to \pi/2^+}\frac{x\cos x}{1-\sin x}.$

▮ $\displaystyle\lim_{x\to \pi/2^+}\frac{x\cos x}{1-\sin x} = \lim_{x\to \pi/2^+}\frac{-x\sin x + \cos x}{-\cos x}.$ But $\displaystyle\frac{-x\sin x + \cos x}{-\cos x} = x\tan x - 1.$ Since $\displaystyle\lim_{x\to \pi/2^+}\tan x = $
$+\infty,$ our limit is $+\infty.$

25.35 $\displaystyle\lim_{x\to +\infty}\frac{x^6 - 3\sin x}{3x^2 + 4x^6}.$

▮ It is simplest to divide the numerator and denominator by x^6, obtaining $\displaystyle\frac{1 - (3\sin x/x^6)}{3/x^4 + 4},$ which approaches
$\frac{1}{4}.$ $\displaystyle\left(\lim_{x\to +\infty}\frac{\sin x}{x^6} = 0,$ since $|\sin x|\le 1.\right)$ Use of L'Hôpital's rule would have been very tedious in this case.

25.36 $\displaystyle\lim_{x\to 0}\frac{1-\cos 2x}{\sin 3x}.$

▮ $\displaystyle\lim_{x\to 0}\frac{1-\cos 2x}{\sin 3x} = \lim_{x\to 0}\frac{2\sin 2x}{3\cos 3x} = \frac{0}{3} = 0.$

25.37 $\displaystyle\lim_{x\to 0}\frac{1-\cos x}{x^2}.$

▮ $\displaystyle\lim_{x\to 0}\frac{1-\cos x}{x^2} = \lim_{x\to 0}\frac{\sin x}{2x} = \frac{1}{2}\lim_{x\to 0}\frac{\sin x}{x} = \frac{1}{2}\cdot 1 = \frac{1}{2}.$

25.38 $\lim\limits_{x \to 1} \dfrac{x^4 - 1}{x - 1}$.

▮ $\lim\limits_{x \to 1} \dfrac{x^4 - 1}{x - 1} = \lim\limits_{x \to 1} \dfrac{4x^3}{1} = 4$.

25.39 $\lim\limits_{y \to 0} \dfrac{\sin 2y}{3y}$.

▮ $\lim\limits_{y \to 0} \dfrac{\sin 2y}{3y} = \lim\limits_{y \to 0} \dfrac{2\cos 2y}{3} = \dfrac{2}{3}$.

25.40 $\lim\limits_{x \to 0} \dfrac{\sin 2x}{2x^2 + 3x}$.

▮ $\lim\limits_{x \to 0} \dfrac{\sin 2x}{2x^2 + 3x} = \lim\limits_{x \to 0} \dfrac{2\cos 2x}{4x + 3} = \dfrac{2}{3}$.

25.41 $\lim\limits_{x \to 0} \dfrac{1 + x - e^x}{x^2}$.

▮ $\lim\limits_{x \to 0} \dfrac{1 + x - e^x}{x^2} = \lim\limits_{x \to 0} \dfrac{1 - e^x}{2x} = \lim\limits_{x \to 0} \dfrac{-e^x}{2} = -\dfrac{1}{2}$.

25.42 $\lim\limits_{x \to 0} (\csc x - \cot x)$.

▮ $\csc x - \cot x = \dfrac{1}{\sin x} - \dfrac{\cos x}{\sin x} = \dfrac{1 - \cos x}{\sin x}$.   Hence, $\lim\limits_{x \to 0} (\csc x - \cot x) = \lim\limits_{x \to 0} \dfrac{1 - \cos x}{\sin x} = \lim\limits_{x \to 0} \dfrac{\sin x}{\cos x} = \lim\limits_{x \to 0} \tan x = 0$.

25.43 $\lim\limits_{x \to +\infty} \left(\cos \dfrac{1}{x}\right)^x$.

▮ Let $y = \left(\cos \dfrac{1}{x}\right)^x$. Then $\ln y = x \ln\left(\cos \dfrac{1}{x}\right)$. So $\lim\limits_{x \to +\infty} x \ln\left(\cos \dfrac{1}{x}\right) = \lim\limits_{x \to +\infty} \dfrac{\ln(\cos 1/x)}{1/x} = \lim\limits_{u \to 0^+} \dfrac{\ln(\cos u)}{u} = \lim\limits_{u \to 0^+} \dfrac{(1/\cos u)\cdot(-\sin u)}{1} = \lim\limits_{u \to 0^+} (-\tan u) = 0$. Hence, $\lim\limits_{x \to +\infty} y = \lim\limits_{x \to +\infty} e^{\ln y} = e^0 = 1$.

25.44 $\lim\limits_{x \to 0} \dfrac{\sec x - 1}{x^2}$.

▮ $\lim\limits_{x \to 0} \dfrac{\sec x - 1}{x^2} = \lim\limits_{x \to 0} \dfrac{\sec x \tan x}{2x} = \lim\limits_{x \to 0} \dfrac{1}{2} \sec x \cdot \dfrac{\tan x}{x} = \dfrac{1}{2} \cdot 1 \cdot 1 = \dfrac{1}{2}$,   since $\lim\limits_{x \to 0} \dfrac{\tan x}{x} = 1$ by   Problem 25.11.

25.45 $\lim\limits_{x \to 0^+} (\sin x)^{\tan x}$.

▮ Let $y = (\sin x)^{\tan x}$. Then $\ln y = \tan x \cdot \ln(\sin x) = \dfrac{\ln(\sin x)}{\cot x}$. But $\lim\limits_{x \to 0^+} \dfrac{\ln(\sin x)}{\cot x} = \lim\limits_{x \to 0^+} \dfrac{(1/\sin x)(\cos x)}{-\csc^2 x} = -\lim\limits_{x \to 0^+} \cos x \sin x = 0$. Then $\lim\limits_{x \to 0^+} y = \lim\limits_{x \to 0^+} e^{\ln y} = e^0 = 1$.

25.46 $\lim\limits_{x \to 1} \dfrac{\sin(x^2 - 1)}{x - 1}$.

▮ $\lim\limits_{x \to 1} \dfrac{\sin(x^2 - 1)}{x - 1} = \lim\limits_{x \to 1} \dfrac{\cos(x^2 - 1)\cdot 2x}{1} = 1 \cdot 2 = 2$.

25.47 $\lim\limits_{x \to 0} \dfrac{x + \sin 2x}{x - \sin x}$.

▮ $\lim\limits_{x \to 0} \dfrac{x + \sin 2x}{x - \sin 2x} = \lim\limits_{x \to 0} \dfrac{1 + 2\cos 2x}{1 - 2\cos 2x} = \dfrac{3}{-1} = -3$.

25.48 $\displaystyle\lim_{x\to 0}\frac{\cos 2x-\cos x}{\sin^2 x}$.

▮ $\displaystyle\lim_{x\to 0}\frac{\cos 2x-\cos x}{\sin^2 x}=\lim_{x\to 0}\frac{-2\sin 2x+\sin x}{2\sin x\cos x}=\lim_{x\to 0}\frac{-2\sin 2x+\sin x}{\sin 2x}=\lim_{x\to 0}\frac{-4\cos 2x+\cos x}{2\cos 2x}=\frac{-3}{2}$.

25.49 $\displaystyle\lim_{x\to 1}\csc \pi x\cdot\ln x$.

▮ $\displaystyle\lim_{x\to 1}\csc \pi x\cdot\ln x=\lim_{x\to 1}\frac{\ln x}{\sin \pi x}=\lim_{x\to 1}\frac{1/x}{\pi\cos \pi x}=\frac{1}{\pi\cdot(-1)}=-\frac{1}{\pi}$.

25.50 $\displaystyle\lim_{x\to +\infty}x^n e^{-x^{1/n}}$ for any positive integer n.

▮ Let $t=x^{1/n}$. Then $\displaystyle\lim_{x\to +\infty}x^n e^{-x^{1/n}}=\lim_{t\to +\infty}\frac{t^{n^2}}{e^t}=0$, by Problem 24.75.

25.51 $\displaystyle\lim_{x\to +\infty}\frac{(\ln x)^n}{x}$ for any positive integer n.

▮ Let $u=\ln x$, $x=e^u$. Then $\displaystyle\lim_{x\to +\infty}\frac{(\ln x)^n}{x}=\lim_{u\to +\infty}\frac{u^n}{e^u}=0$ by Problem 24.75. This result generalizes that of Problem 23.43.

25.52 $\displaystyle\lim_{x\to +\infty}\frac{\ln x}{\sqrt[n]{x}}$ for any positive integer n.

▮ Let $u=\sqrt[n]{x}$, $u^n=x$, $\ln x=n\ln u$. Hence, $\displaystyle\lim_{x\to +\infty}\frac{\ln x}{\sqrt[n]{x}}=\lim_{u\to +\infty}\frac{n\ln u}{u}=n\lim_{u\to +\infty}\frac{\ln u}{u}=n\cdot 0=0$.

25.53 $\displaystyle\lim_{x\to +\infty}\left(1+\frac{4}{x}\right)^{2x}$.

▮ $\displaystyle\lim_{x\to +\infty}\left(1+\frac{4}{x}\right)^{2x}=\left[\lim_{x\to +\infty}\left(1+\frac{4}{x}\right)^x\right]^2=(e^4)^2=e^8$, since $\displaystyle\lim_{x\to +\infty}\left(1+\frac{4}{x}\right)^x=e^4$ by Problem 24.74.

25.54 Sketch the graph $y=(\ln x)^n/x$ when n is an even positive integer.

▮ See Fig. 25-1. $y'=[n(\ln x)^{n-1}-(\ln x)^n]/x^2=[(\ln x)^{n-1}(n-\ln x)]/x^2$. Setting $y'=0$, we find $\ln x=0$ or $n=\ln x$, that is, $x=1$ and $x=e^n$ are the critical numbers. By the first derivative test, there is a relative maximum at $x=e^n$, $y=(n/e)^n$ and a relative minimum at $(1,0)$. As $x\to +\infty$, $y\to 0$ by Problem 25.51. As $x\to 0^+$, $y\to +\infty$.

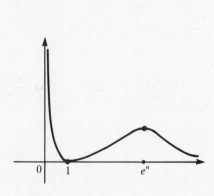

Fig. 25-1

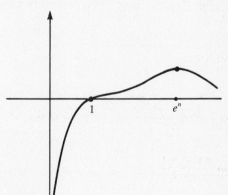

Fig. 25-2

25.55 Sketch the graph of $y=(\ln x)^n/x$ for positive odd integers n.

▮ As in Problem 25.54, $x=e^n$ yields a relative maximum. For $n>1$, $x=1$ is a critical number, but yields only an inflection point. When $n>1$, there are two other inflection points. As $x\to +\infty$, $y\to 0$ by Problem 25.51. As $x\to 0^+$, $y\to -\infty$. Figure 25-2 gives the graph for $n>1$, the graph for $n=1$ is given in Fig. 24-4.

25.56   Graph $y = x^n e^{-x}$ for positive even integers n.

▌ See Fig. 25-3. $y' = -x^n e^{-x} + nx^{n-1}e^{-x} = x^{n-1}e^{-x}(n - x)$.   Hence, $x = n$ and $x = 0$ are critical numbers.   The first derivative test tells us that there is a relative maximum at $x = n$ and a relative minimum at $x = 0$.   As $x \to +\infty$, $y \to 0$ by Problem 24.75.   As $x \to -\infty$, $y \to +\infty$.

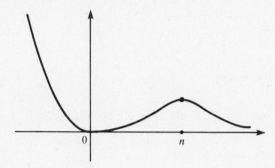

Fig. 25-3

25.57   Graph $y = x^n e^{-x}$ for odd positive integers n.

▌ As in Problem 25.56, there is a relative maximum at $x = n$.   For $n > 1$ [Fig. 25-4(a)], calculation of the second derivative yields $x^{n-2}e^{-x}[x^2 - 2nx + n(n - 1)]$.   Then, there is an inflection point at $x = 0$,   and two other inflection points in the first quadrant.   For the special case $n = 1$ [Fig. 25-4(b)], $y'' = e^{-x}(2 - x)$,   and there is only one inflection point, at $x = 2$.   In either case, as $x \to +\infty$, $y \to +\infty$,   and, as $x \to -\infty$, $y \to -\infty$.

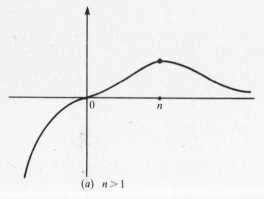 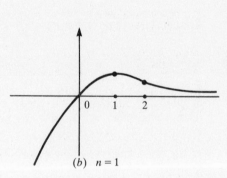

Fig. 25-4

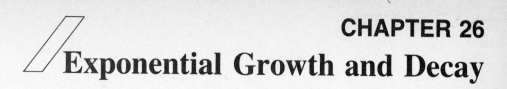

CHAPTER 26
Exponential Growth and Decay

26.1 A quantity y is said to grow or decay exponentially in time if $D_t y \overset{①}{=} Ky$ for some constant K. (K is called the growth constant or decay constant, depending on whether it is positive or negative.) Show that $y = y_0 e^{Kt}$, where y_0 is the value of y at time $t = 0$.

$$\frac{dy}{dt} \overset{①}{=} Ky, \quad \int \frac{dy}{y} = \int K\,dt, \quad \ln|y| = Kt + c, \quad |y| = e^{Kt+c}$$
$$y = \pm e^{Kt+c}, \quad y_0 = \pm e^c$$

▮ $D_T\left(\dfrac{y}{e^{Kt}}\right) = \dfrac{e^{Kt} \cdot D_t y - y(Ke^{Kt})}{e^{2Kt}} = \dfrac{e^{Kt}(D_t y - Ky)}{e^{2Kt}} = 0.$ Hence, y/e^{Kt} is a constant C, $y = Ce^{Kt}$. When $t = 0$, $y_0 = Ce^0 = C$. Thus, $y = y_0 e^{Kt}$.

26.2 A bacteria culture grows exponentially so that the initial number has doubled in 3 hours. How many times the initial number will be present after 9 hours?

▮ Let y be the number of bacteria. Then $y = y_0 e^{Kt}$. By the given information, $2y_0 = y_0 e^{3K}$, $2 = e^{3K}$, $\ln 2 = \ln e^{3K} = 3K$, $K = (\ln 2)/3$. When $t = 9$, $y = y_0 e^{9K} = y_0 e^{3\ln 2} = y_0(e^{\ln 2})^3 = y_0 \cdot 2^3 = 8y_0$. Thus, the initial number has been multiplied by 8.

26.3 A certain chemical decomposes exponentially. Assume that 200 grams becomes 50 grams in 1 hour. How much will remain after 3 hours?

▮ Let y be the number of grams present at time t. Then $y = y_0 e^{Kt}$. The given information tells us that $50 = 200 e^K$, $e^K = \frac{1}{4}$, $K = \ln\frac{1}{4} = -\ln 4$. When $t = 3$, $y = 200 e^{3K} = 200(e^K)^3 = 200(\frac{1}{4})^3 = \frac{200}{64} = \boxed{\frac{25}{8}}$ 3.125 grams.

26.4 Show that, when a quantity grows or decays exponentially, the rate of increase over a fixed time interval is constant (that is, it depends only on the time interval, not on the time at which the interval begins).

▮ The formula for the quantity is $y = y_0 e^{Kt}$. Let Δ be a fixed time interval, and let t be any time. Then $y(t+\Delta)/y(t) = y_0 e^{K(t+\Delta)}/y_0 e^{Kt} = e^{K\Delta}$, which does not depend on t. $e.g.\ 5{:}45\ PM = 17.75$

26.5 If the world population in 1980 was 4.5 billion and if it is growing exponentially with a growth constant $K = 0.04 \ln 2$, find the population in the year 2030.

▮ Let y be the population in billions in year t, with $t = 0$ in 1980. Then $y = 4.5 e^{0.04(\ln 2)t}$. In 2030, when $t = 50$, $y = 4.5 e^{0.04(\ln 2)\cdot 50} = 4.5(e^{\ln 2})^2 = 4.5(2)^2 = \underline{18\ \text{billion people}}$.

26.6 If a quantity y grows exponentially with a growth constant K and if during each unit of time there is an increase in y of r percent, find the relationship between K and r.

▮ $y = y_0 e^{Kt}$. When $t = 1$, $y = (1 + r/100)y_0$. Hence, $(1 + r/100)y_0 = y_0 e^K$, $1 + r/100 = e^K$. So $K = \ln(1 + r/100)$ and $r = 100(e^K - 1)$.

26.7 If a population is increasing exponentially at the rate of 2 percent per year, what will be the percentage increase over a period of 10 years?

▮ In the notation of Problem 26.6, $r = 2$, $K = \ln(1.02) \approx 0.0198$ (by a table of logarithms). Hence, after 10 years, $y = y_0 e^{Kt} = y_0 e^{(0.0198)10} = y_0 e^{0.198} \approx (1.219)y_0$ (using a table for the exponential function). Hence, over 10 years, there will be an increase of about 21.9 percent.

26.8 If an amount of money y_0 is invested at a rate of r percent per year, compounded n times per year, what is the amount of money that will be available after k years?

▮ After the first period of interest ($\frac{1}{n}$th of a year), the amount will be $y_0(1 + r/100n)$; after the second period, $y_0(1 + r/100n)^2$, etc. The interval of k years contains kn periods of interest, and, therefore, the amount present after k years will be $y_0(1 + r/100n)^{kn}$.

26.9 An amount of money y_0 earning r percent per year is compounded continuously (that is, assume that it is compounded n times per year, and then let n approach infinity). How much is available after k years?

❚ By Problem 26.8, $y = y_0(1 + r/100n)^{kn}$ if the money is compounded n times per year. If we let n approach infinity, we get $\lim_{n \to \infty} y_0(1 + r/100n)^{kn} = y_0 \lim_{n \to +\infty} (1 + r/100n)^{nk} = y_0 [\lim_{h \to +\infty} (1 + r/100n)^n]^k = y_0(e^{r/100})^k = y_0 e^{0.01rk}$. (Here we have used Problem 24.74.) Thus, the money grows exponentially, with growth constant $K = 0.01r$.

26.10 If an amount of money earning 8 percent per year is compounded quarterly, what is the equivalent yearly rate of return?

❚ By Problem 26.8, the amount present after 1 year will be $y_0(1 + \frac{8}{400})^4 = y_0(1.02)^4 \approx 1.0824 y_0$. Thus, the equivalent yearly rate is 8.24 percent.

26.11 If an amount of money earning 8 percent per year is compounded 10 times per year, what is the equivalent yearly rate of return?

❚ By Problem 26.8, the amount after 1 year will be $y_0(1 + \frac{8}{1000})^{10} = y_0(1.008)^{10} \approx 1.0829 y$. Thus, the equivalent yearly rate is 8.29 percent.

26.12 If an amount of money receiving interest of 8 percent per year is compounded continuously, what is the equivalent yearly rate of return?

❚ By Problem 26.9, the amount after 1 year will be $y_0 e^{0.08}$, which, by a table of values of e^x, is approximately $1.0833 y_0$. Hence, the equivalent yearly rate is about 8.33 percent.

26.13 A sum of money, compounded continuously, is multiplied by 5 in 8 years. If it amounts to $10,000 after 24 years, what was the initial sum of money?

❚ By the formula of Problem 26.9, $y = y_0 e^{0.01rk}$, where r is the rate of interest and k is the number of years. Hence, $5y_0 = y_0 e^{0.08r}$, $5 = e^{0.08r}$. After 24 years, $10,000 = y_0 e^{0.024r} = y_0(e^{0.08r})^3 = y_0(5)^3 = 125 y_0$. Hence, the initial quantity y_0 was $\frac{10,000}{125} = 80$ dollars.

26.14 If a quantity of money, earning interest compounded continuously, is worth 55 times the initial amount after 100 years, what was the yearly rate of interest?

❚ By Problem 26.9, $y = y_0 e^{0.01rk}$, r is the percentage rate of interest, y_0 is the initial amount, and k is the number of years. Then, $55 y_0 = y_0 e^r$, $55 = e^r$. Hence, $r = \ln 55$, and, by a table of logarithms, $r = \ln(5.5 \times 10) = \ln(5.5) + \ln 10 \approx 1.7047 + 2.3026 = 4.0073$. Thus, the rate is approximately 4 percent per year.

26.15 Assume that a quantity y decays exponentially, with a decay constant K. The half-life T is defined to be the time interval after which half of the original quantity remains. Find the relationship between K and T.

❚ $y = y_0 e^{Kt}$. By definition, $\frac{1}{2} y_0 = y_0 e^{KT}$. So $\frac{1}{2} = e^{KT}$, $KT = \ln \frac{1}{2} = -\ln 2$.

26.16 The half-life of radium is 1690 years. If 10 percent of an original quantity of radium remains, how long ago was the radium created?

❚ Let y be the number of grams of radium t years after the radium was created. Then $y = y_0 e^{Kt}$, where $1690K = -\ln 2$, by Problem 26.15. If at the present time t, $y = \frac{1}{10} y_0$, then $\frac{1}{10} y_0 = y_0 e^{Kt}$, $\frac{1}{10} = e^{Kt}$, $Kt = \ln \frac{1}{10} = -\ln 10$. Hence, $-(\ln 2/1690)t = -\ln 10$, $t = 1690 \ln 10 / \ln 2 \approx 5614.477$. Thus, the radium was created about 5614 years ago.

26.17 If radioactive carbon-14 has a half-life of 5750 years, what will remain of 1 gram after 3000 years?

❚ The amount of carbon is $y = y_0 e^{Kt}$. We know that $5750K = -\ln 2$. Since $y_0 = 1$ and $t = 3000$, $y = e^{3000K} = e^{-\frac{3000}{5750} \ln 2} \approx e^{-0.3616} \approx 0.69657$ (from a table for e^{-x}). Thus, about 0.7 gram will remain.

26.18 If 20 percent of a radioactive element disappears in 1 year, compute its half-life.

❚ Let y_0 be the original amount, and let T be the half-life. Then $0.8 y_0$ remains when $t = 1$. Thus, $0.8 y_0 = y_0 e^K$, $0.8 = e^K$, $K = \ln 0.8 \approx -0.2231$ (from a table of logarithms). But $KT = -\ln 2 \approx -0.6931$. So $-0.2231 T \approx -0.6931$, $T \approx 3.1067$ years.

26.19 Fruit flies are being bred in an enclosure that can hold a maximum of 640 flies. If the flies multiply exponentially, with a growth constant $K = 0.05$ and where time is measured in days, how long will it take an initial population of 20 to fill the enclosure?

❚ If y is the number of flies $y = 20e^{0.05t}$. When the enclosure is full, $y = 640$. Hence, $640 = 20e^{0.05t}$, $32 = e^{0.05t}$, $0.05t = \ln 32 = \ln (2^5) = 5 \ln 2 \approx 3.4655$. Hence, $t \approx 69.31$. Thus, it will take a little more than 69 days to fill the enclosure.

26.20 The number of bacteria in a certain culture is growing exponentially. If 100 bacteria are present initially and 400 are present after 1 hour, how many bacteria are present after $3\frac{1}{2}$ hours?

❚ The growth equation is $y = 100e^{Kt}$. The given information tells us that $400 = 100e^K$, $4 = e^K$. At $t = 3.5$, $y = 100e^{3.5K} = 100(e^K)^{3.5} = 100(4)^{7/2} = 100(2)^7 = 100(128) = 12,800$.

26.21 A certain radioactive substance has a half-life of 3 years. If 10 grams are present initially, how much of the substance remains after 9 years?

❚ Since 9 years is three half-lives, $(\frac{1}{2})^3 10 = 1.25$ grams will remain.

26.22 If y represents the amount by which the temperature of a body exceeds that of the surrounding air, then the rate at which y decreases is proportional to y (Newton's law of cooling). If y was initially 8 degrees and was 7 degrees after 1 minute, what will it be after 2 minutes?

❚ Since $D_t y = Ky$, $y = y_0 e^{Kt} = 8e^{Kt}$. The given facts tell us that $7 = 8e^K$, $e^K = \frac{7}{8}$. When $t = 2$, $y = 8e^{2K} = 8(e^K)^2 = 8(\frac{7}{8})^2 = 6.125$ degrees.

26.23 When a condenser is discharging electricity, the rate at which the voltage V decreases is proportional to V. If the decay constant is $K = -0.025$, per second, how long does it take before V has decreased to one-quarter of its initial value?

❚ $V = V_0 e^{-0.025t}$. When V is one-quarter of its initial value, $\frac{1}{4}V_0 = V_0 e^{-0.025t}$, $\frac{1}{4} = e^{-0.025t}$, $-0.025t = \ln \frac{1}{4} = -\ln 4 = -2 \ln 2 \approx -1.3862$. Hence, $t \approx 55.448$.

26.24 The mass y of a growing substance is $7(5)^t$ grams after t minutes. Find the initial quantity and the growth constant K.

❚ $y = 7(5)^t = 7(e^{\ln 5})^t = 7e^{(\ln 5)t}$. When $t = 0$, $y = 7$ grams is the initial quantity. The growth constant K is $\ln 5$.

26.25 If the population of Latin America has a doubling time of 27 years, by what percent does it grow per year?

❚ The population $y = y_0 e^{Kt}$. By the given information, $2y_0 = y_0 e^{27K}$, $2 = e^{27K}$, $e^K = \sqrt[27]{2} \approx 1.0234$. By the solution to Problem 26.6, the percentage increase per year $r = 100(e^K - 1) \approx 100(1.0234 - 1) = 2.34$.

26.26 If in 1980 the population of the United States was 225 million and increasing exponentially with a growth constant of 0.007, and the population of Mexico was 62 million and increasing exponentially with a growth constant of 0.024, when will the two populations be equal if they continue to grow at the same rate?

❚ The United States' population $y_U = 225e^{0.007t}$, and Mexico's population $y_M = 62e^{0.024t}$. When they are the same, $225e^{0.007t} = 62e^{0.024t}$, $3.6290 \approx e^{0.017t}$, $0.017t \approx \ln 3.629 \approx 1.2890$, $t \approx 75.82$. Hence, the populations would be the same in the year 2055.

26.27 A bacterial culture, growing exponentially, increases from 100 to 400 grams in 10 hours. How much was present after 3 hours?

❚ $y = 100e^{Kt}$. Hence, $400 = 100e^{10K}$, $4 = e^{10K}$, $2 = e^{5K}$, $5K = \ln 2 \approx 0.6931$, $K \approx 0.13862$. After 3 hours, $y = 100e^{3K} = 100e^{0.41586} \approx 100(1.5156) = 151.56$.

26.28 The population of Russia in 1980 was 255 million and growing exponentially with a growth constant of 0.012. The population of the United States in 1980 was 225 million and growing exponentially with a growth constant of 0.007. When will the population of Russia be twice as large as that of the United States?

❚ The population of Russia is $y_R = 225e^{0.012t}$ and that of the United States is $y_U = 225e^{0.007t}$. When the population of Russia is twice that of the United States, $255e^{0.012t} = 450e^{0.007t}$, $e^{0.005t} \approx 1.7647$, $0.005t \approx \ln 1.7647 \approx 0.56798$. $t \approx 113.596$ years; i.e., in the year 2093.

26.29 A bacterial culture, growing exponentially, increases from 200 to 500 grams in the period from 6 a.m. to 9 a.m. How many grams will be present at noon?

▌ $y = 200e^{Kt}$, where $t = 0$ at 6 a.m. Then, $500 = 200e^{3K}$, $e^{3K} = \frac{5}{2}$. At noon, $t = 6$ and $y = 200e^{6K} = 200(e^{3K})^2 = 200(\frac{5}{2})^2 = 1250$ grams.

26.30 A radioactive substance decreases from 8 grams to 7 grams in 1 hour. Find its half-life.

▌ $y = 8e^{Kt}$. Then $7 = 8e^K$, $e^K = 0.875$, $K = \ln 0.875 \approx -0.1335$. The half-life $T = -\ln 2/K \approx 0.6931/0.1335 \approx 5.1918$ hours.

26.31 A doomsday equation is an equation of the form $\dfrac{dP}{dt} = KP^{1.01}$, with $K > 0$. Solve this equation and show that $\lim\limits_{t \to t_1} P = +\infty$ for some t_1.

▌ $\displaystyle\int \frac{dP}{P^{1.01}} = \int K\, dt = Kt + C$. But, $\displaystyle\int \frac{dP}{P^{1.01}} = -\frac{100}{P^{0.01}}$. Setting $t = 0$, we find that $C = -100/P_0^{0.01}$. Now, $P^{0.01} = -100/(Kt + C)$. As $t \to -C/K$ from below, $P \to +\infty$.

26.32 Cobalt-60, with a half-life of 5.3 years, is extensively used in medical radiology. How long does it take for 90 percent of a given quantity to decay?

▌ $y = y_0 e^{Kt}$. Since $T = 5.3$ and $KT = -\ln 2$, $K = -\ln 2/5.3 \approx -0.1308$. When 90 percent of y_0 has decayed, $y = 0.1y_0 = y_0 e^{Kt}$, $0.1 = e^{Kt}$, $Kt = \ln \frac{1}{10} = -\ln 10 \approx -2.3026$. So, $t \approx -2.3026/(-0.1308) \approx 17.604$ years.

26.33 In a chemical reaction, a compound decomposes exponentially. If it is found by experiment that 8 grams diminish to 4 grams in 2 hours, when will 1 gram be left?

▌ The half-life is 2 hours. $(\frac{1}{2})^n 8 = 1$, $n = 3$. Thus 1 gram remains after three half-lifes, or 6 hours.

26.34 A tank initially contains 400 gallons of brine in which 100 pounds of salt are dissolved. Pure water is running into the tank at the rate of 20 gallons per minute, and the mixture (which is kept uniform by stirring) is drained off at the same rate. How many pounds of salt remain in the tank after 30 minutes?

▌ Let y be the number of pounds of salt in the mixture at time t. Since the concentration of salt at any given time is $y/400$ pounds per gallon, and 20 gallons flow out per minute, the rate at which y is diminishing is $20 \cdot y/400 = 0.05y$ pounds per minute. Hence, $D_t y = -0.05y$, and, thus, y is decaying exponentially with a decay constant of -0.05. Hence, $y = 100e^{-0.05t}$. So, after 30 minutes, $y = 100e^{-1.5} \approx 100(0.2231) = 22.31$ pounds.

26.35 Solve Problem 26.34 with the modification that instead of pure water, brine containing $\frac{1}{10}$ pound per gallon is run into the tank at 20 gallons per minute, the mixture being drained off at the same rate.

▌ As before, the tank is losing salt at the rate of $0.05y$ pounds per minute. However, it is also gaining salt at the rate of $20(\frac{1}{10}) = 2$ pounds per minute. Hence, $\dfrac{dy}{dt} = 2 - 0.05y$, $\displaystyle\int \frac{1}{2 - 0.05y}\, dy = \int dt$, $-20 \ln |2 - 0.05y| = t + C$. When $t = 0$, $-20 \ln |2 - 5| = C$, $C = -20 \ln 3$. Hence, $-20 \ln |2 - 0.05y| = t - 20 \ln 3$, $\ln |2 - 0.05y| = -0.05t + \ln 3$, $\ln \left|\dfrac{2 - 0.05y}{3}\right| = -0.05t$, $\left|\dfrac{2 - 0.05y}{3}\right| = e^{-0.05t}$. Note that y is always > 40, since $y(0) = 100$ and $y = 40$ is impossible. [$\ln |2 - 0.05(40)| = \ln 0$ is undefined.] Hence, $0.05y > 0.05(40) = 2$, and $|2 - 0.05y| = 0.05y - 2$. Thus, $(0.05y - 2)/3 = e^{-0.05t}$. When $t = 30$, $(0.05y - 2)/3 = e^{-1.5} \approx 0.2231$, $0.05y \approx 2.6693$, $y \approx 53.386$ pounds.

26.36 A country has 5 billion dollars of paper money in circulation. Each day 30 million dollars is brought into the banks for deposit and the same amount is paid out. The government decides to issue new paper money; whenever the old money comes into the banks, it is destroyed and replaced by the new money. How long will it take for the paper money in circulation to become 90 percent new?

▌ Let y be the number of millions of dollars in old money. Each day, $(y/5000) \cdot 30 = 0.006y$ millions of dollars of old money is turned in at the banks. Hence, $D_t y = -0.006y$, and, thus, y is decreasing exponentially, with a decay constant of -0.006. Hence, $y = 5000e^{-0.006t}$. When 90 percent of the money is new, $y = 500$, $500 = 5000e^{-0.006t}$, $0.1 = e^{-0.006t}$, $-0.006t = \ln \frac{1}{10} = -\ln 10$, $0.006t = \ln 10 \approx 2.3026$, $t \approx 383.77$ days.

26.37 The number of bacteria in a culture doubles every hour. How long does it take for a thousand bacteria to produce a billion?

❙ $y = y_0 e^{Kt}$. We are told that $y = 2y_0$ when $t = 1$. Hence, $2y_0 = y_0 e^K$, $2 = e^K$. If we start with 1000, we obtain a billion when $10^9 = 10^3 e^{Kt}$, $10^6 = (e^K)^t = 2^t$, $\ln(10^6) = t \ln 2$, $6 \ln 10 = t \ln 2$, $t = (6 \ln 10)/\ln 2 \approx 6(2.3026)/0.6931 \approx 19.9$ hours.

26.38 The world population at the beginning of 1970 was 3.6 billion. The weight of the earth is 6.586×10^{21} tons. If the population continues to increase exponentially, with a growth constant $K = 0.02$ and with time measured in years, in what year will the weight of all people equal the weight of the earth, if we assume that the average person weighs 120 pounds?

❙ The earth weighs $6.586 \times 10^{21} \times 2000$ pounds. When this is equal to the weight of y billion people, $120 \times 10^9 y = 6.586 \times 10^{21} \times 2000$, or $y \approx 1.1 \times 10^{14}$. Thus we must solve $1.1 \times 10^{14} = 3.6 e^{0.02t}$ for t. Taking logarithms, $\ln 1.1 + 14 \ln 10 = \ln 3.6 + 0.02t$, $0.02t \approx 31.05$, $t \approx 1552.5$ years. The date would be $1970 + 1552 = 3522$.

26.39 An object cools from 120 to 95°F in half an hour when surrounded by air whose temperature is 70°F. Use Newton's law of cooling (Problem 26.22) to find its temperature at the end of another half an hour.

❙ Let y be the difference in temperature between the object and the air. By Newton's law, $y = y_0 e^{Kt}$. Since $y_0 = 120 - 70 = 50$, $y = 50 e^{Kt}$. At $t = \frac{1}{2}$, $25 = 50 e^{K/2}$, $\frac{1}{2} = e^{K/2} = (e^K)^{1/2}$, $e^K = \frac{1}{4}$. When $t = 1$, $y = 50 e^K = 50 \cdot \frac{1}{4} = 12.5$. Hence, the temperature of the object is $70 + 12.5 = 82.5$°F.

26.40 What is the present value of a sum of money which if invested at 5 percent interest, continuously compounded, will become \$1000 in 10 years?

❙ Let y be the value of the money at time t, and let y_0 be its present value. Then, by Problem 26.9, $y = y_0 e^{0.05t}$. In 10 years, $1000 = y_0 e^{0.05(10)} = y_0 e^{0.5}$. So $y_0 = 1000/e^{0.5} \approx 1000/1.64872 \approx 606.53$. Hence, the present value is about \$606.53.

26.41 Show that if interest is compounded continuously at an annual percentage r the effective annual percentage rate of interest is $100(e^{0.01r} - 1)$.

❙ Let s be the effective annual percentage rate. Then $y_0 e^{0.01r} = y_0(1 + 0.01s)$. So $e^{0.01r} = 1 + 0.01s$, $0.01s = e^{0.01r} - 1$, $s = 100(e^{0.01r} - 1)$.

26.42 If money is invested at 5 percent, compounded continuously, in how many years will it double in value?

❙ $2y_0 = y_0 e^{0.05t}$, by Problem 26.9. So $2 = e^{0.05t}$, $0.05t = \ln 2$, $t = 20 \ln 2 \approx 20(0.6931) = 13.862$. Thus, the money will double in a little less than 13 years and 315 days.

26.43 A radioactive substance decays exponentially. What is the average quantity present over the first half-life?

❙ $y = y_0 e^{Kt}$, so $y_{avg} = \frac{1}{T} \int_0^T y_0 e^{Kt} \, dt = \frac{y_0}{KT}(e^{KT} - 1)$. But by Problem 26.15, $KT = -\ln 2$. Hence, $y_{avg} = \frac{y_0}{-\ln 2}\left(\frac{1}{2} - 1\right) = \frac{y_0}{2 \ln 2} = \frac{y_0}{\ln 4}$.

CHAPTER 27
Inverse Trigonometric Functions

27.1 Draw the graph of $y = \sin^{-1} x$.

▮ By definition, as x varies from -1 to 1, y varies from $-\pi/2$ to $\pi/2$. The graph of $y = \sin^{-1} x$ is obtained from the graph of $y = \sin x$ [Fig. 27-1(a)] by reflection in the line $y = x$. See Fig. 27-1(b).

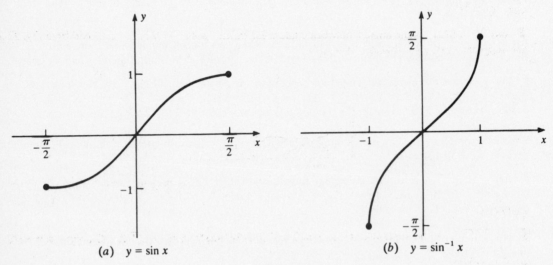

(a) $y = \sin x$ (b) $y = \sin^{-1} x$

Fig. 27-1

27.2 Show that $D_x(\sin^{-1} x) = 1/\sqrt{1 - x^2}$.

▮ Let $y = \sin^{-1} x$. Then $\sin y = x$. By implicit differentiation, $\cos y \cdot D_x y = 1$, $D_x y = 1/\cos y$. But $\cos y = \pm\sqrt{1 - \sin^2 y} = \pm\sqrt{1 - x^2}$. Since, by definition, $-\pi/2 \le y \le \pi/2$, $\cos y \ge 0$, and, therefore, $\cos y = \sqrt{1 - x^2}$ and $D_x y = 1/\sqrt{1 - x^2}$.

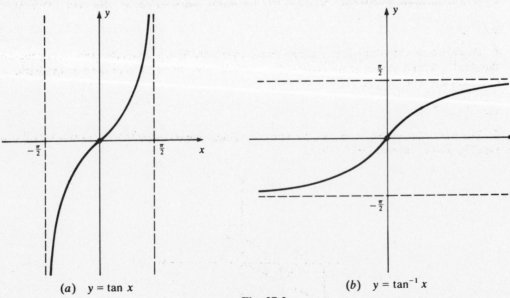

(a) $y = \tan x$ (b) $y = \tan^{-1} x$

Fig. 27-2

27.3 Draw the graph of $y = \tan^{-1} x$.

❚ As x varies from $-\infty$ to $+\infty$, $\tan^{-1} x$ varies from $-\pi/2$ to $\pi/2$. The graph of $y = \tan^{-1} x$ is obtained from that of $y = \tan x$ [Fig. 27-2(a)] by reflection in the line $y = x$. See Fig. 27-2(b).

27.4 Show that $D_x(\tan^{-1} x) = 1/(1 + x^2)$.

❚ Let $y = \tan^{-1} x$. Then $\tan y = x$. By implicit differentiation, $\sec^2 y \cdot D_x y = 1$, $D_x y = 1/\sec^2 y = 1/(1 + \tan^2 y) = 1/(1 + x^2)$.

In Problems 27.5–27.13, find the indicated number.

27.5 $\cos^{-1}(-\sqrt{3}/2)$.

❚ $\cos^{-1}(-\sqrt{3}/2)$ is the angle θ between 0 and π for which $\cos \theta = -\sqrt{3}/2$. It is seen from Fig. 27-3 that θ is the supplement of $\pi/6$, that is, $\theta = 5\pi/6$.

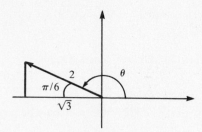

Fig. 27-3

27.6 $\sin^{-1}(\sqrt{2}/2)$.

❚ $\sin^{-1}\sqrt{2}/2$ is the angle θ between $-\pi/2$ and $\pi/2$ for which $\sin \theta = \sqrt{2}/2$. Clearly, $\theta = \pi/4$.

27.7 $\sin^{-1}(-\sqrt{2}/2)$.

❚ $\sin^{-1}(-\sqrt{2}/2)$ is the angle θ between $-\pi/2$ and $\pi/2$ for which $\sin \theta = -\sqrt{2}/2$. Clearly, $\theta = -\pi/4$.

27.8 $\tan^{-1} 1$.

❚ $\tan^{-1} 1$ is the angle θ between $-\pi/2$ and $\pi/2$ for which $\tan \theta = 1$, that is, $\theta = \pi/4$.

27.9 $\tan^{-1}(\sqrt{3}/3)$.

❚ $\pi/6$ is the angle θ between $-\pi/2$ and $\pi/2$ for which $\tan \theta = (\sqrt{3}/3)$. So $\tan^{-1}(\sqrt{3}/3) = \pi/6$.

27.10 $\sec^{-1}\sqrt{2}$.

❚ By definition, the value of $\sec^{-1} x$ is either an angle in the first quadrant (for positive arguments) or an angle in the third quadrant (for negative arguments). In this case, the value is in the first quadrant, so $\sec^{-1}\sqrt{2} = \cos^{-1} 1/\sqrt{2} = \pi/4$.

27.11 $\sec^{-1}(-2\sqrt{3}/3)$.

❚ The value θ must be in the third quadrant. Since $\sec \theta = -2\sqrt{3}/3$, $\cos \theta = -3/2\sqrt{3} = -\sqrt{3}/2$. Thus (see Fig. 27-4), $\theta = \pi + \pi/6 = 7\pi/6$.

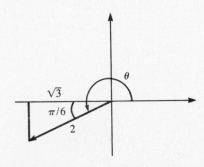

Fig. 27-4

27.12 $\csc^{-1}(-\sqrt{2})$.

▮ By definition, the value θ must be an angle in either the first or third quadrant. Since $\csc\theta = -\sqrt{2}$, θ is in the third quadrant, $\sin\theta = -1/\sqrt{2}$, and $\theta = \pi + \pi/4 = 5\pi/4$.

27.13 $\cot^{-1}(-1)$.

▮ The value θ is, by definition, in $(0, \pi)$. Since the argument is negative, θ is in the second quadrant, $\cot\theta = -1$, $\tan\theta = 1/(-1) = -1$, $\theta = \pi/2 + \pi/4 = 3\pi/4$.

27.14 Let $\theta = \cos^{-1}\frac{1}{3}$. Find $\sin\theta$.

▮ Since $\frac{1}{3} > 0$, θ is in $(0, \pi/2)$. $\cos\theta = \frac{1}{3}$, and $\sin\theta = \pm\sqrt{1 - \cos^2\theta} = \pm\sqrt{\frac{8}{9}} = \pm 2\sqrt{2}/3$. Since θ is in the first quadrant, the $+$ sign applies and $\sin\theta = 2\sqrt{2}/3$.

27.15 Let $\theta = \sin^{-1}(-\frac{1}{4})$. Find $\cos\theta$.

▮ Since $-\frac{1}{4} < 0$, $-\pi/2 < \theta < 0$. $\cos\theta = \pm\sqrt{1 - \sin^2\theta} = \pm\sqrt{\frac{15}{16}} = \pm\sqrt{15}/4$. But $\cos\theta > 0$, since $-\pi/2 < \theta < 0$. Hence, $\cos\theta = \sqrt{15}/4$.

In Problems 27.16–27.20, compute the indicated function value.

27.16 $\sin(\cos^{-1}\frac{4}{5})$.

▮ Let $\theta = \cos^{-1}\frac{4}{5}$. Since $\frac{4}{5} > 0$, $0 < \theta < \pi/2$ and $\sin\theta$ must be positive. $\sin\theta = \sqrt{1 - \cos^2\theta} = \sqrt{\frac{9}{25}} = \frac{3}{5}$.

27.17 $\tan(\sec^{-1}\frac{13}{5})$.

▮ Let $\theta = \sec^{-1}\frac{13}{5}$. Since $\frac{13}{5} > 0$, $0 < \theta < \pi/2$, and therefore, $\tan\theta > 0$. But $\tan\theta = \sqrt{\sec^2\theta - 1} = \sqrt{\frac{144}{25}} = \frac{12}{5}$.

27.18 $\cos(\sin^{-1}\frac{3}{5} + \sec^{-1} 3)$.

▮ Let $\theta_1 = \sin^{-1}\frac{3}{5}$ and $\theta_2 = \sec^{-1} 3$. Since $\frac{3}{5}$ and 3 are positive, θ_1 and θ_2 are in the first quadrant. Then $\cos(\sin^{-1}\frac{3}{5} + \sec^{-1} 3) = \cos(\theta_1 + \theta_2) = \cos\theta_1\cos\theta_2 - \sin\theta_1\sin\theta_2 = \frac{4}{5}\cdot\frac{1}{3} - \frac{3}{5}(2\sqrt{2}/3) = \frac{4}{15} - 6\sqrt{2}/15 = (4 - 6\sqrt{2})/15$.

27.19 $\sin(\cos^{-1}\frac{1}{5} - \tan^{-1} 2)$.

▮ Let $\theta_1 = \cos^{-1}\frac{1}{5}$ and $\theta_2 = \tan^{-1} 2$. Then θ_1 and θ_2 are in the first quadrant. $\sin(\cos^{-1}\frac{1}{5} - \tan^{-1} 2) = \sin(\theta_1 - \theta_2) = \sin\theta_1\cos\theta_2 - \cos\theta_1\sin\theta_2 = (2\sqrt{6}/5)(1/\sqrt{5}) - \frac{1}{5}(2/\sqrt{5}) = (2\sqrt{6} - 2)/5\sqrt{5}$.

27.20 $\sin^{-1}(\sin\pi)$.

▮ $\sin\pi = 0$. Hence, $\sin^{-1}(\sin\pi) = \sin^{-1} 0 = 0$. Note that $\sin^{-1}(\sin x)$ is not necessarily equal to x.

27.21 Find the domain and range of the function $f(x) = \cos(\tan^{-1} x)$.

▮ Since the domain of $\tan^{-1} x$ is the whole set $\mathcal{R}$ of real numbers and $\cos u$ is defined for all u, the domain of f is $\mathcal{R}$. The values of $\tan^{-1} x$ form the interval $(-\pi/2, \pi/2)$, and the cosines of angles in that interval form the interval $(0, 1]$, which is, therefore, the range of f.

In Problems 27.22–27.25, differentiate the given function.

27.22 $\sin^{-1} x + \cos^{-1} x$.

▮ $D_x(\sin^{-1} x + \cos^{-1} x) = \dfrac{1}{\sqrt{1 - x^2}} + \dfrac{-1}{\sqrt{1 - x^2}} = 0$.

27.23 $\tan^{-1} x + \cot^{-1} x$.

▮ $D_x(\tan^{-1} x + \cot^{-1} x) = \dfrac{1}{1 + x^2} + \dfrac{-1}{1 + x^2} = 0$.

27.24 $\sec^{-1} x + \csc^{-1} x$.

▮ $D_x(\sec^{-1} x + \csc^{-1} x) = \dfrac{1}{x\sqrt{x^2 - 1}} + \dfrac{-1}{x\sqrt{x^2 - 1}} = 0$.

27.25 $\tan^{-1} x + \tan^{-1} \dfrac{1}{x}$.

▮ $D_x\left(\tan^{-1} x + \tan^{-1} \dfrac{1}{x}\right) = \dfrac{1}{1 + x^2} + \dfrac{1}{1 + (1/x)^2} \cdot \left(-\dfrac{1}{x^2}\right) = \dfrac{1}{1 + x^2} + \dfrac{-1}{x^2 + 1} = 0$.

27.26 Explain the answers to Problems 27.22–27.25.

▮ In each case, since the derivative is 0, the function has to be a constant. Consider, for example, $\sin^{-1} x + \cos^{-1} x$. When $x > 0$, $\theta_1 = \sin^{-1} x$ and $\theta_2 = \cos^{-1} x$ are acute angles whose sum is $\pi/2$ [see Fig. 27-5(a)]. For $-x$, $\sin^{-1}(-x) = -\theta_1$ and $\cos^{-1}(-x) = \pi - \theta_2$, and, therefore, $\sin^{-1}(-x) + \cos^{-1}(-x) = -\theta_1 + \pi - \theta_2 = \pi - (\theta_1 + \theta_2) = \pi - \pi/2 = \pi/2$. [See Fig. 27.5(b).] Hence, in all cases, $\sin^{-1} x + \cos^{-1} x = \pi/2$.

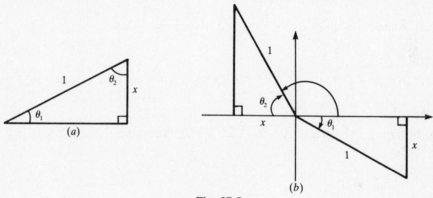

Fig. 27-5

In Problems 27.27–27.37, find the derivative of the given function.

27.27 $y = x \tan^{-1} x$.

▮ $y' = x \cdot \dfrac{1}{1 + x^2} + \tan^{-1} x = \dfrac{x}{1 + x^2} + \tan^{-1} x$.

27.28 $y = \sin^{-1} \sqrt{x}$.

▮ By the chain rule, $y' = \dfrac{1}{\sqrt{1 - (\sqrt{x})^2}} \cdot \dfrac{1}{2\sqrt{x}} = \dfrac{1}{2\sqrt{x}\sqrt{1 - x}}$.

27.29 $y = \tan^{-1}(\cos x)$.

▮ $y' = \dfrac{1}{1 + (\cos x)^2}(-\sin x) = \dfrac{-\sin x}{1 + \cos^2 x}$.

27.30 $y = \ln(\cot^{-1} 3x)$.

▮ $y' = \dfrac{1}{\cot^{-1} 3x}\left[-\dfrac{1}{1 + (3x)^2} \cdot 3\right] = \dfrac{-3}{(\cot^{-1} 3x)(1 + 9x^2)}$.

27.31 $y = e^x \cos^{-1} x$.

▮ $y' = e^x\left(\dfrac{-1}{\sqrt{1 - x^2}}\right) + \cos^{-1} x \cdot e^x = e^x\left(\cos^{-1} x - \dfrac{1}{\sqrt{1 - x^2}}\right)$.

27.32 $y = \ln(\tan^{-1} x)$.

❚ $y' = \dfrac{1}{\tan^{-1} x} \cdot \dfrac{1}{1 + x^2}$.

27.33 $y = \csc^{-1} \dfrac{1}{x}$.

❚ $y' = \dfrac{-1}{(1/x)\sqrt{(1/x)^2 - 1}}\left(-\dfrac{1}{x^2}\right) = \dfrac{1}{x\sqrt{1/x^2 - 1}}\,\dfrac{|x|}{x}\,\dfrac{1}{\sqrt{1 - x^2}}$.

27.34 $y = x\sqrt{a^2 - x^2} + a^2 \sin^{-1}(x/a)$, where $a > 0$.

❚ $y' = x \cdot \dfrac{-x}{\sqrt{a^2 - x^2}} + \sqrt{a^2 - x^2} + \dfrac{a^2}{\sqrt{1 - (x/a)^2}} \cdot \dfrac{1}{a} = \dfrac{-x^2}{\sqrt{a^2 - x^2}} + \sqrt{a^2 - x^2} + \dfrac{a^2}{\sqrt{a^2 - x^2}} = \dfrac{2(a^2 - x^2)}{\sqrt{a^2 - x^2}}$

$= 2\sqrt{a^2 - x^2}$.

27.35 $y = \tan^{-1}\left(\dfrac{a + x}{1 - ax}\right)$.

❚ $y' = \dfrac{1}{1 + [(a + x)/(1 - ax)]^2} \cdot \dfrac{(1 - ax) - (a + x)(-a)}{(1 - ax)^2} = \dfrac{1 - ax + a^2 + ax}{(1 - ax)^2 + (a + x)^2} = \dfrac{1 + a^2}{1 + a^2 x^2 + a^2 + x^2}$

$= \dfrac{1 + a^2}{(1 + a^2)(1 + x^2)} = \dfrac{1}{1 + x^2}$.

27.36 $y = \sin^{-1}\dfrac{1}{x} + \sec^{-1} x$.

❚ $y' = \dfrac{1}{\sqrt{1 - (1/x)^2}}\left(-\dfrac{1}{x^2}\right) + \dfrac{1}{x\sqrt{x^2 - 1}} = \dfrac{|x|}{\sqrt{x^2 - 1}}\left(-\dfrac{1}{x^2}\right) + \dfrac{1}{x\sqrt{x^2 - 1}} = \dfrac{-1}{|x|\sqrt{x^2 - 1}} + \dfrac{1}{x\sqrt{x^2 - 1}}$,

and therefore $\dfrac{1}{\sqrt{x^2 - 1}}\left(\dfrac{1}{x} - \dfrac{1}{|x|}\right) = \begin{cases} 0 & \text{for } x > 1 \\ \dfrac{2}{x\sqrt{x^2 - 1}} & \text{for } x < -1 \end{cases}$

27.37 $y = \tan^{-1}\dfrac{2}{x}$.

❚ $y' = \dfrac{1}{1 + (2/x)^2}\left(-\dfrac{2}{x^2}\right) = \dfrac{-2}{x^2 + 4}$.

27.38 What identity is implied by the result of Problem 27.35?

❚ Two functions with the same derivative differ by a constant. Hence, $\tan^{-1}[(a + x)/(1 - ax)] = \tan^{-1} x + C$. When $x = 0$, $\tan^{-1} a = \tan^{-1} 0 + C = 0 + C = C$. Thus, the identity is $\tan^{-1}[(a + x)/(1 - ax)] = \tan^{-1} x + \tan^{-1} a$.

In Problems 27.39–27.57, evaluate the indicated antiderivative.

27.39 $\displaystyle\int \dfrac{1}{4 + x^2}\, dx$.

❚ Let $x = 2u$, $dx = 2\,du$. Then $\displaystyle\int \dfrac{dx}{4 + x^2} = \dfrac{1}{2}\int \dfrac{du}{1 + u^2} = \dfrac{1}{2}\tan^{-1} u + C = \dfrac{1}{2}\tan^{-1}\dfrac{x}{2} + C$. This is a special case of the formula $\displaystyle\int \dfrac{dx}{a^2 + x^2} = \dfrac{1}{a}\tan^{-1}\dfrac{x}{a} + C$ for $a > 0$.

27.40 $\int \dfrac{dx}{4 + 9x^2}$.

▮ $\int \dfrac{dx}{4 + 9x^2} = \int \dfrac{dx}{9(\frac{4}{9} + x^2)} = \dfrac{1}{9} \int \dfrac{dx}{(\frac{2}{3})^2 + x^2}$. By the formula given in Problem 27.39, this is equal to $\dfrac{1}{9} \cdot \dfrac{1}{\frac{2}{3}} \cdot \tan^{-1} \dfrac{x}{\frac{2}{3}} + C = \dfrac{1}{6} \tan^{-1} \dfrac{3x}{2} + C$.

27.41 $\int \dfrac{dx}{\sqrt{25 - x^2}}$.

▮ Let $x = 5u$, $dx = 5\,du$. Then $\int \dfrac{dx}{\sqrt{25 - x^2}} = \int \dfrac{5\,du}{\sqrt{25 - 25u^2}} = \int \dfrac{du}{\sqrt{1 - u^2}} = \sin^{-1} u + C = \sin^{-1} \dfrac{x}{5} + C$. This is a special case of the formula $\int \dfrac{dx}{\sqrt{a^2 - x^2}} = \sin^{-1} \dfrac{x}{a} + C$ for $a > 0$.

27.42 $\int \dfrac{dx}{\sqrt{25 - 16x^2}}$.

▮ $\int \dfrac{dx}{\sqrt{25 - 16x^2}} = \int \dfrac{dx}{\sqrt{16(\frac{25}{16} - x^2)}} = \dfrac{1}{4} \int \dfrac{dx}{\sqrt{(\frac{5}{4})^2 - x^2}}$. By the formula given in Problem 27.41, this is equal to $\dfrac{1}{4} \sin^{-1} \dfrac{x}{\frac{5}{4}} + C = \dfrac{1}{4} \sin^{-1} \dfrac{4x}{5} + C$.

27.43 $\int \dfrac{dx}{(x - 3)\sqrt{x^2 - 6x + 8}}$.

▮ Let $u = x - 3$, $du = dx$. Then $x^2 - 6x + 8 = (x - 3)^2 - 1 = u^2 - 1$. So $\int \dfrac{dx}{(x - 3)\sqrt{x^2 - 6x + 8}} = \int \dfrac{du}{u\sqrt{u^2 - 1}} = \sec^{-1} u + C = \sec^{-1} (x - 3) + C$.

27.44 $\int \dfrac{dx}{\sqrt{3 - 2x^2}}$.

▮ $\int \dfrac{dx}{\sqrt{3 - 2x^2}} = \int \dfrac{dx}{\sqrt{2(\frac{3}{2} - x^2)}} = \dfrac{1}{\sqrt{2}} \int \dfrac{dx}{\sqrt{\frac{3}{2} - x^2}}$. By the formula in Problem 27.41, we obtain

$\dfrac{1}{\sqrt{2}} \sin^{-1} \dfrac{x}{\sqrt{\frac{3}{2}}} + C = \dfrac{1}{\sqrt{2}} \sin^{-1} \dfrac{\sqrt{2}x}{\sqrt{3}} + C$.

27.45 $\int \dfrac{dx}{2 + 7x^2}$.

▮ $\int \dfrac{dx}{2 + 7x^2} = \int \dfrac{dx}{7(\frac{2}{7} + x^2)}$. By the formula in Problem 27.39, we obtain $\dfrac{1}{7} \cdot \dfrac{1}{\sqrt{\frac{2}{7}}} \tan^{-1} \dfrac{x}{\sqrt{\frac{2}{7}}} + C = \dfrac{1}{\sqrt{14}} \tan^{-1} \dfrac{\sqrt{14}x}{2} + C$.

27.46 $\int \dfrac{dx}{x\sqrt{x^2 - 4}}$.

▮ Let $x = 2u$, $dx = 2\,du$, $\sqrt{x^2 - 4} = \sqrt{4u^2 - 4} = \sqrt{4(u^2 - 1)} = 2\sqrt{u^2 - 1}$. Then $\int \dfrac{dx}{x\sqrt{x^2 - 4}} = \int \dfrac{2\,du}{2u \cdot 2\sqrt{u^2 - 1}} = \dfrac{1}{2} \int \dfrac{du}{u\sqrt{u^2 - 1}} = \dfrac{1}{2} \sec^{-1} u + C = \dfrac{1}{2} \sec^{-1} (x/2) + C$. This is a special case of the formula $\int \dfrac{dx}{x\sqrt{x^2 - a^2}} = \dfrac{1}{a} \sec^{-1} \dfrac{x}{a} + C$ for $a > 0$.

27.47 $\int \dfrac{dx}{x\sqrt{9x^2 - 16}}$.

▮ $\displaystyle\int \frac{dx}{x\sqrt{9x^2-16}} = \int \frac{dx}{x\sqrt{9(x^2-\frac{16}{9})}} = \frac{1}{3}\int \frac{dx}{x\sqrt{x^2-(\frac{4}{3})^2}}$. From the formula in Problem 27.46, we get

$\displaystyle\frac{1}{3}\cdot\frac{1}{(\frac{4}{3})}\sec^{-1}\left(\frac{x}{\frac{4}{3}}\right)+C = \frac{1}{4}\sec^{-1}\frac{3x}{4}+C$.

27.48 $\displaystyle\int \frac{dx}{x\sqrt{3x^2-2}}$.

▮ $\displaystyle\int \frac{dx}{x\sqrt{3x^2-2}} = \int \frac{dx}{x\sqrt{3(x^2-\frac{2}{3})}} = \frac{1}{\sqrt{3}}\int \frac{dx}{x\sqrt{x^2-\frac{2}{3}}}$. From the formula in Problem 27.46, we obtain

$\displaystyle\frac{1}{\sqrt{3}}\cdot\frac{1}{\sqrt{\frac{2}{3}}}\sec^{-1}\frac{x}{\sqrt{\frac{2}{3}}}+C = \frac{\sqrt{2}}{2}\sec^{-1}\frac{\sqrt{6}x}{2}+C$.

27.49 $\displaystyle\int \frac{dx}{(1+x)\sqrt{x}}$.

▮ Let $x=u^2$, $dx=2u\,du$. Then $\displaystyle\int \frac{dx}{(1+x)\sqrt{x}} = \int \frac{2u\,du}{(1+u^2)u} = 2\int \frac{du}{1+u^2} = 2\tan^{-1}u+C = 2\tan^{-1}\sqrt{x}+C$.

27.50 $\displaystyle\int \frac{x\,dx}{x^4+9}$.

▮ Let $u=x^2$, $du=2x\,dx$. Then $\displaystyle\int \frac{x\,dx}{x^4+9} = \frac{1}{2}\int \frac{du}{u^2+9}$. From the formula in Problem 27.39, we get

$\displaystyle\frac{1}{2}\cdot\frac{1}{3}\tan^{-1}\frac{u}{3}+C = \frac{1}{6}\tan^{-1}\frac{x^2}{3}+C$.

27.51 $\displaystyle\int \frac{dx}{\sqrt{6x-x^2}}$.

▮ By completing the square, $x^2-6x = (x-3)^2-9$. So $6x-x^2 = 9-(x-3)^2$. Let $u=x-3$, $du=dx$.

Then $\displaystyle\int \frac{dx}{\sqrt{6x-x^2}} = \int \frac{du}{\sqrt{9-u^2}}$. From the formula in Problem 27.41, we get $\sin^{-1}\frac{u}{3}+C = \sin^{-1}\frac{x-3}{3}+C$.

27.52 $\displaystyle\int \frac{2x\,dx}{\sqrt{6x-x^2}}$.

▮ Note that $D_x(6x-x^2) = 6-2x$. Now,

$\displaystyle\int \frac{2x\,dx}{\sqrt{6x-x^2}} = \int \frac{6-(6-2x)}{\sqrt{6x-x^2}}\,dx = -\int \frac{6-2x}{\sqrt{6x-x^2}}\,dx + 6\int \frac{dx}{\sqrt{6x-x^2}} = -2(6x-x^2)^{1/2}+6\sin^{-1}\frac{x-3}{3}+C$.

(The second integral was evaluated in Problem 27.51; the first integral was found by the formula of Problem 19.1.)

27.53 $\displaystyle\int \frac{x\,dx}{x^2+8x+20}$.

▮ By completing the square, $x^2+8x+20 = (x+4)^2+4$. Let $u=x+4$, $du=dx$. Then

$$\int \frac{x\,dx}{x^2+8x+20} = \int \frac{(u-4)\,du}{u^2+4} = \int \frac{u\,du}{u^2+4} - 4\int \frac{du}{u^2+4} = \frac{1}{2}\ln(u^2+4)-4\cdot\frac{1}{2}\tan^{-1}\frac{u}{2}+C$$

$$= \frac{1}{2}\ln(x^2+8x+20)-2\tan^{-1}\frac{x+4}{2}+C.$$

27.54 $\displaystyle\int \frac{x^3\,dx}{x^2-2x+4}$.

▮ Dividing x^3 by x^2-2x+4, we obtain $x+2-\dfrac{8}{x^2-2x+4}$. Hence,

$$\int \frac{x^3\,dx}{x^2-2x+4} = \int\left(x+2-\frac{8}{x^2-2x+4}\right)dx = \frac{1}{2}x^2+2x-8\int \frac{1}{(x-1)^2+3}\,dx$$

$$= \frac{1}{2}x^2+2x-8\cdot\frac{1}{\sqrt{3}}\tan^{-1}\frac{x-1}{\sqrt{3}}+C = \frac{1}{2}x^2+2x-\frac{8}{\sqrt{3}}\tan^{-1}\frac{x-1}{\sqrt{3}}+C.$$

27.55 $\displaystyle\int \frac{x\,dx}{\sqrt{4-x^4}}.$

▮ Let $u = x^2$, $du = 2x\,dx$. Then $\displaystyle\int \frac{x\,dx}{\sqrt{4-x^4}} = \frac{1}{2}\int \frac{du}{\sqrt{4-u^2}} = \frac{1}{2}\sin^{-1}\frac{u}{2} + C = \frac{1}{2}\sin^{-1}\frac{x^2}{2} + C.$

27.56 $\displaystyle\int \frac{e^x\,dx}{4+e^{2x}}.$

▮ Let $u = e^x$, $du = e^x\,dx$. Then $\displaystyle\int \frac{e^x\,dx}{4+e^{2x}} = \int \frac{du}{4+u^2} = \frac{1}{2}\tan^{-1}\frac{u}{2} + C = \frac{1}{2}\tan^{-1}\frac{e^x}{2} + C.$

27.57 $\displaystyle\int \frac{\cos x\,dx}{5+\sin^2 x}.$

▮ Let $u = \sin x$, $du = \cos x\,dx$. Then $\displaystyle\int \frac{\cos x\,dx}{5+\sin^2 x} = \int \frac{du}{5+u^2} = \frac{1}{\sqrt{5}}\tan^{-1}\frac{u}{\sqrt{5}} + C = \frac{1}{\sqrt{5}}\tan^{-1}\frac{\sin x}{\sqrt{5}} + C.$

27.58 Find an equation of the tangent line to the graph of $y = \sin^{-1}\dfrac{x}{3}$ at the origin.

▮ $y' = \dfrac{1}{\sqrt{1-(x/3)^2}}\left(\dfrac{1}{3}\right)$, which is $\frac{1}{3}$ when $x = 0$. Hence, the slope of the tangent line is $\frac{1}{3}$, and, since it goes through the origin, its equation is $y = \frac{1}{3}x$.

27.59 A ladder that is 13 feet long leans against a wall. The bottom of the ladder is sliding away from the base of the wall at the rate of 5 ft/s. How fast is the radian measure of the angle between the ladder and the ground changing at the moment when the bottom of the ladder is 12 feet from the base of the wall?

▮ Let x be the distance of the bottom of the ladder from the base of the wall, and let θ be the angle between the ladder and the ground. We are told that $D_t x = 5$, and $\theta = \cos^{-1}\dfrac{x}{13}$. Hence, $D_t\theta = \dfrac{-1}{\sqrt{1-(x/13)^2}}\cdot\dfrac{1}{13}\cdot D_t x = \dfrac{-5}{13\sqrt{1-(x/13)^2}}$. When $x = 12$, $D_t\theta = \dfrac{-5}{13\sqrt{\frac{25}{169}}} = \dfrac{-5}{13(\frac{5}{13})} = -1$ rad/s.

27.60 The beam from a lighthouse 3 miles from a straight coastline turns at the rate of 5 revolutions per minute. How fast is the point P at which the beam hits the shore moving when that point is 4 miles from point A on the shore directly opposite the lighthouse?

▮ Let x be the distance from P to A, and let θ be the angle between the beam and the line PA. We are told $D_t\theta = 10\pi$ rad/min. Clearly, $\theta = \tan^{-1}\dfrac{x}{3}$. So $10\pi = D_t\theta = \{1/[1+(x/3)^2]\}\cdot\frac{1}{3}\cdot D_t x$. When $x = 4$, $10\pi = (1/\frac{25}{9})\cdot\frac{1}{3}\cdot D_t x = \frac{3}{25}D_t x$. Hence, $D_t x = 250\pi/3$ mi/min $= 5000\pi$ mi/h $\approx 15{,}708$ mi/h.

27.61 Find the area under the curve $y = 1/(1+x^2)$, above the x-axis, and between the lines $x = 0$ and $x = 1$.

▮ $A = \displaystyle\int_0^1 \frac{dx}{1+x^2} = \tan^{-1}x\,]_0^1 = \tan^{-1}1 - \tan^{-1}0 = \pi/4 - 0 = \pi/4.$

27.62 Find the area under the curve $y = 1/\sqrt{1-x^2}$, above the x-axis, and between the lines $x = 0$ and $x = \frac{1}{2}$.

▮ $A = \displaystyle\int_0^{1/2} \frac{dx}{\sqrt{1-x^2}} = \sin^{-1}x\,]_0^{1/2} = \sin^{-1}\frac{1}{2} - \sin^{-1}0 = \pi/6 - 0 = \pi/6.$

27.63 The region $\mathcal{R}$ under the curve $y = 1/x^2\sqrt{x^2-1}$, above the x-axis and between the lines $x = 2/\sqrt{3}$ and $x = 2$, is revolved around the y-axis. Find the volume of the resulting solid.

▮ By the cylindrical shell formula, $V = 2\pi\int_{2/\sqrt{3}}^2 xy\,dx = 2\pi\displaystyle\int_{2/\sqrt{3}}^2 \frac{dx}{x\sqrt{x^2-1}} = 2\pi\sec^{-1}x\,]_{2/\sqrt{3}}^2 = 2\pi(\sec^{-1}2 - \sec^{-1}2/\sqrt{3}) = 2\pi(\pi/3 - \pi/6) = 2\pi(\pi/6) = \frac{1}{3}\pi^2.$

27.64 Use integration to show that the circumference of a circle of radius r is $2\pi r$.

▌ Find the arc length of the part of the circle $x^2 + y^2 = r^2$ in the first quadrant and multiply it by 4. Since $y = \sqrt{r^2 - x^2}$, $y' = -x/\sqrt{r^2 - x^2}$ and $(y')^2 = x^2/(r^2 - x^2)$. So $1 + (y')^2 = 1 + x^2/(r^2 - x^2) = (r^2 - x^2 + x^2)/(r^2 - x^2) = r^2/(r^2 - x^2)$. Thus, $L = 4\int_0^r \sqrt{1 + (y')^2}\, dx = 4\int_0^r \dfrac{r}{\sqrt{r^2 - x^2}}\, dx = 4r \sin^{-1} \dfrac{x}{r}\Big]_0^r = 4r(\sin^{-1} 1 - \sin^{-1} 0) = 4r(\pi/2 - 0) = 2\pi r$.

27.65 A person is viewing a painting hung high on a wall. The vertical dimension of the painting is 2 feet and the bottom of the painting is 2 feet above the eye level of the viewer. Find the distance x that the viewer should stand from the wall in order to maximize the angle θ subtended by the painting.

▌ From Fig. 27-6, $\theta = \tan^{-1} 4/x - \tan^{-1} 2/x$. So

$$D_x\theta = \frac{1}{1 + (4/x)^2} \cdot \frac{-4}{x^2} - \frac{1}{1 + (2/x)^2} \cdot \frac{-2}{x^2} = \frac{-4}{x^2 + 16} + \frac{2}{x^2 + 4}$$

$$= \frac{-4x^2 - 16 + 2x^2 + 32}{(x^2 + 16)(x^2 + 4)} = \frac{2(8 - x^2)}{(x^2 + 16)(x^2 + 4)}$$

Thus, the only positive critical number is $x = \sqrt{8} = 2\sqrt{2}$, which, by the first derivative test, yields a relative (and, therefore, an absolute) maximum.

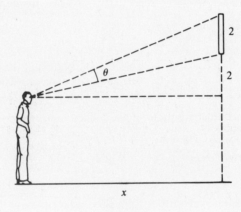

Fig. 27-6

27.66 For what values of x is the equation $\sin^{-1}(\sin x) = x$ true? (Recall Problem 27.20.)

▌ The range of $\sin^{-1} u$ is $[-\pi/2, \pi/2]$, and, in fact, for each x in $[-\pi/2, \pi/2]$, $\sin^{-1}(\sin x) = x$.

27.67 For what values of x is the equation $\cos^{-1}(\cos x) = x$ true?

▌ The range of $\cos^{-1} u$ is $[0, \pi]$. For each x in $[0, \pi]$, $\cos^{-1}(\cos x) = x$.

27.68 For what values of x does the equation $\sin^{-1}(-x) = -\sin^{-1} x$ hold?

▌ $\sin^{-1} x$ is defined only for x in $[-1, 1]$. Consider any such x. Let $\theta = \sin^{-1} x$. Then $\sin \theta = x$ and $-\pi/2 \le \theta \le \pi/2$. Note that $\sin(-\theta) = -\sin \theta = -x$ and $-\pi/2 \le -\theta \le \pi/2$. Hence, $\sin^{-1}(-x) = -\theta = -\sin^{-1} x$. Thus, the set of solutions of $\sin^{-1}(-x) = -\sin^{-1} x$ is $[-1, 1]$.

27.69 If $x^2 - x \tan^{-1} y = \ln y$, find y'.

▌ Use implicit differentiation. $2x - x[1/(1 + y^2)] \cdot y' - \tan^{-1} y = (1/y)y'$, $2x - \tan^{-1} y = y'[1/y + x/(1 + y^2)]$,
$$y' = \frac{y(1 + y^2)(2x - \tan^{-1} y)}{1 + y^2 + xy}.$$

27.70 If $\cos^{-1} xy = e^{2y}$, find y'.

▐ Use implicit differentiation. $\dfrac{-1}{\sqrt{1-(xy)^2}} \cdot (xy' + y) = e^{2y} \cdot 2y'$, $xy' + y = -2e^{2y} y'\sqrt{1 - x^2 y^2}$,

$y'(x + 2e^{2y}\sqrt{1 - x^2 y^2}) = -y$, $y' = \dfrac{-y}{x + 2e^{2y}\sqrt{1 - x^2 y^2}}$.

27.71 Sketch the graph of $y = \tan^{-1} x - \ln\sqrt{1 + x^2}$.

▐ See Fig. 27-7.

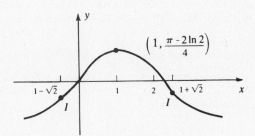

Fig. 27-7

$$y = \tan^{-1} x - \tfrac{1}{2} \ln(1 + x^2)$$

$$y' = \frac{1}{1 + x^2} - \frac{1}{2}\frac{1}{1 + x^2} \cdot 2x = \frac{1 - x}{1 + x^2}$$

$$y'' = \frac{(1 + x^2)(-1) - (1 - x)(2x)}{(1 + x^2)^2} = \frac{-1 - x^2 - 2x + 2x^2}{(1 + x^2)^2} = \frac{x^2 - 2x - 1}{(1 + x^2)^2}$$

For the only critical number, $x = 1$, $y'' = -\tfrac{1}{2} < 0$, and, therefore, there is a relative maximum at $x = 1$, $y = \pi/4 - \tfrac{1}{2}\ln 2 \approx 0.4$. Note that $y(0) = 0$. Also, as $x \to \pm\infty$, $y \to -\infty$. Setting $y'' = 0$, we find two inflection points at $x = 1 \pm \sqrt{2}$.

27.72 Find the derivative of $y = \sin(\sin^{-1} x^2)$.

▐ Since $\sin(\sin^{-1} x^2) = x^2$, $y' = 2x$. Note that y is defined only for $-1 \le x \le 1$ (and y' only for $-1 < x < 1$).

27.73 If $y = \tan^{-1}(e^{\sin x})$, find y'.

▐ $y' = \dfrac{1}{1 + (e^{\sin x})^2} \cdot e^{\sin x} \cdot \cos x = \dfrac{e^{\sin x} \cos x}{1 + e^{2 \sin x}}$.

27.74 If $y = \dfrac{1}{ab} \tan^{-1}\left(\dfrac{b}{a} \tan x\right)$, find y'.

▐ $y' = \dfrac{1}{ab} \cdot \dfrac{1}{1 + [(b/a) \tan x]^2} \cdot \dfrac{b}{a} \sec^2 x = \dfrac{\sec^2 x}{a^2 + b^2 \tan^2 x} = \dfrac{1}{a^2 \cos^2 x + b^2 \sin^2 x}$.

27.75 Refer to Fig. 27-8. In a circular arena of radius r, there is a light at L. A boy starting from B runs toward the center O at the rate of 10 ft/s. At what rate will his shadow be moving along the side when he is halfway from B to O?

▐ Let P be the boy's position, x the distance of P from B, θ the angle OLP, and s the arc intercepted by θ. We are given that $D_t x = 10$. Then $s = r(2\theta)$, $\theta = \tan^{-1}(r - x)/r$. $D_t s = 2rD_t \theta = 2r \cdot \dfrac{1}{1 + [(r - x)/r]^2} \cdot \left(-\dfrac{1}{r}\right) \cdot D_t x = \dfrac{-2r^2}{r^2 + (r - x)^2} \cdot 10 = \dfrac{-20r^2}{r^2 + (r - x)^2}$. When $x = \dfrac{r}{2}$, $D_t s = \dfrac{-20r^2}{r^2 + r^2/4} = -\dfrac{20}{\frac{5}{4}} = -16$. Hence, the shadow is moving at the rate of 16 ft/s.

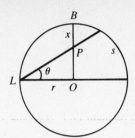

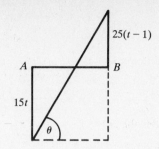

Fig. 27-8   **Fig. 27-9**

27.76 See Fig. 27-9. Two ships sail from A at the same time. One sails south at 15 mi/h; the other sails east at 25 mi/h for 1 hour until it reaches point B, and then sails north. Find the rate of rotation of the line joining them, after 3 hours.

▌ Let θ be the angle between the line joining the ships and the line parallel to AB. The distance AB is 25 miles. Then $\theta = \tan^{-1} \dfrac{15t + 25(t-1)}{25} = \tan^{-1} \dfrac{40t - 25}{25}$. Hence, $D_t\theta = \dfrac{1}{1 + [(40t-25)/25]^2}\left(\dfrac{40}{25}\right)$. When $t = 3$, $D_t\theta = \dfrac{40}{25 + (19)^2} = \dfrac{20}{193}$ radian per hour.

27.77 A balloon is released at eye level and rises at the rate of 5 ft/s. An observer 50 ft away watches the balloon rise. How fast is the angle of elevation increasing 6 seconds after the moment of release?

▌ Let x be the height of the balloon above the observer, and let θ be the angle of elevation. Then $\theta = \tan^{-1}(x/50)$ and $D_t\theta = \{1/[1 + (x/50)^2]\} \cdot \frac{1}{50} \cdot D_t x = \frac{1}{10} \cdot 1/[1 + (x/50)^2]$. When $t = 6$, $x = 30$, and $D_t\theta = \frac{5}{68} \approx 0.07$ rad/s.

27.78 A billboard, 54 feet wide, is perpendicular to a straight road and is 18 feet from the nearest point A on the road. As a motorist approaches the billboard along the road, at what point does she see the billboard in the widest angle?

▌ In Fig. 27-10, let x be the distance of the motorist from A, and let θ be her angle of vision of the billboard. Then $\theta = \cot^{-1}(x/72) - \cot^{-1}(x/18)$. Then $D_x\theta = -1/[1 + (x/72)^2] \cdot \frac{1}{72} + 1/[1 + (x/18)^2] \cdot \frac{1}{18} = 18/[(18)^2 + x^2] - 72/[(72)^2 + x^2]$. Setting $D_x\theta = 0$, we obtain $(72)^2 + x^2 = 4(18)^2 + 4x^2$, $x^2 = 81 \cdot 16$, $x = 36$. The first derivative test will verify that this yields a maximum value of θ.

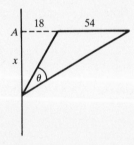

Fig. 27-10

27.79 A person walking along a straight path at the rate of 6 ft/s is followed by a spotlight that comes from a point 30 feet from the path. How fast is the spotlight turning when the person is 40 feet past the point A on the path nearest the light?

▌ Let x be the distance of the person from A, and let θ be the angle between the spotlight and the line to A. Then $\theta = \tan^{-1}(x/30)$, and $D_x\theta = \{1/[1 + (x/30)^2]\} \cdot \frac{1}{30} \cdot D_t x = 180/(900 + x^2)$. When $x = 40$, $D_x\theta = \frac{18}{250} = 0.072$ radian per second.

27.80 Show geometrically that $\sin^{-1}(x/\sqrt{x^2 + 1}) = \tan^{-1}x$ for $x \geq 0$.

▌ Consider (Fig. 27-11) a right triangle with legs of length 1 and x, and let θ be the angle opposite the side of length x. Then $\tan \theta = x$ and $\sin \theta = x/\sqrt{x^2 + 1}$. Hence, $\tan^{-1}x = \theta = \sin^{-1}(x/\sqrt{x^2 + 1})$.

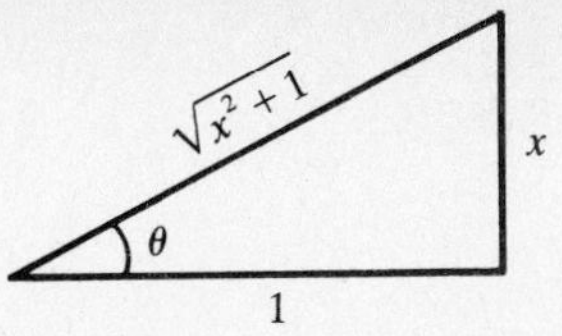

Fig. 27-11

27.81 Find the area under the curve $y = 1/(x\sqrt{x^2 - 1})$ and above the segment $[\sqrt{2}, 2]$ on the x-axis.

▮ $A = \int_{\sqrt{2}}^{2} \dfrac{1}{x\sqrt{x^2 - 1}} \, dx = \sec^{-1} x \,]_{\sqrt{2}}^{2} = \sec^{-1} 2 - \sec^{-1} \sqrt{2} = \dfrac{\pi}{3} - \dfrac{\pi}{4} = \dfrac{\pi}{12}.$

27.82 Find the area under $y = 1/(1 + 3x^2)$ and above the segment $[0, 1]$.

▮ $A = \int_{0}^{1} \dfrac{dx}{1 + 3x^2} = \dfrac{1}{3} \int_{0}^{1} \dfrac{dx}{\frac{1}{3} + x^2} = \dfrac{1}{3} \cdot \sqrt{3} \tan^{-1} \sqrt{3}x \,]_{0}^{1} = \dfrac{\sqrt{3}}{3} (\tan^{-1} \sqrt{3} - \tan^{-1} 0) = \dfrac{\sqrt{3}}{3} \left(\dfrac{\pi}{3} - 0 \right) = \dfrac{\pi\sqrt{3}}{9}.$

CHAPTER 28
Integration by Parts

In Problems 28.1–28.24, find the indicated antiderivative.

28.1 $\int x^2 e^{-x}\, dx.$

I We use integration by parts: $\int u\, dv = uv - \int v\, du$. In this case, let $u = x^2$, $dv = e^{-x}\, dx$. Then $du = 2x\, dx$, $v = -e^{-x}$. Hence, $\int x^2 e^{-x}\, dx = -x^2 e^{-x} + 2\int xe^{-x}\, dx$. [To calculate the latter, we use another integration by parts: $u = x$, $dv = e^{-x}\, dx$; $du = dx$, $v = -e^{-x}$. Then $\int xe^{-x}\, dx = -xe^{-x} + \int e^{-x}\, dx = -xe^{-x} - e^{-x} = -e^{-x}(x + 1)$.] Hence, $\int x^2 e^{-x}\, dx = -x^2 e^{-x} + 2[-e^{-x}(x + 1)] + C = -e^{-x}(x^2 + 2x + 2) + C.$

28.2 $\int e^x \sin x\, dx.$

I Let $u = \sin x$, $dv = e^x\, dx$, $du = \cos x\, dx$, $v = e^x$. Then

$$\int e^x \sin x\, dx = e^x \sin x - \int e^x \cos x\, dx. \qquad (1)$$

We use integration by parts again for the latter integral: let $u = \cos x$, $dv = e^x\, dx$, $du = -\sin x\, dx$, $v = e^x$. Then $\int e^x \cos x\, dx = e^x \cos x + \int e^x \sin x\, dx$. Substituting in (1), $\int e^x \sin x\, dx = e^x \sin x - (e^x \cos x + \int e^x \sin x\, dx) = e^x \sin x - e^x \cos x - \int e^x \sin x\, dx$. Thus, $2\int e^x \sin x\, dx = e^x (\sin x - \cos x) + C$, $\int e^x\, dx = \frac{1}{2}e^x(\sin x - \cos x) + C_1$.

28.3 $\int x^3 e^x\, dx.$

I Let $u = x^3$, $dv = e^x\, dx$, $du = 3x^2$, $v = e^x$. Then $\int x^3 e^x\, dx = x^3 e^x - 3\int x^2 e^x\, dx$. But Problem 28.1 gives, with x replaced by $-x$, $\int x^2 e^x\, dx = e^x(x^2 - 2x + 2) + C$. Hence, $\int x^3 e^x\, dx = e^x(x^3 - 3x^2 + 6x - 6) + C.$

28.4 $\int \sin^{-1} x\, dx.$

I Let $u = \sin^{-1} x$, $dv = dx$, $du = (1/\sqrt{1 - x^2})\, dx$, $v = x$. Then $\int \sin^{-1} x\, dx = x \sin^{-1} x - \int (x/\sqrt{1 - x^2})\, dx = x \sin^{-1} x + \frac{1}{2}\int (1 - x^2)^{-1/2}(-2x)\, dx = x \sin^{-1} x + \frac{1}{2}\cdot 2(1 - x^2)^{1/2} + C = x \sin^{-1} x + \sqrt{1 - x^2} + C.$

28.5 $\int x \sin x\, dx.$

I Let $u = x$, $dv = \sin x\, dx$, $du = dx$, $v = -\cos x$. Then $\int x \sin x\, dx = -x \cos x + \int \cos x\, dx = -x \cos x + \sin x + C.$

28.6 $\int x^2 \cos x\, dx.$

I Let $u = x^2$, $dv = \cos x\, dx$, $du = 2x\, dx$, $v = \sin x$. Then, using Problem 28.5, $\int x^2 \cos x\, dx = x^2 \sin x - 2\int x \sin x\, dx = x^2 \sin x - 2(-x \cos x + \sin x) + C = (x^2 - 2) \sin x + 2x \cos x + C.$

28.7 $\int \cos (\ln x)\, dx.$

I Let $x = e^{y - \pi/2}$, $\cos (\ln x) = \sin y$, $dx = e^{y - \pi/2}\, dy$, and use Problem 28.2: $\int \cos (\ln x)\, dx = e^{-\pi/2}\int e^y \sin y\, dy = e^{-\pi/2}[\frac{1}{2}e^y(\sin y - \cos y)] + C = \frac{1}{2}x[\cos (\ln x) + \sin (\ln x)] + C.$

28.8 $\int x \cos (5x - 1)\, dx.$

I Let $u = x$, $dv = \cos (5x - 1)\, dx$, $du = dx$, $v = \frac{1}{5}\sin (5x - 1)$. Then $\int x \cos (5x - 1)\, dx = \frac{1}{5}x \sin (5x - 1) - \frac{1}{5}\int \sin (5x - 1)\, dx = \frac{1}{5}x \sin (5x - 1) + \frac{1}{5}\cdot\frac{1}{5}\cos (5x - 1) + C = \frac{1}{25}[5x \sin (5x - 1) + \cos (5x - 1)] + C.$

232

28.9 $\int e^{ax} \cos bx \, dx$.

▮ Let $u = \cos bx$, $dv = e^{ax} \, dx$, $du = -b \sin bx$, $v = (1/a)e^{ax}$. Then $\int e^{ax} \cos bx \, dx = (1/a)e^{ax} \cos bx + (b/a) \int e^{ax} \sin bx \, dx$. Apply integration by parts to the latter: $u = \sin bx$, $dv = e^{ax} \, dx$, $du = b \cos bx$, $v = (1/a)e^{ax}$.   So $\int e^{ax} \sin bx \, dx = (1/a)e^{ax} \sin bx - (b/a) \int e^{ax} \cos bx \, dx$.   Hence, by substitution, $\int e^{ax} \cos bx \, dx = (1/a)e^{ax} \cos bx + (b/a)[(1/a)e^{ax} \sin bx - (b/a) \int e^{ax} \cos bx \, dx] = (1/a)e^{ax} \cos bx + (b/a^2)e^{ax} \sin bx - (b^2/a^2) \int e^x \cos bx \, dx$. Thus, $(1 + b^2/a^2) \int e^{ax} \cos bx \, dx = (e^{2x}/a^2)(a \cos bx + b \sin bx) + C$, $\int e^{ax} \cos bx \, dx = [e^{ax}/(a^2 + b^2)](a \cos bx + b \sin bx) + C_1$.

28.10 $\int \sin^2 x \, dx$.

▮ Let $u = \sin x$, $dv = \sin x \, dx$, $du = \cos x \, dx$, $v = -\cos x$. Then $\int \sin^2 x \, dx = -\sin x \cos x + \int \cos^2 x \, dx = -\sin x \cos x + \int (1 - \sin^2 x) \, dx = -\sin x \cos x + x - \int \sin^2 x \, dx$.   So $2 \int \sin^2 x \, dx = x - \sin x \cos x + C$, $\int \sin^2 x \, dx = \frac{1}{2}(x - \sin x \cos x) + C_1$.

28.11 $\int \cos^3 x \, dx$.

▮ $\int \cos^3 x \, dx = \int \cos x \, (1 - \sin^2 x) \, dx = \int \cos x \, dx - \int \sin^2 x \cos x \, dx = \sin x - \frac{1}{3} \sin^3 x + C$.

28.12 $\int \cos^4 x \, dx$.

▮ $\int \cos^4 x \, dx = \int \left(\frac{1 + \cos 2x}{2}\right)^2 dx = \frac{1}{4} \int (1 + 2 \cos 2x + \cos^2 2x) \, dx = \frac{1}{4}\left(x + \sin 2x + \int \frac{1 + \cos 4x}{2} \, dx\right)$
$= \frac{1}{4}[x + \sin 2x + \frac{1}{2}(x + \frac{1}{4} \sin 4x)] + C = \frac{3}{8}x + \frac{1}{4} \sin 2x + \frac{1}{32} \sin 4x + C$.

28.13 $\int xe^{3x} \, dx$.

▮ Let $u = x$, $dv = e^{3x} \, dx$, $du = dx$, $v = \frac{1}{3}e^{3x}$. Then $\int xe^{3x} \, dx = \frac{1}{3}xe^{3x} - \frac{1}{3} \int e^{3x} \, dx = \frac{1}{3}xe^x - \frac{1}{3} \cdot \frac{1}{3}e^{3x} + C = \frac{1}{9}e^{3x}(3x - 1) + C$.

28.14 $\int x \sec^2 x \, dx$.

▮ Let $u = x$, $dv = \sec^2 x \, dx$, $du = dx$, $v = \tan x$. Then $\int x \sec^2 x \, dx = x \tan x - \int \tan x \, dx = x \tan x - \ln |\sec x| + C$.

28.15 $\int x \cos^2 x \, dx$.

▮ Let $u = x$, $dv = \cos^2 x \, dx$, $du = dx$, $v = \int \frac{1 + \cos 2x}{2} \, dx = \frac{1}{2}(x + \frac{1}{2} \sin 2x)$. Then $\int x \cos^2 x \, dx = \frac{1}{2}x(x + \frac{1}{2} \sin 2x) - \frac{1}{2} \int (x + \frac{1}{2} \sin 2x) \, dx = \frac{1}{2}x^2 + \frac{1}{4} \sin 2x - \frac{1}{2}(\frac{1}{2}x^2 - \frac{1}{2} \cdot \frac{1}{2} \cos 2x) + C = \frac{1}{4}x^2 + \frac{1}{8}(2 \sin 2x + \cos 2x) + C$.

28.16 $\int (\ln x)^2 \, dx$.

▮ Let $x = e^{-t}$ and use Problem 28.1: $\int (\ln x)^2 \, dx = -\int t^2 e^{-t} \, dt = e^{-t}(t^2 + 2t + 2) + C = x[(\ln x)^2 - 2 \ln x + 2] + C$.

28.17 $\int x \sin 2x \, dx$.

▮ Let $2x = y$ and use Problem 28.5: $\int x \sin 2x \, dx = \frac{1}{4} \int y \sin y \, dy = \frac{1}{4}(-y \cos y + \sin y) + C = \frac{1}{4}(-2x \cos 2x + \sin 2x) + C$.

28.18 $\int x \sin (x^2) \, dx$.

▮ Use a substitution $u = x^2$, $du = 2x \, dx$.   Then $\int x \sin x^2 \, dx = \frac{1}{2} \int \sin u \, du = -\frac{1}{2} \cos u + C = -\frac{1}{2} \cos x^2 + C$.

28.19 $\int \dfrac{\ln x}{x^2}\, dx.$

▌ Let $u = \ln x$, $dv = (1/x^2)\, dx$, $du = (1/x)\, dx$, $v = -1/x$. Then $\int \dfrac{\ln x}{x^2}\, dx = \dfrac{-\ln x}{x} + \int \dfrac{1}{x^2}\, dx = \dfrac{-\ln x}{x} - \dfrac{1}{x} + C = \dfrac{-1}{x}(\ln x + 1) + C.$

28.20 $\int x^2 e^{3x}\, dx.$

▌ Let $3x = -y$ and use Problem 28.1: $\int x^2 e^{3x}\, dx = -\dfrac{1}{27}\int y^2 e^{-y}\, dy = \frac{1}{27} e^{-y}(y^2 + 2y + 2) + C = \frac{1}{27} e^{3x}(9x^2 - 6x + 2) + C.$

28.21 $\int x^2 \tan^{-1} x\, dx.$

▌ Let $u = \tan^{-1} x$, $dv = x^2\, dx$, $du = [1/(1+x^2)]\, dx$, $v = \frac{1}{3} x^3$. Then $\int x^2 \tan^{-1} x\, dx = \frac{1}{3} x^3 \tan^{-1} x - \dfrac{1}{3}\int \dfrac{x^3}{1+x^2}\, dx = \frac{1}{3} x^3 \tan^{-1} x - \dfrac{1}{3}\int \left(x - \dfrac{x}{1+x^2}\right) dx = \frac{1}{3} x^3 \tan^{-1} x - \frac{1}{3}[\frac{1}{2} x^2 - \frac{1}{2}\ln(1+x^2)] + C = \frac{1}{3} x^3 \tan^{-1} x - \frac{1}{6} x^2 + \frac{1}{6}\ln(1+x^2) + C.$

28.22 $\int \ln(x^2 + 1)\, dx.$

▌ Let $u = \ln(x^2 + 1)$, $dv = dx$, $du = [2x/(x^2+1)]\, dx$, $v = x$. So $\int \ln(x^2 + 1)\, dx = x \ln(x^2 + 1) - 2\int \dfrac{x^2}{x^2+1}\, dx = x \ln(x^2 + 1) - 2\int \left(1 - \dfrac{1}{x^2+1}\right) dx = x \ln(x^2 + 1) - 2(x - \tan^{-1} x) + C = x \ln(x^2 + 1) - 2x + 2\tan^{-1} x + C.$

28.23 $\int \dfrac{x^3}{\sqrt{1+x^2}}\, dx.$

▌ A simple substitution works: let $u = 1 + x^2$, $du = 2x\, dx$. $\int \dfrac{x^3}{\sqrt{1+x^2}}\, dx = \dfrac{1}{2}\int \dfrac{u-1}{\sqrt{u}}\, du = \dfrac{1}{2}\int (u^{1/2} - u^{-1/2})\, du = \frac{1}{2}(\frac{2}{3} u^{3/2} - 2u^{1/2}) + C = \frac{1}{3} u^{1/2}(u - 3) + C = \frac{1}{3}\sqrt{1+x^2}(x^2 - 2) + C.$

28.24 $\int x^2 \ln x\, dx.$

▌ Let $u = \ln x$, $dv = x^2\, dx$, $du = (1/x)\, dx$, $v = \frac{1}{3} x^3$. Then $\int x^2 \ln x\, dx = \frac{1}{3} x^3 \ln x - \frac{1}{3}\int x^2\, dx = \frac{1}{3} x^3 \ln x - \frac{1}{3}\cdot\frac{1}{3} x^3 + C = \frac{1}{9} x^3(3\ln x - 1) + C.$

28.25 Let $\mathcal{R}$ be the region bounded by the curve $y = \ln x$, the x-axis, and the line $x = e$. Find the area of $\mathcal{R}$.

▌ $A = \int_1^e \ln x\, dx = x \ln x \,]_1^e - \int_1^e 1\, dx = (e - 0) - (e - 1) = 1.$

28.26 Find the volume of the solid obtained by revolving the region $\mathcal{R}$ of Problem 28.25 about the x-axis.

▌ By the disk formula, $v = \pi \int_1^e (\ln x)^2\, dx$. By Problem 28.16, $v = \pi x[(\ln x)^2 - 2\ln x + 2] \,]_1^e = \pi[e(1 - 2 + 2) - 2] = \pi(e - 2).$

28.27 Find the volume of the solid obtained by revolving the region $\mathcal{R}$ of Problem 28.25 about the y-axis.

▌ We use the cylindrical shell formula: $V = 2\pi \int_1^e x \ln x\, dx = 2\pi(x^2/2)\ln x \,]_1^e - 2\pi \int_1^e \dfrac{x}{2}\, dx = 2\pi[e^2/2 - (e^3/4 - \frac{1}{4})] = (\pi/2)(e^2 + 1).$

28.28 Let $\mathcal{R}$ be the region bounded by the curve $y = \ln x/x$, the x-axis, and the line $x = e$. Find the area of $\mathcal{R}$.

▌ $A = \int_1^e \dfrac{\ln x}{x}\, dx = \frac{1}{2}(\ln x)^2 \,]_1^e = \frac{1}{2}(1 - 0) = \frac{1}{2}.$

28.29 Find the volume of the solid obtained by revolving the region $\mathcal{R}$ of Problem 28.28 about the y-axis.

▌ By the cylindrical shell formula, $V = 2\pi \int_1^e xy\, dx = 2\pi \int_1^e \ln x\, dx = 2\pi x(\ln x - 1) \,]_1^e = 2\pi[e(1 - 1) - (0 - 1)] = 2\pi.$

28.30   Find the volume of the solid obtained by revolving the region $\mathcal{R}$ of Problem 28.28 about the x-axis.

❚   By the disk formula, $V = \pi \int_1^e (\ln x / x)^2 \, dx$.   Change the variable to $t = \ln x$ and use Problem 28.1:
$V = \pi \int_0^1 t^2 e^{-t} \, dt = -\pi e^{-t}(t^2 + 2t + 2) \big]_0^1 = \pi(2 - 5/e)$.

28.31   Use the solution to Problem 28.30 to establish the following bounds on e: $2.5 \le e \le 2.823$.

❚   In Problem 28.30, it was shown that $\displaystyle\int_1^e \left(\frac{\ln x}{x}\right)^2 dx = 2 - 5/e$.   By Problem 24.59, $1/e$ is the maximum value of $\ln x / x$.   Hence, $\displaystyle\int_1^e \left(\frac{\ln x}{x}\right)^2 dx \le \frac{e - 1}{e^2}$.   {In general, $\int_a^b f(x) \, dx \le M(b - a)$ if M is an upper bound of $f(x)$ on $[a, b]$.}   Thus, $0 \le 2 - 5/e \le (e - 1)/e^2$.   The left-hand inequality gives $e \ge 2.5$.   The right-hand inequality gives $2e^2 - 5e \le e - 1$, $2e^2 - 6e + 1 \le 0$.   Since the roots of $2x^2 - 6x + 1 = 0$ are $(3 \pm \sqrt{7})/2$, $e \le (3 + \sqrt{7})/2 < 2.823$ (since $\sqrt{7} < 2.646$).

28.32   Let $\mathcal{R}$ be the region under one arch of the curve $y = \sin x$, above the x-axis, between $x = 0$ and $x = \pi$. Find the volume of the solid obtained by revolving $\mathcal{R}$ about the y-axis.

❚   By the cylindrical shell formula and Problem 28.5, $V = 2\pi \int_0^\pi x \sin x \, dx = 2\pi[(\pi + 0) - (0 + 0)] = 2\pi^2$.

28.33   If n is a positive integer, find $\int_0^{2\pi} x \cos nx \, dx$.

❚   Let $u = x$, $dv = \cos nx \, dx$, $du = dx$, $v = (1/n) \sin nx$.   Then $\int x \cos nx \, dx = \frac{1}{n} x \sin nx - \frac{1}{n} \int \sin nx \, dx = \frac{1}{n} x \sin nx + \frac{1}{n^2} \cos nx = \frac{1}{n^2}(nx \sin nx + \cos nx)$.   Hence, $\int_0^{2\pi} x \cos nx \, dx = \frac{1}{n^2}(nx \sin nx + \cos nx) \big]_0^{2\pi} = \frac{1}{n^2}(1 - 1) = 0$.

28.34   If n is a positive integer, find $\int_0^{2\pi} x \sin nx \, dx$.

❚   Let $u = x$, $dv = \sin nx \, dx$, $du = dx$, $v = -(1/n) \cos nx$.   Then $\int x \sin nx \, dx = -\frac{1}{n} x \cos nx + \frac{1}{n} \int \cos nx \, dx = -\frac{1}{n} x \cos nx + \frac{1}{n^2} \sin nx$.   Hence, $\int_0^{2\pi} x \sin nx \, dx = \frac{1}{n^2}(\sin nx - nx \cos nx) \big]_0^{2\pi} = \frac{1}{n^2}[(0 - 2\pi n) - (0 - 0)] = -\frac{2\pi}{n}$.

28.35   Find a reduction formula for $\int \cos^n x \, dx$ for $n \ge 2$.

❚   Let $u = \cos^{n-1} x$, $dv = \cos x \, dx$, $du = -(n - 1) \cos^{n-2} x \sin x \, dx$, $v = \sin x$.   Then $\int \cos^n x \, dx = \sin x \cos^{n-1} x + (n - 1) \int \cos^{n-2} x \sin^2 x \, dx$ $= \sin x \cos^{n-1} x + (n - 1) \int \cos^{n-2} x (1 - \cos^2 x) \, dx$ $= \sin x \cos^{n-1} x + (n - 1) \int \cos^{n-2} x \, dx - (n - 1) \int \cos^n x \, dx$.   Hence, $n \int \cos^n x \, dx = \sin x \cos^{n-1} x + (n - 1) \int \cos^{n-2} x \, dx$, $\int \cos^n x \, dx = \frac{\sin x \cos^{n-1} x}{n} + \frac{n - 1}{n} \int \cos^{n-2} x \, dx$.

28.36   Apply the reduction formula of Problem 28.35 to find $\int \cos^6 x \, dx$,   using the result of Problem 28.12.

❚ $\int \cos^6 x \, dx = \frac{\sin x \cos^5 x}{6} + \frac{5}{6} \int \cos^4 x \, dx = \frac{\sin x \cos^5 x}{6} + \frac{5}{6}\left(\frac{3}{8}x + \frac{1}{4}\sin 2x + \frac{1}{32} \sin 4x\right) + C$.

28.37   Find a reduction formula for $\int \sin^n x \, dx$ for $n \ge 2$.

❚   Let $u = \sin^{n-1} x$, $dv = \sin x \, dx$, $du = (n - 1) \sin^{n-2} x \cos x \, dx$, $v = -\cos x$.   Then $\int \sin^n x \, dx = -\cos x \sin^{n-1} x + (n - 1) \int \sin^{n-2} x \cos^2 x \, dx$ $= -\cos x \sin^{n-1} x + (n - 1) \int \sin^{n-2} x (1 - \sin^2 x) \, dx$ $= -\cos x \sin^{n-1} x + (n - 1) \int \sin^{n-2} x \, dx - (n - 1) \int \sin^n x \, dx$.   Hence, $n \int \sin^n x \, dx = -\cos x \sin^{n-1} x + (n - 1) \int \sin^{n-2} x \, dx$, $\int \sin^n x \, dx = -\frac{\cos x \sin^{n-1} x}{n} + \frac{n - 1}{n} \int \sin^{n-2} x \, dx$.

28.38   Use the reduction formula of Problem 28.37 to find $\int \sin^4 x \, dx$,   using the result of Problem 28.10.

❚ $\int \sin^4 x \, dx = -\frac{\cos x \sin^3 x}{4} + \frac{3}{4} \int \sin^2 dx = -\frac{\cos x \sin^3 x}{4} + \frac{3}{4} \cdot \frac{1}{2}(x - \sin x \cos x) + C$

$$= -\frac{\cos x \sin^3 x}{4} + \frac{3}{8}(x - \sin x \cos x) + C.$$

28.39 Find a reduction formula for $\int \sec^n x \, dx$ for $n \geq 2$.

▌ Let $u = \sec^{n-2} x$, $dv = \sec^2 x \, dx$, $du = (n-2) \sec^{n-3} x \sec x \tan x \, dx$, $v = \tan x$. Then $\int \sec^n x \, dx =$ $\tan x \sec^{n-2} x - (n-2) \int \sec^{n-2} x \tan^2 x \, dx = \tan x \sec^{n-2} x - (n-2) \int \sec^{n-2} x (\sec^2 x - 1) \, dx =$ $\tan x \sec^{n-2} x - (n-2) \int \sec^n x \, dx + (n-2) \int \sec^{n-2} x \, dx$. Hence, $(n-1) \int \sec^n x \, dx = \tan x \sec^{n-2} x + (n-2) \int \sec^{n-2} x \, dx$. Thus, $\int \sec^n x \, dx = \dfrac{\tan x \sec^{n-2} x}{n-1} + \dfrac{n-2}{n-1} \int \sec^{n-2} x \, dx$.

28.40 Use the reduction formula of Problem 28.39 to find $\int \sec^3 x \, dx$.

▌ $\int \sec^3 x \, dx = \dfrac{\tan x \sec x}{2} + \dfrac{1}{2} \int \sec x \, dx = \dfrac{1}{2} (\tan x \sec x + \ln |\sec x + \tan x|) + C$.

28.41 Use the reduction formula of Problem 28.39 to find $\int \sec^4 x \, dx$.

▌ $\int \sec^4 x \, dx = \dfrac{\tan x \sec^2 x}{3} + \dfrac{2}{3} \int \sec^2 x \, dx = \dfrac{\tan x \sec^2 x}{3} + \dfrac{2}{3} \tan x + C = \dfrac{1}{3} (\tan x)(\sec^2 x + 2) + C$.

28.42 Establish a reduction formula for $\int x^n e^{ax} \, dx$ for $n \geq 1$.

▌ Let $u = x^n$, $dv = e^{ax} \, dx$, $du = nx^{n-1} \, dx$, $v = (1/a)e^{ax}$. So $\int x^n e^{ax} \, dx = \dfrac{1}{a} x^n e^{ax} - \dfrac{n}{a} \int x^{n-1} e^{ax} \, dx$.

28.43 Apply the reduction formula of Problem 28.42 and the result of Problem 28.20 to find $\int x^3 e^{3x} \, dx$.

▌ $\int x^3 e^{3x} \, dx = \frac{1}{3} x^3 e^{3x} - \int x^2 e^{3x} \, dx = \frac{1}{3} x^3 e^{3x} - \frac{1}{27} e^{3x}(9x^2 - 6x + 2) + C = \frac{1}{27} e^{3x}(9x^3 - 9x^2 + 6x - 2) + C$.

28.44 Find a reduction formula for $\int x^n \sin ax \, dx$ for $n \geq 1$.

▌ Let $u = x^n$, $dv = \sin ax \, dx$, $du = nx^{n-1} \, dx$, $v = -(1/a) \cos ax$. Then $\int x^n \sin ax \, dx = -\dfrac{1}{a} x^n \cos ax + \dfrac{n}{a} \int x^{n-1} \cos ax \, dx$.

28.45 Use the reduction formula of Problem 28.44 and the solution of Problem 19.33 to find $\int x^2 \sin x \, dx$.

▌ $\int x^2 \sin x \, dx = -x^2 \cos x + 2 \int x \cos x \, dx = -x^2 \cos x + 2(x \sin x + \cos x) + C = 2x \sin x + \cos x \, (2 - x^2) + C$.

28.46 Find $\int x\sqrt{1 + x} \, dx$.

▌ Let $u = x$, $dv = \sqrt{1 + x} \, dx$, $du = dx$, $v = \frac{2}{3}(1 + x)^{3/2}$. Then $\int x\sqrt{1 + x} \, dx = \dfrac{2}{3} x(1 + x)^{3/2} - \dfrac{2}{3} \int (1 + x)^{3/2} \, dx = \frac{2}{3} x(1 + x)^{3/2} - \frac{2}{3} \cdot \frac{2}{5}(1 + x)^{5/2} + C = \frac{2}{15}(1 + x)^{3/2}[5x - 2(1 + x)] + C = \frac{2}{15}(1 + x)^{3/2}(3x - 2) + C$.

28.47 Prove the reduction formula $\displaystyle\int \dfrac{x^2 \, dx}{(a^2 + x^2)^n} = \dfrac{-x}{(2n-2)(a^2 + x^2)^{n-1}} + \dfrac{1}{2n-2} \int \dfrac{dx}{(a^2 + x^2)^{n-1}}$.

▌ Let $u = x$, $dv = \dfrac{x \, dx}{(a^2 + x^2)^n}$, $du = dx$, $v = \dfrac{-1}{(2n-2)(a^2 + x^2)^{n-1}}$. Then

$$\int \dfrac{x^2 \, dx}{(a^2 + x^2)^n} = \dfrac{-x}{(2n-2)(a^2 + x^2)^{n-1}} + \dfrac{1}{2n-2} \int \dfrac{dx}{(a^2 + x^2)^{n-1}}$$

28.48 Apply the reduction formula of Problem 28.47 to find $\displaystyle\int \dfrac{x^2 \, dx}{(a^2 + x^2)^2}$.

▌ $\displaystyle\int \dfrac{x^2 \, dx}{(a^2 + x^2)^2} = \dfrac{-x}{2(a^2 + x^2)} + \dfrac{1}{2} \int \dfrac{dx}{a^2 + x^2} = \dfrac{1}{2} \left(\dfrac{-x}{a^2 + x^2} + \dfrac{1}{a} \tan^{-1} \dfrac{x}{a} \right) + C$.

28.49 Find $\int \cos x^{1/3} \, dx$, using Problem 28.6.

▌ First make the substitution $x = w^3$, $dx = 3w^2 \, dw$. Then $\int \cos x^{1/3} \, dx = 3 \int w^2 \cos w \, dw = 3[(w^2 - 2) \sin w + 2w \cos w] + C = 3(x^{2/3} - 2) \sin x^{1/3} + 6x^{1/3} \cos x^{1/3} + C$.

28.50 Find $\int e^{\sqrt{x}}\, dx$.

$\blacksquare$ First make the substitution $x = w^2$, $dx = 2w\, dw$. Then $\int e^{\sqrt{x}}\, dx = 2 \int we^w\, dw$. By the reduction formula in Problem 28.42, $\int we^w\, dw = we^w - \int e^w\, dw = we^w - e^w = e^w(w - 1)$. So we obtain $2e^w(w - 1) + C = 2e^{\sqrt{x}}(\sqrt{x} - 1) + C$.

28.51 Evaluate $\int x(ax + b)^3\, dx$ by integration by parts.

$\blacksquare$ Let $u = x$, $dv = (ax + b)^3\, dx$, $du = dx$, $v = \dfrac{1}{4a}(ax + b)^4$. Then $\int x(ax + b)^3\, dx = \dfrac{1}{4a} x(ax + b)^4 - \dfrac{1}{4a} \int (ax + b)^4\, dx = \dfrac{1}{4a} x(ax + b)^4 - \dfrac{1}{4a} \cdot \dfrac{1}{5a}(ax + b)^5 + C = \dfrac{(ax + b)^4}{20a^2}[5ax - (ax + b)] + C = \dfrac{(ax + b)^4}{20a^2}(4ax - b) + C$.

28.52 Do Problem 28.51 by means of a substitution.

$\blacksquare$ Let $u = ax + b$, $du = a\, dx$. Then $\int x(ax + b)^3\, dx = \dfrac{1}{a} \int \dfrac{u - b}{a} u^3\, du = \dfrac{1}{a^2}(\int u^4\, du - \int bu^3\, du) = \dfrac{1}{a^2}\left(\dfrac{1}{5} u^5 - \dfrac{b}{4} u^4\right) + C = \dfrac{u^4}{20a^2}(4u - 5b) + C = \dfrac{(ax + b)^4}{20a^2}(4ax - b) + C$.

28.53 The region under the curve $y = \cos x$ between $x = 0$ and $x = \pi/2$ is revolved about the y-axis. Find the volume of the resulting solid.

$\blacksquare$ By the cylindrical shell formula, $V = 2\pi \int_0^{\pi/2} xy\, dx = 2\pi \int_0^{\pi/2} x \cos x\, dx$. By the solution of Problem 19.33, we obtain $2\pi(x \sin x + \cos x)\,]_0^{\pi/2} = 2\pi[(\pi/2 - 0) - (0 + 1)] = 2\pi(\pi/2 - 1) = \pi(\pi - 2)$.

28.54 Find $\int (\sin^{-1} x)^2\, dx$ with the aid of Problem 28.6.

$\blacksquare$ Make the substitution $y = \sin^{-1} x$, $dy = \dfrac{1}{\sqrt{1 - x^2}}\, dx$. Then $\int (\sin^{-1} x)^2\, dx = \int y^2 \sqrt{1 - \sin^2 y}\, dy = \int y^2 \cos y\, dy = (y^2 - 2) \sin y + 2y \cos y + C = [(\sin^{-1} x)^2 - 2] x + 2(\sin^{-1} x)\sqrt{1 - x^2} + C$.

28.55 Find $\int x^n \ln x\, dx$ for $n \neq -1$.

$\blacksquare$ Let $u = \ln x$, $dv = x^n\, dx$, $du = \dfrac{1}{x}\, dx$, $v = \dfrac{x^{n+1}}{n + 1}$. Then $\int x^n \ln x\, dx = \dfrac{x^{n+1}}{n + 1} \ln x - \dfrac{1}{n + 1} \int x^n\, dx = \dfrac{x^{n+1}}{n + 1} \ln x - \dfrac{1}{n + 1} \dfrac{x^{n+1}}{n + 1} + C = \dfrac{x^{n+1}}{(n + 1)^2}[(n + 1) \ln x - 1] + C$.

28.56 Derive the reduction formula $\int x^m (\ln x)^n\, dx = \dfrac{x^{m+1}(\ln x)^n}{m + 1} - \dfrac{n}{m + 1} \int x^m (\ln x)^{n-1}\, dx$.

$\blacksquare$ Let $u = (\ln x)^n$, $dv = x^m\, dx$, $du = \dfrac{n(\ln x)^{n-1}}{x}$, $v = \dfrac{1}{m + 1} x^{m+1}$. Then $\int x^m (\ln x)^n\, dx = \dfrac{x^{m+1}(\ln x)^n}{m + 1} - \dfrac{n}{m + 1} \int x^m (\ln x)^{n-1}\, dx$.

28.57 Find $\int x^5 (\ln x)^2\, dx$.

$\blacksquare$ By the reduction formula of Problem 28.56, $\int x^5 (\ln x)^2\, dx = \frac{1}{6} x^6 (\ln x)^2 - \frac{2}{6} \int x^5 \ln x\, dx = \frac{1}{6} x^6 (\ln x)^2 - \frac{1}{3}(\frac{1}{6} x^6 \ln x - \frac{1}{6} \int x^5\, dx) = \frac{1}{6} x^6 (\ln x)^2 - \frac{1}{18} x^6 \ln x + \frac{1}{18} \cdot \frac{1}{6} x^6 + C = \frac{1}{108} x^6 [18(\ln x)^2 - 6 \ln x + 1] + C$.

CHAPTER 29
Trigonometric Integrands and Substitutions

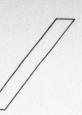

29.1 Find $\int \cos^2 ax \, dx$. *[handwritten: $\cos 2u = \cos(u+u) = \cos u \cos u - \sin u \sin u = \cos^2 u - \sin^2 u = \cos^2 u - 1 + \cos^2 u = 2\cos^2 u - 1$ $\therefore \cos^2 u = (1+\cos 2u)/2$; $(1/2a)\sin(ax+ax) = (1/2a) \cdot 2 \sin ax \cos ax = (1/a) \sin ax \cos ax$]*

▮ $\int \cos^2 ax \, dx = \int \dfrac{1 + \cos 2ax}{2} \, dx = \dfrac{1}{2}\left(x + \dfrac{1}{2a}\sin 2ax\right) + C = \dfrac{1}{2}\left(x + \dfrac{1}{a}\sin ax \cos ax\right) + C$.

29.2 Find $\int \sin^2 ax \, dx$.

▮ Using Problem 29.1, $\int \sin^2 ax \, dx = \int (1 - \cos^2 ax)\, dx = x - \dfrac{1}{2}\left(x + \dfrac{1}{a}\sin ax \cos ax\right) + C = \dfrac{1}{2}\left(x - \dfrac{1}{a}\sin ax \cos ax\right) + C$.

In Problems 29.3–29.16, find the indicated antiderivative.

29.3 $\int \sin x \cos^2 x \, dx$.

▮ Let $u = \cos x$, $du = -\sin x \, dx$. Then $\int \sin x \cos^2 dx = -\int u^2 \, du = -\frac{1}{3}u^3 + C = -\frac{1}{3}\cos^3 x + C$.

29.4 $\int \sin^4 x \cos^5 x \, dx$.

▮ Since the power of $\cos x$ is odd, let $u = \sin x$, $du = \cos x \, dx$. Then $\int \sin^4 x \cos^5 x \, dx = \int \sin^4 x(1 - \sin^2 x)^2 \cos x \, dx = \int u^4(1 - u^2)^2 \, du = \int u^4(1 - 2u^2 + u^4) \, du = \int (u^4 - 2u^6 + u^8) \, du = \frac{1}{5}u^5 - \frac{2}{7}u^7 + \frac{1}{9}u^9 + C = u^5(\frac{1}{5} - \frac{2}{7}u^2 + \frac{1}{9}u^4) + C = \sin^5 x(\frac{1}{5} - \frac{2}{7}\sin^2 x + \frac{1}{9}\sin^4 x) + C$.

29.5 $\int \cos^6 x \, dx$.

▮ $\int \cos^6 x \, dx = \int (\cos^2 x)^3 \, dx = \int \left(\dfrac{1 + \cos 2x}{2}\right)^3 dx = \dfrac{1}{8}\int (1 + 3\cos 2x + 3\cos^2 2x + \cos^3 2x) \, dx = \dfrac{1}{8}[x + \dfrac{3}{2}\sin 2x + 3\int \dfrac{1 + \cos 4x}{2} \, dx + \int (1 - \sin^2 2x)\cos 2x \, dx]$. Now, $\int \dfrac{1 + \cos 4x}{2} \, dx = \frac{1}{2}(x + \frac{1}{4}\sin 4x) = \frac{1}{2}(x + \frac{1}{4}\sin 2x \cos 2x)$. Also, in $\int (1 - \sin^2 2x)\cos 2x \, dx$, let $u = \sin 2x$, $du = 2\cos 2x \, dx$. So we get $\frac{1}{2}\int (1 - u^2) \, du = \frac{1}{2}(u - \frac{1}{3}u^3) = \frac{1}{6}u(3 - 2u^3) = \frac{1}{6}\sin 2x(3 - 2\sin^3 2x)$. Hence, the entire answer is $\frac{1}{8}[x + \frac{3}{2}\sin 2x + \frac{3}{2}(x + \frac{1}{4}\sin 2x \cos 2x) + \frac{1}{6}\sin 2x(3 - 2\sin^3 2x)] + C = \frac{1}{8}[\frac{5}{2}x + 2\sin 2x + \frac{3}{4}\sin 2x \cos 2x - \frac{1}{3}\sin^4 2x] + C$. [Can you show that this answer agrees with Problem 28.36?]

29.6 $\int \cos^4 x \sin^2 x \, dx$.

▮ $\int \cos^4 x \sin^2 x \, dx = \int \left(\dfrac{1 + \cos 2x}{2}\right)^2 \left(\dfrac{1 - \cos 2x}{2}\right) dx = \dfrac{1}{8}\int (1 + 2\cos 2x + \cos^2 2x)(1 - \cos 2x) \, dx = \dfrac{1}{8}\int (1 + \cos 2x - \cos^2 2x - \cos^3 2x) \, dx = \frac{1}{8}(x + \frac{1}{2}\sin 2x - \int \cos^2 2x \, dx - \int \cos^3 2x \, dx)$. Now, $\int \cos^2 2x \, dx = \frac{1}{2}(x + \frac{1}{2}\sin 2x \cos 2x)$ by Problem 29.1. Also, $\int \cos^3 2x \, dx = \int (1 - \sin^2 2x)\cos 2x \, dx = \int \cos 2x \, dx - \int \sin^2 2x \cos 2x \, dx = \frac{1}{2}\sin 2x - \frac{1}{6}\sin^3 2x$. Hence, we get $\frac{1}{8}[x + \frac{1}{2}\sin 2x - \frac{1}{2}(x + \frac{1}{2}\sin 2x \cos 2x) + \frac{1}{2}\sin 2x - \frac{1}{6}\sin^3 2x] + C = \frac{1}{8}[(x/2) + \sin 2x - \frac{1}{4}\sin 2x \cos 2x - \frac{1}{6}\sin^3 2x] + C$.

29.7 $\int \tan^2 \dfrac{x}{2} \, dx$.

▮ Let $x = 2u$, $dx = 2 \, du$. Then $\int \tan^2 \dfrac{x}{2} \, dx = 2\int \tan^2 u \, du = 2\int (\sec^2 u - 1) \, du = 2(\tan u - u) + C = 2\left(\tan \dfrac{x}{2} - \dfrac{x}{2}\right) + C = 2\tan \dfrac{x}{2} - x + C$.

29.8 $\int \tan^4 x \, dx$.

▮ $\int \tan^4 x \, dx = \int \tan^2 x(\sec^2 x - 1) \, dx = \int \tan^2 x \sec^2 x \, dx - \int \tan^2 x \, dx = \frac{1}{3}\tan^3 x - \int (\sec^2 x - 1) \, dx$

$= \frac{1}{3}\tan^3 x - \tan x + x + C$

29.9 $\int \sec^5 x \, dx$.

   ▌ Use Problem 28.39:

$$\int \sec^5 x \, dx = \frac{\tan x \sec^3 x}{4} + \frac{3}{4} \int \sec^3 x \, dx = \frac{\tan x \sec^3 x}{4} + \frac{3}{4}\left(\frac{\tan x \sec x}{2} + \frac{1}{3} \int \sec x \, dx \right)$$

$$= \frac{\tan x \sec^3 x}{4} + \frac{3}{8} \tan x \sec x + \frac{1}{4} \ln|\sec x + \tan x| + C$$

29.10 $\int \tan^2 x \sec^4 x \, dx$.

   ▌ Since the exponent of $\sec x$ is even, $\int \tan^2 x \sec^4 x \, dx = \int \tan^2 x \, (1 + \tan^2 x) \sec^2 x \, dx = \int (\tan^2 x \sec^2 x + \tan^4 x \sec^2 x) \, dx = \frac{1}{3} \tan^3 x + \frac{1}{5} \tan^5 x + C$.

29.11 $\int \tan^3 x \sec^3 x \, dx$.

   ▌ Since the exponent of $\tan x$ is odd, $\int \tan^3 x \sec^3 dx = \int (\sec^2 x - 1) \sec^2 x \sec x \tan x \, dx = \int (\sec^4 x \sec x \tan x - \sec^2 x \sec x \tan x) \, dx = \frac{1}{5} \sec^5 x - \frac{1}{3} \sec^3 x + C$.

29.12 $\int \tan^4 x \sec x \, dx$.

   ▌ $\int \tan^4 x \sec x \, dx = \int (\sec^2 x - 1)^2 \sec x \, dx = \int (\sec^4 x - 2 \sec^2 x + 1) \sec x \, dx = \int (\sec^5 x - 2 \sec^3 x + \sec x) \, dx$. By Problems 28.39 and 28.40, $\int \sec^5 x \, dx = \frac{\tan x \sec^3 x}{4} + \frac{3}{4} \int \sec^3 x \, dx$,   and $\int \sec^3 x \, dx = \frac{1}{2} (\tan x \sec x + \ln|\sec x + \tan x|)$.   Thus, we get $\frac{\tan x \sec^3 x}{4} + \frac{3}{4} \int \sec^3 x \, dx - 2 \int \sec^3 x \, dx + \ln|\sec x + \tan x| = \frac{\tan x \sec^3 x}{4} - \frac{5}{4} \int \sec^3 x \, dx + \ln|\sec x + \tan x| = \frac{\tan x \sec^3 x}{4} - \frac{5}{4}\left(\frac{1}{2} \tan x \sec x + \ln|\sec x + \tan x|\right) + \ln|\sec x + \tan x| + C = \frac{\tan x \sec^3 x}{4} - \frac{5}{8} \tan x \sec x + \frac{3}{8} \ln|\sec x + \tan x| + C$.

29.13 $\int \sin 2x \cos 2x \, dx$.

   ▌ $\int \sin 2x \cos 2x \, dx = \frac{1}{2} \int \sin 2x \cdot 2 \cos 2x \, dx = \frac{1}{2} \cdot \frac{1}{2} \sin^2 2x + C = \frac{1}{4} \sin^2 2x + C = \frac{1}{4}(2 \sin x \cos x)^2 + C$

$$= \sin^2 x \cos^2 x + C.$$

29.14 $\int \sin \pi x \cos 3\pi x \, dx$.

   ▌ Use the formula $\sin Ax \cos Bx = \frac{1}{2}[\sin (A + B)x + \sin (A - B)x]$.   Then $\int \sin \pi x \cos 3\pi x \, dx = \frac{1}{2} \int [\sin 4\pi x + \sin (-2\pi x)] \, dx = \frac{1}{2} \int (\sin 4\pi x - \sin 2\pi x) \, dx = \frac{1}{2}\left(- \frac{\cos 4\pi x}{4\pi} + \frac{\cos 2\pi x}{2\pi}\right) + C = \frac{1}{8\pi}(2 \cos 2\pi x - \cos 4\pi x) + C$.

29.15 $\int \sin 5x \sin 7x \, dx$.

   ▌ Recall $\sin Ax \sin Bx = \frac{1}{2}[\cos (A - B)x - \cos (A + B)x]$.   So $\int \sin 5x \sin 7x \, dx = \frac{1}{2} \int [\cos (-2x) - \cos 12x] \, dx = \frac{1}{2} \int (\cos 2x - \cos 12x) \, dx = \frac{1}{2}\left(\frac{\sin 2x}{2} - \frac{\sin 12x}{12}\right) + C = \frac{1}{24}(6 \sin 2x - \sin 12x) + C$.

29.16 $\int \cos 4x \cos 9x \, dx$.

   ▌ Recall $\cos Ax \cos Bx = \frac{1}{2}[\cos (A - B)x + \cos (A + B)x]$.   So $\int \cos 4x \cos 9x \, dx = \frac{1}{2} \int [\cos (-5x) + \cos 13x] \, dx = \frac{1}{2} \int (\cos 5x + \cos 13x) \, dx = \frac{1}{12}\left(\frac{\sin 5x}{5} + \frac{\sin 13x}{13}\right) + C$.

29.17   Calculate $\int_0^\pi \sin nx \sin kx \, dx$ when n and k are distinct positive integers.

   ▌ $\sin nx \sin kx = \frac{1}{2}[\cos (n - k)x - \cos (n + k)x]$. So $\int_0^\pi \sin nx \sin kx \, dx = \frac{1}{2} \int_0^\pi [\cos (n - k)x - \cos (n + k)x] \, dx = \frac{1}{2}\left[\frac{\sin (n - k)x}{n - k} - \frac{\sin (n + k)x}{n + k}\right]\Big|_0^\pi = \frac{1}{2}[(0 - 0) - (0 - 0)] = 0$.

29.18 Calculate $\int_0^\pi \sin^2 nx \, dx$ when n is a positive integer (the exceptional case in Problem 29.17).

❚ By Problem 29.2, $\int_0^\pi \sin^2 nx \, dx = \frac{1}{2}\left(x - \frac{\sin nx \cos nx}{n}\right)\Big]_0^\pi = \frac{1}{2}[(\pi - 0) - (0 - 0)] = \frac{\pi}{2}$.

In Problems 29.19–29.29, evaluate the given antiderivative.

29.19 $\int \frac{\sqrt{x^2 - 1}}{x} \, dx$.

❚ Since $\sqrt{x^2 - 1}$ is in the integrand, we let $x = \sec \theta$, $dx = \sec \theta \tan \theta \, d\theta$. Then (Fig. 29-1), $\sqrt{x^2 - 1} = \tan \theta$. Hence, $\int \frac{\sqrt{x^2 - 1}}{x} \, dx = \int \frac{\tan \theta}{\sec \theta} \sec \theta \tan \theta \, d\theta = \int \tan^2 \theta \, d\theta = \int (\sec^2 \theta - 1) \, d\theta = \tan \theta - \theta + C = \sqrt{x^2 - 1} - \sec^{-1} x + C$.

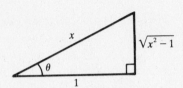
Fig. 29-1

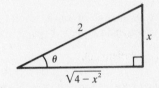
Fig. 29-2

29.20 $\int \frac{1}{\sqrt{4 - x^2}} \, dx$.

❚ Since $\sqrt{4 - x^2}$ is present, we let $x = 2 \sin \theta$, $dx = 2 \cos \theta \, d\theta$. Then (Fig. 29-2), $\sqrt{4 - x^2} = 2 \cos \theta$. So $\int \frac{1}{\sqrt{4 - x^2}} \, dx = \int \frac{4 \sin^2 \theta}{2 \cos \theta} 2 \cos \theta \, d\theta = 4 \int \sin^2 \theta \, d\theta = 4 \int \frac{1 - \cos 2\theta}{2} \, d\theta = 2\left(\theta - \frac{\sin 2\theta}{2}\right) + C = 2(\theta - \sin \theta \cos \theta) + C = 2\left(\sin^{-1} \frac{x}{2} - \frac{x}{2} \cdot \frac{\sqrt{4 - x^2}}{2}\right) + C = 2 \sin^{-1} \frac{x}{2} - \frac{x\sqrt{4 - x^2}}{2} + C$.

29.21 $\int \frac{\sqrt{1 + x^2}}{x} \, dx$.

❚ Since $\sqrt{1 + x^2}$ is present, we let $x = \tan \theta$, $dx = \sec^2 \theta \, d\theta$. Then (Fig. 29-3), $\sqrt{1 + x^2} = \sec \theta$. So $\int \frac{\sqrt{1 + x^2}}{x} \, dx = \int \frac{\sec \theta}{\tan \theta} \sec^2 \theta \, d\theta = \int \frac{\sec \theta}{\tan \theta} (1 + \tan^2 \theta) \, d\theta = \int (\csc \theta + \sec \theta \tan \theta) \, d\theta = \ln |\csc \theta - \cot \theta| + \sec \theta + C = \ln \left| \frac{\sqrt{1 + x^2} - 1}{x} \right| + \sqrt{1 + x^2} + C$.

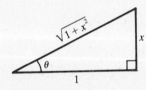
Fig. 29-3

29.22 $\int \frac{x}{\sqrt{2 - x^2}} \, dx$.

❚ A trigonometric substitution is not required here. $\int \frac{x}{\sqrt{2 - x^2}} \, dx = -\frac{1}{2} \int (2 - x^2)^{-1/2}(-2x) \, dx = -\frac{1}{2} \cdot 2(2 - x^2)^{1/2} + C = -\sqrt{2 - x^2} + C$.

29.23 $\int \frac{dx}{x^2 \sqrt{x^2 - 9}}$.

❚ Since $\sqrt{x^2 - 9}$ is present, let $x = 3 \sec \theta$, $dx = 3 \sec \theta \tan \theta \, d\theta$, and (Fig. 29-4) $\sqrt{x^2 - 9} = 3 \tan \theta$. Then $\int \frac{dx}{x^2 \sqrt{x^2 - 9}} = \int \frac{3 \sec \theta \tan \theta \, d\theta}{9 \sec^2 \theta \cdot 3 \tan \theta} = \frac{1}{9} \int \cos \theta \, d\theta = \frac{1}{9} \sin \theta + C = \frac{1}{9} \frac{\sqrt{x^2 - 9}}{x} + C$.

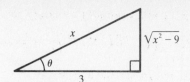

Fig. 29-4

29.24 $\int \dfrac{dx}{(4 - x^2)^{3/2}}$.

▌ Since $\sqrt{4 - x^2}$ is present, let $x = 2 \sin \theta$, $dx = 2 \cos \theta \, d\theta$, $\sqrt{4 - x^2} = 2 \cos \theta$. Then $\int \dfrac{dx}{(4 - x^2)^{3/2}} = \int \dfrac{2 \cos \theta \, d\theta}{8 \cos^3 \theta} = \dfrac{1}{4} \int \sec^2 \theta \, d\theta = \dfrac{1}{4} \tan \theta + C = \dfrac{1}{4} \dfrac{x}{\sqrt{4 - x^2}} + C$.

29.25 $\int \dfrac{dx}{(x^2 + 9)^2}$.

▌ Since $x^2 + 9$ appears, let $x = 3 \tan \theta$, $dx = 3 \sec^2 \theta \, d\theta$, and (Fig. 29-5) $x^2 + 9 = 9 \sec^2 \theta$. Then, using Problem 29.1, $\int \dfrac{dx}{(x^2 + 9)^2} = \int \dfrac{3 \sec^2 \theta \, d\theta}{(9 \sec^2 \theta)^2} = \dfrac{1}{27} \int \dfrac{d\theta}{\sec^2 \theta} = \dfrac{1}{27} \int \cos^2 \theta \, d\theta = \dfrac{1}{27} \cdot \dfrac{1}{2} (\theta + \sin \theta \cos \theta) + C = \dfrac{1}{54} \left(\tan^{-1} \dfrac{x}{3} + \dfrac{x}{\sqrt{x^2 + 9}} \dfrac{3}{\sqrt{x^2 + 9}} \right) + C = \dfrac{1}{54} \left(\tan^{-1} \dfrac{x}{3} + \dfrac{3x}{x^2 + 9} \right) + C$.

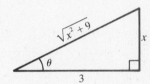

Fig. 29-5

29.26 $\int \dfrac{dx}{x \sqrt{16 - 9x^2}}$.

▌ Since $\sqrt{16 - 9x^2}$ is present, let $x = \frac{4}{3} \sin \theta$, $dx = \frac{4}{3} \cos \theta \, d\theta$, and $\sqrt{16 - 9x^2} = \sqrt{16 - 16 \sin^2 \theta} = 4 \cos \theta$. Then $\int \dfrac{dx}{x \sqrt{16 - 9x^2}} = \int \dfrac{\frac{4}{3} \cos \theta \, d\theta}{\frac{4}{3} \sin \theta \cdot 4 \cos \theta} = \dfrac{1}{4} \int \dfrac{d\theta}{\sin \theta} = \dfrac{1}{4} \int \csc \theta \, d\theta = \dfrac{1}{4} \ln |\csc \theta - \cot \theta| + C = \dfrac{1}{4} \ln \left| \dfrac{4}{3x} - \dfrac{\sqrt{16 - 9x^2}}{3x} \right| + C = \dfrac{1}{4} \ln \left| \dfrac{4 - \sqrt{16 - 9x^2}}{3x} \right| + C$.

29.27 $\int x^2 \sqrt{1 - x^2} \, dx$.

▌ Let $x = \sin \theta$, $dx = \cos \theta \, d\theta$, $\sqrt{1 - x^2} = \cos \theta$. Then, using Problem 29.2, $\int x^2 \sqrt{1 - x^2} \, dx = \int \sin^2 \theta \cos \theta \cos \theta \, d\theta = \int \sin^2 \theta \cos^2 \theta \, d\theta = \int \left(\dfrac{\sin 2\theta}{2} \right)^2 d\theta = \dfrac{1}{4} \int \sin^2 2\theta \, d\theta = \frac{1}{4} \cdot \frac{1}{2} (\theta - \frac{1}{2} \sin 2\theta \cos 2\theta) + C = \frac{1}{8} [\theta - \sin \theta \cos \theta (1 - 2 \sin^2 \theta)] + C = \frac{1}{8} [\sin^{-1} x - x \sqrt{1 - x^2} (1 - 2x^2)] + C$.

29.28 $\int e^{3x} \sqrt{1 - e^{2x}} \, dx$.

▌ Let $x = \ln v$, and use Problem 29.27: $\int e^{3x} \sqrt{1 - e^{2x}} \, dx = \int v^2 \sqrt{1 - v^2} \, dv = \frac{1}{8} [\sin^{-1} v - v \sqrt{1 - v^2} (1 - 2v^2)] + C = \frac{1}{8} [\sin^{-1} e^x - e^x \sqrt{1 - e^{2x}} (1 - 2e^{2x})] + C$.

29.29 $\int \dfrac{dx}{(x^2 - 6x + 13)^2}$.

▌ By completing the square, $x^2 - 6x + 13 = (x - 3)^2 + 4$. Let $x - 3 = 2 \tan \theta$, $dx = 2 \sec^2 \theta \, d\theta$, $x^2 - 6x + 13 = 4 \sec^2 \theta$ (see Fig. 29-6). Then $\int \dfrac{dx}{(x^2 - 6x + 13)^2} = \int \dfrac{2 \sec^2 \theta \, d\theta}{(4 \sec^2 \theta)^2} = \dfrac{1}{8} \int \dfrac{d\theta}{\sec^2 \theta} = \dfrac{1}{8} \int \cos^2 \theta \, d\theta$. By Problem 29.1, we have $\dfrac{1}{8} \cdot \dfrac{1}{2} (\theta + \sin \theta \cos \theta) + C = \dfrac{1}{16} \left(\tan^{-1} \dfrac{x - 3}{2} + \dfrac{x - 3}{\sqrt{x^2 - 6x + 13}} \dfrac{2}{\sqrt{x^2 - 6x + 13}} \right) + C = \dfrac{1}{16} \left[\tan^{-1} \dfrac{x - 3}{2} + \dfrac{2(x - 3)}{x^2 - 6x + 13} \right] + C$.

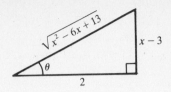

Fig. 29-6

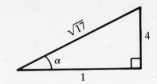

Fig. 29-7

29.30 Find the arc length of the parabola $y = x^2$ from $(0, 0)$ to $(2, 4)$.

❙ $y' = 2x$. So $L = \int_0^2 \sqrt{1 + (y')^2}\,dx = \int_0^2 \sqrt{1 + 4x^2}\,dx$. Let $x = \frac{1}{2}\tan\theta$, $dx = \frac{1}{2}\sec^2\theta\,d\theta$, $\sqrt{1 + 4x^2} = \sqrt{1 + \tan^2\theta} = \sec\theta$. As x increases from 0 to 2, θ increases from 0 to $\alpha = \tan^{-1} 4$ (Fig. 29-7). So, by Problem 28.40, $\int_0^2 \sqrt{1 + 4x^2}\,dx = \frac{1}{2}\int_0^\alpha \sec^3\theta\,d\theta = \frac{1}{4}(\tan\theta\sec\theta + \ln|\sec\theta + \tan\theta|)]_0^\alpha = \frac{1}{4}\{[4\sqrt{17} + \ln(\sqrt{17} + 4)] - (0 + 0)\} = \sqrt{17} + \frac{1}{4}\ln(\sqrt{17} + 4)$.

29.31 Find the arc length of the curve $y = \ln x$ from $(1, 0)$ to $(e, 1)$.

❙ $y' = \frac{1}{x}$, $1 + (y')^2 = 1 + \frac{1}{x^2} = \frac{x^2 + 1}{x^2}$. Then, using Problem 29.21, $L = \int_1^e \sqrt{1 + (y')^2}\,dx = \int_1^e \frac{\sqrt{x^2 + 1}}{x}\,dx$

$= \left(\ln\left|\frac{\sqrt{1 + x^2} - 1}{x}\right| + \sqrt{1 + x^2}\right)\Big]_1^e = \left(\ln\left|\frac{\sqrt{1 + e^2} - 1}{e}\right| + \sqrt{1 + e^2}\right) - [\ln(\sqrt{2} - 1) + \sqrt{2}] = \sqrt{e^2 + 1} - (1 + \sqrt{2}) +$

$\ln\frac{\sqrt{e^2 + 1} - 1}{\sqrt{2} - 1}$.

29.32 Find the arc length of the curve $y = e^x$ from $(0, 1)$ to $(1, e)$.

❙ This has the same answer as Problem 29.31, since the two arcs are mirror images of each other in the line $y = x$.

29.33 Find the arc length of the curve $y = \ln\cos x$ from $(0, 0)$ to $(\pi/3, -\ln 2)$.

❙ $y' = \frac{1}{\cos x}(-\sin x) = -\tan x$, $1 + (y')^2 = \sec^2 x$. So $L = \int_0^{\pi/3} \sec x\,dx = \ln|\sec x + \tan x|]_0^{\pi/3} = \ln|2 + \sqrt{3}| - \ln 1 = \ln(2 + \sqrt{3})$.

29.34 Find $\int \sqrt{1 - \cos x}\,dx$.

❙ From the identity $\sin^2 x = \frac{1 - \cos 2x}{2}$, we obtain $\sin^2\frac{x}{2} = \frac{1 - \cos x}{2}$. Thus, $\int \sqrt{1 - \cos x}\,dx = \sqrt{2}\int \sin\frac{x}{2}\,dx = \sqrt{2}\cdot(-2)\cos\frac{x}{2} + C = -2\sqrt{2}\cos\frac{x}{2} + C$.

29.35 Find $\int (1 + \cos ax)^{3/2}\,dx$.

❙ From the identity $\cos^2 x = \frac{1 + \cos 2x}{2}$, we get $\cos^2\frac{ax}{2} + \frac{1 + \cos ax}{2}$. Hence, $\int (1 + \cos ax)^{3/2}\,dx = 2\sqrt{2}\int \cos^3\frac{ax}{2}\,dx = 2\sqrt{2}\left(\int \cos\frac{ax}{2}\,dx - \int \sin^2\frac{ax}{2}\cos\frac{ax}{2}\,dx\right) = 2\sqrt{2}\left(\frac{2}{a}\sin\frac{ax}{2} - \frac{2}{a}\cdot\frac{1}{3}\sin^3\frac{ax}{2} + C\right) = \frac{4\sqrt{2}}{a}\sin\frac{ax}{2}\left(1 - \frac{1}{3}\sin^2\frac{ax}{2}\right) + C$.

29.36 Find $\int \frac{1}{\sqrt{1 - \sin 2x}}\,dx$.

❙ From the identity in the solution of Problem 29.34, $\sqrt{1 - \sin 2x} = \sqrt{1 - \cos(\pi/2 - 2x)} = \sqrt{2}\sin(\pi/4 - x)$. Hence, $\int \frac{dx}{\sqrt{1 - \sin 2x}} = \frac{\sqrt{2}}{2}\int \frac{dx}{\sin(\pi/4 - x)} = \frac{\sqrt{2}}{2}\int \csc(\pi/4 - x)\,dx = \frac{\sqrt{2}}{2}\ln|\csc(\pi/4 - x) - \cot(\pi/4 - x)| + C$.

29.37 Find $\int \dfrac{\cos^3 x}{1 - \sin x}\, dx$.

$\blacksquare\quad \dfrac{\cos^3 x}{1 - \sin x} = \dfrac{\cos^3 x}{1 - \sin x} \cdot \dfrac{1 + \sin x}{1 + \sin x} = \dfrac{\cos^3 x(1 + \sin x)}{\cos^2 x} = \cos x(1 + \sin x) = \cos x + \cos x \sin x.$ Thus,

$\int \dfrac{\cos^3 x}{1 - \sin x}\, dx = \int \cos x\, dx + \int \cos x \sin x\, dx = \sin x + \tfrac{1}{2}\sin^2 x + C.$

29.38 Find $\int \dfrac{x^2\, dx}{(4 - x^2)^{5/2}}$.

$\blacksquare\quad$ Refer to Fig. 29-8. Let $x = 2\sin\theta$, $dx = 2\cos\theta\, d\theta$. So $\int \dfrac{x^2\, dx}{(4 - x^2)^{5/2}} = \int \dfrac{4\sin^2\theta \cdot 2\cos\theta\, d\theta}{32\cos^5\theta} =$
$\dfrac{1}{4}\int \dfrac{\sin^2\theta}{\cos^4\theta}\, d\theta = \dfrac{1}{4}\int \tan^2\theta \sec^2\theta\, d\theta = \dfrac{1}{4} \cdot \dfrac{1}{3}\tan^3\theta + C = \dfrac{1}{12} \dfrac{x^3}{(4 - x^2)^{3/2}} + C.$

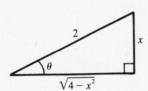

Fig. 29-8

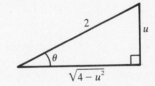

Fig. 29-9

29.39 Find $\int \dfrac{dx}{(4x - x^2)^{3/2}}$.

$\blacksquare\quad$ By completing the square, $x^2 - 4x = (x - 2)^2 - 4$. So $4x - x^2 = 4 - (x - 2)^2$. Let $u = x - 2$,
$du = dx$. Then $\int \dfrac{dx}{(4x - x^2)^{3/2}} = \int \dfrac{du}{(4 - u^2)^{3/2}}$. Let $u = 2\sin\theta$ (Fig. 29-9), $du = 2\cos\theta\, d\theta$. Then
$\int \dfrac{du}{(4 - u^2)^{3/2}} = \int \dfrac{2\cos\theta\, d\theta}{8\cos^3\theta} = \dfrac{1}{4}\int \sec^2\theta\, d\theta = \dfrac{1}{4}\tan\theta + C = \dfrac{1}{4} \dfrac{u}{\sqrt{4 - u^2}} + C = \dfrac{1}{4} \dfrac{x - 2}{\sqrt{4x - x^2}} + C.$

29.40 Find $\int x \sin^{-1} x\, dx$.

$\blacksquare\quad$ First, use integration by parts. Let $u = \sin^{-1} x$, $dv = x\, dx$, $du = \dfrac{1}{\sqrt{1 - x^2}}\, dx$, $v = \dfrac{1}{2}x^2$. Then
$\int x \sin^{-1} x\, dx = \dfrac{1}{2}x^2 \sin^{-1} x - \dfrac{1}{2}\int \dfrac{x^2}{\sqrt{1 - x^2}}\, dx.$ For the latter integral, use a trigonometric substitution. Let
$x = \sin\theta$, $dx = \cos\theta\, d\theta$. Then $\int \dfrac{x^2}{\sqrt{1 - x^2}}\, dx = \int \dfrac{\sin^2\theta \cos\theta\, d\theta}{\cos\theta} = \int \sin^2\theta\, d\theta = \tfrac{1}{2}(\theta - \sin\theta \cos\theta)$ (by
Problem 29.2). Hence, the answer is $\tfrac{1}{2}x^2 \sin^{-1} x - \tfrac{1}{4}(\sin^{-1} x - x\sqrt{1 - x^2}) + C = \tfrac{1}{4}[(2x^2 - 1)\sin^{-1} x +$
$x\sqrt{1 - x^2}] + C.$

29.41 Find $\int \dfrac{\sqrt{a^2 - x^2}}{x}\, dx$.

$\blacksquare\quad$ Refer to Fig. 29-10. Let $x = a\sin\theta$, $dx = a\cos\theta\, d\theta$. Then $\int \dfrac{\sqrt{a^2 - x^2}}{x}\, dx = \int \dfrac{a\cos\theta \cdot a\cos\theta\, d\theta}{a\sin\theta} =$
$a\int \dfrac{\cos^2\theta}{\sin\theta}\, d\theta = a\int \dfrac{1 - \sin^2\theta}{\sin\theta}\, d\theta = a(\int \csc\theta\, d\theta - \int \sin\theta\, d\theta) = a(\ln|\csc\theta - \cot\theta| + \cos\theta) + C =$
$a\left(\ln\left|\dfrac{a}{x} - \dfrac{\sqrt{a^2 - x^2}}{x}\right| + \dfrac{\sqrt{a^2 - x^2}}{a}\right) + C = a\ln\left|\dfrac{a - \sqrt{a^2 - x^2}}{x}\right| + \sqrt{a^2 - x^2} + C.$

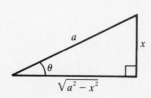

Fig. 29-10

29.42 Assume that an object A, starting at $(a, 0)$, is connected by a string of constant length a to an object B, starting at $(0, 0)$. As B moves up the y-axis, A traces out a curve called a *tractrix*. Find its equation.

▌ Let A be at (x, y). The string must be tangent to the curve. From Fig. 29-11, the slope of the tangent line is $y' = -\dfrac{\sqrt{a^2 - x^2}}{x}$. So, by Problem 29.41, $y = -\int \dfrac{\sqrt{a^2 - x^2}}{x}\, dx = -a \ln \left| \dfrac{a - \sqrt{a^2 - x^2}}{x} \right| - \sqrt{a^2 - x^2} + C$. Since $y = 0$ when $x = a$, $C = 0$.

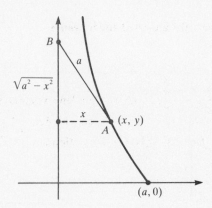

Fig. 29-11

29.43 In a disk of radius a, a chord b units from the center cuts off a region of the disk called a *segment*. Find a formula for the area of the segment.

▌ From a diagram, the area $A = 2 \int_b^a \sqrt{a^2 - x^2}\, dx$. Let $x = a \sin \theta$, $dx = a \cos \theta\, d\theta$. Then, $\int \sqrt{a^2 - x^2}\, dx = \int a \cos \theta \cdot a \cos \theta\, d\theta = a^2 \int \cos^2 \theta\, d\theta$. By Problem 29.1, this is $a^2 \cdot \dfrac{1}{2}(\theta + \sin \theta \cos \theta) = \dfrac{1}{2} a^2 \left(\sin^{-1} \dfrac{x}{a} + \dfrac{x}{a} \dfrac{\sqrt{a^2 - x^2}}{a} \right)$. Hence, $A = a^2 \left(\sin^{-1} \dfrac{x}{a} + \dfrac{x}{a} \dfrac{\sqrt{a^2 - x^2}}{a} \right) \Big]_b^a = a^2 \left[\left(\dfrac{\pi}{2} + 0 \right) - \left(\sin^{-1} \dfrac{b}{a} + \dfrac{b\sqrt{a^2 - b^2}}{a^2} \right) \right] = a^2 \left(\dfrac{\pi}{2} - \sin^{-1} \dfrac{b}{a} \right) - b\sqrt{a^2 - b^2}$.

29.44 The region under $y = 1/(x^2 + 1)$, above the x-axis, between $x = 0$ and $x = 1$, is revolved about the x-axis. Find the volume of the resulting solid.

▌ By the disk formula, $V = \pi \int_0^1 y^2\, dx = \pi \int_0^1 \dfrac{dx}{(x^2 + 1)^2}$. Let $x = \tan \theta$, $dx = \sec^2 \theta\, d\theta$. Then $V = \pi \int_0^{\pi/4} \dfrac{\sec^2 \theta\, d\theta}{\sec^4 \theta} = \pi \int_0^{\pi/4} \cos^2 \theta\, d\theta = \dfrac{\pi}{2} (\theta + \sin \theta \cos \theta)]_0^{\pi/4} = \dfrac{\pi}{2} \left[\left(\dfrac{\pi}{4} + \dfrac{1}{2} \right) - (0 + 0) \right] = \dfrac{\pi(\pi + 2)}{8}$.

29.45 Find $\displaystyle\int \dfrac{dx}{1 - \sin x}$.

▌ Let $z = \tan \dfrac{x}{2}$, $x = 2 \tan^{-1} z$, $dx = \dfrac{2}{1 + z^2}\, dz$. Then $\sin x = 2 \sin \dfrac{x}{2} \cos \dfrac{x}{2} = 2 \dfrac{\tan (x/2)}{\sec^2 (x/2)} = \dfrac{2z}{1 + z^2}$.

Thus,

$$\int \frac{dx}{1 - \sin x} = \int \frac{1}{1 - 2z/(1 + z^2)} \cdot \frac{2}{1 + z^2}\, dz = 2 \int \frac{dz}{1 + z^2 - 2z} = 2 \int \frac{dz}{(z - 1)^2} = -\frac{2}{z - 1} + C = \frac{2}{1 - \tan (x/2)} + C$$

[More generally, the above change of variable, $z = \tan (x/2)$, converts the indefinite integral of any rational function of $\sin x$ and $\cos x$ into the indefinite integral of a rational function of z.]

CHAPTER 30
Integration of Rational Functions: The Method of Partial Fractions

In Problems 30.1–30.21, evaluate the indicated antiderivative.

30.1 $\int \dfrac{dx}{x^2 - 9}$.

▌ $\dfrac{1}{x^2 - 9} = \dfrac{1}{(x-3)(x+3)} = \dfrac{A}{x-3} + \dfrac{B}{x+3}$. Clear the denominators by multiplying both sides by $(x-3)(x+3)$: $1 = A(x+3) + B(x-3)$. Let $x = 3$. Then $1 = 6A$, $A = \frac{1}{6}$. Let $x = -3$. Then $1 = -6B$, $B = -\frac{1}{6}$. So $1/(x^2 - 9) = \frac{1}{6}[1/(x-3)] - \frac{1}{6}[1/(x+3)]$. Hence, $\int \dfrac{dx}{x^2 - 9} = \frac{1}{6} \ln |x-3| - \frac{1}{6} \ln |x+3| + C = \frac{1}{6} \ln |(x-3)/(x+3)| + C$.

30.2 $\int \dfrac{x \, dx}{(x+2)(x+3)}$.

▌ $\dfrac{x}{(x+2)(x+3)} = \dfrac{A}{x+2} + \dfrac{B}{x+3}$. Then $x = A(x+3) + B(x+2)$. Let $x = -3$. Then $-3 = -B$, $B = 3$. Let $x = -2$. Then $-2 = A$. So $\dfrac{x}{(x+2)(x+3)} = -\dfrac{2}{x+2} + \dfrac{3}{x+3}$. Hence, $\int \dfrac{x \, dx}{(x+2)(x+3)} = -2 \ln |x+2| + 3 \ln |x+3| + C = \ln |(x+3)^3/(x+2)^2| + C$.

30.3 $\int \dfrac{x^4 - 4x^2 + x + 1}{x^2 - 4} \, dx$.

▌ Since the degree of the numerator is at least as great as that of the denominator, carry out the long division, obtaining $\dfrac{x^4 - 4x^2 + x + 1}{x^2 - 4} = x^2 + \dfrac{x+1}{x^2 - 4}$. Thus, $\int \dfrac{x^4 - 4x^2 + x + 1}{x^2 - 4} \, dx = \frac{1}{3}x^3 + \int \dfrac{x+1}{x^2 - 4} \, dx$. But $\dfrac{x+1}{x^2 - 4} = \dfrac{x+1}{(x+2)(x-2)} = \dfrac{A}{x+2} + \dfrac{B}{x-2}$. Then $x + 1 = A(x-2) + B(x+2)$. Let $x = 2$. Then $3 = 4B$, $B = \frac{3}{4}$. Let $x = -2$. Then $-1 = -4A$, $A = \frac{1}{4}$. Thus, $\dfrac{x+1}{(x+2)(x-2)} = \dfrac{1}{4} \dfrac{1}{x+2} + \dfrac{3}{4} \dfrac{1}{x-2}$, and $\int \dfrac{x+1}{(x+2)(x-2)} \, dx = \frac{1}{4} \ln |x+2| + \frac{3}{4} \ln |x-2| = \frac{1}{4} \ln |(x+2)(x-2)^3|$. Hence, the complete answer is $\frac{1}{3}x^3 + \frac{1}{4} \ln |(x+2)(x-2)^3| + C$.

30.4 $\int \dfrac{2x^2 + 1}{(x-1)(x-2)(x-3)} \, dx$.

▌ $\dfrac{2x^2 + 1}{(x-1)(x-2)(x-3)} = \dfrac{A}{x-1} + \dfrac{B}{x-2} + \dfrac{C}{x-3}$. Then $2x^2 + 1 = A(x-2)(x-3) + B(x-1)(x-3) + C(x-1)(x-2)$. Let $x = 1$. Then $3 = 2A$, $A = \frac{3}{2}$. Let $x = 2$. Then $9 = -B$, $B = -9$. Let $x = 3$. Then $19 = 2C$, $C = \frac{19}{2}$. Thus, $\dfrac{2x^2 + 1}{(x-1)(x-2)(x+3)} = \dfrac{3}{2} \dfrac{1}{x-1} - \dfrac{9}{x-2} + \dfrac{19}{2} \dfrac{1}{x-3}$. Hence, $\int \dfrac{2x^2 + 1}{(x-1)(x-2)(x-3)} \, dx = \frac{3}{2} \ln |x-1| - 9 \ln |x-2| + \frac{19}{2} \ln |x-3| + C_1 = \frac{1}{2} \ln \left| \dfrac{(x-1)^3(x-3)^{19}}{(x-2)^{18}} \right| + C_1$.

30.5 $\int \dfrac{x^2 - 4}{x^3 - 3x^2 - x + 3} \, dx$.

▌ We must factor the denominator. Clearly $x = 1$ is a root. Dividing the denominator by $x - 1$ we obtain $x^2 - 2x - 3 = (x-3)(x+1)$. Hence, the denominator is $(x-1)(x-3)(x+1)$. $\dfrac{x^2 - 4}{(x-1)(x-3)(x+1)} = \dfrac{A}{x-1} + \dfrac{B}{x-3} + \dfrac{C}{x+1}$. Then $x^2 - 4 = A(x-3)(x+1) + B(x-1)(x+1) + C(x-1)(x-3)$. Let $x = 1$. Then $-3 = -4A$, $A = \frac{3}{4}$. Let $x = 3$. Then $5 = 8B$, $B = \frac{5}{8}$. Let $x = -1$.

245

Then $\quad -3 = 8C, \quad\quad C = -\frac{3}{8}.$ $\quad$ Thus, $\quad \dfrac{x^2 - 4}{(x - 1)(x - 3)(x + 1)} = \dfrac{3}{4}\dfrac{1}{x - 1} + \dfrac{5}{8}\dfrac{1}{x - 3} - \dfrac{3}{8}\dfrac{1}{x + 1},$ $\quad$ and $\displaystyle\int \dfrac{x^2 - 4}{(x - 1)(x - 3)(x + 1)}\,dx = \frac{3}{4}\ln|x - 1| + \frac{5}{8}\ln|x - 3| - \frac{3}{8}\ln|x + 1| + C_1 = \frac{1}{8}(6\ln|x - 1| + 5\ln|x - 3| - 3\ln|x + 1|) + C_1 = \dfrac{1}{8}\ln\left|\dfrac{(x - 1)^6(x - 3)^5}{(x + 1)^3}\right| + C_1.$

30.6 $\displaystyle\int \dfrac{x^3 + 1}{x(x + 3)(x + 2)(x - 1)}\,dx.$

▌ $\dfrac{x^3 + 1}{x(x + 3)(x + 2)(x - 1)} = \dfrac{A}{x} + \dfrac{B}{x + 3} + \dfrac{C}{x + 2} + \dfrac{D}{x - 1}.$ $\quad$ So $\quad x^3 + 1 = A(x + 3)(x + 2)(x - 1) + Bx(x + 2)(x - 1) + Cx(x + 3)(x - 1) + Dx(x + 3)(x + 2).$ $\quad$ Let $\quad x = 0.$ $\quad$ Then $\quad 1 = -6A, \quad A = -\frac{1}{6}.$ $\quad$ Let $\quad x = -3.$ $\quad$ Then $-26 = -12B, \quad B = \frac{13}{6}.$ $\quad$ Let $\quad x = -2.$ $\quad$ Then $\quad -7 = 6C, \quad C = -\frac{7}{6}.$ $\quad$ Let $\quad x = 1.$ $\quad$ Then $\quad 2 = 12D, \quad D = \frac{1}{6}.$ Thus, $\quad \dfrac{x^3 + 1}{x(x + 3)(x + 2)(x - 1)} = -\dfrac{1}{6}\dfrac{1}{x} + \dfrac{13}{6}\dfrac{1}{x + 3} - \dfrac{7}{6}\dfrac{1}{x + 2} + \dfrac{1}{6}\dfrac{1}{x - 1},$ $\quad$ and $\displaystyle\int \dfrac{x^3 + 1}{x(x + 3)(x + 2)(x - 1)}\,dx = -\frac{1}{6}\ln|x| + \frac{13}{6}\ln|x + 3| - \frac{7}{6}\ln|x + 2| + \frac{1}{6}\ln|x - 1| + C_1 = \frac{1}{6}(-\ln|x| + 13\ln|x + 3| - 7\ln|x + 2| + \ln|x - 1|) + C_1 = \frac{1}{6}\ln\left|\dfrac{(x + 3)^{13}(x - 1)}{x(x + 2)^7}\right| + C_1.$

30.7 $\displaystyle\int \dfrac{x\,dx}{x^4 - 13x^2 + 36}.$

▌ $x^4 - 13x^2 + 36 = (x^2 - 9)(x^2 - 4) = (x - 3)(x + 3)(x + 2)(x - 2).$ $\quad$ Let $\quad \dfrac{x}{(x - 3)(x + 3)(x + 2)(x - 2)} = \dfrac{A}{x - 3} + \dfrac{B}{x + 3} + \dfrac{C}{x + 2} + \dfrac{D}{x - 2}.$ $\quad$ Then $\quad x = A(x + 3)(x + 2)(x - 2) + B(x - 3)(x + 2)(x - 2) + C(x - 3)(x + 3)(x - 2) + D(x - 3)(x + 3)(x + 2).$ $\quad$ Let $\quad x = 3.$ $\quad$ Then $\quad 3 = 30A, \quad A = \frac{1}{10}.$ $\quad$ Let $\quad x = -3.$ $\quad$ Then $-3 = -30B, \quad B = \frac{1}{10}.$ $\quad$ Let $\quad x = -2.$ $\quad$ Then $\quad -2 = 20C, \quad C = -\frac{1}{10}.$ $\quad$ Let $\quad x = 2.$ $\quad$ Then $\quad 2 = -20D, \quad D = -\frac{1}{10}.$ So $\quad \dfrac{x}{x^4 - 13x^2 + 36} = \dfrac{1}{10}\dfrac{1}{x - 3} + \dfrac{1}{10}\dfrac{1}{x + 3} - \dfrac{1}{10}\dfrac{1}{x + 2} - \dfrac{1}{10}\dfrac{1}{x - 2},$ $\quad$ and $\quad\displaystyle\int \dfrac{x}{x^4 - 13x^2 + 36}\,dx = \frac{1}{10}\ln|x - 3| + \frac{1}{10}\ln|x + 3| - \frac{1}{10}\ln|x + 2| - \frac{1}{10}\ln|x - 2| + C_1 = \frac{1}{10}(\ln|x - 3| + \ln|x + 3| - \ln|x + 2| - \ln|x - 2|) + C_1 = \dfrac{1}{10}\ln\left|\dfrac{(x - 3)(x + 3)}{(x + 2)(x - 2)}\right| + C_1.$

30.8 $\displaystyle\int \dfrac{x - 5}{x^2(x + 1)}\,dx.$

▌ $\dfrac{x - 5}{x^2(x + 1)} = \dfrac{A}{x} + \dfrac{B}{x^2} + \dfrac{C}{x + 1}.$ $\quad$ Multiply both sides by $x^2(x + 1)$, obtaining $\quad x - 5 = Ax(x + 1) + B(x + 1) + Cx^2.$ $\quad$ Let $\quad x = 0.$ $\quad$ Then $\quad -5 = B.$ $\quad$ Let $\quad x = -1.$ $\quad$ Then $\quad -6 = C.$ $\quad$ To find A, compare coefficients of x^2 on both sides of the equation: $\quad 0 = A + C, \quad A = -C = 6.$ $\quad$ Thus, $\quad \dfrac{x - 5}{x^2(x + 1)} = \dfrac{6}{x} - \dfrac{5}{x^2} - \dfrac{6}{x + 1},$ $\quad$ and $\displaystyle\int \dfrac{x - 5}{x^2(x + 1)}\,dx = 6\ln|x| + \frac{5}{x} - 6\ln|x + 1| + C_1 = 6\ln\left|\dfrac{x}{x + 1}\right| + \dfrac{5}{x} + C_1.$

30.9 $\displaystyle\int \dfrac{2x\,dx}{(x - 2)^2(x + 2)}.$

▌ $\dfrac{2x}{(x - 2)^2(x + 2)} = \dfrac{A}{x - 2} + \dfrac{B}{(x - 2)^2} + \dfrac{C}{x + 2}.$ $\quad$ Then $\quad 2x = A(x - 2)(x + 2) + B(x + 2) + C(x - 2)^2.$ $\quad$ Let $\quad x = 2.$ $\quad$ Then $\quad 4 = 4B, \quad B = 1.$ $\quad$ Let $\quad x = -2.$ $\quad$ Then $\quad -4 = 16C, \quad C = -\frac{1}{4}.$ $\quad$ To find A, equate coefficients of x^2: $0 = A + C, \quad A = -C = \frac{1}{4}.$ $\quad$ Thus, $\quad \dfrac{2x}{(x - 2)^2(x + 2)} = \dfrac{1}{4}\dfrac{1}{x - 2} + \dfrac{1}{(x - 2)^2} - \dfrac{1}{4}\dfrac{1}{x + 2},$ $\quad$ and $\displaystyle\int \dfrac{2x\,dx}{(x - 2)^2(x + 2)} = \frac{1}{4}\ln|x - 2| - \dfrac{1}{x - 2} - \frac{1}{4}\ln|x + 2| + C_1 = \dfrac{1}{4}\ln\left|\dfrac{x - 2}{x + 2}\right| - \dfrac{1}{x - 2} + C_1.$

30.10 $\displaystyle\int \dfrac{x + 4}{x^3 + 6x^2 + 9x}\,dx.$

▮ $\dfrac{x+4}{x^3+6x^2+9x} = \dfrac{x+4}{x(x+3)^2} = \dfrac{A}{x} + \dfrac{B}{x+3} + \dfrac{C}{(x+3)^2}$. Then $\quad x+4 = A(x+3)^2 + Bx(x+3) + Cx$. Let $x=0$. Then $4=9A$, $A=\frac{4}{9}$. Let $x=-3$. Then $1=-3C$, $C=-\frac{1}{3}$. To find B, equate coefficients of x^2: $\quad 0=A+B$, $\quad B=-A=-\frac{4}{9}$. Thus, $\quad \dfrac{x+4}{x^3+6x^2+9x} = \dfrac{4}{9}\dfrac{1}{x} - \dfrac{4}{9}\dfrac{1}{x+3} - \dfrac{1}{3}\dfrac{1}{(x+3)^2}$, and

$$\int \frac{x+4}{x^3+6x^2+9x}\, dx = \frac{4}{9}\ln|x| - \frac{4}{9}\ln|x+3| + \frac{1}{3}\frac{1}{x+3} + C_1 = \frac{4}{9}\ln\left|\frac{x}{x+3}\right| + \frac{1}{3}\frac{1}{x+3} + C_1.$$

30.11 $\displaystyle\int \frac{x^4\, dx}{x^3-2x^2-7x-4}$.

▮ First we divide the numerator by the denominator, obtaining $x+2+\dfrac{11x^2+18x+8}{x^3-2x^2-7x-4}$. Hence, $\displaystyle\int \frac{x^4\, dx}{x^3-2x^2-7x-4} = \frac{1}{2}x^2 + 2x + \int \frac{11x^2+18x+8}{x^3-2x^2-7x-4}\, dx$. Now we seek to factor the denominator. Looking for its roots, we examine the integral factors of the constant term, -4. We find that $x=-1$ is a root of the denominator, and, dividing the latter by $x+1$ yields $x^2-3x-4=(x-4)(x+1)$. Thus, the denominator is $(x-4)(x+1)^2$. Now, $\dfrac{11x^2+18x+8}{(x-4)(x+1)^2} = \dfrac{A}{x-4} + \dfrac{B}{x+1} + \dfrac{C}{(x+1)^2}$. Hence, $11x^2+18x+8 = A(x+1)^2 + B(x-4)(x+1) + C(x-4)$. Let $x=4$. Then $256=25A$, $A=\frac{256}{25}$. Let $x=-1$. Then $1=-5C$, $C=-\frac{1}{5}$. To find B, equate coefficients of x^2: $11=A+B$, $B=11-A=\frac{19}{25}$. Thus, $\dfrac{11x^2+18x+8}{(x-4)(x+1)^2} = \dfrac{256}{25}\dfrac{1}{x-4} + \dfrac{19}{25}\dfrac{1}{x+1} - \dfrac{1}{5}\dfrac{1}{(x+1)^2}$, and $\displaystyle\int \frac{11x^2+18x+8}{(x-4)(x+1)^2}\, dx = \frac{256}{25}\ln|x-4| + \frac{19}{25}\ln|x+1| + \frac{1}{5}\frac{1}{x+1} + C_1$. Thus, the complete answer is $\frac{1}{2}x^2 + 2x + \frac{256}{25}\ln|x-4| + \frac{19}{25}\ln|x+1| + \frac{1}{5}\frac{1}{x+1} + C_1$.

30.12 $\displaystyle\int \frac{dx}{x(x^2+5)}$.

▮ $\dfrac{1}{x(x^2+5)} = \dfrac{A}{x} + \dfrac{Bx+C}{x^2+5}$. So $\quad 1 = A(x^2+5) + Bx^2 + Cx$. Let $x=0$. Then $1=5A$, $A=\frac{1}{5}$. Equate coefficients of x^2: $\quad 0=A+B$, $B=-A=-\frac{1}{5}$. Equate coefficients of x: $\quad 0=C$. Hence, $\dfrac{1}{x(x^2+5)} = \dfrac{1}{5}\dfrac{1}{x} - \dfrac{1}{5}\dfrac{x}{x^2+5}$, and $\displaystyle\int \frac{dx}{x(x^2+5)} = \frac{1}{5}\left(\ln|x| - \frac{1}{2}\ln|x^2+5|\right) + C_1 = \frac{1}{5}\ln\left|\frac{x}{\sqrt{x^2+5}}\right| + C_1$.

30.13 $\displaystyle\int \frac{x^2\, dx}{(x-1)(x^2+4x+5)}$.

▮ x^2+4x+5 is irreducible, since its discriminant $b^2-4ac=-4<0$. Thus, $\dfrac{x^2}{(x-1)(x^2+4x+5)} = \dfrac{A}{x-1} + \dfrac{Bx+C}{x^2+4x+5}$. $\quad x^2 = A(x^2+4x+5) + (Bx+C)(x-1)$. Let $x=1$. Then $1=10A$, $A=\frac{1}{10}$. Equate coefficients of x^2: $\quad 1=A+B$, $B=1-A=\frac{9}{10}$. Equate constant coefficients: $\quad 0=5A-C$, $C=5A=\frac{5}{10}$. So $\dfrac{x^2}{(x-1)(x^2+4x+5)} = \dfrac{1}{10}\dfrac{1}{x-1} + \dfrac{\frac{9}{10}x+\frac{5}{10}}{x^2+4x+5} = \dfrac{1}{10}\dfrac{1}{x-1} + \dfrac{1}{10}\dfrac{9x+5}{x^2+4x+5}$. Hence, $\displaystyle\int \frac{x^2\, dx}{(x-1)(x^2+4x+5)} = \frac{1}{10}\ln|x-1| + \frac{1}{10}\int \frac{9x+5}{x^2+4x+5}\, dx$. To compute the latter integral, complete the square: $\quad x^2+4x+5 = (x+2)^2+1$. Let $\quad x+2=\tan\theta$, $\quad dx=\sec^2\theta\, d\theta$, $\quad 9x+5=9\tan\theta-13$. Thus (Fig. 30-1), $\displaystyle\int \frac{9x+5}{x^2+4x+5}\, dx = \int \frac{(9\tan\theta-13)\sec^2\theta\, d\theta}{\sec^2\theta} = \int (9\tan\theta-13)\, d\theta = 9\ln|\sec\theta| - 13\theta = 9\ln|\sqrt{x^2+4x+5}| - 13\tan^{-1}(x+2) = \frac{9}{2}\ln(x^2+4x+5) - 13\tan^{-1}(x+2)$. Hence, the complete answer is $\frac{1}{10}\ln|x-1| + \frac{9}{20}\ln(x^2+4x+5) - \frac{13}{10}\tan^{-1}(x+2) + C_1$.

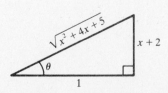

Fig. 30-1

30.14 $\int \dfrac{dx}{(x^2+1)(x^2+4)}$.

▌ Neither x^2+1 nor x^2+4 factors. Then $\dfrac{1}{(x^2+1)(x^2+4)} = \dfrac{Ax+B}{x^2+1} + \dfrac{Cx+D}{x^2+4}$. Hence, $1 = (Ax + B)(x^2+4) + (Cx+D)(x^2+1)$. Equate coefficients of x^3: (*) $0 = A + C$. Equate coefficients of x: $0 = 4A + C$. Subtracting (*) from this equation, we get $3A = 0$, $A = 0$, $C = 0$. Equate coefficients of x^2: (**) $0 = B + D$. Equate constant coefficients: $1 = 4B + D$. Subtracting (**) from this equation, we get $3B = 1$, $B = \frac{1}{3}$, $D = -\frac{1}{3}$. Thus, $\dfrac{1}{(x^2+1)(x^2+4)} = \dfrac{1}{3}\dfrac{1}{x^2+1} - \dfrac{1}{3}\dfrac{1}{x^2+4}$. Hence, $\int \dfrac{dx}{(x^2+1)(x^2+4)} = \frac{1}{3}\tan^{-1} x - \frac{1}{3}\frac{1}{2}\tan^{-1}(x/2) + C_1 = \frac{1}{3}\tan^{-1} x - \frac{1}{6}\tan^{-1}(x/2) + C_1$.

30.15 $\int \dfrac{x^4+1}{x^3+9x}\,dx$.

▌ Dividing x^4+1 by x^3+9x, we obtain $x + \dfrac{1-9x^2}{x^3+9x}$. So $\int \dfrac{x^4+1}{x^3+9x}\,dx = \frac{1}{2}x^2 + \int \dfrac{1-9x^2}{x^3+9x}\,dx$. Now, $\dfrac{1-9x^2}{x^3+9x} = \dfrac{1-9x^2}{x(x^2+9)} = \dfrac{A}{x} + \dfrac{Bx+C}{x^2+9}$. Then $1-9x^2 = A(x^2+9) + x(Bx+C)$. Let $x=0$. Then $1=9A$, $A = \frac{1}{9}$. Equate coefficients of x^2: $-9 = A + B$, $B = -9 - A = -\frac{82}{9}$. Equate coefficients of x: $0 = C$. Thus, $\dfrac{1-9x^2}{x(x^2+9)} = \dfrac{1}{9}\dfrac{1}{x} - \dfrac{82}{9}\dfrac{x}{x^2+9}$, and $\int \dfrac{1-9x^2}{x(x^2+9)}\,dx = \frac{1}{9}\ln|x| - \frac{41}{9}\ln(x^2+9) + C_1$. The complete answer is, therefore, $\frac{1}{2}x^2 + \frac{1}{9}\ln|x| - \frac{41}{9}\ln(x^2+9) + C_1$.

30.16 $\int \dfrac{dx}{x(x^2+1)^2}$.

▌ $\dfrac{1}{x(x^2+1)^2} = \dfrac{A}{x} + \dfrac{Bx+C}{x^2+1} + \dfrac{Dx+E}{(x^2+1)^2}$. Then $1 = A(x^2+1)^2 + x(x^2+1)(Bx+C) + x(Dx+E)$. Let $x=0$. Then $1 = A$. Equate coefficients of x^4: $0 = A + B$, $B = -A = -1$. Equate coefficients of x^3: $0 = C$. Equate coefficients of x^2: $0 = 2A + B + D$, $D = -2A - B = -1$. Equate coefficients of x: $0 = C + E$, $E = -C = 0$. Thus, $\dfrac{1}{x(x^2+1)^2} = \dfrac{1}{x} - \dfrac{1}{x^2+1} - \dfrac{1}{(x^2+1)^2}$. As in the solution of Problem 29.25, $\int \dfrac{dx}{(x^2+1)^2} = \frac{1}{2}\left(\tan^{-1} x + \dfrac{x}{x^2+1}\right)$. Hence, $\int \dfrac{dx}{x(x^2+1)^2} = \ln|x| - \dfrac{3}{2}\tan^{-1} x - \dfrac{1}{2}\dfrac{x}{x^2+1} + C$.

30.17 $\int \dfrac{x^2\,dx}{(x-1)(x^2+4)^2}$.

▌ $\dfrac{x^2}{(x-1)(x^2+4)^2} = \dfrac{A}{x-1} + \dfrac{Bx+C}{x^2+4} + \dfrac{Dx+E}{(x^2+4)^2}$. Then $x^2 = A(x^2+4)^2 + (x-1)(x^2+4)(Bx+C) + (x-1)(Dx+E)$. Let $x=1$. Then $1 = 25A$, $A = \frac{1}{25}$. Equate coefficients of x^4: $0 = A + B$, $B = -A = -\frac{1}{25}$. Equate coefficients of x^3: $0 = -B + C$, $C = B = -\frac{1}{25}$. Equate coefficients of x^2: $1 = 8A + 4B - C + D$. Then $D = \frac{20}{25} = \frac{4}{5}$. Equate constant terms: $0 = 16A - 4C - E$. Then $E = \frac{20}{25} = \frac{4}{5}$. Thus, $\dfrac{x^2}{(x-1)(x^2+4)^2} = \dfrac{1}{25}\dfrac{1}{x-1} - \dfrac{1}{25}\dfrac{x+1}{x^2+4} + \dfrac{4}{5}\dfrac{x+1}{(x^2+4)^2}$. Hence, $\int \dfrac{x^2\,dx}{(x-1)(x^2+4)^2} = \dfrac{1}{25}\ln|x-1| - \dfrac{1}{25}\int \dfrac{x+1}{x^2+4}\,dx + \dfrac{4}{5}\int \dfrac{x+1}{(x^2+4)}\,dx$. Now

$$\int \dfrac{x+1}{x^2+4}\,dx = \int \dfrac{x\,dx}{x^2+4} + \int \dfrac{dx}{x^2+4} = \frac{1}{2}\ln(x^2+4) + \frac{1}{2}\tan^{-1}\frac{x}{2} \tag{I}$$

$$\int \dfrac{(x+1)\,dx}{(x^2+4)^2} = \int \dfrac{x\,dx}{(x^2+4)^2} + \int \dfrac{dx}{(x^2+4)^2} = -\frac{1}{2}\dfrac{1}{x^2+4} + \frac{1}{16}\left(\tan^{-1}\frac{x}{2} + \dfrac{2x}{x^2+4}\right) \tag{II}$$

where the last integration is performed as in Problem 29.25. Hence, the complete answer is

$$\frac{1}{25}\ln|x-1| - \frac{1}{25}\left[\frac{1}{2}\ln(x^2+4) + \frac{1}{2}\tan^{-1}\frac{x}{2}\right] + \frac{4}{5}\left(\frac{1}{8}\dfrac{x-4}{x^2+4} + \frac{1}{16}\tan^{-1}\frac{x}{2}\right) + C_1$$

$$= \frac{1}{25}\ln\dfrac{|x-1|}{\sqrt{x^2+4}} + \frac{3}{100}\tan^{-1}\frac{x}{2} + \frac{1}{10}\dfrac{x-4}{x^2+4} + C_1$$

30.18 $\int \dfrac{x^3+1}{x(x^2+x+1)^2}\,dx$.

▮ $x^2 + x + 1$ is irreducible $(b^2 - 4ac = -3 < 0)$. $\dfrac{x^3 + 1}{x(x^2 + x + 1)^2} = \dfrac{A}{x} + \dfrac{Bx + C}{x^2 + x + 1} + \dfrac{Dx + E}{(x^2 + x + 1)^2}$. Then $x^3 + 1 = A(x^2 + x + 1)^2 + x(x^2 + x + 1)(Bx + C) + x(Dx + E)$. Let $x = 0$. Then $1 = 9A$, $A = \frac{1}{9}$. Equate coefficients of x^4: $0 = A + B$, $B = -A = -\frac{1}{9}$. Equate coefficients of x^3: $1 = 2A + C + B$. Then $C = 1 - 2A - B = \frac{8}{9}$. Equate coefficients of x^2: $0 = 3A + B + C + D$. Then $D = -3A - B - C = -\frac{10}{9}$. Equate coefficients of x: $0 = 2A + C + E$, $E = 2A - C = -\frac{10}{9}$. Thus, $\dfrac{x^3 + 1}{x(x^2 + x + 1)^2} = \dfrac{1}{9}\dfrac{1}{x} + \dfrac{1}{9}\dfrac{-x + 8}{x^2 + x + 1} - \dfrac{10}{9}\dfrac{x + 1}{(x^2 + x + 1)^2}$. Then $\displaystyle \int \dfrac{x^3 + 1}{x(x^2 + x + 1)^2}\, dx = \dfrac{1}{9}\ln|x| + \dfrac{1}{9}\int \dfrac{8 - x}{x^2 + x + 1}\, dx - \dfrac{10}{9}\int \dfrac{(x + 1)\, dx}{(x^2 + x + 1)^2}$.

$$\int \frac{8 - x}{x^2 + x + 1}\, dx = \int \frac{(8 - x)\, dx}{(x + \frac{1}{2})^2 + \frac{3}{4}} = \int \frac{(-x - \frac{1}{2}) + \frac{17}{2}}{(x + \frac{1}{2})^2 + \frac{3}{4}}\, dx$$

$$= -\int \frac{x + \frac{1}{2}}{(x + \frac{1}{2})^2 + \frac{3}{4}}\, dx + \frac{17}{2}\int \frac{dx}{(x + \frac{1}{2})^2 + \frac{3}{4}}$$

$$= -\frac{1}{2}\ln\left[\left(x + \frac{1}{2}\right)^2 + \frac{3}{4}\right] + \frac{17}{2} \cdot \frac{2}{\sqrt{3}}\tan^{-1}\frac{x + \frac{1}{2}}{\sqrt{3}/2}$$

$$= -\frac{1}{2}\ln(x^2 + x + 1) + \frac{17}{\sqrt{3}}\tan^{-1}\frac{2x + 1}{\sqrt{3}} \qquad (I)$$

$$\int \frac{(x + 1)\, dx}{(x^2 + x + 1)^2} = \int \frac{(x + 1)\, dx}{[(x + \frac{1}{2})^2 + \frac{3}{4}]^2} = \int \frac{(x + \frac{1}{2}) + \frac{1}{2}}{[(x + \frac{1}{2})^2 + \frac{3}{4}]^2}\, dx$$

$$= \int \frac{(x + \frac{1}{2})\, dx}{[(x + \frac{1}{2})^2 + \frac{3}{4}]^2} + \frac{1}{2}\int \frac{dx}{[(x + \frac{1}{2})^2 + \frac{3}{4}]^2}$$

$$= -\frac{1}{2}\frac{1}{(x + \frac{1}{2})^2 + \frac{3}{4}} + \frac{1}{2} \cdot \frac{1}{2(\sqrt{3}/2)^3}\left[\tan^{-1}\frac{x + \frac{1}{2}}{\sqrt{3}/2} + \frac{(\sqrt{3}/2)x}{(x + \frac{1}{2})^2 + \frac{3}{4}}\right]$$

$$= -\frac{1}{2}\frac{1}{x^2 + x + 1} + \frac{2}{3\sqrt{3}}\tan^{-1}\frac{2x + 1}{\sqrt{3}} + \frac{1}{3}\frac{x}{x^2 + x + 1}$$

$$= \frac{1}{6}\frac{2x - 3}{x^2 + x + 1} + \frac{2}{3\sqrt{3}}\tan^{-1}\frac{2x + 1}{\sqrt{3}} \qquad (II)$$

Hence, the complete answer is $\dfrac{1}{9}\ln|x| + \dfrac{1}{9}\left[-\dfrac{1}{2}\ln(x^2 + x + 1) + \dfrac{17}{\sqrt{3}}\tan^{-1}\dfrac{2x + 1}{\sqrt{3}}\right] - \dfrac{10}{9}\left(\dfrac{1}{6}\dfrac{2x - 1}{x^2 + x + 1} + \right.$ $\left. \dfrac{2}{3\sqrt{3}}\tan^{-1}\dfrac{2x + 1}{\sqrt{3}}\right) + C_1 = \dfrac{1}{18}\ln\dfrac{x^2}{x^2 + x + 1} + \dfrac{31}{27\sqrt{3}}\tan^{-1}\dfrac{2x + 1}{\sqrt{3}} - \dfrac{5}{27}\dfrac{2x - 1}{x^2 + x + 1} + C_1$.

30.19 $\displaystyle \int \frac{x - 1}{x^3 + 2x^2 - x - 2}\, dx$.

▮ $x = 1$ is a root of the denominator. When the latter is divided by $x - 1$, we obtain $x^2 + 3x + 2 = (x + 2)(x + 1)$. Since $x - 1$ cancels out, we are left with $\displaystyle \int \frac{dx}{(x + 2)(x + 1)}$. Now, $\dfrac{1}{(x + 2)(x + 1)} = \dfrac{A}{x + 2} + \dfrac{B}{x + 1}$. Then $1 = A(x + 1) + B(x + 2)$. Let $x = -1$. Then $1 = B$. Let $x = -2$. Then $1 = -A$, $A = -1$. Thus, $\dfrac{1}{(x + 2)(x + 1)} = -\dfrac{1}{x + 2} + \dfrac{1}{x + 1}$, and $\displaystyle \int \frac{dx}{(x + 2)(x + 1)} = -\ln|x + 2| + \ln|x + 1| + C_1 = \ln\left|\dfrac{x + 1}{x + 2}\right| + C_1$.

30.20 $\displaystyle \int \frac{x^2 + 2}{x(x^2 + 5x + 6)}\, dx$.

▮ Since $x^2 + 5x + 6 = (x + 2)(x + 3)$, we have $\dfrac{x^2 + 2}{x(x + 2)(x + 3)} = \dfrac{A}{x} + \dfrac{B}{x + 2} + \dfrac{C}{x + 3}$. Then $x^2 + 2 = A(x + 2)(x + 3) + Bx(x + 3) + Cx(x + 2)$. Let $x = 0$. Then $2 = 6A$, $A = \frac{1}{3}$. Let $x = -2$. Then $6 = -2B$, $B = -3$. Let $x = -3$. Then $11 = 3C$, $C = \frac{11}{3}$. Thus, $\dfrac{x^2 + 2}{x(x + 2)(x + 3)} = \dfrac{1}{3}\dfrac{1}{x} - \dfrac{3}{x + 2} + \dfrac{11}{3}\dfrac{1}{x + 3}$. Hence, $\displaystyle \int \frac{x^2 + 2}{x(x + 2)(x + 3)}\, dx = \frac{1}{3}\ln|x| - 3\ln|x + 2| + \frac{11}{3}\ln|x + 3| + C_1$.

30.21 $\int \dfrac{dx}{1+e^x}$.

▮ Let $u = 1 + e^x$, $du = e^x\,dx$. Then $\int \dfrac{dx}{1+e^x} = \int \dfrac{du}{u(u-1)}$. Let $\dfrac{1}{u(u-1)} = \dfrac{A}{u} + \dfrac{B}{u-1}$. Then $1 = A(u-1) + Bu$. Let $u = 0$. Then $1 = -A$, $A = -1$. Let $u = 1$. Then $1 = B$. So $\dfrac{1}{u(u-1)} = -\dfrac{1}{u} + \dfrac{1}{u-1}$, and $\int \dfrac{du}{u(u-1)} = -\ln|u| + \ln|u-1| + C = -\ln(1+e^x) + \ln e^x + C = -\ln(1+e^x) + x + C$.

30.22 Find the area of the region in the first quadrant under the curve $y = \dfrac{1}{x^3+27}$ and to the left of the line $x = 3$.

▮ $A = \int_0^3 \dfrac{dx}{x^3+27}$. Note that $x = -3$ is a root of $x^3 + 27$, and, dividing the latter by $x - 3$, we obtain $x^2 - 3x + 9$, which is irreducible. Let $\dfrac{1}{(x+3)(x^2-3x+9)} = \dfrac{A}{x+3} + \dfrac{Bx+C}{x^2-3x+9}$. Then $1 = A(x^2-3x+9) + (x+3)(Bx+C)$. Let $x = -3$. Then $1 = 27A$, $A = \frac{1}{27}$. Equate the coefficients of x^2: $0 = A + B$, $B = -A = -\frac{1}{27}$. Equate the constant coefficients: $1 = 9A + 3C$, $C = \frac{6}{27}$. Thus, $\dfrac{1}{x^3+27} = \dfrac{1}{27}\dfrac{1}{x+3} - \dfrac{1}{27}\dfrac{x-6}{x^2-3x+9}$. Hence, $\int \dfrac{dx}{x^3+27} = \dfrac{1}{27}\ln|x+3| - \dfrac{1}{27}\int \dfrac{x-6}{x^2-3x+9}\,dx$. But $\int \dfrac{x-6}{x^2-3x+9}\,dx = \int \dfrac{(x-6)\,dx}{(x-\frac{3}{2})^2 + \frac{27}{4}} = \int \dfrac{(x-\frac{3}{2})\,dx}{(x-\frac{3}{2})^2 + \frac{27}{4}} + \int \dfrac{-\frac{9}{2}\,dx}{(x-\frac{3}{2})^2 + \frac{27}{4}} = \dfrac{1}{2}\ln[(x-\frac{3}{2})^2 + \frac{27}{4}] - \dfrac{9}{2}\cdot\dfrac{1}{(3\sqrt{3}/2)}\tan^{-1}\dfrac{x-\frac{3}{2}}{3\sqrt{3}/2} = \dfrac{1}{2}\ln(x^2-3x+9) - \sqrt{3}\tan^{-1}\dfrac{2x-3}{3\sqrt{3}}$. Thus, the complete integral is $\left[\dfrac{1}{27}\ln|x+3| - \dfrac{1}{54}\ln(x^2-3x+9) + \dfrac{\sqrt{3}}{27}\tan^{-1}\dfrac{2x-3}{3\sqrt{3}}\right]_0^3 = \left(\dfrac{1}{27}\ln 6 - \dfrac{1}{54}\ln 9 + \dfrac{\sqrt{3}}{27}\tan^{-1}\dfrac{1}{\sqrt{3}}\right) - \left[\dfrac{1}{27}\ln 3 - \dfrac{1}{54}\ln 9 + \dfrac{\sqrt{3}}{27}\tan^{-1}\left(-\dfrac{1}{\sqrt{3}}\right)\right] = \dfrac{1}{81}(3\ln 2 + \pi\sqrt{3})$. $\left(\text{Note that }\tan^{-1}\dfrac{1}{\sqrt{3}} = \dfrac{\pi}{6}.\right)$

30.23 Evaluate $\int \dfrac{dx}{\sin x + \cos x}$.

▮ Use the substitution of Problem 29.45, followed by the method of partial fractions. Thus, let $z = \tan(x/2)$, $x = 2\tan^{-1}z$, $dx = \dfrac{2}{1+z^2}\,dz$, $\sin x = \dfrac{2z}{1+z^2}$, $\cos x = \sqrt{1-\sin^2 x} = \dfrac{1-z^2}{1+z^2}$. Then

$$\int \dfrac{dx}{\sin x + \cos x} = \int \dfrac{-2}{z^2-2z-1}\,dz = \dfrac{1}{\sqrt{2}}\int \dfrac{dz}{z-(1-\sqrt{2})} - \dfrac{1}{\sqrt{2}}\int \dfrac{dz}{z-(1+\sqrt{2})}$$

$$= \dfrac{1}{\sqrt{2}}\ln|z-(1-\sqrt{2})| - \dfrac{1}{\sqrt{2}}\ln|z-(1+\sqrt{2})| + C$$

$$= \dfrac{1}{\sqrt{2}}\ln\left|\dfrac{z-(1-\sqrt{2})}{z-(1+\sqrt{2})}\right| + C$$

$$= \dfrac{1}{\sqrt{2}}\ln\left|\dfrac{\tan(x/2)-(1-\sqrt{2})}{\tan(x/2)-(1+\sqrt{2})}\right| + C$$

30.24 Find $\int \dfrac{\cos x}{\sin x - 1}\,dx$.

▮ Using the same approach as in Problem 30.23, we have: $\int \dfrac{\cos x\,dx}{\sin x - 1} = 2\int \dfrac{z+1}{(1+z^2)(z-1)}\,dz$. Let $\dfrac{z+1}{(1+z^2)(z-1)} = \dfrac{A}{z-1} + \dfrac{Bz+C}{1+z^2}$. Then $z+1 = A(1+z^2) + (z-1)(Bz+C)$. Let $z = 1$. Then $2 = 2A$, $A = 1$. Equate coefficients of z^2: $0 = A + B$, $B = -A = -1$. Equate constant coefficients: $1 = A + C$, $C = 1 - A = 0$. Thus, $\dfrac{z+1}{(1+z^2)(z-1)} = \dfrac{A}{z-1} - \dfrac{z}{1+z^2}$, and $2\int \dfrac{z+1}{(1+z^2)(z-1)}\,dz = 2[\ln|z-1| - \frac{1}{2}\ln(1+z^2)] + C_1 = \ln|z-1|^2 - \ln(1+z^2) + C_1 = \ln\dfrac{(z-1)^2}{1+z^2} + C_1 = \ln(1-\sin x) + C_1$.

30.25 Find $\int \dfrac{\cos x \, dx}{\sin x - 1}$ by a suitable substitution.

▌ Let $u = \sin x - 1$, $du = \cos x \, dx$. Then $\int \dfrac{\cos x \, dx}{\sin x - 1} = \int \dfrac{du}{u} = \ln|u| + C = \ln|\sin x - 1| + C = \ln(1 - \sin x) + C$, since $\sin x \le 1$. This is the same result as in Problem 30.24.

30.26 Find $\int \dfrac{x^2 + 3}{(x-1)^3(x+1)} \, dx$.

▌ $\dfrac{x^2 + 3}{(x-1)^3(x+1)} = \dfrac{A}{x+1} + \dfrac{B}{x-1} + \dfrac{C}{(x-1)^2} + \dfrac{D}{(x-1)^3}$. Then $x^2 + 3 = A(x-1)^3 + B(x+1)(x-1)^2 + C(x+1)(x-1) + D(x+1)$. Let $x = -1$. Then $4 = -8A$, $A = -\frac{1}{2}$. Let $x = 1$. Then $4 = 2D$, $D = 2$. Equate coefficients of x^3: $0 = A + B$, $B = -A = \frac{1}{2}$. Equate constant coefficients: $3 = -A + B - C + D$, $C = -A + B + D - 3 = 0$. Thus, $\dfrac{x^2 + 3}{(x-1)^3(x+1)} = -\dfrac{1}{2}\dfrac{1}{x+1} + \dfrac{1}{2}\dfrac{1}{x-1} + \dfrac{2}{(x-1)^3}$, and $\int \dfrac{(x^2+3)\,dx}{(x-1)^3(x+1)} = -\dfrac{1}{2}\ln|x+1| + \dfrac{1}{2}\ln|x-1| - \dfrac{1}{(x-1)^2} + C_1 = \dfrac{1}{2}\ln\left|\dfrac{x-1}{x+1}\right| - \dfrac{1}{(x-1)^2} + C_1$.

In Problems 30.27–30.33, evaluate the given antiderivative.

30.27 $\int \dfrac{dx}{\sqrt[3]{x} - x}$.

▌ Make a substitution to eliminate the radical. Let $x = z^3$, $dx = 3z^2 \, dz$. Then $\int \dfrac{dx}{\sqrt[3]{x} - x} = \int \dfrac{3z^2 \, dz}{z - z^3} = 3\int \dfrac{z \, dz}{1 - z^2} = 3 \cdot (-\frac{1}{2}) \ln|1 - z^2| + C = -\frac{3}{2} \ln|1 - z^2| + C = -\frac{3}{2} \ln|1 - x^{2/3}| + C$.

30.28 $\int \dfrac{dx}{1 + \sqrt[4]{x - 1}}$.

▌ Let $x - 1 = z^4$, $dx = 4z^3 \, dz$. Then $\int \dfrac{dx}{1 + \sqrt[4]{x-1}} = \int \dfrac{4z^3 \, dz}{1 + z} = 4\int \left(z^2 - z + 1 - \dfrac{1}{1+z}\right) dz = 4(\frac{1}{3}z^3 - \frac{1}{2}z^2 + z - \ln|1 + z|) + C = \frac{4}{3}(\sqrt[4]{x-1})^3 - 2\sqrt{x-1} + \sqrt[4]{x-1} - \ln(1 + \sqrt[4]{x-1}) + C$.

30.29 $\int \dfrac{dx}{x\sqrt{1 + 3x}}$.

▌ Let $1 + 3x = z^2$, $3 \, dx = 2z \, dz$. Then $\int \dfrac{dx}{x\sqrt{1+3x}} = \dfrac{2}{3}\int \dfrac{z \, dz}{[(z^2-1)/3] \cdot z} = 2\int \dfrac{dz}{(z-1)(z+1)} = \int \left(\dfrac{1}{z-1} - \dfrac{1}{z+1}\right) dz = \ln|z - 1| - \ln|z + 1| + C = \ln\left|\dfrac{z-1}{z+1}\right| + C = \ln\left|\dfrac{(z-1)^2}{z^2-1}\right| + C = \ln\left|\dfrac{z^2 - 2z + 1}{3x}\right| + C = \ln\left|\dfrac{3x + 2 - 2\sqrt{1+3x}}{x}\right| + C$.

30.30 $\int \dfrac{dx}{\sqrt[3]{x} + \sqrt{x}}$.

▌ Let $x = z^6$. (In general, let $x = z^m$, where m is the least common multiple of the radicals.) $dx = 6z^5 \, dz$. Then $\int \dfrac{dx}{\sqrt[3]{x} + \sqrt{x}} = \int \dfrac{6z^5 \, dz}{z^2 + z^3} = 6\int \dfrac{z^3 \, dz}{1 + z}$. From Problem 30.28, we get $6(\frac{1}{3}z^3 - \frac{1}{2}z^2 + z - \ln|1 + z|) + C = 2\sqrt{x} - 3\sqrt[3]{x} + \sqrt[6]{x} - \ln(1 + \sqrt[6]{x}) + C$.

30.31 $\int \dfrac{dx}{\sqrt{\sqrt{x} + 1}}$.

▌ Let $\sqrt{x} + 1 = z^2$, $-\dfrac{1}{2\sqrt{x}} \, dx = 2z \, dz$, $-\dfrac{1}{2(z^2 - 1)} \, dx = 2z \, dz$, $dx = -4z(z^2 - 1) \, dz$. Then $\int \dfrac{dx}{\sqrt{\sqrt{x} + 1}} = -4\int \dfrac{z(z^2 - 1) \, dz}{z} = -4\int (z^2 - 1) \, dz = -4(\frac{1}{3}z^3 - z) + C = -\frac{4}{3}z(z^2 - 3) + C = -\frac{4}{3}\sqrt{\sqrt{x} + 1}(\sqrt{x} - 2) + C$.

30.32 $\int \sqrt{1 + e^x}\, dx$

▮ Let $1 + e^x = z^2$, $e^x\, dx = 2z\, dz$, $(z^2 - 1)\, dx = 2z\, dz$, $dx = \dfrac{2z}{z^2 - 1}\, dz$. Then $\int \sqrt{1 + e^x}\, dx =$
$\int z \cdot \dfrac{2z}{z^2 - 1}\, dz = 2 \int \dfrac{z^2\, dz}{z^2 - 1} = 2 \int \dfrac{(z^2 - 1) + 1}{z^2 - 1}\, dz = \int \left(2 + \dfrac{2}{z^2 - 1}\right) dz = 2z + \ln \left|\dfrac{z - 1}{z + 1}\right| + C$ [from Problem
30.29] $= 2z + \ln \left|\dfrac{z^2 - 1}{(z + 1)^2}\right| + C = 2z + \ln \left|\dfrac{e^x}{(z + 1)^2}\right| + C = 2z + \ln e^x - 2 \ln |z + 1| + C = 2\sqrt{1 + e^x} + x -$
$2 \ln (\sqrt{1 + e^x} + 1) + C$.

30.33 $\int \dfrac{x^{2/3}}{x + 1}\, dx$.

▮ Let $x = z^3$, $dx = 3z^2\, dz$. So $\int \dfrac{x^{2/3}}{x + 1}\, dx = \int \dfrac{z^2 \cdot 3z^2\, dz}{z^3 + 1} = 3 \int \dfrac{z^4}{z^3 + 1}\, dz = 3 \int \left(z - \dfrac{z}{z^3 + 1}\right) dz =$
$3\left(\dfrac{1}{2} z^2 - \int \dfrac{z}{z^3 + 1}\, dz\right)$. Now, $z^3 + 1 = (z + 1)(z^2 - z + 1)$. So, $\int \dfrac{z}{z^3 + 1}\, dz = \int \dfrac{z\, dz}{(z + 1)(z^2 - z + 1)}$. Let
$\dfrac{z}{(z + 1)(z^2 - z + 1)} = \dfrac{A}{z + 1} + \dfrac{Bz + C}{z^2 - z + 1}$. Then $z = A(z^2 - z + 1) + (z + 1)(Bz + C)$. Let $z = -1$. Then
$-1 = 3A$, $A = -\frac{1}{3}$. Equate coefficients of z^2: $0 = A + B$, $B = -A = \frac{1}{3}$. Equate constant coefficients:
$0 = A + C$, $C = -A = \frac{1}{3}$. Then $\dfrac{z}{(z + 1)(z^2 - z + 1)} = -\dfrac{1}{3} \dfrac{1}{z + 1} + \dfrac{1}{3} \dfrac{z + 1}{z^2 - z + 1}$. Hence,
$\int \dfrac{z\, dz}{(z + 1)(z^2 - z + 1)} = -\dfrac{1}{3} \ln |z + 1| + \dfrac{1}{3} \int \dfrac{z + 1}{z^2 - z + 1}\, dz$. Now, $\int \dfrac{z + 1}{z^2 - z + 1}\, dz = \dfrac{1}{2} \int \dfrac{2z + 2}{z^2 - z + 1}\, dz =$
$\dfrac{1}{2} \int \dfrac{2z - 1}{z^2 - z + 1}\, dz + \dfrac{1}{2} \int \dfrac{3}{z^2 - z + 1}\, dz = \dfrac{1}{2} \ln (z^2 - z + 1) + \dfrac{3}{2} \int \dfrac{dz}{(z - \frac{1}{2})^2 + \frac{3}{4}} = \dfrac{1}{2} \ln (z^2 - z + 1) +$
$\dfrac{3}{2} \cdot \dfrac{2}{\sqrt{3}} \tan^{-1} \dfrac{2z - 1}{\sqrt{3}}$. Hence, our complete answer is $\dfrac{3}{2} z^2 - 3\left[-\dfrac{1}{3} \ln |z + 1| + \dfrac{1}{3} \dfrac{1}{2} \ln (z^2 - z + 1) +\right.$
$\left.\dfrac{\sqrt{3}}{3} \tan^{-1} \dfrac{2z - 1}{\sqrt{3}}\right] + C = \dfrac{3}{2} z^2 + \ln |z + 1| - \dfrac{1}{2} \ln (z^2 - z + 1) - \sqrt{3} \tan^{-1} \dfrac{2z - 1}{\sqrt{3}} + C = \dfrac{3}{2} x^{2/3} + \ln |x^{1/3} + 1| -$
$\dfrac{1}{2} \ln \dfrac{x + 1}{x^{1/3} + 1} - \sqrt{3} \tan^{-1} \dfrac{2x^{1/3} - 1}{\sqrt{3}} + C$.

CHAPTER 31
Integrals for Surface Area, Work, Centroids

SURFACE AREA OF A SOLID OF REVOLUTION

31.1 If the region under a curve $y = f(x)$, above the x-axis, and between $x = a$ and $x = b$, is revolved about the x-axis, state a formula for the surface area S of the resulting solid.

$\blacksquare$ $S = 2\pi \int y \, ds = 2\pi \int_a^b y\sqrt{1 + \left(\dfrac{dy}{dx}\right)^2} \, dx = 2\pi \int_{f(a)}^{f(b)} y\sqrt{1 + \left(\dfrac{dx}{dy}\right)^2} \, dy$.

[For revolution about the y-axis, change y to x in either integrand.]

31.2 Find the surface area of a sphere of radius r.

$\blacksquare$ Revolve the upper semicircle $y = \sqrt{r^2 - x^2}$ about the x-axis. Since $x^2 + y^2 = r^2$, $2x + 2yy' = 0$, $y' = -x/y$, $(y')^2 = x^2/y^2$, $1 + (y')^2 = 1 + x^2/y^2 = (y^2 + x^2)/y^2 = r^2/y^2$. Hence, the surface area $S = 2\pi \int_{-r}^r y \cdot \dfrac{r}{y} \, dx = 2\pi \int_{-r}^r r \, dx = 2\pi r x \,]_{-r}^r = 4\pi r^2$.

In Problems 31.3–31.13, find the surface area generated when the given arc is revolved about the given axis.

31.3 $y = x^2$, $0 \le x \le \frac{1}{2}$; about the x-axis.

$\blacksquare$ $y' = 2x$. Hence, the surface area $S = 2\pi \int_0^{1/2} x^2 \sqrt{1 + 4x^2} \, dx$. Let $x = \frac{1}{2}\tan\theta$, $dx = \frac{1}{2}\sec^2\theta \, d\theta$. So we get $2\pi \int_0^{\pi/4} \frac{1}{4}\tan^2\theta \sec\theta \cdot \frac{1}{2}\sec^2\theta \, d\theta = \dfrac{\pi}{4}\int_0^{\pi/4}\tan^2\theta \sec^3\theta \, d\theta = \dfrac{\pi}{4}\int_0^{\pi/4}(\sec^5\theta - \sec^3\theta) \, d\theta$. By the reduction formula of Problem 29.39, $\int \sec^5\theta \, d\theta = \dfrac{1}{4}\sec^3\theta\tan\theta + \dfrac{3}{4}\int \sec^3\theta \, d\theta$. By Problem 29.40, $\int \sec^3\theta \, d\theta = \dfrac{1}{2}(\tan\theta\sec\theta + \ln|\sec\theta + \tan\theta|)$. Thus we get $\dfrac{\pi}{4}\left(\dfrac{1}{4}\sec^3\theta\tan\theta - \dfrac{1}{8}\sec\theta\tan\theta - \dfrac{1}{8}\ln|\sec\theta + \tan\theta|\right)\Big]_0^{\pi/4} = \dfrac{\pi}{4}\left(\dfrac{2\sqrt{2}}{4} - \dfrac{1}{8}\sqrt{2} - \dfrac{1}{8}\ln|\sqrt{2} + 1|\right) = \dfrac{\pi}{32}[3\sqrt{2} - \ln(1 + \sqrt{2})]$.

31.4 The same arc as in Problem 31.3, but about the y-axis.

$\blacksquare$ Use $S = 2\pi \int_0^{1/4} x\sqrt{1 + \left(\dfrac{dx}{dy}\right)^2} \, dy$. Since $x = \sqrt{y}$, $\dfrac{dx}{dy} = \dfrac{1}{2\sqrt{y}}$, $1 + \left(\dfrac{dx}{dy}\right)^2 = 1 + \dfrac{1}{4y} = \dfrac{4y + 1}{4y}$. So $S = 2\pi \int_0^{1/4} \sqrt{y}\,\dfrac{\sqrt{4y+1}}{2\sqrt{y}} \, dy = \pi \int_0^{1/4}\sqrt{4y + 1} \, dy = \pi\,\frac{1}{4}(\frac{2}{3})(4y + 1)^{3/2}\,]_0^{1/4} = \dfrac{\pi}{6}(2^{3/2} - 1) = \dfrac{\pi}{6}(2\sqrt{2} - 1)$.

31.5 $y = x^3$, $0 \le x < 1$; about the x-axis.

$\blacksquare$ $y' = 3x^2$, so $S = 2\pi \int_0^1 x^3\sqrt{1 + 9x^4} \, dx = 2\pi \cdot \frac{1}{36}(\frac{2}{3})(1 + 9x^4)^{3/2}\,]_0^1 = \dfrac{\pi}{27}[(10)^{3/2} - 1] = \dfrac{\pi}{27}(10\sqrt{10} - 1)$.

31.6 $y = \frac{1}{3}\sqrt{x}(3 - x)$, $0 \le x \le 3$; about the x-axis.

$\blacksquare$ $y' = \dfrac{1}{2\sqrt{x}} - \dfrac{1}{2}\sqrt{x} = \dfrac{1}{2}\left(\dfrac{1-x}{\sqrt{x}}\right)$. So $1 + (y')^2 = \dfrac{(x+1)^2}{4x}$. $S = 2\pi \int_0^3 \frac{1}{3}\sqrt{x}(3 - x)\cdot\dfrac{x+1}{2\sqrt{x}} \, dx = \dfrac{\pi}{3}\int_0^3 (3 + 2x - x^2) \, dx = \dfrac{\pi}{3}(3x + x^2 - \frac{1}{3}x^3)\,]_0^3 = \dfrac{\pi}{3}(9 + 9 - 9) = 3\pi$.

31.7 $x^{2/3} + y^{2/3} = a^{2/3}$, in the first quadrant; about the x-axis.

$\blacksquare$ $\dfrac{2}{3}x^{-1/3} + \dfrac{2}{3}y^{-1/3}y' = 0$, $y' = -\dfrac{y^{1/3}}{x^{1/3}}$, $1 + (y')^2 = 1 + \dfrac{y^{2/3}}{x^{2/3}} = \dfrac{x^{2/3} + y^{2/3}}{x^{2/3}} = \dfrac{a^{2/3}}{x^{2/3}}$. So $S = 2\pi \int_0^a a^{1/3}\dfrac{(a^{2/3} - x^{2/3})^{3/2}}{x^{1/3}} \, dx = 2\pi a^{1/3}(-\frac{3}{2})(\frac{2}{5})(a^{2/3} - x^{2/3})^{5/2}\,]_0^a = -\dfrac{6\pi}{5}a^{1/3}(-a^{5/3}) = \dfrac{6\pi}{5}a^{6/3} = \dfrac{6\pi}{5}a^2$.

31.8 $y = 2\sqrt{15 - x}$, $0 \le x \le 15$; about the x-axis.

▋ $y' = \dfrac{-1}{\sqrt{15-x}}$, $1 + (y')^2 = \dfrac{16-x}{15-x}$. Then $S = 2\pi \int_0^{15} 2\sqrt{15-x} \cdot \dfrac{\sqrt{16-x}}{\sqrt{15-x}}\, dx = 4\pi \int_0^{15} \sqrt{16-x}\, dx = 4\pi(-\tfrac{2}{3})(16-x)^{3/2} \,\big]_0^{15} = -\dfrac{8\pi}{3}(1-64) = 168\pi$.

31.9 $y = \dfrac{1}{4}x^4 + \dfrac{1}{8x^2}$, $1 \le x \le 2$; about the y-axis.

▋ Use $S = 2\pi \int_1^2 x\sqrt{1+(y')^2}\, dx$. $y' = x^3 - \dfrac{1}{4x^3}$, $(y')^2 = x^6 - \dfrac{1}{2} + \dfrac{1}{16x^6}$. Then $1 + (y')^2 = x^6 + \dfrac{1}{2} + \dfrac{1}{16x^6} = \left(x^3 + \dfrac{1}{4x^3}\right)^2$. So $S = 2\pi \int_1^2 x\left(x^3 + \dfrac{1}{4x^3}\right)dx = \pi \int_1^2 \left(x^4 + \dfrac{1}{4x^2}\right)dx = 2\pi\left(\dfrac{1}{5}x^5 - \dfrac{1}{4x}\right)\Big]_1^2 = 2\pi\left[\left(\dfrac{32}{5} - \dfrac{1}{8}\right) - \left(\dfrac{1}{5} - \dfrac{1}{4}\right)\right] = \dfrac{253\pi}{20}$.

31.10 $y^2 = 12x$, $0 \le x \le 3$; about the x-axis.

▋ $S = 2\pi \int_0^3 y\sqrt{1+(y')^2}\, dx$. Now, $2yy' = 12$, $4y^2(y')^2 = 144$, $(y')^2 = \dfrac{36}{y^2} = \dfrac{3}{x}$, $1 + (y')^2 = \dfrac{y^2+36}{y^2}$. Hence, $S = 2\pi \int_0^3 y \dfrac{\sqrt{y^2+36}}{y}\, dx = 2\pi \int_0^3 \sqrt{12x+36}\, dx = 4\pi\sqrt{3} \int_0^3 \sqrt{x+3}\, dx = 4\pi\sqrt{3}(\tfrac{2}{3})(x+3)^{3/2}\,\big]_0^3 = \dfrac{8\pi\sqrt{3}}{3}(6^{3/2} - 3^{3/2}) = \dfrac{8\pi\sqrt{3}}{3}[3^{3/2}(2\sqrt{2}-1)] = 24\pi(2\sqrt{2}-1)$.

31.11 $y^2 + 4x = 2\ln y$, $0 \le y \le 3$; about the x-axis.

▋ $S = 2\pi \int_1^3 y\sqrt{1+\left(\dfrac{dx}{dy}\right)^2}\, dy$. We have $2y + 4\dfrac{dx}{dy} = \dfrac{2}{y}$, $2\dfrac{dx}{dy} = \dfrac{1}{y} - y = \dfrac{1-y^2}{y}$, $\left(\dfrac{dx}{dy}\right)^2 = \dfrac{(1-y^2)^2}{4y^2}$, $1 + \left(\dfrac{dx}{dy}\right)^2 = \dfrac{4y^2+(1-y^2)^2}{4y^2} = \dfrac{(1+y^2)^2}{4y^2}$ Thus, $S = 2\pi \int_1^3 y \dfrac{1+y^2}{2y}\, dy = \pi \int_1^3 (1+y^2)\, dy = \pi\left(y + \dfrac{1}{3}y^3\right)\Big]_1^3 = \pi\left(12 - \dfrac{4}{3}\right) = \dfrac{32\pi}{3}$.

31.12 $y = \dfrac{x^3}{6} + \dfrac{1}{2x}$, $1 \le x \le 2$; about the y-axis.

▋ $S = 2\pi \int_1^2 x\sqrt{1+(y')^2}\, dx$. Now, $y' = \dfrac{1}{2}x^2 - \dfrac{1}{2x^2}$, $(y')^2 = \dfrac{1}{4}x^4 - \dfrac{1}{2} + \dfrac{1}{4x^4}$, $1+(y')^2 = \dfrac{1}{4}x^4 + \dfrac{1}{2} + \dfrac{1}{4x^4} = \left(\dfrac{1}{2}x^2 + \dfrac{1}{2x^2}\right)^2$. So $S = 2\pi \int_1^2 x\left(\dfrac{1}{2}x^2 + \dfrac{1}{2x^2}\right)dx = 2\pi\left(\dfrac{1}{8}x^4 + \dfrac{1}{2}\ln|x|\right)\Big]_1^2 = 2\pi\left(2 + \dfrac{1}{2}\ln 2 - \dfrac{1}{8}\right) = \pi\left(\dfrac{15}{4} + \ln 2\right)$.

31.13 $y = \ln x$, $1 \le x \le 7$; about the y-axis.

▋ $S = 2\pi \int_1^7 x\sqrt{1+(y')^2}\, dx$. Now, $y' = \dfrac{1}{x}$, $1+(y')^2 = \dfrac{x^2+1}{x^2}$. Thus, $S = 2\pi \int_1^7 \sqrt{x^2+1}\, dx$. Let $x = \tan\theta$, $dx = \sec^2\theta\, d\theta$. Then $\int \sqrt{x^2+1}\, dx = \int \sec\theta \sec^2\theta\, d\theta = \int \sec^3\theta\, d\theta$. By Problem 28.40, this is equal to $\tfrac{1}{2}(\sec\theta\tan\theta + \ln|\sec\theta + \tan\theta|) = \tfrac{1}{2}(x\sqrt{x^2+1} + \ln|\sqrt{x^2+1}+x|)$. So we arrive at $2\pi \cdot \tfrac{1}{2}(x\sqrt{x^2+1} + \ln|\sqrt{x^2+1}+x|)\big]_1^7 = \pi[7\sqrt{50} + \ln(\sqrt{50}+7) - \sqrt{2} - \ln(\sqrt{2}+1)] = \pi\left(34\sqrt{2} + \ln\dfrac{5\sqrt{2}+7}{\sqrt{2}+1}\right) = \pi[34\sqrt{2} + \ln(3 + 2\sqrt{2})]$. $\left(\text{We used the fact that } \dfrac{5\sqrt{2}+7}{\sqrt{2}+1} = \dfrac{5\sqrt{2}+7}{\sqrt{2}+1} \cdot \dfrac{1-\sqrt{2}}{1-\sqrt{2}} = 3 + 2\sqrt{2}.\right)$

31.14 Find the surface area of a right circular cone of height h and radius of base r.

▋ As is shown in Fig. 31-1, the cone is obtained by revolving about the x-axis the region in the first quadrant under the line $y = \dfrac{r}{h}(h - x)$. $y' = -\dfrac{r}{h}$, $1+(y')^2 = \dfrac{r^2+h^2}{h^2}$. Hence, $S = 2\pi \int_0^h \dfrac{r}{h}(h-x) \dfrac{\sqrt{r^2+h^2}}{h}\, dx = \dfrac{2\pi r\sqrt{r^2+h^2}}{h^2}\left(hx - \dfrac{1}{2}x^2\right)\Big]_0^h = \dfrac{2\pi r\sqrt{r^2+h^2}}{h^2}\left(h^2 - \dfrac{1}{2}h^2\right) = \pi r\sqrt{r^2+h^2}$.

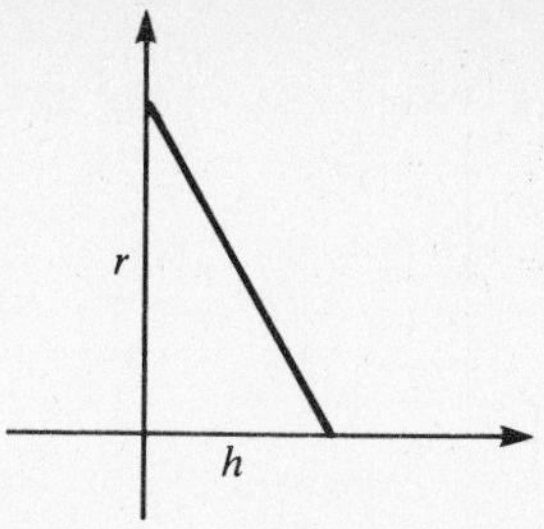

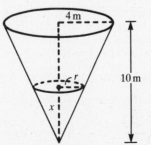

Fig. 31-1

31.15 Find the surface area of a cap of a sphere with radius a determined by a plane at a distance b from the center.

▮ The cap is generated by revolving about the x-axis the region in the first quadrant under $y = \sqrt{a^2 - x^2}$,
between $x = b$ and $x = a$. Since $x^2 + y^2 = a^2$, $2x + 2yy' = 0$, $y' = -\dfrac{x}{y}$, $1 + (y')^2 = \dfrac{y^2 + x^2}{y^2} = \dfrac{a^2}{y^2}$.
Hence, $S = 2\pi \int_b^a y\sqrt{1 + (y')^2}\, dx = 2\pi \int_b^a a\, dx = 2\pi a(a - b)$.

WORK

31.16 A spring with a natural length of 10 inches is stretched $\frac{1}{2}$ inch by a 12-pound force. Find the work done in stretching the spring from 10 to 18 inches.

▮ We use Hooke's law: The spring pulls back with a restoring force of $F = kx$ pounds, where the spring is stretched x inches beyond its natural length, and k is a constant. Then, $12 = \frac{1}{2}k$, $k = 24$. $F = 24x$, and the work $W = \int_0^8 F\, dx = \int_0^8 24x\, dx = 12x^2\,]_0^8 = 12(64) = 768$ in · lb $= 64$ ft · lb.

31.17 A spring supporting a railroad car has a natural length of 12 inches, and a force of 8000 pounds compresses it $\frac{1}{2}$ inch. Find the work done in compressing it from 12 to 9 inches.

▮ Hooke's law also holds for compression. Then $F = kx$, $8000 = \frac{1}{2}k$, $k = 16,000$. So the work $W = \int_0^3 16,000\, x\, dx = 8000x^2\,]_0^3 = 8000(9) = 72,000$ in · lb $= 6000$ ft · lb.

31.18 A bucket, weighing 5 pounds when empty, is loaded with 60 pounds of sand, and then lifted (at constant speed) 10 feet. Sand leaks out of a hole in the bucket at a uniform rate, and a third of the sand is lost by the end of the lifting. Find the work done in the lifting process.

▮ Let x be the height above the initial position. The sand leaks out at the uniform rate of 2 pounds per foot. The force being exerted when the bucket is at position x is $65 - 2x$, the weight of the load. Hence, the work $W = \int_0^{10} (65 - 2x)\, dx = (65x - x^2)\,]_0^{10} = 650 - 100 = 550$ ft · lb.

31.19 A 5-lb monkey is attached to the end of a 30-ft hanging rope that weighs 0.2 lb/ft. The monkey climbs the rope to the top. How much work has it done?

▮ At height x above its initial position, the monkey must exert a force $5 + 0.2x$ to balance its own weight and the weight of rope below that point. Hence, the work $W = \int_0^{30} (5 + 0.2x)\, dx = 5x + 0.1x^2\,]_0^{30} = 150 + 90 = 240$ ft · lb.

31.20 A conical tank, 10 meters deep and 8 meters across at the top, is filled with water to a depth of 5 meters. The tank is emptied by pumping the water over the top edge. How much work is done in the process?

Fig. 31-2

▌ See Fig. 31-2.   Let x be the distance from the bottom of the tank.   Consider a slab of water of thickness Δx at depth x.   By similar triangles, we see that $r/x = \frac{4}{10}$,   where r is the radius of the slab.   Then $r = 2x/5$.   The weight of the slab is approximately $w\pi r^2 \Delta x$, where w is the weight-density of water (about 9.8 kN/m³) and the slab is raised a distance of $10 - x$.   Hence, the work is $\int_0^5 w\pi r^2 (10 - x)\, dx = w\pi \int_0^5 \frac{4x^2}{25}(10 - x)\, dx =$
$\frac{4\pi w}{25}\left(\frac{10}{3} x^3 - \frac{1}{4} x^4\right)\Big]_0^5 = \frac{4\pi w}{25}\left(\frac{1250}{3} - \frac{625}{4}\right) = \frac{4\pi w}{25}\left(\frac{3125}{12}\right) = \frac{125}{3} \pi w \approx 1283 \text{ kJ}.$

31.21   A 100-ft cable weighing 5 lb/ft supports a safe weighing 500 lb.   Find the work done in winding 80 ft of the cable on a drum.

▌ Let x denote the length of cable that has been wound on the drum.   The total weight (unwound cable and safe) is $500 + 5(100 - x) = 1000 - 5x$,   and the work done in raising the safe a distance Δx is $(1000 - 5x)\,\Delta x$. Hence, the work $W = \int_0^{80} (1000 - 5x)\, dx = (1000x - \frac{5}{2}x^2)\,]_0^{80} = 64{,}000 \text{ ft} \cdot \text{lb}.$

31.22   A particle carrying a positive electrical charge $+q$ is released at a distance d from an immovable positive charge $+Q$.   How much work is done on the particle as its distance increases to $2d$?

▌ By Coulomb's law, the repulsive force on the particle when at distance $d + x$ is $kQq/(d + x)^2$,   where k is a universal constant.   Then the work is $W = \int_0^d \frac{kQq}{(d + x)^2}\, dx = kQq\,\frac{-1}{d + x}\,\Big]_0^d = kQq\left(-\frac{1}{2d} + \frac{1}{d}\right) = \frac{kQq}{2d}.$

31.23   The expansion of a gas in a cylinder causes a piston to move so that the volume of the enclosed gas increases from 15 to 25 cubic inches.   Assuming the relationship between the pressure p (lb/in²) and the volume y (in³) is $pv^{1.4} = 60$,   find the work done.

▌ If A is the area of a cross section of the cylinder, pA is the force exerted by the gas.   A volume increase Δv causes the piston to move a distance $\Delta v/A$, and the corresponding work is $pA \cdot \frac{\Delta v}{A} = \frac{60}{v^{1.4}} \Delta v.$ Then the work
$W = 60 \int_{15}^{25} \frac{dv}{v^{1.4}} = -\frac{60}{0.4} v^{-0.4}\,\Big]_{15}^{25} = -150\left(\frac{1}{25^{0.4}} - \frac{1}{15^{0.4}}\right) \approx 9.39 \text{ in} \cdot \text{lb}.$

CENTROID OF A PLANAR REGION

In Problems 31.24–31.32, find the centroid of the given planar region.

31.24   The region bounded by $y = x^2$, $y = 0$,   and $x = 1$ (Fig. 31-3).

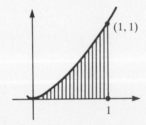

Fig. 31-3

▌ The area $A = \int_0^1 x^2\, dx = \frac{1}{3}x^3\,]_0^1 = \frac{1}{3}$.   The moment about the y-axis is $M_y = \int_0^1 x \cdot x^2\, dx = \frac{1}{4}x^4\,]_0^1 = \frac{1}{4}$. Hence, the x-coordinate of the centroid is $\bar{x} = M_y/A = \frac{1}{4}/\frac{1}{3} = \frac{3}{4}$.   The moment about the x-axis is $M_x = \int_0^1 y(1 - \sqrt{y})\, dy = \frac{1}{2}y^2 - \frac{2}{5}y^{5/2}\,]_0^1 = \frac{1}{2} - \frac{2}{5} = \frac{1}{10}$.   Hence, the y-coordinate of the centroid is $\bar{y} = M_x/A = (\frac{1}{10})/(\frac{1}{3}) = \frac{3}{10}$.   Thus, the centroid is $(\frac{3}{4}, \frac{3}{10})$.

31.25   The region bounded by the semicircle $y = \sqrt{a^2 - x^2}$,   and $y = 0$.

▌ By symmetry, $\bar{x} = 0$.   (In general, the centroid lies on any line of symmetry of the region.)   The area $A = \frac{\pi a^2}{2}$.   The moment about the x-axis $M_x = 2\int_0^a y\sqrt{a^2 - y^2}\, dy = -\frac{2}{3}(a^2 - y^2)^{3/2}\,\Big]_0^a = \frac{2}{3}(a^2)^{3/2} = \frac{2}{3}a^3.$
Thus, $\bar{y} = \left(\frac{2}{3}a^3\right)\Big/(\pi a^2/2) = \frac{4a}{3\pi}$.   Thus, the centroid is $\left(0, \frac{4a}{3\pi}\right)$.

31.26 The region bounded by $y = \sin x$, $y = 0$, from $x = 0$ to $x = \pi$ (Fig. 31-4).

$\blacksquare$ By symmetry, $\bar{x} = \dfrac{\pi}{2}$. The area $A = \int_0^\pi \sin x \, dx = -\cos x \,]_0^\pi = -(-1-1) = 2$. The moment about the x-axis $M_x = \int_0^1 y \cdot 2\left(\dfrac{\pi}{2} - \sin^{-1} y\right) dy = 2\int_0^1 \left(\dfrac{\pi}{2} y - y\sin^{-1} y\right) dy$. By Problem 29.40, $\int y\sin^{-1} y \, dy = \dfrac{1}{4}[(2y^2 - 1)\sin^{-1} y + y\sqrt{1-y^2}]$. Thus, $M_x = 2\left\{\dfrac{\pi}{4}y^2 - \dfrac{1}{4}((2y^2-1)\sin^{-1} y + y\sqrt{1-y^2})\right\}\Big]_0^1 = 2\left(\dfrac{\pi}{4} - \dfrac{\pi}{8}\right) = \dfrac{\pi}{4}$. Hence, $\bar{y} = \dfrac{\pi/4}{2} = \dfrac{\pi}{8}$.

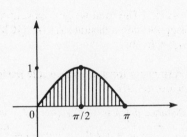

Fig. 31-4

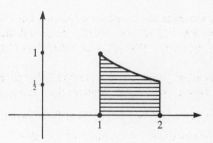

Fig. 31-5

31.27 The region bounded by $y = \dfrac{1}{x}$, $y = 0$, $x = 1$, $x = 2$ (Fig. 31-5).

$\blacksquare$ The area $A = \int_1^2 \dfrac{1}{x} \, dx = \ln x \,]_1^2 = \ln 2$. The moment about the y-axis is $M_y = \int_1^2 x \cdot \dfrac{1}{x} \, dx = \int_1^2 1 \, dx = 1$. Thus, $\bar{x} = \dfrac{1}{\ln 2}$. The moment about the x-axis is $M_x = \int_0^{1/2} y \, dy + \int_{1/2}^1 y\left(\dfrac{1}{y} - 1\right) dy = \dfrac{1}{2}y^2\Big]_0^{1/2} + \left(y - \dfrac{1}{2}y^2\right)\Big]_{1/2}^1 = \dfrac{1}{8} + \left[\left(1 - \dfrac{1}{2}\right) - \left(\dfrac{1}{2} - \dfrac{1}{8}\right)\right] = \dfrac{1}{4}$. Thus, $\bar{y} = \dfrac{\frac{1}{4}}{\ln 2}$. So the centroid is $\left(\dfrac{1}{\ln 2}, \dfrac{1}{4\ln 2}\right)$.

31.28 A right triangle with legs r and h.

$\blacksquare$ Let the right angle be at the origin, and let the legs r and h be along the positive x-axis and y-axis, respectively (Fig. 31-6). The hypotenuse is along the line $y = -\dfrac{h}{r}x + h$. The area A is $\dfrac{1}{2}hr$. The moment about the y-axis is $M_y = \int_0^r x\left(h - \dfrac{h}{r}x\right) dx = \left(\dfrac{1}{2}hx^2 - \dfrac{h}{3r}x^3\right)\Big]_0^r = \dfrac{1}{2}hr^2 - \dfrac{h}{3}r^2 = \dfrac{1}{6}hr^2$. Hence, $\bar{x} = \dfrac{\frac{1}{6}hr^2}{\frac{1}{2}hr} = \dfrac{1}{3}r$. In similar manner, $\bar{y} = \frac{1}{3}h$.

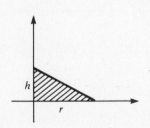

Fig. 31-6

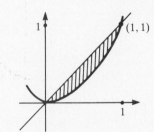

Fig. 31-7

31.29 The region bounded by $y = x^2$ and $y = x$ (Fig. 31-7).

$\blacksquare$ The area $A = \int_0^1 (x - x^2) \, dx = \dfrac{1}{2}x^2 - \dfrac{1}{3}x^3\Big]_0^1 = \dfrac{1}{2} - \dfrac{1}{3} = \dfrac{1}{6}$. The moment about y-axis is $M_y = \int_0^1 x(x - x^2) \, dx = \left(\dfrac{1}{3}x^3 - \dfrac{1}{4}x^4\right)\Big]_0^1 = \dfrac{1}{3} - \dfrac{1}{4} = \dfrac{1}{12}$. Thus, $\bar{x} = \dfrac{\frac{1}{12}}{\frac{1}{6}} = \dfrac{1}{2}$. The moment about the x-axis $M_x = \int_0^1 y(\sqrt{y} - y) \, dy = \left(\dfrac{2}{5}y^{5/2} - \dfrac{1}{3}y^3\right)\Big]_0^1 = \dfrac{2}{5} - \dfrac{1}{3} = \dfrac{1}{15}$. Thus, $\bar{y} = \dfrac{\frac{1}{15}}{\frac{1}{6}} = \dfrac{2}{5}$.

31.30 The region bounded by $y = e^x$, $y = e^{-x}$, and $x = 1$ (Fig. 31-8).

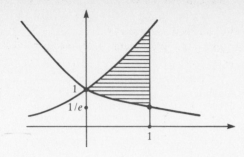

Fig. 31-8

∎ The area $A = \int_0^1 (e^x - e^{-x})\, dx = (e^x + e^{-x})\,]_0^1 = e + e^{-1} - 2 = (e-1)^2/e$. The moment about the y-axis is $M_y = \int_0^1 x(e^x - e^{-x})\, dx = \int_0^1 xe^x\, dx - \int_0^1 xe^{-x}\, dx = e^x(x-1)\,]_0^1 + e^{-x}(x+1)\,]_0^1 = 1 + (2/e - 1) = 2/e$. Hence, $\bar{x} = 2/(e-1)^2$. The moment about the x-axis is $M_x = \int_{1/e}^1 y(1 + \ln y)\, dy + \int_1^e y(1 - \ln y)\, dy = \int_{1/e}^e y\, dy + \int_{1/e}^1 y \ln y\, dy - \int_1^e y \ln y\, dy$. To compute $\int y \ln y\, dy$, we use integration by parts. Let $u = \ln y$, $dv = y\, dy$, $du = \dfrac{1}{y}\, dy$, $v = \frac{1}{2}y^2$. Then $\int y \ln dy = \frac{1}{2}y^2 \ln y - \frac{1}{4}y^2 = \frac{1}{4}y^2(2 \ln y - 1)$. Thus, $M_x = \frac{1}{2}y^2\,]_{1/e}^e + \frac{1}{4}y^2(2 \ln y - 1)\,]_{1/e}^1 - \frac{1}{4}y^2(2 \ln y - 1)\,]_1^e = \frac{1}{2}(e^2 - e^{-2}) + (-\frac{1}{4} + \frac{3}{4}e^{-2}) - (-\frac{1}{4}e^2 + \frac{1}{4}) = \frac{1}{4}(3e^2 - 2 + e^{-2})$. Thus, $\bar{y} = \dfrac{e(3e^2 - 2 + e^{-2})}{4(e-1)^2}$.

31.31 The region under $y = 4 - x^2$ in the first quadrant (Fig. 31-9).

∎ The area $A = \int_0^2 (4 - x^2)\, dx = (4x - \frac{1}{3}x^3)\,]_0^2 = \frac{16}{3}$. The moment about the y-axis is $M_y = \int_0^2 x(4 - x^2)\, dx = 2x^2 - \frac{1}{4}x^4\,]_0^2 = 8 - 4 = 4$. Thus, $\bar{x} = \dfrac{4}{\frac{16}{3}} = \dfrac{3}{4}$. The moment about the x-axis is $M_x = \int_0^4 y\sqrt{4 - y}\, dy$. Let $u = 4 - y$, $du = -dy$. Then $M_x = -\int_4^0 (4 - u)\sqrt{u}\, du = \int_0^4 (4\sqrt{u} - u^{3/2})\, du = (\frac{8}{3}u^{3/2} - \frac{2}{5}u^{5/2})\,]_0^4 = \frac{64}{3} - \frac{64}{5} = \frac{128}{15}$. Thus, $\bar{y} = \dfrac{\frac{128}{15}}{\frac{16}{3}} = \dfrac{8}{5}$.

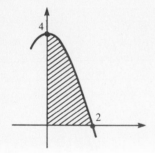

Fig. 31-9

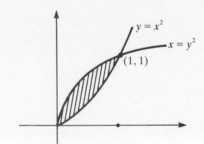

Fig. 31-10

31.32 The region between $y = x^2$ and $x = y^2$ (Fig. 31-10).

∎ The area $A = \int_0^1 (\sqrt{x} - x^2)\, dx = (\frac{2}{3}x^{3/2} - \frac{1}{3}x^3)\,]_0^1 = \frac{2}{3} - \frac{1}{3} = \frac{1}{3}$. The moment about the y-axis is $M_y = \int_0^1 x(\sqrt{x} - x^2)\, dx = (\frac{2}{5}x^{5/2} - \frac{1}{4}x^4)\,]_0^1 = \frac{2}{5} - \frac{1}{4} = \frac{3}{20}$. Hence, $\bar{x} = \dfrac{\frac{3}{20}}{\frac{1}{3}} = \dfrac{9}{20}$. By symmetry about the line $y = x$, $\bar{y} = \dfrac{9}{20}$.

31.33 Use Pappus's theorem to find the volume of a torus obtained by revolving a circle of radius a about a line in its plane at a distance b from its center $(b > a)$.

∎ Pappus's theorem states that the volume of a solid generated by revolving a region $\mathcal{R}$ about a line $\mathcal{L}$ not passing through the region is equal to the product of the area A of $\mathcal{R}$ and the distance d traveled around the line by its centroid. In this case, $A = \pi a^2$; the centroid is the center of the circle (by symmetry), so that $d = 2\pi b$. Hence, the volume $V = \pi a^2 \cdot 2\pi b = 2\pi^2 a^2 b$.

31.34 Use Pappus's theorem to find the volume of a right circular cone of height h and radius of base b.

❚ The cone is obtained by revolving a right triangle with legs r and h around the side of length h (Fig. 31-11). The area of the triangle is $A = \frac{1}{2}hr$ and, by Problem 31.28, the centroid is located at $(\frac{1}{3}r, \frac{1}{3}h)$. Therefore, $d = 2\pi(\frac{1}{3}r) = \frac{2}{3}\pi r$ and $V = (\frac{1}{2}hr) \cdot (\frac{2}{3}\pi r) = \frac{1}{3}\pi r^2 h$.

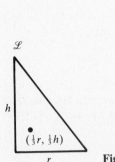

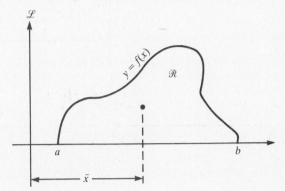

Fig. 31-11   **Fig. 31-12**

31.35 Establish Pappus's theorem in the important special case where the axis of revolution $\mathcal{L}$ is the y-axis and the region $\mathcal{R}$ lies completely in the first quadrant, being bounded by the x-axis and the curve $y = f(x)$. (See Fig. 31-12.)

❚ By the cylindrical shell method, the volume of revolution is $V = 2\pi \int_a^b xy\,dx$. But the x-coordinate of the centroid of $\mathcal{R}$ is defined as $\bar{x} = \dfrac{M_y}{A} = \dfrac{1}{A}\int_a^b xy\,dx$. Hence, $V = 2\pi A\bar{x} = A(2\pi\bar{x}) = Ad$.

CHAPTER 32
Improper Integrals

32.1 Determine whether the area in the first quadrant under the curve $y = 1/x$, for $x \geq 1$, is finite.

❚ This is equivalent to determining whether the improper integral $\int_1^\infty (1/x)\, dx$ is convergent. $\int_1^\infty (1/x)\, dx = \lim_{v \to +\infty} \int_1^v (1/x)\, dx = \lim_{v \to +\infty} \ln v\,]_1^v = \lim_{v \to +\infty} \ln v = +\infty$. Thus, the integral diverges and the area is infinite.

$$= \lim_{v \to +\infty} (\ln v - \ln 1)$$
$$0$$

32.2 Determine whether $\int_1^\infty (1/x^2)\, dx$ converges.

❚ $\int_1^\infty \frac{1}{x^2}\, dx = \lim_{v \to +\infty} \int_1^v \frac{1}{x^2}\, dx = \lim_{v \to +\infty} -\frac{1}{x}\Big]_1^v = \lim_{v \to +\infty} \left(-\frac{1}{v} + 1\right) = 1.$ Thus, the integral converges.

32.3 For what values of p is $\int_1^\infty (1/x)^p\, dx$ convergent?

❚ By Problem 32.1, we know that the integral is divergent when $p = 1$.

$$\int_1^\infty \frac{1}{x^p}\, dx = \lim_{v \to +\infty} \int_1^v \frac{1}{x^p}\, dx = \lim_{v \to +\infty} \left(-\frac{1}{p-1} \cdot \frac{1}{x^{p-1}}\right)\Big]_1^v = \lim_{v \to +\infty} \left[-\frac{1}{p-1}\left(\frac{1}{v^{p-1}} - 1\right)\right]$$

The last limit is $1/(p-1)$ if $p > 1$, and $+\infty$ if $p < 1$. Thus, the integral converges if and only if $p > 1$.

32.4 For $p > 1$, is $\int_1^\infty \frac{\ln x}{x^p}\, dx$ convergent?

❚ First we evaluate $\int [(\ln x)/x^p]\, dx$ by integration by parts. Let $u = \ln x$, $dv = (1/x^p)\, dx$, $du = (1/x)\, dx$, $v = \frac{1}{1-p}\frac{1}{x^{p-1}}$. Thus, $\int \frac{\ln x}{x^p}\, dx = \frac{1}{1-p}\frac{\ln x}{x^{p-1}} - \int \frac{1}{1-p}\frac{1}{x^p}\, dx = \frac{1}{1-p}\frac{\ln x}{x^{p-1}} - \frac{1}{(1-p)^2}\frac{1}{x^{p-1}}$. Hence,

$$\int_1^\infty \frac{\ln x}{x^p}\, dx = \lim_{v \to +\infty} \int_1^v \frac{\ln x}{x^p}\, dx = \lim_{v \to +\infty} \frac{1}{1-p}\frac{\ln x}{x^{p-1}} - \frac{1}{(1-p)^2}\frac{1}{x^{p-1}}\Big]_1^v$$

$$= \lim_{v \to +\infty}\left[\frac{1}{1-p}\frac{\ln v}{v^{p-1}} - \frac{1}{(1-p)^2}\frac{1}{v^{p-1}}\right] - \left[-\frac{1}{(1-p)^2}\right] = \frac{1}{(1-p)^2}$$

$\Big[$In the last step, we used L'Hôpital's rule to evaluate $\lim_{v \to +\infty} \frac{\ln v}{v^{p-1}} = \lim_{v \to +\infty} \frac{1/v}{(p-1)v^{p-2}} = \lim_{v \to +\infty} \frac{1}{p-1} \times \frac{1}{v^{p-1}} = 0.\Big]$ Thus, the integral converges for all $p > 1$.

32.5 For $p \leq 1$, is $\int_1^\infty \frac{\ln x}{x^p}\, dx$ convergent?

❚ $\frac{\ln x}{x^p} \geq \frac{1}{x^p}$ for $x \geq e$. Hence, $\int_e^v \frac{\ln x}{x^p}\, dx \geq \int_e^v \frac{1}{x^p}\, dx \to +\infty$ by Problem 32.3. Hence, $\int_1^\infty \frac{\ln x}{x^p}\, dx$ is divergent for $p \leq 1$.

32.6 Evaluate $\int_a^\infty xe^{-x}\, dx$.

❚ By integration by parts, we find $\int xe^{-x}\, dx = -e^{-x}(x+1)$. Hence, $\int_a^\infty xe^{-x}\, dx = \lim_{v \to +\infty} (-e^{-x}(x+1))\,]_a^v = \lim_{v \to +\infty}[-e^{-v}(v+1) + e^{-a}(a+1)] = e^{-a}(a+1)$. [In the last step, we used L'Hôpital's rule to evaluate $\lim_{v \to +\infty} (v+1)/e^v = \lim_{v \to +\infty} (1/e^v) = 0.$]

32.7 For positive p, show that $\int_1^\infty x^p e^{-x}\, dx$ converges.

▌ By Problem 32.6, $\int_1^\infty xe^{-x}\, dx = 2e^{-1}$. For $0 < p < 1$, $\int_1^v x^p e^{-x}\, dx \le \int_1^v xe^{-x}\, dx$. Hence, $\int_1^\infty x^p e^{-x}\, dx$ converges. Now let us consider $p > 1$. By the reduction formula of Problem 28.42, $\int x^p e^{-x}\, dx = -x^p e^{-x} + p \int x^{p-1} e^{-x}\, dx$. Hence, $\int_1^\infty x^p e^{-x}\, dx = \lim_{v \to +\infty} \int_1^v x^p e^{-x}\, dx = \lim_{v \to +\infty} (-x^p e^{-x}]_1^v + p \int_1^v x^{p-1} e^{-x}\, dx) = \lim_{v \to +\infty} (-v^p e^{-v} + e^{-1}) + p \int_1^v x^{p-1} e^{-x}\, dx = e^{-1} + p \int_1^\infty x^{p-1} e^{-x}\, dx$. (Note that we used L'Hôpital's rule to show that $\lim_{v \to +\infty} v^p e^{-v} = 0$.) Hence, the question eventually reduces to the case of $p \le 1$. Thus, we have convergence for all positive p.

32.8 Is $\displaystyle\int_e^\infty \frac{dx}{(\ln x)^p}$ convergent when $p \ge 1$?

▌ By successive applications of L'Hôpital's rule, we see that $\lim_{x \to +\infty} (\ln x)^p / x = 0$. Hence, $(\ln x)^p / x < 1$ for sufficiently large x. Thus, for some x_0, if $x \ge x_0$, $(\ln x)^p < x$, $1/(\ln x)^p > 1/x$. So, $\displaystyle\int_{x_0}^v \frac{dx}{(\ln x)^p} > \int_{x_0}^v \frac{dx}{x} \to +\infty$. Hence, the integral must be divergent for arbitrary $p \ge 1$.

32.9 If $\int_a^\infty f(x)\, dx = +\infty$ and $g(x) \ge f(x)$ for all $x \ge x_0$, show that $\int_a^\infty g(x)\, dx$ is divergent.

▌ $\int_a^v g(x)\, dx = \int_a^{x_0} g(x)\, dx + \int_{x_0}^v g(x)\, dx \ge \int_a^{x_0} g(x)\, dx + \int_{x_0}^v f(x)\, dx \to +\infty$.

32.10 Show that $\displaystyle\int_e^\infty \frac{dx}{(\ln x)^p}$ is divergent for $p < 1$.

▌ For $x \ge e$, $(\ln x)^p \le \ln x$, and, therefore, $1/(\ln x)^p \ge 1/\ln x$. Now apply Problems 32.8 and 32.9.

32.11 Evaluate $\displaystyle\int_0^\infty \frac{dx}{x^2 + a^2}$.

▌ $\displaystyle\int_0^v \frac{dx}{x^2 + a^2} = \frac{1}{\sqrt{a}} \tan^{-1} \frac{x}{a} \Big]_0^v = \frac{1}{\sqrt{a}} \tan^{-1} \frac{v}{a}$. But, $\lim_{v \to +\infty} \tan^{-1} \frac{v}{a} = \frac{\pi}{2}$. Hence, $\displaystyle\int_0^\infty \frac{dx}{x^2 + a^2} = \frac{\pi}{2\sqrt{a}}$.

32.12 Evaluate $\displaystyle\int_0^\infty \frac{dx}{e^x + e^{-x}}$.

▌ $\displaystyle\frac{1}{e^x + e^{-x}} = \frac{e^x}{e^{2x} + 1}$. Let $u = e^x$, $du = e^x\, dx$. Then $\displaystyle\int \frac{e^x\, dx}{e^{2x} + 1} = \int \frac{du}{u^2 + 1} = \tan^{-1} u = \tan^{-1} e^x$. Hence,

$$\int_0^\infty \frac{dx}{e^x + e^{-x}} = \lim_{v \to +\infty} \int_0^v \frac{e^x\, dx}{e^{2x} + 1} = \lim_{v \to +\infty} (\tan^{-1} e^v - \tan^{-1} 1) = \frac{\pi}{2} - \frac{\pi}{4} = \frac{\pi}{4}$$

32.13 Evaluate $\int_0^\infty e^{-x} \cos x\, dx$.

▌ By Problem 28.9, $\int e^{-x} \cos x\, dx = \frac{1}{2} e^{-x}(\sin x - \cos x)$. Hence, $\int_0^\infty e^{-x} \cos x\, dx = \lim \frac{1}{2} e^{-x}(\sin x - \cos x)]_0^v = \lim_{v \to +\infty} \frac{1}{2}[e^{-v}(\cos v - \sin v) - (-1)] = \frac{1}{2}$, since $\lim_{v \to +\infty} \frac{\cos v}{e^v} = \lim_{v \to +\infty} \frac{\sin v}{e^v} = 0$. ($|\cos v| \le 1$ and $|\sin v| \le 1$.)

32.14 Evaluate $\int_0^\infty e^{-x}\, dx$.

▌ $\int_0^\infty e^{-x}\, dx = \lim_{v \to +\infty} \int_0^v e^{-x}\, dx = \lim_{v \to +\infty} (-e^{-x})]_0^v = \lim_{v \to +\infty} (1 - e^{-v}) = 1 - 0 = 1$.

32.15 Evaluate $\displaystyle\int_2^\infty \frac{dx}{x \ln^2 x}$.

▌ $\displaystyle\int \frac{1}{x \ln^2 x}\, dx = \int (\ln x)^{-2} \cdot \left(\frac{1}{x}\right) dx = -(\ln x)^{-1}$. Hence,

$$\int_2^\infty \frac{dx}{x \ln^2 x} = \lim_{v \to +\infty} \int_2^v \frac{1}{x \ln^2 x}\, dx = \lim_{v \to +\infty} -(\ln x)^{-1}]_2^v = \lim_{v \to +\infty} \left(-\frac{1}{\ln v} + \frac{1}{\ln 2}\right) = \frac{1}{\ln 2}$$

32.16   Evaluate $\displaystyle\int_2^\infty \frac{1}{x \ln^k x}\,dx$.

  ▮ $\displaystyle\int_2^\infty \frac{1}{x \ln^k x}\,dx = \lim_{v \to +\infty} \int_2^v \frac{1}{x \ln^k x}\,dx = \lim_{v \to +\infty} \frac{(\ln x)^{-k+1}}{-k+1}\Big]_2^v = \lim_{v \to +\infty} \frac{1}{k-1}\left[\frac{-1}{(\ln v)^{k-1}} + \frac{1}{(\ln 2)^{k-1}}\right] = $
$\displaystyle\frac{1}{k-1}\frac{1}{(\ln 2)^{k-1}}$.

32.17   Evaluate $\displaystyle\int_1^\infty \frac{e^{-\sqrt{x}}}{\sqrt{x}}\,dx$.

  ▮ Let $u = \sqrt{x}$, $du = \dfrac{1}{2\sqrt{x}}\,dx$, $2u\,du = dx$. Then $\displaystyle\int_1^\infty \frac{e^{-\sqrt{x}}}{\sqrt{x}}\,dx = 2\int_1^\infty e^{-u}\,du = \lim_{v \to +\infty} -2e^{-u}\,]_1^v = $
$\displaystyle\lim_{v \to +\infty}\left(-\frac{2}{e^v} + \frac{2}{e}\right) = \frac{2}{e}$.

32.18   Evaluate $\displaystyle\int_{-\infty}^0 xe^x\,dx$.

  ▮ Let $x = -y$. Then $\displaystyle\int_{-\infty}^0 xe^x\,dx = -\int_0^\infty ye^{-y}\,dy = -1$ [by Problem 32.6].

32.19   Evaluate $\displaystyle\int_0^\infty e^{-\sqrt{x}}\,dx$.

  ▮ Let $x = t^2$. Then $\displaystyle\int_0^\infty e^{-\sqrt{x}}\,dx = 2\int_0^\infty te^{-t}\,dt = 2$ [by Problem 32.6].

32.20   Evaluate $\displaystyle\int_0^\infty x^2 e^{-x}\,dx$.

  ▮ By Problem 28.1, $\displaystyle\int_0^\infty x^2 e^{-x}\,dx = \lim_{v \to +\infty} \int_0^v x^2 e^{-x}\,dx = \lim_{v \to +\infty} (-e^{-x}(x^2 + 2x + 2))\,]_0^v = \lim_{v \to +\infty}\{-[e^{-v}(v^2 + 2v + $
$2) - 2]\} = 2$.   [Here, we used L'Hôpital's rule to see that $\displaystyle\lim_{v \to +\infty} e^{-v}(v^2 + 2v + 1) = 0$.]

32.21   Find $\displaystyle\int_0^\infty x^3 e^{-x}\,dx$.

  ▮ By the reduction formula of Problem 28.42 and the result of Problem 32.20, $\displaystyle\int x^3 e^{-x}\,dx = -x^3 e^{-x} + $
$3\int x^2 e^{-x}\,dx$. So, $\displaystyle\int_0^\infty x^3 e^{-x}\,dx = \lim_{v \to +\infty} (-x^3 e^{-x})\,]_0^v + \lim_{v \to +\infty} 3\int_0^v x^2 e^{-x}\,dx = 0 + 3 \cdot 2 = 3!$.

32.22   Show that $\displaystyle\int_0^\infty x^n e^{-x}\,dx = n!$ for all natural numbers n.

  ▮ By Problem 32.14, we know that the formula holds for $n = 0$. Assume now, for the sake of induction, that
the formula holds for $n - 1$. By the reduction formula of Problem 28.42, $\displaystyle\int_0^\infty x^n e^{-x}\,dx = \lim_{v \to +\infty} (-x^n e^{-x})\,]_0^v + $
$n\int_0^\infty x^{n-1} e^{-x}\,dx = n \cdot (n - 1)! = n!$. [The gamma function $\Gamma(u)$ is defined as $\Gamma(u) = \int_0^\infty x^{u-1} e^{-x}\,dx$. This
problem shows that $\Gamma(n + 1) = n!$.]

32.23   Investigate $\displaystyle\int_0^1 \frac{1}{x}\,dx$.

  ▮ $\displaystyle\int_0^1 \frac{1}{x}\,dx = \lim_{u \to 0^+} \int_u^1 \frac{1}{x}\,dx = \lim_{u \to 0^+} (\ln 1 - \ln u) = +\infty$.   Thus, the integral diverges.

32.24   Investigate $\displaystyle\int_0^1 \frac{1}{\sqrt{x}}\,dx$.

  ▮ $\displaystyle\int_0^1 \frac{1}{\sqrt{x}}\,dx = \lim_{u \to 0^+} \int_u^1 \frac{1}{\sqrt{x}}\,dx = \lim_{u \to 0^+} 2\sqrt{x}\,]_u^1 = \lim_{u \to 0^+} 2(1 - \sqrt{u}) = 2$.

32.25   Investigate $\displaystyle\int_0^1 \frac{1}{x^2}\,dx$.

  ▮ $\displaystyle\int_0^1 \frac{1}{x^2}\,dx = \lim_{u \to 0^+} \int_u^1 \frac{1}{x^2}\,dx = \lim_{u \to 0^+} \left(-\frac{1}{x}\right)\Big]_u^1 = \lim_{u \to 0^+} \left(-1 + \frac{1}{u}\right) = +\infty$.   Thus, the integral diverges.

32.26 For what values of k, with $k \neq 1$ and $k > 0$, does $\int_0^1 \frac{1}{x^k}\, dx$ converge?

▮ $\int_0^1 \frac{1}{x^k}\, dx = \lim_{u \to 0^+} \int_u^1 \frac{1}{x^k}\, dx = \lim_{u \to 0^+} \frac{1}{-k+1}\, x^{-k+1}\,\Big]_u^1 = \lim_{u \to 0^+} \frac{1}{1-k}\left(1 - \frac{1}{u^{k-1}}\right)$. If $k > 1$, this limit is $+\infty$, whereas, if $k < 1$, the limit is $1/(1-k)$.

32.27 Evaluate $\int_0^a \frac{dx}{a^2 - x^2}$, where $a > 0$.

▮ $\int \frac{dx}{a^2 - x^2} = \frac{1}{2a} \int \frac{dx}{a-x} + \frac{1}{2a} \int \frac{dx}{a+x} = -\frac{1}{2a} \ln|a-x| + \frac{1}{2a} \ln|a+x|$. Thus,

$$\int_0^a \frac{dx}{a^2 - x^2} = \lim_{v \to a^-} \int_0^v \frac{dx}{a^2 - x^2} = \lim_{v \to a^-} \left(-\frac{1}{2a} \ln|a-x| + \frac{1}{2a} \ln|a+x|\right)\Big]_0^v$$

$$= \lim_{v \to a^-} \left(-\frac{1}{2a} \ln|a-v| + \frac{1}{2a} \ln|a+v| + \frac{\ln a}{2a}\right) = +\infty + 2a \ln 2a + \frac{\ln a}{2a} = +\infty$$

Thus, the integral diverges.

32.28 Evaluate $\int_0^a \frac{dx}{\sqrt{a^2 - x^2}}$ for $a > 0$.

▮ $\int_0^a \frac{dx}{\sqrt{a^2 - x^2}} = \lim_{v \to a^-} \int_0^v \frac{dx}{\sqrt{a^2 - x^2}} = \lim_{v \to a^-} \sin^{-1} \frac{x}{a}\,\Big]_0^v = \lim_{v \to a^-} \sin^{-1} \frac{v}{a} = \frac{\pi}{2}$.

33.29 Evaluate $\int_1^3 \frac{dx}{x - 2}$.

▮ There is a discontinuity at $x = 2$. So, $\int_1^3 \frac{dx}{x-2} = \lim_{v \to 2^-} \int_1^v \frac{dx}{x-2} + \lim_{w \to 2^+} \int_w^3 \frac{dx}{x-2} = \lim_{v \to 2^-} \ln|x-2|\,]_1^v + \lim_{w \to 2^+} \ln|x-2|\,]_w^3 = \lim_{v \to 2^-} \ln|v-2| + \lim_{w \to 2^+} -\ln|w-2|$. Neither limit exists. Therefore, the integral diverges.

32.30 Evaluate $\int_1^5 \frac{dx}{\sqrt[3]{x-2}}$.

▮ There is a discontinuity at $x = 2$. Thus, $\int_1^5 \frac{dx}{\sqrt[3]{x-2}} = \int_1^2 \frac{dx}{\sqrt[3]{x-2}} + \int_2^5 \frac{dx}{\sqrt[3]{x-2}} = \lim_{v \to 2^-} \int_1^v \frac{dx}{\sqrt[3]{x-2}} + \lim_{u \to 2^+} \int_u^5 \frac{dx}{\sqrt[3]{x-2}} = \lim_{v \to 2^-} \frac{3}{2}(x-2)^{2/3}\,]_1^v + \lim_{u \to 2^+} \frac{3}{2}(x-2)^{2/3}\,]_u^5 = \lim_{v \to 2^-} [\frac{3}{2}(v-2)^{2/3} - \frac{3}{2}] + \lim_{u \to 2^+} [\frac{3}{2}(3)^{2/3} - \frac{3}{2}(u-2)^{2/3}] = -\frac{3}{2} + \frac{3}{2}(3)^{2/3} = \frac{3}{2}(\sqrt[3]{9} - 1)$.

32.31 Evaluate $\int_0^{\pi/2} \sec x\, dx$.

▮ $\int_0^{\pi/2} \sec x\, dx = \lim_{v \to \pi/2^-} \int_0^v \sec x\, dx = \lim_{v \to \pi/2^-} \ln|\sec x + \tan x|\,]_0^v = \lim_{v \to \pi/2^-} |\sec v + \tan v| = +\infty$. Thus, the integral diverges.

32.32 Evaluate $\int_0^{\pi/2} \frac{\sin x\, dx}{\sqrt{1 - \cos x}}$.

▮ $\int_0^{\pi/2} \frac{\sin x\, dx}{\sqrt{1 - \cos x}} = \lim_{u \to 0^+} \int_u^{\pi/2} \frac{\sin x\, dx}{\sqrt{1 - \cos x}} = \lim_{u \to 0^+} 2(1 - \cos x)^{1/2}\,]_u^{\pi/2} = \lim_{u \to 0^+} 2[1 - (1 - \cos u)^{1/2}] = 2$.

32.33 Find the area under the curve $y = x/\sqrt{1 - x^2}$ for $0 \leq x \leq 1$.

▮ $A = \int_0^1 \frac{x}{\sqrt{1 - x^2}}\, dx = \lim_{v \to 1^-} \int_0^v \frac{x}{\sqrt{1 - x^2}}\, dx = \lim_{v \to 1^-} (-(1 - x^2)^{1/2})\,]_0^v = \lim_{v \to 1^-} [-(1 - v^2)^{1/2} + 1] = 1$.

32.34 Find the area under $y = 1/(x^2 - a^2)$ for $x \ge a + 1$.

▮ From Problem 32.27,

$$A = \int_{a+1}^{\infty} \frac{dx}{x^2 - a^2} = \lim_{u \to +\infty} \int_{a+1}^{u} \frac{dx}{x^2 - a^2} = \frac{1}{2a} \lim_{v \to +\infty} \ln \left| \frac{a - x}{a + x} \right| \Big]_{a+1}^{v}$$

$$= \frac{1}{2a} \lim_{v \to +\infty} \left(\ln \frac{v - a}{v + a} - \ln \frac{1}{|2a + 1|} \right) = \frac{1}{2a} \ln |2a + 1|$$

32.35 Evaluate $\int_{-1}^{8} \frac{1}{x^{1/3}}\, dx$.

▮ There is a discontinuity at $x = 0$. So, $\int_{-1}^{8} \frac{1}{x^{1/3}}\, dx = \int_{-1}^{0} \frac{dx}{x^{1/3}} + \int_{0}^{8} \frac{dx}{x^{1/3}}$. For the first integral, $\int_{-1}^{0} \frac{dx}{x^{1/3}} = \lim_{u \to 0^-} \frac{3}{2} x^{2/3} \Big]_{-1}^{u} = \lim_{u \to 0^-} \frac{3}{2}(u^{2/3} - 1) = -\frac{3}{2}$. Also, $\int_{0}^{8} \frac{dx}{x^{1/3}} = \lim_{v \to 0^+} \frac{3}{2} x^{2/3} \Big]_{v}^{8} = \lim_{v \to 0^+} \frac{3}{2}(4 - v^{2/3}) = 6$. Thus, the value is $-\frac{3}{2} + 6 = \frac{9}{2}$.

32.36 Evaluate $\int_0^1 \ln x\, dx$.

▮ By integration by parts, $\int \ln x\, dx = x(\ln x - 1)$. Thus, $\int_0^1 \ln x\, dx = \lim_{v \to 0^+} x(\ln x - 1) \Big]_{v}^{1} = \lim_{v \to 0^+} [-1 - v(\ln v - 1)] = -1 - 0 = -1$. [The limit $\lim_{v \to 0^+} v(\ln v - 1) = 0$ is obtained by L'Hôpital's rule.]

32.37 Evaluate $\int_0^1 x \ln x\, dx$.

▮ By integration by parts, $\int x \ln x\, dx = \frac{1}{4}x^2(2 \ln x - 1)$ (Take $u = \ln x$, $v = x\, dx$.) Then $\int_0^1 x \ln x\, dx = \lim_{u \to 0^+} \frac{1}{4}x^2(2 \ln x - 1) \Big]_{u}^{1} = \lim_{u \to 0^+} [-\frac{1}{4} - \frac{1}{4}u^2(2 \ln u - 1)] = -\frac{1}{4}$.

32.38 Find the first quadrant area under $y = e^{-2x}$.

▮ $A = \int_0^{\infty} e^{-2x}\, dx = \lim_{v \to +\infty} \int_0^{v} e^{-2x}\, dx = \lim_{v \to +\infty} (-\frac{1}{2}e^{-2x}) \Big]_0^{v} = \lim_{v \to +\infty} [-\frac{1}{2}(e^{-2v} - 1)] = \frac{1}{2}$.

32.39 Find the volume of the solid obtained by revolving the region of Problem 32.38 about the x-axis.

▮ By the disk formula, $V = \pi \int_0^{\infty} (e^{-2x})^2\, dx = \pi \int_0^{\infty} e^{-4x}\, dx = \pi \lim_{v \to +\infty} \int_0^{v} e^{-4x}\, dx = \pi \lim_{v \to +\infty} (-\frac{1}{4}e^{-4x}) \Big]_0^{v} = -\frac{\pi}{4} \lim_{v \to +\infty} (e^{-4v} - 1) = \pi/4$.

32.40 Let $\mathcal{R}$ be the region in the first quadrant under $xy = 9$ and to the right of $x = 1$. Find the volume generated by revolving $\mathcal{R}$ about the x-axis.

▮ By the disk formula, $V = \pi \int_1^{\infty} \left(\frac{9}{x} \right)^2 dx = 81\pi \int_1^{\infty} \frac{dx}{x^2} = 81\pi \lim_{v \to +\infty} \left(-\frac{1}{v} + 1 \right) = 81\pi$.

32.41 Find the surface area of the volume in Problem 32.40.

▮ $S = 2\pi \int_1^{\infty} y\sqrt{1 + (y')^2}\, dx$. Note that $y = 9/x$, $y' = -9/x^2$,

$$S = 2\pi \int_1^{\infty} \frac{9}{x} \frac{\sqrt{x^4 + 81}}{x^2}\, dx = 18\pi \int_1^{\infty} \frac{\sqrt{x^4 + 81}}{x^3}\, dx$$

But

$$\frac{\sqrt{x^4 + 81}}{x^3} > \frac{\sqrt{x^4}}{x^3} = \frac{1}{x}$$

so by Problem 32.9, the integral diverges.

32.42 Investigate $\int_0^1 \dfrac{dx}{1-x^4}$.

▮ $\int_0^1 dx/(1-x^4) = \lim_{v\to 1^-} \int_0^v dx/(1-x^4)$. For $0 \le x < 1$, $1-x^4 = (1-x)(1+x)(1+x^2) < 4(1-x)$. Hence, $1/(1-x^4) > 1/4(1-x)$. But, $\int_0^1 dx/4(1-x) = \lim_{v\to 1^-} -\frac{1}{4}\ln|1-x|\,]_0^v = \lim_{v\to 1^-} -\frac{1}{4}(\ln|1-v|) = +\infty$. Thus, $\int_0^1 dx/(1-x^4) = +\infty$.

32.43 Determine whether $\int_0^1 \dfrac{dx}{x^2 + \sqrt{x}}$ converges.

▮ For $0 < x \le 1$, $1/(x^2 + \sqrt{x}) < 1/\sqrt{x}$ and (Problem 32.24) $\int_0^1 dx/\sqrt{x}$ converges. Hence $\int_0^1 dx/(x^2 + \sqrt{x})$ converges.

32.44 Determine whether $\int_0^\infty \cos x\, dx$ converges.

▮ $\int_0^\infty \cos x\, dx = \lim_{v\to +\infty} \int_0^v \cos x\, dx = \lim_{v\to +\infty} \sin x\,]_0^v = \lim_{v\to +\infty} \sin v$. Since the latter limit does not exist, $\int_0^\infty \cos x\, dx$ is not convergent.

32.45 Evaluate $\int_{1/\pi}^\infty \dfrac{1}{x^2} \sin \dfrac{1}{x}\, dx$.

▮ $\int_{1/\pi}^\infty \dfrac{1}{x^2} \sin \dfrac{1}{x}\, dx = \lim_{v\to +\infty} \int_{1/\pi}^v \dfrac{1}{x^2} \sin \dfrac{1}{x}\, dx = \lim_{v\to +\infty} \cos \dfrac{1}{x}\,\Big]_{1/\pi}^v = \lim_{v\to +\infty} \left(\cos \dfrac{1}{v} + 1\right) = 1 + 1 = 2$.

32.46 $\int_1^\infty \left(\dfrac{1}{\sqrt{x}} - \dfrac{1}{\sqrt{x+3}}\right) dx$.

▮ $\int_1^\infty \left(\dfrac{1}{\sqrt{x}} - \dfrac{1}{\sqrt{x+3}}\right) dx = \lim_{v\to +\infty} \int_1^v \left(\dfrac{1}{\sqrt{x}} - \dfrac{1}{\sqrt{x+3}}\right) dx = \lim_{v\to +\infty} (2\sqrt{x} - 2\sqrt{x+3})\,]_1^v = \lim_{v\to +\infty} 2[(\sqrt{v} - \sqrt{v+3}) - (1-2)] = \lim_{v\to +\infty} 2(\sqrt{v} - \sqrt{v+3}) + 2$. Now, $\sqrt{v} - \sqrt{v+3} = (\sqrt{v} - \sqrt{v+3}) \cdot \dfrac{\sqrt{v} + \sqrt{v+3}}{\sqrt{v} + \sqrt{v+3}} = \dfrac{-3}{\sqrt{v} + \sqrt{v+3}} \to 0$ as $v \to +\infty$. Hence, the improper integral has the value 2.

32.47 Show that the region in the first quadrant under the curve $y = 1/(x+1)^2$ has a finite area but does not have a centroid.

▮ $A = \int_0^\infty \dfrac{dx}{(x+1)^2} = \lim_{v\to +\infty} -\dfrac{1}{x+1}\,\Big]_0^v = \lim_{v\to +\infty} \left(-\dfrac{1}{v+1} + 1\right) = 1$. However,

$$M_y = \int_0^\infty \dfrac{x\, dx}{(x+1)^2} = \int_0^\infty \left[\dfrac{1}{x+1} - \dfrac{1}{(x+1)^2}\right] dx = \lim_{v\to +\infty} \left(\ln|x+1| + \dfrac{1}{x+1}\right)\Big]_0^v = +\infty$$

Hence, the x-coordinate of the centroid is infinite.

32.48 For what positive values of p is $\int_0^1 \dfrac{dx}{(1-x)^p}$ convergent?

▮ Let $u = 1 - x$, $du = -dx$. Then $\int_0^1 \dfrac{dx}{(1-x)^p} = -\int_1^0 \dfrac{du}{u^p}$. By Problem 32.26, the latter converges when and only when $p < 1$.

32.49 Evaluate $\int_0^\infty \dfrac{x\, dx}{x^4 + 1}$.

▮ Let $u = x^2$, $du = 2x\, dx$. Then $\int_0^\infty \dfrac{x\, dx}{x^4 + 1} = \dfrac{1}{2} \int_0^\infty \dfrac{du}{u^2 + 1} = \dfrac{1}{2} \lim_{v\to +\infty} \tan^{-1} u\,]_0^v = \dfrac{1}{2} \lim_{v\to +\infty} (\tan^{-1} v) = \dfrac{1}{2} \cdot (\pi/2) = \pi/4$.

32.50 Evaluate $\displaystyle\int_e^\infty \frac{dx}{x \ln x\sqrt{\ln x}}$.

▮ $\displaystyle\int_e^\infty \frac{dx}{x \ln x\sqrt{\ln x}} = \int_e^\infty (\ln x)^{-3/2} \cdot \frac{1}{x}\, dx = \lim_{v\to+\infty} -2(\ln x)^{-1/2}\,]_e^v = \lim_{v\to\infty}\left(-\frac{2}{\sqrt{\ln v}} + 2\right) = 2.$

32.51 Evaluate $\displaystyle\int_0^{\pi/2} \frac{dx}{1-\sin x}$.

▮ $\displaystyle\int_0^{\pi/2} \frac{dx}{1-\sin x} = \int_0^{\pi/2} \frac{1}{1-\sin x}\cdot\frac{1+\sin x}{1+\sin x}\,dx = \int_0^{\pi/2}\frac{1+\sin x}{\cos^2 x}\,dx = \int_0^{\pi/2} (\sec^2 x + \tan x\sec x)\,dx$

$\displaystyle = \lim_{v\to\pi/2^-} (\tan x + \sec x)\,]_0^v = \lim_{v\to\pi/2^-} [(\tan v + \sec v) - 1] = +\infty.$

32.52 Evaluate $\displaystyle\int_0^9 \frac{dx}{(x-1)^{2/3}}$.

▮ There is a discontinuity at $x=1$. Then

$\displaystyle\int_0^9 \frac{dx}{(x-1)^{2/3}} = \int_0^1 \frac{dx}{(x-1)^{2/3}} + \int_1^9 \frac{dx}{(x-1)^{2/3}} = \lim_{v\to1^-}\int_0^v \frac{dx}{(x-1)^{2/3}} + \lim_{w\to1^+}\int_w^9 \frac{dx}{(x-1)^{2/3}}$

$\displaystyle = \lim_{v\to1^-} 3(x-1)^{1/3}\,\Big]_0^v + \lim_{w\to1^+} 3(x-1)^{1/3}\,\Big]_w^9 = \lim_{v\to1^-} 3[(v-1)^{1/3} - (-1)] + \lim_{w\to1^+} 3[2 - (w-1)^{1/3}]$

$= 3 + 6 = 9$

32.53 Evaluate $\displaystyle\int_0^{\pi/2} \cot x\, dx$.

▮ $\displaystyle\int_0^{\pi/2} \cot x\, dx = \lim_{u\to0^+}\int_u^{\pi/2} \cot x\, dx = \lim_{u\to0^+} \ln|\sin x|\,]_u^{\pi/2} = \lim_{u\to0^+} -\ln(\sin u) = +\infty.$

32.54 Evaluate $\displaystyle\int_0^\infty \tan^{-1} x\, dx$.

▮ $\tan^{-1} x > \pi/4$ for $x>1$. Hence by Problem 32.9, $\displaystyle\int_0^\infty \tan^{-1} x\, dx = +\infty$.

32.55 Find the area between the curves $y = 1/x$ and $y = 1/(x+1)$ to the right of the line $x = 1$.

▮ The area $\displaystyle A = \int_1^\infty \left(\frac{1}{x} - \frac{1}{x+1}\right)dx = \lim_{v\to+\infty}\int_1^v \left(\frac{1}{x} - \frac{1}{x+1}\right)dx = \lim_{v\to+\infty} (\ln|x| - \ln|x+1|)\,]_1^v$

$\displaystyle = \lim_{v\to+\infty} \ln\left|\frac{v}{v+1}\right| - (-\ln 2) = \ln 2$

32.56 Find the area in the first quadrant under the curve $y = 1/(x^2 + 6x + 10)$.

▮ $\displaystyle A = \int_0^\infty \frac{dx}{x^2 + 6x + 10} = \lim_{v\to+\infty}\int_0^v \frac{dx}{(x+3)^2 + 1} = \lim_{v\to+\infty} \tan^{-1}(x+3)\,]_0^v = \lim_{v\to+\infty} [\tan^{-1}(v+3) - \tan^{-1} 3]$

$= \pi/2 - \tan^{-1} 3$

Problems 32.57–32.60 refer to the *Laplace transform* $L\{f\} \equiv \int_0^\infty e^{-st}f(t)\, dt$ of a function $f(t)$, where $s>0$. ($L\{f\}$ may not be defined at some or all $s>0$.) It is assumed that $\lim_{t\to+\infty} e^{-st}f(t) = 0$.

32.57 Calculate $L\{t\}$.

▮ $\displaystyle L\{t\} = \int_0^\infty e^{-st}t\, dt = \lim_{w\to+\infty}\int_0^w e^{-st}t\, dt = \lim_{w\to+\infty}\left(-\frac{1}{s}te^{-st} - \frac{1}{s^2}e^{-st}\right)\Big]_0^w$

$\displaystyle = \lim_{w\to+\infty}\left(-\frac{1}{s}we^{-sw} - \frac{1}{s^2}e^{-sw} + \frac{1}{s^2}\right) = 0 + 0 + \frac{1}{s^2}$

(Here, the integration was performed by parts: $u=t$, $dv = e^{-st}\, dt$.) Thus, $L\{t\} = 1/s^2$.

32.58 Calculate $L\{e^t\}$.

$\blacksquare$
$$L\{e^t\} = \int_0^\infty e^{-st} e^t \, dt = \int_0^\infty e^{(1-s)t} \, dt = \lim_{v \to +\infty} \int_0^v e^{(1-s)t} \, dt = \lim_{v \to +\infty} \frac{1}{1-s} e^{(1-s)t} \bigg]_0^v$$

$$= \lim_{v \to +\infty} \frac{1}{1-s} e^{(1-s)v} - \frac{1}{1-s} = \frac{1}{s-1}$$

The last limit is valid when $s > 1$. Thus, $L\{e^t\} = 1/(s-1)$ (defined for $s > 1$).

32.59 Calculate $L\{\cos t\}$.

$\blacksquare$ $L\{\cos t\} = \int_0^\infty e^{-st} \cos t \, dt = \lim_{v \to +\infty} \int_0^v e^{-st} \cos t \, dt$. By integration by parts (see Problem 28.9), we obtain

$$\lim_{v \to +\infty} \frac{e^{-st}}{s^2 + 1} (-s \cos t + \sin t) \bigg]_0^v = \lim_{v \to +\infty} \frac{e^{-sv}}{s^2 + 1} (\sin v - s \cos v) + \frac{s}{s^2 + 1} = \frac{s}{s^2 + 1}$$

Thus, $L\{\cos t\} = s/(s^2 + 1)$.

32.60 If $L\{f\}$ and $L\{f'\}$ are defined, show that $L\{f'\} = -f(0) + sL\{f\}$.

$\blacksquare$ For $L\{f'\}$, we use integration by parts with $u = e^{-st}$, $dv = f'(t) \, dt$. Then $L\{f'\} = \int_0^\infty e^{-st} f'(t) \, dt = \lim_{v \to +\infty} [e^{-st} f(t)]_0^v + s \int_0^v e^{-st} f(t) \, dt] = \lim_{v \to +\infty} [e^{-sv} f(v) - f(0) + s \int_0^v e^{-st} f(t) \, dt] = -f(0) + sL\{f\}$. [Here, we have used the basic hypothesis that $\lim_{v \to +\infty} e^{-sv} f(v) = 0$.]

CHAPTER 33
Planar Vectors

33.1   Find the vector from the point $A(1, -2)$ to the point $B(3, 7)$.

   ▌ The vector $\overrightarrow{AB} = (3 - 1, 7 - (-2)) = (2, 9)$. In general, the vector $\overrightarrow{P_1 P_2}$ from $P_1(x_1, y_1)$ to $P_2(x_2, y_2)$ is $(x_2 - x_1, \ y_2 - y_1)$.

33.2   Given vectors $\mathbf{A} = (2, 4)$ and $\mathbf{C} = (-3, 8)$, find $\mathbf{A} + \mathbf{C}$, $\mathbf{A} - \mathbf{C}$, and $3\mathbf{A}$.

   ▌ By componentwise addition, subtraction, and scalar multiplication, $\mathbf{A} + \mathbf{C} = (2 + (-3), 4 + 8) = (-1, 12)$, $\mathbf{A} - \mathbf{C} = (2 - (-3), \ 4 - 8) = (5, -4)$, and $3\mathbf{A} = (3 \cdot 2, 3 \cdot 4) = (6, 12)$.

33.3   Given $\mathbf{A} = 3\mathbf{i} + 4\mathbf{j}$ and $\mathbf{C} = 2\mathbf{i} - \mathbf{j}$, find the magnitude and direction of $\mathbf{A} + \mathbf{C}$.

   ▌ $\mathbf{A} + \mathbf{C} = 5\mathbf{i} + 3\mathbf{j}$. Therefore, $|\mathbf{A} + \mathbf{C}| = \sqrt{(5)^2 + (3)^2} = \sqrt{34}$. If θ is the angle made by $\mathbf{A} + \mathbf{C}$ with the positive x-axis, $\tan \theta = \frac{3}{5}$. From a table of tangents, $\theta \approx 30° 58'$.   *TI-84 : $\tan^{-1}(3/5) = 30.96376$ Deg. mode) Ans ▸ DMS $30° 57' 49.524''$*

33.4   Describe a method for resolving a vector $\mathbf{A}$ into components $\mathbf{A}_1$ and $\mathbf{A}_2$ that are, respectively, parallel and perpendicular to a given nonzero vector $\mathbf{B}$.

   ▌ $\mathbf{A} = \mathbf{A}_1 + \mathbf{A}_2$, $\mathbf{A}_1 = c\mathbf{B}$, $\mathbf{A}_2 \cdot \mathbf{B} = 0$. So, $\mathbf{A}_2 = \mathbf{A} - \mathbf{A}_1 = \mathbf{A} - c\mathbf{B}$, $0 = \mathbf{A}_2 \cdot \mathbf{B} = (\mathbf{A} - c\mathbf{B}) \cdot \mathbf{B} = \mathbf{A} \cdot \mathbf{B} - c|\mathbf{B}|^2$. Hence, $c = (\mathbf{A} \cdot \mathbf{B}) / |\mathbf{B}|^2$. Therefore, $\mathbf{A}_1 = \dfrac{\mathbf{A} \cdot \mathbf{B}}{|\mathbf{B}|^2} \mathbf{B}$, and $\mathbf{A}_2 = \mathbf{A} - c\mathbf{B} = \mathbf{A} - \dfrac{\mathbf{A} \cdot \mathbf{B}}{|\mathbf{B}|^2} \mathbf{B}$. Here, $(\mathbf{A} \cdot \mathbf{B}) / |\mathbf{B}|$ is the *scalar projection* of $\mathbf{A}$ on $\mathbf{B}$, and $\left(\dfrac{\mathbf{A} \cdot \mathbf{B}}{|\mathbf{B}|} \right) \dfrac{\mathbf{B}}{|\mathbf{B}|} = \mathbf{A}_1$ is the *vector projection* of $\mathbf{A}$ on $\mathbf{B}$.

33.5   Resolve $\mathbf{A} = (4, 3)$ into components $\mathbf{A}_1$ and $\mathbf{A}_2$ that are, respectively, parallel and perpendicular to $\mathbf{B} = (3, 1)$.

   ▌ From Problem 33.4 $c = (\mathbf{A} \cdot \mathbf{B}) / |\mathbf{B}|^2 = [(4 \cdot 3) + (3 \cdot 1) / 10] = \frac{3}{2}$. So, $\mathbf{A}_1 = c\mathbf{B} = \frac{3}{2}(3, 1) = (\frac{9}{2}, \frac{3}{2})$ and $\mathbf{A}_2 = \mathbf{A} - \mathbf{A}_1 = (4, 3) - (\frac{9}{2}, \frac{3}{2}) = (-\frac{1}{2}, \frac{3}{2})$.

33.6   Show that the vector $\mathbf{A} = (a, b)$ is perpendicular to the line $ax + by + c = 0$.

   ▌ Let $P_1(x_1, y_1)$ and $P_2(x_2, y_2)$ be two points on the line. Then $ax_1 + by_1 + c = 0$ and $ax_2 + by_2 + c = 0$. By subtraction, $a(x_1 - x_2) + b(y_1 - y_2) = 0$, or $(a, b) \cdot (x_1 - x_2, y_1 - y_2) = 0$. Thus, $(a, b) \cdot \overrightarrow{P_2 P_1} = 0$, $(a, b) \perp \overrightarrow{P_2 P_1}$. (Recall that two nonzero vectors are perpendicular to each other if and only if their dot product is 0.) Hence, (a, b) is perpendicular to the line.

33.7   Use vector methods to find an equation of the line M through the point $P_1(2, 3)$ that is perpendicular to the line $L: x + 2y + 5 = 0$.

   ▌ By Problem 33.6, $\mathbf{A} = (1, 2)$ is perpendicular to the line L. Let $P(x, y)$ be any point on the line M. $\overrightarrow{P_1 P} = (x - 2, y - 3)$ is parallel to M. So, $(x - 2, y - 3) = c(1, 2)$ for some scalar c. Hence, $x - 2 = c$, $y - 3 = 2c$. So, $y - 3 = 2(x - 2)$, $y = 2x - 1$.

33.8   Use vector methods to find an equation of the line N through the points $P_1(1, 4)$ and $P_2(3, -2)$.

   ▌ Let $P(x, y)$ be any point on N. Then $\overrightarrow{P_1 P} = (x - 1, y - 4)$ and $\overrightarrow{P_1 P_2} = (3 - 1, -2 - 4) = (2, -6)$. Clearly, $(6, 2)$ is perpendicular to $\overrightarrow{P_1 P_2}$, and, therefore, to $\overrightarrow{P_1 P}$. Thus, $0 = (6, 2) \cdot (x - 1, y - 4) = 6(x - 1) + 2(y - 4) = 6x + 2y - 14$. Hence, $3x + y - 7 = 0$ is an equation of N.

33.9   Use vector methods to find the distance from $P(2, 3)$ to the line $3x + 4y - 12 = 0$. See Fig. 33-1.

   ▌ At any convenient point on the line, say $A(4, 0)$, construct the vector $\mathbf{B} = (3, 4)$, which is perpendicular to the line. The required distance d is the magnitude of the scalar projection of $\overrightarrow{AP}$ on $\mathbf{B}$:

$$\frac{|\overrightarrow{AP} \cdot \mathbf{B}|}{|\mathbf{B}|} = \frac{|(-2, 3) \cdot (3, 4)|}{\sqrt{3^2 + 4^2}} = \frac{6}{5} \quad \text{[by Problem 33.4]}$$

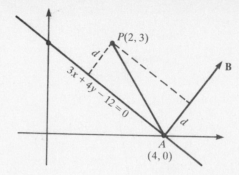

Fig. 33-1

33.10 Generalize the method of Problem 33.9 to find a formula for the distance from a point $P(x_1, y_1)$ to the line $ax + by + c = 0$.

▌ Take the point $A(-c/a, 0)$ on the line. The vector $\mathbf{B} = (a, b)$ is perpendicular to the line. As in Problem 33.9,

$$d = \frac{|\overrightarrow{AP} \cdot \mathbf{B}|}{|\mathbf{B}|} = \frac{|(x_1 + c/a, y_1 - 0) \cdot (a, b)|}{\sqrt{a^2 + b^2}} = \frac{|ax_1 + by_1 + c|}{\sqrt{a^2 + b^2}}$$

This derivation assumes $a \neq 0$. If $a = 0$, a similar derivation can be given, taking A to be $(0, -c/b)$.

33.11 If $\mathbf{A}, \mathbf{B}, \mathbf{C}, \mathbf{D}$ are consecutive sides of an oriented quadrilateral $PQRS$ (Fig. 33-2), show that $\mathbf{A} + \mathbf{B} + \mathbf{C} + \mathbf{D} = \mathbf{0}$. [$\mathbf{0}$ is $(0, 0)$, the *zero vector*.]

▌ $\overrightarrow{PR} = \overrightarrow{PQ} + \overrightarrow{QR} = \mathbf{A} + \mathbf{B}$. $\overrightarrow{PR} = \overrightarrow{PS} + \overrightarrow{SR} = -\mathbf{D} - \mathbf{C}$. Hence, $\mathbf{A} + \mathbf{B} = -\mathbf{D} - \mathbf{C}$, $\mathbf{A} + \mathbf{B} + \mathbf{C} + \mathbf{D} = \mathbf{0}$.

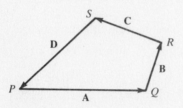

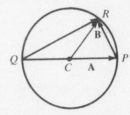

Fig. 33-2   **Fig. 33-3**

33.12 Prove by vector methods that an angle inscribed in a semicircle is a right angle.

▌ Let $\measuredangle QRP$ be inscribed on a diameter of a circle with center C and radius r (Fig. 33-3). Let $\mathbf{A} = \overrightarrow{CP}$ and $\mathbf{B} = \overrightarrow{CR}$. Then $\overrightarrow{QR} = \mathbf{A} + \mathbf{B}$ and $\overrightarrow{PR} = \mathbf{B} - \mathbf{A}$. $\overrightarrow{QR} \cdot \overrightarrow{PR} = (\mathbf{A} + \mathbf{B}) \cdot (\mathbf{B} - \mathbf{A}) = \mathbf{A} \cdot \mathbf{B} - \mathbf{A} \cdot \mathbf{A} + \mathbf{B} \cdot \mathbf{B} - \mathbf{B} \cdot \mathbf{A} = -r^2 + r^2 = 0$ (since $\mathbf{A} \cdot \mathbf{A} = \mathbf{B} \cdot \mathbf{B} = r^2$). Hence, $\overrightarrow{QR} \perp \overrightarrow{PR}$ and $\measuredangle QRP$ is a right angle.

33.13 Find the length of $\mathbf{A} = \mathbf{i} + \sqrt{3}\mathbf{j}$ and the angle it makes with the positive x-axis.

▌ $|\mathbf{A}| = \sqrt{1 + (\sqrt{3})^2} = \sqrt{4} = 2$. $\tan \theta = \sqrt{3}$, $\theta = \pi/3$.

33.14 Write the vector from $P_1(7, 5)$ to $P_2(6, 8)$ in the form $a\mathbf{i} + b\mathbf{j}$.

▌ $\overrightarrow{P_1 P_2} = (6 - 7, \ 8 - 5) = (-1, 3) = -\mathbf{i} + 3\mathbf{j}$.

33.15 Write the unit vector in the direction of $(5, 12)$ in the form $a\mathbf{i} + b\mathbf{j}$.

▌ $|(5, 12)| = \sqrt{25 + 144} = 13$. So, the required vector is $\frac{1}{13}(5, 12) = \frac{5}{13}\mathbf{i} + \frac{12}{13}\mathbf{j}$.

33.16 Write the vector of length 2 and direction $150°$ in the form $a\mathbf{i} + b\mathbf{j}$.

▌ In general, the vector of length r obtained by a counterclockwise rotation θ from the positive axis is given by $r(\cos \theta \, \mathbf{i} + \sin \theta \, \mathbf{j})$. In this case, we have $2\left(-\dfrac{\sqrt{3}}{2}\mathbf{i} + \dfrac{1}{2}\mathbf{j}\right) = -\sqrt{3}\mathbf{i} + \mathbf{j}$.

33.17 Given $O(0,0)$, $A(3,1)$, and $B(1,5)$ as vertices of the parallelogram $OAPB$, find the coordinates of P (see Fig. 33-4).

▮ Let $\mathbf{A} = (3,1)$ and $\mathbf{B} = (1,5)$. Then, by the parallelogram law, $\overrightarrow{OP} = \mathbf{A} + \mathbf{B} = (3,1) + (1,5) = (4,6)$. Hence, P has coordinates $(4,6)$.

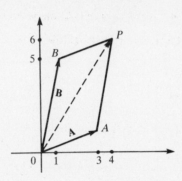

Fig. 33-4

33.18 Find k so that $\mathbf{A} = (3,-2)$ and $\mathbf{B} = (1,k)$ are perpendicular.

▮ We must have $0 = \mathbf{A} \cdot \mathbf{B} = 3 \cdot 1 + (-2) \cdot k = 3 - 2k$. Hence, $2k = 3$, $k = \frac{3}{2}$.

33.19 Find a vector perpendicular to the vector $(2,5)$.

▮ In general, given a vector (a,b), a perpendicular vector is $(b,-a)$, since $(a,b) \cdot (b,-a) = ab - ab = 0$. In this case, take $(5,-2)$.

33.20 Find the vector projection of $\mathbf{A} = (2,7)$ on $\mathbf{B} = (-3,1)$.

▮ By Problem 33.4, the projection is $\dfrac{\mathbf{A} \cdot \mathbf{B}}{|\mathbf{B}|^2} \mathbf{B} = \dfrac{(-6) + 7}{(-3)^2 + (1)^2} (-3,1) = \frac{1}{10}(-3,1) = (-\frac{3}{10}, \frac{1}{10})$.

33.21 Show that $\mathbf{A} = (3,-6)$, $\mathbf{B} = (4,2)$, and $\mathbf{C} = (-7,4)$ are the sides of a right triangle.

▮ $\mathbf{A} + \mathbf{B} + \mathbf{C} = 0$. Hence, $\mathbf{A}, \mathbf{B}, \mathbf{C}$ form a triangle. In addition, $\mathbf{A} \cdot \mathbf{B} = 3 \cdot 4 + (-6) \cdot 2 = 0$. Hence, $\mathbf{A} \perp \mathbf{B}$.

33.22 In Fig. 33-5, the ratio of segment PQ to segment PR is a certain number t, with $0 < t < 1$. Express $\mathbf{C}$ in terms of $\mathbf{A}$, $\mathbf{B}$, and t.

▮ $\overrightarrow{PQ} = t\,\overrightarrow{PR}$. But, $\overrightarrow{PR} = \mathbf{B} - \mathbf{A}$. So, $\mathbf{C} = \mathbf{A} + \overrightarrow{PQ} = \mathbf{A} + t\overrightarrow{PR} = \mathbf{A} + t(\mathbf{B} - \mathbf{A}) = (1-t)\mathbf{A} + t\mathbf{B}$.

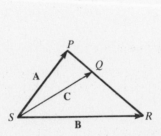

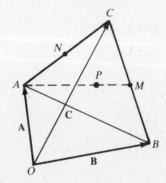

Fig. 33-5   Fig. 33-6

33.23 Prove by vector methods that the three medians of a triangle intersect at a point that is two-thirds of the way from any vertex to the opposite side.

▮ See Fig. 33-6. Let O be a point outside the given triangle $\triangle ABC$, and let $\mathbf{A} = \overrightarrow{OA}$, $\mathbf{B} = \overrightarrow{OB}$, $\mathbf{C} = \overrightarrow{OC}$. Let M be the midpoint of side BC. By Problem 33.22, $\overrightarrow{OM} = \frac{1}{2}(\mathbf{B} + \mathbf{C})$. So, $\overrightarrow{AM} = \overrightarrow{OM} - \mathbf{A} = \frac{1}{2}(\mathbf{B} + \mathbf{C}) - \mathbf{A}$. Let P be the point two-thirds of the way from A to M. Then $\overrightarrow{OP} = \mathbf{A} + \frac{2}{3}\overrightarrow{AM} = \mathbf{A} + \frac{2}{3}[\frac{1}{2}(\mathbf{B} + \mathbf{C}) - \mathbf{A}] = \frac{1}{3}(\mathbf{A} + \mathbf{B} + \mathbf{C})$. Similarly, if N is the midpoint of AC and Q is the point two-thirds of the way from B to N, $\overrightarrow{OQ} = \frac{1}{3}(\mathbf{A} + \mathbf{B} + \mathbf{C}) = \overrightarrow{OP}$. Hence, $P = Q$.

33.24 Find the two unit vectors that are parallel to the vector $7\mathbf{i} - \mathbf{j}$.

▮ $|7\mathbf{i} - \mathbf{j}| = \sqrt{(7)^2 + (-1)^2} = \sqrt{50} = 5\sqrt{2}$. Hence, $\mathbf{A} = \dfrac{1}{5\sqrt{2}}(7\mathbf{i} - \mathbf{j}) = \dfrac{7}{5\sqrt{2}}\mathbf{i} - \dfrac{1}{5\sqrt{2}}\mathbf{j}$ is a unit vector in the same direction as $7\mathbf{i} - \mathbf{j}$, and $-\mathbf{A} = -\dfrac{7}{5\sqrt{2}}\mathbf{i} + \dfrac{1}{5\sqrt{2}}\mathbf{j}$ is the unit vector in the opposite direction.

33.25 Find a vector of length 5 that has the direction opposite to the vector $\mathbf{B} = 7\mathbf{i} + 24\mathbf{j}$.

▮ $|\mathbf{B}| = \sqrt{49 + 576} = 25$. Hence, the unit vector in the direction of $\mathbf{B}$ is $\mathbf{C} = \frac{7}{25}\mathbf{i} + \frac{24}{25}\mathbf{j}$. Thus, the desired vector is $-5\mathbf{C} = -\frac{7}{5}\mathbf{i} - \frac{24}{5}\mathbf{j}$.

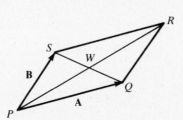

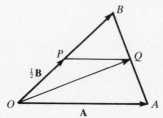

Fig. 33-7　　　　　　　　　　　　　　**Fig. 33-8**

33.26 Use vector methods to show that the diagonals of a parallelogram bisect each other.

▮ Let the diagonals of parallelogram $PQRS$ intersect at W (Fig. 33-7). Let $\mathbf{A} = \overrightarrow{PQ}$, $\mathbf{B} = \overrightarrow{PS}$. Then $\overrightarrow{PR} = \mathbf{A} + \mathbf{B}$, $\overrightarrow{SQ} = \mathbf{A} - \mathbf{B}$. Now, $\mathbf{B} = \overrightarrow{PW} + \overrightarrow{WS} = \overrightarrow{PW} - \overrightarrow{SW} = x\overrightarrow{PR} - y\overrightarrow{SQ} = x(\mathbf{A} + \mathbf{B}) - y(\mathbf{A} - \mathbf{B}) = (x - y)\mathbf{A} + (x + y)\mathbf{B}$, where x and y are certain numbers between 0 and 1. Hence, $x - y = 0$ and $x + y = 1$. So, $x = y = \frac{1}{2}$. Therefore, $\overrightarrow{PW} = \frac{1}{2}\overrightarrow{PR}$ and $\overrightarrow{SW} = \frac{1}{2}\overrightarrow{SQ}$, and the diagonals bisect each other.

33.27 Use vector methods to show that the line joining the midpoints of two sides of a triangle is parallel to and one-half the length of the third side.

▮ Let P and Q be the midpoints of sides OB and AB of $\triangle OAB$ (Fig. 33-8). Let $\mathbf{A} = \overrightarrow{OA}$ and $\mathbf{B} = \overrightarrow{OB}$. By Problem 33.22, $\overrightarrow{OQ} = \frac{1}{2}(\mathbf{A} + \mathbf{B})$. Also, $\overrightarrow{OP} = \frac{1}{2}\mathbf{B}$. Hence, $\overrightarrow{PQ} = \overrightarrow{OQ} - \overrightarrow{OP} = \frac{1}{2}(\mathbf{A} + \mathbf{B}) - \frac{1}{2}\mathbf{B} = \frac{1}{2}\mathbf{A}$. Thus, $\overrightarrow{PQ}$ is parallel to $\mathbf{A}$ and is half its length.

33.28 Prove Cauchy's inequality: $|\mathbf{A} \cdot \mathbf{B}| \le |\mathbf{A}|\,|\mathbf{B}|$.

▮ **Case 1.** $\mathbf{A} = \mathbf{0}$. Then $|\mathbf{A} \cdot \mathbf{B}| = |\mathbf{0} \cdot \mathbf{B}| = 0 = 0 \cdot |\mathbf{B}| = |\mathbf{0}|\,|\mathbf{B}|$. **Case 2.** $\mathbf{A} \ne \mathbf{0}$. Let $u = \mathbf{A} \cdot \mathbf{A}$ and $v = \mathbf{A} \cdot \mathbf{B}$, and let $\mathbf{C} = u\mathbf{B} - v\mathbf{A}$. Then, $\mathbf{C} \cdot \mathbf{C} = u^2(\mathbf{B} \cdot \mathbf{B}) - 2uv(\mathbf{A} \cdot \mathbf{B}) + v^2(\mathbf{A} \cdot \mathbf{A}) = u^2(\mathbf{B} \cdot \mathbf{B}) - uv^2 = u[u(\mathbf{B} \cdot \mathbf{B}) - v^2]$. Since $\mathbf{A} \ne \mathbf{0}$, $u > 0$. In addition, $\mathbf{C} \cdot \mathbf{C} \ge 0$. Thus, $u(\mathbf{B} \cdot \mathbf{B}) - v^2 \ge 0$, $u(\mathbf{B} \cdot \mathbf{B}) \ge v^2$, $\sqrt{u}\sqrt{\mathbf{B} \cdot \mathbf{B}} \ge |v|$, $|\mathbf{A}|\,|\mathbf{B}| \ge |\mathbf{A} \cdot \mathbf{B}|$.

33.29 Prove the triangle inequality: $|\mathbf{A} + \mathbf{B}| \le |\mathbf{A}| + |\mathbf{B}|$.

▮ By the Cauchy inequality, $|\mathbf{A} + \mathbf{B}|^2 = (\mathbf{A} + \mathbf{B}) \cdot (\mathbf{A} + \mathbf{B}) = \mathbf{A} \cdot \mathbf{A} + 2\mathbf{A} \cdot \mathbf{B} + \mathbf{B} \cdot \mathbf{B} \le |\mathbf{A}|^2 + 2|\mathbf{A}|\,|\mathbf{B}| + |\mathbf{B}|^2 = (|\mathbf{A}| + |\mathbf{B}|)^2$. Therefore, $|\mathbf{A} + \mathbf{B}| \le |\mathbf{A}| + |\mathbf{B}|$.

33.30 Prove $|\mathbf{A} + \mathbf{B}|^2 + |\mathbf{A} - \mathbf{B}|^2 = 2(|\mathbf{A}|^2 + |\mathbf{B}|^2)$, and interpret it geometrically.

▮ $|\mathbf{A} + \mathbf{B}|^2 + |\mathbf{A} - \mathbf{B}|^2 = (\mathbf{A} + \mathbf{B}) \cdot (\mathbf{A} + \mathbf{B}) + (\mathbf{A} - \mathbf{B}) \cdot (\mathbf{A} - \mathbf{B}) = \mathbf{A} \cdot \mathbf{A} + 2\mathbf{A} \cdot \mathbf{B} + \mathbf{B} \cdot \mathbf{B} + \mathbf{A} \cdot \mathbf{A} - 2\mathbf{A} \cdot \mathbf{B} + \mathbf{B} \cdot \mathbf{B} = 2\mathbf{A} \cdot \mathbf{A} + 2\mathbf{B} \cdot \mathbf{B} = 2(|\mathbf{A}|^2 + |\mathbf{B}|^2)$. Thus, the sum of the squares of the diagonals of a parallelogram is equal to twice the sum of the squares of the sides.

33.31 Show that, if $\mathbf{A} \cdot \mathbf{B} = \mathbf{A} \cdot \mathbf{C}$ and $\mathbf{A} \ne \mathbf{0}$, we cannot necessarily conclude that $\mathbf{B} = \mathbf{C}$.

▮ If $\mathbf{A} \cdot \mathbf{B} = \mathbf{A} \cdot \mathbf{C}$, then $\mathbf{A} \cdot (\mathbf{B} - \mathbf{C}) = 0$. So, $\mathbf{B} - \mathbf{C}$ can be any vector perpendicular to $\mathbf{A}$; $\mathbf{B} - \mathbf{C}$ need not be $\mathbf{0}$.

33.32 Prove by vector method that the diagonals of a rhombus are perpendicular.

▮ Let $PQRS$ be a rhombus, $\mathbf{A} = \overrightarrow{PQ}$, $\mathbf{B} = \overrightarrow{PS}$ (see Fig. 33-9). Then $|\mathbf{A}| = |\mathbf{B}|$. The diagonal vectors are $\overrightarrow{PR} = \mathbf{A} + \mathbf{B}$ and $\overrightarrow{SQ} = \mathbf{A} - \mathbf{B}$. Then $\overrightarrow{PR} \cdot \overrightarrow{SQ} = (\mathbf{A} + \mathbf{B}) \cdot (\mathbf{A} - \mathbf{B}) = \mathbf{A} \cdot \mathbf{A} + \mathbf{B} \cdot \mathbf{A} - \mathbf{A} \cdot \mathbf{B} - \mathbf{B} \cdot \mathbf{B} = |\mathbf{A}|^2 - |\mathbf{B}|^2 = 0$. Hence, $\overrightarrow{PR} \perp \overrightarrow{SQ}$.

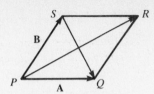

Fig. 33-9

33.33 Find the cosine of the angle between $\mathbf{A} = (1, 2)$ and $\mathbf{B} = (3, -4)$.

❚ $\mathbf{A} \cdot \mathbf{B} = |\mathbf{A}| \, |\mathbf{B}| \cos \theta$. So, $(1, 2) \cdot (3, -4) = \sqrt{5}\sqrt{25} \cos \theta$, $3 - 8 = 5\sqrt{5} \cos \theta$, $-1 = \sqrt{5} \cos \theta$, $\cos \theta = -1/\sqrt{5} = -\sqrt{5}/5$. Since $\cos \theta < 0$, θ is an obtuse angle.

33.34 Find the distance between the point $(2, 3)$ and the line $5x - 12y + 3 = 0$.

❚ By Problem 33.10, the distance is $\dfrac{|5 \cdot 2 - 12 \cdot 3 + 3|}{\sqrt{25 + 144}} = \dfrac{23}{13}$.

33.35 Find k so that the angle between $\mathbf{A} = (3, -2)$ and $\mathbf{B} = (1, k)$ is $60°$.

❚ $\mathbf{A} \cdot \mathbf{B} = |\mathbf{A}| \, |\mathbf{B}| \cos \theta$, $3 - 2k = \sqrt{13}\sqrt{1 + k^2} \cdot \frac{1}{2}$, $9 - 12k + 4k^2 = \frac{13}{4}(1 + k^2)$, $36 - 48k + 16k^2 = 13 + 13k^2$, $3k^2 - 48k + 23 = 0$, $k = \dfrac{48 \pm \sqrt{2028}}{6} = \dfrac{24 \pm \sqrt{507}}{3}$.

33.36 Find k so that $\mathbf{A} = (3, -2)$ and $\mathbf{B} = (1, k)$ are parallel.

❚ Let $\mathbf{A} = c\mathbf{B}$, $(3, -2) = c(1, k)$, $3 = c$ and $-2 = ck$. Hence, $-2 = 3k$, $k = -\frac{2}{3}$.

33.37 Prove that, if $\mathbf{A}$ is perpendicular to both $\mathbf{B}$ and $\mathbf{C}$, then $\mathbf{A}$ is perpendicular to any vector of the form $u\mathbf{B} + v\mathbf{C}$.

❚ $\mathbf{A} \cdot \mathbf{B} = 0$ and $\mathbf{A} \cdot \mathbf{C} = 0$. Hence, $\mathbf{A} \cdot (u\mathbf{B} + v\mathbf{C}) = u(\mathbf{A} \cdot \mathbf{B}) + v(\mathbf{A} \cdot \mathbf{C}) = u \cdot 0 + v \cdot 0 = 0$.

33.38 Let $\mathbf{A}$ and $\mathbf{B}$ be nonzero vectors, and let $a = |\mathbf{A}|$ and $b = |\mathbf{B}|$. Show that $\mathbf{C} = b\mathbf{A} + a\mathbf{B}$ bisects the angle between $\mathbf{A}$ and $\mathbf{B}$.

❚ Since $a > 0$ and $b > 0$, $\mathbf{C} = (a + b)\left(\dfrac{b}{a + b}\mathbf{A} + \dfrac{a}{a + b}\mathbf{B}\right) \equiv (a + b)\mathbf{C}^*$ lies between $\mathbf{A}$ and $\mathbf{B}$ (see Fig. 33-10). Let θ_1 be the angle between $\mathbf{A}$ and $\mathbf{C}$, and θ_2 the angle between $\mathbf{B}$ and $\mathbf{C}$. Now, $\mathbf{A} \cdot \mathbf{C} = \mathbf{A} \cdot (b\mathbf{A} + a\mathbf{B}) = b\mathbf{A} \cdot \mathbf{A} + a\mathbf{A} \cdot \mathbf{B}$ and $\mathbf{B} \cdot \mathbf{C} = \mathbf{B} \cdot (b\mathbf{A} + a\mathbf{B}) = b\mathbf{A} \cdot \mathbf{B} + a\mathbf{B} \cdot \mathbf{B}$. Then,

$$\cos \theta_1 = \frac{\mathbf{A} \cdot \mathbf{C}}{|\mathbf{A}| \, |\mathbf{C}|} = \frac{b|\mathbf{A}|}{|\mathbf{C}|} + \frac{\mathbf{A} \cdot \mathbf{B}}{|\mathbf{C}|} = \frac{ab + \mathbf{A} \cdot \mathbf{B}}{C}$$

Likewise,

$$\cos \theta_2 = \frac{\mathbf{A} \cdot \mathbf{B}}{|\mathbf{C}|} + \frac{a|\mathbf{B}|}{|\mathbf{C}|} = \frac{ab + \mathbf{A} \cdot \mathbf{B}}{|\mathbf{C}|} = \cos \theta_1$$

Hence, $\theta_1 = \theta_2$.

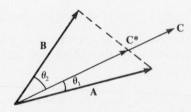

Fig. 33-10

33.39 Write the vector $\mathbf{A} = (7, 3)$ as the sum of a vector $\mathbf{C}$ parallel to $\mathbf{B} = (5, -12)$ and another vector $\mathbf{D}$ that is perpendicular to $\mathbf{C}$.

❚ The projection of $\mathbf{A}$ on $\mathbf{B}$ is $\mathbf{C} = \dfrac{\mathbf{A} \cdot \mathbf{B}}{|\mathbf{B}|^2} \mathbf{B} = -\frac{1}{169}(5, -12) = (-\frac{5}{169}, \frac{12}{169})$. Let $\mathbf{D} = \mathbf{A} - \mathbf{C} = (7, 3) - (-\frac{5}{169}, \frac{12}{169}) = (\frac{1188}{169}, \frac{495}{169})$. Note that $\mathbf{C} \cdot \mathbf{D} = \dfrac{-5940}{(169)^2} + \dfrac{5940}{(169)^2} = 0$.

33.40 For nonzero vectors **A** and **B**, find a necessary and sufficient condition that $\mathbf{A} \cdot \mathbf{B} = |\mathbf{A}| \, |\mathbf{B}|$.

▌ $\mathbf{A} \cdot \mathbf{B} = |\mathbf{A}| \, |\mathbf{B}| \cos \theta$, where θ is the angle between **A** and **B**. Hence, $\mathbf{A} \cdot \mathbf{B} = |\mathbf{A}| \, |\mathbf{B}|$ if and only if $\cos \theta = 1$, that is, if and only if $\theta = 0$, which is equivalent to **A** and **B** having the same direction. (In other words, $\mathbf{A} = \theta \mathbf{B}$ for some positive scalar u.)

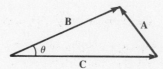

Fig. 33-11

33.41 Derive the low of cosines by vector methods.

▌ In the triangle of Fig. 33-11, let $|\mathbf{A}| = a$, $|\mathbf{B}| = b$, $|\mathbf{C}| = c$. Then $a^2 = \mathbf{A} \cdot \mathbf{A} = (\mathbf{B} - \mathbf{C}) \cdot (\mathbf{B} - \mathbf{C}) = \mathbf{B} \cdot \mathbf{B} + \mathbf{C} \cdot \mathbf{C} - 2\mathbf{B} \cdot \mathbf{C} = b^2 + c^2 - 2bc \cos \theta$.

CHAPTER 34
Parametric Equations, Vector Functions, Curvilinear Motion

PARAMETRIC EQUATIONS OF PLANE CURVES

34.1 Sketch the curve given by the parametric equations $x = a \cos \theta$, $y = a \sin \theta$.

▮ Note that $x^2 + y^2 = a^2 \cos^2 \theta + a^2 \sin^2 \theta = a^2(\cos^2 \theta + \sin^2 \theta) = a^2$. Thus, we have a circle of radius a with center at the origin. As shown in Fig. 34-1, the parameter θ can be thought of as the angle between the positive x-axis and the vector from the origin to the curve.

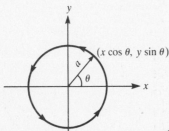

Fig. 34-1

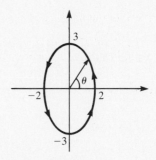

Fig. 34-2

34.2 Sketch the curve with the parametric equations $x = 2 \cos \theta$, $y = 3 \sin \theta$.

▮ $\dfrac{x^2}{4} + \dfrac{y^2}{9} = 1$. Hence, the curve is an ellipse with semimajor axis of length 3 along the y-axis and semiminor axis of length 2 along the x-axis (Fig. 34-2).

34.3 Sketch the curve with the parametric equations $x = t$, $y = t^2$.

▮ $y = t^2 = x^2$. Hence, the curve is a parabola with vertex at the origin and the y-axis as its axis of symmetry (Fig. 34-3).

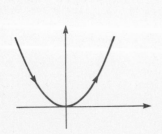

Fig. 34-3

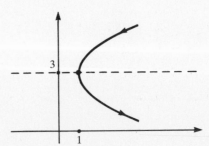

Fig. 34-4

34.4 Sketch the curve with the parametric equations $x = t$, $y = t^2$.

▮ $x = 1 + (3 - y)^2$, $x - 1 = (y - 3)^2$. Hence, the curve is a parabola with vertex at $(1, 3)$ and axis of symmetry $y = 3$ (Fig. 34-4).

34.5 Sketch the curve with the parametric equations $x = \sin t$, $y = -3 + 2 \cos t$.

▮ $x^2 + \left(\dfrac{y + 3}{2}\right)^2 = \sin^2 t + \cos^2 t = 1$. Thus, we have an ellipse with center $(0, -3)$, semimajor axis of length 2 along the y-axis, and semiminor axis of length 1 along the line $y = -3$ (Fig. 34-5).

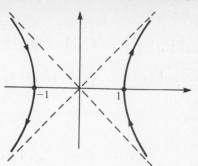

Fig. 34-5

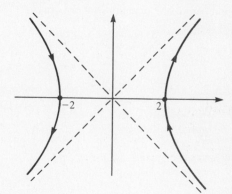

Fig. 34-6

34.6   Sketch the curve with the parametric equations $x = \sec t,\quad y = \tan t$.

▮ $x^2 = y^2 + 1$. Hence, $x^2 - y^2 = 1$. Thus, the curve is a rectangular hyperbola with the perpendicular asymptotes $y = \pm x$. See Fig. 34-6.

34.7   Sketch the curve with the parametric equations $x = \sin t,\quad y = \cos 2t$.

▮ $y = \cos 2t = 1 - 2\sin^2 t = 1 - 2x^2$, defined for $|x| \le 1$. Thus, the curve is an arc of a parabola, with vertex at $(0, 1)$, opening downward, and with the y-axis as axis of symmetry (Fig. 34-7).

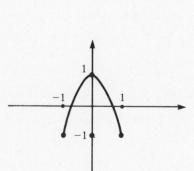

Fig. 34-7

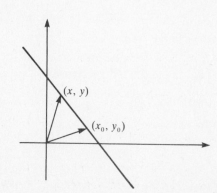

Fig. 34-8

34.8   Sketch the curve with the parametric equations $x = t + 1/t,\quad y = t - 1/t$.

▮ $x^2 = t^2 + 2 + 1/t^2$, $y^2 = t^2 - 2 + 1/t^2$. Subtracting the second equation from the first, we obtain the hyperbola $x^2 - y^2 = 4$ (Fig. 34-8).

34.9   Sketch the curve with the parametric equations $x = 1 + t,\quad y = 1 - t$.

▮ $x + y = 2$. Thus, we have a straight line, going through the point $(1, 1)$ and parallel to the vector $(1, -1)$; see Fig. 34-9.

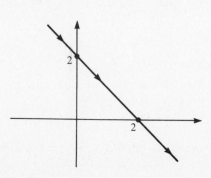

Fig. 34-9

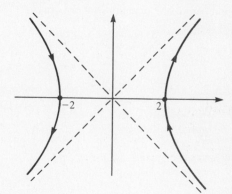

Fig. 34-10

34.10 Sketch the curve with the parametric equations $x = x_0 + at$, $y = y_0 + bt$, where a and b are not both 0.

▋ $bx = bx_0 + abt$, $ay = ay_0 + abt$. Subtracting the second equation from the first, we get $bx - ay = bx_0 - ay_0$. This is a line through the point (x_0, y_0) and parallel to the vector (a, b), since $(x, y) - (x_0, y_0) = t(a, b)$. See Fig. 34-10.

34.11 Find parametric equations for the ellipse $\dfrac{x^2}{25} + \dfrac{y^2}{144} = 1$.

▋ Let $x = 5 \cos \theta$, $y = 12 \sin \theta$. Then $\dfrac{x^2}{25} + \dfrac{y^2}{144} = \cos^2 \theta + \sin^2 \theta = 1$.

34.12 Find parametric equations for the hyperbola $\dfrac{x^2}{a^2} - \dfrac{y^2}{b^2} = 1$.

▋ Let $x = at + a/4t$, $y = bt - b/4t$. Then $(x/a)^2 = t^2 + \frac{1}{2} + 1/16t^2$, $(y/b)^2 = t^2 - \frac{1}{2} + 1/16t^2$. Hence, $(x/a)^2 - (y/b)^2 = 1$. Another possibility (cf. Problem 34.6) would be $x = a \sec u$, $y = b \tan u$.

34.13 Find parametric equations for $x^{2/3} + y^{2/3} = a^{2/3}$.

▋ It suffices to have $x^{2/3} = a^{2/3} \cos^2 \theta$ and $y^{2/3} = a^{2/3} \sin^2 \theta$. So, let $x = a \cos^3 \theta$, $y = a \sin^3 \theta$.

34.14 Find parametric equations for the circle $x^2 + y^2 - 4y = 0$.

▋ Complete the square: $x^2 + (y - 2)^2 = 4$. It suffices to have $x = 2 \cos \theta$, $y - 2 = 2 \sin \theta$. So, let $x = 2 \cos \theta$, $y = 2 + 2 \sin \theta$.

34.15 Sketch the curve given by the parametric equations $x = \cosh t$, $y = \sinh t$.

▋ We know that $\cosh^2 t - \sinh^2 t = 1$. Hence, we have $x^2 - y^2 = 1$. Since $x = \cosh t > 0$, we have only one branch of the hyperbola (Fig. 34-11).

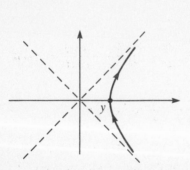

Fig. 34-11

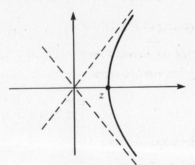

Fig. 34-12

34.16 Sketch the curve given by the parametric equations $x = 2 \cosh t$, $y = 3 \sinh t$.

▋ Since $\cosh^2 t - \sinh^2 t = 1$, $\left(\dfrac{x}{2}\right)^2 - \left(\dfrac{y}{3}\right)^2 = 1$, $\dfrac{x^2}{4} - \dfrac{y^2}{9} = 1$. Thus, we have one branch of a hyperbola, as shown in Fig. 34-12.

34.17 Find dy/dx and d^2y/dx^2 for the circle $x = r \cos \theta$, $y = r \sin \theta$.

▋ Recall that $\dfrac{dy}{dx} = \dfrac{dy/d\theta}{dx/d\theta}$. Since $dx/d\theta = -r \sin \theta$ and $dy/d\theta = r \cos \theta$, we have $dy/dx = r \cos \theta / (-r \sin \theta) = -\cot \theta = -x/y$. Remember also that $d^2y/dx^2 = \dfrac{d}{dx}\left(\dfrac{dy}{dx}\right) = \dfrac{d}{d\theta}\left(\dfrac{dy}{dx}\right) \Big/ \left(\dfrac{dx}{d\theta}\right)$. Hence,

$$\frac{d^2y}{dx^2} = \frac{\dfrac{d}{d\theta}(-\cot \theta)}{-r \sin \theta} = -\frac{\csc^2 \theta}{r \sin \theta} = -\frac{1}{r} \csc^3 \theta = -\frac{r^2}{y^3}$$

34.18 Find dy/dx and d^2y/dx^2 for $x = t^3 + t$, $y = t^7 + t + 1$.

▮ $dx/dt = 3t^2 + 1$, $dy/dt = 7t^6 + 1$. Then $\dfrac{dy}{dx} = \dfrac{dy/dt}{dx/dt} = \dfrac{7t^6 + 1}{3t^2 + 1}$. Further,

$$\frac{d^2y}{dx^2} = \frac{d}{dx}\left(\frac{dy}{dx}\right) = \frac{d}{dt}\left(\frac{dy}{dx}\right) \bigg/ \frac{dx}{dt}$$

But, $\dfrac{d}{dt}\left(\dfrac{dy}{dx}\right) = \dfrac{(3t^2 + 1)(42t^5) - (7t^6 + 1)6t}{(3t^2 + 1)^2} = \dfrac{t(84t^6 + 42t^4 - 6)}{(3t^2 + 1)^2} = \dfrac{6t(14t^6 + 7t^4 - 1)}{(3t^2 + 1)^2}$

Hence $\dfrac{d^2y}{dx^2} = \dfrac{6t(14t^6 + 7t^4 - 1)}{(3t^2 + 1)^3}$.

34.19 Find dy/dx and d^2y/dx^2 along the general curve $x = x(t)$, $y = y(t)$.

▮ Using dot notation for t-derivatives, we have

$$\frac{dy}{dx} = \frac{\dot{y}}{\dot{x}}$$

$$\frac{d^2y}{dx^2} = \frac{d}{dt}\left(\frac{\dot{y}}{\dot{x}}\right) \bigg/ \dot{x} = \frac{\dot{x}\ddot{y} - \dot{y}\ddot{x}}{\dot{x}^3}$$

34.20 Find the angle at which the cycloid $x = a\theta - a\sin\theta$, $y = a - a\cos\theta$ meets the x-axis at the origin.

▮ $dx/d\theta = a - a\cos\theta$, $dy/d\theta = a\sin\theta$. Hence, $dy/dx = \sin\theta/(1 - \cos\theta) = \cot\theta/2$, $\lim\limits_{\theta \to 0^+} \cot(\theta/2) = +\infty$. Therefore, the cycloid comes in vertically at the origin.

34.21 Find the slope of the curve $x = t^5 + \sin 2\pi t$, $y = t + e^t$ at $t = 1$.

▮ $\dfrac{dx}{dt} = t^4 + 2\pi\cos 2\pi t$, $\dfrac{dy}{dt} = 1 + e^t$. Hence, for $t = 1$, $\dfrac{dy}{dx} = \dfrac{dy/dt}{dx/dt} = \dfrac{1 + e^t}{t^4 + 2\pi\cos 2\pi t} = \dfrac{1 + e}{1 + 2\pi}$.

34.22 Find the slope of the curve $x = t^2 + e^t$, $y = t + e^t$ at the point $(1, 1)$.

▮ $dx/dt = 2t + e^t$, $dy/dt = 1 + e^t$. Hence, $\dfrac{dy}{dx} = \dfrac{1 + e^t}{2t + e^t}$. The point $(1, 1)$ corresponds to the parameter value $t = 0$. So, the slope is $dy/dx = \frac{2}{1} = 2$.

34.23 Find dy/dx and d^2y/dx^2 for $x = a\cos^3\theta$, $y = a\sin^3\theta$.

▮ $dx/d\theta = -3a\cos^2\theta\sin\theta$, $dy/d\theta = 3a\sin^2\theta\cos\theta$. So, $\dfrac{dy}{dx} = \dfrac{dy/d\theta}{dx/d\theta} = -\dfrac{\sin\theta}{\cos\theta} = -\tan\theta$. Further, $\dfrac{d}{d\theta}\left(\dfrac{dy}{dx}\right) = -\sec^2\theta$, and

$$\frac{d^2y}{dx^2} = \frac{d}{d\theta}\left(\frac{dy}{dx}\right) \bigg/ \frac{dx}{d\theta} = \frac{-\sec^2\theta}{-3a\cos^2\theta\sin\theta} = \frac{1}{3a}\sec^4\theta\csc\theta$$

34.24 Find the slope of $x = e^{-t}\cos 2t$, $y = e^{-2t}\sin 2t$ at $t = 0$.

▮ $dx/dt = -2e^{-t}\sin 2t - e^{-t}\cos 2t$, $dy/dt = 2e^{-2t}\cos 2t - 2e^{-2t}\sin 2t$. At $t = 0$, $dx/dt = -1$, $dy/dt = 2$. So, $\dfrac{dy}{dx} = \dfrac{dy/dt}{dx/dt} = -2$.

34.25 Find the coordinates of the highest point of the curve $x = 96t$, $y = 96t - 16t^2$.

▮ We must maximize y. $dy/dt = 96 - 32t$, $d^2y/dt^2 = -32$. So, the only critical number is $t = 3$, and, by the second-derivative test, we have a relative (and, therefore, an absolute) maximum. When $t = 3$, $x = 288$, $y = 144$.

34.26 Find an equation of the tangent line to the curve $x = 3e^t$, $y = 5e^{-t}$ at $t = 0$.

▮ $dx/dt = 3e^t$, $dy/dt = -5e^{-t}$, $\dfrac{dy}{dx} = \dfrac{dy/dt}{dx/dt} = -\frac{5}{3}e^{-2t}$. At $t = 0$, $dy/dx = -\frac{5}{3}$, $x = 3$, $y = 5$. Hence, the tangent line is $y - 5 = -\frac{5}{3}(x - 3)$, $3y - 15 = -5x + 15$, $5x + 3y - 30 = 0$. **Another method.** At $t = 0$, the tangent vector $(dx/dt, dy/dt) = (3, -5)$, so the normal vector is $(5, 3)$. Then, by Problem 33.6, the tangent line is given by $5x + 3y + c = 0$, where c is determined by the condition that the point $(x, y)_{t=0} = (3, 5)$ lies on the line.

34.27 Find an equation of the normal line to the curve $x = a \cos^4 \theta$, $y = a \sin^4 \theta$ at $\theta = \pi/4$.

▮ $dx/d\theta = 4a \cos^3 \theta (-\sin \theta)$, $dy/d\theta = 4a \sin^3 \theta (\cos \theta)$. At $\theta = \pi/4$, $dx/dt = -a$, $dy/d\theta = a$, giving as tangent vector $(-a, a) = -a(1, -1)$. So the normal line has equation $x - y + c = 0$. To find c, substitute the values of x and y corresponding to $\theta = \pi/4$: $\dfrac{a}{4} - \dfrac{a}{4} + c = 0$ or $c = 0$.

34.28 Find the slope of the curve $x = 3t - 1$, $y = 9t^2 - 3t$ when $t = 1$.

▮ $dx/dt = 3$, $dy/dt = 18t - 3$. Hence, $dy/dt = 6t - 1 = 5$ when $t = 1$.

34.29 For the curve of Problem 34.28, determine where it is concave upward.

▮ A curve is concave upward where $d^2y/dx^2 > 0$. In this case, $\dfrac{d^2y}{dx^2} = \dfrac{\dfrac{d}{dt}\left(\dfrac{dy}{dx}\right)}{dx/dt} = \dfrac{6}{3} = 2 > 0$. Hence, the curve is concave upward everywhere.

34.30 Where is the curve $x = \ln t$, $y = e^t$ concave upward?

▮ $dx/dt = 1/t$, $dy/dt = e^t$. So, $dy/dx = te^t$, $\dfrac{d^2y}{dx^2} = \dfrac{\dfrac{d}{dt}(te^t)}{1/t} = e^t t(t + 1)$. Thus, $d^2y/dx^2 > 0$ if and only if $t(t + 1) > 0$. Since $t > 0$ (in order for $x = \ln t$ to be defined), the curve is concave upward everywhere.

34.31 Where does the curve $x = 2t^2 - 5$, $y = t^3 + t$ have a tangent line that is perpendicular to the line $x + y + 3 = 0$?

▮ $dx/dt = 4t$, $dy/dt = 3t^2 + 1$. So, $dy/dx = (3t^2 + 1)/4t$. The slope of the line $x + y + 3 = 0$ is -1, and, therefore, the slope of a line perpendicular to it is 1. Thus, we must have $dy/dx = 1$, $(3t^2 + 1)/4t = 1$, $3t^2 + 1 = 4t$, $3t^2 - 4t + 1 = 0$, $(3t - 1)(t - 1) = 0$, $t = \frac{1}{3}$ or $t = 1$. Hence, the required points are $(-\frac{43}{9}, \frac{10}{27})$ and $(-3, 2)$.

34.32 Find the arc length of the circle $x = a \cos \theta$, $y = a \sin \theta$, $0 \le \theta \le 2\pi$.

▮ Recall that the arc length $s = \int_a^b \sqrt{(dx/du)^2 + (dy/du)^2}\, du$, where u is the parameter. In this case, $dx/d\theta = -a \sin \theta$, $dy/d\theta = a \cos \theta$, and $s = \int_0^{2\pi} \sqrt{a^2 \sin^2 \theta + a^2 \cos^2 \theta}\, d\theta = a \int_0^{2\pi} d\theta = a(2\pi - 0) = 2\pi a$, the standard formula for the circumference of a circle of radius a.

34.33 Find the arc length of the curve $x = e^t \cos t$, $y = e^t \sin t$, from $t = 0$ to $t = \pi$.

▮ $dx/dt = e^t(-\sin t) + e^t(\cos t) = e^t(\cos t - \sin t)$, $(dx/dt)^2 = e^t(\cos^2 t - 2 \sin t \cos t + \sin^2 t) = e^{2t}(1 - 2 \sin t \cos t)$. $dy/dt = e^t \cos t + e^t \sin t = e^t(\cos t + \sin t)$, $(dy/dt)^2 = e^{2t}(\cos^2 t + 2 \sin t \cos t + \sin^2 t) = e^{2t}(1 + 2 \sin t \cos t)$. So, $s = \int_0^\pi \sqrt{e^{2t}(1 - 2 \sin t \cos t) + e^{2t}(1 + \sin t \cos t)}\, dt = \int_0^\pi \sqrt{2}e^t\, dt = \sqrt{2}e^t\,]_0^\pi = \sqrt{2}(e^\pi - 1)$.

34.34 Find the arc length of the curve $x = \frac{1}{2} \ln (1 + t^2)$, $y = \tan^{-1} t$, from $t = 0$ to $t = 1$.

▮ $dx/dt = t/(1 + t^2)$, $(dx/dt)^2 = t^2/(1 + t^2)^2$. $dy/dt = 1/(1 + t^2)$, $(dy/dt)^2 = 1/(1 + t^2)^2$. Hence,

$$s = \int_0^1 \sqrt{\frac{t^2}{(1 + t^2)^2} + \frac{1}{(1 + t^2)^2}}\, dt = \int_0^1 \frac{1}{\sqrt{1 + t^2}}\, dt$$

The substitution $t = \tan \theta$, $dt = \sec^2 \theta\, d\theta$ yields $\int_0^{\pi/4} \sec \theta\, d\theta = \ln |\sec \theta + \tan \theta|\,]_0^{\pi/4} = \ln |\sqrt{2} + 1| - \ln 1 = \ln (\sqrt{2} + 1)$.

34.35 Find the arc length of $x = 2 \cos \theta + \cos 2\theta + 1$, $y = 2 \sin \theta + \sin 2\theta$, for $0 \le \theta \le 2\pi$.

▮ $dx/d\theta = -2 \sin \theta - 2 \sin 2\theta$, $(dx/d\theta)^2 = 4(\sin^2 \theta + 2 \sin \theta \sin 2\theta + \sin^2 2\theta)$. $dy/d\theta = 2 \cos \theta + 2 \cos 2\theta$, $(dy/d\theta)^2 = 4(\cos^2 \theta + 2 \cos \theta \cos 2\theta + \cos^2 2\theta)$. So,

$$s = \int_0^{2\pi} \sqrt{(dx/d\theta)^2 + (dy/d\theta)^2}\, d\theta = \int_0^{2\pi} 2\sqrt{2 + 2 \sin \theta \sin 2\theta + 2 \cos \theta \cos 2\theta}\, d\theta = 2\sqrt{2} \int_0^{2\pi} \sqrt{1 + \cos \theta}\, d\theta.$$

[Note that $\sin \theta \sin 2\theta + \cos \theta \cos 2\theta = \cos (2\theta - \theta) = \cos \theta$.] Since $1 + \cos \theta = 2 \cos^2(\theta/2)$, $\sqrt{1 + \cos \theta} = \sqrt{2}\,|\cos (\theta/2)|$. Thus, we have: $2\sqrt{2}[\int_0^\pi \sqrt{2} \cos (\theta/2)\, d\theta + \int_\pi^{2\pi} -\sqrt{2} \cos (\theta/2)\, d\theta] = 4\{2 \sin (\theta/2)\,]_0^\pi - 2 \sin (\theta/2)\,]_\pi^{2\pi}\} = 8[(1 - 0) - (0 - 1)] = 16$.

34.36 Find the arc length of $x = \frac{1}{2}t^2$, $y = \frac{1}{9}(6t + 9)^{3/2}$, from $t = 0$ to $t = 4$.

$dx/dt = t$, $dy/dt = (6t + 9)^{1/2}$, $(dx/dt)^2 = t^2$, $(dy/dt)^2 = 6t + 9$. So, $s = \int_0^4 \sqrt{t^2 + 6t + 9}\, dt = \int_0^4 (t + 3)\, dt = (\frac{1}{2}t^2 + 3t)\,]_0^4 = 8 + 12 = 20$.

34.37 Find the arc length of $x = a\cos^3\theta$, $y = a\sin^3\theta$, from $\theta = 0$ to $\theta = \pi/2$.

$dx/d\theta = -3a\cos^2\theta\sin\theta$, $(dx/d\theta)^2 = 9a^2\cos^4\theta\sin^2\theta$, $dy/d\theta = 3a\sin^2\theta\cos\theta$, $(dy/d\theta)^2 = 9a^2\sin^4\theta\cos^2\theta$. Thus, $s = 3a\int_0^{\pi/2}\sqrt{\cos^4\theta\sin^2\theta + \sin^4\theta\cos^2\theta}\, d\theta = 3a\int_0^{\pi/2}\sqrt{\cos^2\theta\sin^2\theta(\cos^2\theta + \sin^2\theta)}\, d\theta = 3a\int_0^{\pi/2}\cos\theta\sin\theta\, d\theta = 3a(\frac{1}{2}\sin^2\theta)\,]_0^{\pi/2} = \dfrac{3a}{2}(1 - 0) = \dfrac{3a}{2}$.

34.38 Find the arc length of $x = \cos t + t\sin t$, $y = \sin t - t\cos t$, from $t = \pi/6$ to $t = \pi/4$.

$dx/dt = -\sin t + \sin t + t\cos t = t\cos t$, $dy/dt = \cos t - \cos t + t\sin t = t\sin t$. Hence,

$$s = \int_{\pi/6}^{\pi/4}\sqrt{t^2\cos^2 t + t^2\sin^2 t}\, dt = \int_{\pi/6}^{\pi/4} t\, dt = \frac{1}{2}t^2\,\Big]_{\pi/6}^{\pi/4} = \frac{1}{2}\left(\frac{\pi^2}{16} - \frac{\pi^2}{36}\right) = \frac{\pi^2}{288}(9 - 4) = \frac{5\pi^2}{288}$$

34.39 Find the length of one arch of the cycloid $x = a(\theta - \sin\theta)$, $y = a(1 - \cos\theta)$, $0 \le \theta \le 2\pi$.

$dx/d\theta = a(1 - \cos\theta)$, $dy/d\theta = a\sin\theta$. So, $s = \int_0^{2\pi} a\sqrt{(1 - \cos\theta)^2 + \sin^2\theta}\, d\theta = \int_0^{2\pi} a\sqrt{2 - 2\cos\theta}\, d\theta = \int_0^{2\pi} 2a\sin(\theta/2)\, d\theta = 4a(-\cos(\theta/2))\,]_0^{2\pi} = -4a(-1 - 1) = 8a$.

34.40 Find the arc length of $x = u$, $y = u^{3/2}$, $0 \le u \le \frac{4}{3}$.

$dx/du = 1$, $dy/du = \frac{3}{2}u^{1/2}$. Hence, $s = \int_0^{4/3}\sqrt{1 + \frac{9}{4}u}\, du = \frac{4}{9}\cdot\frac{2}{3}(1 + \frac{9}{4}u)^{3/2}\,]_0^{4/3} = \frac{8}{27}(8 - 1) = \frac{56}{27}$.

34.41 Find the arc length of $x = \ln\sin\theta$, $y = \theta$, $\pi/6 \le \theta \le \pi/2$.

$dx/d\theta = \cot\theta$, $dy/d\theta = 1$. Hence, $s = \int_{\pi/6}^{\pi/2}\sqrt{\cot^2\theta + 1}\, d\theta = \int_{\pi/6}^{\pi/2}\csc\theta\, d\theta = \ln|\csc\theta - \cot\theta|\,]_{\pi/6}^{\pi/2} = [\ln 1 - \ln(2 - \sqrt{3})] = -\ln(2 - \sqrt{3}) = \ln(2 + \sqrt{3})$.

34.42 Rework Problem 34.33 in polar coordinates (r, θ), where $r = \sqrt{x^2 + y^2}$, $\tan\theta = y/x$.

On the given curve, $\sqrt{x^2 + y^2} = \sqrt{e^{2t}(\cos^2 t + \sin^2 t)} = e^t$ and $y/x = \tan t$. Thus, replacing t by θ, we have as the equation of the curve in polar coordinates: $r = e^\theta$ or $\theta = \ln r$ (a logarithmic spiral). Using the arc length formula $s = \int_a^b\sqrt{(dr/d\theta)^2 + r^2}\, d\theta$ we retrieve $s = \int_0^\pi\sqrt{e^{2\theta} + e^{2\theta}}\, d\theta = \int_0^\pi\sqrt{2}\, e^\theta\, d\theta$.

VECTOR-VALUED FUNCTIONS

34.43 If $\mathbf{F}(u) = (f(u), g(u))$ is a **two-dimensional** vector function, $\lim_{u\to a}\mathbf{F}(u) = (\lim_{u\to a} f(u), \lim_{u\to a} g(u))$, where the limit on the left exists if and only if the limits on the right exist. Taking this as the definition of vector convergence, show that $\mathbf{F}'(u) = (f'(u), g'(u))$.

$\mathbf{F}'(u) = \lim_{\Delta u\to 0}\dfrac{\mathbf{F}(u + \Delta u) - \mathbf{F}(u)}{\Delta u} = \lim_{\Delta u\to 0}\dfrac{(f(u + \Delta u)g(u + \Delta u)) - (f(u)g(u))}{\Delta u}$

$= \lim_{\Delta u\to 0}\dfrac{(f(u + \Delta u) - f(u),\ g(u + \Delta u) - g(u))}{\Delta u} = \lim_{\Delta u\to 0}\left(\dfrac{f(u + \Delta u) - f(u)}{\Delta u},\ \dfrac{g(u + \Delta u) - g(u)}{\Delta u}\right)$

This last limit is, by the definition, equal to $\left(\lim_{\Delta u\to 0}\dfrac{f(u + \Delta u) - f(u)}{\Delta u},\ \lim_{\Delta u\to 0}\dfrac{g(u + \Delta u) - g(u)}{\Delta u}\right) = (f'(u), g'(u))$.

34.44 If $\mathbf{R}(\theta) = (r\cos\theta, r\sin\theta)$, with fixed $r > 0$, show that $\mathbf{R}'(\theta) \perp \mathbf{R}(\theta)$.

By Problem 34.43, $\mathbf{R}'(\theta) = (-r\sin\theta, r\cos\theta)$. Then $\mathbf{R}(\theta)\cdot\mathbf{R}'(\theta) = (r\cos\theta, r\sin\theta)\cdot(-r\sin\theta, r\cos\theta) = -r^2\cos\theta\sin\theta + r^2\sin\theta\cos\theta = 0$.

34.45 Show that, if $\mathbf{R}(u)$ traces out a curve, then $\mathbf{R}'(u)$ is a tangent vector pointing in the direction of motion along the curve.

Refer to Fig. 34.13. Let $\overrightarrow{OP} = \mathbf{R}(u)$ and $\overrightarrow{OQ} = \mathbf{R}(u + \Delta u)$. Then $\overrightarrow{PQ} = \mathbf{R}(u + \Delta u) - \mathbf{R}(u + \Delta u) - \mathbf{R}(u)$ and $\dfrac{1}{\Delta u}\overrightarrow{PQ} = \dfrac{\mathbf{R}(u + \Delta u) - \mathbf{R}(u)}{\Delta u}$. As $\Delta u\to 0$, Q approaches P, and the direction of $\overrightarrow{PQ}$ (which is the direction of $\overrightarrow{PQ}/\Delta u$) approaches the direction of $\mathbf{R}'(u)$, which is thus a tangent vector at P.

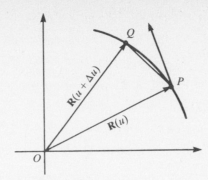

Fig. 34-13

34.46 Show that, if $\mathbf{R}(t)$ traces out a curve and the parameter t represents time, then $\mathbf{R}(t)$ is the velocity vector, that is, its direction is the direction of motion and its length is the speed.

▌ By Problem 34.45, we already know that $\mathbf{R}'(t)$ has the direction of the tangent vector along the curve. By Problem 34.43, $\mathbf{R}'(t) = (dx/dt, dy/dt)$, since $\mathbf{R}(t) = (x(t), y(t))$. Hence, $|\mathbf{R}'(t)| = \sqrt{(dx/dt)^2 + (dy/dt)^2} = ds/dt$, where s is the arc length along the curve (measured from some fixed point on the curve). But ds/dt is the speed. [The "speed" is how fast the end of the position vector $\mathbf{R}(t)$ is moving, which is the rate of change of its position s along the curve.]

34.47 If $\mathbf{R}(s)$ is a vector function tracing out a curve and the parameter s is the arc length, show that the tangent vector $\mathbf{R}'(s)$ is a unit vector, that is, it has constant length 1.

▌ $|\mathbf{R}'(s)| = \left|\left(\dfrac{dx}{ds}, \dfrac{dy}{ds}\right)\right| = \sqrt{\left(\dfrac{dx}{ds}\right)^2 + \left(\dfrac{dy}{ds}\right)^2} = \dfrac{ds}{ds} = 1.$

34.48 For the curve $\mathbf{R}(t) = (t, t^2)$, find the tangent vector $\mathbf{v} \equiv \mathbf{R}'(t)$, the speed, the unit tangent vector $\mathbf{T}$, and the acceleration vector $\mathbf{a} \equiv \mathbf{R}''(t)$.

▌ $\mathbf{v} = (1, 2t)$ and $\mathbf{a} = (0, 2)$. The speed is $|\mathbf{v}| = \sqrt{1 + (2t)^2} = \sqrt{1 + 4t^2}$, and the unit tangent vector $\mathbf{T} = \mathbf{v}/|\mathbf{v}| = \left(\dfrac{1}{\sqrt{1 + 4t^2}}, \dfrac{2t}{\sqrt{1 + 4t^2}}\right).$

34.49 Find the unit tangent vector for the circle $\mathbf{R} = (a \cos \theta, a \sin \theta)$.

▌ $\mathbf{R}'(\theta) = (-a \sin \theta, a \cos \theta)$. $|\mathbf{R}'(\theta)| = \sqrt{a^2 \sin^2 \theta + a^2 \cos^2 \theta} = a$. Hence, $\mathbf{T} = \mathbf{R}'(\theta)/|\mathbf{R}'(\theta)| = (-\sin \theta, \cos \theta)$.

34.50 Find the unit tangent vector for the curve $\mathbf{R}(\theta) = (e^\theta, e^{-\theta})$.

▌ $\mathbf{R}'(\theta) = (e^\theta, -e^{-\theta})$. $|\mathbf{R}'(\theta)| = \sqrt{e^{2\theta} + e^{-2\theta}} = \sqrt{e^{4\theta} + 1}/e^\theta$. Hence, the unit tangent vector $\mathbf{T} = \dfrac{1}{|\mathbf{R}'(\theta)|} \mathbf{R}'(\theta) = \dfrac{e^\theta}{\sqrt{e^{4\theta} + 1}} (e^\theta, -e^{-\theta}) = \dfrac{1}{\sqrt{e^{4\theta} + 1}} (e^{2\theta}, -1).$

34.51 Find the tangent vector $\mathbf{v}$ and acceleration vector $\mathbf{a}$ for the ellipse $\mathbf{R}(t) = (a \cos t, b \sin t)$, and show that $\mathbf{a}$ is opposite in direction to $\mathbf{R}(t)$ and of the same length.

▌ $\mathbf{v} = \mathbf{R}'(t) = (-a \sin t, b \cos t)$, $\mathbf{a} = \mathbf{R}''(t) = (-a \cos t, -b \sin t) = -(a \cos t, b \sin t) = -\mathbf{R}(t).$

34.52 Find the magnitude and direction of the velocity vector for $\mathbf{R}(t) = (e^t, e^{2t} - 4e^t + 3)$ at $t = 0$.

▌ $\mathbf{R}'(t) = (e^t, 2e^{2t} - 4e^t) = e^t(1, 2e^t + 4)$. When $t = 0$, $\mathbf{R}'(0) = (1, -2)$, $|\mathbf{R}'(0)| = \sqrt{5}$, and the vector $\mathbf{R}'(0)$ is in the fourth quadrant, with an angle $\theta = \tan^{-1}(-2) \approx -63° 26'$.

34.53 Find the velocity and acceleration vectors for $\mathbf{R}(t) = (2 - t, 2t^3 - t)$ at $t = 1$.

▌ $\mathbf{R}'(t) = (-1, 6t^2)$, $\mathbf{R}''(t) = (0, 12t)$. Hence, $\mathbf{R}'(1) = (-1, 6)$ and $\mathbf{R}''(1) = (0, 12)$.

34.54 If $\mathbf{R}(u) = f(u)\mathbf{F}(u)$, show that $\mathbf{R}'(u) = f(u)\mathbf{F}'(u) + f'(u)\mathbf{F}(u)$, analogous to the product formula for ordinary derivatives.

▮ $\mathbf{R}(u + \Delta u) - \mathbf{R}(u) = f(u + \Delta u)\mathbf{F}(u + \Delta u) - f(u)\mathbf{F}(u) = [f(u + \Delta u) - f(u)]\mathbf{F}(u + \Delta u) + f(u)[\mathbf{F}(u + \Delta u) - \mathbf{F}(u)]$. Hence,

$$\frac{\mathbf{R}(u + \Delta u) - \mathbf{R}(u)}{\Delta u} = \frac{f(u + \Delta u) - f(u)}{\Delta u}\mathbf{F}(u + \Delta u) + f(u)\left[\frac{\mathbf{F}(u + \Delta u) - \mathbf{F}(u)}{\Delta u}\right] \rightarrow f'(u)\mathbf{F}(u) + f(u)\mathbf{F}'(u) \text{ as } \Delta u \rightarrow 0$$

34.55 If $\mathbf{R}(t) = t^2(\ln t, \sin t)$, calculate $\mathbf{R}'(t)$.

▮ By Problem 34.54, $\mathbf{R}'(t) = t^2(1/t, \cos t) + 2t(\ln t, \sin t) = (1 + 2t \ln 2, t(\cos t + 2 \sin t))$.

34.56 If $\mathbf{R}(t) = (\sin t)(e^t, t)$, calculate $\mathbf{R}''(t)$.

▮ Applying Problem 34.54 twice, $\mathbf{R}'(t) = (\sin t)(e^t, 1) + (\cos t)(e^t, t)$, $\mathbf{R}''(t) = (\sin t)(e^t, 0) + (\cos t)(e^t, 1) + (\cos t)(e^t, 1) - (\sin t)(e^t, t) = (2e^t \cos t, 2 \cos t - t \sin t)$.

34.57 If $h(u) = \mathbf{F}(u) \cdot \mathbf{G}(u)$, show that $h'(u) = \mathbf{F}(u) \cdot \mathbf{G}'(u) + \mathbf{F}'(u) \cdot \mathbf{G}(u)$, another analogue of the product formula for derivatives.

▮
$$\frac{h(u + \Delta u) - h(u)}{\Delta u} = \frac{\mathbf{F}(u + \Delta u) \cdot \mathbf{G}(u + \Delta u) - \mathbf{F}(u) \cdot \mathbf{G}(u)}{\Delta u}$$

$$= \frac{[\mathbf{F}(u + \Delta u) - \mathbf{F}(u)]\mathbf{G}(u + \Delta u) + \mathbf{F}(u) \cdot [\mathbf{G}(u + \Delta u) - \mathbf{G}(u)]}{\Delta u}$$

$$= \frac{\mathbf{F}(u + \Delta u) - \mathbf{F}(u)}{\Delta u} \cdot \mathbf{G}(u + \Delta u) + \mathbf{F}(u) \cdot \frac{\mathbf{G}(u + \Delta u) - \mathbf{G}(u)}{\Delta u}$$

$$\rightarrow \mathbf{F}'(u) \cdot \mathbf{G}(u) + \mathbf{F}(u) \cdot \mathbf{G}'(u) \quad \text{as} \quad \Delta u \rightarrow 0$$

34.58 If $\mathbf{F}(t) = (t, \ln t)$ and $\mathbf{G}(t) = (e^t, t^2)$, find $\dfrac{d}{dt}[\mathbf{F}(t) \cdot \mathbf{G}(t)]$.

▮ By Problem 34.57, $\dfrac{d}{dt}[\mathbf{F}(t) \cdot \mathbf{G}(t)] = (t, \ln t) \cdot (e^t, 2t) + (1, 1/t) \cdot (e^t, t^2) = te^t + 2t \ln t + e^t + t$.

34.59 If $|\mathbf{R}(t)|$ is a constant $c > 0$, show that the tangent vector $\mathbf{R}'(t)$ is perpendicular to the position vector $\mathbf{R}(t)$.

▮ $\mathbf{R}(t) \cdot \mathbf{R}(t) = |\mathbf{R}(t)|^2 = c^2$. So, $\dfrac{d}{dt}[\mathbf{R}(t) \cdot \mathbf{R}(t)] = 0$. But, by Problem 34.57, $\dfrac{d}{dt}[\mathbf{R}(t) \cdot \mathbf{R}(t)] = \mathbf{R}(t) \cdot \mathbf{R}'(t) + \mathbf{R}'(t) \cdot \mathbf{R}(t) = 2\mathbf{R}(t) \cdot \mathbf{R}'(t)$. Hence, $\mathbf{R}(t) \cdot \mathbf{R}'(t) = 0$.

34.60 Give a geometric argument for the result of Problem 34.59.

▮ If $|\mathbf{R}(t)| = c \neq 0$, then the endpoint of $\mathbf{R}(t)$ moves on the circle of radius c with center at the origin. At each point of a circle, the tangent line is perpendicular to the radius vector.

34.61 For any vector function $\mathbf{F}(u)$ and scalar function $h(u)$, prove a chain rule: $\dfrac{d}{du}\mathbf{F}(h(u)) = h'(u)\mathbf{F}'(h(u))$.

▮ Let $\mathbf{F}(u) = (f(u), g(u))$. Then $\mathbf{F}(h(u)) = (f(h(u)), g(h(u)))$. Hence, by Problem 34.43 and the regular chain rule, $\dfrac{d}{du}[\mathbf{F}(h(u)] = \left(\dfrac{d}{du} f(h(u)), \dfrac{d}{du} g(h(u))\right) = (f'(h(u))h'(u), g'(h(u))h'(u)) = h'(u)(f'(h(u)), g'(h(u))) = h'(u)\mathbf{F}'(h(u))$.

34.62 Let $\mathbf{F}(u) = (\cos u, \sin 2u)$ and let $\mathbf{G}(t) = \mathbf{F}(t^2)$. Find $\mathbf{G}'(t)$.

▮ $\mathbf{F}'(u) = (-\sin u, 2 \cos 2u)$. By Problem 34.61, $\mathbf{G}'(t) = \dfrac{d}{dt}(t^2)\mathbf{F}'(t^2) = 2t(-\sin t^2, 2 \cos 2t^2)$.

34.63 Let $\mathbf{F}(u) = (u^3, u^4)$ and let $\mathbf{G}(t) = \mathbf{F}(e^{-t})$. Find $\mathbf{G}'(t)$.

▮ $\mathbf{F}'(u) = (3u^2, 4u^3)$. By Problem 34.61, $\mathbf{G}'(t) = \dfrac{d}{dt}(e^{-t})\mathbf{F}'(e^{-t}) = -e^{-t}(3e^{-2t}, 4e^{-3t}) = -e^{-4t}(3e^t, 4)$.

34.64 At $t = 2$, $\mathbf{F}(t) = \mathbf{i} + \mathbf{j}$ and $\mathbf{F}'(t) = 2\mathbf{i} - 3\mathbf{j}$. Find $\dfrac{d}{dt}[t^2\mathbf{F}(t)]$ at $t = 2$.

▮ By Problem 34.54, $\dfrac{d}{dt}[t^2\mathbf{F}(t)] = t^2\mathbf{F}'(t) + 2t\mathbf{F}(t)$. At $t = 2$, $\dfrac{d}{dt}[t^2\mathbf{F}(t)] = 4(2\mathbf{i} - 3\mathbf{j}) + 4(\mathbf{i} + \mathbf{j}) = 12\mathbf{i} - 8\mathbf{j}$.

34.65 At $t = -1$, $\mathbf{G}(t) = (3, 0)$ and $\mathbf{G}'(t) = (2, 3)$. Find $\dfrac{d}{dt}[t^3\mathbf{G}(t)]$ at $t = -1$.

▮ By Problem 34.54, $\dfrac{d}{dt}[t^3\mathbf{G}(t)] = t^3\mathbf{G}'(t) + 3t^2\mathbf{G}(t)$. At $t = -1$, $\dfrac{d}{dt}[t^3\mathbf{G}(t)] = -(2, 3) + 3(3, 0) = (7, -3)$.

34.66 Prove the converse of Problem 34.59.

▮ By Problem 34.57, $\dfrac{d}{dt}[\mathbf{R}(t) \cdot \mathbf{R}(t)] = 2\mathbf{R}(t) \cdot \mathbf{R}'(t)$. Hence, $\mathbf{R}(t) \cdot \mathbf{R}'(t) = 0$ implies $\mathbf{R}(t) \cdot \mathbf{R}(t) = c^2$ for

some positive constant c. Then $|\mathbf{R}(t)| = \sqrt{\mathbf{R}(t) \cdot \mathbf{R}(t)} = c$.

34.67 Give an example to show that if $\mathbf{R}(t)$ is a unit vector, then $\mathbf{R}'(t)$ need not be a unit vector (nor even a vector of constant length).

▮ Consider $\mathbf{R}(t) = (\cos t^2, \sin t^2)$. Then

$|\mathbf{R}(t)| = 1$. However, $\mathbf{R}'(t) = (-2t \sin t^2, 2t \cos t^2)$ and $|\mathbf{R}'(t)| = \sqrt{4t^2 \sin^2 t^2 + 4t^2 \cos^2 t^2} = 2|t|$.

34.68 If $\mathbf{F}(u) = \mathbf{A}$ for all u, show that $\mathbf{F}'(u) = \mathbf{0}$, and, conversely, if $\mathbf{F}'(u) = \mathbf{0}$ for all u, then $\mathbf{F}(u)$ is a constant vector.

▮ Let $\mathbf{F}(u) = (f(u), g(u))$. If $f(u) = a_1$ and $g(u) = a_2$ for all u, then $\mathbf{F}'(u) = (f'(u), g'(u)) = (0, 0) = \mathbf{0}$. Conversely, if $\mathbf{F}'(u) = \mathbf{0}$ for all u, then $f'(u) = 0$ and $g'(u) = 0$ for all u, and, therefore, $f(u)$ and $g(u)$ are constants, and, thus, $\mathbf{F}(u)$ is constant.

34.69 Show that, if $\mathbf{B} \neq \mathbf{0}$, then $\mathbf{R}(t) = \mathbf{A} + t\mathbf{B}$ represents a straight line and has constant velocity vector and zero acceleration vector.

▮ It is clear from the parallelogram law (Fig. 34-14) that $\mathbf{R}(t)$ generates the line $\mathscr{L}$ that passes through the endpoint of $\mathbf{A}$ and is parallel to $\mathbf{B}$. We have:

$$\mathbf{R}'(t) = \frac{d}{dt}(\mathbf{A}) + \frac{d}{dt}(t\mathbf{B}) = \mathbf{0} + t\frac{d\mathbf{B}}{dt} + \mathbf{B} = \mathbf{0} + t\mathbf{0} + \mathbf{B} = \mathbf{B}$$

Further, $\mathbf{R}''(t) = d\mathbf{B}/dt = \mathbf{0}$.

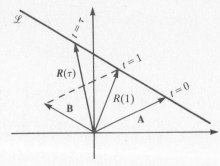

Fig. 34-14

34.70 As a converse to Problem 34.69, show that if $\mathbf{R}'(u) = \mathbf{B} \neq \mathbf{0}$ for all u, then $\mathbf{R}(u) = \mathbf{A} + u\mathbf{B}$, a straight line.

▮ Let $\mathbf{R}(u) = (f(u), g(u))$ and $\mathbf{B} = (b_1, b_2)$. Then $f'(u) = b_1$ and $g'(u) = b_2$. Hence, $f(u) = b_1 u + a_1$ and $g(u) = b_2 u + a_2$. So, $\mathbf{R}(u) = (b_1 u + a_1, b_2 u + a_2) = \mathbf{A} + u\mathbf{B}$.

34.71 If $\mathbf{R}''(u) = \mathbf{0}$ for all u, show that $\mathbf{R}(u) = \mathbf{A} + u\mathbf{B}$, either a constant function or a straight line.

▮ By Problem 34.68, $\mathbf{R}'(u)$ is a constant, $\mathbf{B}$. If $\mathbf{B} = \mathbf{0}$, then $\mathbf{R}(u)$ is a constant, again by Problem 34.68. If $\mathbf{B} \neq \mathbf{0}$, then $\mathbf{R}(u)$ has the form $\mathbf{A} + u\mathbf{B}$, by Problem 34.70.

34.72 Show that the angle θ between the position vector $\mathbf{R}(t) = (e^t \cos t, e^t \sin t)$ and the velocity vector $\mathbf{R}'(t)$ is $\pi/4$.

▮ $\mathbf{R}(t) = e^t(\cos t, \sin t)$. By the product formula, $\mathbf{R}'(t) = e^t(-\sin t, \cos t) + e^t(\cos t, \sin t) = e^t(\cos t - \sin t, \cos t + \sin t)$. Therefore, $|\mathbf{R}(t)| = e^t|(\cos t \sin t)| = e^t$ and $|\mathbf{R}'(t)| = e^t|(\cos t - \sin t, \cos t + \sin t)| = e^t\sqrt{\cos^2 t - 2\cos t \sin t + \sin^2 t + \cos^2 t + 2\cos t \sin t + \sin^2 t} = e^t\sqrt{2}$ (in agreement with $ds/dt = \sqrt{2}e^t$ as calculated in Problem 34.33). Now, $\mathbf{R}(t) \cdot \mathbf{R}'(t) = e^t(\cos t, \sin t) \cdot e^t(\cos t - \sin t, \cos t + \sin t) = e^{2t}(\cos^2 t - \cos t \sin t + \sin t \cos t + \sin^2 t) = e^{2t}$. But, $\mathbf{R}(t) \cdot \mathbf{R}'(t) = |\mathbf{R}(t)||\mathbf{R}'(t)|\cos\theta$, $e^{2t} = e^t \cdot \sqrt{2}e^t \cdot \cos\theta$, $\cos\theta = 1/\sqrt{2}$, $\theta = \pi/4$.

34.73 Derive the following quotient rule: $\dfrac{d}{dt}\left[\dfrac{G(u)}{f(u)}\right] = \dfrac{f(u)G'(u) - f'(u)G(u)}{[f(u)]^2}$.

▌ By the product rule, $\dfrac{d}{dt}\left[\dfrac{G(u)}{f(u)}\right] = \dfrac{d}{dt}\left[\dfrac{1}{f(u)}G(u)\right] = \dfrac{1}{f(u)}G'(u) - \dfrac{f'(u)}{[f(u)]^2}G(u) = \dfrac{f(u)G'(u) - f'(u)G(u)}{[f(u)]^2}$.

34.74 Assume that an object moves on a circle of radius r with constant speed $v > 0$. Show that the acceleration vector is directed toward the center of the circle and has length v^2/r.

▌ The position vector $\mathbf{R}(t)$ satisfies $|\mathbf{R}(t)| = r$ and $|\mathbf{R}'(t)| = v$. Let θ be the angle from the positive x-axis to $\mathbf{R}(t)$ and let s be the corresponding arc length on the circle. Then $s = r\theta$, and, since the object moves with constant speed v, $s = vt$. Hence, $\theta = vt/r$. We can write $\mathbf{R}(t) = r(\cos\theta, \sin\theta)$. By the chain rule, $\mathbf{R}'(t) = r(-\sin\theta, \cos\theta)\dfrac{d\theta}{dt} = r(-\sin\theta, \cos\theta)\dfrac{v}{r} = v(-\sin\theta, \cos\theta)$. Again by the chain rule, $\mathbf{R}''(t) = v(-\cos\theta - \sin\theta)\dfrac{d\theta}{dt} = -v(\cos\theta, \sin\theta)\dfrac{v}{r} = -\dfrac{v^2}{r^2}(r\cos\theta, r\sin\theta) = -\dfrac{v^2}{r^2}\mathbf{R}(t)$. Hence, the acceleration vector $\mathbf{R}''(t)$ points in the opposite direction to $\mathbf{R}(t)$, that is, toward the center of the circle, and $|\mathbf{R}''(t)| = \dfrac{v^2}{r^2}|\mathbf{R}(t)| = \dfrac{v^2}{r^2}\cdot r = v^2/r$.

34.75 Let $\mathbf{R}(t) = (\cos(\pi e^t/2), \sin(\pi e^t/2))$. Determine $\mathbf{R}(0)$, $\mathbf{v}(0)$, and $\mathbf{a}(0)$, and show these vectors in a diagram.

▌ Let $\theta = \pi e^t/2$; thus, $\mathbf{R}(t) = (\cos\theta, \sin\theta)$. By the chain rule, $\mathbf{v}(t) = \mathbf{R}'(t) = (-\sin\theta, \cos\theta)\dfrac{d\theta}{dt} = \theta(-\sin\theta, \cos\theta)$, and

$$\mathbf{a}(t) = \dfrac{d}{dt}\mathbf{v} = \theta(-\cos\theta, -\sin\theta)\dfrac{d\theta}{dt} + \dfrac{d\theta}{dt}(-\sin\theta, \cos\theta) = -\theta^2(\cos\theta, \sin\theta) + \theta(-\sin\theta, \cos\theta) = -\theta^2\mathbf{R}(t) + \theta\mathbf{v}(t)$$

When $t = 0$, $\theta = \pi/2$, and $\mathbf{R}(0) = (0, 1)$, $\mathbf{v}(0) = \dfrac{\pi}{2}(-1, 0)$, $\mathbf{a}(0) = -\dfrac{\pi^2}{4}(0, 1) + \dfrac{\pi}{4}\cdot\dfrac{\pi}{2}(-1, 0) = -\dfrac{\pi^2}{8}(1, 2)$.

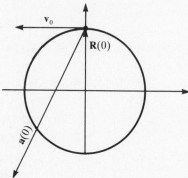

Fig. 34-15

34.76 Let $\mathbf{R}(t) = (10\cos 2\pi t, 10\sin 2\pi t)$. Find the acceleration vector.

▌ Let $\theta = 2\pi t$. Then $d\theta/dt = 2\pi$ and $\mathbf{R}(t) = 10(\cos\theta, \sin\theta)$. By the chain rule, $\mathbf{v}(t) = \mathbf{R}'(t) = 10(-\sin\theta, \cos\theta)\dfrac{d\theta}{dt} = 20\pi(-\sin\theta, \cos\theta)$. By the chain rule again, the acceleration vector $\mathbf{a}(t) = d\mathbf{v}/dt = 20\pi(-\cos\theta, -\sin\theta)\dfrac{d\theta}{dt} = -40\pi^2(\cos\theta, \sin\theta) = -4\pi^2\mathbf{R}(t)$. Note that this is a special case of Problem 34.74.

34.77 Let $\mathbf{R}(t) = (t, 1/t)$. Find $\mathbf{v}(t)$, the speed $|\mathbf{v}(t)|$, and $\mathbf{a}(t)$. Describe what happens as $t \to +\infty$.

▌ $\mathbf{v}(t) = (1, -1/t^2)$, $|\mathbf{v}(t)| = \sqrt{1 + (1/t^4)}$, and $\mathbf{a}(t) = (0, 2/t^3)$. As $t \to +\infty$, the direction of $\mathbf{R}(t)$ approaches that of the positive x-axis and its length approaches $+\infty$. The velocity vector approaches the unit vector $\mathbf{i} = (1, 0)$, and the speed approaches 1. The acceleration vector always points along the positive y-axis, and its length approaches 0.

34.78 Let $\mathbf{R}(t) = (t + \cos t, t - \sin t)$. Show that the acceleration vector has constant length.

▌ $\mathbf{R}'(t) = (1 - \sin t, 1 - \cos t)$, and $\mathbf{R}''(t) = (-\cos t, \sin t)$. Hence, $|\mathbf{R}''(t)| = \sqrt{\cos^2 t + \sin^2 t} = 1$.

34.79 Prove that, if the acceleration vector is always perpendicular to the velocity vector, then the speed is constant.

▌ This follows from Problem 34.66, substituting $\mathbf{R}'(t)$ for $\mathbf{R}(t)$.

34.80 Prove the converse of Problem 34.79: If the speed is constant, then the velocity and acceleration vectors are perpendicular.

▌ This is a special case of Problem 34.59 if we substitute $\mathbf{R}'(t)$ for $\mathbf{R}(t)$.

34.81 Let $\mathbf{T}(t)$ be the unit tangent vector to a curve $\mathbf{R}(t)$. Show that, wherever $\mathbf{T}'(t) \neq \mathbf{0}$, the *principal unit normal vector* $\mathbf{N} = \mathbf{T}'/|\mathbf{T}'|$ is perpendicular to $\mathbf{T}$.

▌ The argument of Problem 34.59 applies to any differentiable vector function. Therefore, $\mathbf{T} \cdot \mathbf{T}' = 0$, which in turn implies $\mathbf{T} \cdot \mathbf{N} = 0$ wherever $\mathbf{N}$ is defined (i.e., wherever $|\mathbf{T}'| > 0$).

34.82 Show that the principal unit normal vector $\mathbf{N}(t)$ (Problem 34.81) points in the direction in which $\mathbf{T}(t)$ is turning as t increases.

▌ Since $\mathbf{N}(t)$ has the same direction as $\mathbf{T}'$, we need prove the result only for $\mathbf{T}'$. $\mathbf{T}'(t) = \lim\limits_{\Delta t \to 0} \dfrac{\mathbf{T}(t + \Delta t) - \mathbf{T}(t)}{\Delta t} = \lim\limits_{\Delta t \to 0} \dfrac{\Delta \mathbf{T}}{\Delta t}$. For small Δt, $\Delta \mathbf{T}/\Delta t$ has approximately the same direction as $\mathbf{T}'(t)$, and, when $\Delta t > 0$, $\Delta \mathbf{T}$ and $\Delta \mathbf{T}/\Delta t$ have the same direction. If we place $\mathbf{T}(t + \Delta t)$ and $\mathbf{T}(t)$ with their tails at the origin, $\Delta \mathbf{T}$ is the vector from the head of $\mathbf{T}(t)$ to the head of $\mathbf{T}(t + \Delta t)$. Figure 34-16 shows that $\Delta \mathbf{T}$ points in the direction in which $\mathbf{T}$ is turning (to the right in this case).

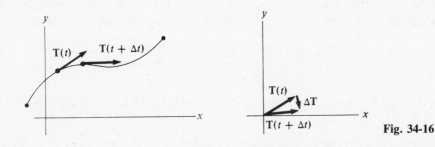

Fig. 34-16

34.83 If $\mathbf{R}(t) = (r \cos t, r \sin t)$, where t represents time, show that the principal unit normal vector $\mathbf{N}(t)$ has the same direction as the acceleration vector.

▌ $\mathbf{R}'(t) = r(-\sin t, \cos t)$ and $|\mathbf{R}'(t)| = r$. So, $\mathbf{T}(t) = \mathbf{R}'/|\mathbf{R}'| = (-\sin t, \cos t)$. Hence, $\mathbf{T}'(t) = (-\cos t, -\sin t)$ and $|\mathbf{T}'(t)| = 1$. Thus, $\mathbf{N}(t) = \mathbf{T}'(t) = -(\cos t, \sin t) = -\dfrac{1}{r}\mathbf{R}$. In this case, by Problem 34.74, $\mathbf{N}(t)$ has the same direction as the acceleration vector. Note that $\mathbf{N}(t)$ is pointing in the direction in which $\mathbf{T}(t)$ is turning, that is, "inside" the curve.

34.84 Compute $\mathbf{N}(t)$ for the curve $\mathbf{R}(t) = (2 \cos t^2, 2 \sin t^2)$.

▌ $\mathbf{R}'(t) = (-4t \sin t^2, 4t \cos t^2) = 4t(-\sin t^2, \cos t^2)$. So, $|\mathbf{R}'(t)| = 4|t|$. Thus, $\mathbf{T}(t) = \dfrac{t}{|t|}(-\sin t^2, \cos t^2)$. Hence, $\mathbf{T}'(t) = \dfrac{t}{|t|}(-2t \cos t^2, -2t \sin t^2) = -\dfrac{2t^2}{|t|}(\cos t^2, \sin t^2)$. So, $|\mathbf{T}'(t)| = 2|t|$, and, therefore, $\mathbf{N}(t) = \mathbf{T}'(t)/|\mathbf{T}'(t)| = -(\cos t^2, \sin t^2)$. Again, $\mathbf{N}(t)$ has the same direction as the acceleration vector, that is, it points toward the center of the circle traced out by $\mathbf{R}(t)$.

34.85 Show that when arc length s is chosen as the curve parameter, the principal unit normal vector has the simple expression $\mathbf{N}(s) = \mathbf{R}''(s)/|\mathbf{R}''(s)|$.

▌ By Problem 34.47, $\mathbf{T}(s) = \mathbf{R}'(s)$, and so $\mathbf{T}'(s) = \mathbf{R}''(s)$ and $\mathbf{N}(s) = \mathbf{T}'(s)/|\mathbf{T}'(s)| = \mathbf{R}''(s)/|\mathbf{R}''(s)|$ [except where $\mathbf{R}''(s) = \mathbf{0}$].

34.86 Show that, for the cycloid $\mathbf{R}(t) = a(t - \sin t, 1 - \cos t)$, where t is the time, the acceleration vector is not parallel to the principal unit normal vector.

▌ $\mathbf{R}'(t) = a(1 - \cos t, \sin t)$, $|\mathbf{R}'(t)| = a\sqrt{(1 - \cos t)^2 + \sin^2 t} = a\sqrt{2 - 2\cos t} = a\sqrt{2}\sqrt{1 - \cos t}$. Then

$$\mathbf{T}(t) = \frac{\sqrt{2}}{2}\frac{1}{\sqrt{1 - \cos t}}(1 - \cos t, \sin t)$$

and
$$\mathbf{T}'(t) = \frac{\sqrt{2}}{2} \left[\frac{1}{\sqrt{1 - \cos t}} (\sin t, \cos t) - \frac{1}{2} \frac{\sin t}{(1 - \cos t)^{3/2}} (1 - \cos t, \sin t) \right]$$

$$= \frac{\sqrt{2}}{2} \frac{1}{\sqrt{1 - \cos t}} [(\sin t, \cos t) - \tfrac{1}{2} (\sin t, 1 + \cos t)] = \frac{\sqrt{2}}{4} \frac{1}{\sqrt{1 - \cos t}} (\sin t, \cos t - 1)$$

Thus $\mathbf{N}(t)$ has the direction of $(\sin t, \cos t - 1)$. However, the acceleration vector $\mathbf{R}''(t) = a(\sin t, \cos t)$.

34.87 At a given point on a curve, does the unit tangent vector depend on the direction in which the curve is being swept out?

❚ We let $\mathbf{S}(t) = \mathbf{R}(-t)$. Then $\mathbf{S}(t)$ traces the same curve as $\mathbf{R}(t)$, but in the reverse direction. $\mathbf{S}'(t) = -\mathbf{R}'(-t)$ and $|\mathbf{S}'(t)| = |\mathbf{R}'(-t)|$. Let $\mathbf{T}^*(t)$ be the unit tangent vector for the curve $\mathbf{S}(t)$. Then $\mathbf{T}^*(t) = \mathbf{S}'(t)/|\mathbf{S}'(t)| = -\mathbf{R}'(-t)/|\mathbf{R}'(-t)| = -\mathbf{T}(-t)$. Hence, at each point, the direction of the unit tangent vector has been reversed.

34.88 At a given point on a curve, does the principal unit normal vector depend on the direction in which the curve is being swept out?

❚ Use the same notation as in Problem 34.87. Let $\mathbf{N}^*(t)$ be the principal unit normal vector on the reversed curve $\mathbf{S}(t)$. Since $\mathbf{T}^*(t) = -\mathbf{T}(-t)$, $\dfrac{d}{dt} \mathbf{T}^*(t) = \mathbf{T}'(-t)$, by the chain rule. Hence,

$$\mathbf{N}^*(t) = \frac{1}{\left| \dfrac{d}{dt} \mathbf{T}^*(t) \right|} \frac{d}{dt} \mathbf{T}^*(t) = \frac{1}{|\mathbf{T}'(-t)|} \mathbf{T}'(-t) = \mathbf{N}(-t)$$

Thus, the principal unit vector is not changed by reversing the direction along a curve.

34.89 Let ϕ be the angle between the velocity vector and the positive x-axis. Show that $|d\mathbf{T}/d\phi| = 1$.

❚ Since $\mathbf{T}$ is a unit vector in the same direction as the velocity vector, $\mathbf{T} = (\cos \phi, \sin \phi)$. Hence, $d\mathbf{T}/d\phi = (-\sin \phi, \cos \phi)$ and $|d\mathbf{T}/d\phi| = 1$.

34.90 Show that there is a scalar function $f(t)$ such that $\mathbf{N}'(t) = f(t)\mathbf{T}(t)$, and find a formula for $f(t)$.

❚ Parameterize the curve as $\mathbf{R}(t) = (t, t^3)$.

Then $\qquad \mathbf{R}'(t) = (1, 3t^2), \qquad |\mathbf{R}'(t)| = \sqrt{1 + 9t^4}, \qquad \mathbf{T}(t) = \dfrac{1}{\sqrt{1 + 9t^4}} (1, 3t^2).$

$$\mathbf{T}'(t) = \frac{1}{\sqrt{1 + 9t^4}} (0, 6t) - \frac{18t^3}{(1 + 9t^4)^{3/2}} (1, 3t^2)$$

So, $\mathbf{T}'(0) = \mathbf{0}$ and, therefore, $\mathbf{N}(0)$ is not defined.

34.91 Show that there is a scalar function $f(t)$ such that $\mathbf{N}'(t) = f(t)\mathbf{T}(t)$, and find a formula for $f(t)$.

❚ Since $\mathbf{N}(t)$ is a unit vector, Problem 34.59 shows that $\mathbf{N}'(t) \perp \mathbf{N}(t)$. Since $\mathbf{T}(t) \perp \mathbf{N}(t)$, $\mathbf{N}'(t)$ and $\mathbf{T}(t)$ must be parallel (remember that we are in two dimensions). Hence, there is a scalar function $f(t)$ such that $\mathbf{N}'(t) = f(t)\mathbf{T}(t)$. Since $\mathbf{N}(t) \cdot \mathbf{T}(t) = 0$, the product rule yields $\mathbf{N} \cdot \mathbf{T}' + \mathbf{N}' \cdot \mathbf{T} = 0$, $\mathbf{N}' \cdot \mathbf{T} = -\dfrac{\mathbf{T}'}{|\mathbf{T}'|} \cdot \mathbf{T}' = -|\mathbf{T}'|$. Hence, $-|\mathbf{T}'| = \mathbf{N}' \cdot \mathbf{T} = f(t)\mathbf{T} \cdot \mathbf{T} = f(t)$, since $\mathbf{T} \cdot \mathbf{T} = |\mathbf{T}|^2 = 1$. When $t = s = $ arc length, the scalar f measures the curvature of $\mathbf{R}(s)$; see Problem 34.105.

34.92 Show that $\mathbf{T}'(t) = \dfrac{v\mathbf{a} - (dv/dt)\mathbf{v}}{v^2}$.

❚ $\mathbf{T}(t) = (1/v)\mathbf{R}'(t)$. By the product rule,

$$\mathbf{T}'(t) = \frac{1}{v} \mathbf{R}''(t) - \frac{1}{v^2} \frac{dv}{dt} \mathbf{R}'(t) = \frac{1}{v^2} \left[v\mathbf{R}''(t) - \frac{dv}{dt} \mathbf{R}'(t) \right] = \frac{v\mathbf{a} - (dv/dt)\mathbf{v}}{v^2}$$

34.93 Generalize the result of Problem 34.83 to any motion at constant speed.

❚ If $dv/dt = 0$, Problem 34.92 gives $\mathbf{T}' = (1/v)\mathbf{a}$ or $\mathbf{N} = (1/v)|\mathbf{T}'|\mathbf{a}$.

34.94 Define the *curvature* κ and *radius of curvature* ρ of a curve $\mathbf{R}(t)$.

❚ As in Problem 34.89, let ϕ denote the angle between the velocity vector $\mathbf{R}'(t)$ and the positive x-axis. The curvature κ is defined as $d\phi/ds$, where s is the arc length. The curvature measures how fast the tangent vector turns as a point moves along the curve. The radius of curvature is defined as $\rho = |1/\kappa|$.

34.95 For a circle of radius a, traced out in the counterclockwise direction, show that the curvature is $1/a$, and the radius of curvature is a, the radius of the circle.

❚ If (x_0, y_0) is the center of the circle, then $\mathbf{R}(t) = (x_0 + a \cos t, y_0 + a \sin t)$ traces out the circle, where t is the angle from the positive x-axis to $\mathbf{R}(t) - (x_0, y_0)$. Then $\mathbf{R}'(t) = a(-\sin t, \cos t)$, and $ds/dt = a$. The angle ϕ made by the positive x-axis with $\mathbf{R}'(t)$ is $\tan^{-1}\left(\dfrac{a \cos t}{-a \sin t}\right) = \tan^{-1}(-\cot t)$, or that angle $+ \pi$. Hence,

$$\frac{d\phi}{ds} = \frac{1}{1 + \cot^2 t}\,(\csc^2 t) \cdot \frac{dt}{ds} = \frac{dt}{ds} = \frac{1}{a}$$

So, $\kappa = 1/a$, and by definition, the radius of curvature is a.

34.96 Find the curvature of a straight line $\mathbf{R}(t) = \mathbf{A} + t\mathbf{B}$.

❚ $\mathbf{R}'(t) = \mathbf{B}$. Since $\mathbf{R}'(t)$ is constant, ϕ is constant, and, therefore, $\kappa = d\phi/ds = 0$.

34.97 Show that, for a curve $y = f(x)$, the curvature is given by the formula $\kappa = y''/[1 + (y')^2]^{3/2}$. (We assume that $ds/dx > 0$, that is, the arc length increases with x.)

❚ Since y' is the slope of the tangent line, $\tan \phi = y'$. Hence, differentiating with respect to s, $\sec^2 \phi\,(d\phi/ds) = y''/(ds/dx)$. But, $\sec^2 \phi = 1 + \tan^2 \phi = 1 + (y')^2$, and $ds/dx = [1 + (y')^2]^{1/2}$. Thus,

$$\kappa = \frac{d\phi}{ds} = \frac{y''/[1 + (y')^2]}{[1 + (y')^2]^{1/2}} = \frac{y''}{[1 + (y')^2]^{3/2}}$$

34.98 Find the curvature of the parabola $y = x^2$ at the point $(0, 0)$.

❚ $y' = 2x$ and $y'' = 2$. Hence, by Problem 34.97, $\kappa = 2/(1 + 4x^2)^{3/2}$. When $x = 0$, $\kappa = 2$.

34.99 Find the curvature of the hyperbola $xy = 1$ at $(1, 1)$.

❚ $y' = -1/x^2$ and $y'' = 2/x^3$. By Problem 34.97,

$$\kappa = \frac{2/x^3}{[1 + (1/x^4)]^{3/2}} = \frac{2}{x^3}\,\frac{x^6}{(x^4 + 1)^{3/2}} = \frac{2x^3}{(x^4 + 1)^{3/2}}$$

When $x = 1$, $\kappa = 2/2\sqrt{2} = \sqrt{2}/2$.

34.100 Given a curve in parametric form $\mathbf{R}(t) = (x(t), y(t))$, show that $\kappa = \dfrac{(\dot{x}\ddot{y} - \dot{y}\ddot{x})}{[(\dot{x})^2 + (\dot{y})^2]^{3/2}}$. (Here, the dots indicate differentiation with respect to t, and we assume that $dx/dt > 0$ and $ds/dt > 0$.)

❚ Substitute the results of Problem 34.19 into the formula of Problem 34.97.

34.101 Find the curvature at the point $\theta = \pi$ of the cycloid $\mathbf{R}(\theta) = a(\theta - \sin \theta, 1 - \cos \theta)$.

❚ Use the formula of Problem 34.100, taking t to be θ. Then $\dot{x} = a(1 - \cos \theta)$, $\dot{y} = a \sin \theta$, $\ddot{x} = a \sin \theta$, $\ddot{y} = a \cos \theta$, $(\dot{x})^2 + (\dot{y})^2 = a^2(1 - \cos \theta)^2 + a^2 \sin^2 \theta = a^2(2 - 2 \cos \theta)$. Then

$$\kappa = \frac{a(1 - \cos \theta)a \cos \theta - a \sin \theta\, a \sin \theta}{2\sqrt{2}a^3(1 - \cos \theta)^{3/2}} = \frac{1}{2\sqrt{2}a}\,\frac{\cos \theta - 1}{(1 - \cos \theta)^{3/2}} = -\frac{1}{2\sqrt{2}a}\,\frac{1}{\sqrt{1 - \cos \theta}}$$

When $\theta = \pi$, $\kappa = -1/4a$.

34.102 Find the radius of curvature of $y = \ln x$ at $x = e$.

▮ $y' = 1/x$, $y'' = -1/x^2$, and $[1 + (y')^2]^{3/2} = [1 + (1/x^2)]^{3/2} = (x^2 + 1)^{3/2}/x^3$. By Problem 34.97,

$$\kappa = \frac{-(1/x^2)}{(x^2 + 1)^{3/2}/x^3} = -\frac{x}{(x^2 + 1)^{3/2}}$$

When $x = e$, $\kappa = -e/(e^2 + 1)^{3/2}$, and $\rho = (e^2 + 1)^{3/2}/e$.

34.103 Find the curvature of $\mathbf{R}(t) = (\cos^3 t, \sin^3 t)$ at $t = \pi/4$.

▮ Use Problem 34.100. $\dot{x} = -3 \cos^2 t \sin t$, $\dot{y} = 3 \sin^2 t \cos t$, $\ddot{x} = -3(\cos^3 t - 2 \sin^2 t \cos t) = -3 \cos t (3 \cos^2 t - 2)$. $\ddot{y} = 3(-\sin^3 t + 2 \cos^2 t \sin t) + 3 \sin t (3 \cos^2 t - 1)$. $[(\dot{x})^2 + (\dot{y})^2]^{3/2} = (9 \cos^4 t \sin^2 t + 9 \sin^4 t \cos^2 t)^{3/2} = (9 \cos^2 t \sin^2 t)^{3/2} = 27 \cos^3 t \sin^3 t$. Now, $\dot{x}\ddot{y} - \dot{y}\ddot{x} = -3 \cos^2 t \sin t (3) \sin t (3 \cos^2 t - 1) - 3 \sin^2 t \cos t (-3) \cos t (3 \cos^2 t - 2) = -9 \cos^2 t \sin^2 t$. $\kappa = -9 \cos^2 t \sin^2 t / 27 \cos^3 t \sin^3 t = -1/(3 \cos t \sin t)$. When $t = \pi/4$, $\kappa = -\frac{2}{3}$.

34.104 For what value of x is the radius of curvature of $y = e^x$ smallest?

▮ $y' = y'' = e^x$. By Problem 34.97, $\kappa = e^x/(1 + e^{2x})^{3/2}$, and the radius of curvature ρ is $(1 + e^{2x})^{3/2}/e^x$. Then

$$\frac{d\rho}{dx} = \frac{e^x \cdot \frac{3}{2}(1 + e^{2x})^{1/2}(2e^{2x}) - e^x(1 + e^{2x})^{3/2}}{e^{2x}} = \frac{(1 + e^{2x})^{1/2}[3e^{2x} - (1 + e^{2x})]}{e^x} = \frac{(1 + e^{2x})^{1/2}(2e^{2x} - 1)}{e^x}$$

Setting $d\rho/dx = 0$, we find $2e^{2x} = 1$, $2x = \ln \frac{1}{2} = -\ln 2$, $x = -(\ln 2)/2$. The first-derivative test shows that this yields a relative (and, therefore, an absolute) minimum.

34.105 For a given curve $\mathbf{R}(t)$, show that $\mathbf{T}'(t) = (v/\rho)\mathbf{N}(t)$.

▮ By definition of $\mathbf{N}(t)$, $\mathbf{T}'(t) = |\mathbf{T}'(t)|\mathbf{N}(t)$. We must show that $|\mathbf{T}'(t)| = v/\rho$. By the chain rule, $\mathbf{T}'(t) = \dfrac{d\mathbf{T}}{d\phi}\dfrac{d\phi}{dt}$. By Problem 34.89, $|d\mathbf{T}/d\phi| = 1$. Hence, $|\mathbf{T}'(t)| = |d\phi/dt|$. But, $\dfrac{d\phi}{dt} = \dfrac{d\phi}{ds}\dfrac{ds}{dt}$, and, therefore, $\left|\dfrac{d\phi}{dt}\right| = \left|\dfrac{d\phi}{ds}\right|\left|\dfrac{ds}{dt}\right|$. Since $\kappa = d\phi/ds$, $|d\phi/ds| = 1/\rho$. Thus, $|\mathbf{T}'(t)| = v/\rho$.

34.106 Assume that, for a curve $\mathbf{R}(t)$, $v = ds/dt > 0$. Then the acceleration vector $\mathbf{a}$ can be represented as the following linear combination of the perpendicular vectors $\mathbf{T}$ and $\mathbf{N}$: $\mathbf{a} = (d^2s/dt^2)\mathbf{T} + (v^2/\rho)\mathbf{N}$. The coefficients of $\mathbf{T}$ and $\mathbf{N}$ are called, respectively, the *tangential* and *normal* components of the acceleration vector. (Since the normal component v^2/ρ is positive, the acceleration vector points "inside" the curve, just as $\mathbf{N}$ does.)

▮ By definition of $\mathbf{T}$, $\mathbf{T} = \mathbf{R}'(t)/v$. So, $\mathbf{R}'(t) = v\mathbf{T}$. By the product rule, $\mathbf{R}''(t) = v\mathbf{T}' + (dv/dt)\mathbf{T}$. By Problem 34.105, $\mathbf{T}' = (v/\rho)\mathbf{N}$. Hence, $\mathbf{a} = \mathbf{R}''(t) = (d^2s/dt^2)\mathbf{T} + (v^2/\rho)\mathbf{N}$.

34.107 Find the tangential and normal components of the acceleration vector for the curve $\mathbf{R}(t) = (e^t, e^{2t})$ (a motion along the parabola $y = x^2$).

▮ $\mathbf{R}' = (e^t, 2e^{2t})$, $\mathbf{R}'' = (e^t, 4e^{2t})$. $v = |\mathbf{R}'| = \sqrt{e^{2t} + 4e^{4t}} = e^t\sqrt{1 + 4e^{2t}}$. Hence,

$$\frac{dv}{dt} = \frac{4e^t \cdot e^{2t}}{\sqrt{1 + 4e^{2t}}} + e^t\sqrt{1 + 4e^{2t}} = \frac{e^t(8e^{2t} + 1)}{\sqrt{1 + 4e^{2t}}} \qquad |\mathbf{R}''| = \sqrt{e^{2t} + 16e^{4t}} = e^t\sqrt{1 + 16e^{2t}}$$

Since $\mathbf{R}'' = (dv/dt)\mathbf{T} + (v^2/\rho)\mathbf{N}$, and $\mathbf{T}$ and $\mathbf{N}$ are perpendicular, the Pythagorean theorem yields $|\mathbf{R}''|^2 = (dv/dt)^2 + (v^2/\rho)^2$. Hence,

$$e^{2t}(1 + 16e^{2t}) = \frac{e^{2t}(8e^{2t} + 1)^2}{1 + 4e^{2t}} + \left(\frac{v^2}{\rho}\right)^2$$

So, $\left(\dfrac{v^2}{\rho}\right)^2 = e^{2t}(1 + 16e^{2t}) - \dfrac{e^{2t}(8e^{2t} + 1)^2}{1 + 4e^{2t}} = \dfrac{e^{2t}}{1 + 4e^{2t}}[(1 + 16e^{2t})(1 + 4e^{2t}) - (64e^{4t} + 16e^{2t} + 1)]$

$$= \frac{e^{2t}}{1 + 4e^{2t}}(4e^{2t}) = \frac{4e^{4t}}{1 + 4e^{2t}}$$

Thus, $v^2/\rho = 2e^{2t}/\sqrt{1 + 4e^{2t}}$. This is the normal component, and we already found the tangential component as $e^t(8e^{2t} + 1)/\sqrt{1 + 4e^{2t}}$. Note that we avoided a direct computation of v^2/ρ, which is usually tedious.

34.108 Find the tangential and normal components of the acceleration vector for the curve $\mathbf{R} = (t^2, t^3)$.

❚ $\mathbf{R}' = (2t, 3t^2)$, $\mathbf{R}'' = (2, 6t) = 2(1, 3t)$. Then $v = |\mathbf{R}'| = \sqrt{4t^2 + 9t^4}$. Hence, $\dfrac{dv}{dt} = \dfrac{1}{2} \dfrac{8t + 36t^3}{\sqrt{4t^2 + 9t^4}} =$

$\dfrac{2t(2 + 9t^2)}{\sqrt{4t^2 + 9t^4}}$. $|\mathbf{R}''| = 2\sqrt{1 + 9t^2}$. By the Pythagorean theorem, $|\mathbf{R}''|^2 = |dv/dt|^2 + (v^2/\rho)^2$. Hence,

$$\left(\frac{v^2}{\rho}\right)^2 = 4(1 + 9t^2) - \frac{4t^2(2 + 9t^2)^4}{4t^2 + 9t^4} = \frac{4}{4t^2 + 9t^4}[(1 + 9t^2)(4t^2 + 9t^4) - t^2(2 + 9t^2)^4] = \frac{36t^2}{4 + 9t^2}$$

So, $v^2/\rho = 6|t|/\sqrt{4 + 9t^2}$ is the normal component of the acceleration vector, and we already found the tangential component as $2t(2 + 9t^2)/\sqrt{4t^2 + 9t^4}$.

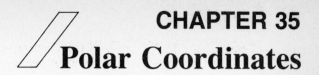

35.1 Write the relations between polar coordinates (r, θ) and rectangular coordinates (x, y).

▌ $x = r \cos \theta$, $y = r \sin \theta$; or, inversely, $r^2 = x^2 + y^2$, $\tan \theta = y/x$. See Fig. 35-1. Note that, because $\cos(\theta + \pi) = -\cos\theta$ and $\sin(\theta + \pi) = -\sin\theta$, (r, θ) and $(-r, \theta + \pi)$ represent the same point (x, y).

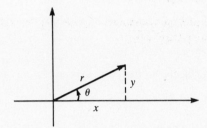

Fig. 35-1

35.2 Give all possible polar representations of the point with rectangular coordinates $(1, 0)$.

▌ $(1, 2\pi n)$ for all integers n, and $(-1, (2n+1)\pi)$ for all integers n.

35.3 Give all possible polar representations of the point with rectangular coordinates $(1, 1)$.

▌ $\left(\sqrt{2}, \dfrac{\pi}{4} + 2\pi n\right)$ for all integers n, and $\left(-\sqrt{2}, \dfrac{5\pi}{4} + 2\pi n\right)$ for all integers n. $\;=\left(-\sqrt{2}, \; {}^{\pi}\!/_{4} + (2n+1)\pi\right)$

35.4 Find the rectangular coordinates of the point with polar coordinates $(2, \pi/6)$.

▌ $x = 2\cos(\pi/6) = 2 \cdot (\sqrt{3}/2) = \sqrt{3}$, $y = 2\sin(\pi/6) = 2 \cdot \frac{1}{2} = 1$. Thus, in rectangular coordinates, the point is $(\sqrt{3}, 1)$.

35.5 Find the rectangular coordinates of the point with polar coordinates $(-4, \pi/3)$.

▌ $x = -4\cos(\pi/3) = -4 \cdot \frac{1}{2} = -2$. $y = -4\sin(\pi/3) = -4(\sqrt{3}/2) = -2\sqrt{3}$. Thus, in rectangular coordinates, the point is $(-2, -2\sqrt{3})$.

35.6 Find the rectangular coordinates of the point with polar coordinates $(3, 3\pi/4)$.

▌ $x = 3\cos(3\pi/4) = 3 \cdot (-\sqrt{2}/2) = -3\sqrt{2}/2$. $y = 3\sin(3\pi/4) = 3 \cdot (\sqrt{2}/2) = 3\sqrt{2}/2$. Thus, in rectangular coordinates, the point is $(-3\sqrt{2}/2, 3\sqrt{2}/2)$.

35.7 Describe the graph of the polar equation $r = 2$.

▌ $x^2 + y^2 = r^2 = 4$. Thus, the graph is the circle of radius 2 with center at the pole.

35.8 Describe the graph of the polar equation $r = -2$.

▌ $x^2 + y^2 = r^2 = 4$. Thus, the graph is the circle of radius 2 with center at the pole.

35.9 Describe the graph of the polar equation $r = a$.

▌ $x^2 + y^2 = r^2 = a^2$. Hence, the graph is the circle of radius $|a|$ with center at the pole.

35.10 Describe the graph of the polar equation $\theta = \pi/4$.

▌ The graph is the line through the pole making an angle of $\pi/4$ radian with the polar axis (Fig. 35-2). Note that we obtain the points on that line below the x-axis because r can assume negative values.

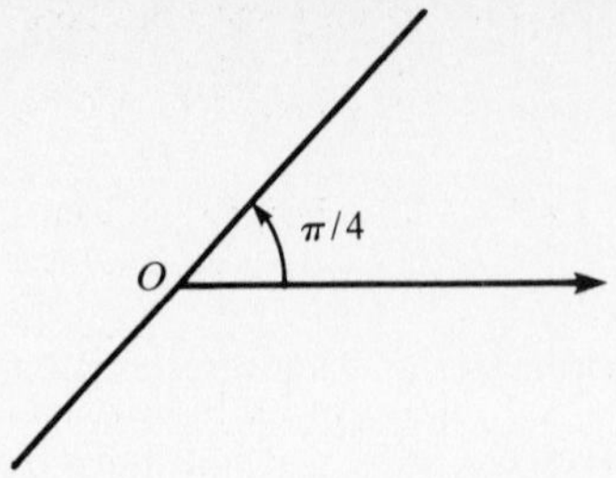

Fig. 35-2

35.11 Describe the graph of the polar equation $\theta = 0$.

▌ This is simply the line through the polar axis, or the x-axis in rectangular coordinates.

35.12 Write a polar equation for the y-axis.

▌ $\theta = \pi/2$ yields the line perpendicular to the polar axis and going through the pole, which is the y-axis.

35.13 Describe the graph of the polar equation $r = 2 \sin \theta$.

▌ Multiplying both sides by r, we obtain $r^2 = 2r \sin \theta$, $x^2 + y^2 = 2y$, $x^2 + y^2 - 2y = 0$, $x^2 + (y - 1)^2 = 1$. Thus, the graph is the circle with center at $(0, 1)$ and radius 1.

35.14 Describe the graph of the polar equation $r = 4 \cos \theta$.

▌ Multiplying both sides by r, we obtain $r^2 = 4r \cos \theta$, $x^2 + y^2 = 4x$, $x^2 - 4x + y^2 = 0$, $(x - 2)^2 + y^2 = 4$. Thus, the graph is the circle with center at $(2, 0)$ and radius 2.

35.15 Describe the graph of the polar equation $r = \tan \theta \sec \theta$.

▌ $r = \sin \theta / \cos^2 \theta$, $r \cos^2 \theta = \sin \theta$, $r^2 \cos^2 \theta = r \sin \theta$, $x^2 = y$. Thus, the graph is a parabola.

35.16 Describe the graph of the polar equation $r = 6/\sqrt{9 - 5 \sin^2 \theta}$.

▌ $r^2 = 36/(9 - 5 \sin^2 \theta)$, $r^2(9 - 5 \sin^2 \theta) = 36$, $9r^2 - 5r^2 \sin^2 \theta = 36$, $9(x^2 + y^2) - 5y^2 = 36$, $9x^2 + 4y^2 = 36$, $\dfrac{x^2}{4} + \dfrac{y^2}{9} = 1$. The graph is an ellipse.

35.17 Describe the graph of the polar equation $r = -10 \cos \theta$.

▌ Multiply both sides by r: $r^2 = -10r \cos \theta$, $x^2 + y^2 = -10x$, $x^2 + 10x + y^2 = 0$, $(x + 5)^2 + y^2 = 25$. Thus, the graph is the circle with center $(-5, 0)$ and radius 5.

35.18 Describe the graph of the polar equation $r = 6(\sin \theta + \cos \theta)$.

▌ Multiply both sides by r: $r^2 = 6r \sin \theta + 6r \cos \theta$, $x^2 + y^2 = 6y + 6x$, $x^2 - 6x + y^2 - 6y = 0$, $(x - 3)^2 + (y - 3)^2 = 18$. Thus, the graph is the circle with center at $(3, 3)$ and radius $3\sqrt{2}$.

35.19 Describe the graph of the polar equation $r = 5 \csc \theta$.

▌ $r = 5/\sin \theta$, $r \sin \theta = 5$, $y = 5$. Thus, the graph is a horizontal line.

35.20 Describe the graph of the polar equation $r = -3 \sec \theta$.

▌ $r = -3/\cos \theta$, $r \cos \theta = -3$, $x = -3$. Hence, the graph is a vertical line.

35.21 Change the rectangular equation $x^2 + y^2 = 16$ into a polar equation.

▌ $r^2 = 16$, $r = 4$ (or $r = -4$).

35.22 Change the rectangular equation $x^2 - y^2 = 1$ into a polar equation.

▌ $r^2 \cos^2 \theta - r^2 \sin^2 \theta = 1$, $r^2(\cos^2 \theta - \sin^2 \theta) = 1$, $r^2 \cos 2\theta = 1$, $r^2 = \sec 2\theta$.

35.23 Transform the rectangular equation $xy = 4$ into a polar equation.

▌ $r\cos\theta \cdot r\sin\theta = 4$, $r^2\sin\theta\cos\theta = 4$, $r^2(\sin 2\theta)/2 = 4$, $r^2 = 8\csc 2\theta$.

35.24 Transform the rectangular equation $x = 3$ into a polar equation.

▌ $r\cos\theta = 3$, $r = 3\sec\theta$.

35.25 Transform the rectangular equation $x + 2y = 3$ into a polar equation.

▌ $r\cos\theta + 2r\sin\theta = 3$, $r(\cos\theta + 2\sin\theta) = 3$, $r = 3/(\cos\theta + 2\sin\theta)$.

35.26 Find a rectangular equation equivalent to the polar equation $\theta = \pi/3$.

▌ $\tan\theta = \tan(\pi/3) = \sqrt{3}$. Hence, $y/x = \sqrt{3}$, $y = \sqrt{3}x$.

35.27 Find a rectangular equation equivalent to the polar equation $r = \tan\theta$.

▌ $\sqrt{x^2 + y^2} = y/x$, $x^2 + y^2 = y^2/x^2$, $y^2 = x^2(x^2 + y^2)$, $y^2 = x^4 + x^2y^2$, $y^2(1 - x^2) = x^4$, $y^2 = x^4/(1 - x^2)$.

35.28 Show that the point with polar coordinates $(3, 3\pi/4)$ lies on the curve $r = 3\sin 2\theta$.

▌ Observe that $r = 3$, $\theta = 3\pi/4$ do not satisfy the equation $r = 3\sin 2\theta$. However (Problem 35.1), the point with polar coordinates $(3, 3\pi/4)$ also has polar coordinates $(-3, 7\pi/4)$, and $r = -3$, $\theta = 7\pi/4$ satisfy the equation $r = 3\sin 2\theta$, since $3\sin 2(7\pi/4) = 3\sin(7\pi/2) = 3(-1) = -3$.

35.29 Show that the point with polar coordinates $(3, 3\pi/2)$ lies on the curve with the polar equation $r^2 = 9\sin\theta$.

▌ Notice that $r = 3$, $\theta = 3\pi/2$ do not satisfy the equation $r^2 = 9\sin\theta$. However, the point with polar coordinates $(3, 3\pi/2)$ also has polar coordinates $(-3, \pi/2)$, and $r = -3$, $\theta = \pi/2$ satisfy the equation $r^2 = 9\sin\theta$, since $(-3)^2 = 9 \cdot 1$.

35.30 Sketch the graph of $r = 1 + \cos\theta$.

▌ See Fig. 35-3. At $\theta = 0$, $r = 2$. As θ increases to $\pi/2$, r decreases to 1. As θ increases to π, r decreases to 0. Then, as θ increases to $3\pi/2$, r increases to 1, and finally, as θ increases to 2π, r increases to 2. After $\theta = 2\pi$, the curve repeats itself. The graph is called a cardioid.

θ	0	$\pi/2$	π	$3\pi/2$	2π
r	2	1	0	1	2

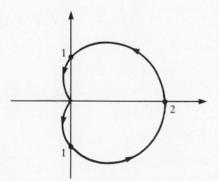

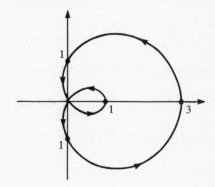

Fig. 35-3

Fig. 35-4

35.31 Sketch the graph of $r = 1 + 2\cos\theta$.

▌ See Fig. 35-4. As θ goes from 0 to $\pi/2$, r decreases from 3 to 1. As θ increases further to $2\pi/3$, r decreases to 0. As θ goes on to π, r decreases to -1, and then, as θ moves up to $4\pi/3$, r goes back up to 0. As θ moves on to $3\pi/2$, r goes up to 1, and, finally, as θ increases to 2π, r grows to 3. This kind of graph is called a limaçon.

θ	0	$\pi/2$	$2\pi/3$	π	$4\pi/3$	$3\pi/2$	2π
r	3	1	0	-1	0	1	3

35.32 Sketch the graph of $r^2 = \cos 2\theta$.

▌ The construction of the graph, Fig. 35-5, is indicated in the table of values. Note that some values of θ yield two values of r, and some yield none at all (when $\cos 2\theta$ is negative). The graph repeats from $\theta = \pi$ to $\theta = 2\pi$. The graph is called a lemniscate.

θ	0	$\pi/4$	$\cdots$	$3\pi/4$	π
r	± 1	0	$\cdots$	0	± 1

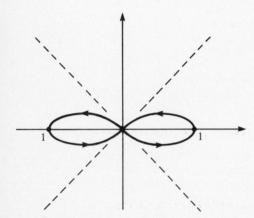

Fig. 35-5

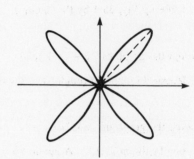

Fig. 35-6

35.33 Sketch the graph of $r = \sin 2\theta$.

▌ The accompanying table of values yields Fig. 35-6. The graph is called a four-leaved rose.

θ	0	$\pi/4$	$\pi/2$	$3\pi/4$	π	$5\pi/4$	$3\pi/2$	$7\pi/4$	2π
r	0	1	0	-1	0	1	0	-1	0

35.34 Sketch the graph of $r = 1 - \cos \theta$.

▌ The graph of $r = f(\theta - \alpha)$ is the graph of $r = f(\theta)$ rotated counterclockwise through α radians. Thus, a rotation of Fig. 35-3 through π radians gives the graph of $r = 1 + \cos (\theta - \pi) = 1 - \cos \theta$.

35.35 Sketch the graph of $r = 1 - \sin \theta$.

▌ Rotate Fig. 35-3 through $3\pi/2$ radians (or $-\pi/2$ radians): $r = 1 + \cos (\theta - 3\pi/2) = 1 - \sin \theta$.

35.36 Sketch the graph of $r = 1 + \sin \theta$.

▌ Rotate Fig. 35-3 through $\pi/2$ radians: $r = 1 + \cos (\theta - \pi/2) = 1 + \sin \theta$.

35.37 Sketch the graph of $r^2 = \sin 2\theta$.

▌ Rotate Fig. 35-5 through $\pi/4$ radians: $r^2 = \cos 2(\theta - \pi/4) = \cos (2\theta - \pi/2) = \sin 2\theta$.

35.38 Sketch the graph of $r = 4 + 2 \cos \theta$.

▌ See Fig. 35-7. This figure is also called a limaçon (but with a "dimple" instead of a loop).

θ	0	$\pi/2$	π	$3\pi/2$	2π
r	6	4	2	4	6

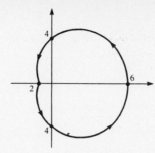

Fig. 35-7

35.39 Sketch the graph of $r = 2 + 4\cos\theta$.

▮ Scale up Fig. 35-4 by the factor 2.

35.40 Sketch the graph of $r = 4 - 2\sin\theta$.

▮ Rotate Fig. 35-7 through $3\pi/2$ (or $-\pi/2$) radians: $r = 4 + 2\cos(\theta - 3\pi/2) = 4 - 2\sin\theta$.

35.41 Sketch the graph of $r = 2a\sin\theta$.

▮ See Problem 35.13. A geometrical construction is shown in Fig. 35-8, for the case $a > 0$.

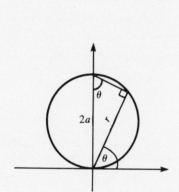

Fig. 35-8

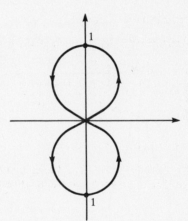

Fig. 35-9

35.42 Sketch the graph of $r^2 = \sin\theta$.

▮ As shown in Fig. 35-9, the curve is a lemniscate.

θ	0	$\pi/2$	π	$\cdots$	2π
r	0	± 1	0	$\cdots$	0

35.43 Sketch the graph of $r = 2 + \cos 2\theta$.

▮ r is never negative. The graph, Fig. 35-10, is symmetric with respect to the pole, since $\cos 2(\theta + \pi) = \cos 2\theta$.

θ	0	$\pi/4$	$\pi/2$	$3\pi/4$	π	$5\pi/4$	$3\pi/2$	$7\pi/4$	2π
r	3	2	1	2	3	2	1	2	3

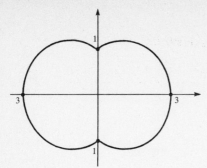

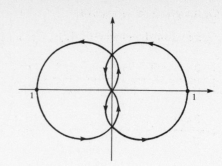

Fig. 35-10   Fig. 35-11

35.44   Sketch the graph of $r = \sin(\theta/2)$.

▌ See Fig. 35-11. From $\theta = 0$ to $\theta = 2\pi$,   we obtain a "cardioid", and then, from $\theta = 2\pi$ to $\theta = 4\pi$, the reflection of the first "cardioid" in the y-axis.

θ	0	$\pi/2$	π	$3\pi/2$	2π	$5\pi/2$	3π	$7\pi/2$	4π
r	0	$\sqrt{2}/2$	1	$\sqrt{2}/2$	0	$-\sqrt{2}/2$	-1	$-\sqrt{2}/2$	0

35.45   Sketch the graph of $r = \cos^2(\theta/2)$.

▌ $r = \cos^2(\theta/2) = (1 + \cos\theta)/2$.   Hence, the graph is a contraction toward the pole by a factor of $\frac{1}{2}$, of Fig. 35-3.

35.46   Sketch the graph of $r = \tan\theta$.

▌ The behavior of $\tan\theta$ is shown in the table of values.   Recall that $\tan\theta \to +\infty$ as $\theta \to (\pi/2)$,   and $\tan\theta \to -\pi$ as $\theta \to (\pi/2)^+$.   See Fig. 35-12.

θ	0	$\pi/4 \to \pi/2 \leftarrow 3\pi/4$	π	$5\pi/4 \to 3\pi/2 \leftarrow 7\pi/4$	2π
r	0	$1 \to +\infty, -\infty \leftarrow -1$	0	$1 \to +\infty, -\infty \leftarrow -1$	0

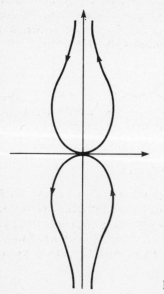

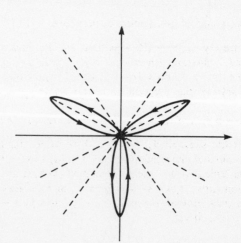

Fig. 35-12   Fig. 35-13

35.47   Sketch the graph of $r = \sin 3\theta$.

▌ It is convenient to use increments in θ of $\pi/6$ to construct Fig. 35-13.   Note that the graph repeats itself after $\theta = \pi$.   The result is a three-leaved rose.

θ	0	$\pi/6$	$\pi/3$	$\pi/2$	$2\pi/3$	$5\pi/6$	π
r	0	1	0	-1	0	1	0

35.48 Sketch the graph of $r = \sin 4\theta$.

▐ Figure 35-14 shows an eight-leaved rose. Use increments of $\pi/8$ in θ.

θ	0	$\pi/8$	$\pi/4$	$3\pi/8$	$\pi/2$	$5\pi/8$	$3\pi/4$	$7\pi/8$	π	$9\pi/8$	$5\pi/4$	$11\pi/8$
r	0	1	0	−1	0	1	0	−1	0	1	0	−1

θ	$3\pi/2$	$13\pi/8$	$7\pi/4$	$15\pi/8$	2π
r	0	1	0	−1	0

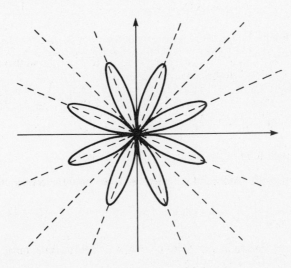

Fig. 35-14

35.49 Find the largest value of y on the cardioid $r = 2(1 + \cos \theta)$.

▐ $y = r \sin \theta = 2(1 + \cos \theta) \sin \theta$. Hence, $dy/d\theta = 2[(1 + \cos \theta) \cos \theta - \sin^2 \theta]$. Setting $dy/d\theta = 0$, we find $(1 + \cos \theta) \cos \theta = \sin^2 \theta$, $\cos \theta + \cos^2 \theta = 1 - \cos^2 \theta$, $2 \cos^2 \theta + \cos \theta - 1 = 0$, $(2 \cos \theta - 1)(\cos \theta + 1) = 0$, $\cos \theta = \frac{1}{2}$ or $\cos \theta = -1$. Hence, we get the critical numbers π, $\pi/3$, $5\pi/3$. If we calculate y for these values and for the endpoints 0 and 2π, the largest value $3\sqrt{3}/2$ is assumed when $\theta = \pi/3$.

35.50 Find all points of intersection of the curves $r = 1 + \sin^2 \theta$ and $r = -1 - \sin^2 \theta$.

▐ If we try to solve the equations simultaneously, we obtain $2 \sin^2 \theta = -2$, $\sin^2 \theta = -1$, which is never satisfied. However, there are, in fact, infinitely many points of intersection because the curves are identical. Assume (r, θ) satisfies $r = 1 + \sin^2 \theta$. The point (r, θ) is identical with the point $(-1 - \sin^2 \theta, \theta + \pi)$, which satisfies the equation $r = -1 - \sin^2 \theta$ [because $\sin^2 (\theta + \pi) = \sin^2 \theta$].

35.51 Find the points of intersection of the curves $r = 4 \cos \theta$ and $r = 4\sqrt{3} \sin \theta$.

▐ For the same (r, θ), $4\sqrt{3} \sin \theta = 4 \cos \theta$, $\tan \theta = 1/\sqrt{3}$, and, therefore, $\theta = \pi/6$ or $\theta = 7\pi/6$. So, these points of intersection are $(2\sqrt{3}, \pi/6)$ and $(-2\sqrt{3}, 7\pi/6)$. However, notice that these points are identical. So, we have just one point of intersection thus far. Considering intersection points where (r, θ) satisfies one equation and $(-r, \theta + \pi)$ satisfies the other equation, we obtain no new solutions. However, observe that both curves pass through the pole. Hence, this is a second point of intersection. (The curves are actually two intersecting circles.)

35.52 Find the points of intersection of the curves $r = \sqrt{2} \sin \theta$ and $r^2 = \cos 2\theta$.

▐ Solving simultaneously, we have $2 \sin^2 \theta = \cos 2\theta$, $2 \sin^2 \theta = 1 - 2 \sin^2 \theta$, $4 \sin^2 \theta = 1$, $\sin^2 \theta = \frac{1}{4}$, $\sin \theta = \pm \frac{1}{2}$. Hence, we obtain the solutions $\theta = \pi/6, 5\pi/6, 7\pi/6, 11\pi/6$, and the corresponding points $(\sqrt{2}/2, \pi/6), (\sqrt{2}/2, 5\pi/6), (-\sqrt{2}/2, 7\pi/6), (-\sqrt{2}/2, 11\pi/6)$. However, the first and third of these points are identical, as are the second and fourth. So, we have obtained two intersection points. If we use (r, θ) for the first equation and $(-r, \theta + \pi)$ for the second equation, we obtain the same pair of equations as before. Notice that both curves pass through the pole (when $r = 0$), which is a third point of intersection.

35.53 Find the intersection points of the curves $r^2 = 4 \cos 2\theta$ and $r^2 = 4 \sin 2\theta$.

❚ From $4 \cos 2\theta = 4 \sin 2\theta$, we obtain $\tan 2\theta = 1$, $2\theta = \pi/4$ or $5\pi/4$. The latter is impossible, since $4 \cos (5\pi/4) < 0$ and cannot equal r^2. Hence, $\theta = \pi/8$, $r^2 = 2\sqrt{2}$, $r = \pm 2^{3/4}$. Putting $(-r, \ \theta + \pi)$ in the second equation, we obtain the same equation as before and no new points are found. However, both curves pass through the pole $r = 0$, which is a third point of intersection.

35.54 Find the points of intersection of the curves $r = 1$ and $r = -1$.

❚ As $r = 1$ and $r = -1$ both represent the circle $x^2 + y^2 = 1$, there are an infinity of intersection points.

35.55 Find the area enclosed by the cardioid $r = 1 + \cos \theta$.

❚ As was shown in Problem 35.30, the curve is traced out for $\theta = 0$ to $\theta = 2\pi$. The general formula for the area is $\frac{1}{2} \int_{\theta_1}^{\theta_2} r^2 \, d\theta$. In this case, we have

$$A = \frac{1}{2} \int_0^{2\pi} (1 + \cos \theta)^2 \, d\theta = \frac{1}{2} \int_0^{2\pi} (1 + 2 \cos \theta + \cos^2 \theta) \, d\theta = \frac{1}{2} \int_0^{2\pi} [1 + 2 \cos \theta + \frac{1}{2}(1 + \cos 2\theta)] \, d\theta$$

$$= \frac{1}{2}(\theta + 2 \sin \theta + \frac{1}{2}(\theta + \frac{1}{2} \sin 2\theta)) \,]_0^{2\pi} = \frac{1}{2}(3\pi) = \frac{3\pi}{2}$$

35.56 Find the area inside the inner loop of the limaçon $r = 1 + 2 \cos \theta$.

❚ Problem 35.31 shows that the inner loop is traced out from $\theta = 5\pi/6$ to $\theta = 7\pi/6$. So, the area

$$A = \frac{1}{2} \int_{5\pi/6}^{7\pi/6} r^2 \, d\theta = \frac{1}{2} \int_{5\pi/6}^{7\pi/6} (1 + 4 \cos \theta + 4 \cos^2 \theta) \, d\theta = \frac{1}{2} \int_{5\pi/6}^{7\pi/6} [1 + 4 \cos \theta + 2(1 + \cos 2\theta)] \, d\theta$$

$$= \frac{1}{2}(3\theta + 4 \sin \theta + \sin 2\theta) \,\Big]_{5\pi/6}^{7\pi/6} = \frac{1}{2}\left[\left(\frac{7\pi}{2} - 2 + \frac{\sqrt{3}}{2} \right) - \left(\frac{5\pi}{2} + 2 - \frac{\sqrt{3}}{2} \right) \right] = \frac{1}{2}(\pi + \sqrt{3}) - 2$$

35.57 Find the area inside one loop of the lemniscate $r^2 = \cos 2\theta$.

❚ From Problem 35.32, we see that the area of one loop is double the area in the first quadrant. The latter area is swept out from $\theta = 0$ to $\theta = \pi/4$. Hence, the required area is $2 \cdot \frac{1}{2} \int_0^{\pi/4} \cos 2\theta \, d\theta = \frac{1}{2} \sin 2\theta \,]_0^{\pi/4} = \frac{1}{2}(1 - 0) = \frac{1}{2}$.

35.58 Find the area inside one petal of the four-leaved rose $r = \sin 2\theta$.

❚ From Problem 35.33, we see that the area of one petal is swept out from $\theta = 0$ to $\theta = \pi/2$. Hence, the area is

$$\frac{1}{2} \int_0^{\pi/2} \sin^2 2\theta \, d\theta = \frac{1}{2} \int_0^{\pi/2} \frac{1 - \cos 4\theta}{2} \, d\theta = \frac{1}{4}(\theta - \frac{1}{4} \sin 4\theta) \,\Big]_0^{\pi/2} = \frac{1}{4}\left(\frac{\pi}{2} \right) = \frac{\pi}{8}$$

35.59 Find the area inside the limaçon $r = 4 + 2 \cos \theta$.

❚ From Problem 35.38, we see that the area is swept out from $\theta = 0$ to $\theta = 2\pi$. So, the area is $\frac{1}{2} \int_0^{2\pi} (16 + 16 \cos \theta + 4 \cos^2 \theta) \, d\theta = \frac{1}{2} \int_0^{2\pi} [16 + 16 \cos \theta + 2(1 + \cos 2\theta)] \, d\theta = \frac{1}{2}(18\theta + 16 \sin \theta + \sin 2\theta) \,]_0^{2\pi} = \frac{1}{2}(36\pi) = 18\pi$.

35.60 Find the area inside the limaçon $r = 1 + 2 \cos \theta$, but outside its inner loop.

❚ From Problem 35.31, we see that it suffices to double the area above the x-axis. The latter can be obtained by subtracting the area above the x-axis and inside the loop from the area outside the loop. Since the desired area outside the loop is swept out from $\theta = 0$ to $\theta = 2\pi/3$ and the desired area inside the loop is swept out from $\theta = \pi$ to $\theta = 4\pi/3$, we obtain $2(\frac{1}{2} \int_0^{2\pi/3} r^2 \, d\theta - \frac{1}{2} \int_\pi^{4\pi/3} r^2 \, d\theta) = \int_0^{2\pi/3}(1 + 4 \cos \theta + \cos^2 \theta) \, d\theta - \int_\pi^{4\pi/3}(1 + 4 \cos \theta + 4 \cos^2 \theta) \, d\theta = \int_0^{2\pi/3}[1 + 4 \cos \theta + 2(1 + \cos 2\theta)] \, d\theta - \int_\pi^{4\pi/3}[1 + 4 \cos \theta + 2(1 + \cos 2\theta)] \, d\theta = (3\theta + 4 \sin \theta + \sin 2\theta) \,]_0^{2\pi/3} - (3\theta + 4 \sin \theta + \sin 2\theta) \,]_\pi^{4\pi/3} = (2\pi + 2\sqrt{3} - \frac{1}{2}\sqrt{3}) - [(4\pi - 2\sqrt{3} + \frac{1}{2}\sqrt{3}) - (3\pi)] = \pi + 3\sqrt{3}$.

35.61 Find the area inside $r = 2 + \cos 2\theta$.

❚ From Problem 35.43, we see that the area is swept out from $\theta = 0$ to $\theta = 2\pi$. Hence, the area is $\frac{1}{2} \int_0^{2\pi} [4 + 4 \cos 2\theta + \cos^2 (2\theta)] \, d\theta = \frac{1}{2} \int_0^{2\pi} [4 + 4 \cos \theta + \frac{1}{2}(1 + \cos 4\theta)] \, d\theta = \frac{1}{2}(\frac{9}{2}\theta + 2 \sin 2\theta + \frac{1}{8} \sin 4\theta) \,]_0^{2\pi} = \frac{1}{2}(9\pi) = 9\pi/2$.

35.62 Find the area inside $r = \cos^2(\theta/2)$.

▍ By Problem 35.45, we see that the area is swept out from $\theta = 0$ to $\theta = 2\pi$. Hence, the area is

$$\frac{1}{2}\int_0^{2\pi} \cos^4 \frac{\theta}{2}\, d\theta = \frac{1}{2}\int_0^{2\pi}\left(\frac{1+\cos\theta}{2}\right)^2 d\theta = \frac{1}{8}\int_0^{2\pi}(1 + 2\cos\theta + \cos^2\theta)\, d\theta$$

$$= \frac{1}{8}\int_0^{2\pi}[1 + 2\cos\theta + \tfrac{1}{2}(1 + \cos 2\theta)]\, d\theta = \tfrac{1}{8}(\tfrac{3}{2}\theta + 2\sin\theta + \tfrac{1}{4}\sin 2\theta)\big]_0^{2\pi} = \tfrac{1}{8}(3\pi) = \frac{3\pi}{8}$$

35.63 Find the area swept out by $r = \tan\theta$ from $\theta = 0$ to $\theta = \pi/4$.

▍ See Fig. 35-12. The area is $\frac{1}{2}\int_0^{\pi/4}\tan^2\theta\, d\theta = \frac{1}{2}\int_0^{\pi/4}(\sec^2\theta - 1)\, d\theta = \frac{1}{2}(\tan\theta - \theta)\big]_0^{\pi/4} = \frac{1}{2}[1 - (\pi/4)]$.

35.64 Find the area of one petal of the three-leaved rose $r = \sin 3\theta$.

▍ By Problem 35.47, one petal is swept out from $\theta = 0$ to $\theta = \pi/3$. Hence, the area is $\frac{1}{2}\int_0^{\pi/3}\sin^2(3\theta)\, d\theta = \frac{1}{2}\int_0^{\pi/3}\frac{1}{2}(1 - \cos 6\theta)\, d\theta = \frac{1}{4}(\theta - \frac{1}{6}\sin 6\theta)\big]_0^{\pi/3} = \frac{1}{4}(\pi/3) = \pi/12$.

35.65 Find the area inside one petal of the eight-leaved rose $r = \sin 4\theta$.

▍ From Problem 35.48, one petal is swept out from $\theta = 0$ to $\theta = \pi/4$. Hence, the area is $\frac{1}{2}\int_0^{\pi/4}\sin^2(4\theta)\, d\theta = \frac{1}{2}\int_0^{\pi/4}\frac{1}{2}(1 - \cos 8\theta)\, d\theta = \frac{1}{4}(\theta - \frac{1}{8}\sin 8\theta)\big]_0^{\pi/4} = \frac{1}{4}(\pi/4) = \pi/16$.

35.66 Find the area inside the cardioid $r = 1 + \cos\theta$ and outside the circle $r = 1$.

▍ In Fig. 35-15, area $ABC = $ area $OBC - $ area OAC is one-half the required area. Thus, the area is $2 \cdot \frac{1}{2}\int_0^{\pi/2}[(1+\cos\theta)^2 - (1)^2]\, d\theta = \int_0^{\pi/2}(2\cos\theta + \cos^2\theta)\, d\theta = \int_0^{\pi/2}[2\cos\theta + \frac{1}{2}(1 + \cos 2\theta)]\, d\theta = (\frac{1}{2}\theta + 2\sin\theta + \frac{1}{4}\sin 2\theta)\big]_0^{\pi/2} = (\pi/4) + 2$.

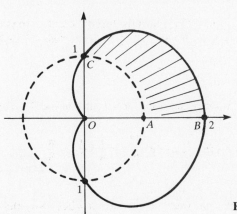

Fig. 35-15

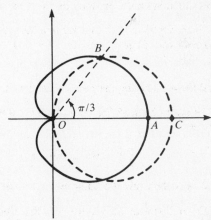

Fig. 35-16

35.67 Find the area common to the circle $r = 3\cos\theta$ and the cardioid $r = 1 + \cos\theta$.

▍ In Fig. 35-16, area AOB consists of two parts, one swept out by the radius vector $r = 1 + \cos\theta$ as θ varies from 0 to $\pi/3$ and the other swept out by $r = 3\cos\theta$ as θ varies from $\pi/3$ to $\pi/2$. Hence, the area is

$$2 \cdot \frac{1}{2}\int_0^{\pi/3}(1 + \cos\theta)^2\, d\theta + 2 \cdot \frac{1}{2}\int_{\pi/3}^{\pi/2} 9\cos^2\theta\, d\theta = \int_0^{\pi/3}[1 + 2\cos\theta + \tfrac{1}{2}(1 + \cos 2\theta)]\, d\theta + 9\int_{\pi/3}^{\pi/2}\tfrac{1}{2}(1 + \cos 2\theta)\, d\theta$$

$$= (\tfrac{3}{2}\theta + 2\sin\theta + \tfrac{1}{4}\sin 2\theta)\big]_0^{\pi/3} + \tfrac{9}{2}(\theta + \tfrac{1}{2}\sin 2\theta)\big]_{\pi/3}^{\pi/2}$$

$$= \left(\frac{\pi}{2} + \sqrt{3} + \frac{\sqrt{3}}{8}\right) + \frac{9}{2}\left[\frac{\pi}{2} - \left(\frac{\pi}{3} + \frac{\sqrt{3}}{4}\right)\right] = \frac{5\pi}{4}$$

35.68 Find the area bounded by the curve $r = 2\cos\theta$, $0 \le \theta \le \pi$.

▍ $A = \frac{1}{2}\int_0^{\pi} 4\cos^2\theta\, d\theta = 2\int_0^{\pi}\frac{1}{2}(1 + \cos 2\theta)\, d\theta = (\theta + \frac{1}{2}\sin 2\theta)\big]_0^{\pi} = \pi$. This could have been obtained more easily by noting that the region is a circle of radius 1.

35.69 Find the area inside the circle $r = \sin \theta$ and outside the cardioid $r = 1 - \cos \theta$.

▮ For $0 \le \theta \le \pi/2$, $\sin^2 \theta = 1 - \cos^2 \theta \ge 1 - \cos \theta \ge (1 - \cos \theta)^2$, so that $\sin \theta \ge 1 - \cos \theta$ and the desired area is as shown in Fig. 35-17.

$$A = \frac{1}{2} \int_0^{\pi/2} [\sin^2 \theta - (1 - \cos \theta)^2] \, d\theta = \frac{1}{2} \int_0^{\pi/2} (2 \cos \theta - 1 - \cos 2\theta) \, d\theta = \frac{1}{2} (2 \sin \theta - \theta - \tfrac{1}{2} \sin 2\theta) \Big]_0^{\pi/2}$$

$$= \tfrac{1}{2} \left(2 - \frac{\pi}{2} \right) = 1 - \frac{\pi}{4}$$

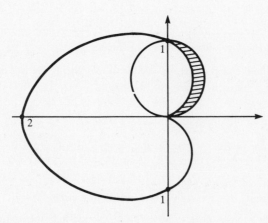

Fig. 35-17

35.70 Find the area swept out by the Archimedean spiral $r = \theta$ from $\theta = 0$ to $\theta = 2\pi$.

▮ The area A is $\frac{1}{2} \int_0^{2\pi} \theta^2 \, d\theta = \tfrac{1}{6} \theta^3 \big]_0^{2\pi} = \tfrac{1}{6} (2\pi)^3 = \tfrac{4}{3} \pi^3$.

35.71 Find the area swept out by the equiangular spiral $r = e^\theta$ from $\theta = 0$ to $\theta = 2\pi$.

▮ $A = \tfrac{1}{2} \int_0^{2\pi} e^{2\theta} \, d\theta = \tfrac{1}{4} e^{2\theta} \big]_0^{2\pi} = \tfrac{1}{4} (e^{4\pi} - 1)$.

35.72 Find the centroid of the region inside the first-quadrant loop of the rose $r = \sin 2\theta$ (Fig. 35-6).

▮ The area A is $\frac{1}{2} \int_0^{\pi/2} \sin^2 2\theta \, d\theta = \frac{1}{2} \int_0^{\pi/2} \tfrac{1}{2} (1 - \cos 4\theta) \, d\theta = \tfrac{1}{4} (\theta - \tfrac{1}{4} \sin 4\theta) \big]_0^{\pi/2} = \pi/8$. The general formulas for the centroid $(\bar{x}, \bar{y})$ are $A\bar{x} = \frac{1}{3} \int_{\theta_1}^{\theta_2} r^3 \cos \theta \, d\theta$ and $A\bar{y} = \frac{1}{3} \int_{\theta_1}^{\theta_2} r^3 \sin \theta \, d\theta$, where A is the area. In this case, we have $(\pi/8)\bar{x} = \frac{1}{3} \int_0^{\pi/2} \sin^3 (2\theta) \cos \theta \, d\theta = \tfrac{8}{3} \int_0^{\pi/2} \sin^3 \theta \cos^4 \theta \, d\theta = \tfrac{8}{3} \int_0^{\pi/2} (1 - \cos^2 \theta) \cdot \cos^4 \theta \sin \theta \, d\theta = \tfrac{8}{3} \int_0^{\pi/2} (\cos^4 \theta \sin \theta - \cos^6 \theta \sin \theta) \, d\theta = \tfrac{8}{3} (-\tfrac{1}{5} \cos^5 \theta + \tfrac{1}{7} \cos^7 \theta) \big]_0^{\pi/2} = -\tfrac{8}{3} (-\tfrac{1}{5} + \tfrac{1}{7}) = \tfrac{16}{105}$. Hence, $\bar{x} = 128/105\pi$. By symmetry, $\bar{y} = 128/105\pi$.

35.73 Find the centroid of the right half of the cardioid $r = 1 + \sin \theta$. (See Problem 35.36.)

▮ By Problem 35.55 (and a rotation), the area A is $3\pi/4$. The region in question is swept out from $\theta = -\pi/2$ to $\theta = \pi/2$. Hence, $(3\pi/4)\bar{x} = \frac{1}{3} \int_{-\pi/2}^{\pi/2} (1 + \sin \theta)^3 \cos \theta \, d\theta = \frac{1}{3} \cdot \tfrac{1}{4} (1 + \sin \theta)^4 \big]_{-\pi/2}^{\pi/2} = \tfrac{1}{12} (16) = \tfrac{4}{3}$. Thus, $\bar{x} = 16/9\pi$. In a similar manner, one finds $\bar{y} = \tfrac{5}{6}$.

35.74 Sketch the graph of $r^2 = 1 + \sin \theta$.

▮ See Fig. 35-18; there are two loops inside a larger oval.

θ	0	$\pi/2$	π	$3\pi/2$	2π
r	± 1	$\pm \sqrt{2}$	± 1	0	± 1

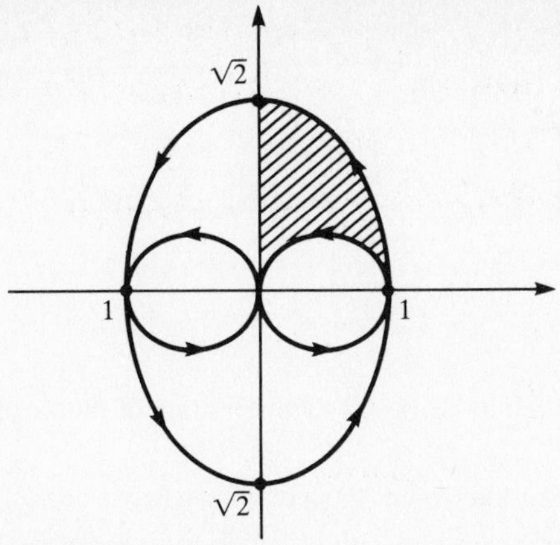

Fig. 35-18

35.75 Find the area between the inner and outer ovals of $\quad r^2 = 1 + \sin\theta\quad$ in the first quadrant (see Fig. 35-18).

❙ The area inside the outer oval is

$$A_o = \frac{1}{2}\int_0^{\pi/2}(1+\sin\theta)\,d\theta = \tfrac{1}{2}(\theta - \cos\theta)\,]_0^{\pi/2} = \frac{1}{2}\left[\frac{\pi}{2} - (-1)\right] = \frac{\pi + 2}{4}$$

The area inside the inner oval is the same as

$$A_i = \frac{1}{2}\int_\pi^{3\pi/2}(1+\sin\theta)\,d\theta = \tfrac{1}{2}(\theta - \cos\theta)\,]_\pi^{3\pi/2} = \frac{1}{2}\left[\frac{3\pi}{2} - (\pi + 1)\right] = \frac{\pi - 2}{4}$$

Hence $$A_o - A_i = \frac{\pi + 2}{4} - \frac{\pi - 2}{4} = 1$$

35.76 Find the arc length of the spiral $\quad r = \theta\quad$ from $\quad\theta = 0\quad$ to $\quad\theta = 1$.

❙ The general arc length formula is $\int_{\theta_1}^{\theta_2}\sqrt{r^2 + (dr/d\theta)^2}\,d\theta$. In this case, $dr/d\theta = 1$, and we have $L = \int_0^1\sqrt{\theta^2 + 1}\,d\theta$. Letting $\theta = \tan u$, $d\theta = \sec^2 u\,du$, we obtain $\int_0^{\pi/4}\sec^3 u\,du = \frac{1}{2}(\sec u\tan u + \ln|\sec u + \tan u|)\,]_0^{\pi/4} = \frac{1}{2}[\sqrt{2} + \ln(\sqrt{2} + 1)]$.

35.77 Find the arc length of the spiral $\quad r = e^\theta\quad$ from $\quad\theta = 0\quad$ to $\quad\theta = \ln 2$.

❙ $dr/d\theta = e^\theta$. Hence, $L = \int_0^{\ln 2}\sqrt{e^{2\theta} + e^{2\theta}}\,d\theta = \int_0^{\ln 2}\sqrt{2}\,e^\theta\,d\theta = \sqrt{2}\,e^\theta\,]_0^{\ln 2} = \sqrt{2}\,(2 - 1) = \sqrt{2}$.

35.78 Find the length of the cardioid $\quad r = 1 - \cos\theta\quad$ (Problem 35.34).

❙ $dr/d\theta = \sin\theta$. So, $L = \int_0^{2\pi}\sqrt{(1 - \cos\theta)^2 + \sin^2\theta}\,d\theta = \int_0^{2\pi}\sqrt{2 - 2\cos\theta}\,d\theta = \sqrt{2}\int_0^{2\pi}\sqrt{1 - \cos\theta}\,d\theta = \sqrt{2}\int_0^{2\pi}\sqrt{2}\sin(\theta/2)\,d\theta = 2(-2\cos(\theta/2))\,]_0^{2\pi} = -4(-1 - 1) = 8$.

35.79 Find the length of $\quad r = \theta^2\quad$ from $\quad\theta = 0\quad$ to $\quad\theta = \sqrt{5}$.

❙ $dr/d\theta = 2\theta$. $L = \int_0^{\sqrt{5}}\sqrt{\theta^4 + 4\theta^2}\,d\theta = \int_0^{\sqrt{5}}\theta\sqrt{\theta^2 + 4}\,d\theta = \frac{1}{2}\cdot\frac{2}{3}(\theta^2 + 4)^{3/2}\,]_0^{\sqrt{5}} = \frac{1}{3}(27 - 8) = \frac{19}{3}$.

35.80 Find the length of $\quad r = e^{\theta/2}\quad$ from $\quad\theta = 0\quad$ to $\quad\theta = 2\pi$.

❙ $dr/d\theta = \frac{1}{2}e^{\theta/2}$. Then

$$L = \int_0^{2\pi}\sqrt{e^\theta + \tfrac{1}{4}e^\theta}\,d\theta = \int_0^{2\pi}\frac{\sqrt{5}}{2}\,e^{\theta/2}\,d\theta = \frac{\sqrt{5}}{2}\,(2e^{\theta/2})\,\Big]_0^{2\pi} = \sqrt{5}\,(e^\pi - 1)$$

35.81 Find the arc length of $\quad r = 1/\theta\quad$ from $\quad\theta = 1\quad$ to $\quad\theta = \sqrt{3}$.

❙ $dr/d\theta = -1/\theta^2$. Then $L = \int_1^{\sqrt{3}}\sqrt{\frac{1}{\theta^2} + \frac{1}{\theta^4}}\,d\theta = \int_1^{\sqrt{3}}\frac{\sqrt{\theta^2 + 1}}{\theta^2}\,d\theta$. Let $\theta = \tan u$, $d\theta = \sec^2 u\,du$. We get

$$\int_{\pi/4}^{\pi/3}\sec u\csc^2 u\,du = \int_{\pi/4}^{\pi/3}(\sec u + \csc u\cot u)\,du = (\ln|\sec u + \tan u| - \csc u)\,\Big]_{\pi/4}^{\pi/3}$$

$$= \left[\ln(2 + \sqrt{3}) - \frac{2}{\sqrt{3}}\right] - [\ln(\sqrt{2} + 1) - \sqrt{2}] = \ln\frac{2 + \sqrt{3}}{1 + \sqrt{2}} + \sqrt{2} - \frac{2}{\sqrt{3}}$$

35.82 Find the arc length of $r = \cos^2(\theta/2)$ (Problem 35.45).

▮ $dr/d\theta = -\cos(\theta/2)\sin(\theta/2)$. Then $L = \int_0^{2\pi} \sqrt{\cos^4(\theta/2) + \cos^2(\theta/2)\sin^2(\theta/2)} \, d\theta = \int_0^{2\pi} |\cos(\theta/2)| \, d\theta = \int_0^{\pi} \cos(\theta/2) \, d\theta - \int_{\pi}^{2\pi} \cos(\theta/2) \, du = 2\sin(\theta/2) \,]_0^{\pi} - 2\sin(\theta/2) \,]_{\pi}^{2\pi} = 2[(1-0) - (0-1)] = 4$.

35.83 Find the arc length of $r = \sin^3(\theta/3)$ from $\theta = 0$ to $\theta = 3\pi/2$.

▮ $dr/d\theta = \sin^2(\theta/3)\cos(\theta/3)$. Then $L = \int_0^{3\pi/2} \sqrt{\sin^6(\theta/3) + \sin^4(\theta/3)\cos^2\theta} \, d\theta = \int_0^{3\pi/2} \sin^2(\theta/3) \, d\theta = \int_0^{3\pi/2} \frac{1}{2}[1 - \cos(2\theta/3)] \, d\theta = \frac{1}{2}(\theta - \frac{3}{2}\sin(2\theta/3)) \,]_0^{3\pi/2} = \frac{1}{2}(3\pi/2) = 3\pi/4$.

35.84 Find the area of the surface generated by revolving the upper half of the cardioid $r = 1 - \cos\theta$ about the polar axis.

▮ The general formula for revolution about the polar axis is $S = 2\pi \int_{\theta_1}^{\theta_2} r\sin\theta \sqrt{r^2 + (dr/d\theta)^2} \, d\theta$ $(= \int 2\pi y \, ds$, the disk formula). In this case, the calculation in Problem 35.78 shows that $\sqrt{r^2 + (dr/d\theta)^2} = 2\sin(\theta/2)$. Hence, $S = 2\pi \int_0^{\pi} (1 - \cos\theta)\sin\theta \cdot 2\sin(\theta/2) \, d\theta = 4\pi \int_0^{\pi} 2\sin^2(\theta/2) \cdot 2\sin(\theta/2)\cos(\theta/2) \cdot \sin(\theta/2) \, d\theta = 16\pi \int_0^{\pi} \sin^4(\theta/2)\cos(\theta/2) \, d\theta = 16\pi(\frac{2}{5}\sin^5(\theta/2)) \,]_0^{\pi} = 16\pi(\frac{2}{5}) = 32\pi/5$.

35.85 Find the surface area generated by revolving the lemniscate $r^2 = \cos 2\theta$ (Fig. 35-5) about the polar axis.

▮ The required area is twice that generated by revolving the first-quadrant arc. $r^2 + (dr/d\theta)^2 = \cos 2\theta + [(-\sin 2\theta)/r]^2 = 1/r^2$. So, the surface area $S = 2 \cdot 2\pi \int_0^{\pi/4} r\sin\theta \cdot (1/r) \, d\theta = 4\pi \int_0^{\pi/4} \sin\theta \, d\theta = 2\pi(2 - \sqrt{2})$.

35.86 Find the surface area generated by revolving the lemniscate $r^2 = \cos 2\theta$ about the 90°-line.

▮ The general form for revolutions about the 90°-line is $S = \int 2\pi x \, ds = 2\pi \int_{\theta_1}^{\theta_2} r\cos\theta \sqrt{r^2 + (dr/d\theta)^2} \, d\theta$. As in Problem 35.85, the required area is twice that generated by revolving the first-quadrant arc. Thus, $S = 2 \cdot 2\pi \int_0^{\pi/4} r\cos\theta \cdot (1/r) \, d\theta = 4\pi \int_0^{\pi/4} \cos\theta \, d\theta = 2\sqrt{2}\pi$.

35.87 Describe the graph of $r = 2\sin\theta + 4\cos\theta$.

▮ Multiply both sides by r: $r^2 = 2r\sin\theta + 4r\cos\theta$, $x^2 + y^2 = 2y + 4x$, $x^2 - 4x + y^2 - 2y = 0$, $(x-2)^2 + (y-1)^2 = 5$. Thus, the graph is a circle with center $(2, 1)$ and radius $\sqrt{5}$. Because $(-2)^2 + (-1)^2 = 5$, the circle passes through the pole.

35.88 Find the centroid of the arc of the circle $r = 2\sin\theta + 4\cos\theta$ from $\theta = 0$ to $\theta = \pi/2$.

▮ The general formulas for the centroid of an arc are $\bar{x}L = \int_{\theta_1}^{\theta_2} r\cos\theta \sqrt{r^2 + (dr/d\theta)^2} \, d\theta$ and $\bar{y}L = \int_{\theta_1}^{\theta_2} r\sin\theta \sqrt{r^2 + (dr/d\theta)^2} \, d\theta$, where L is the arc length. In this case, the arc happens to be half of a circle of radius $\sqrt{5}$, and, thus, its arc length $L = \pi\sqrt{5}$. Since $dr/d\theta = 2\cos\theta - 4\sin\theta$ and $r^2 + (dr/d\theta)^2 = 20$, we have: $\pi\sqrt{5}\bar{x} = 4\sqrt{5} \int_0^{\pi/2} (\sin\theta\cos\theta + 2\cos^2\theta) \, d\theta = 4\sqrt{5}(\frac{1}{2}\sin^2\theta + \frac{1}{2}\sin 2\theta) \,]_0^{\pi/2} = 2\sqrt{5}(\pi + 1)$, and $\bar{x} = 2(\pi + 1)/\pi$. $\pi\sqrt{5}\bar{y} = 4\sqrt{5} \int_0^{\pi/2} (\sin^2\theta + 2\sin\theta\cos\theta) \, d\theta = 4\sqrt{5}((\theta/2) - \frac{1}{4}\sin 2\theta + \sin^2\theta) \,]_0^{\pi/2} = 4\sqrt{5}[(\pi/4) + 1]$, and $\bar{y} = (\pi + 4)/4$.

35.89 Derive an expression for $\tan\phi$, where ϕ is the angle made by the tangent line to the curve $r = f(\theta)$ with the positive x-axis.

▮ Let a prime denote differentiation with respect to θ. Then $\tan\phi = dy/dx = y'/x'$. Since $y = r\sin\theta$, $y' = r\cos\theta + r'\sin\theta$. Since $x = r\cos\theta$, $x' = -r\sin\theta + r'\cos\theta$. Thus,

$$\tan\phi = \frac{r\cos\theta + r'\sin\theta}{-r\sin\theta + r'\cos\theta} = \frac{r + r'\tan\theta}{-r\tan\theta + r'}$$

35.90 Find the slope of the tangent line to the spiral $r = \theta$ at $\theta = \pi/3$.

▮ $r' = 1$. Hence, by Problem 35.89, with $\tan\theta = \sqrt{3}$,

$$m = \tan\phi = \frac{\pi/3 + \sqrt{3}}{-(\pi/3)\sqrt{3} + 1} = \frac{\pi + 3\sqrt{3}}{-\pi\sqrt{3} + 3}$$

35.91 Find the equation of the tangent line to the cardioid $r = 1 - \cos \theta$ at $\theta = \pi/2$.

▍ $r' = \sin \theta$. Hence, by Problem 35.89, the slope $m = -r'/r = -\frac{1}{1} = -1$. When $\theta = \pi/2$, $x = 0$ and $y = 1$. So, an equation of the line is $y - 1 = -x$, $x + y = 1$; in polar coordinates, the line is $r(\cos \theta + \sin \theta) = 1$ or $r = 1/(\cos \theta + \sin \theta)$.

35.92 Show that, if θ_1 is such that $r = f(\theta_1) = 0$, then the direction of the tangent line to the curve $r = f(\theta)$ at the pole $(0, \theta_1)$ is θ_1.

▍ At $(0, \theta_1)$, $r = 0$ and $r' = f'(\theta_1)$. If $r' \neq 0$, then, by the formula of Problem 35.89,

$$\tan \phi = \frac{0 + \sin \theta_1 \cdot f'(\theta_1)}{0 + \cos \theta_1 \cdot f'(\theta_1)} = \tan \theta_1$$

and so, $\phi = \theta_1$. If $r' = 0$,

$$\tan \phi = \lim_{\theta \to \theta_1} \frac{\sin \theta \cdot f'(\theta)}{\cos \theta \cdot f'(\theta)} = \tan \theta_1$$

and again $\phi = \theta_1$.

35.93 Find the slope of the three-leaved rose $r = \cos 3\theta$ at the pole (see Fig. 35-19).

▍ When $r = 0$, $\cos 3\theta = 0$. Then $3\theta = \pi/2, 3\pi/2$, or $5\pi/2$, and $\theta = \pi/6, \pi/2$, or $5\pi/6$. By Problem 35.92, $\tan \phi = 1/\sqrt{3}, \infty$, or $-1/\sqrt{3}$, respectively.

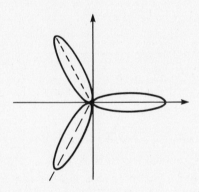

Fig. 35-19

35.94 Find the slope of $r = 1/\theta$ when $\theta = \pi/3$.

▍ For $r = 3/\pi$ and $r' = -1/\theta^2 = -9/\pi^2$, Problem 35.89 gives

$$\tan \phi = \frac{(3/\pi) - (9/\pi^2)\sqrt{3}}{-(3/\pi)\sqrt{3} - (9/\pi^2)} = \frac{3\sqrt{3} - \pi}{\pi\sqrt{3} + 3}$$

35.95 Investigate $r = 1 + \sin \theta$ for horizontal and vertical tangents.

▍ $$\tan \phi = \frac{(1 + \sin \theta) \cos \theta + \cos \theta \sin \theta}{-(1 + \sin \theta) \sin \theta + \cos^2 \theta} = \frac{\cos \theta (1 + 2 \sin \theta)}{(\sin \theta + 1)(2 \sin \theta - 1)}$$

For horizontal tangents, set $\tan \phi = 0$ and solve: $\cos \theta = 0$ or $1 + 2 \sin \theta = 0$. Hence, θ is $\pi/2, 3\pi/2$, $7\pi/6$, or $11\pi/6$. For $\theta = \pi/2$, there is a horizontal tangent at $(2, \pi/2)$. For $\theta = 7\pi/6$ and $11\pi/6$, there are horizontal tangents at $(\frac{1}{2}, 7\pi/6)$ and $(\frac{1}{2}, 11\pi/6)$. For $\theta = 3\pi/2$, the denominator of $\tan \phi$ also is 0, and, by Problem 35.92, there is a vertical tangent at the pole. For the other vertical tangents, we set the denominator $(\sin \theta + 1)(2 \sin \theta - 1) = 0$ and obtain the additional cases $\theta = \pi/6$ and $5\pi/6$. These yield vertical tangents at $(\frac{3}{2}, \pi/6)$ and $(\frac{3}{2}, 5\pi/6)$. See Fig. 35-20.

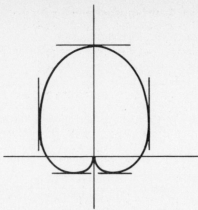

Fig. 35-20

35.96 Find the slope of $r = 2 + \sin \theta$ when $\theta = \pi/6$.

❚ By Problem 35.89, the slope $m = \dfrac{\frac{5}{2} + (\sqrt{3}/2)(1/\sqrt{3})}{-\frac{5}{2}(1/\sqrt{3}) + (\sqrt{3}/2)} = -3\sqrt{3}$.

35.97 Find the slope of $r = \sin^3 (\theta/3)$ when $\theta = \pi/2$.

❚ When $\theta = \pi/2$, $r' = \sin^2 (\theta/3) = \frac{1}{4}$ and $r = \frac{1}{8}$. Hence, by Problem 35.89, $\tan \phi = -r'/r = -2$.

35.98 Show that the angle ψ from the radius vector $\overrightarrow{OP}$ to the tangent line at a point $P(r, \theta)$ (Fig. 35-21) is given by $\tan \psi = r/r'$, where $r' = dr/d\theta$.

❚ By Problem 35.89,

$$\tan \psi = \tan (\phi - \theta) = \frac{\tan \phi - \tan \theta}{1 + \tan \phi \tan \theta} = \frac{(r + r' \tan \theta)/(-r \tan \theta + r') - \tan \theta}{1 + [(r + r' \tan \theta)/(-r \tan \theta + r')] \cdot \tan \theta} = \frac{r(1 + \tan^2 \theta)}{r'(1 + \tan^2 \theta)} = \frac{r}{r'}$$

provided $\cos \theta \neq 0$. A limiting process yields the same result as $\cos \theta \to 0$.

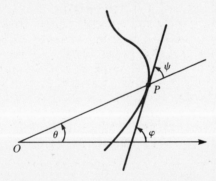

Fig. 35-21

35.99 Find $\tan \psi$ (see Problem 35.98) for $r = 2 + \cos \theta$ at $\theta = \pi/3$.

❚ At $\theta = \pi/3$, $r = 2 + \frac{1}{2} = \frac{5}{2}$, and $r' = -\sin \theta = -\sqrt{3}/2$, so $\tan \psi = r/r' = -5/\sqrt{3}$.

35.100 Find $\tan \psi$ (see Problem 35.98) for $r = 2 \sin 3\theta$ at $\theta = \pi/4$.

❚ At $\theta = \pi/4$, $r = 2(1/\sqrt{2}) = \sqrt{2}$, and $r' = 6 \cos 3\theta = 6(-1/\sqrt{2}) = -3\sqrt{2}$. Hence, $\tan \psi = r/r' = -\frac{1}{3}$.

35.101 Show that, at each point of $r = ae^{c\theta}$, the radius vector makes a fixed angle with the tangent line. (That is why the curve $r = ae^{c\theta}$ is called an equiangular spiral.)

❚ $r' = ace^{c\theta}$. Hence, by Problem 35.98, $\tan \psi = r/r' = 1/c$. So, $\psi = \tan^{-1}(1/c)$.

35.102 Show that the angle ψ that the radius vector to any point of the cardioid $r = a(1 - \cos\theta)$ makes with the curve is one-half the angle θ that the radius vector makes with the polar axis.

▮ $r' = a\sin\theta$.   By Problem 35.98, $\tan\psi = r/r' = (1 - \cos\theta)/\sin\theta = 2\sin^2(\theta/2)/[2\sin(\theta/2)\cos(\theta/2)] = \tan(\theta/2)$.   Hence $\psi = \theta/2$.

35.103 For the spiral of Archimedes, $r = a\theta\ (\theta \ge 0)$, show that the angle ψ between the radius vector and the tangent line is $\pi/4$ when $\theta = 1$ and $\psi \to \pi/2$ as $\theta \to +\infty$.

▮ $r' = a$. By Problem 35.98, $\tan\psi = r/r' = \theta$. When $\theta = 1$, $\tan\psi = 1$ and $\psi = \pi/4$. As $\theta \to +\infty$, $\tan\psi \to +\infty$ and $\psi \to \pi/2$.

35.104 Prove the converse of Problem 35.102.

▮ If $\psi = \frac{1}{2}\theta$, so that $r/r' = \tan\frac{1}{2}\theta = (1 - \cos\theta)/\sin\theta$, then $\displaystyle\int \frac{dr}{r} = \int \frac{\sin\theta}{1 - \cos\theta}\, d\theta$, $\ln r = \ln a(1 - \cos\theta)$ [$a = $ const.], $r = a(1 - \cos\theta)$.

35.105 Find the intersection points of the curves $r = \sin\theta$ and $r = \cos\theta$.

▮ Setting $\sin\theta = \cos\theta$, we obtain the intersection point $(\sqrt{2}/2, \pi/4)$. Substituting $(-r, \theta + \pi)$ for (r, θ) in either equation yields no additional points.   However, both curves pass through the pole, which is, therefore, a second intersection point. [In fact, the curves are two circles of radius $\frac{1}{2}$, with centers $(0, \frac{1}{2})$ and $(\frac{1}{2}, 0)$, respectively.]

35.106 Find the angle ζ at which the curves $r = \sin\theta$ and $r = \cos\theta$ intersect at the point $(\sqrt{2}/2, \pi/4)$.

▮ Clearly, $\zeta = \psi_1 - \psi_2$, where ψ_1 and ψ_2 are the angles between the common radius vector and the two tangent lines. Hence, $\tan\zeta = \tan(\psi_1 - \psi_2) = (\tan\psi_1 - \tan\psi_2)/(1 + \tan\psi_1 \tan\psi_2)$.   For $r = \sin\theta$, $r' = \cos\theta$ and, therefore, by Problem 35.98, $\tan\psi_1 = r/r' = \tan(\pi/4) = 1$. For $r = \cos\theta$, $r' = -\sin\theta$ and $\tan\psi_2 = r/r' = -\tan(\pi/4) = -1$. Hence, $\tan\zeta = [1 - (-1)]/[1 + 1(-1)] = \infty$. Therefore, $\zeta = \pi/2$

35.107 Find the angles of intersection of the curves $r = 3\cos\theta$ and $r = 1 + \cos\theta$.

▮ Solving the two equations simultaneously yields $\cos\theta = \frac{1}{2}$, and, therefore, $\theta = \pi/3, 5\pi/3$, with $r = \frac{3}{2}$. No other points lie on both curves except the pole. For $r = 3\cos\theta$, $r' = -3\sin\theta$ and $\tan\psi_1 = -\cot\theta$. For $r = 1 + \cos\theta$, $r' = -\sin\theta$ and $\tan\psi_2 = -(1 + \cos\theta)/\sin\theta$. Here, ψ_1 and ψ_2 are, as usual, the angles between the radius vector and the tangent lines. The angle ζ between the curves is $\psi_1 - \psi_2$, and, therefore, $\tan\zeta = (\tan\psi_1 - \tan\psi_2)/(1 + \tan\psi_1 \tan\psi_2)$.   Now, at $\theta = \pi/3$, $\tan\psi_1 = -1/\sqrt{3}$, $\tan\psi_2 = \sqrt{3}$, and $\tan\zeta = [(-1/\sqrt{3}) + \sqrt{3}]/[1 + (-1/\sqrt{3})(-\sqrt{3})] = 1/\sqrt{3}$. Hence, $\zeta = \pi/6$. By symmetry, the angle at $(\frac{3}{2}, 5\pi/3)$ also is $\pi/6$. The circle $r = 3\cos\theta$ passes through the pole when $\theta = \pi/2$, and the cardioid passes through the pole when $\theta = \pi$. Hence, by Problem 35.92, those are the directions of the tangent lines, and, therefore, the curves are orthogonal at the pole.

35.108 For a curve $r = f(\theta)$, show that the curvature $\kappa = [r^2 + 2(r')^2 - rr'']/(r^2 + r'^2)^{3/2}$, where $r' = dr/d\theta$ and $r'' = d^2r/d\theta^2$.

▮ By definition, $\kappa = d\phi/ds$. But, $\phi = \theta + \psi$ (Fig. 35-21) and

$$\frac{d\phi}{ds} = \frac{d\theta}{ds} + \frac{d\psi}{ds} = \frac{d\theta}{ds} + \frac{d\psi}{d\theta}\frac{d\theta}{ds} = \frac{d\theta}{ds}\left(1 + \frac{d\psi}{d\theta}\right)$$

We know by Problem 35.98 that $\tan\psi = r/r'$. Hence, by differentiation,

$$\sec^2\psi \frac{d\psi}{d\theta} = \frac{(r')^2 - rr''}{(r')^2} \qquad \left[1 + \left(\frac{r}{r'}\right)^2\right]\frac{d\psi}{d\theta} = \frac{(r'')^2 - rr''}{(r')^2}$$

$$\frac{d\psi}{d\theta} = \frac{(r')^2 - rr''}{(r')^2 + r^2} \qquad 1 + \frac{d\psi}{d\theta} = \frac{r^2 + 2(r')^2 - rr''}{(r')^2 + r^2}$$

But, we know that $ds/d\theta = \sqrt{r^2 + (r')^2}$. Hence,

$$\kappa = \left(1 + \frac{d\psi}{d\theta}\right)\frac{ds}{d\theta} = \frac{r^2 + 2(r')^2 - rr''}{[r^2 + (r')^2]^{3/2}}$$

35.109 Compute the curvature of $r = 2 + \sin \theta$.

▮ $r' = \cos \theta$, $r'' = -\sin \theta$. By the formula of Problem 35.108,

$$\kappa = \frac{(2 + \sin \theta)^2 + 2\cos^2 \theta - (2 + \sin \theta)(-\sin \theta)}{[(2 + \sin \theta)^2 + \cos^2 \theta]^{3/2}} = \frac{6(1 + \sin \theta)}{(5 + 4\sin \theta)^{3/2}}$$

35.110 Show that the spirals $r = \theta$ and $r = 1/\theta$ intersect orthogonally at $(1, 1)$.

▮ For $r = \theta$, $r' = 1$, and $\tan \psi_1 = r/r' = \theta = 1$. For $r = 1/\theta$, $r' = -1/\theta^2 = -1$, and $\tan \psi_2 = -1$. Hence, $\tan (\psi_1 - \psi_2) = [1 - (-1)]/[1 + (1)(-1)] = \infty$. Hence, $\psi_1 - \psi_2 = \pi/2$.

35.111 Prove the converse of Problem 35.101.

▮ If $r/r' = 1/C$, then $\int (dr/r) = \int C \, d\theta$, $\ln r = c\theta + \ln a$ [$a = $ const.], $r = ae^{C\theta}$. Note that $C = 0$ gives a circle, a degenerate spiral that maintains the fixed angle $\pi/2$ with its radii.

35.112 Show that, if a point moves at a constant speed v along the equiangular spiral $r = ae^{c\theta}$, then the radius r changes at a constant rate.

▮
$$v = \frac{ds}{dt} = \sqrt{\left(\frac{dr}{dt}\right)^2 + r^2 \left(\frac{d\theta}{dt}\right)^2}$$

But, along the curve, $dr/dt = ace^{c\theta}(d\theta/dt) = cr(d\theta/dt)$; hence

$$v = \sqrt{\left(\frac{dr}{dt}\right)^2 + \frac{1}{c^2} \left(\frac{dr}{dt}\right)^2} = \frac{dr}{dt} \sqrt{1 + \left(\frac{1}{c^2}\right)}$$

or $dr/dt = v/\sqrt{1 + (1/c^2)} = $ const.

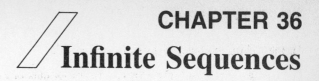

CHAPTER 36
Infinite Sequences

In Problems 36.1–36.18, write a formula for the nth term a_n of the sequence and determine its limit (if it exists). It is understood that $n = 1, 2, 3, \ldots$.

36.1 $1, \frac{1}{2}, \frac{1}{4}, \frac{1}{8}, \frac{1}{16}, \ldots$.

▮ $a_n = \dfrac{1}{2^{n-1}}$. Clearly, $\displaystyle\lim_{n \to +\infty} \dfrac{1}{2^{n-1}} = 0$.

36.2 $1, -1, 1, -1, \ldots$.

▮ $a_n = (-1)^{n+1}$. There is no limit.

36.3 $1, \frac{1}{2}, \frac{1}{3}, \frac{1}{4}, \ldots$.

▮ $a_n = \dfrac{1}{n}$. $\displaystyle\lim_{n \to +\infty} \dfrac{1}{n} = 0$.

36.4 $1, 0, 1, 0, 1, 0, \ldots$.

▮ $a_n = \frac{1}{2}[1 + (-1)^{n+1}]$. Clearly there is no limit.

36.5 $\frac{1}{2}, \frac{1}{4}, \frac{1}{6}, \frac{1}{8}, \ldots$.

▮ $a_n = \dfrac{1}{2n}$. $\displaystyle\lim_{n \to +\infty} \dfrac{1}{2n} = 0$.

36.6 $1, \frac{1}{3}, \frac{1}{5}, \frac{1}{7}, \ldots$.

▮ $a_n = \dfrac{1}{2n-1}$. $\displaystyle\lim_{n \to +\infty} \dfrac{1}{2n-1} = 0$.

36.7 $\frac{1}{2}, \frac{2}{3}, \frac{3}{4}, \frac{4}{5}, \frac{5}{6}, \ldots$.

▮ $a_n = \dfrac{n}{n+1}$. $\displaystyle\lim_{n \to +\infty} \dfrac{n}{n+1} = \lim_{n \to +\infty} \left(1 - \dfrac{1}{n+1}\right) = 1$.

36.8 $1, -\frac{1}{2}, \frac{1}{3}, -\frac{1}{4}, \frac{1}{5}, \ldots$.

▮ $a_n = (-1)^{n+1} \dfrac{1}{n}$. $\displaystyle\lim_{n \to +\infty} a_n = 0$.

36.9 $\frac{2}{1}, \left(\frac{3}{2}\right)^2, \left(\frac{4}{3}\right)^3, \left(\frac{5}{4}\right)^4, \ldots$.

▮ $a_n = \left(\dfrac{n+1}{n}\right)^n$. $\displaystyle\lim_{n \to +\infty} a_n = \lim_{n \to +\infty} \left(1 + \dfrac{1}{n}\right)^n = e$, since $\displaystyle\lim_{x \to +\infty} \left(1 + \dfrac{1}{x}\right)^x = e$.

36.10 $\ln \frac{2}{1}, \ln \frac{3}{2}, \ln \frac{4}{3}, \ldots$.

▮ $a_n = \ln \dfrac{n+1}{n}$. $\displaystyle\lim_{n \to +\infty} \ln\left(1 + \dfrac{1}{n}\right) = \ln 1 = 0$. Note that this depends on the continuity of $\ln x$ at $x = 1$.

36.11 0.9, 0.99, 0.999, 0.9999, ...

▮ $a_n = 1 - \dfrac{1}{10^n}$. $\displaystyle\lim_{n \to +\infty} a_n = 1$.

36.12 $\dfrac{2}{1}, \dfrac{4}{2}, \dfrac{8}{6}, \dfrac{16}{24}, \dfrac{32}{120}, \ldots$.

▮ $a_n = \dfrac{2^n}{n!}$. For $n > 4$, $\dfrac{2}{1} \cdot \dfrac{2}{2} \cdot \dfrac{2}{3} \cdot \dfrac{2}{4} \cdot \dfrac{2}{5} \cdot \ldots \cdot \dfrac{2}{n} < \dfrac{2}{1} \cdot \dfrac{2}{2} \cdot \dfrac{2}{3} \cdot \dfrac{2}{4} \cdot \dfrac{1}{2} \cdot \dfrac{1}{2} \cdot \ldots \cdot \dfrac{1}{2} = \dfrac{2}{3} \cdot \dfrac{1}{2^{n-4}}$. Since $\displaystyle\lim_{n \to +\infty} \dfrac{2}{3} \cdot \dfrac{1}{2^{n-4}} = 0$, $\displaystyle\lim_{n \to +\infty} a_n = 0$.

36.13 $\dfrac{a}{1}, \dfrac{a^2}{2}, \dfrac{a^3}{6}, \dfrac{a^4}{24}, \dfrac{a^5}{120}, \ldots$.

▮ $a_n = \dfrac{a^n}{n!}$. Let κ be the least integer $\geq 2|a|$. For $n > \kappa$, $\dfrac{a^n}{n!} = \dfrac{a}{1} \cdot \dfrac{a}{2} \cdot \dfrac{a}{3} \cdot \ldots \cdot \dfrac{a}{\kappa + 1} \cdot \dfrac{a}{\kappa + 2} \cdot \ldots \cdot \dfrac{a}{n} = \dfrac{a^\kappa}{\kappa!} \cdot \dfrac{a}{\kappa + 1} \cdot \dfrac{a}{\kappa + 2} \cdot \ldots \cdot \dfrac{a}{n}$. Each of $\left|\dfrac{a}{\kappa + 1}\right|, \left|\dfrac{a}{\kappa + 2}\right|, \ldots, \left|\dfrac{a}{n}\right|$ is $< \dfrac{1}{2}$. Therefore, $\left|\dfrac{a^n}{n!}\right| < \dfrac{|a|^\kappa}{\kappa!} \cdot \dfrac{1}{2^{n-\kappa}} \to 0$ as $n \to \infty$. So $\displaystyle\lim_{n \to +\infty} a_n = 0$.

36.14 $\dfrac{1}{1}, \dfrac{2}{4}, \dfrac{6}{27}, \dfrac{24}{256}, \dfrac{120}{3125}, \ldots$.

▮ $a_n = \dfrac{n!}{n^n} = \dfrac{1}{n} \cdot \dfrac{2}{n} \cdot \ldots \cdot \dfrac{n}{n} \leq \dfrac{1}{n} \to 0$. Hence, $\displaystyle\lim_{n \to +\infty} a_n = 0$.

36.15 $1, \sqrt{2}, \sqrt[3]{3}, \sqrt[4]{4}, \ldots$.

▮ $a_n = \sqrt[n]{n} = n^{1/n} = e^{(\ln n)/n} \to e^0 = 1$. Here we have used the fact that $\displaystyle\lim_{n \to +\infty} (\ln n)/n = 0$, which follows by L'Hôpital's rule.

36.16 $\dfrac{2}{4}, \dfrac{4}{5}, \dfrac{6}{6}, \dfrac{8}{7}, \dfrac{10}{8}, \ldots$.

▮ $a_n = \dfrac{2n}{n + 3} = \dfrac{2}{1 + 3/n} \to \dfrac{2}{1} = 2$.

36.17 $1, -\dfrac{1}{4}, \dfrac{1}{9}, -\dfrac{1}{16}, \dfrac{1}{25}, \ldots$.

▮ $a_n = \dfrac{(-1)^{n+1}}{n^2} \to 0$.

36.18 $\cos \pi, \cos (\pi/2), \cos (\pi/3), \cos (\pi/4), \ldots$.

▮ $a_n = \cos (\pi/n) \to \cos 0 = 1$.

In Problems 36.19–36.45, determine whether the given sequence converges, and, if it does, find the limit.

36.19 $a_n = \sin (n\pi/4)$.

▮ The sequence takes on the values $\sqrt{2}/2, 1, \sqrt{2}/2, 0, -\sqrt{2}/2, -1, -\sqrt{2}/2, 0$, and then keeps repeating in this manner. Hence, there is no limit.

36.20 $a_n = n/e^n$.

▮ As L'Hôpital's rule yields $\displaystyle\lim_{x \to +\infty} x/e^x = 0$, $\displaystyle\lim_{n \to +\infty} n/e^n = 0$.

36.21 $a_n = (\ln n)/n$.

▮ The sequence converges to 0 (see Problem 36.15).

36.22 $a_n = \dfrac{3n-2}{7n+5}$.

▌ $\displaystyle\lim_{n\to+\infty}\frac{3n-2}{7n+5} = \lim_{n\to+\infty}\frac{3-2/n}{7+5/n} = \frac{3}{7}$.

36.23 $a_n = \dfrac{4n+5}{n^3-2n+3}$.

▌ In general, the same method that was used for rational functions $\left(\text{like } \dfrac{4x+5}{x^3-2x+5}\right)$ can be used here. Divide by the highest power in the denominator. $\displaystyle\lim_{n\to+\infty}\frac{4n+5}{n^3-2n+3} = \lim_{n\to+\infty}\frac{4/n^2+5/n^3}{1-2/n^2+3/n^3} = \frac{0}{1} = 0$.

36.24 $a_n = \dfrac{2n^5-3n+20}{5n^4+2}$.

▌ $\displaystyle\lim_{n\to+\infty}\frac{2n^5-3n+20}{5n^4+2} = \lim_{n\to+\infty}\frac{2n-3/n^3+20/n^4}{5+2/n^4} = +\infty$.

36.25 $a_n = \sqrt[n]{1/n^k}$, where $k>0$.

▌ $a_n = \dfrac{1}{n^{k/n}} = \dfrac{1}{e^{(k/n)\ln n}} \to \dfrac{1}{e^0} = 1$, since $\dfrac{\ln n}{n} \to 0$.

36.26 $a_n = \dfrac{\sin n}{n}$.

▌ Since $|\sin n| \le 1, |a_n| \le \dfrac{1}{n}$. Therefore, $\displaystyle\lim_{n\to+\infty} a_n = 0$.

36.27 $a_n = \sqrt{n+1} - \sqrt{n}$.

▌ $\sqrt{n+1} - \sqrt{n} = (\sqrt{n+1} - \sqrt{n})\dfrac{\sqrt{n+1}+\sqrt{n}}{\sqrt{n+1}+\sqrt{n}} = \dfrac{1}{\sqrt{n+1}+\sqrt{n}} \to 0$.

36.28 $a_n = \dfrac{n^2+5}{\sqrt{4n^4+n}}$.

▌ $\dfrac{n^2+5}{\sqrt{4n^4+n}} = \dfrac{1+5/n^2}{\sqrt{4+1/n^3}} \to \dfrac{1}{\sqrt{4}} = \dfrac{1}{2}$.

36.29 $a_n = \dfrac{\sqrt[3]{2n^5-n^2+4}}{n^2+1}$.

▌ $\dfrac{\sqrt[3]{2n^5-n^2+4}}{n^2+1} = \dfrac{\sqrt[3]{2/n-1/n^4+4/n^6}}{1+1/n^2} \to \dfrac{0}{1} = 0$.

36.30 $a_n = \ln(n+1) - \ln n$.

▌ $\ln(n+1) - \ln n = \ln\dfrac{n+1}{n} = \ln\left(1+\dfrac{1}{n}\right) \to \ln 1 = 0$.

36.31 $a_n = \dfrac{2^n}{3^n-5}$.

▌ $\dfrac{2^n}{3^n-5} = \dfrac{2^n/3^n}{1-5/3^n} = \dfrac{\left(\frac{2}{3}\right)^n}{1-5/3^n} \to \dfrac{0}{1} = 0$.

36.32 $a_n = 2n\sin\dfrac{\pi}{n}$.

▌ $\displaystyle\lim_{n\to+\infty} 2n\sin\frac{\pi}{n} = 2\pi\lim_{n\to+\infty}\frac{\sin(\pi/n)}{\pi/n} = 2\pi\lim_{\theta\to 0^+}\frac{\sin\theta}{\theta} = 2\pi\cdot 1 = 2\pi$.

36.33 $a_n = \dfrac{\sqrt{n}\,\sin{(n!\,e^n)}}{n+1}$.

∎ $|a_n| \le \dfrac{\sqrt{n}}{n+1} = \dfrac{1/\sqrt{n}}{1+1/n} \to 0$. Hence, $\lim\limits_{n \to +\infty} a_n = 0$.

36.34 $a_n = \dfrac{\ln{(n+100)}}{n}$.

∎ $\dfrac{\ln{(n+100)}}{n} = \dfrac{\ln{(n+100)}}{n+100} \cdot \dfrac{n+100}{n} = \dfrac{\ln{(n+100)}}{n+100} \cdot \left(1 + \dfrac{100}{n}\right) \to 0 \cdot 1 = 0$, since $\lim\limits_{x \to +\infty} \dfrac{\ln{x}}{x} = 0$.

36.35 $a_n = \dfrac{n^{10}}{2^n}$.

∎ $a_n = \left(\dfrac{n}{e^{n(\ln 2)/10}}\right)^{10} = \left(\dfrac{10}{\ln 2}\right)^{10} \left[\dfrac{n(\ln 2)/10}{e^{n(\ln 2)/10}}\right]^{10} \to 0$, because $\lim\limits_{x \to +\infty} \dfrac{x}{e^x} = 0$.

36.36 $a_n = \sqrt{n}(\sqrt{n+a} - \sqrt{n})$.

∎ $\sqrt{n}(\sqrt{n+a} - \sqrt{n}) = \sqrt{n}(\sqrt{n+a} - \sqrt{n}) \cdot \dfrac{\sqrt{n+a} + \sqrt{n}}{\sqrt{n+a} + \sqrt{n}} = \dfrac{\sqrt{n}}{\sqrt{n+a} + \sqrt{n}} \cdot a = \dfrac{a}{\sqrt{1+a/n} + 1} \to \dfrac{a}{2}$.

36.37 $a_n = n\left[\left(a + \dfrac{1}{n}\right)^4 - a^4\right]$.

∎ $a_n = \dfrac{(a + 1/n)^4 - a^4}{1/n} \to$ the derivative of x^4 at $x = a$, that is, $4a^3$.

36.38 $a_n = \sqrt[n]{a^n + b^n}$, where $0 < a < b$.

∎ $a_n = \sqrt[n]{b^n(a^n/b^n + 1)} = b\sqrt[n]{(a/b)^n + 1} \to b$.

36.39 $a_n = \dfrac{(n+1)^n}{n^{n+1}}$.

∎ $a_n = \left(\dfrac{n+1}{n}\right)^n \cdot \dfrac{1}{n} = \left(1 + \dfrac{1}{n}\right)^n \cdot \dfrac{1}{n} \to e \cdot 0 = 0$.

36.40 $a_n = \dfrac{(n+1)\ln{n} - n\ln{(n+1)}}{\ln{n}}$.

∎ $a_n = \dfrac{n\ln{n} - n\ln{(n+1)} + \ln{n}}{\ln{n}} = \dfrac{\ln{[n/(n+1)]^n}}{\ln{n}} + 1 = \dfrac{\ln{1/(1+1/n)^n}}{\ln{n}} + 1 \to 0 + 1 = 1$, since $\lim\limits_{n \to +\infty} \left(1 + \dfrac{1}{n}\right)^n = e$.

36.41 $a_n = \left(1 + \dfrac{1}{n^2}\right)^n$.

∎ $a_n = \left[\left(1 + \dfrac{1}{n^2}\right)^{n^2}\right]^{1/n} \to e^0 = 1$.

36.42 $a_n = \left(1 + \dfrac{1}{2n}\right)^n$.

∎ $a_n = \left[\left(1 + \dfrac{1}{2n}\right)^{2n}\right]^{1/2} \to e^{1/2}$.

36.43 $a_n = \tanh{n}$.

∎ $a_n = \dfrac{e^n - e^{-n}}{e^n + e^{-n}} = \dfrac{1 - e^{-2n}}{1 + e^{-2n}} \to \dfrac{1}{1} = 1$.

36.44 $a_n = \dfrac{1}{\sqrt{n^2 + 1} - n}$.

   ▋ $a_n = \dfrac{1}{\sqrt{n^2 + 1} - n} \cdot \dfrac{\sqrt{n^2 + 1} + n}{\sqrt{n^2 + 1} + n} = \dfrac{\sqrt{n^2 + 1} + n}{1} \to +\infty$.

36.45 $a_n = nr^n$ where $|r| < 1$.

   ▋ By the method of Problem 36.35, $|a_n| \to 0$; hence, $a_n \to 0$.

36.46 Give a rigorous proof that $\lim\limits_{n \to +\infty} \dfrac{3n + 5}{7n - 4} = \dfrac{3}{7}$.

   ▋ Let $\varepsilon > 0$. We must find an integer k such that, if $n \ge k$, then $\left| \dfrac{3n + 5}{7n - 4} - \dfrac{3}{7} \right| < \varepsilon$. Now, $\left| \dfrac{3n + 5}{7n - 4} - \dfrac{3}{7} \right| = \left| \dfrac{21n + 35 - 21n + 12}{7(7n - 4)} \right| = \dfrac{47}{7(7n - 4)}$. But $\dfrac{47}{7(7n - 4)} < \varepsilon \Leftrightarrow \dfrac{7\varepsilon}{47} > \dfrac{1}{7n - 4} \Leftrightarrow 7n - 4 > \dfrac{47}{7\varepsilon} \Leftrightarrow$ $7n > \dfrac{47}{7\varepsilon} + 4 \Leftrightarrow n > \dfrac{1}{7}\left(\dfrac{47}{7\varepsilon} + 4 \right)$. So choose k to be the least integer that exceeds $\dfrac{1}{7}\left(\dfrac{47}{7\varepsilon} + 4 \right)$.

36.47 Prove that a convergent sequence must be bounded.

   ▋ Assume $\lim\limits_{n \to +\infty} a_n = L$. Then there exists an integer k such that, if $n \ge k$, then $|a_n - L| < 1$. By the triangle inequality, $|a_n| = |(a_n - L) + L| \le |a_n - L| + |L| < |L| + 1$. Let $M =$ the maximum of the numbers $|a_1|, |a_2|, \ldots, |a_k|, |L| + 1$. Then $|a_n| \le M$ for all n.

36.48 Give an example to show that the converse of the theorem in Problem 36.47 is false.

   ▋ See Problem 36.2.

36.49 If $\lim\limits_{n \to +\infty} a_n = L$ and $\lim\limits_{n \to +\infty} b_n = K$, prove that $\lim\limits_{n \to +\infty} (a_n + b_n) = L + K$.

   ▋ Assume $\varepsilon > 0$. Since $\lim a_n = L$, there must be an integer n_1 such that, if $n \ge n_1$, $|a_n - L| < \varepsilon/2$. Likewise, there exists an integer n_2 such that, if $n \ge n_2$, then $|b_n - K| < \varepsilon/2$. Let n_0 be the maximum of n_1 and n_2. If $n \ge n_0$, we conclude by the triangle inequality that $|(a_n + b_n) - (L + K)| = |(a_n - L) + (b_n - K)| \le |a_n - L| + |b_n - K| < \varepsilon/2 + \varepsilon/2 = \varepsilon$.

36.50 If $\lim\limits_{n \to +\infty} a_n = L$ and $\lim\limits_{n \to +\infty} b_n = K$, prove that $\lim\limits_{n \to +\infty} a_n b_n = LK$.

   ▋ By Problem 36.47, there exists a positive number M such that $|a_n| \le M$ for all n. There exists an integer n_1 such that, if $n \ge n_1$, then $|a_n - L| < \varepsilon/2|K|$. This applies when $K \ne 0$; when $K = 0$, let $n_1 = 1$. In either case, if $n \ge n_1$, $|K||a_n - L| < \varepsilon/2$. Also choose n_2 so that, if $n \ge n_2$, then $|b_n - K| < \varepsilon/2M$ (and, therefore, $M|b_n - K| < \varepsilon/2$). Let n_0 be the maximum of n_1 and n_2. If $n \ge n_0$, $|a_n b_n - LK| = |a_n(b_n - K) + K(a_n - L)| \le |a_n(b_n - K)| + |K(a_n - L)| = |a_n||b_n - K| + |K||a_n - L| \le M|b_n - K| + |K||a_n - L| < \varepsilon/2 + \varepsilon/2 = \varepsilon$.

36.51 If $\lim\limits_{n \to +\infty} a_n = L$, show that $\lim\limits_{n \to +\infty} \dfrac{a_1 + \cdots + a_n}{n} = L$.

   ▋ Assume $\varepsilon > 0$. There exists an integer n_1 such that, if $n \ge n_1$, then $|a_n - L| < \dfrac{\varepsilon}{2}$. Let n_2 be such that $\dfrac{|(a_1 - L) + \cdots + (a_{n_1} - L)|}{n_2} < \dfrac{\varepsilon}{2}$. Let n_0 be the maximum of n_1 and n_2. If $n > n_0$,

$$\left| \frac{a_1 + \cdots + a_n}{n} - L \right| = \left| \frac{a_1 + \cdots + a_n - nL}{n} \right| = \frac{|(a_1 - L) + \cdots + (a_n - L)|}{n}$$

$$\le \frac{|(a_1 - L) + \cdots + (a_{n_1} - L)|}{n} + \frac{|a_{n_1 + 1} - L| + \cdots + |a_n - L|}{n}$$

$$\le \frac{|(a_1 - L) + \cdots + (a_{n_1} - L)|}{n_2} + \frac{n(\varepsilon/2)}{n} < \frac{\varepsilon}{2} + \frac{\varepsilon}{2} = \varepsilon.$$

36.52 Show that a sequence may converge in the mean (Problem 36.51) without converging in the ordinary sense.

▮ See Problem 36.2.

36.53 If $\lim\limits_{n \to +\infty} a_n = L$ and each $a_n > 0$ (so that $L \geq 0$), prove that $\lim\limits_{n \to +\infty} \sqrt[n]{a_1 a_2 \cdots a_n} = L$.

▮ Let $b_n = \ln a_n$. Then $\lim\limits_{n \to +\infty} b_n = \ln L$. So $\ln \sqrt[n]{a_1 a_2 \cdots a_n} = \dfrac{1}{n} \ln a_1 a_2 \cdots a_n = \dfrac{1}{n} \sum b_n \to \ln L$, by Problem 36.51. Hence, $\sqrt[n]{a_1 a_2 \cdots a_n} \to L$. (The proof assumed only that $L > 0$. The result also can be proved when $L = 0$ by a slight variation in the argument.)

36.54 Show that $a_n = 2n/(3n+1)$ is an increasing sequence.

▮ $a_{n+1} = \dfrac{2(n+1)}{3(n+1)+1} = \dfrac{2n+2}{3n+4}$. Then, $a_n < a_{n+1} \Leftrightarrow \dfrac{2n}{3n+1} < \dfrac{2n+2}{3n+4} \Leftrightarrow 6n^2 + 8n < 6n^2 + 8n + 2$. The last inequality is obvious.

36.55 Determine whether $a_n = (5n-2)/(4n+1)$ is increasing, decreasing, or neither.

▮ $a_{n+1} = \dfrac{5(n+1)-2}{4(n+1)+1} = \dfrac{5n+3}{4n+5}$. Then $a_n < a_{n+1} \Leftrightarrow \dfrac{5n-2}{4n+1} < \dfrac{5n+3}{4n+1} \Leftrightarrow 20n^2 - 3n - 2 < 20n^2 + 17n + 3 \Leftrightarrow -5 < 20n$. The last inequality is obvious, and, therefore, the sequence is increasing.

36.56 Determine whether $a_n = 3^n/(1+3^n)$ is increasing, decreasing, or neither.

▮ $a_{n+1} = \dfrac{3^{n+1}}{1+3^{n+1}}$. Then, $a_n < a_{n+1} \Leftrightarrow \dfrac{3^n}{1+3^n} < \dfrac{3^{n+1}}{1+3^{n+1}} \Leftrightarrow \dfrac{1}{1+3^n} < \dfrac{3}{1+3^{n+1}} \Leftrightarrow 1 + 3^{n+1} < 3 + 3^{n+1}$. Since the last inequality is true, the sequence is increasing.

36.57 Determine whether the sequence $a_n = n!/2^n$ is increasing, decreasing, or neither.

▮ $a_{n+1} = \dfrac{(n+1)!}{2^{n+1}}$. Then, $a_n < a_{n+1} \Leftrightarrow \dfrac{n!}{2^n} < \dfrac{(n+1)!}{2^{n+1}} \Leftrightarrow 1 < \dfrac{n+1}{2} \Leftrightarrow 2 < n+1$. Thus, the sequence is increasing for $n > 1$.

36.58 Determine whether the sequence $a_n = n/2^n$ is increasing, decreasing, or neither.

▮ $a_{n+1} = \dfrac{n+1}{2^{n+1}}$. Then, $a_n < a_{n+1} \Leftrightarrow \dfrac{n}{2^n} < \dfrac{n+1}{2^{n+1}} \Leftrightarrow n < \dfrac{n+1}{2} \Leftrightarrow 2n < n+1 \Leftrightarrow n < 1$. Thus, the sequence is decreasing.

36.59 Determine whether the sequence $a_n = (n^2-1)/n$ is increasing, decreasing, or neither.

▮ $a_{n+1} = \dfrac{(n+1)^2-1}{n+1} = \dfrac{n^2+2n}{n+1}$. Then, $a_n < a_{n+1} \Leftrightarrow \dfrac{n^2-1}{n} < \dfrac{n^2+2n}{n+1} \Leftrightarrow n^3 + n^2 - n - 1 < n^3 + 2n^2 \Leftrightarrow -n - 1 < n^2$. Hence, the sequence is increasing.

36.60 Determine whether the sequence $a_n = n^n/n!$ is increasing, decreasing, or neither.

▮ $a_{n+1} = \dfrac{(n+1)^{n+1}}{(n+1)!}$. Then $a_n < a_{n+1} \Leftrightarrow \dfrac{n^n}{n!} < \dfrac{(n+1)^{n+1}}{(n+1)!} \Leftrightarrow n^n < (n+1)^n$. Since the last inequality holds, the sequence is increasing.

36.61 Determine whether the sequence $a_n = \dfrac{1 \cdot 3 \cdot 5 \cdot \; \cdots \; \cdot (2n-1)}{2 \cdot 4 \cdot 6 \cdot \; \cdots \; \cdot (2n)}$ is increasing, decreasing, or neither.

▮ $a_{n+1} = \dfrac{1 \cdot 3 \cdot 5 \cdot \; \cdots \; \cdot (2n+1)}{2 \cdot 4 \cdot 6 \cdot \; \cdots \; \cdot (2n+2)}$. Then, $\dfrac{a_{n+1}}{a_n} = \dfrac{2n+1}{2n+2} < 1$. So $a_{n+1} < a_n$ and the sequence is decreasing.

36.62 Show that the sequence of Problem 36.61 is convergent.

▮ Since the sequence is decreasing and bounded below by 0, it must converge. (In general, any bounded monotonic sequence converges.)

36.63 Determine whether the sequence $a_n = \dfrac{1 \cdot 3 \cdot 5 \cdot \cdots \cdot (2n-1)}{2^n \cdot n!}$ is increasing, decreasing, or neither.

▮ $\dfrac{a_{n+1}}{a_n} = \dfrac{2n+1}{2(n+1)} = \dfrac{2n+1}{2n+2} < 1.$ So $a_{n+1} < a_n$ and the sequence is decreasing.

36.64 We know that $\lim\limits_{n \to +\infty} (0.99)^n = 0$. How large must n be taken so that $(0.99)^n < 0.001$?

▮ We must have $(\frac{99}{100})^n < 1/10^3$, $n \log \frac{99}{100} < \log (1/10^3)$, $n(\log 99 - 2) < -3$, $n(2 - \log 99) > 3$. From a table of common logarithms, $\log 99 \approx 1.9956$. So $n(0.0044) > 3$, $n > 3/0.0044 \approx 681.7$. So n should be at least 682.

36.65 Let $f(x) = \lim\limits_{n \to +\infty} \dfrac{x^{2n}}{1 + x^{2n}}$. For which x is $f(x)$ defined?

▮
$$\lim_{n \to +\infty} \frac{x^{2n}}{1 + x^{2n}} = \lim_{n \to +\infty} \frac{1}{1/x^{2n} + 1} = \begin{cases} 1 & \text{if } |x| > 1 \\ \frac{1}{2} & \text{if } |x| = 1 \\ 0 & \text{if } |x| < 1 \end{cases}$$

CHAPTER 37
Infinite Series

37.1 Prove that, if $\Sigma\, a_n$ converges, then $\displaystyle\lim_{n\to+\infty} a_n = 0$.

▮ Let $\displaystyle S = \sum_{n=0}^{\infty} a_n$. Then $\displaystyle a_n = \sum_{k=1}^{n} a_k - \sum_{k=1}^{n-1} a_k \to S - S = 0$.

37.2 Show that the harmonic series $\Sigma\, 1/n = 1 + \frac{1}{2} + \frac{1}{3} + \cdots$ diverges.

▮ $1 > \frac{1}{2}$, $\quad \frac{1}{2} \geq \frac{1}{2}$, $\quad \frac{1}{3} + \frac{1}{4} > \frac{2}{4} = \frac{1}{2}$, $\quad \frac{1}{5} + \frac{1}{6} + \frac{1}{7} + \frac{1}{8} > \frac{4}{8} = \frac{1}{2}$, $\quad \frac{1}{9} + \frac{1}{10} + \cdots + \frac{1}{16} > \frac{8}{16} = \frac{1}{2}$, $\quad$ etc. $\quad$ Therefore, $1 + \frac{1}{2} + \frac{1}{3} + \frac{1}{4} + \cdots > \frac{1}{2} + \frac{1}{2} + \frac{1}{2} + \frac{1}{2} + \cdots \to +\infty$. $\left[\text{Alternatively, by the integral test, } \int_1^{\infty} \frac{1}{x}\, dx = \right.$
$\left. \displaystyle\lim_{u\to+\infty} \int_1^{u} \frac{1}{x}\, dx = \lim_{u\to+\infty} (\ln x\,]_1^{u}) = \lim_{u\to+\infty} \ln u = +\infty. \right]$

37.3 Does $\displaystyle\lim_{n\to+\infty} a_n = 0$ imply that $\Sigma\, a_n$ converges?

▮ No. The harmonic series $\Sigma\, 1/n$ (Problem 37.2) is a counterexample.

37.4 Let $S_n = a + ar + \cdots + ar^{n-1}$, with $r \neq 1$. Show that $S_n = \dfrac{a(r^n - 1)}{r - 1}$.

▮ $rS_n = ar + ar^2 + \cdots + ar^{n-1} + ar^n$. $S_n = a + ar + ar^2 + \cdots + ar^{n-1}$. Hence, $(r-1)S_n = ar^n - a = a(r^n - 1)$.
Thus, $S_n = \dfrac{a(r^n - 1)}{r - 1}$.

37.5 Let $a \neq 0$. Show that the infinite geometric series $\displaystyle\sum_{n=0}^{\infty} ar^n = \dfrac{a}{1 - r}$ if $|r| < 1$ and diverges if $|r| \geq 1$.

▮ By Problem 37.4, $S_n = \dfrac{a(r^n - 1)}{r - 1}$. If $|r| < 1$, $S_n \to \dfrac{a(-1)}{r - 1} = \dfrac{a}{1 - r}$, since $r^n \to 0$; if $|r| > 1$,
$|S_n| \to +\infty$, since $|r|^n \to +\infty$. If $r = 1$, the series is $a + a + a + \cdots$, which diverges since $a \neq 0$. If
$r = -1$, the series is $a - a + a - a + \cdots$, which oscillates between a and 0.

37.6 Evaluate $\displaystyle\sum_{n=0}^{\infty} \dfrac{1}{2^n} = 1 + \frac{1}{2} + \frac{1}{4} + \cdots$.

▮ By Problem 37.5, with $r = \dfrac{1}{2}$, $\displaystyle\sum_{n=0}^{\infty} \dfrac{1}{2^n} = \dfrac{1}{1 - \frac{1}{2}} = 2$.

37.7 Evaluate $\displaystyle\sum_{n=0}^{\infty} \dfrac{(-1)^n}{3^n} = 1 - \frac{1}{3} + \frac{1}{9} - \frac{1}{27} + \cdots$.

▮ By Problem 37.5, with $r = -\dfrac{1}{3}$, $\displaystyle\sum_{n=0}^{\infty} \dfrac{(-1)^n}{3^n} = \dfrac{1}{1 - (-\frac{1}{3})} = \dfrac{3}{4}$.

37.8 Show that the infinite decimal $0.9999\cdots$ is equal to 1.

▮ $0.999\cdots = \dfrac{9}{10} + \dfrac{9}{10^2} + \dfrac{9}{10^3} + \cdots = \dfrac{\frac{9}{10}}{1 - \frac{1}{10}} = 1$, by Problem 37.5, with $r = \frac{1}{10}$.

37.9 Evaluate the infinite repeating decimal $d = 0.215626262\cdots$.

▮ $d = 0.215 + \dfrac{62}{10^5} + \dfrac{62}{10^7} + \dfrac{62}{10^9} + \cdots$. By Problem 37.5, with $r = \dfrac{1}{10^2}$, $\dfrac{62}{10^5} + \dfrac{62}{10^7} + \dfrac{62}{10^9} + \cdots =$
$\dfrac{62/10^5}{1 - 1/10^2} = \dfrac{62}{99,000}$. Hence, $d = \dfrac{215}{1000} + \dfrac{62}{99,000} = \dfrac{21,347}{99,000}$.

37.10 Investigate the series $\dfrac{1}{1\cdot2}+\dfrac{1}{2\cdot3}+\dfrac{1}{3\cdot4}+\cdots+\dfrac{1}{n(n+1)}+\cdots$.

▌ $\dfrac{1}{n(n+1)}=\dfrac{1}{n}-\dfrac{1}{n+1}$. Hence, the partial sum

$$S_n=\left(1-\frac{1}{2}\right)+\left(\frac{1}{2}-\frac{1}{3}\right)+\left(\frac{1}{3}-\frac{1}{4}\right)+\cdots+\left(\frac{1}{n}-\frac{1}{n+1}\right)=1-\frac{1}{n+1}\to1.$$

The series converges to 1. (The method used here is called "telescoping.")

37.11 Study the series $\dfrac{1}{1\cdot3}+\dfrac{1}{3\cdot5}+\dfrac{1}{5\cdot7}+\cdots+\dfrac{1}{(2n-1)(2n+1)}+\cdots$.

▌ $\dfrac{1}{(2n-1)(2n+1)}=\dfrac{1}{2}\left(\dfrac{1}{2n-1}-\dfrac{1}{2n+1}\right)$. So $S_n=\dfrac{1}{2}\left(\dfrac{1}{1}-\dfrac{1}{3}\right)+\dfrac{1}{2}\left(\dfrac{1}{3}-\dfrac{1}{5}\right)+\cdots+\dfrac{1}{2}\left(\dfrac{1}{2n-1}-\dfrac{1}{2n+1}\right)=$
$\dfrac{1}{2}\left(1-\dfrac{1}{2n+1}\right)\to\dfrac{1}{2}$. Thus, the series converges to $\dfrac{1}{2}$.

37.12 Find the sum of the series $4-1+\frac{1}{4}-\frac{1}{16}+\cdots$.

▌ This is a geometric series with ratio $r=-\frac{1}{4}$ and first term $a=4$. Hence, it converges to $\dfrac{4}{1-(-\frac{1}{4})}=\frac{16}{5}$.

37.13 Test the convergence of $1+\frac{3}{2}+\frac{9}{4}+\frac{27}{8}+\cdots$.

▌ This is a geometric series with ratio $r=\frac{3}{2}>1$. Hence, it is divergent.

37.14 Test the convergence of $3+\frac{5}{2}+\frac{7}{3}+\frac{9}{4}+\cdots$.

▌ The series has the general term $a_n=\dfrac{2n+3}{n+1}$ (starting with $n=0$), but $\lim a_n=\lim\dfrac{2+3/n}{1+1/n}=2\neq0$. Hence, by Problem 37.1, the series diverges.

37.15 Investigate the series $\dfrac{1}{1\cdot3}+\dfrac{1}{2\cdot4}+\dfrac{1}{3\cdot5}+\dfrac{1}{4\cdot6}+\cdots$.

▌ Rewrite the series as $\left(\dfrac{1}{1\cdot3}+\dfrac{1}{3\cdot5}+\cdots\right)+\dfrac{1}{2^2}\left(\dfrac{1}{1\cdot2}+\dfrac{1}{2\cdot3}+\cdots\right)=\dfrac{1}{2}+\dfrac{1}{4}(1)=\dfrac{3}{4}$, by Problems 37.11 and 37.10.

37.16 Test the convergence of $3+\sqrt3+\sqrt[3]3+\sqrt[4]3+\cdots$.

▌ $a_n=\sqrt[n]3=3^{1/n}=e^{(\ln3)/n}\to e^0=1\neq0$. Hence, by Problem 37.1, the series diverges.

37.17 Study the series $\displaystyle\sum_{n=1}^\infty\dfrac{1}{n(n+4)}=\dfrac{1}{1\cdot5}+\dfrac{1}{2\cdot6}+\dfrac{1}{3\cdot7}+\cdots$.

▌ $\dfrac{1}{n(n+4)}=\dfrac{1}{4}\left(\dfrac{1}{n}-\dfrac{1}{n+4}\right)$. So the partial sum $S_n=\dfrac{1}{4}\left(\dfrac{1}{1}-\dfrac{1}{5}\right)+\dfrac{1}{4}\left(\dfrac{1}{2}-\dfrac{1}{6}\right)+\cdots+\dfrac{1}{4}\left(\dfrac{1}{n}-\dfrac{1}{n+4}\right)=$
$\dfrac{1}{4}\left(1+\dfrac{1}{2}+\dfrac{1}{3}+\dfrac{1}{4}\right)+\dfrac{1}{4}\left(-\dfrac{1}{n+1}-\dfrac{1}{n+2}-\dfrac{1}{n+3}-\dfrac{1}{n+4}\right)\to\dfrac{1}{4}\left(1+\dfrac{1}{2}+\dfrac{1}{3}+\dfrac{1}{4}\right)=\dfrac{25}{48}$.

37.18 Study the series $\displaystyle\sum_{n=1}^\infty\dfrac{1}{n(n+1)(n+2)}=\dfrac{1}{1\cdot2\cdot3}+\dfrac{1}{2\cdot3\cdot4}+\dfrac{1}{3\cdot4\cdot5}+\cdots$.

▌ $\dfrac{1}{n(n+1)(n+2)}=\dfrac{1}{n(n+2)}-\dfrac{1}{(n+1)(n+2)}=\dfrac{1}{2}\left(\dfrac{1}{n}-\dfrac{1}{n+2}\right)-\left(\dfrac{1}{n+1}-\dfrac{1}{n+2}\right)=\dfrac{1}{2}\left(\dfrac{1}{n}+\dfrac{1}{n+2}\right)-$
$\dfrac{1}{n+1}$. Thus, $\dfrac{1}{1\cdot2\cdot3}=\dfrac{1}{2}\left(\dfrac{1}{1}+\dfrac{1}{3}\right)-\dfrac{1}{2}$, $\dfrac{1}{2\cdot3\cdot4}=\dfrac{1}{2}\left(\dfrac{1}{2}+\dfrac{1}{4}\right)-\dfrac{1}{3}$, $\dfrac{1}{3\cdot4\cdot5}=\dfrac{1}{2}\left(\dfrac{1}{3}+\dfrac{1}{5}\right)-\dfrac{1}{4},\ldots$,
$\dfrac{1}{n(n+1)(n+2)}=\dfrac{1}{2}\left(\dfrac{1}{n}+\dfrac{1}{n+2}\right)-\dfrac{1}{n+1}$. The partial sum $\dfrac{1}{2}-\dfrac{1}{2}\left(\dfrac{1}{2}\right)+\dfrac{1}{n+1}\left(-\dfrac{1}{2}\right)+\dfrac{1}{n+2}\left(\dfrac{1}{2}\right)\to\dfrac{1}{2}-$
$\dfrac{1}{4}=\dfrac{1}{4}$.

37.19 Study the series $\displaystyle\sum_{n=1}^{\infty} \frac{n}{(n+1)!} = \frac{1}{2!} + \frac{2}{3!} + \frac{3}{4!} + \cdots$.

▮ $\dfrac{n}{(n+1)!} = \dfrac{1}{n!} - \dfrac{1}{(n+1)!}$. Hence, the partial sum $\left(\dfrac{1}{1!} - \dfrac{1}{2!}\right) + \left(\dfrac{1}{2!} - \dfrac{1}{3!}\right) + \cdots + \left[\dfrac{1}{n!} - \dfrac{1}{(n+1)!}\right] =$
$1 - \dfrac{1}{(n+1)!} \to 1$.

37.20 Study the series $\displaystyle\sum_{n=1}^{\infty} \frac{1}{(4n-3)(4n+1)} = \frac{1}{1\cdot5} + \frac{1}{5\cdot9} + \frac{1}{9\cdot13} + \cdots$.

▮ $\dfrac{1}{(4n-3)(4n+1)} = \dfrac{1}{4}\left(\dfrac{1}{4n-3} - \dfrac{1}{4n+1}\right)$. So

$$S_n = \frac{1}{4}\left(\frac{1}{1} - \frac{1}{5}\right) + \frac{1}{4}\left(\frac{1}{5} - \frac{1}{9}\right) + \cdots + \frac{1}{4}\left(\frac{1}{4n-3} - \frac{1}{4n+1}\right) = \frac{1}{4}\left(1 - \frac{1}{4n+1}\right) \to \frac{1}{4}.$$

37.21 Evaluate $\displaystyle\sum_{n=1}^{\infty} (-1)^{n+1}\frac{2n+1}{n(n+1)}$.

▮ $\dfrac{2n+1}{n(n+1)} = \dfrac{1}{n} + \dfrac{1}{n+1}$. So the partial sum $\left(\dfrac{1}{1} + \dfrac{1}{2}\right) - \left(\dfrac{1}{2} + \dfrac{1}{3}\right) + \left(\dfrac{1}{3} + \dfrac{1}{4}\right) - \left(\dfrac{1}{4} + \dfrac{1}{5}\right) + \cdots$ is either
$1 - \dfrac{1}{n+1}$ or $1 - \dfrac{1}{n+1} + \left(\dfrac{1}{n+1} + \dfrac{1}{n+2}\right)$. In either case, the partial sum approaches 1.

37.22 Evaluate $\displaystyle\sum_{n=1}^{\infty} \frac{2n+1}{n^2(n+1)^2}$.

▮ $\dfrac{2n+1}{n^2(n+1)^2} = \dfrac{1}{n^2} - \dfrac{1}{(n+1)^2}$. The partial sum $S_n = \left(\dfrac{1}{1} - \dfrac{1}{2^2}\right) + \left(\dfrac{1}{2^2} - \dfrac{1}{3^2}\right) + \cdots + \left[\dfrac{1}{n^2} - \dfrac{1}{(n+1)^2}\right] =$
$1 - \dfrac{1}{(n+1)^2} \to 1$.

37.23 Evaluate $\displaystyle\sum_{n=1}^{\infty} \frac{e^n}{n^3} = e + \frac{e^2}{8} + \frac{e^3}{27} + \frac{e^4}{64} + \cdots$.

▮ $\dfrac{e^n}{n^3} \to +\infty$; so, by Problem 37.1, the series diverges.

37.24 Evaluate $\displaystyle\sum_{n=1}^{\infty} \frac{1}{\sqrt{n} + \sqrt{n-1}}$.

▮ $\dfrac{1}{\sqrt{n} + \sqrt{n-1}} = \dfrac{1}{\sqrt{n} + \sqrt{n-1}} \cdot \dfrac{\sqrt{n} - \sqrt{n-1}}{\sqrt{n} - \sqrt{n-1}} = \sqrt{n} - \sqrt{n-1}$. The partial sum $(\sqrt{1} - \sqrt{0}) + (\sqrt{2} - \sqrt{1}) + \cdots + (\sqrt{n} - \sqrt{n-1}) = \sqrt{n} \to +\infty$. The series diverges.

37.25 Find the sum $S = \displaystyle\sum_{n=0}^{\infty} \frac{1}{3^n}$, and show that it is correct by exhibiting a formula which, for each $\varepsilon > 0$, specifies an integer m for which $|S_n - S| < \varepsilon$ holds for all $n > m$ (where S_n is the nth partial sum).

▮ $\displaystyle\sum_{n=0}^{\infty} \frac{1}{3^n}$ is a geometric series with ratio $r = \frac{1}{3}$ and first term $a = 1$. So the sum $S = \dfrac{1}{1 - \frac{1}{3}} = \dfrac{3}{2}$. In fact,
assume $\varepsilon > 0$. Then, by Problem 37.4, $S_n = 1 + \dfrac{1}{3} + \cdots + \dfrac{1}{3^{n-1}} = \dfrac{(\frac{1}{3})^n - 1}{\frac{1}{3} - 1} = \dfrac{3}{2}\left(1 - \dfrac{1}{3^n}\right)$. Now, $|S_n - S| =$
$\left|\dfrac{3}{2} - \dfrac{3}{2}\left(1 - \dfrac{1}{3^n}\right)\right| = \dfrac{1}{2 \cdot 3^{n-1}}$. We want $\dfrac{1}{2 \cdot 3^{n-1}} < \varepsilon$, $\dfrac{1}{2\varepsilon} < 3^{n-1}$, $-\ln 2\varepsilon < (n-1)\ln 3$, $-\dfrac{\ln 2\varepsilon}{\ln 3} < n - 1$.
Choose m to be the least positive integer that exceeds $-\dfrac{\ln 2\varepsilon}{\ln 3}$.

37.26 Determine the value of the infinite decimal $0.666 + \cdots$.

▮ $0.666\cdots = \dfrac{6}{10} + \dfrac{6}{10^2} + \dfrac{6}{10^3} + \cdots$ is a geometric series with ratio $r = \frac{1}{10}$ and first term $a = \frac{6}{10}$. Hence,
the sum is $\dfrac{\frac{6}{10}}{1 - \frac{1}{10}} = \dfrac{2}{3}$.

37.27 Evaluate $\displaystyle\sum_{n=0}^{\infty} \frac{1}{n+100} = \frac{1}{100} + \frac{1}{101} + \cdots$.

▮ $\displaystyle\sum_{n=0}^{\infty} \frac{1}{n+100}$ is the harmonic series minus the first 99 terms. However, convergence or divergence is not affected by deletion or addition of any finite number of terms. Since the harmonic series is divergent (by Problem 37.2), so is the given series.

37.28 Evaluate $\displaystyle\sum_{n=1}^{\infty} \frac{1}{3n} = \frac{1}{3} + \frac{1}{6} + \frac{1}{9} + \cdots$.

▮ $\displaystyle\sum_{n=1}^{\infty} \frac{1}{3n} = \frac{1}{3}\sum_{n=1}^{\infty} \frac{1}{n}$. Since the harmonic series is divergent, so is the given series.

37.29 Evaluate $\displaystyle\sum_{n=1}^{\infty} \ln \frac{n}{n+1}$.

▮ $\ln[n/(n+1)] = \ln n - \ln(n+1)$, and $S_n = (\ln 1 - \ln 2) + (\ln 2 - \ln 3) + \cdots + [\ln n - \ln(n+1)] = -\ln(n+1) \to -\infty$. Thus, the given series diverges.

37.30 Evaluate $\displaystyle\sum_{n=0}^{\infty} e^{-n} = 1 + e^{-1} + e^{-2} + \cdots$.

▮ This is a geometric series with ratio $r = 1/e < 1$ and first term $a = 1$. Hence, it converges to $\dfrac{1}{1 - 1/e} = \dfrac{e}{e-1}$.

37.31 Evaluate $\displaystyle\sum_{n=0}^{\infty} \left(\frac{1}{2^n} + \frac{1}{5^n} \right)$.

▮ This series converges because it is the sum of two convergent series, $\displaystyle\sum_{n=0}^{\infty} \frac{1}{2^n}$ and $\displaystyle\sum_{n=0}^{\infty} \frac{1}{5^n}$ (both are geometric series with ratio $r < 1$). $\displaystyle\sum_{n=0}^{\infty} \frac{1}{2^n} = \frac{1}{1 - \frac{1}{2}} = 2$, and $\displaystyle\sum_{n=0}^{\infty} \frac{1}{5^n} = \frac{1}{1 - \frac{1}{5}} = \frac{5}{4}$. Hence, the sum of the given series is $2 + \frac{5}{4} = \frac{13}{4}$.

37.32 (Zeno's paradox) Achilles (A) and a tortoise (T) have a race. T gets a 1000-ft head start, but A runs at 10 ft/s while the tortoise only does 0.01 ft/s. When A reaches T's starting point, T has moved a short distance ahead. When A reaches that point, T again has moved a short distance ahead, etc. Zeno claimed that A would never catch T. Show that this is not so.

▮ When A reaches T's starting point, 100 s have passed and T has moved $0.01 \times 100 = 1$ ft. A covers that additional 1 ft in 0.1 s, but T has moved $0.01 \times 0.1 = 0.001$ ft further. A needs 0.0001 s to cover that distance, but T meanwhile has moved $0.01 \times 0.0001 = 0.000001$ ft; and so on. The limit of the distance between A and T approaches 0. The time involved is $100 + 0.1 + 0.0001 + 0.0000001 + \cdots$, which is a geometric series with first term $a = 100$ and ratio $r = \frac{1}{1000}$. Its sum is $100/(1 - \frac{1}{1000})$. Thus, Achilles catches up with (and then passes) the tortoise in a little over 100 s, just as we knew he would. The seeming paradox arises from the artificial division of the event into infinitely many shorter and shorter steps.

37.33 A rubber ball falls from an initial height of 10 m; whenever it hits the ground, it bounces up two-thirds of the previous height. What is the total distance covered by the ball before it comes to rest?

▮ The distance is $10 + 10(\frac{2}{3}) + 10(\frac{2}{3})^2 + 10(\frac{2}{3})^3 + \cdots$. This is a geometric series with ratio $\frac{2}{3}$ and first term $a = 10$. Hence, the sum is $10/(1 - \frac{2}{3}) = 30$ m.

37.34 Investigate $\displaystyle\sum_{n=1}^{\infty} \frac{3}{5^n + 2}$.

▮ $3/(5^n + 2) < 5/5^n = 1/5^{n-1}$. So this series of positive terms is term by term less than the convergent geometric series $\displaystyle\sum_{n=1}^{\infty} \frac{1}{5^{n-1}}$. Hence, by the comparison test, the given series is convergent. However, we cannot directly compute the sum of the series. We can only say that the sum is less than $\displaystyle\sum_{n=1}^{\infty} \frac{1}{5^{n-1}} = \frac{1}{1 - \frac{1}{5}} = \frac{5}{4}$.

37.35 Determine whether $\displaystyle\sum_{n=1}^{\infty} \frac{1}{2^n - 3}$ is convergent.

▌ $1/(2^n - 3) < 1/2^{n-1}$ for all $n \geq 3$ (since $2^{n-1} < 2^n - 3 \Leftrightarrow 3 < 2^n - 2^{n-1} = 2^{n-1}$). Hence, beginning with the third term, the given series is term by term less than the convergent series $\displaystyle\sum_{n=3}^{\infty} \frac{1}{2^{n-1}}$, and, therefore, converges.

37.36 If $0 < p \leq 1$, show that the series $\displaystyle\sum_{n=1}^{\infty} \frac{1}{n^p} = 1 + \frac{1}{2^p} + \frac{1}{3^p} + \cdots$ is divergent.

▌ $1/n^p \geq 1/n$ since $n^p \leq n$. Therefore, by the comparison test and the fact that $\displaystyle\sum_{n=1}^{\infty} \frac{1}{n}$ is divergent, $\displaystyle\sum_{n=1}^{\infty} \frac{1}{n^p}$ is divergent.

37.37 Determine whether $\displaystyle\sum_{n=0}^{\infty} \frac{1}{n!}$ is convergent.

▌ For $n \geq 1$, $1/n! = 1/(1 \cdot 2 \cdot \,\cdots\, \cdot n) \leq 1/(1 \cdot 2 \cdot 2 \cdot \,\cdots\, \cdot 2) = 1/2^{n-1}$. Hence, $\displaystyle\sum_{n=0}^{\infty} \frac{1}{n!}$ is convergent, by comparison with the convergent series $\displaystyle\sum_{n=0}^{\infty} \frac{1}{2^{n-1}}$. The sum $(= e)$ of the given series is $< 1 + \displaystyle\sum_{n=1}^{\infty} \frac{1}{2^{n-1}} = 1 + 2 = 3$.

37.38 Determine whether $\displaystyle\sum_{n=1}^{\infty} \frac{1}{(n^3 + 10)^{1/4}}$ is convergent.

▌ $1/(n^3 + 10) \geq 1/2n^3$ for $n \geq 3$ (since $n^3 > 10$ for $n \geq 3$). Therefore, $1/(n^3 + 10)^{1/4} \geq 1/2^{1/4} n^{3/4}$. Thus, the given series is divergent by comparison with $\dfrac{1}{2^{1/4}} \displaystyle\sum_{n=1}^{\infty} \frac{1}{n^{3/4}}$ (see Problem 37.36).

37.39 State the integral test.

▌ Let $\sum a_n$ be a series of positive terms such that there is a continuous, decreasing function $f(x)$ for which $f(n) = a_n$ for all positive integers $n \geq n_0$. Then $\Sigma\, a_n$ converges if and only if the improper integral $\int_{n_0}^{\infty} f(x)\, dx$ converges.

37.40 For $p > 1$, show that the so-called p-series $\displaystyle\sum_{n=1}^{\infty} \frac{1}{n^p}$ converges. (Compare with Problem 37.36.)

▌ Use the integral test (Problem 37.39), with $f(x) = 1/x^p$. $\displaystyle\int_1^{\infty} \frac{dx}{x^p} = \lim_{u \to +\infty} \int_1^u \frac{dx}{x^p} = \lim_{u \to +\infty} \frac{1}{1-p} \left.\frac{1}{x^{p-1}}\right]_1^u = \lim_{u \to +\infty} \frac{1}{1-p} \left(\frac{1}{u^{p-1}} - 1\right) = \frac{1}{p-1}$. Hence, $\displaystyle\sum_{n=1}^{\infty} \frac{1}{n^p}$ converges.

37.41 Determine whether $\displaystyle\sum_{n=2}^{\infty} \frac{1}{n \ln n}$ converges.

▌ Use the integral test with $f(x) = \dfrac{1}{x \ln x}$.

$$\int_2^{\infty} \frac{1}{x \ln x}\, dx = \lim_{u \to +\infty} \int_1^u \frac{1}{x \ln x}\, dx = \lim_{u \to \infty} \ln(\ln x) \Big]_1^u = \lim_{u \to +\infty} [\ln(\ln u) - \ln 0] = +\infty.$$

Hence, $\displaystyle\sum_{n=2}^{\infty} \frac{1}{n \ln n}$ diverges.

37.42 State the limit comparison test.

▌ Let $\Sigma\, a_n$ and $\Sigma\, b_n$ be a series of positive terms.
Case I. If $\displaystyle\lim_{n \to +\infty} a_n/b_n = L > 0$, then $\Sigma\, a_n$ converges if and only if $\Sigma\, b_n$ converges.
Case II. If $\displaystyle\lim_{n \to +\infty} a_n/b_n = 0$ and $\Sigma\, b_n$ converges, then $\Sigma\, a_n$ converges.
Case III. If $\displaystyle\lim_{n \to +\infty} a_n/b_n = +\infty$ and Σb_n diverges, then $\Sigma\, a_n$ diverges.

37.43 If $\Sigma\, a_n$ and $\Sigma\, b_n$ are series of positive terms and $\displaystyle\lim_{n\to+\infty} a_n/b_n = 0$, show by example that $\Sigma\, a_n$ may converge and $\Sigma\, b_n$ not converge.

❚ Let $a_n = 1/n^2$ and $b_n = 1/n$. $\displaystyle\lim_{n\to+\infty} a_n/b_n = \lim_{n\to+\infty} 1/n = 0$. But, $\Sigma\, 1/n^2$ converges and $\Sigma\, 1/n$ diverges.

37.44 Determine whether $\displaystyle\sum_{n=1}^{\infty} \frac{1}{\sqrt{n^3 + 3}}$ converges.

❚ Use the limit comparison test with the convergent p-series $\displaystyle\sum \frac{1}{n^{3/2}}$. Then $\displaystyle\lim_{n\to+\infty} \frac{1/\sqrt{n^3 + 3}}{1/n^{3/2}} =$ $\displaystyle\lim_{n\to\infty} \sqrt{\frac{n^3}{n^3 + 3}} = \lim_{n\to+\infty} \sqrt{\frac{1}{1 + 3/n^3}} = 1$. Therefore, the given series is convergent.

37.45 Determine whether $\displaystyle\sum_{n=1}^{\infty} \frac{1}{n3^n}$ is convergent.

❚ Use the comparison test: $\displaystyle \frac{1}{n3^n} \le \frac{1}{3^n}$. The geometric series $\displaystyle\sum_{n=1}^{\infty} \frac{1}{3^n}$ is convergent. Hence, the given series is convergent.

37.46 Determine whether $\displaystyle\sum_{n=1}^{\infty} \frac{n^3}{2n^4 + 1}$ is convergent.

❚ Intuitively, we ignore the 1 in the denominator, so we use the limit comparison test with the divergent series $\displaystyle\sum 1/n$. $\displaystyle\lim_{n\to+\infty} \frac{n^3/(2n^4 + 1)}{1/n} = \lim_{n\to+\infty} \frac{n^4}{2n^4 + 1} = \lim_{n\to+\infty} \frac{1}{2 + 1/n^4} = \frac{1}{2}$. Hence, the given series is divergent.

37.47 Determine whether $\displaystyle\sum_{n=2}^{\infty} \frac{\ln n}{n^2 + 1}$ is convergent.

❚ Use the limit comparison test with the convergent p-series $\displaystyle\sum 1/n^{3/2}$. $\displaystyle\lim_{n\to+\infty} \frac{(\ln n)/(n^2 + 1)}{1/n^{3/2}} =$ $\displaystyle\lim_{n\to+\infty} \frac{n^{3/2} \ln n}{n^2 + 1} = \lim_{n\to+\infty} \frac{1}{1 + 1/n^2} \frac{\ln n}{n^{1/2}} = 0$ $\left(\text{since, by L'Hôpital's rule, } \displaystyle\lim_{n\to+\infty} \frac{\ln n}{n^{1/2}} = \lim_{n\to+\infty} \frac{1/n}{\frac{1}{2} n^{-1/2}} = \lim_{n\to+\infty} \frac{2}{\sqrt{n}} = \right.$ $0\Big)$. Therefore, the given series is convergent.

37.48 Determine whether $\displaystyle\sum_{n=1}^{\infty} \frac{1}{n^n}$ is convergent.

❚ $\displaystyle \frac{1}{n^n} = \frac{1}{e^{n \ln n}} < \frac{1}{e^n}$, for $n \ge 3$. Hence, the series converges, by comparison with the convergent geometric series $\displaystyle\sum_{n=1}^{\infty} \frac{1}{e^n}$.

37.49 Determine whether $\displaystyle\sum_{n=1}^{\infty} \frac{\ln n}{n}$ converges.

❚ Use the integral test with $\displaystyle f(x) = \frac{\ln x}{x}$. $\left[\text{Note that } f(x) \text{ is decreasing, since } f'(x) = \frac{1 - \ln x}{x^2} < 0 \text{ for } x > e.\right]$ $\displaystyle\int_1^{\infty} \frac{\ln x}{x}\, dx = \lim_{u\to+\infty} \int_1^u \frac{\ln x}{x}\, dx = \lim_{u\to+\infty} \{\tfrac{1}{2}(\ln x)^2\,]_1^u\} = \lim_{u\to+\infty} \tfrac{1}{2}(\ln u)^2 = +\infty$. Therefore, the given series diverges.

37.50 Determine whether $\displaystyle\sum_{n=2}^{\infty} \frac{1}{n(\ln n)^2}$ converges.

❚ Use the integral test with $f(x) = 1/x(\ln x)^2$. $\displaystyle\int_2^{\infty} \frac{dx}{x(\ln x)^2} = \lim_{u\to+\infty} \int_2^u \frac{dx}{x(\ln x)^2} = \lim_{u\to+\infty} \left(-\frac{1}{\ln x}\, \right]_2^u \Big) =$ $\displaystyle\lim_{u\to+\infty} \left[-\left(\frac{1}{\ln u} - \frac{1}{\ln 2}\right)\right] = \frac{1}{\ln 2}$. Hence, the series converges.

37.51 Give an example of a series that is conditionally convergent (that is, convergent but not absolutely convergent).

❚ $\displaystyle\sum_{n=1}^{\infty} \frac{(-1)^{n+1}}{n} = 1 - \frac{1}{2} + \frac{1}{3} - \frac{1}{4} + \cdots$ is convergent by the alternating series test (the terms are alternately positive and negative, and their magnitudes decrease to zero). But $\displaystyle\sum_{n=1}^{\infty} \left|\frac{(-1)^{n+1}}{n}\right| = \sum_{n=1}^{\infty} \frac{1}{n}$ diverges.

37.52 State the ratio test for a series $\Sigma\, a_n$.

▮ Assume $a_n \neq 0$ for $n \geq n_0$.
Case I. If $\lim_{n \to +\infty} |a_{n+1}/a_n| < 1$, the series is absolutely convergent.
Case II. If $\lim_{n \to +\infty} |a_{n+1}/a_n| > 1$, the series is divergent.
Case III. If $\lim_{n \to +\infty} |a_{n+1}/a_n| = 1$, nothing can be said about convergence or divergence.

37.53 Determine whether $\displaystyle\sum_{n=1}^{\infty} \frac{n^2}{e^n}$ is convergent.

▮ Apply the ratio test. $\displaystyle\lim_{n \to +\infty} |a_{n+1}/a_n| = \lim_{n \to +\infty} \frac{(n+1)^2/e^{n+1}}{n^2/e^n} = \lim_{n \to +\infty} \frac{(n+1)^2}{n^2} \frac{1}{e} = \lim_{n \to +\infty} \left(1 + \frac{1}{n}\right)^2 \cdot \frac{1}{e} = \frac{1}{e} < 1$.
Hence, the series converges. (The integral test is also applicable.)

37.54 Determine whether $\displaystyle\sum_{n=1}^{\infty} \frac{\tan^{-1} n}{n^2 + 1}$ is convergent.

▮ Use the comparison test with the convergent p-series $\dfrac{\pi}{2} \Sigma \dfrac{1}{n^2}$. Clearly, $\dfrac{\tan^{-1} n}{n^2 + 1} < \dfrac{\pi/2}{n^2}$. Hence, the given series converges. (The integral test is also applicable.)

37.55 Determine whether $1 - \dfrac{2}{3} + \dfrac{3}{3^2} - \dfrac{4}{3^3} + \cdots = \displaystyle\sum_{n=0}^{\infty} \frac{(-1)^n(n+1)}{3^n}$ is convergent.

▮ By the ratio test, the series is absolutely convergent; so it is certainly convergent.

37.56 Determine whether $\displaystyle\sum_{n=0}^{\infty} (-1)^n \frac{n+1}{2n+1}$ converges.

▮ Although the terms are alternatively positive and negative, the alternating series test does not apply, since $\displaystyle\lim_{n \to +\infty} \frac{n+1}{2n+1} = \lim_{n \to +\infty} \frac{1 + 1/n}{2 + 1/n} = \frac{1}{2} \neq 0$. Since the nth term does not approach 0, the series is not convergent.

37.57 Find the error if the sum of the first three terms is used as an approximation to the sum of the alternating series $\displaystyle\sum_{n=1}^{\infty} (-1)^{n+1} \frac{1}{n}$.

▮ The error is less than the magnitude of the first term omitted. Thus, the approximation is $1 - \frac{1}{2} + \frac{1}{3} = \frac{5}{6}$, and the error is less than 0.25. Hence, the actual value V satisfies $\frac{7}{12} < V < \frac{13}{12}$. [N.B. It can be shown that $V = \ln 2 \approx 0.693$.]

37.58 Find the sum of the infinite series $\displaystyle\sum_{n=1}^{\infty} \frac{(-1)^n}{n!} = -1 + \frac{1}{2} - \frac{1}{6} + \frac{1}{24} - \cdots$ correct to three decimal places.

▮ We want an error < 0.0005. Hence, we must find the least n so that $1/n! < 0.0005 = \frac{1}{2000}$; that is, $n! > 2000$. Since $6! = 720$ and $7! = 5040$, the desired value of n is 7. So we must use $-1 + \frac{1}{2} - \frac{1}{6} + \frac{1}{24} - \frac{1}{120} + \frac{1}{720} = -\frac{455}{720} \approx -0.628$. [The actual value is $e^{-1} - 1$.]

37.59 Approximate the sum $\displaystyle\sum_{n=0}^{\infty} \frac{(-1)^n}{4^n} = 1 - \frac{1}{4} + \frac{1}{16} - \frac{1}{64} + \cdots$ to two decimal places using the method based on the alternating series test. Check your answer by finding the actual sum.

▮ We must find the least n such that $1/4^n < 0.005 = \frac{1}{200}$, that is, $200 < 4^n$, or $n \geq 4$. So, if we use $1 - \frac{1}{4} + \frac{1}{16} - \frac{1}{64} = \frac{51}{64}$, the error will be less than 0.0005. Since $\frac{51}{64} = 0.796\cdots$, our approximation is 0.80. To check, note that the given series is a geometric series with ratio $-\frac{1}{4}$, and, therefore, the sum of the series is $\dfrac{1}{1 - (-\frac{1}{4})} = \dfrac{4}{5} = 0.8$. Thus, our approximation is actually the exact value of the sum.

37.60 Estimate the error when $\displaystyle\sum_{n=1}^{\infty} \frac{(-1)^{n+1}}{n^2} = 1 - \frac{1}{4} + \frac{1}{9} - \frac{1}{16} + \cdots$ is approximated by its first 10 terms.

▮ The error is less than the magnitude of the first term omitted, which is $1/11^2 \approx 0.0083$.

37.61 How many terms must be used to approximate the sum of the series in Problem 37.60 correctly to one decimal place?

▮ We must have $1/n^2 < 0.05 = \frac{1}{20}$, or $20 < n^2$, $n \geq 5$. Thus, we must use the first four terms $1 - \frac{1}{4} + \frac{1}{9} - \frac{1}{16} = \frac{115}{144} \approx 0.79$. Hence, correct to one decimal place, the sum is 0.8.

37.62 Estimate the error when the series $1 - \frac{1}{3} + \frac{1}{5} - \frac{1}{7} + \cdots$ is approximated by its first 50 terms.

▮ $a_n = (-1)^{n+1}/(2n - 1)$. The error will be less than the magnitude of the first term omitted: $1/[2(51) - 1] = \frac{1}{101}$. Notice how poor the guaranteed approximation is for such a large number of terms.

37.63 Estimate the number of terms of the series $1 - \frac{1}{2^4} + \frac{1}{3^4} - \frac{1}{4^4} + \cdots$ required for an approximation of the sum which will be correct to four decimal places.

▮ We must have $1/n^4 < 0.00005 = \frac{1}{20,000}$, $n^4 > 20,000$, $n \geq 12$. Hence, we should use the first 11 terms.

37.64 Show that, if $\Sigma\, a_n$ converges by the integral test for a function $f(x)$, the error R_n, if we use the first n terms, satisfies $\int_{n+1}^{\infty} f(x)\, dx < R_n < \int_n^{\infty} f(x)\, dx$.

▮ $R_n = a_{n+1} + a_{n+2} + \cdots$. If we approximate by the lower rectangles in Fig. 37-1, then $R_n < \int_n^{n+1} f(x)\, dx + \int_{n+1}^{n+2} f(x)\, dx + \cdots = \int_n^{\infty} f(x)\, dx$. If we use the upper rectangles, $R_n > \int_{n+1}^{n+2} f(x)\, dx + \int_{n+2}^{n+3} f(x)\, dx + \cdots = \int_{n+1}^{\infty} f(x)\, dx$.

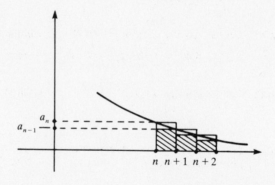

Fig. 37-1

37.65 Estimate the error when $\displaystyle\sum_{n=1}^{\infty} \frac{1}{4n^2}$ is approximated by the first 10 terms.

▮ By Problem 37.64, $R_{10} < \int_{10}^{\infty} \frac{dx}{4x^2} = \lim_{u \to +\infty} \left(-\frac{1}{4x}\right]_{10}^u\right) = \lim_{u \to +\infty} -\left(\frac{1}{4u} - \frac{1}{40}\right) = \frac{1}{40} = 0.025$. In addition, $R_{10} > \int_{11}^{\infty} \frac{dx}{4x^2} = \lim_{u \to +\infty} \left(-\frac{1}{4x}\right]_{11}^u\right) = \lim_{u \to +\infty} -\left(\frac{1}{4u} - \frac{1}{44}\right) = \frac{1}{44} \approx 0.0227$. Hence, the error lies between 0.023 and 0.025.

37.66 How many terms are necessary to approximate $\displaystyle\sum_{n=1}^{\infty} \frac{1}{n^3}$ correctly to three decimal places?

▮ By Problem 37.64, the error $R_n < \int_n^{\infty} \frac{dx}{x^3} = \lim_{u \to +\infty} \left(\frac{-1}{2x^2}\right]_n^u\right) = \lim_{u \to +\infty} -\frac{1}{2}\left(\frac{1}{u^2} - \frac{1}{n^2}\right) = \frac{1}{2n^2}$. To get $1/2n^2 < 0.005 = \frac{1}{200}$, we need $100 < n^2$, $n > 10$. So at least 11 terms are required.

37.67 What is the error if we approximate a convergent geometric series $\displaystyle\sum_{n=0}^{\infty} ar^n$ by the first n terms $a + ar + \cdots + ar^{n-1}$?

▮ The error is $ar^n + ar^{n+1} + \cdots$, a geometric series with sum $ar^n/(1 - r)$.

37.68 If we approximate the geometric series $\sum_{n=0}^{\infty} \frac{1}{5^n} = 1 + \frac{1}{5} + \frac{1}{25} + \cdots$ by means of the first 10 terms, what will the error be?

▮ By Problem 37.67, the error is $\dfrac{(\frac{1}{5})^{10}}{1 - \frac{1}{5}} = \dfrac{1}{4} \cdot \dfrac{1}{5^9}$.

37.69 For the convergent p-series $\sum_{n=1}^{\infty} \frac{1}{n^p}$ ($p > 1$), show that the error R_n after n terms is less than $\dfrac{1}{(p-1)n^{p-1}}$.

▮ By Problem 37.64, $R_n < \int_n^{\infty} \dfrac{dx}{x^p} = \lim_{u \to +\infty} \int_n^u \dfrac{dx}{x^p} = \lim_{u \to +\infty} \left\{ \dfrac{1}{1-p} x^{-p+1} \Big]_n^u \right\} = \lim_{u \to +\infty} \dfrac{1}{1-p} \left(\dfrac{1}{u^{p-1}} - \dfrac{1}{n^{p-1}} \right) = \dfrac{1}{p-1} \dfrac{1}{n^{p-1}}$.

37.70 Study the convergence of $\sum_{n=0}^{\infty} (-1)^{n+1} \dfrac{2^n}{n!}$.

▮ Use the ratio test. $\lim_{n \to +\infty} |a_{n+1}/a_n| = \lim_{n \to +\infty} \dfrac{2^{n+1}/(n+1)!}{2^n/n!} = \lim_{n \to +\infty} \dfrac{2}{n+1} = 0$. Therefore, the series is absolutely convergent.

37.71 Determine whether $\sum_{n=0}^{\infty} \dfrac{n!}{(2n)!}$ converges.

▮ Use the ratio test.

$$\lim_{n \to +\infty} \left| \dfrac{a_{n+1}}{a_n} \right| = \lim_{n \to +\infty} \dfrac{(n+1)!/[2(n+1)]!}{n!/(2n)!} = \lim_{n \to +\infty} \dfrac{n+1}{(2n+1)(2n+2)} = \lim_{n \to +\infty} \dfrac{1/n + 1/n^2}{(2+1/n)(2+2/n)} = \dfrac{0}{4} = 0.$$

Hence, the series converges.

37.72 Determine whether $\sum_{n=1}^{\infty} n(\frac{3}{4})^n$ converges.

▮ Use the ratio test. $\lim_{n \to +\infty} \left| \dfrac{a_{n+1}}{a_n} \right| = \lim_{n \to +\infty} \dfrac{(n+1)(\frac{3}{4})^{n+1}}{n(\frac{3}{4})^n} = \lim_{n \to +\infty} \dfrac{n+1}{n} \cdot \dfrac{3}{4} = \lim_{n \to +\infty} \dfrac{3}{4}\left(1 + \dfrac{1}{n}\right) = \dfrac{3}{4} < 1$. Hence, the series converges.

37.73 Is $\sum_{n=1}^{\infty} n^n e^{-n^2}$ convergent?

▮ Yes. Since $\ln x/x \leq 1/e < \frac{1}{2}$, we have $n^n e^{-n^2} = 1/e^{n^2[1-(\ln n)/n]} < 1/e^{n(1-1/2)} = (1/\sqrt{e})^n$, so the given series is dominated by a convergent geometric series.

37.74 Determine whether $\sum_{n=1}^{\infty} (-1)^{n+1} \dfrac{\sin(\pi/n)}{n^2}$ converges.

▮ $\left| \dfrac{\sin(\pi/n)}{n^2} \right| \leq \dfrac{1}{n^2}$. Hence, the series is absolutely convergent by comparison with the convergent p-series $\sum_{n=1}^{\infty} \dfrac{1}{n^2}$.

37.75 Study the convergence of $1 - \dfrac{1}{2^2} + \dfrac{1}{3^3} - \dfrac{1}{4^4} + \cdots$.

▮ The series is $\sum_{n=1}^{\infty} \dfrac{(-1)^{n+1}}{n^n}$. Note that, for $n \geq 2$, $\dfrac{1}{n^n} \leq \dfrac{1}{2^n}$. So, by comparison with the convergent geometric series $\sum_{n=1}^{\infty} \dfrac{1}{2^n}$, the series is absolutely convergent.

37.76 Study the convergence of $1 - \dfrac{2^2 + 1}{2^3 + 1} + \dfrac{3^2 + 1}{3^3 + 1} - \dfrac{4^2 + 1}{4^3 + 1} + \cdots$.

▌ This is the series $\displaystyle\sum_{n=1}^{\infty} (-1)^{n+1} \dfrac{n^2 + 1}{n^3 + 1}$. By the alternating series test, it is convergent. By the limit comparison test with $\Sigma\, 1/n$, $\displaystyle\lim_{n \to +\infty} \dfrac{(n^2 + 1)/(n^3 + 1)}{1/n} = \lim_{n \to +\infty} \dfrac{n(n^2 + 1)}{n^3 + 1} = \lim_{n \to +\infty} \dfrac{1 + 1/n^2}{1 + 1/n^3} = 1$. Since $\displaystyle\sum_{n=1}^{\infty} \dfrac{1}{n}$ is divergent, $\displaystyle\sum_{n=1}^{\infty} \dfrac{n^2 + 1}{n^3 + 1}$ diverges. Hence, the given series is conditionally convergent.

37.77 Prove the following special case of the ratio test: A series of positive terms $\Sigma\, a_n$ is convergent if $\displaystyle\lim_{n \to +\infty} a_{n+1}/a_n = L < 1$.

▌ Choose r so that $L < r < 1$. There exists an integer k such that, if $n \geq k$, then $|a_{n+1}/a_n - L| < r - L$, and, therefore, $a_{n+1}/a_n = |L + (a_{n+1}/a_n - L)| \leq L + |a_{n+1}/a_n - L| < L + (r - L) = r$. So, if $n \geq k$, $a_{n+1}/a_n < r$. Hence, $a_{k+1}/a_k < r$, $a_{k+1} < ra_k$; $a_{k+2}/a_{k+1} < r$, $a_{k+2} < ra_{k+1} < r^2 a_k$; ...; $a_{k+m}/a_{k+m-1} < r$, $a_{k+m} < r^m a_k$. Hence, by comparison with the convergent geometric series $ra_k + r^2 a_k + r^3 a_k + \cdots$, the series $\displaystyle\sum_{i=k+1}^{\infty} a_i = a_{k+1} + a_{k+2} + \cdots$ is convergent, and, therefore, the given series is convergent (since it is obtained from a convergent sequence by addition of a finite number of terms).

37.78 Test $\displaystyle\sum_{n=1}^{\infty} \dfrac{n}{(n+1)(n+2)(n+3)}$ for convergence.

▌ Use the limit comparison test with the convergent p-series $\displaystyle\sum_{n=1}^{\infty} \dfrac{1}{n^2}$. $\dfrac{n/(n+1)(n+2)(n+3)}{1/n^2} = \dfrac{n^3}{(n+1)(n+2)(n+3)} = \dfrac{1}{(1+1/n)(1+2/n)(1+3/n)} \to 1$. Hence, the given series converges.

37.79 Test $\displaystyle\sum_{n=1}^{\infty} \dfrac{n^3}{(\ln 2)^n}$ for convergence.

▌ Use the ratio test. $\displaystyle\lim_{n \to +\infty} \dfrac{a_{n+1}}{a_n} = \lim_{n \to +\infty} \dfrac{(n+1)^3/(\ln 2)^{n+1}}{n^3/(\ln 2)^n} = \lim_{n \to +\infty} \left(1 + \dfrac{1}{n}\right)^3 \cdot \dfrac{1}{\ln 2} = \dfrac{1}{\ln 2} > 1$, since $\ln 2 < \ln e = 1$. Therefore, the series diverges.

37.80 Test $\displaystyle\sum_{n=1}^{\infty} \dfrac{n^3}{(\ln 3)^n}$ for convergence.

▌ Use the ratio test. $\displaystyle\lim_{n \to +\infty} \dfrac{a_{n+1}}{a_n} = \lim_{n \to +\infty} \dfrac{(n+1)^{n+1}/(n+1)!}{n^n/n!} = \lim_{n \to +\infty} \left(1 + \dfrac{1}{n}\right)^n = e > 1$. Therefore, the series converges.

37.81 Test $\displaystyle\sum_{n=1}^{\infty} \dfrac{n^n}{n!}$ for convergence.

▌ Use the ratio test. $\displaystyle\lim_{n \to +\infty} \dfrac{a_{n+1}}{a_n} = \lim_{n \to +\infty} \dfrac{(n+1)^{n+1}/(n+1)!}{n^n/n!} = \lim_{n \to +\infty} \left(1 + \dfrac{1}{n}\right)^n = e > 1$. Therefore, the series diverges.

37.82 Determine the nth term of and test for convergence the series $\dfrac{1}{4^2} + \dfrac{1}{7^2} + \dfrac{1}{10^2} + \dfrac{1}{13^2} + \cdots$.

▌ The nth term is $1/(3n + 1)^2$. Use the limit comparison test with the convergent p-series $\Sigma\, 1/n^2$. $\displaystyle\lim_{n \to +\infty} \dfrac{1/(3n+1)^2}{1/n^2} = \lim_{n \to +\infty} \left(\dfrac{n}{3n+1}\right)^2 = \lim_{n \to +\infty} \left(\dfrac{1}{3 + 1/n}\right)^2 = \dfrac{1}{9}$. Therefore, the given series converges.

37.83 Determine the nth term of and test for convergence the series $\dfrac{1}{2} + \dfrac{1}{3 \cdot 4} + \dfrac{1}{4 \cdot 5 \cdot 6} + \dfrac{1}{5 \cdot 6 \cdot 7 \cdot 8} + \cdots$.

▌ The nth term is $1/(n+1)(n+2) \cdots (2n)$. The nth term is less than $1/2 \cdot 2 \cdots 2 = 1/2^n$. Hence, the series is convergent by comparison with the convergent geometric series $\displaystyle\sum_{n=1}^{\infty} \dfrac{1}{2^n}$.

37.84 Determine the nth term of and test for convergence the series $\dfrac{2}{1\cdot 3}+\dfrac{3}{2\cdot 4}+\dfrac{4}{3\cdot 5}+\dfrac{5}{4\cdot 6}+\cdots$.

▌ The nth term is $(n+1)/n(n+2)$. Use the limit comparison test with the divergent series $\Sigma\, 1/n$.
$\displaystyle\lim_{n\to +\infty}\frac{n+1/n(n+2)}{1/n}=\lim_{n\to +\infty}\frac{n+1}{n+2}=\lim_{n\to +\infty}\frac{1+1/n}{1+2/n}=1$. Hence, the given series diverges.

37.85 Determine the nth term of and test for convergence the series $\dfrac{1}{2}+\dfrac{2}{3^2}+\dfrac{3}{4^3}+\dfrac{4}{5^4}+\cdots$.

▌ The nth term is $n/(n+1)^n$. Observe that $\dfrac{n}{(n+1)^n}=\dfrac{n}{(n+1)(n+1)\cdots(n+1)}<\dfrac{1}{n^2}$. Hence, the series
converges by comparison with the convergent p-series $\Sigma\, 1/n^2$.

37.86 Determine the nth term of and test for convergence the series $1+\frac{3}{5}+\frac{4}{10}+\frac{5}{17}+\cdots$.

▌ The nth term is $(n+1)/(n^2+1)$. Use the limit comparison test with the divergent series $\Sigma\, 1/n$.
$\displaystyle\lim_{n\to +\infty}\frac{(n+1)/(n^2+1)}{1/n}=\lim_{n\to +\infty}\frac{n(n+1)}{n^2+1}=\lim_{n\to +\infty}\frac{1+1/n}{1+1/n^2}=\frac{1}{1}=1$. Therefore, the given series is divergent.

37.87 Determine the nth term of and test for convergence the series $\dfrac{2}{5}+\dfrac{2\cdot 4}{5\cdot 8}+\dfrac{2\cdot 4\cdot 6}{5\cdot 8\cdot 11}+\dfrac{2\cdot 4\cdot 6\cdot 8}{5\cdot 8\cdot 11\cdot 14}+\cdots$.

▌ The nth term is $\dfrac{2\cdot 4\cdot\,\cdots\,\cdot(2n)}{5\cdot 8\cdot\,\cdots\,1\cdot(3n+2)}$. The ratio test yields

$$\lim_{n\to +\infty}\frac{2\cdot 4\cdot\,\cdots\,\cdot(2n)(2n+2)/5\cdot 8\cdot\,\cdots\,\cdot(3n+2)(3n+5)}{2\cdot 4\cdot\,\cdots\,\cdot(2n)/5\cdot 8\cdot\,\cdots\,\cdot(3n+2)}=\lim_{n\to +\infty}\frac{2n+2}{3n+5}=\lim_{n\to +\infty}\frac{2+2/n}{3+5/n}=\frac{2}{3}<1$$

Hence, the series converges.

37.88 Determine the nth term of and test for convergence the series $\frac{3}{2}+\frac{5}{10}+\frac{7}{30}+\frac{9}{68}+\cdots$.

▌ The nth term is $(2n+1)/(n^3+n)$. Use the limit comparison test with the convergent p-series $\Sigma\, 1/n^2$.
$\displaystyle\lim_{n\to +\infty}\frac{(2n+1)/(n^3+n)}{1/n^2}=\lim_{n\to +\infty}\frac{n^2(2n+1)}{n(n^2+1)}=\lim_{n\to +\infty}\frac{n(2n+1)}{n^2+1}=\lim_{n\to\infty}\frac{2+1/n}{1+1/n^2}=2$. Hence, the series is convergent.

37.89 Determine the nth term of and test for convergence the series $\frac{3}{2}+\frac{5}{24}+\frac{7}{108}+\frac{9}{320}+\cdots$.

▌ The nth term is $(2n+1)/(n+1)n^3$. Use the limit comparison test with the convergent p-series $\Sigma\, 1/n^3$.
$\displaystyle\lim_{n\to +\infty}\frac{(2n+1)/(n+1)n^3}{1/n^3}=\lim_{n\to +\infty}\frac{2n+1}{n+1}=\lim_{n\to +\infty}\frac{2+1/n}{1+1/n}=2$. Hence, the given series converges.

37.90 Determine the nth term of and test for convergence the series $\dfrac{1}{2^2-1}+\dfrac{2}{3^2-2}+\dfrac{3}{4^2-3}+\dfrac{4}{5^2-4}+\cdots$.

▌ The nth term is $n/[(n+1)^2-n]$. Use the limit comparison test with the divergent series $\Sigma\, 1/n$.
$\displaystyle\lim_{n\to +\infty}\frac{n/[(n+1)^2-n]}{1/n}=\lim_{n\to +\infty}\frac{n^2}{n^2+n+1}=\lim_{n\to +\infty}\frac{1}{1+1/n+1/n^2}=1$. Hence, the given series is divergent.

37.91 Prove the *root test*: A series of positive terms $\Sigma\, a_n$ converges if $\displaystyle\lim_{n\to +\infty}\sqrt[n]{a_n}<1$ and diverges if $\displaystyle\lim_{n\to +\infty}\sqrt[n]{a_n}>1$.

▌ Assume $\displaystyle\lim_{n\to +\infty}\sqrt[n]{a_n}=L<1$. Choose r so that $L<r<1$. Then there exists an integer k such that, if $n\geq k$, $\sqrt[n]{a_n}<r$, and, therefore, $a_n<r^n$. Hence, the series $a_k+a_{k+1}+\cdots$ is convergent by comparison with the convergent geometric series $\displaystyle\sum_{\infty} r^n$. So the given series is convergent. Assume now that $\displaystyle\lim_{n\to +\infty}\sqrt[n]{a_n}=L>1$. Choose r so that $L>r>1$. Then there exists an integer k such that, if $n\geq k$, $\sqrt[n]{a_n}>r$, and, therefore, $a_n>r^n$. Thus, by comparison with the divergent geometric series $\displaystyle\sum_{n=k} r^n$, the series $a_k+a_{k+1}+\cdots$ is divergent, and, therefore, the given series is divergent.

37.92 Test $\displaystyle\sum_{n=2}^{\infty}\frac{1}{(\ln n)^n}$ for convergence.

▮ Use the root test (Problem 37.91). $\displaystyle\lim\sqrt[n]{a_n}=\lim_{n\to+\infty}\frac{1}{\ln n}=0$. Therefore, the series converges.

37.93 Test $\displaystyle\sum_{n=1}^{\infty}\frac{2^{n+1}}{n^n}$ for convergence.

▮ Use the root test (Problem 37.91). $\displaystyle\lim\sqrt[n]{a_n}=\lim_{n\to+\infty}\sqrt[n]{2(2^n/n^n)}=\lim_{n\to+\infty}\sqrt[n]{2}(2/n)=1\cdot 0=0$. Therefore, the series converges.

37.94 Test $\displaystyle\sum_{n=1}^{\infty}\left(1+\frac{1}{n}\right)^n$ for convergence.

▮ The root test gives no information, since $\displaystyle\lim_{n\to+\infty}\sqrt[n]{a_n}=1$. However, $\displaystyle\lim_{n\to+\infty}a_n=e\neq 0$. Hence, the series diverges, by Problem 37.1.

37.95 Determine the nth term of and test for convergence the series $\displaystyle\frac{1}{2}-\frac{2}{3}\cdot\frac{1}{2^3}+\frac{3}{4}\cdot\frac{1}{3^3}-\frac{4}{5}\cdot\frac{1}{4^3}+\cdots$.

▮ $a_n=(-1)^{n+1}[n/(n+1)]/(1/n^3)=(-1)^{n+1}[1/n^2(n+1)]$. Since $|a_n|=1/n^2(n+1)<1/n^3$, the given series is absolutely convergent by comparison with the convergent p-series $\Sigma\, 1/n^3$.

37.96 Determine the nth term of and test for convergence the series $\displaystyle\frac{2}{3}-\frac{3}{4}\cdot\frac{1}{2}+\frac{4}{5}\cdot\frac{1}{3}-\frac{5}{6}\cdot\frac{1}{4}+\cdots$.

▮ $a_n=(-1)^{n+1}[(n+1)/(n+2)](1/n)$. The alternating series test implies that the series is convergent. However, it is only conditionally convergent. By the limit comparison test with $\Sigma\, 1/n$, $\displaystyle\lim_{n\to+\infty}\frac{(n+1)/(n+2)n}{1/n}=\lim_{n\to+\infty}\frac{n+1}{n+2}=\lim_{n\to+\infty}\frac{1+1/n}{1+2/n}=1$. Hence, $\Sigma\, |a_n|$ diverges.

37.97 Determine the nth term of and test for convergence the series $\displaystyle 2-\frac{2^3}{3!}+\frac{2^5}{5!}-\frac{2^7}{7!}+\cdots$.

▮ $a_n=2^{2n-1}/(2n-1)!$. Use the ratio test. $\displaystyle\lim_{n\to+\infty}\frac{a_{n+1}}{a_n}=\lim_{n\to+\infty}\frac{2^{2n+1}/(2n+1)!}{2^{2n-1}/(2n-1)!}=\lim_{n\to+\infty}\frac{4}{2n(2n+1)}=0$. Therefore, the series is absolutely convergent.

37.98 Determine the nth term of and test for convergence the series $\displaystyle\frac{1}{2}-\frac{4}{2^3+1}+\frac{9}{3^3+1}-\frac{16}{4^3+1}+\cdots$.

▮ $a_n=(-1)^{n+1}n^2/(n^3+1)$. The series converges by the alternating series test. It is only conditionally convergent, since $\displaystyle\Sigma\, |a_n|=\sum_{n=1}^{\infty}\frac{n^2}{n^3+1}$ diverges, by the limit comparison test with the divergent series $\Sigma\, 1/n$. $\displaystyle\lim_{n\to+\infty}\frac{n^2/(n^3+1)}{1/n}=\lim_{n\to+\infty}\frac{n^3}{n^3+1}=\lim_{n\to+\infty}\frac{1}{1+1/n^3}=1$.

37.99 Determine the nth term of and test for convergence the series $2-\frac{3}{2}+\frac{4}{3}-\frac{5}{4}+\cdots$.

▮ $a_n=(-1)^{n+1}(n+1)/n$. This is divergent, since $\lim|a_n|=1\neq 0$.

37.100 Test the series $\displaystyle\sum_{n=1}^{\infty}(-1)^{n+1}\frac{1}{\ln(n+1)}$ for convergence.

▮ This series converges by the alternating series test. However, it is only conditionally convergent, since $\displaystyle\sum_{n=1}^{\infty}\frac{1}{\ln(n+1)}$ is divergent. To see this, note that $1/\ln n>1/n$ and use the comparison test with the divergent series $\Sigma\, 1/n$.

37.101 Determine the nth term of and test for convergence the series $\dfrac{1}{\sqrt{2}} - \dfrac{1}{\sqrt{6}} + \dfrac{1}{\sqrt{12}} - \dfrac{1}{\sqrt{20}} + \dfrac{1}{\sqrt{30}} + \cdots$.

▮ $a_n = (-1)^{n+1}/\sqrt{n(n+1)}$. Convergence follows by the alternating series test. However, applying the limit comparison test with $\Sigma\, 1/n$ we see that $\quad \lim\limits_{n \to +\infty} \dfrac{1/\sqrt{n(n+1)}}{1/n} = \lim\limits_{n \to +\infty} \dfrac{n}{\sqrt{n(n+1)}} = \lim\limits_{n \to +\infty} \sqrt{\dfrac{n}{n+1}} =$ $\lim\limits_{n \to +\infty} \sqrt{\dfrac{1}{1+1/n}} = 1$. Therefore, $\Sigma\, \dfrac{1}{\sqrt{n(n+1)}}$ diverges, and the given series is conditionally convergent.

37.102 Determine the nth term of and test for convergence the series $\quad 1 - \dfrac{1}{2^3} + \dfrac{1}{6^3} - \dfrac{1}{24^3} + \dfrac{1}{120^3} + \cdots$.

▮ $a_n = (-1)^{n+1}/(n!)^3$. Use the ratio test $\quad \lim\limits_{n \to +\infty} \dfrac{a_{n+1}}{a_n} = \lim\limits_{n \to +\infty} \dfrac{1/[(n+1)!]^3}{1/(n!)^3} = \lim\limits_{n \to +\infty} \dfrac{1}{(n+1)^3} = 0$. Hence, the series is absolutely convergent.

37.103 Show by example that the sum of two divergent series can be convergent.

▮ One trivial example is $\Sigma\, 1/n + \Sigma\, (-1/n) = 0$. Another example is $\Sigma\, 1/n + \Sigma\, (1-n)/n^2 = \Sigma\, 1/n^2$. Of course, the sum of two divergent series of nonnegative terms must be divergent.

37.104 Show how to rearrange the terms of the conditionally convergent series $\quad 1 - \frac{1}{2} + \frac{1}{3} - \frac{1}{4} + \cdots \quad$ so as to obtain a series whose sum is 1.

▮ Use the first n_1 positive terms until the sum is >1. Then use the first n_2 negative terms until the sum becomes <1. Then repeat with more positive terms until the sum becomes >1, then more negative terms until the sum becomes <1, etc. Since the difference between the partial sums and 1 is less than the last term used, the new series $\quad 1 + \frac{1}{3} - \frac{1}{2} + \frac{1}{5} - \frac{1}{4} + \frac{1}{7} + \frac{1}{9} - \cdots \quad$ converges to 1. (Note that the series of positive terms $\quad 1 + \frac{1}{3} + \frac{1}{5} + \cdots \quad$ and the series of negative terms $\quad \frac{1}{2} + \frac{1}{4} + \frac{1}{6} + \cdots \quad$ are both divergent, so the described procedure always can be carried out.)

37.105 Test $\quad \sum\limits_{n=1}^{\infty} \dfrac{5^n}{n^{n+1}} \quad$ for convergence.

▮ Use the root test. $\lim\limits_{n \to +\infty} \sqrt[n]{a_n} = \lim\limits_{n \to +\infty} \dfrac{1}{\sqrt[n]{n}} \cdot \dfrac{5}{n} = \lim\limits_{n \to +\infty} \dfrac{1}{\sqrt[n]{n}} \cdot \lim\limits_{n \to +\infty} \dfrac{5}{n} = 1 \cdot 0 = 0$. (We know that $\lim\limits_{n \to +\infty} \sqrt[n]{n} = 1$ by Problem 36.15.) Hence, the series converges.

37.106 Show that the root test gives no information when $\quad \lim\limits_{n \to +\infty} \sqrt[n]{a_n} = 1$.

▮ Let $a_n = 1/n$. $\Sigma\, 1/n$ is divergent and $\quad \lim\limits_{n \to +\infty} \sqrt[n]{a_n} = \lim\limits_{n \to +\infty} 1/\sqrt[n]{n} = 1$. On the other hand, let $a_n = 1/n^2$. Then $\Sigma\, 1/n^2$ is convergent and $\quad \lim\limits_{n \to +\infty} \sqrt[n]{a_n} = \lim\limits_{n \to +\infty} 1/(\sqrt[n]{n})^2 = 1$.

37.107 Show that the ratio test gives no information when $\quad \lim |a_{n+1}/a_n| = 1$.

▮ Let $a_n = 1/n$. Then $\Sigma\, 1/n$ is divergent, but $\quad \lim\limits_{n \to +\infty} \left| \dfrac{a_{n+1}}{a_n} \right| = \lim\limits_{n \to +\infty} \dfrac{1/(n+1)}{1/n} = \lim\limits_{n \to +\infty} \dfrac{n}{n+1} =$ $\lim\limits_{n \to +\infty} \dfrac{1}{1+1/n} = 1$. On the other hand, let $a_n = 1/n^2$. Then $\Sigma\, 1/n^2$ converges, but $\quad \lim\limits_{n \to +\infty} \left| \dfrac{a_{n+1}}{a_n} \right| =$ $\lim \dfrac{1/(n+1)^2}{1/n^2} = \lim\limits_{n \to +\infty} \left(\dfrac{n}{n+1} \right)^2 = \lim\limits_{n \to +\infty} \left(\dfrac{1}{1+1/n} \right)^2 = 1$.

37.108 Determine whether $\quad \sum\limits_{n=1}^{\infty} \dfrac{(\ln n)^2}{n^2} \quad$ converges.

▮ Use the limit comparison test with $\Sigma\, 1/n^{3/2}$. $\lim\limits_{n \to +\infty} \dfrac{(\ln n)^2/n^2}{1/n^{3/2}} = \lim\limits_{n \to +\infty} \dfrac{(\ln n)^2}{n^{1/2}} = \lim\limits_{n \to +\infty} \dfrac{2 \ln n \cdot (1/n)}{\frac{1}{2}(1/\sqrt{n})} =$ $\lim\limits_{n \to +\infty} 4 \dfrac{\ln n}{n^{1/2}} = 4 \lim\limits_{n \to +\infty} \dfrac{1/n}{\frac{1}{2} n^{-1/2}} = 8 \lim\limits_{n \to +\infty} \dfrac{1}{n^{1/2}} = 0$. (We have used L'Hôpital's rule twice.) Since $\Sigma\, 1/n^{3/2}$ converges, so does the given series.

37.109 Determine whether $\quad \sum\limits_{n=1}^{\infty} \dfrac{(-3)^n(1+n^2)}{n!} \quad$ converges.

▮ Use the ratio test. $\lim \left| \dfrac{a_{n+1}}{a_n} \right| = \lim_{n \to +\infty} \dfrac{3^{n+1}[1+(n+1)^2]/(n+1)!}{3^n(1+n^2)/n!} = \lim_{n \to +\infty} \dfrac{3}{n+1} \cdot \dfrac{n^2+2n+2}{1+n^2} =$ $\lim_{n \to +\infty} \dfrac{3}{n+1} \cdot \dfrac{1+2/n+2/n^2}{1/n^2+1} = 0 \cdot 1 = 0$. Hence, the series is absolutely convergent.

37.110 Show that, in Fig. 37-2, the areas in the rectangles and above $y = 1/x$ add up to a number γ between $\frac{1}{2}$ and 1. (γ is called *Euler's constant*.)

▮ The area in question is less than the sum S of the indicated rectangles. $S = \frac{1}{2} + (\frac{1}{2} - \frac{1}{3}) + (\frac{1}{3} - \frac{1}{4}) + \cdots = 1$. So the area is finite and <1. On the other hand, the area is greater than the sum of the triangles (half the rectangles), which is $\frac{1}{2}$. Note that $\gamma = \lim_{n \to +\infty} \left(\sum_{k=1}^{n} \dfrac{1}{k} - \ln n \right)$. It is an unsolved problem as to whether γ is rational.

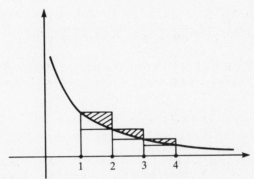

Fig. 37-2

37.111 If $\Sigma\, a_n$ is divergent and $\Sigma\, b_n$ is convergent, show that $\Sigma\, (a_n - b_n)$ is divergent.

▮ Assume $\Sigma\, (a_n - b_n)$ is convergent. Then, $\Sigma\, a_n = \Sigma\, b_n + \Sigma\, (a_n - b_n)$ is convergent, contrary to hypothesis.

37.112 Determine whether $\displaystyle\sum_{n=1}^{\infty} \left[\dfrac{1}{n} - \left(\dfrac{2}{3} \right)^n \right]$ converges.

▮ The given series is the difference of a divergent and a convergent series, and is, therefore, by Problem 37.111, divergent.

37.113 Find the values of x for which the series $1 + x + x^2 + \cdots$ converges, and express the sum as a function of x.

▮ This is a geometric series with ratio x. Therefore, it converges for $|x| < 1$. The sum is $1/(1-x)$. Thus, for $|x| < 1$, $1/(1-x) = 1 + x + x^2 + \cdots = \displaystyle\sum_{n=0}^{\infty} x^n$.

37.114 Find the values of x for which the series $x + x^3 + x^5 + \cdots$ converges, and express the sum as a function of x.

▮ This is a geometric series with ratio x^2. Hence, it converges for $|x^2| < 1$, that is, for $|x| < 1$. By the formula $a/(1-r)$ for the sum of a geometric series, the sum is $x/(1-x^2)$.

37.115 Find the values of x for which the series $1/x + 1/x^2 + 1/x^3 + \cdots$ converges and express the sum as a function of x.

▮ This is a geometric series with ratio $1/x$. It converges for $|1/x| < 1$, that is, for $|x| > 1$. The sum is $\dfrac{1/x}{1 - 1/x} = \dfrac{1}{x-1}$.

37.116 Find the values of x for which the series $\ln x + (\ln x)^2 + (\ln x)^3 + \cdots$ converges and express the sum as a function of x.

▮ This is a geometric series with ratio $\ln x$. It converges for $|\ln x| < 1$, $-1 < \ln x < 1$, $1/e < x < e$. The sum is $(\ln x)/(1 - \ln x)$.

CHAPTER 38
Power Series

In Problems 38.1–38.24, find the interval of convergence of the given power series. Use the ratio test, unless otherwise instructed.

38.1 $\sum x^n/n$.

■ $\lim\limits_{n \to +\infty} \dfrac{|x|^{n+1}/(n+1)}{|x|^n/n} = \lim\limits_{n \to +\infty} |x| \dfrac{n}{n+1} = \lim\limits_{n \to +\infty} |x| \dfrac{1}{1+1/n} = |x|$. Therefore, the series converges absolutely for $|x| < 1$, and diverges for $|x| > 1$. When $x = 1$, we have the divergent harmonic series $\sum 1/n$. When $x = -1$, the series is $\sum (-1)^n/n$, which converges by the alternating series test. Hence, the series converges for $-1 \le x < 1$.

38.2 $\sum x^n/n^2$.

■ $\lim\limits_{n \to +\infty} \dfrac{|x|^{n+1}/(n+1)^2}{|x|^n/n^2} = \lim\limits_{n \to +\infty} |x| \left(\dfrac{n}{n+1}\right)^2 = \lim\limits_{n \to +\infty} |x| \left(\dfrac{1}{1+1/n}\right)^2 = |x|$. Thus, the series converges absolutely for $|x| < 1$, and diverges for $|x| > 1$. When $x = 1$, we have the convergent p-series $\sum 1/n^2$. When $x = -1$, the series converges by the alternating series test. Hence, the power series converges for $-1 \le x \le 1$.

38.3 $\sum x^n/n!$.

■ $\lim\limits_{n \to +\infty} \dfrac{|x|^{n+1}/(n+1)!}{|x|^n/n!} = \lim\limits_{n \to +\infty} \dfrac{|x|}{n+1} = 0$. Therefore, the series converges for all x.

38.4 $\sum n! x^n$.

■ $\lim\limits_{n \to +\infty} \dfrac{(n+1)!|x|^{n+1}}{n!|x|^n} = \lim\limits_{n \to +\infty} |x|(n+1) = +\infty$ (except when $x = 0$). Thus, the series converges only for $x = 0$.

38.5 $\sum x^n/2^n$.

■ This is a geometric series with ratio $x/2$. Hence, we have convergence for $|x/2| < 1$, $|x| < 2$, and divergence for $|x| > 2$. When $x = 2$, we have $\sum 1$, which diverges. When $x = -2$, we have $\sum (-1)^n$, which is divergent. Hence, the power series converges for $-2 < x < 2$.

38.6 $\sum x^n/(n \cdot 2^n)$.

■ $\lim\limits_{n \to +\infty} \dfrac{|x|^{n+1}/(n+1)2^{n+1}}{|x|^n/n2^n} = \lim\limits_{n \to +\infty} \dfrac{|x|}{2} \dfrac{n}{n+1} = \lim\limits_{n \to +\infty} \dfrac{|x|}{2} \dfrac{1}{1+1/n} = \dfrac{|x|}{2}$. Thus, we have convergence for $|x| < 2$, and divergence for $|x| > 2$. When $x = 2$, we obtain the divergent harmonic series. When $x = -2$, we have the convergent alternating series $\sum (-1)^n/n$. Therefore, the power series converges for $-2 \le x < 2$.

38.7 $\sum nx^n$.

■ $\lim\limits_{n \to +\infty} \dfrac{(n+1)|x|^{n+1}}{n|x|^n} = \lim\limits_{n \to +\infty} |x|\left(1 + \dfrac{1}{n}\right) = |x|$. So we have convergence for $|x| < 1$, and divergence for $|x| > 1$. When $x = 1$, the divergent series $\sum n$ arises. When $x = -1$, we have the divergent series $\sum (-1)^n n$. Therefore, the series converges for $-1 < x < 1$.

38.8 $\sum 3^n x^n/n4^n$.

■ $\lim\limits_{n \to +\infty} \dfrac{3^{n+1}|x|^{n+1}/(n+1)4^{n+1}}{3^n|x|^n/n \cdot 4^n} = \lim\limits_{n \to +\infty} \dfrac{3}{4} \dfrac{n}{n+1} |x| = \lim\limits_{n \to +\infty} \dfrac{3}{4} \dfrac{1}{1+1/n} |x| = \dfrac{3}{4} |x|$. Thus, we have convergence for $|x| < \frac{4}{3}$, and divergence for $|x| > \frac{4}{3}$. For $x = \frac{4}{3}$, we obtain the divergent series $\sum 1/n$, and, for $x = -\frac{4}{3}$, we obtain the convergent alternating series $\sum (-1)^n/n$. Therefore, the power series converges for $-\frac{4}{3} \le x < \frac{4}{3}$.

38.9 $\Sigma (ax)^n$, $a > 0$.

▐ $\lim\limits_{n \to +\infty} \dfrac{|ax|^{n+1}}{|ax|^n} = a|x|$. So we have convergence for $|x| < 1/a$, and divergence for $|x| > 1/a$. When $x = 1/a$, we obtain the divergent series $\Sigma 1$, and, when $x = -1/a$, we obtain the divergent series $\Sigma (-1)^n$. Therefore, the power series converges for $-1/a < x < 1/a$.

38.10 $\Sigma n(x - 1)^n$.

▐ A translation of Problem 38.7 shows that the power series converges for $0 < x < 2$.

38.11 $\Sigma \dfrac{x^n}{(n^2 + 1)}$.

▐ $\lim\limits_{n \to +\infty} \dfrac{|x|^{n+1}/[(n+1)^2 + 1]}{|x|^n/(n^2 + 1)} = \lim\limits_{n \to +\infty} |x| \dfrac{n^2 + 1}{n^2 + 2n + 2} = \lim\limits_{n \to +\infty} |x| \dfrac{1 + 1/n^2}{1 + 2/n + 2/n^2} = |x|$. Thus, we have convergence for $|x| < 1$, and divergence for $|x| > 1$. When $x = 1$, we get the convergent series $\Sigma 1/(n^2 + 1)$ (by comparison with the convergent p-series $\Sigma 1/n^2$); when $x = -1$, we have the convergent alternating series $\Sigma (-1)^n/(n^2 + 1)$. Therefore, the power series converges for $-1 \le x \le 1$.

38.12 $\Sigma (x + 2)^n/\sqrt{n}$.

▐ $\lim\limits_{n \to +\infty} \dfrac{|x + 2|^{n+1}/\sqrt{n+1}}{|x + 2|^n/\sqrt{n}} = \lim\limits_{n \to +\infty} |x + 2| \sqrt{\dfrac{n}{n+1}} = \lim\limits_{n \to +\infty} |x + 2| \sqrt{\dfrac{1}{1 + 1/n}} = |x + 2|$. So we have convergence for $|x + 2| < 1$, $-1 < x + 2 < 1$, $-3 < x < -1$, and divergence for $x < -3$ or $x > -1$. For $x = -1$, we have the divergent series $\Sigma 1/\sqrt{n}$ (Problem 37.36), and, for $x = -3$, we have the convergent alternating series $\Sigma (-1)^n(1/\sqrt{n})$. Hence, the power series converges for $-3 \le x < -1$.

38.13 $\sum\limits_{n=0}^{\infty} (-1)^n \dfrac{x^{2n}}{(2n)!}$.

▐ $\lim\limits_{n \to +\infty} \dfrac{|x|^{2n+2}/(2n + 2)!}{|x|^{2n}/(2n)!} = \lim\limits_{n \to +\infty} \dfrac{|x|^2}{(2n + 1)(2n + 2)} = 0$. Thus, the power series converges for all x.

38.14 $\sum\limits_{n=0}^{\infty} (-1)^n \dfrac{x^{2n+1}}{(2n + 1)!}$.

▐ $\lim\limits_{n \to +\infty} \dfrac{|x|^{2n+3}/(2n + 3)!}{|x|^{2n+1}/(2n + 1)!} = \lim\limits_{n \to +\infty} \dfrac{|x|^2}{(2n + 2)(2n + 3)} = 0$. Hence, the power series converges for all x.

38.15 $\sum\limits_{n=1}^{\infty} \dfrac{x^n}{\ln (n + 1)}$.

▐ $$\lim\limits_{n \to +\infty} \dfrac{|x|^{n+1}/\ln (n + 2)}{|x|^n/\ln (n + 1)} = \lim\limits_{n \to +\infty} |x| \dfrac{\ln (n + 1)}{\ln (n + 2)} = |x|$$

$\left[\text{By L'Hôpital's rule } \lim\limits_{x \to +\infty} \dfrac{\ln (x + 1)}{\ln (x + 2)} = \lim\limits_{x \to +\infty} \dfrac{1/(x + 1)}{1/(x + 2)} = \lim\limits_{x \to +\infty} \dfrac{x + 2}{x + 1} = \lim\limits_{x \to +\infty} \dfrac{1 + 2/x}{1 + 1/x} = 1.\right]$ Hence, we have convergence for $|x| < 1$, and divergence for $|x| > 1$. For $x = 1$, $\Sigma 1/\ln (n + 1)$ is divergent (Problem 37.100). For $x = -1$, $\Sigma (-1)^n/\ln (n + 1)$ converges by the alternating series test. Therefore, the power series converges for $-1 \le x < 1$.

38.16 $\Sigma x^n/n(n + 1)$.

▐ $\lim\limits_{n \to +\infty} \dfrac{|x|^{n+1}/(n + 1)(n + 2)}{|x|^n/n(n + 1)} = \lim\limits_{n \to +\infty} |x| \dfrac{n}{n + 2} = \lim\limits_{n \to +\infty} |x| \dfrac{1}{1 + 2/n} = |x|$. Thus, we have convergence for $|x| < 1$ and divergence for $|x| > 1$. When $x = \pm 1$, the series is convergent (by Problem 37.10). Hence, the power series converges for $-1 \le x \le 1$.

38.17 $\sum\limits_{n=1}^{\infty} \dfrac{x^n}{n^n}$.

▐ $\lim\limits_{n \to +\infty} \dfrac{|x|^{n+1}/(n + 1)^{n+1}}{|x|^n/n^n} = \lim\limits_{n \to +\infty} |x| \cdot \dfrac{1}{(1 + 1/n)^n} \cdot \dfrac{1}{n + 1} = |x| \cdot \dfrac{1}{e} \cdot 0 = 0$. Hence, the series converges for all x.

38.18 $\Sigma \, x^n/n5^n$.

❚ $\lim\limits_{n \to +\infty} \dfrac{|x|^{n+1}/(n+1)5^{n+1}}{|x|^n/n5^n} = \lim\limits_{n \to +\infty} |x| \cdot \dfrac{1}{5} \cdot \dfrac{n}{n+1} = |x|/5$. Thus, the series converges for $|x| < 5$ and diverges for $|x| > 5$. For $x = 5$, we get the divergent series $\Sigma \, 1/n$, and, for $x = -5$, we get the convergent alternating series $\Sigma \, (-1)^n/n$. Hence, the power series converges for $-5 \le x < 5$.

38.19 $\Sigma \, x^{2n}/(n+1)(n+2)(n+3)$.

❚ $\lim\limits_{n \to +\infty} \dfrac{|x|^{2n+2}/(n+2)(n+3)(n+4)}{|x|^{2n}/(n+1)(n+2)(n+3)} = \lim\limits_{n \to +\infty} |x|^2 \dfrac{n+1}{n+4} = \lim\limits_{n \to +\infty} |x|^2 \dfrac{1+1/n}{1+4/n} = |x|^2$. Hence, we have convergence for $|x| < 1$ and divergence for $|x| > 1$. For $x = \pm 1$, we get absolute convergence by Problem 37.18. Hence, the series converges for $-1 \le x \le 1$.

38.20 $\displaystyle\sum_{n=2}^{\infty} \dfrac{x^n}{(\ln n)^n}$; use the root test for absolute convergence.

❚ $\lim\limits_{n \to +\infty} \dfrac{|x|}{\ln n} = 0$; the series converges (absolutely) for all x.

38.21 $\Sigma \, x^n/(1+n^3)$.

❚ $\lim\limits_{n \to +\infty} \dfrac{|x|^{n+1}/[1+(n+1)^3]}{|x|^n/(1+n^3)} = \lim |x| \dfrac{1+n^3}{1+(n+1)^3} = \lim\limits_{n \to +\infty} |x| \dfrac{1/n^3+1}{1/n^3+(1+1/n)^3} = |x|$. Hence, the series converges for $|x| < 1$, and diverges for $|x| > 1$. For $x = \pm 1$, the series is absolutely convergent by limit comparison with the convergent p-series $\Sigma \, 1/n^3$. Therefore, the series converges for $-1 \le x \le 1$.

38.22 $\Sigma \, (x+3)^n/n$.

❚ $\lim\limits_{n \to +\infty} \dfrac{|x+3|^{n+1}/(n+1)}{|x+3|^n/n} = \lim\limits_{n \to +\infty} |x+3| \dfrac{n}{n+1} = |x+3|$. Thus, the series converges for $|x+3| < 1$, $-1 < x+3 < 1$, $-4 < x < -2$ and diverges for $x < -4$ or $x > -2$. For $x = -2$, we get the divergent series $\Sigma \, 1/n$. For $x = -4$, we have the convergent alternating series $\Sigma \, (-1)^n/n$. Hence, the power series converges for $-4 \le x < -2$.

38.23 $\displaystyle\sum_{n=1}^{\infty} \dfrac{x^n}{n^n}$; use the root test for absolute convergence.

❚ $\lim\limits_{n \to +\infty} \dfrac{|x|}{n} = 0$; the series converges (absolutely) for all x.

38.24 $\displaystyle\sum_{n=2}^{\infty} \dfrac{x^n}{n(\ln n)^2}$.

❚ $\qquad \lim\limits_{n \to +\infty} \dfrac{|x|^{n+1}/(n+1)[\ln (n+1)]^2}{|x|^n/n(\ln n)^2} = \lim\limits_{n \to +\infty} |x| \dfrac{n}{n+1} \left[\dfrac{\ln n}{\ln (n+1)} \right]^2 = |x|$

$\left[\text{By L'Hôpital's rule,} \ \lim\limits_{n \to +\infty} \dfrac{\ln n}{\ln (n+1)} = 1. \right]$ Hence, we have convergence for $|x| < 1$ and divergence for $|x| > 1$. For $x = \pm 1$, we have absolute convergence by Problem 37.50. Hence, the power series converges for $-1 \le x \le 1$.

38.25 Find the radius of convergence of the power series $\displaystyle\sum_{n=1}^{\infty} \dfrac{(n!)^2}{(2n)!} x^n$.

❚ $\lim\limits_{n \to +\infty} \dfrac{[(n+1)!]^2 |x|^{n+1}/(2n+2)!}{(n!)^2 |x|^n/(2n)!} = \lim\limits_{n \to +\infty} |x| \dfrac{(n+1)^2}{(2n+1)(2n+2)} = \dfrac{|x|}{4}$. Therefore, the series converges for $|x| < 4$ and diverges for $|x| > 4$. Hence, the radius of convergence is 4.

38.26 Prove that, if a power series $\Sigma \, a_n x^n$ converges for $x = b$, then it converges absolutely for all x such that $|x| < |b|$.

❚ Since $\Sigma \, a_n b^n$ converges, $\lim\limits_{n \to +\infty} |a_n b^n| = 0$. Since a convergent sequence is bounded, there exists an M such that $|a_n b^n| \le M$ for all n. Let $|x/b| = r < 1$. Then $|a_n x^n| = |a_n b^n| \cdot |x^n/b^n| \le Mr^n$. Therefore, by comparison with the convergent geometric series $\Sigma \, Mr^n$, $\Sigma \, |a_n x^n|$ is convergent.

38.27 Prove that, if the radius of convergence of $\Sigma\, a_n x^n$ is R, then the radius of convergence of $\Sigma\, a_n x^{2n}$ is $\sqrt{R}$.

▌ Assume $|u| < \sqrt{R}$. Then $u^2 < R$. Hence, $\Sigma\, a_n (u^2)^n$ converges, and, therefore, $\Sigma\, a_n u^{2n}$ converges. Now, assume $|u| > \sqrt{R}$. Then $u^2 > R$, and, therefore, $\Sigma\, a_n (u^2)^n$ diverges. Thus, $\Sigma\, a_n u^{2n}$ diverges.

38.28 Find the radius of convergence of $\displaystyle\sum_{n=1}^{\infty} \frac{(n!)^2}{(2n)!} x^{2n}$.

▌ By Problem 38.25, the radius of convergence of $\displaystyle\sum_{n=1}^{\infty} \frac{(n!)^2}{(2n)!} x^n$ is 4. Hence, by Problem 38.27, the radius of convergence of $\displaystyle\sum_{n=1}^{\infty} \frac{(n!)^2}{(2n)!} x^{2n}$ is $\sqrt{4} = 2$.

38.29 If $\displaystyle\lim_{n \to +\infty} \sqrt[n]{|a_n|} = L > 0$, show that the radius of convergence of $\Sigma\, a_n x^n$ is $1/L$.

▌ Assume $|x| < 1/L$. Then $L < 1/|x|$. Choose r so that $L < r < 1/|x|$. Then $|rx| < 1$. Since $\displaystyle\lim_{n \to +\infty} \sqrt[n]{|a_n|} = L$, there exists an integer k such that, if $n \geq k$, then $\sqrt[n]{|a_n|} < r$, and, therefore, $|a_n| < r^n$. Hence, for $n \geq k$, $|a_n x^n| < r^n |x|^n = |rx|^n$. Thus, eventually, $\Sigma\, |a_n x^n|$ is term by term less than the convergent geometric series $\Sigma\, |rx|^n$ and is convergent by the comparison test. Now, on the other, hand, assume $|x| > 1/L$. Then $L > 1/|x|$. Choose r so that $L > r > 1/|x|$. Then $|rx| > 1$. Since $\displaystyle\lim_{n \to +\infty} \sqrt[n]{|a_n|} = L$, there exists an integer k such that, if $n \geq k$, then $\sqrt[n]{|a_n|} > r$, and, therefore, $|a_n| > r^n$. Hence, for $n \geq k$, $|a_n x^n| > r^n |x|^n = |rx|^n > 1$. Thus, we cannot have $\displaystyle\lim_{n \to +\infty} a_n x^n = 0$, and, therefore, $\Sigma\, a_n x^n$ cannot converge. (The result also holds when $L = 0$. Then the series converges for all x.)

38.30 Find the radius of convergence of $\displaystyle\sum_{n=1}^{\infty} \left(1 + \frac{1}{n}\right)^{n^2} x^n$.

▌ $\displaystyle\lim_{n \to +\infty} \sqrt[n]{|a_n|} = \lim_{n \to +\infty} \left(1 + \frac{1}{n}\right)^n = e$. Therefore, by Problem 38.29, the radius of convergence is $1/e$.

38.31 Find the radius of convergence of the binomial series $\displaystyle 1 + \frac{m}{1} x + \frac{m(m-1)}{1\cdot2} x^2 + \frac{m(m-1)(m-2)}{1\cdot2\cdot3} x^3 + \cdots$.

▌ Use the ratio test.

$$\lim_{n \to +\infty} \frac{|m(m-1)\cdots(m-n)|\,|x|^{n+1}/(n+1)!}{|m(m-1)\cdots(m-n+1)|\,|x|^n/n!} = \lim_{n \to +\infty} \frac{|x|}{n+1} \cdot |m-n| = \lim_{n \to +\infty} |x|\, \frac{|m/n - 1|}{1 + 1/n} = |x|$$

Hence, the radius of convergence is 1.

38.32 If $\Sigma\, a_n x^n$ has a radius of convergence r_1 and if $\Sigma\, b_n x^n$ has a radius of convergence $r_2 > r_1$, what is the radius of convergence of the sum $\Sigma\, (a_n + b_n) x^n$?

▌ For $|x| < r_1$, both $\Sigma\, a_n x^n$ and $\Sigma\, b_n x^n$ are convergent, and, therefore, so is $\Sigma\, (a_n + b_n) x^n$. Now, take x so that $r_1 < |x| < r_2$. Then $\Sigma\, a_n x^n$ diverges and $\Sigma\, b_n x^n$ converges. Hence, $\Sigma\, (a_n + b_n) x^n$ diverges (by Problem 37.111). Thus, the radius of convergence of $\Sigma\, (a_n + b_n) x^n$ is r_1.

38.33 Let k be a fixed positive integer. Find the radius of convergence of $\displaystyle\sum \frac{(n!)^k}{(kn)!} x^n$.

▌ Use the ratio test. $\displaystyle\lim_{n \to +\infty} \frac{[(n+1)!]^k |x|^{n+1}/(kn+k)!}{(n!)^k |x|^n/(kn)!} = \lim_{n \to +\infty} |x|\, \frac{(n+1)^k}{(kn+1)(kn+2)\cdots(kn+k)} =$
$\displaystyle\lim_{n \to +\infty} |x|\, \frac{(1+1/n)^k}{(k+1/n)(k+2/n)\cdots(k+k/n)} = \frac{|x|}{k^k}$. Hence, the radius of convergence is k^k. (Problem 38.25 is a special case of this result.)

38.34 Show that $\displaystyle \frac{1}{1+x} = 1 - x + x^2 - \cdots = \sum_{n=0}^{\infty} (-1)^n x^n$ for $|x| < 1$.

▌ Substitute $-x$ for x in Problem 37.113.

38.35 Show that $\displaystyle \frac{1}{1+x^2} = 1 - x^2 + x^4 - \cdots = \sum_{n=0}^{\infty} (-1)^n x^{2n}$ for $|x| < 1$.

▌ Substitute x^2 for x in the series of Problem 38.34.

38.36 Show that $\tan^{-1} x = x - \dfrac{x^3}{3} + \dfrac{x^5}{5} - \cdots = \displaystyle\sum_{n=0}^{\infty} \dfrac{(-1)^n x^{2n+1}}{2n+1}$ for $|x| < 1$.

▮ Integrate the power series of Problem 38.35 term by term, and note that $\tan^{-1} 0 = 0$.

38.37 Find a power series representation for $\dfrac{1}{(1-x)^2}$.

▮ *Method 1.* By Problem 37.113, for $|x| < 1$, $\dfrac{1}{1-x} = 1 + x + x^2 + \cdots = \displaystyle\sum_{n=0}^{\infty} x^n$. Differentiate this series

term by term. Then, for $|x| < 1$, $\dfrac{1}{(1-x)^2} = 1 + 2x + 3x^2 + \cdots = \displaystyle\sum_{n=1}^{\infty} nx^{n-1} = \displaystyle\sum_{n=0}^{\infty} (n+1)x^n$.

Method 2. $\dfrac{1}{1-x} \cdot \dfrac{1}{1-x} = \left(\displaystyle\sum_{n=0}^{\infty} x^n\right)\left(\displaystyle\sum_{m=0}^{\infty} x^m\right) = \displaystyle\sum_{k=0}^{\infty} \left(\displaystyle\sum_{n+m=k} 1\right)x^k = \displaystyle\sum_{k=0}^{\infty} (k+1)x^k$.

38.38 Find a power series representation for $\ln(1+x)$ for $|x| < 1$.

▮ By Problem 38.34, for $|x| < 1$, $\dfrac{1}{1+x} = 1 - x + x^2 - \cdots = \displaystyle\sum_{n=0}^{\infty} (-1)^n x^n$. Integrate term by term:

$\ln(1+x) = x - \dfrac{x^2}{2} + \dfrac{x^3}{3} - \cdots = \displaystyle\sum_{n=0}^{\infty} \dfrac{(-1)^n}{n+1} x^{n+1} = \displaystyle\sum_{n=1}^{\infty} \dfrac{(-1)^{n+1}}{n} x^n$.

38.39 Show that $e^x = \displaystyle\sum_{n=0}^{\infty} \dfrac{x^n}{n!}$ for all x.

▮ Let $f(x) = \displaystyle\sum_{n=0}^{\infty} \dfrac{x^n}{n!} = 1 + x + \dfrac{x^2}{2!} + \dfrac{x^3}{3!} + \cdots$. By Problem 38.3, $f(x)$ is defined for all x. Differentiate term

by term: $f'(x) = \displaystyle\sum_{n=1}^{\infty} \dfrac{x^{n-1}}{(n-1)!} = \displaystyle\sum_{n=0}^{\infty} \dfrac{x^n}{n!} = f(x)$. Moreover, $f(0) = 1$. Hence, by Problem 24.72, $f(x) = e^x$.

38.40 Find a power series representation for e^{-x}.

▮ By Problem 38.39, $e^x = \displaystyle\sum_{n=0}^{\infty} \dfrac{x^n}{n!}$. Substitute $-x$ for x, obtaining $e^{-x} = \displaystyle\sum_{n=0}^{\infty} \dfrac{(-1)^n}{n!} x^n = 1 - x + \dfrac{x^2}{2!} - \dfrac{x^3}{3!} +$

38.41 Find a power series representation for $e^{-x^2/2}$.

▮ By Problem 38.40, $e^{-x} = \displaystyle\sum_{n=0}^{\infty} \dfrac{(-1)^n}{n!} x^n$. Substitute $\dfrac{x^2}{2}$ for x. Then $e^{-x^2/2} = \displaystyle\sum_{n=0}^{\infty} \dfrac{(-1)^n x^{2n}}{2^n \cdot n!}$.

38.42 Approximate $1/e$ correctly to two decimal places.

▮ By Problem 38.40, $e^{-x} = 1 - x + x^2/2! - x^3/3! + \cdots$. Let $x = 1$. Then $1/e = 1 - 1 + 1/2! - 1/3! + \cdots$.
Let us use the alternating series theorem here. We must find the least n for which $1/n! < 0.005 = \frac{1}{200}$,
$200 < n!$, $n \geq 6$. So we can use $1 - 1 + \frac{1}{2} - \frac{1}{6} + \frac{1}{24} - \frac{1}{120} = \frac{44}{120} = 0.3666 \cdots$. So $1/e \approx 0.37$, correct to two decimal places.

38.43 Find a power series representation for $\cosh x$.

▮ Using the series found in Problems 38.39 and 38.40, we have $\cosh x = \dfrac{e^x + e^{-x}}{2} = \displaystyle\sum_{n=0}^{\infty} \dfrac{1 + (-1)^n}{2} \dfrac{x^n}{n!} =$
$\displaystyle\sum_{n=0}^{\infty} \dfrac{x^{2n}}{(2n)!}$.

38.44 Find a power series representation for $\sinh x$.

▮ Since $D_x(\cosh x) = \sinh x$, we can differentiate the power series of Problem 38.43 to get $\sinh x =$
$\displaystyle\sum_{n=1}^{\infty} \dfrac{x^{2n-1}}{(2n-1)!} = \displaystyle\sum_{n=0}^{\infty} \dfrac{x^{2n+1}}{(2n+1)!}$.

38.45 Find a power series representation for the normal distribution function $\int_0^x e^{-t^2/2} \, dt$.

▮ By Problem 38.41, $e^{-t^2/2} = \displaystyle\sum_{n=0}^{\infty} \dfrac{(-1)^n t^{2n}}{n!2^n}$. Integrate: $\int_0^x e^{-t^2/2} \, dt = \displaystyle\sum_{n=0}^{\infty} \dfrac{(-1)^n}{n!2^n} \int_0^x t^{2n} \, dt = \displaystyle\sum_{n=0}^{\infty} \dfrac{(-1)^n}{n!2^n} \dfrac{x^{2n+1}}{2n+1}$.

38.46 Approximate $\int_0^1 e^{-t^2/2}\, dt$ correctly to three decimal places.

▮ By Problem 38.45, $\int_0^1 e^{-t^2/2}\, dt = \sum_{n=0}^{\infty} \dfrac{(-1)^n}{n!\, 2^n (2n+1)}$. This is an alternating series. Hence, we must find the least n such that $1/n!\,2^n(2n+1) < 0.0005 = \frac{1}{2000}$, $2000 < n!\,2^n(2n+1)$, $n \geq 4$. So we may use $1 - \frac{1}{6} + \frac{1}{40} - \frac{1}{336} = \frac{1437}{1680} \approx 0.855$.

38.47 For a fixed integer $k > 1$, evaluate $\sum_{n=1}^{\infty} \dfrac{n}{k^n} = \dfrac{1}{k} + \dfrac{2}{k^2} + \dfrac{3}{k^3} + \cdots$.

▮ In Problem 38.37, we found that, for $|x| < 1$, $\dfrac{1}{(1-x)^2} = 1 + 2x + 3x^2 + \cdots = \sum_{n=1}^{\infty} n x^{n-1}$. Hence, $\dfrac{x}{(1-x)^2} = x + 2x^2 + 3x^3 + \cdots = \sum_{n=1}^{\infty} n x^n$ for $|x| < 1$. Let $x = \dfrac{1}{k}$. Then $\dfrac{1/k}{(1-1/k)^2} = \dfrac{1}{k} + \dfrac{2}{k^2} + \dfrac{3}{k^3} + \cdots$, and the desired value is $\dfrac{k}{(k-1)^2}$.

38.48 Evaluate $1 \cdot \frac{1}{2} + 2 \cdot \frac{1}{4} + 3 \cdot \frac{1}{8} + 4 \cdot \frac{1}{16} + \cdots$.

▮ In Problem 38.47, let $k = 2$. Then $\sum_{n=1}^{\infty} \dfrac{n}{2^n} = \dfrac{2}{(2-1)^2} = 2$.

38.49 Use power series to solve the differential equation $y' = -xy$ under the boundary condition that $y = 1$ when $x = 0$.

▮ Let $y = f(x) = \sum_{n=0}^{\infty} a_n x^n$. Differentiate: $y' = \sum_{n=1}^{\infty} n a_n x^{n-1} = \sum_{n=0}^{\infty} (n+1) a_{n+1} x^n$. So $\sum_{n=0}^{\infty} (n+1) a_{n+1} x^n = -xy = -x \sum_{n=0}^{\infty} a_n x^n = \sum_{n=0}^{\infty} (-a_n) x^{n+1} = \sum_{n=1}^{\infty} (-a_{n-1}) x^n$. Comparing coefficients, we get $a_1 = 0$, and $(n+1) a_{n+1} = -a_{n-1}$. Since $y = 1$ when $x = 0$, we know that $a_0 = 1$. Now, $a_3 = a_5 = a_7 = \cdots = 0$. Also, $2a_2 = -a_0 = -1$, $a_2 = -\dfrac{1}{2}$. Then, $4a_4 = -a_2$, $a_4 = \dfrac{1}{2 \cdot 4}$. Further, $6a_6 = -a_4$, $a_6 = -\dfrac{1}{2 \cdot 4 \cdot 6}$. Similarly, $8a_8 = -a_6$, $a_8 = \dfrac{1}{2 \cdot 4 \cdot 6 \cdot 8}$, and, in general, $a_{2n} = \dfrac{(-1)^n}{2 \cdot 4 \cdot 6 \cdot \cdots \cdot (2n)} = \dfrac{(-1)^n}{2^n \cdot n!}$. So $f(x) = \sum_{n=0}^{\infty} (-1)^n \dfrac{x^{2n}}{2^n \cdot n!}$. By Problem 38.41, $f(x) = e^{-x^2/2}$.

38.50 Find a power series representation for $\ln(1-x)$.

▮ Substitute $-x$ for x in the series of Problem 38.38: $\ln(1-x) = \sum_{n=1}^{\infty} (-1)^{n+1} \dfrac{(-x)^n}{n} = \sum_{n=1}^{\infty} (-1)^{2n+1} \dfrac{x^n}{n} = -\sum_{n=1}^{\infty} \dfrac{x^n}{n}$, for $|x| < 1$.

38.51 Find a power series representation for $\ln \dfrac{1+x}{1-x}$.

▮ By Problems 38.38 and 38.50, for $|x| < 1$, $\ln \dfrac{1+x}{1-x} = \ln(1+x) - \ln(1-x) = \sum_{n=1}^{\infty} (-1)^{n+1} \dfrac{x^n}{n} + \sum_{n=1}^{\infty} \dfrac{x^n}{n} = \sum_{n=1}^{\infty} [(-1)^{n+1} + 1] \dfrac{x^n}{n} = 2 \sum_{k=0}^{\infty} \dfrac{x^{2k+1}}{2k+1}$.

38.52 Use the power series for $\ln \dfrac{1+x}{1-x}$ to approximate $\ln 2$.

▮ By Problem 38.51, $\ln \dfrac{1+x}{1-x} = 2 \sum_{k=0}^{\infty} \dfrac{x^{2k+1}}{2k+1}$. When $2 = \dfrac{1+x}{1-x}$, $x = \dfrac{1}{3}$. So $\ln 2 = 2 \sum_{k=0}^{\infty} \dfrac{1}{2k+1} \left(\dfrac{1}{3}\right)^{2k+1} = 2[\frac{1}{3} + \frac{1}{3}(\frac{1}{3})^3 + \frac{1}{5}(\frac{1}{3})^5 + \cdots]$. Using the first three terms, we get $\frac{842}{1215} \approx 0.6930$. (The correct value to four decimal places is 0.6931.)

38.53 Use the power series for $\ln \dfrac{1+x}{1-x}$, to approximate $\ln 3$.

▮ When $3 = \dfrac{1+x}{1-x}$, $x = \dfrac{1}{2}$. As in Problem 38.52, $\ln 3 = 2 \sum_{k=0}^{\infty} \dfrac{1}{2k+1} \left(\dfrac{1}{2}\right)^{2k+1} = 2[\frac{1}{2} + \frac{1}{3}(\frac{1}{2})^3 + \frac{1}{5}(\frac{1}{2})^5 + \cdots] \approx 2(\frac{1}{2} + \frac{1}{24} + \frac{1}{160}) = \frac{263}{240} \approx 1.0958$. (The correct value to four places is 1.0986.)

38.54 Approximate $\int_0^{1/3} \dfrac{dx}{1+x^3}$ to three decimal places.

▮ For $|x| < 1$, $\dfrac{1}{1+x} = 1 - x + x^2 - \cdots = \sum\limits_{n=0}^{\infty} (-1)^n x^n$. Hence, $\dfrac{1}{1+x^3} = \sum\limits_{n=0}^{\infty} (-1)^n x^{3n}$. Integrate term-wise: $\int_0^{1/3} \dfrac{dx}{1+x^3} = \sum\limits_{n=0}^{\infty} \dfrac{(-1)^n}{3n+1} \left(\dfrac{1}{3}\right)^{3n+1}$. Since this is an alternating series, we look for the least n such that $\dfrac{1}{3n+1} \cdot \dfrac{1}{3^{3n+1}} < 0.0005 = \dfrac{1}{2000}$, $2000 < (3n+1)3^{3n+1}$. We get $n=2$. Thus, we may use $\frac{1}{3} - \frac{1}{324} = \frac{107}{324} \approx 0.330$.

38.55 Approximate $\tan^{-1} \frac{1}{2}$ to two decimal places.

▮ By Problem 38.36, $\tan^{-1} x = x - \dfrac{x^3}{3} + \dfrac{x^3}{5} - \cdots = \sum\limits_{n=0}^{\infty} (-1)^n \dfrac{x^{2n+1}}{2n+1}$, for $|x| < 1$. Therefore, $\tan^{-1} \frac{1}{2} = \sum\limits_{n=0}^{\infty} \dfrac{(-1)^n}{2n+1} \dfrac{1}{2^{2n+1}}$. By the alternating series theorem, we seek the least n for which $\dfrac{1}{(2n+1)\cdot 2^{2n+1}} < 0.005 = \dfrac{1}{200}$, $200 < (2n+1)2^{2n+1}$. We obtain $n=3$. Thus, we can use $\frac{1}{2} - \frac{1}{24} + \frac{1}{160} = \frac{223}{480} \approx 0.46$.

38.56 Use power series to solve the differential equation $y'' = 4y$ with the boundary conditions $y=0$, $y'=1$ when $x=0$.

▮ Let $y = \sum\limits_{n=0}^{\infty} a_n x^n$. Since $y=0$ when $x=0$, $a_0 = 0$. Differentiate: $y' = \sum\limits_{n=1}^{\infty} na_n x^{n-1}$. Since $y'=1$ when $x=0$, $a_1 = 1$. Differentiate again: $y'' = \sum\limits_{n=2}^{\infty} n(n-1)a_n x^{n-2} = \sum\limits_{n=0}^{\infty} (n+2)(n+1)a_{n+2} x^n$. So $\sum\limits_{n=0}^{\infty} (n+2)(n+1)a_{n+2}x^n = 4y = 4 \sum\limits_{n=0}^{\infty} a_n x^n$. Therefore, $(n+2)(n+1)a_{n+2} = 4a_n$. Since $a_0 = 0$, we get $a_2 = a_4 = a_6 = \cdots = 0$. For odd subscripts, $a_3 = \dfrac{4}{3\cdot 2} a_1 = \dfrac{4}{3\cdot 2} = \dfrac{4}{3!}$. Then, $a_5 = \dfrac{4}{5\cdot 4} a_3 = \dfrac{4^2}{5!}$. In general, $a_{2k+1} = \dfrac{4^k}{(2k+1)!} = \dfrac{2^{2k}}{(2k+1)!}$. Hence, $y = \sum\limits_{k=0}^{\infty} \dfrac{2^{2k} x^{2k+1}}{(2k+1)!} = \dfrac{1}{2} \sum\limits_{k=0}^{\infty} \dfrac{(2x)^{2k+1}}{(2k+1)!}$. By Problem 38.44, $\sinh u = \sum\limits_{k=0}^{\infty} \dfrac{u^{2k+1}}{(2k+1)!}$. Hence, $y = \frac{1}{2} \sinh 2x$.

38.57 Show directly that, if $y'' = -y$, and $y'=1$ and $y=0$ when $x=0$, then $y = \sin x$.

▮ Let $z = dy/dx$. Then $y'' = \dfrac{dz}{dx} = \dfrac{dz}{dy} \cdot \dfrac{dy}{dx} = \dfrac{dz}{dy} z$. Hence, $\dfrac{dz}{dy} z = -y$, $\int z \, dz = -\int y \, dy$, $\dfrac{1}{2} z^2 = -\dfrac{1}{2} y^2 + C$, $z^2 = -y^2 + K$. Since $z=1$ and $y=0$ when $x=0$, $K=1$. Thus, $z^2 = 1 - y^2$, $\dfrac{dy}{dx} = \pm\sqrt{1-y^2}$, $\int \dfrac{dy}{\sqrt{1-y^2}} = \pm \int dx$, $\sin^{-1} y = \pm x + C_1$. Since $y=0$ when $x=0$, $C_1 = 0$. So $\sin^{-1} y = \pm x$, $y = \sin(\pm x) = \pm \sin x$. Then $y = \sin x$. (If $y = -\sin x$, then $y' = -\cos x$, and $y' = -1$ when $x=0$.)

38.58 Show that $\sin x = \sum\limits_{n=0}^{\infty} (-1)^n \dfrac{x^{2n+1}}{(2n+1)!}$.

▮ Let $y = \sum\limits_{n=0}^{\infty} (-1)^n \dfrac{x^{2n+1}}{(2n+1)!}$. When $x=0$, $y=0$. By differentiation, $y' = \sum\limits_{n=0}^{\infty} (-1)^n \dfrac{x^{2n}}{(2n)!}$. Hence, $y'=1$ when $x=0$. Further, $y'' = \sum\limits_{n=1}^{\infty} (-1)^n \dfrac{x^{2n-1}}{(2n-1)!} = \sum\limits_{n=0}^{\infty} (-1)^{n+1} \dfrac{x^{2n+1}}{(2n+1)!} = -\sum\limits_{n=0}^{\infty} (-1)^n \dfrac{x^{2n+1}}{(2n+1)!} = -y$. Hence, by Problem 38.57, $y = \sin x$.

38.59 Show that $\cos x = \sum\limits_{n=0}^{\infty} (-1)^n \dfrac{x^{2n}}{(2n)!}$.

▮ By Problem 38.58, $\sin x = \sum\limits_{n=0}^{\infty} (-1)^n \dfrac{x^{2n+1}}{(2n+1)!}$. Differentiate: $\cos x = \sum\limits_{n=0}^{\infty} (-1)^n \dfrac{x^{2n}}{(2n)!}$.

38.60 Show that $\ln 2 = 1 - \dfrac{1}{2} + \dfrac{1}{3} - \dfrac{1}{4} + \cdots = \displaystyle\sum_{n=1}^{\infty} \dfrac{(-1)^{n+1}}{n}$.

▌ By Problem 38.38, $\ln(r + x) = \displaystyle\sum_{n=1}^{\infty} (-1)^{n+1} \dfrac{x^n}{n}$ for $|x| < 1$. By the alternating series theorem, the series converges when $x = 1$. By Abel's theorem, $\ln 2 = \displaystyle\sum_{n=1}^{\infty} \dfrac{(-1)^{n+1}}{n}$. $\left[\text{Abel's theorem reads: Let } f(x) = \displaystyle\sum_{n=0}^{\infty} a_n x^n \text{ for } -r < x < r. \text{ If the series converges for } x = r, \text{ then } \lim f(x) \text{ exists and is equal to } a_n r^n.\right]$

38.61 Show that $\dfrac{\pi}{4} = 1 - \dfrac{1}{3} + \dfrac{1}{5} - \dfrac{1}{7} + \cdots = \displaystyle\sum_{n=0}^{\infty} \dfrac{(-1)^n}{2n + 1}$.

▌ By Problem 38.36, $\tan^{-1} x = x - \dfrac{x^3}{3} + \dfrac{x^5}{5} - \dfrac{x^7}{7} + \cdots = \displaystyle\sum_{n=0}^{\infty} (-1)^n \dfrac{x^{2n+1}}{2n + 1}$, for $|x| < 1$. The series converges for $x = 1$, by the alternating series test. Hence, by Abel's theorem (Problem 38.60), $\dfrac{\pi}{4} = \tan^{-1} 1 = 1 - \dfrac{1}{3} + \dfrac{1}{5} - \dfrac{1}{7} + \cdots = \displaystyle\sum_{n=0}^{\infty} \dfrac{(-1)^n}{2n + 1}$.

38.62 Show that the converse of Abel's theorem fails; that is, show that if $f(x) = \displaystyle\sum_{n=0}^{\infty} a_n x^n$ for $|x| < r$, the series has radius of convergence r, and $\displaystyle\lim_{x \to r} f(x) = b$, then $\displaystyle\sum_{n=0}^{\infty} a_n r^n$ does not necessarily converge.

▌ Consider $f(x) = \dfrac{1}{1 + x} = \displaystyle\sum_{n=0}^{\infty} (-1)^n x^n$, with radius of convergence 1. $\displaystyle\lim_{x \to 1^-} f(x) = \dfrac{1}{2}$, but $\displaystyle\sum_{n=0}^{\infty} (-1)^n$ is not convergent.

38.63 Find a power series for $\sin^2 x$.

▌ $\sin^2 x = \dfrac{1 - \cos 2x}{2}$. By Problem 38.59, $\cos x = \displaystyle\sum_{n=0}^{\infty} (-1)^n \dfrac{x^{2n}}{(2n)!}$. Hence, $\cos 2x = \displaystyle\sum_{n=0}^{\infty} (-1)^n \dfrac{2^{2n} x^{2n}}{(2n)!}$, $-\cos 2x = \displaystyle\sum_{n=0}^{\infty} (-1)^{n+1} \dfrac{2^{2n} x^{2n}}{(2n)!}$. Adding 1 eliminates the constant term -1, yielding $1 - \cos 2x = \displaystyle\sum_{n=1}^{\infty} (-1)^{n+1} \dfrac{2^{2n} x^{2n}}{(2n)!}$. So $\sin^2 x = \dfrac{1 - \cos 2x}{2} = \displaystyle\sum_{n=1}^{\infty} (-1)^{n+1} \dfrac{2^{2n-1} x^{2n}}{(2n)!}$.

38.64 Find a power series for $\dfrac{1}{3 - x}$.

▌ $\dfrac{1}{3 - x} = \dfrac{\frac{1}{3}}{1 - x/3}$ is the sum of the geometric series with first term $a = \dfrac{1}{3}$ and ratio $\dfrac{x}{3}$: $\dfrac{1}{3} + \dfrac{1}{3} \dfrac{x}{3} + \dfrac{1}{3} \left(\dfrac{x}{3}\right)^2 + \cdots = \displaystyle\sum_{n=0}^{\infty} \dfrac{x^n}{3^{n+1}}$ for $|x| < 3$.

38.65 Find a power series for $\dfrac{4x}{1 + 2x - 3x^2}$.

▌ $1 + 2x - 3x^2 = (1 - x)(1 + 3x)$, and

$$\dfrac{4x}{1 + 2x - 3x^2} = \dfrac{1}{1 - x} - \dfrac{1}{1 + 3x} = \displaystyle\sum_{n=0}^{\infty} x^n - \displaystyle\sum_{n=0}^{\infty} (-3x)^n = \displaystyle\sum_{n=0}^{\infty} [1 - (-3)^n] x^n \quad \text{for } |x| < \tfrac{1}{3}$$

38.66 Find power series solutions of the differential equation $xy'' + y' - y = 0$.

▌ Let $y = \displaystyle\sum_{n=0}^{\infty} a_n x^n$. Then $y' = \displaystyle\sum_{n=1}^{\infty} n a_n x^{n-1} = \displaystyle\sum_{n=0}^{\infty} (n + 1) a_{n+1} x^n$. Further, $y'' = \displaystyle\sum_{n=2}^{\infty} n(n - 1) a_n x^{n-2}$, and $xy'' = \displaystyle\sum_{n=2}^{\infty} n(n - 1) a_n x^{n-1} = \displaystyle\sum_{n=1}^{\infty} (n + 1)n a_{n+1} x^n$. Now, $0 = xy'' + y' - y = (a_1 - a_0) + \displaystyle\sum_{n=1}^{\infty} [(n + 1)n a_{n+1} + (n + 1) a_{n+1} - a_n] x^n$. Therefore, $a_1 = a_0$ and $(n + 1)^2 a_{n+1} = a_n$. Hence, $a_2 = \dfrac{a_1}{2^2} = \dfrac{a_0}{2^2}$, $a_3 = \dfrac{a_2}{3^2} = \dfrac{a_0}{2^2 \cdot 3^2}$, $a_4 = \dfrac{a_3}{4^2} = \dfrac{a_0}{2^2 \cdot 3^2 \cdot 4^2}$, and, in general, $a_n = \dfrac{a_0}{(n!)^2}$. Thus, $y = \displaystyle\sum_{n=0}^{\infty} \dfrac{a_0}{(n!)^2} x^n = a_0 \displaystyle\sum_{n=0}^{\infty} \dfrac{x^n}{(n!)^2}$, where a_0 is an arbitrary constant.

38.67 Find the interval of convergence of $\displaystyle\sum_{n=0}^{\infty} \dfrac{(-1)^n x^{2n}}{(n!)^2 2^{2n}}$.

▌ Use the ratio test. $\displaystyle\lim_{n \to +\infty} \dfrac{|x|^{2n+2}/[(n + 1)!]^2 2^{2n+2}}{|x|^{2n}/(n!)^2 2^{2n}} = \displaystyle\lim_{n \to +\infty} \dfrac{|x|^2}{4(n + 1)^2} = 0$. Hence, the series converges for all x. The function defined by this series is denoted $J_0(x)$ and is called a Bessel function of the first kind of order zero.

38.68 Find the interval of convergence of $\displaystyle\sum_{n=0}^{\infty} (-1)^n \frac{x^{2n+1}}{(n!)(n+1)!2^{2n+1}}$.

❚ Use the ratio test. $\displaystyle\lim_{n\to+\infty} \frac{|x|^{2n+3}/(n+1)!(n+2)!2^{2n+3}}{|x|^{2n+1}/n!(n+1)!2^{2n+1}} = \lim_{n\to+\infty} \frac{|x|^2}{4(n+1)(n+2)} = 0$. Therefore, we have convergence for all x. The function defined by this series is denoted $J_1(x)$ and is called a Bessel function of the first kind of order one.

38.69 Show that $\dfrac{d}{dx} J_0(x) = -J_1(x)$, where $J_0(x)$ and $J_1(x)$ are the Bessel functions defined in Problems 38.67 and 38.68.

❚ $J_0(x) = \displaystyle\sum_{n=0}^{\infty} \frac{(-1)^n x^{2n}}{(n!)^2 2^{2n}}$. Differentiate:

$$J_0'(x) = \sum_{n=1}^{\infty} \frac{(-1)^n (2n) x^{2n-1}}{(n!)^2 2^{2n}} = \sum_{n=1}^{\infty} \frac{(-1)^n x^{2n-1}}{(n-1)! n! 2^{2n-1}} = \sum_{n=0}^{\infty} \frac{(-1)^{n+1} x^{2n+1}}{n!(n+1)! 2^{2n+1}} = -J_1(x)$$

38.70 Find an ordinary differential equation of second order satisfied by $J_0(t)$.

❚ Let $x = -t^2/4$ in the series of Problem 38.66; the result is, by Problem 38.67, $J_0(t)$. Thus, the above change of variable must take the differential equation $x \dfrac{d^2y}{dx^2} + \dfrac{dy}{dx} - y = 0$ into a differential equation for $y = J_0(t)$. Explicitly, $\dfrac{dy}{dx} = \dfrac{dy/dt}{dz/dt} = -\dfrac{2}{t} \dfrac{dy}{dt}$ and $\dfrac{d^2y}{dx^2} = -\dfrac{2}{t} \dfrac{d}{dt}\left(-\dfrac{2}{t} \dfrac{dy}{dt}\right) = -\dfrac{2}{t}\left(-\dfrac{2}{t} \dfrac{d^2y}{dt^2} + \dfrac{2}{t^2} \dfrac{dy}{dt}\right)$, and the desired differential equation (*Bessel's equation of order zero*) is $\left(-\dfrac{t^2}{4}\right)\left(-\dfrac{2}{t}\right)\left(-\dfrac{2}{t} \dfrac{d^2y}{dt^2} + \dfrac{2}{t^2} \dfrac{dy}{dt}\right) - \dfrac{2}{t} \dfrac{dy}{dt} - y = 0$ or

$$\frac{d^2y}{dt^2} + \frac{1}{t} \frac{dy}{dt} + y = 0 \tag{1}$$

38.71 Show that $z = J_1(t)$ satisfies *Bessels' equation of order 1*: $\dfrac{d^2z}{dt^2} + \dfrac{1}{t} \dfrac{dz}{dt} + \left(1 - \dfrac{1}{t^2}\right)z = 0$.

❚ By Problem 38.69, we obtain a differential equation for $z = J_1(t)$ by letting $\dfrac{dy}{dt} = -z$ in (1) of Problem 38.70: $-\dfrac{dz}{dt} - \dfrac{1}{t} z + y = 0$. Differentiating once more with respect to t: $-\dfrac{d^2z}{dt^2} - \dfrac{1}{t} \dfrac{dz}{dt} + \dfrac{1}{t^2} z - z = 0$ or $\dfrac{d^2z}{dt^2} + \dfrac{1}{t} \dfrac{dz}{dt} + \left(1 - \dfrac{1}{t^2}\right)z = 0$.

37.72 Find the power series expansion of $\dfrac{1}{1+x^2}$ by division.

❚ Let $\dfrac{1}{1+x^2} = \displaystyle\sum_{n=0}^{\infty} a_n x^n$. Then $1 = (1+x^2) \displaystyle\sum_{n=0}^{\infty} a_n x^n = a_0 + a_1 x + \displaystyle\sum_{k=2}^{\infty} (a_k + a_{k-2}) x^k$. Hence, $a_0 = 1$, $a_1 = 0$, and, for $k \geq 2$, $a_k + a_{k-2} = 0$, that is, $a_k = -a_{k-2}$. Thus, $0 = a_3 = a_5 = \cdots$. For even subscripts, $a_2 = -1$, $a_4 = 1$, $a_6 = -1$, and, in general, $a_{2n} = (-1)^n$. Therefore, $\dfrac{1}{1+x^2} = \displaystyle\sum_{n=0}^{\infty} (-1)^n x^{2n} = 1 - x^2 + x^4 - x^6 + x^8 + \cdots$. (Of course, this is obtained more easily by using a geometric series.)

38.73 Show that if $f(x) = \displaystyle\sum_{n=0}^{\infty} a_n x^n$ for $|x| < r$ and $f(x)$ is an even function [that is, $f(-x) = f(x)$], then all odd-order coefficients $a_{2k+1} = 0$.

❚ $\displaystyle\sum_{n=0}^{\infty} a_n x^n = \displaystyle\sum_{n=0}^{\infty} a_n (-x)^n = \displaystyle\sum_{n=0}^{\infty} (-1)^n a_n x^n$. Equating coefficients, we see that, when n is odd, $a_n = -a_n$, and, therefore, $a_n = 0$.

38.74 Show that if $f(x) = \displaystyle\sum_{n=0}^{\infty} a_n x^n$ for $|x| < r$, and $f(x)$ is an odd function [that is, $f(-x) = -f(x)$], then all even-order coefficients $a_{2k} = 0$.

❚ $\displaystyle\sum_{n=0}^{\infty} a_n x^n = -\displaystyle\sum_{n=0}^{\infty} a_n (-x)^n = \displaystyle\sum_{n=0}^{\infty} (-1)^{n+1} a_n x^n$. Equating coefficients, we see that, when n is even, $a_n = -a_n$, and, therefore, $a_n = 0$.

38.75 For what values of x can $\sin x$ be replaced by x if the allowable error is 0.0005?

▌ By Problem 38.58, $\sin x = x - \dfrac{x^3}{3!} + \dfrac{x^5}{5!} - \dfrac{x^7}{7!} + \cdots = \displaystyle\sum_{n=0}^{\infty} (-1)^n \dfrac{x^{2n+1}}{(2n+1)!}$. Since this is an alternating series, the error is less than the magnitude of the first term omitted. If we only use x, the error is less than $|x|^3/3!$. So we need to have $|x|^3/3! < 0.0005$, $|x|^3 < 0.003$, $|x| < 0.1441\cdots$.

38.76 Use power series to evaluate $\displaystyle\lim_{x \to 0} \dfrac{e^x - e^{-x}}{\sin x}$.

▌ $\displaystyle\lim_{x \to 0} \dfrac{e^x - e^{-x}}{\sin x} = \lim_{x \to 0} \dfrac{(1 + x + x^2/2! + \cdots) - (1 - x + x^2/2! - \cdots)}{x - x^3/3! + x^5/5! - \cdots} = \lim_{x \to 0} \dfrac{2 + x^2/3 + \cdots}{1 - x^2/6 + \cdots} = 2.$

38.77 Approximate $\displaystyle\int_0^1 \sin x/x \, dx$ correctly to six decimal places.

▌ By Problem 38.58, $\sin x = x - \dfrac{x^3}{3!} + \dfrac{x^5}{5!} - \dfrac{x^7}{7!} + \cdots$. So $\displaystyle\int_0^1 \dfrac{\sin x}{x} \, dx = \int_0^1 \left(1 - \dfrac{x^2}{3!} + \dfrac{x^4}{5!} - \dfrac{x^6}{7!} + \cdots\right) dx = x - \dfrac{x^3}{3 \cdot 3!} + \dfrac{x^5}{5 \cdot 5!} - \dfrac{x^7}{7 \cdot 7!} + \cdots \Big]_0^1 = 1 - \dfrac{1}{3 \cdot 3!} + \dfrac{1}{5 \cdot 5!} - \dfrac{1}{7 \cdot 7!} + \cdots$. Note that $9 \cdot 9! = 3,265,920$, and, therefore, the next term, $\dfrac{1}{9 \cdot 9!} < \dfrac{1}{2,000,000}$. Hence, it suffices to calculate $1 - \dfrac{1}{3 \cdot 3!} + \dfrac{1}{5 \cdot 5!} - \dfrac{1}{7 \cdot 7!}$, which yields 0.946083.

38.78 Use the multiplication of power series to verify that $e^x e^{-x} = 1$.

▌ Refer to Problems 38.39 and 38.40.

$$e^x e^{-x} = \left(\sum_{n=0}^{\infty} \dfrac{1}{n!} x^n\right)\left[\sum_{m=0}^{\infty} \dfrac{(-1)^m}{m!} x^m\right]$$

$$= \sum_{k=0}^{\infty} \left[\sum_{n+m=k} \dfrac{(-1)^m}{n!m!}\right] x^k$$

$$= \sum_{k=0}^{\infty} \left[\sum_{n=0}^{k} \dfrac{(-1)^{k-n}}{n!(k-n)!}\right] x^k \qquad (1)$$

But the binomial theorem gives, for $k \geq 1$, $0 = (1-1)^k = \displaystyle\sum_{n=0}^{k} \binom{k}{n}(1)^n(-1)^{k-n} = k! \sum_{n=0}^{k} \dfrac{(-1)^{k-n}}{n!(k-n)!}$. Hence the coefficient of x^k in (1) is zero, for all $k \geq 1$; and we are left with $e^x e^{-x} = 1 \cdot x^0 = 1$.

38.79 Find the first five terms of the power series for $e^x \cos x$ by multiplication of power series.

▌ $e^x = 1 + x + \dfrac{x^2}{2!} + \dfrac{x^3}{3!} + \dfrac{x^4}{4!} + \dfrac{x^5}{5!} + \cdots$ and $\cos x = 1 - \dfrac{x^2}{2!} + \dfrac{x^4}{4!} - \cdots$. Hence, $e^x \cos x = \left(1 + x + \dfrac{x^2}{2!} + \dfrac{x^3}{3!} + \dfrac{x^4}{4!} + \dfrac{x^5}{5!} + \cdots\right)\left(1 - \dfrac{x^2}{2!} + \dfrac{x^4}{4!} - \cdots\right) = 1 + x - \dfrac{1}{3} x^3 - \dfrac{1}{6} x^4 - \dfrac{1}{30} x^5 + \cdots$.

38.80 Find the first five terms of the power series for $e^x \sin x$.

▌ $e^x \sin x = \left(1 + x + \dfrac{x^2}{2!} + \dfrac{x^3}{3!} + \dfrac{x^4}{4!} + \dfrac{x^5}{5!} + \cdots\right)\left(x - \dfrac{x^3}{3!} + \dfrac{x^5}{5!} + \cdots\right) = x + x^2 + \dfrac{1}{3} x^3 - \dfrac{1}{30} x^5 - \dfrac{1}{90} x^6 + \cdots$.

38.81 Find the first four terms of the power series for $\sec x$.

▌ Let $\sec x = \dfrac{1}{\cos x} = \displaystyle\sum_{n=0}^{\infty} a_n x^n$. Then $1 = \cos x \cdot \sum_{n=0}^{\infty} a_n x^n = \left(1 - \dfrac{x^2}{2!} + \dfrac{x^4}{4!} - \dfrac{x^6}{6!} + \cdots\right)\left(a_0 + a_1 x + a_2 x^2 + a_3 x^3 + \cdots\right)$. Now we equate coefficients. From the constant coefficient, $1 = a_0$. From the coefficient of x, $0 = a_1$. From the coefficient of x^2, $0 = a_2 - \frac{1}{2} a_0 = a_2 - \frac{1}{2}$; hence, $a_2 = \frac{1}{2}$. From the coefficient of x^3, $0 = a_3 - \frac{1}{2} a_1 = a_3$. From the coefficient of x^4, $0 = a_4 - \frac{1}{2} a_2 + \frac{1}{24} a_0 = a_4 - \frac{1}{4} + \frac{1}{24}$; hence, $a_4 = \frac{5}{24}$. From the coefficient of x^5, $0 = a_5 - \frac{1}{2} a_3 + \frac{1}{24} a_1 = a_5$. From the coefficient of x^6, $0 = a_6 - \frac{1}{2} a_4 + \frac{1}{24} a_2 - \frac{1}{720} a_0 = a_6 - \frac{5}{48} + \frac{1}{48} - \frac{1}{720}$; hence, $a_6 = \frac{61}{720}$. Thus, $\sec x = 1 + \frac{1}{2} x^2 + \frac{5}{24} x^4 + \frac{61}{720} x^6 + \cdots$.

38.82 Find the first four terms of the power series for $e^x/\cos x$ by long division.

▌ Write the long division as follows:

$$
\begin{array}{r}
1 + x + x^2 + \dfrac{2x^3}{3} + \cdots \\[4pt]
1 + 0\cdot x - \dfrac{x^2}{2} + 0\cdot x^3 + \cdots \overline{\big)\; 1 + x + \dfrac{x^2}{2} + \dfrac{x^3}{6} + \cdots} \\[10pt]
\underline{1 + 0\cdot x - \dfrac{x^2}{2} + 0\cdot x^3 + \cdots} \\[10pt]
x + x^2 + \dfrac{x^3}{6} + \cdots \\[10pt]
\underline{x + 0\cdot x^2 - \dfrac{x^3}{2} + \cdots} \\[10pt]
x^2 + \dfrac{2x^3}{3} + \cdots \\[10pt]
\underline{x^2 + 0\cdot x^3 - \cdots} \\[10pt]
\dfrac{2x^3}{3} + \cdots \\[10pt]
\underline{\dfrac{2x^3}{3} + \cdots} \\[8pt]
\cdots
\end{array}
$$

Hence, the power series for $e^x/\cos x$ begins with $1 + x + x^2 + 2x^3/3$.

38.83 Find the first three terms of the power series for $\tan x$ by long division.

∥ Arrange the division of $\sin x$ by $\cos x$ as follows:

$$
\begin{array}{r}
x + \dfrac{x^3}{3} + \dfrac{2}{15}x^5 + \cdots \\[6pt]
1 + \dfrac{x^2}{2} - \dfrac{x^4}{4!} + \dfrac{x^6}{6!} - \cdots \overline{\big)\; x - \dfrac{x^3}{3!} + \dfrac{x^5}{5!} - \dfrac{x^7}{7!} \cdots} \\[10pt]
\underline{x + \dfrac{x^3}{2} - \dfrac{x^5}{4!} + \dfrac{x^7}{6!} \cdots} \\[10pt]
\dfrac{1}{3}x^3 - \dfrac{1}{30}x^5 + \dfrac{1}{840}x^7 \cdots \\[10pt]
\underline{\dfrac{1}{3}x^3 + \dfrac{1}{6}x^5 - \dfrac{1}{72}x^7 \cdots} \\[10pt]
\dfrac{2}{15}x^5 - \dfrac{4}{315}x^7 \cdots \\[10pt]
\underline{\dfrac{2}{15}x^5 + \dfrac{1}{15}x^7 \cdots}
\end{array}
$$

Hence, the power series for $\tan x$ begins with $x + \tfrac{1}{3}x^3 + \tfrac{2}{15}x^5$.

38.84 Evaluate the power series $x + 2x^2 + 3x^3 + \cdots + nx^n + \cdots$.

∥ By Problem 38.37, $1/(1-x)^2 = 1 + 2x + 3x^2 + \cdots$. Therefore, $x/(1-x)^2 = x + 2x^2 + 3x^3 + \cdots$.

38.85 Evaluate the power series $\displaystyle\sum_{n=1}^{\infty} n^2 x^n$.

∥ Let $f(x) = \displaystyle\sum_{n=1}^{\infty} n^2 x^n = x \sum_{n=1}^{\infty} n^2 x^{n-1} = x g(x)$. Now, $\displaystyle\int g(x)\,dx = \sum_{n=1}^{\infty} n x^n = \frac{x}{(1-x)^2}$, by Problem 38.84. Hence, $g(x) = \dfrac{d}{dx}\dfrac{x}{(1-x)^2} = \dfrac{x+1}{(1-x)^3}$ and $f(x) = \dfrac{x(x+1)}{(1-x)^3}$.

38.86 Write the first three terms of the power series for $\ln \sec x$.

∥ $\dfrac{d}{dx}(\ln \sec x) = \tan x = x + \tfrac{1}{3}x^3 + \tfrac{2}{15}x^5 + \cdots$, by Problem 38.83. Therefore, $\ln \sec x = C + x^2/2 + x^4/12 + x^6/45 + \cdots$. When $x = 0$, $\ln \sec x = 0$. Hence, $C = 0$. Thus, $\ln \sec x = x^2/2 + x^4/12 + x^6/45 + \cdots$.

38.87 Let $f(x) = \sum_{n=0}^{\infty} a_n x^n$. Show that $\dfrac{1}{1-x} f(x) = \sum_{n=0}^{\infty} (a_0 + a_1 + \cdots + a_n)x^n$.

▮ $\dfrac{1}{1-x}(a_0 + a_1 x + \cdots + a_n x^n + \cdots) = (1 + x + x^2 + \cdots + x^n + \cdots)(a_0 + a_1 x + \cdots a_n x^n + \cdots)$. The terms of this product involving x^n are $a_0 x^n + a_1 x \cdot x^{n-1} + \cdots + a_{n-2} x^{n-2} \cdot x^2 + a_{n-1} x^{n-1} \cdot x + a_n x^n$. Hence, the coefficient of x^n will be $a_0 + a_1 + \cdots + a_n$.

38.88 Find a power series for $\dfrac{\ln(1-x)}{x-1}$ by multiplication of power series.

▮ $\ln(1-x) = -(x + x^2/2 + x^3/3 + \cdots)$ by Problem 38.50. Hence, by Problem 38.87, the coefficient of x^n in $\dfrac{\ln(1-x)}{1-x}$ is $-\left(\dfrac{1}{2} + \dfrac{1}{3} + \cdots + \dfrac{1}{n}\right)$. Hence, $\dfrac{\ln(1-x)}{x-1} = \sum_{n=1}^{\infty} \left(1 + \dfrac{1}{2} + \cdots + \dfrac{1}{n}\right)x^n$.

38.89 Write the first four terms of a power series for $(\sin x)/(1-x)$.

▮ $\sin x = x - \dfrac{x^3}{3!} + \dfrac{x^5}{5!} - \cdots + (-1)^k \dfrac{x^{2k+1}}{(2k+1)!} + \cdots$. By Problem 38.87, the coefficient of x^n in $(\sin x)/(1-x)$ will be the sum of the coefficients of the power series for $\sin x$ up through that of x^n. Hence, we get $(\sin x)/(1-x) = x + x^2 + \frac{5}{6}x^3 + \frac{5}{6}x^4 + \cdots$.

38.90 Find the first five terms of a power series for $(\tan^{-1} x)/(1-x)$.

▮ $\tan^{-1} x = x - x^3/3 + x^5/5 - x^7/7 + \cdots$ for $|x| < 1$, by Problem 38.36. Hence, by Problem 38.87, we obtain $(\tan^{-1} x)/(1-x) = x + x^2 + \frac{2}{3}x^3 + \frac{2}{3}x^4 + \frac{13}{15}x^5 + \cdots$.

38.91 Find the first five terms of a power series for $(\cos x)/(1-x)$.

▮ $\cos x = 1 - x^2/2! + x^4/4! - x^6/6! + x^8/8! - \cdots$. Hence, by Problem 38.87, $(\cos x)/(1-x) = 1 + x + \frac{1}{2}x^2 + \frac{1}{2}x^3 + \frac{13}{24}x^4 + \cdots$.

38.92 Use the result of Problem 38.87 to evaluate $\sum_{n=0}^{\infty} (n+1)x^n$.

▮ We want to find $\sum_{n=0}^{\infty} a_n x^n$ so that $a_0 + a_1 + \cdots + a_n = n+1$. A simple choice is $a_i = 1$. Thus, from $\sum_{n=0}^{\infty} x^n = \dfrac{1}{1-x}$ we obtain $\sum_{n=0}^{\infty} (n+1)x^n = \dfrac{1}{(1-x)^2}$.

38.93 If $f(x) = \sum_{n=0}^{\infty} a_n x^n$, show that the even part of $f(x)$, $E(x) = \frac{1}{2}[f(x) + f(-x)]$, is $\sum_{k=0}^{\infty} a_{2k} x^{2k}$, and the odd part of $f(x)$, $O(x) = \frac{1}{2}[f(x) - f(-x)]$, is $\sum_{k=0}^{\infty} a_{2k+1} x^{2k+1}$.

▮ $f(-x) = \sum_{n=0}^{\infty} (-1)^n a_n x^n$. Hence, $f(x) + f(-x) = 2 \sum_{k=0}^{\infty} a_{2k} x^{2k}$, and $f(x) - f(-x) = 2 \sum_{k=0}^{\infty} a_{2k+1} x^{2k+1}$.

38.94 Evaluate $x^2/2 + x^4/4 + \cdots + x^{2k}/2k + \cdots$.

▮ This is the even part (see Problem 38.93) of the series $f(x) = -\ln(1-x) = x + x^2/2 + x^3/3 + x^4/4 + \cdots$. Hence the given series is equal to $\frac{1}{2}[f(x) + f(-x)] = \frac{1}{2}[-\ln(1-x) + \ln(1+x)] = \frac{1}{2}\ln[(1+x)/(1-x)] = \ln\sqrt{(1+x)/(1-x)}$.

38.95 Use Problem 38.93 to evaluate $\sum_{n=0}^{\infty} \dfrac{x^{2n}}{(2n)!}$.

▮ This is the even part of e^x, which is $(e^x + e^{-x})/2 = \cosh x$. (This problem was solved in the reverse direction in Problem 38.43.)

38.96 Find $\lim\limits_{x \to 0} \dfrac{\sin x - x}{x^3}$ by power series methods.

▮ By Problem 38.58, $\sin x - x = -x^3/3! + x^5/5! - x^7/7! + \cdots$. Hence, $(\sin x - x)/x^3 = -1/3! + x^2/5! - x^4/7! + \cdots$, and $\lim\limits_{x \to 0} (\sin x - x)/x^3 = -1/3! = -\frac{1}{6}$.

38.97 Evaluate $x/2! + x^2/3! + x^3/4! + x^4/5! + \cdots$.

▮ Let $f(x) = x/2! + x^2/3! + x^3/4! + x^4/5! + \cdots$. Then $xf(x) = x^2/2! + x^3/3! + x^4/4! + x^5/5! + \cdots = e^x - x - 1$. Hence, $f(x) = (e^x - x - 1)/x$.

38.98 Assume that the coefficients of a power series $\sum\limits_{n=0}^{\infty} a_n x^n$ repeat every k terms, that is, $a_{n+k} = a_n$ for all n. Show that its sum is $\dfrac{a_0 + a_1 x + \cdots + a_{k-1} x^{k-1}}{1 - x^k}$.

▮ Let $g(x) = a_0 + a_1 x + \cdots + a_{k-1} x^{k-1}$. Then, $\sum\limits_{n=0}^{\infty} a_n x^n = g(x) + g(x)x^k + g(x)x^{2k} + \cdots = g(x)(1 + x^k + x^{2k} + \cdots) = \dfrac{g(x)}{1 - x^k}$. The series converges for $|x| < 1$.

38.99 Evaluate $\dfrac{x^2}{1 \cdot 2} - \dfrac{x^3}{2 \cdot 3} + \dfrac{x^4}{3 \cdot 4} - \dfrac{x^5}{4 \cdot 5} + \cdots$.

▮ Let $f(x) = \dfrac{x^2}{1 \cdot 2} - \dfrac{x^3}{2 \cdot 3} + \dfrac{x^4}{3 \cdot 4} - \dfrac{x^5}{4 \cdot 5} + \cdots$. Then $f'(x) = x - \dfrac{x^2}{2} + \dfrac{x^3}{3} - \dfrac{x^4}{4} + \cdots = \ln(1 + x)$, by Problem 38.38. Hence, $f(x) = \int \ln(1 + x)\, dx = (1 + x)[\ln(1 + x) - 1] + C = (1 + x)\ln(1 + x) - x + C_1$. When $x = 0$, $f(x) = 0$, and, therefore, $C_1 = 0$. Thus, $f(x) = (1 + x)\ln(1 + x) - x$.

38.100 Evaluate $x^4/4 + x^8/8 + x^{12}/12 + x^{16}/16 + \cdots$.

▮ By Problem 38.50, $-\ln(1 - u) = u + u^2/2 + u^3/3 + u^4/4 + \cdots$. Hence, $-\ln(1 - x^4) = x^4 + x^8/2 + x^{12}/3 + x^{16}/4 + \cdots$. Thus, the given series is $-\frac{1}{4}\ln(1 - x^4)$.

38.101 Evaluate $\phi(x) \equiv \sum\limits_{n=1}^{\infty} n^3 x^n$.

▮ If $f(x)$ is the function of Problem 38.85, then
$$\phi(x) = xf'(x) = x \frac{d}{dx}\left[\frac{x^2 + x}{(1 - x)^3}\right] = x \frac{(1 - x)^3(2x + 1) + 3(x^2 + x)(1 - x)^2}{(1 - x)^6} = \frac{x(x^2 + 4x + 1)}{(1 - x)^4}.$$

38.102 Evaluate $\psi(x) \equiv \sum\limits_{n=0}^{\infty} (n + 4)x^n$.

▮ $\psi(x) = \sum\limits_{n=0}^{\infty} nx^n + 4\sum\limits_{n=0}^{\infty} x^n = \dfrac{x}{(1 - x)^2} + 4\dfrac{1}{1 - x} = \dfrac{4 - 3x}{(1 - x)^2}$, where we have used Problem 38.84.

38.103 For the binomial series (Problem 38.31), $f(x) = 1 + mx + \dfrac{m(m - 1)}{2!}x^2 + \cdots + \dfrac{m(m - 1)\cdots(m - n + 1)}{n!}x^n + \cdots$, which is convergent for $|x| < 1$, show that $(1 + x)f'(x) = mf(x)$.

▮ $f'(x) = m + m(m - 1)x + \cdots + \dfrac{m(m - 1)\cdots(m - n + 1)}{(n - 1)!}x^{n-1} + \cdots$; $xf'(x) = mx + m(m - 1)x^2 + \cdots + \dfrac{m(m - 1)\cdots(m - n + 1)}{(n - 1)!}x^n + \cdots$. In $(1 + x)f'(x)$, the coefficient of x^n will be
$$\frac{m(m - 1)\cdots(m - n + 1)(m - n)}{n!} + n\frac{m(m - 1)\cdots(m - n + 1)}{n!} = m\frac{m(m - 1)\cdots(m - n + 1)}{n!}$$
Hence, $(1 + x)f'(x) = mf(x)$.

38.104 Prove that the binomial series $f(x)$ of Problem 38.103 is equal to $(1 + x)^m$.

▮ Let $g(x) = \dfrac{f(x)}{(1 + x)^m}$. Then, $g'(x) = \dfrac{(1 + x)^m f'(x) - mf(x)(1 + x)^{m-1}}{(1 + x)^{2m}} = 0$, by Problem 38.103. Hence, $g(x)$ is a constant C. But $f(x) = 1$ when $x = 0$, and, therefore, $C = 1$. Therefore, $f(x) = (1 + x)^m$.

38.105 Show that $\sqrt{1 - x} = 1 - \dfrac{1}{2}x - \dfrac{1}{2 \cdot 4}x^2 - \dfrac{1 \cdot 3}{2 \cdot 4 \cdot 6}x^3 - \cdots - \dfrac{1 \cdot 3 \cdot \cdots \cdot (2n - 3)}{2^n \cdot n!}x^n - \cdots$.

▮ Substitute $-x$ for x and $\frac{1}{2}$ for m in the binomial series of Problems 38.103 and 38.104.

38.106 Derive the series $\dfrac{1}{\sqrt{1 - x}} = 1 + \dfrac{1}{2}x + \dfrac{1 \cdot 3}{2 \cdot 4}x^2 + \dfrac{1 \cdot 3 \cdot 5}{2 \cdot 4 \cdot 6}x^3 + \cdots$.

▮ Substitute $-x$ for x and $-\frac{1}{2}$ for m in the binomial series of Problems 38.103 and 38.104. (Alternatively, take the derivative of the series in Problem 38.105.)

38.107 Obtain the series $\sin^{-1} x = x + \dfrac{1}{2}\dfrac{x^3}{3} + \dfrac{1 \cdot 3}{2 \cdot 4}\dfrac{x^5}{5} + \dfrac{1 \cdot 3 \cdot 5}{2 \cdot 4 \cdot 6}\dfrac{x^7}{7} + \cdots$, $|x| < 1$.

▮ By Problem 38.106, $\dfrac{1}{\sqrt{1-t}} = 1 + \dfrac{1}{2} t + \dfrac{1 \cdot 3}{2 \cdot 4} t^2 + \dfrac{1 \cdot 3 \cdot 5}{2 \cdot 4 \cdot 6} t^3 + \cdots$. Therefore, $\dfrac{1}{\sqrt{1-t^2}} = 1 + \dfrac{1}{2} t^2 +$

$\dfrac{1 \cdot 3}{2 \cdot 4} t^4 + \dfrac{1 \cdot 3 \cdot 5}{2 \cdot 4 \cdot 6} t^6 + \cdots$ and $\sin^{-1} x = \displaystyle\int_0^x \dfrac{1}{\sqrt{1-t^2}}\, dt = x + \dfrac{1}{2} \dfrac{x^3}{3} + \dfrac{1 \cdot 3}{2 \cdot 4} \dfrac{x^5}{5} + \dfrac{1 \cdot 3 \cdot 5}{2 \cdot 4 \cdot 6} \dfrac{x^7}{7} + \cdots$.

38.108 By means of the binomial series, approximate $\sqrt[5]{33}$ correctly to three decimal places.

▮ $\sqrt[5]{33} = \sqrt[5]{32+1} = 2 \cdot \sqrt[5]{1 + \frac{1}{32}}$. By Problem 38.104, $(1 + \frac{1}{32})^{1/5} = 1 + \frac{1}{5}(\frac{1}{32}) + \dfrac{\frac{1}{5}(\frac{1}{5} - 1)}{2}(\frac{1}{32})^2 + \cdots$. Since the series alternates in sign, the error is less than the magnitude of the first term omitted. Now, $\dfrac{\frac{1}{5}(\frac{1}{5} - 1)}{2}\left(\dfrac{1}{32}\right)^2 = \dfrac{1}{12,800} < 0.0005$. Thus, it suffices to use $1 + \frac{1}{5} \frac{1}{32} = \frac{161}{160} \approx 1.006$. Hence, $\sqrt[5]{33} \approx 2.012$.

38.109 Find the radius of convergence of $\displaystyle\sum_{n=0}^{\infty} \dfrac{n!}{10^n} x^n$.

▮ Use the ratio test. $\displaystyle\lim_{n \to +\infty} \dfrac{(n+1)!|x|^{n+1}/10^{n+1}}{n!|x|^n/10^n} = \lim_{n \to +\infty} \dfrac{n+1}{10} |x| \to +\infty$ except for $x = 0$. Hence, the radius of convergence is zero.

38.110 Find the interval of convergence of $\displaystyle\sum_{n=0}^{\infty} \dfrac{2^n}{n!} x^n$.

▮ Use the ratio test. $\displaystyle\lim_{n \to +\infty} \dfrac{2^{n+1}|x|^{n+1}/(n+1)!}{2^n|x|/n!} = \lim_{n \to +\infty} \dfrac{2|x|}{n+1} = 0$. Hence, the series converges for all x.

38.111 Find the interval of convergence of $\displaystyle\sum_{n=1}^{\infty} \dfrac{(-1)^{n+1} x^n}{\sqrt{n} 2^n}$.

▮ Use the ratio test. $\displaystyle\lim_{n \to +\infty} \dfrac{|x|^{n+1}/\sqrt{n+1}\,2^{n+1}}{|x|^n/\sqrt{n}\,2^n} = \lim_{n \to +\infty} \dfrac{|x|}{2} \sqrt{\dfrac{n}{n+1}} = \dfrac{|x|}{2}$. Thus, the series converges for $|x| < 2$ and diverges for $|x| > 2$. For $x = 2$, we obtain a convergent alternating series. For $x = -2$, we get a divergent p-series, $-\Sigma\, 1/\sqrt{n}$.

38.112 Find the interval of convergence of $\displaystyle\sum_{n=1}^{\infty} \dfrac{\ln n}{n} x^n$.

▮ Use the ratio test. $\displaystyle\lim_{n \to +\infty} \dfrac{\ln(n+1)\,|x|^{n+1}/(n+1)}{(\ln n)\,|x|^n/n} = \lim_{n \to +\infty} \dfrac{\ln(n+1)}{\ln n} \dfrac{n}{n+1} |x| = |x|$. Hence, the series converges for $|x| < 1$ and diverges for $|x| > 1$. When $x = 1$, we obtain a divergent series, by the integral test. When $x = -1$, we get a convergent alternating series.

38.113 Find the radius of convergence of the *hypergeometric series*

$$1 + \dfrac{ab}{c} x + \dfrac{a(a+1)b(b+1)}{2!c(c+1)} x^2 + \dfrac{a(a+1)(a+2)b(b+1)(b+2)}{3!c(c+1)(c+2)} x^3 + \cdots$$

▮ Use the ratio test. $\displaystyle\lim_{n \to +\infty} \dfrac{|x|^{n+1} a(a+1)\cdots(a+n)b(b+1)\cdots(b+n)/(n+1)!c(c+1)\cdots(c+n)}{|x|^n a(a+1)\cdots(a+n-1)b(b+1)\cdots(b+n-1)/n!c(c+1)\cdots(c+n-1)} =$

$\displaystyle\lim_{n \to +\infty} \dfrac{(a+n)(b+n)|x|}{n(c+n)} = \lim_{n \to +\infty} \dfrac{(a/n+1)(b/n+1)|x|}{c/n+1} = |x|$. Hence, the radius of convergence is 1.

38.114 If infinitely many coefficients of a power series are nonzero integers, show that the radius of convergence $r \le 1$.

▮ Assume $\Sigma\, a_n x^n$ converges for some $|x| > 1$. Then, $\displaystyle\lim_{n \to +\infty} |a_n|\,|x|^n = 0$. But, for infinitely many values of n, $|a_n|\,|x|^n > 1$, contradicting $\displaystyle\lim_{n \to +\infty} |a_n|\,|x|^n = 0$.

38.115 Denoting the sum of the hypergeometric series (Problem 38.113) by $F(a, b; c; x)$, show that $\tan^{-1} x = xF(1, \frac{1}{2}; \frac{3}{2}; -x^2)$.

▮ By Problem 38.113, $xF(1, \frac{1}{2}; \frac{3}{2}; -x^2) = x - \dfrac{(1)(\frac{1}{2})}{\frac{3}{2}} x^3 + \dfrac{(1)(2)(\frac{1}{2})(\frac{3}{2})}{2!(\frac{3}{2})(\frac{5}{2})} x^5 - \dfrac{(1)(2)(3)(\frac{1}{2})(\frac{3}{2})(\frac{5}{2})}{3!(\frac{3}{2})(\frac{5}{2})(\frac{7}{2})} x^7 + \cdots =$

$x - \dfrac{x^3}{3} + \dfrac{x^5}{5} - \dfrac{x^7}{7} + \cdots = \tan^{-1} x$ (see Problem 38.36).

CHAPTER 39
Taylor and Maclaurin Series

39.1 Find the Maclaurin series of e^x.

    ▌ Let $f(x) = e^x$. Then $f^{(n)}(x) = e^x$ for all $n \geq 0$. Hence, $f^{(n)}(0) = 1$ for all $n \geq 0$. Therefore, the Maclaurin series $\sum_{n=0}^{\infty} \frac{f^{(n)}(0)}{n!} x^n = \sum_{n=0}^{\infty} \frac{x^n}{n!}$.

39.2 Find the Maclaurin series for $\sin x$.

    ▌ Let $f(x) = \sin x$. Then, $f(0) = \sin 0 = 0$, $f'(0) = \cos 0 = 1$, $f''(0) = -\sin 0 = 0$, $f'''(0) = -\cos 0 = -1$, and, thereafter, the sequence of values $0, 1, 0, -1$ keeps repeating. Thus, we obtain $x - \frac{x^3}{3!} + \frac{x^5}{5!} - \frac{x^7}{7!} + \cdots = \sum_{n=0}^{\infty} (-1)^n \frac{x^{2n+1}}{(2n+1)!}$.

39.3 Find the Taylor series for $\sin x$ about $\pi/4$.

    ▌ Let $f(x) = \sin x$. Then $f(\pi/4) = \sin(\pi/4) = \sqrt{2}/2$, $f'(\pi/4) = \cos(\pi/4) = \sqrt{2}/2$, $f''(\pi/4) = -\sin(\pi/4) = -\sqrt{2}/2$, $f'''(\pi/4) = -\cos(\pi/4) = -\sqrt{2}/2$, and, thereafter, this cycle of four values keeps repeating. Thus, the Taylor series for $\sin x$ about $\frac{\pi}{4}$ is $\frac{\sqrt{2}}{2}\left[1 + \left(x - \frac{\pi}{4}\right) - \frac{(x - \pi/4)^2}{2!} - \frac{(x - \pi/4)^3}{3!} + \frac{(x - \pi/4)^4}{4!} + \frac{(x - \pi/4)^5}{5!} - \frac{(x - \pi/4)^6}{6!} - \frac{(x - \pi/4)^7}{7!} + \cdots\right]$.

39.4 Calculate the Taylor series for $1/x$ about 1.

    ▌ Let $f(x) = \frac{1}{x}$. Then, $f'(x) = -\frac{1}{x^2}$, $f''(x) = \frac{2}{x^3}$, $f'''(x) = -\frac{2 \cdot 3}{x^4}$, $f^{(4)}(x) = \frac{2 \cdot 3 \cdot 4}{x^5}$, and, in general, $f^{(n)}(x) = (-1)^n \frac{n!}{x^{n+1}}$. So $f^{(n)}(1) = (-1)^n n!$. Thus, the Taylor series is $\sum_{n=0}^{\infty} \frac{f^{(n)}(1)}{n!} (x-1)^n = \sum_{n=0}^{\infty} (-1)^n (x-1)^n = 1 - (x-1) + (x-1)^2 - (x-1)^3 + \cdots$.

39.5 Find the Maclaurin series for $\ln(1-x)$.

    ▌ Let $f(x) = \frac{1}{x}$. Then, $f'(x) = -\frac{1}{x^2}$, $f''(x) = \frac{2}{x^3}$, $f'''(x) = -\frac{2 \cdot 3}{x^4}$, $f^{(4)}(x) = \frac{2 \cdot 3 \cdot 4}{x^5}$, and, in general, $f^{(n)}(x) = (-1)^n \frac{n!}{x^{n+1}}$. So $f^{(n)}(1) = (-1)^n n!$. Thus, the Taylor series is $\sum_{n=0}^{\infty} \frac{f^{(n)}(1)}{n!} (x-1)^n = \sum_{n=0}^{\infty} (-1)^n (x-1)^n = 1 - (x-1) + (x-1)^2 - (x-1)^3 + \cdots$.

39.6 Find the Taylor series for $\ln x$ around 2.

    ▌ Let $f(x) = \ln x$. Then, $f'(x) = \frac{1}{x}$, $f''(x) = -\frac{1}{x^2}$, $f'''(x) = \frac{2}{x^3}$, $f^{(4)}(x) = -\frac{2 \cdot 3}{x^4}$, and, in general, $f^{(n)}(x) = (-1)^{n+1} \frac{(n-1)!}{x^n}$. So $f(2) = \ln 2$, and, for $n \geq 1$, $f^{(n)}(2) = (-1)^{n+1} \frac{(n-1)!}{2^n}$. Thus, the Taylor series is $\sum_{n=0}^{\infty} \frac{f^{(n)}(2)}{n!} (x-2)^n = \ln 2 + \frac{1}{2}(x-2) - \frac{1}{8}(x-2)^2 + \frac{1}{24}(x-2)^3 - \cdots$.

39.7 Compute the first three nonzero terms of the Maclaurin series for $e^{\cos x}$.

    ▌ Let $f(x) = e^{\cos x}$. Then, $f'(x) = -e^{\cos x} \sin x$, $f''(x) = e^{\cos x}(\sin^2 x - \cos x)$, $f'''(x) = e^{\cos x} \sin x(3 \cos x + 1 - \sin^2 x)$, $f^{(4)}(x) = e^{\cos x}[-\sin^2 x(3 + 2\cos x) + (3\cos x + 1 - \sin^2 x)(\cos x - \sin^2 x)]$. Thus, $f(0) = e$, $f'(0) = 0$, $f''(0) = -e$, $f'''(0) = 0$, $f^{(4)}(0) = 4e$. Hence, the Maclaurin series is $e(1 - \frac{1}{2}x^2 + \frac{1}{6}x^4 + \cdots)$.

39.8 Write the first nonzero terms of the Maclaurin series for $\sec x$.

❚ Let $f(x) = \sec x$. Then, $f'(x) = \sec x \tan x$, $f''(x) = \sec x(1 + 2\tan^2 x)$, $f'''(x) = \sec x \tan x(5 + 6\tan^2 x)$, $f^{(4)}(x) = 12\sec^3 x \tan^2 x + (5 + 6\tan^2 x)(\sec^3 x + \tan^2 x \sec x)$. Thus, $f(0) = 1$, $f'(0) = 0$, $f''(0) = 1$, $f'''(0) = 0$, $f^{(4)} = 5$. The Maclaurin series is $1 + \frac{1}{2}x^2 + \frac{5}{24}x^4 + \cdots$.

39.9 Find the first three nonzero terms of the Maclaurin series for $\tan x$.

❚ Let $f(x) = \tan x$. Then $f'(x) = \sec^2 x$, $f''(x) = 2\tan x \sec^2 x$, $f'''(x) = 2(\sec^4 x + 2\tan^2 x + 2\tan^4 x)$, $f^{(4)}(x) = 8\tan x \sec^2 x(2 + 3\tan^2 x)$, $f^{(5)}(x) = 48\tan^2 x \sec^4 x + 8(2 + 3\tan^2 x)(\sec^4 x + 2\tan^2 x + 2\tan^4 x)$. So $f(0) = 0$, $f'(0) = 1$, $f''(0) = 0$, $f'''(0) = 2$, $f^{(4)}(0) = 0$, $f^{(5)}(0) = 16$. Thus, the Maclaurin series is $x + \frac{1}{3}x^3 + \frac{2}{15}x^3 + \cdots$.

39.10 Write the first three nonzero terms of the Maclaurin series for $\sin^{-1} x$.

❚ Let $f(x) = \sin^{-1} x$. $f'(x) = (1 - x^2)^{-1/2}$, $f''(x) = x(1 - x^2)^{-3/2}$, $f'''(x) = (1 - x^2)^{-5/2}(2x^2 + 1)$. $f^{(4)}(x) = 3x(1 - x^2)^{-7/2}(2x^2 + 3)$, $f^{(5)}(x) = 3(1 - x^2)^{-9/2}(3 + 24x^2 + 8x^4)$. Thus, $f(0) = 0$, $f'(0) = 1$, $f''(0) = 0$, $f'''(0) = 1$, $f^{(4)}(0) = 0$, $f^{(5)}(0) = 9$. Hence, the Maclaurin series is $x + \frac{1}{6}x^3 + \frac{3}{40}x^5 + \cdots$.

39.11 If $f(x) = \sum\limits_{n=0}^{\infty} a_n(x - a)^n$ for $|x - a| < r$, prove that $a_k = \dfrac{f^{(k)}(a)}{k!}$. In other words, if $f(x)$ has a power series expansion about a, that power series must be the Taylor series for $f(x)$ about a.

❚ $f(a) = a_0$. It can be shown that the power series converges uniformly on $|x - a| \le \rho < r$, allowing differentiation term by term: $f'(x) = \sum\limits_{n=1}^{\infty} na_n(x - a)^{n-1}$, $f''(x) = \sum\limits_{n=2}^{\infty} n(n - 1)a_n(x - a)^{n-2}, \ldots$, $f^{(k)}(x) = \sum\limits_{n=k}^{\infty} n(n - 1)\cdots[n - (k - 1)]a_n(x - a)^{n-k}$. If we let $x = a$, $f^{(k)}(a) = k(k - 1)\cdot \cdots \cdot 1 \cdot a_k = k! \cdot a_k$. Hence, $a_k = \dfrac{f^{(k)}(a)}{k!}$.

39.12 Find the Maclaurin series for $\dfrac{1}{1 + x}$.

❚ By Problem 38.34, we know that $\dfrac{1}{1 + x} = 1 - x + x^2 - \cdots = \sum\limits_{n=0}^{\infty} (-1)^n x^n$, for $|x| < 1$. Hence, by Problem 39.11, this must be the Maclaurin series for $\dfrac{1}{1 + x}$.

39.13 Find the Maclaurin series for $\tan^{-1} x$.

❚ By Problem 38.36, we know that $\tan^{-1} x = x - \dfrac{x^3}{3} + \dfrac{x^5}{5} - \cdots = \sum\limits_{n=0}^{\infty} \dfrac{(-1)^n x^{2n+1}}{2n + 1}$, for $|x| < 1$. Hence, by Problem 39.11, this must be the Maclaurin series for $\tan^{-1} x$. A direct calculation of the coefficients is tedious.

39.14 Find the Maclaurin series for $\cosh x$.

❚ By Problem 38.43, $\cosh x = \sum\limits_{n=0}^{\infty} \dfrac{x^{2n}}{(2n)!}$, for all x. Hence, by Problem 39.11, this is the Maclaurin series for $\cosh x$.

39.15 Obtain the Maclaurin series for $\cos^2 x$.

❚ $\cos^2 x = \dfrac{1 + \cos 2x}{2}$. Now, by Problem 38.59, $\cos x = \sum\limits_{n=0}^{\infty} \dfrac{(-1)^n x^{2n}}{(2n)!}$, and, therefore, $\cos 2x = \sum\limits_{n=0}^{\infty} \dfrac{(-1)^n 2^{2n} x^{2n}}{(2n)!}$. Since the latter series has constant term 1, $1 + \cos 2x = 2 + \sum\limits_{n=1}^{\infty} \dfrac{(-1)^n 2^{2n} x^{2n}}{(2n)!}$, and $\dfrac{1 + \cos 2x}{2} = 1 + \sum\limits_{n=1}^{\infty} \dfrac{(-1)^n 2^{2n-1} x^{2n}}{(2n)!}$. By Problem 39.11, this is the Maclaurin series for $\cos^2 x$.

39.16 Find the Taylor series for $\cos x$ about $\frac{\pi}{3}$.

▌ $\cos x = \cos\left[\frac{\pi}{3} + \left(x - \frac{\pi}{3}\right)\right]$ $=$ $\cos\frac{\pi}{3}\cos\left(x - \frac{\pi}{3}\right) - \sin\frac{\pi}{3}\sin\left(x - \frac{\pi}{3}\right)$ $=$ $\frac{1}{2}\sum_{n=0}^{\infty}\frac{(-1)^n(x - \pi/3)^{2n}}{(2n)!} -$

$\frac{\sqrt{3}}{2}\sum_{n=0}^{\infty}\frac{(-1)^n(x - \pi/3)^{2n+1}}{(2n+1)!}$. By Problem 39.11, we have the Taylor series for $\cos x$ about $\frac{\pi}{3}$.

39.17 State Taylor's formula with Lagrange's form of the remainder and indicate how it is used to show that a function is represented by its Taylor series.

▌ If $f(x)$ and its first n derivatives are continuous on an open interval containing a, then, for any x in this interval, there is a number c between a and x such that

$$f(x) = f(a) + f'(a)(x - a) + \frac{f''(a)}{2!}(x - a)^2 + \cdots + \frac{f^{(n-1)}(a)}{(n-1)!}(x - a)^{n-1} + R_n(x)$$

where $R_n(x) = \frac{f^{(n)}(c)}{n!}(x - a)^n$. If $f(x)$ has continuous derivatives of all orders, then, for those x for which $\lim_{n\to+\infty} R_n(x) = 0$, $f(x)$ is equal to its Taylor series.

39.18 Show that e^x is represented by its Maclaurin series.

▌ Let $f(x) = e^x$. Then the Lagrange remainder $R_n(x) = \frac{f^{(n)}(c)}{n!}x^n = e^c\frac{x^n}{n!}$, for some c between x and 0. Thus, $e^c \leq e^{|x|}$ and $|R_n(x)| \leq e^{|x|}\frac{|x|^n}{n!}$. But $\lim_{n\to+\infty}\frac{|x|^n}{n!} = 0$, by Problem 36.13. Therefore, $\lim_{n\to+\infty} R_n(x) = 0$ for all x. So e^x is equal to its Maclaurin series $\sum_{n=0}^{\infty}\frac{x^n}{n!}$ (which was found in Problem 39.1). Note that this was proved in a different manner in Problem 38.39.

39.19 Find the interval on which $\sin x$ may be represented by its Maclaurin series.

▌ Let $f(x) = \sin x$. Note that $f^{(n)}(x)$ is either $\pm\sin x$ or $\pm\cos x$, and, therefore, $|R_n(x)| = \left|\frac{f^{(n)}(c)}{n!}x^n\right| \leq \frac{|x|^n}{n!} \to 0$ (by Problem 36.13). Hence, $\sin x$ is equal to its Maclaurin series for all x. (See Problem 39.2.)

39.20 Show that $\ln(1 - x)$ is equal to its Maclaurin series for $|x| < 1$.

▌ The Maclaurin series for $\ln(1 - x)$ is, by Problem 39.5, $-\sum_{n=1}^{\infty}\frac{1}{n}x^n$, and the latter converges for $|x| < 1$. From the solution of Problem 39.5, $f^{(n)}(x) = -\frac{(n-1)!}{(1-x)^n}$. Hence, $R_n(x) = -\frac{(n-1)!}{(1-c)^n}\frac{x^n}{n!} = -\frac{1}{n}\left(\frac{x}{1-c}\right)^n$. Now, c is between 0 and x. When $0 < c < x$, $|1 - c| > |1 - x|$, and, when $x < c < 0$, $|1 - c| > 1$. In either case, there is a fixed nonzero number k (independent of n) for which $|1 - c| > k$. Thus, $|R_n(x)| = \frac{1}{n}\left|\frac{x}{1-c}\right|^n < \left|\frac{1}{k}\right|^n \cdot \frac{|x|^n}{n} \to 0$.

39.21 Show that $e = \sum_{n=0}^{\infty}\frac{1}{n!}$.

▌ We know that $e^x = \sum_{n=a}^{\infty}\frac{x^n}{n!}$. Now let $x = 1$.

39.22 How large may the angle x be taken if the values of $\cos x$ are to be computed using three terms of the Taylor series about $\pi/3$ and if the computation is to be correct to four decimal places?

▌ Since $f^{(4)}(x) = \sin x$, the Lagrange remainder $|R_3(x)| = \frac{|\sin c|}{3!}|x - \pi/3|^3 \leq \frac{1}{6}|x - \pi/3|^3 = 0.00005$. Then, $|x - \pi/3| \leq \sqrt[3]{0.0003} \approx 0.0669$. Hence, x can lie between $\pi/3 + 0.0669$ and $\pi/3 - 0.0669$. (0.0669 radian is about 3° 50'.)

39.23 For what range of x can $\cos x$ be replaced by the first three nonzero terms of its Maclaurin series to achieve four decimal place accuracy?

$\blacksquare$ The first three nonzero terms of the Maclaurin series for $\cos x$ are $\quad 1 - x^2/2 + x^4/24$. We must have $|R_5(x)| < 0.00005$. Since $|f^{(5)}(c)| = |-\sin c| \le 1$, $\quad |R_5(x)| = \dfrac{|f^{(5)}(c)|}{5!}\,|x|^5 \le |x|^5/120$. Therefore we require $|x|^5/120 < 0.00005$, $\quad |x|^5 < 0.006$, $\quad |x| < \sqrt[5]{0.006} \approx 0.3594$.

39.24 Estimate the error when $\sqrt{e} = e^{1/2}$ is approximated by the first four terms of the Maclaurin series for e^x.

$\blacksquare$ Since $e^x = 1 + x + x^2/2! + x^3/3! + \cdots$, we are approximating $\sqrt{e}$ by $\quad 1 + \frac{1}{2} + \frac{1}{2}(\frac{1}{2})^2 + \frac{1}{6}(\frac{1}{2})^3 = 1 + \frac{1}{2} + \frac{1}{8} + \frac{1}{48} \approx 1.64583$. The error $R_n(x)$ is $\dfrac{f^{(4)}(c)}{4!}\,(\frac{1}{2})^4$ for some c between 0 and $\frac{1}{2}$. Now, $f^{(4)}(x) = e^x$. The error is $e^c/384$, with $0 < c < \frac{1}{2}$. Now, $e^c < e^{1/2} < 2$, since $e < 4$. Hence, the error is less than $\quad \frac{2}{384} = \frac{1}{192} \approx 0.0052$.

39.25 Use the Maclaurin series to estimate e to within two decimal place accuracy.

$\blacksquare$ We have $e = 1 + 1 + 1/2! + 1/3! + 1/4! + \cdots$. Since $f^{(n)}(x) = e^x$, the error $R_n(x) = e^c/n!$ for some number c such that $0 < c < 1$. Since $e^c < e < 3$, we require that $3/n! < 0.005$, that is, $600 < n!$. Hence, we can let $n = 6$. Then e is estimated by $\quad 1 + 1 + \frac{1}{2} + \frac{1}{6} + \frac{1}{24} + \frac{1}{120} = \frac{326}{120} \approx 2.72$ to two decimal places.

39.26 Find the Maclaurin series for $\cos x^2$.

$\blacksquare$ The Maclaurin series for $\cos x$ is $\displaystyle\sum_{n=0}^{\infty} (-1)^n \frac{x^{2n}}{(2n)!}$, which is valid for all x. Hence, the Maclaurin series for $\cos x^2$ is $\displaystyle\sum_{n=0}^{\infty} (-1)^n \frac{x^{4n}}{(2n)!}$, which also holds for all x.

39.27 Estimate $\int_0^{1/2} \cos x^2\, dx$ to three decimal place accuracy.

$\blacksquare$ By Problem 39.26, $\cos x^2 = 1 - x^4/2! + x^8/4! - x^{12}/6! + \cdots$. Integrate termwise: $\int_0^{1/2} \cos x^2\, dx = x - \frac{1}{5}\frac{x^5}{2!} + \frac{1}{9}\frac{x^9}{4!} - \cdots \Big]_0^{1/2} = \frac{1}{2} - \frac{1}{5}\frac{1}{2!}\left(\frac{1}{2}\right)^5 + \frac{1}{9}\frac{1}{4!}\left(\frac{1}{2}\right)^9 - \cdots$. Since this is an alternating series, we must find the first term that is less than 0.0005. Calculation shows that this term is $\frac{1}{9}\frac{1}{4!}\left(\frac{1}{2}\right)^9$. Hence, we need use only $\frac{1}{2} - \frac{1}{320} = \frac{159}{320} \approx 0.497$.

39.28 Estimate $\ln 1.1$ to within three decimal place accuracy.

$\blacksquare$ $\ln(1 + x) = x - x^2/2 + x^3/3 - x^4/4 + \cdots$ for $|x| < 1$. Thus, $\ln 1.1 = (0.1) - \frac{1}{2}(0.1)^2 + \frac{1}{3}(0.1)^3 - \frac{1}{4}(0.1)^4 + \cdots$. This is an alternating series. We must find n so that $(0.1)^n/n = 1/n10^n < 0.0005$, or $2000 < n10^n$. Hence, $n \ge 3$. Therefore, we may use the first two terms: $0.1 - (0.1)^2/2 = 0.1 - 0.005 = 0.095$.

39.29 Estimate $\int_0^1 \sqrt{x}\,\sin x\, dx$ to within two decimal place accuracy.

$\blacksquare$ $\sin x = x - x^3/3! + x^5/5! - x^7/7! + \cdots$. Hence, $\sqrt{x}\,\sin x = x^{3/2} - x^{7/2}/3! + x^{11/2}/5! - x^{15/2}/7! + \cdots$, and, therefore, $\int_0^1 \sqrt{x}\,\sin x\, dx = \frac{2}{5} - \frac{1}{3!}\frac{2}{9} + \frac{1}{5!}\frac{2}{13} - \frac{1}{7!}\frac{2}{17} + \cdots$. This is an alternating series, and, since $\frac{1}{5!}\frac{2}{13} < 0.005$, we may use the first two terms $\frac{2}{5} - \frac{1}{54} = \frac{103}{270} \approx 0.38$.

39.30 If $f(x) = \displaystyle\sum_{n=0}^{\infty} 2^n x^n$, find $f^{(33)}(0)$.

$\blacksquare$ In general, $a_n = \dfrac{f^{(n)}(0)}{n!}$. So $f^{(33)}(0) = 33!\,a_{33} = 33!\,2^{33}$.

39.31 If $f(x) = \tan^{-1} x$, find $f^{(99)}(0)$.

▌ $\tan^{-1} x = \sum_{n=0}^{\infty} (-1)^n \dfrac{x^{2n+1}}{2n+1}$. Since $a_n = \dfrac{f^{(n)}(0)}{n!}$, $f^{(99)}(0) = 99! \cdot a_{99}$. But $a_{99} = (-1)^{49} \frac{1}{99} = -\frac{1}{99}$. Thus, $f^{(99)}(0) = -(99!)/99 = -(98!)$.

39.32 Find $f^{(100)}(0)$ if $f(x) = e^{x^2}$.

▌ $e^x = \sum_{n=0}^{\infty} \dfrac{1}{n!} x^n$. Hence, $f(x) = \sum_{n=0}^{\infty} \dfrac{1}{n!} x^{2n}$. $\dfrac{f^{(100)}(0)}{100!} = a_{100} = \dfrac{1}{(50)!}$. Hence, $f^{(100)}(0) = \dfrac{100!}{50!}$.

39.33 A certain function $f(x)$ satisfies $f(0) = 2$, $f'(0) = 1$, $f''(0) = 4$, $f'''(0) = 12$, and $f^{(n)}(0) = 0$ if $n > 3$. Find a formula for $f(x)$.

▌ The Maclaurin series for $f(x)$ is the polynomial $2 + x + \dfrac{4}{2!} x^2 + \dfrac{12}{3!} x^3$. Thus, $f(x) = 2 + x + 2x^2 + 2x^3$.

39.34 Exhibit the nth nonzero term of the Maclaurin series for $\ln(1 + x^2)$.

▌ $\ln(1 + x) = x - x^2/2 + x^3/3 - \cdots + (-1)^{n+1} x^n/n + \cdots$ for $|x| < 1$. Hence, $\ln(1 + x^2) = x^2 - x^4/2 + x^6/3 + \cdots + (-1)^{n+1} x^{2n}/n + \cdots$. Thus, the nth nonzero term is $(-1)^{n+1} x^{2n}/n$.

39.35 Exhibit the nth nonzero term of the Maclaurin series for e^{-x^2}.

▌ $e^x = \sum_{n=0}^{\infty} \dfrac{x^n}{n!}$. Then, $e^{-x^2} = \sum_{n=0}^{\infty} (-1)^n \dfrac{x^{2n}}{n!}$. The nth nonzero term is $(-1)^{n-1} x^{2n-2} / (n-1)!$.

39.36 Find the Maclaurin series for $x \sin 3x$.

▌ $\sin x = \sum_{n=0}^{\infty} (-1)^n \dfrac{x^{2n+1}}{(2n+1)!}$. Hence, $\sin 3x = \sum_{n=0}^{\infty} (-1)^n \cdot 3^{2n+1} \dfrac{x^{2n+1}}{(2n+1)!}$, and $x \sin 3x = \sum_{n=0}^{\infty} (-1)^n 3^{2n+1} \dfrac{x^{2n+2}}{(2n+1)!}$.

39.37 Find the Taylor series for $f(x) = 2x^2 + 4x - 3$ about 1.

▌ **Method 1.** $f'(x) = 4x + 4$, $f''(x) = 4$, and $f^{(n)}(x) = 0$ for $n > 2$. Thus, $f(1) = 3$, $f'(1) = 8$, $f''(1) = 4$, and $f(x) = 3 + 8(x - 1) + 2(x - 1)^2$.

Method 2. $f(x) = 2[(x - 1) + 1]^2 + 4[(x - 1) + 1] - 3 = 2(x - 1)^2 + 4(x - 1) + 2 + 4(x - 1) + 4 - 3 = 3 + 8(x - 1) + 2(x - 1)^2$. This is the Taylor series, by Problem 39.11.

39.38 Find the Taylor series for x^4 about -3.

▌ $x^4 = [-3 + (x + 3)]^4$

$= (-3)^4 + 4(-3)^3(x + 3) + \dfrac{4 \cdot 3}{1 \cdot 2} (-3)^2 (x + 3)^2 + \dfrac{4 \cdot 3 \cdot 2}{1 \cdot 2 \cdot 3} (-3)(x + 3)^3 + \dfrac{4 \cdot 3 \cdot 2 \cdot 1}{1 \cdot 2 \cdot 3 \cdot 4} (x + 3)^4$

$= 81 - 108(x + 3) + 54(x + 3)^2 - 12(x + 3)^3 + (x + 3)^4$.

39.39 Find the Maclaurin series for $\int \sqrt{1 + x^3} \, dx$.

▌ By Problem 38.104, $\sqrt{1 + x} = 1 + \dfrac{1}{2} x + \cdots + \dfrac{\frac{1}{2}(\frac{1}{2} - 1) \cdots (\frac{1}{2} - n + 1)}{n!} x^n + \cdots$. Then,

$\sqrt{1 + x^3} = 1 + \dfrac{1}{2} x^3 + \cdots + \dfrac{\frac{1}{2}(\frac{1}{2} - 1) \cdots (\frac{1}{2} - n + 1)}{n!} x^{3n} + \cdots$ and $\int \sqrt{1 + x^3} \, dx = x + \dfrac{1}{8} x^4 + \cdots +$

$\dfrac{\frac{1}{2}(\frac{1}{2} - 1) \cdots (\frac{1}{2} - n + 1)}{n!(3n + 1)} x^{3n+1} + \cdots$. The nth term may be rewritten as $(-1)^{n-1} \dfrac{1 \cdot 3 \cdot 5 \cdot \cdots \cdot (2n - 3)}{2^n n!(3n + 1)} x^{3n+1}$

for $n > 1$.

39.40 Estimate $\int_0^{1/2} \dfrac{\ln(1+x)}{x}\,dx$ to within two decimal place accuracy.

▮ For $|x| < 1$, $\dfrac{\ln(1+x)}{x} = \displaystyle\sum_{n=1}^{\infty} (-1)^{n+1} \dfrac{x^{n-1}}{n}$. Hence, $\int_0^{1/2} \dfrac{\ln(1+x)}{x}\,dx = \displaystyle\sum_{n=1}^{\infty}(-1)^{n+1}\dfrac{x^n \big]_0^{1/2}}{n^2} = $
$\displaystyle\sum_{n=1}^{\infty}\dfrac{(-1)^{n+1}}{n^2 2^n}$. This is an alternating series. Therefore, we seek n for which $1/n^2 2^n \le 0.005$, $200 \le n^2 2^n$, $n \ge 4$. Hence, we may use $\frac{1}{2} - \frac{1}{16} + \frac{1}{72} = \frac{65}{144} \approx 0.45$.

39.41 Approximate $\int_0^1 e^{-x^2}\,dx$ to within two decimal place accuracy.

▮ $e^x = \displaystyle\sum_{n=0}^{\infty}\dfrac{x^n}{n!}$. Hence, $e^{-x^2} = \displaystyle\sum_{n=0}^{\infty}\dfrac{(-1)^n x^{2n}}{n!}$, and $\int_0^1 e^{-x^2}\,dx = \displaystyle\sum_{n=0}^{\infty}\left(\dfrac{(-1)^n x^{2n+1}}{(2n+1)n!}\Big]_0^1\right) = \displaystyle\sum_{n=0}^{\infty}\dfrac{(-1)^n}{(2n+1)n!}$.
Since this is an alternating series, we must find n such that $\dfrac{1}{(2n+1)n!} \le 0.005$, $200 \le (2n+1)n!$, $n \ge 4$. Hence, we use $1 - \frac{1}{3} + \frac{1}{10} - \frac{1}{42} = \frac{26}{35} \approx 0.74$.

39.42 Find the Maclaurin series for $f(x) = \dfrac{1}{x^2+x+1}$.

▮ $f(x) = \dfrac{1-x}{1-x^3}$. From $\dfrac{1}{1-x} = \displaystyle\sum_{n=0}^{\infty} x^n$, we obtain $\dfrac{1}{1-x^3} = \displaystyle\sum_{n=0}^{\infty} x^{3n}$, and so $f(x) = \dfrac{1}{1-x^3} - x\dfrac{1}{1-x^3} = $
$\displaystyle\sum_{n=0}^{\infty} x^{3n} - \displaystyle\sum_{n=0}^{\infty} x^{3n+1}$.

39.43 If $f(x) = \dfrac{1}{x^2+x+1}$, find $f^{(36)}(0)$.

▮ From the Maclaurin expansion found in Problem 39.42, $\dfrac{f^{(36)}(0)}{36!} = a_{36} = +1$ (because 36 is divisible by 3); hence, $f^{(36)}(0) = 36!$.

39.44 Prove that e is irrational.

▮ $e^x = \displaystyle\sum_{n=0}^{\infty}\dfrac{x^n}{n!}$. Hence, $\dfrac{1}{e} = \displaystyle\sum_{n=0}^{\infty}\dfrac{(-1)^n}{n!}$. By the alternating series theorem, $0 < \left|\dfrac{1}{e} - \displaystyle\sum_{n=0}^{k-1}\dfrac{(-1)^n}{n!}\right| < \dfrac{1}{k!}$.
Therefore, $0 < (k-1)!\left|\dfrac{1}{e} - \displaystyle\sum_{n=0}^{k-1}\dfrac{(-1)^n}{n!}\right| < \dfrac{1}{k} \le \dfrac{1}{2}$ for $k \ge 2$. So $0 < \left|\dfrac{(k-1)!}{e} - (k-1)!\displaystyle\sum_{n=0}^{k-1}\dfrac{(-1)^n}{n!}\right| < \dfrac{1}{2}$.
But $(k-1)!\displaystyle\sum_{n=0}^{k-1}\dfrac{(-1)^n}{n!}$ is an integer. If e were rational, then k could be chosen large enough so that $(k-1)!/e$ would be an integer. Hence, for k large enough and suitably chosen, $\Big|\dfrac{(k-1)!}{e} - $
$(k-1)!\displaystyle\sum_{n=0}^{k-1}\dfrac{(-1)^n}{n!}\Big|$ would be an integer strictly between 0 and $\frac{1}{2}$, which is impossible.

39.45 Find the Taylor series for $\cos x$ about $\pi/2$.

▮ $\cos x = \sin(\pi/2 - x) = -\sin(x - \pi/2)$. Since $\sin x = \displaystyle\sum_{n=0}^{\infty}(-1)^n\dfrac{x^{2n+1}}{(2n+1)!}$, we have
$\cos x = \displaystyle\sum_{n=0}^{\infty}(-1)^{n+1}\dfrac{(x-\pi/2)^{2n+1}}{(2n+1)!}$.

39.46 Find the Maclaurin series for $\ln(2+x)$.

▮ $\ln(2+x) = \ln[2(1+x/2)] = \ln 2 + \ln(1+x/2)$. But $\ln(1+x) = \displaystyle\sum_{n=1}^{\infty}\dfrac{(-1)^{n+1}x^n}{n}$. Hence, $\ln\left(1+\dfrac{x}{2}\right) = $
$\displaystyle\sum_{n=1}^{\infty}\dfrac{(-1)^{n+1}}{n}\left(\dfrac{x}{2}\right)^n = \displaystyle\sum_{n=1}^{\infty}\dfrac{(-1)^{n+1}}{n2^n}x^n$. Thus, $\ln(2+x) = \ln 2 + \displaystyle\sum_{n=1}^{\infty}\dfrac{(-1)^{n+1}}{n2^n}x^n$.

39.47 Find the Taylor series for $\sqrt{x}$ about 2.

$\sqrt{x} = \sqrt{2 + (x - 2)} = \sqrt{2}\sqrt{1 + [(x - 2)/2]}$. Recall from Problem 38.104 that

$$\sqrt{1 + x} = 1 + \sum_{n=1}^{\infty} \frac{\frac{1}{2}(\frac{1}{2} - 1) \cdots (\frac{1}{2} - n + 1)}{n!} x^n$$

Hence,
$$\sqrt{1 + [(x - 2)/2]} = 1 + \sum_{n=1}^{\infty} \frac{\frac{1}{2}(\frac{1}{2} - 1) \cdots (\frac{1}{2} - n + 1)}{n!} \frac{(x - 2)^n}{2^n}$$

Thus,
$$\sqrt{x} = \sqrt{2} + \sqrt{2} \sum_{n=1}^{\infty} \frac{1 \cdot (1 - 2) \cdot (1 - 4) \cdot \cdots \cdot [1 - 2(n - 1)]}{n! \cdot 2^{2n}} (x - 2)^n$$

$$= \sqrt{2} + \sqrt{2} \sum_{n=1}^{\infty} (-1)^{n-1} \frac{1 \cdot 3 \cdot 5 \cdot \cdots \cdot (2n - 3)}{2^{2n} n!} (x - 2)^n$$

39.48 Find the Maclaurin series for $\sin x \cos x$.

$\sin x \cos x = \frac{1}{2} \sin 2x$. Now, $\sin x = \sum_{n=0}^{\infty} (-1)^n \frac{x^{2n+1}}{(2n + 1)!}$. So $\sin 2x = \sum_{n=0}^{\infty} (-1)^n \frac{2^{2n+1} x^{2n+1}}{(2n + 1)!}$, and,

therefore, $\dfrac{\sin 2x}{2} = \sum_{n=0}^{\infty} (-1)^n \dfrac{2^{2n}}{(2n + 1)!} x^{2n+1}$.

39.49 Show that $1 = \sum_{n=1}^{\infty} \dfrac{n}{(n + 1)!}$.

$e^x = \sum_{n=0}^{\infty} \dfrac{x^n}{n!}$. Hence, $e^x - 1 = \sum_{n=1}^{\infty} \dfrac{x^n}{n!}$, and, therefore, $\dfrac{e^x - 1}{x} = \sum_{n=1}^{\infty} \dfrac{x^{n-1}}{n!} = \sum_{n=0}^{\infty} \dfrac{x^n}{(n + 1)!}$.

Now, let $g(x) = \dfrac{d}{dx}\left(\dfrac{e^x - 1}{x}\right) = \sum_{n=1}^{\infty} \dfrac{nx^{n-1}}{(n + 1)!}$. Then $g(1) = \sum_{n=1}^{\infty} \dfrac{n}{(n + 1)!}$. On the other hand, $g(x) =$
$\dfrac{d}{dx}\left(\dfrac{e^x - 1}{x}\right) = \dfrac{xe^x - (e^x - 1)}{x^2}$, and, thus, $g(1) = \dfrac{e - (e - 1)}{1} = 1$.

39.50 Find the Maclaurin series for 2^x.

$2^x = e^{x \ln 2}$. Now, $e^x = \sum_{n=0}^{\infty} \dfrac{x^n}{n!}$. Therefore, $2^x = e^{x \ln 2} = \sum_{n=0}^{\infty} \dfrac{(\ln 2)^n}{n!} x^n$.

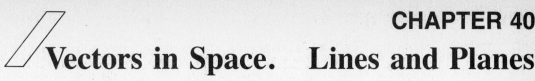

CHAPTER 40

Vectors in Space.   Lines and Planes

40.1 Find the distance between the points $(2, 4, 7)$ and $(1, 5, 10)$.

The distance between any two points (x_1, y_1, z_1) and (x_2, y_2, z_2) is $\sqrt{(x_1 - x_2)^2 + (y_1 - y_2)^2 + (z_1 - z_2)^2}$. Hence, the distance between $(2, 4, 7)$ and $(1, 5, 10)$ is $\sqrt{(2-1)^2 + (4-5)^2 + (7-10)^2} = \sqrt{1+1+9} = \sqrt{11}$.

40.2 Find the distance between $(-1, 2, 0)$ and $(4, 3, -5)$.

The distance is $\sqrt{(-1-4)^2 + (2-3)^2 + [0-(-5)]^2} = \sqrt{25+1+25} = \sqrt{51}$.

40.3 Find the distance between a point (x, y, z) and the origin $(0, 0, 0)$.

The distance is $\sqrt{(x-0)^2 + (y-0)^2 + (z-0)^2} = \sqrt{x^2 + y^2 + z^2}$.

40.4 Find the distance between $(2, -1, 4)$ and the origin.

By Problem 40.3, the distance is $\sqrt{(2)^2 + (-1)^2 + (4)^2} = \sqrt{4+1+16} = \sqrt{21}$.

40.5 Find the midpoint of the line segment connecting the points $(4, 3, 1)$ and $(-2, 5, 7)$.

The coordinates of the midpoint are the averages of the coordinates of the endpoints. In this case, the midpoint is $\left[\dfrac{4+(-2)}{2}, \dfrac{3+5}{2}, \dfrac{1+7}{2} \right] = (1, 4, 4)$.

40.6 Find the equation of a sphere $\mathscr{S}$ of radius r and center (a, b, c).

A point (x, y, z) is on $\mathscr{S}$ if and only if its distance from (a, b, c) is r, that is, if and only if $\sqrt{(x-a)^2 + (y-b)^2 + (z-c)^2} = r$, or, equivalently, $(x-a)^2 + (y-b)^2 + (z-c)^2 = r^2$.

40.7 Find the equation of the sphere with radius 5 and center $(2, -1, 3)$.

By Problem 40.6, the equation is $(x-2)^2 + (y+1)^2 + (z-3)^2 = 25$.

40.8 Describe the surface with the equation $x^2 + y^2 + z^2 = 49$.

By Problem 40.6, this is the sphere with center $(0, 0, 0)$ and radius 7.

40.9 Describe the graph of the equation $x^2 + 4x + y^2 + z^2 - 8z = 5$.

Complete the square in x and in z: $(x+2)^2 + y^2 + (z-4)^2 = 5 + 4 + 16 = 25$. This is the equation of the sphere with center $(-2, 0, 4)$ and radius 5.

40.10 Show that every sphere has an equation of the form $x^2 + y^2 + z^2 + Ax + By + Cz + D = 0$.

The sphere with center (a, b, c) and radius r has the equation $(x-a)^2 + (y-b)^2 + (z-c)^2 = r^2$. Expanding, we obtain $x^2 - 2ax + a^2 + y^2 - 2by + b^2 + z^2 - 2cz + c^2 = r^2$, which is equivalent to $x^2 + y^2 + z^2 - 2ax - 2by - 2cz + (a^2 + b^2 + c^2 - r^2) = 0$.

40.11 When does an equation $x^2 + y^2 + z^2 + Ax + By + Cz + D = 0$ represent a sphere?

Complete the squares: $\left(x + \dfrac{A}{2}\right)^2 + \left(y + \dfrac{B}{2}\right)^2 + \left(z + \dfrac{C}{2}\right)^2 = \dfrac{A^2}{4} + \dfrac{B^2}{4} + \dfrac{C^2}{4} - D$. This is a sphere if and only if the right side is positive; that is, if and only if $A^2 + B^2 + C^2 - 4D > 0$. In that case, the sphere has radius $\frac{1}{2}\sqrt{A^2 + B^2 + C^2 - 4D}$ and center $\left(-\dfrac{A}{2}, -\dfrac{B}{2}, -\dfrac{C}{2}\right)$. When $A^2 + B^2 + C^2 - 4D = 0$, the graph is a single point $\left(-\dfrac{A}{2}, -\dfrac{B}{2}, -\dfrac{C}{2}\right)$. When $A^2 + B^2 + C^2 - 4D < 0$, there are no points on the graph at all.

40.12   Find an equation of the sphere with the points $P(7, -1, -2)$ and $Q(3, 4, 6)$ as the ends of a diameter.

▌ The center is the midpoint of the segment PQ, namely, $(5, \frac{3}{2}, 4)$. The length of the diameter is $\sqrt{(7-3)^2 + (-1-4)^2 + (2-6)^2} = \sqrt{16+25+16} = \sqrt{57}$. So, the radius is $\frac{1}{2}\sqrt{57}$. Thus, an equation of the sphere is $(x-5)^2 + (y - \frac{3}{2})^2 + (z-4)^2 = \frac{57}{4}$.

40.13   Show that the three points $P(1, 2, 3)$, $Q(4, -5, 2)$, and $R(0, 0, 0)$ are the vertices of a right triangle.

▌ $\overline{PR} = \sqrt{1^2 + 2^2 + 3^2} = \sqrt{14}$, $\overline{QR} = \sqrt{4^2 + (-5)^2 + 2^2} = \sqrt{45}$, $\overline{PQ} = \sqrt{(1-4)^2 + [2-(-5)]^2 + (3-2)^2} = \sqrt{59}$. Thus, $\overline{PQ}^2 = \overline{PR}^2 + \overline{QR}^2$. Hence, by the converse of the Pythagorean theorem, $\triangle PQR$ is a right triangle with right angle at R.

40.14   If line L passes through point $(1, 2, 3)$ and is perpendicular to the xy-plane, what are the coordinates of the points on the line that are at a distance 7 from the point $P(3, -1, 5)$?

▌ The points on L have coordinates $(1, 2, z)$. Their distance from P is $\sqrt{(3-1)^2 + (-1-2)^2 + (5-z)^2} = \sqrt{13 + (5-z)^2}$. Set this equal to 7. $\sqrt{13 + (5-z)^2} = 7$, $13 + (5-z)^2 = 49$, $(5-z)^2 = 36$, $5 - z = \pm 6$, $z = -1$ or $z = 11$. So, the required points are $(1, 2, -1)$ and $(1, 2, 11)$.

40.15   Show that the points $P(2, -1, 5)$, $Q(6, 0, 6)$, and $R(14, 2, 8)$ are collinear.

▌ $\overline{PQ} = \sqrt{4^2 + 1^2 + 1^2} = \sqrt{18} = 3\sqrt{2}$. $\overline{QR} = \sqrt{8^2 + 2^2 + 2^2} = \sqrt{72} = 6\sqrt{2}$. $\overline{PR} = \sqrt{12^2 + 3^2 + 3^2} = \sqrt{162} = 9\sqrt{2}$. So, $\overline{PR} = \overline{PQ} + \overline{QR}$. Hence, the points are collinear. (If three points are not collinear, they form a triangle; then the sum of two sides must be greater than the third side.)

40.16   Find an equation of the sphere with center in the xz-plane and passing through the points $P(0, 8, 0)$, $Q(4, 6, 2)$, and $R(0, 12, 4)$.

▌ Let $C(a, 0, c)$ be the center. Then $\overline{PC} = \overline{QC} = \overline{RC}$. So, $\sqrt{a^2 + 64 + c^2} = \sqrt{(a-4)^2 + 36 + (c-2)^2} = \sqrt{a^2 + 144 + (c-4)^2}$. Hence, $a^2 + 64 + c^2 = (a-4)^2 + 36 + (c-2)^2 = a^2 + 144 + (c-4)^2$. Equate the first and third terms: $64 + c^2 = 144 + (c-4)^2$, $c^2 = 80 + c^2 - 8c + 16$, $8c = 96$, $c = 12$. Substitute 12 for c in the first equation: $a^2 + 64 + 144 = (a-4)^2 + 36 + 100$, $a^2 + 72 = a^2 - 8a + 16$, $8a = -56$, $a = -7$. So, the center is $(-7, 0, 12)$. The radius $\overline{PC} = \sqrt{49 + 64 + 144} = \sqrt{257}$.   E.g.: $(x+7)^2 + y^2 + (z-12)^2 = 257$

40.17   If the midpoints of the sides of $\triangle PQR$ are $(5, -1, 3)$, $(4, 2, 1)$, and $(2, 1, 0)$, find the vertices.

▌ Let $P = (x_1, y_1, z_1)$, $Q = (x_2, y_2, z_2)$, $R = (x_3, y_3, z_3)$. Assume $(5, -1, 3)$ is the midpoint of PQ, $(4, 2, 1)$ is the midpoint of QR, and $(2, 1, 0)$ is the midpoint of PR. Then, $(x_1 + x_2)/2 = 5$, $(x_2 + x_3)/2 = 4$, and $(x_1 + x_3)/2 = 2$. So, $x_1 + x_2 = 10$, $x_2 + x_3 = 8$, and $x_1 + x_3 = 4$. Subtract the second equation from the first: $x_1 - x_3 = 2$, and add this to the third equation: $2x_1 = 6$, $x_1 = 3$. Hence, $x_2 = 7$, $x_3 = 1$. Similarly, $y_1 + y_2 = -2$, $y_2 + y_3 = 4$, $y_1 + y_3 = 2$. Then, $y_1 - y_3 = -6$, $2y_1 = -4$, $y_1 = -2$. So, $y_2 = 0$, $y_3 = 4$. Finally, $z_1 + z_2 = 6$, $z_2 + z_3 = 2$, $z_1 + z_3 = 0$. Therefore, $z_1 - z_3 = 4$, $2z_1 = 4$, $z_1 = 2$, $z_2 = 4$, $z_3 = -2$. Hence, $P = (3, -2, 2)$, $Q = (7, 0, 4)$, and $R = (1, 4, -2)$.

40.18   Describe the intersection of the graphs of $x^2 + y^2 = 1$ and $z = 2$.

▌ As shown in Fig. 40-1, $x^2 + y^2 = 1$ is a cylinder of radius 1 with the z-axis as its axis of symmetry. $z = 2$ is a plane two units above and parallel to the xy-plane. Hence, the intersection is a circle of radius 1 with center at $(0, 0, 2)$ in the plane $z = 2$.

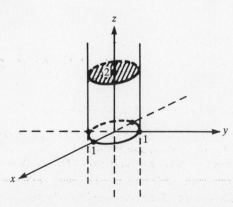

Fig. 40-1

40.19 Describe the intersection of the graphs of $y = x$ and $y = 5$.

▐ See Fig. 40-2. $y = x$ is a plane, obtained by moving the line $y = x$ in the xy-plane in a direction perpendicular to the xy-plane; $y = 5$ is a plane parallel to the xz-plane. The intersection is the line through the point $(5, 5, 0)$ and perpendicular to the xy-plane.

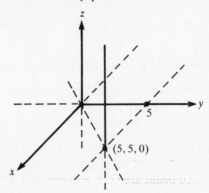

Fig. 40-2

40.20 Find the point on the y-axis equidistant from $(2, 5, -3)$ and $(-3, 6, 1)$.

▐ Let the point be $(0, y, 0)$. Then $\sqrt{(2-0)^2 + (5-y)^2 + (-3-0)^2} = \sqrt{(-3-0)^2 + (6-y)^2 + (1-0)^2}$, $13 + (5-y)^2 = 10 + (6-y)^2$, $y^2 - 10y + 38 = y^2 - 12y + 46$, $2y = 8$, $y = 4$. The desired point is $(0, 4, 0)$.

40.21 Describe the graph of $x^2 + y^2 = 0$.

▐ $x^2 + y^2 = 0$ if and only if $x = 0$ and $y = 0$. The graph consists of all points $(0, 0, z)$, that is, the z-axis.

40.22 Write an equation of the sphere with radius 2 and center on the positive x-axis, and tangent to the yz-plane.

▐ The radius from the center to the point of tangency on the yz-plane must be perpendicular to that plane. Hence, the radius must lie on the x-axis and the point of tangency must be the origin. So, the center must be $(2, 0, 0)$ and the equation of the sphere is $(x-2)^2 + y^2 + z^2 = 4$.

40.23 Find an equation of the sphere with center at $(1, 2, 3)$ and tangent to the yz-plane.

▐ The radius from the center to the yz-plane is perpendicular to the yz-plane and, therefore, cuts the yz-plane at $(0, 2, 3)$. So, the radius is 1, and an equation of the sphere is $(x-1)^2 + (y-2)^2 + (z-3)^2 = 1$.

40.24 Find the locus of all points (x, y, z) that are twice as far from $(3, 2, 0)$ as from $(3, 2, 6)$.

▐ $\sqrt{(x-3) + (y-2)^2 + z^2} = 2\sqrt{(x-3)^2 + (y-2)^2 + (z-6)^2}$, $(x-3)^2 + (y-2)^2 + z^2 = 4[(x-3)^2 + (y-2)^2 + (z-6)^2]$, $3(x-3)^2 + 3(y-2)^2 + 4(z-6)^2 - z^2 = 0$, $3x^2 - 18x + 27 + 3y^2 - 12y + 12 + 3z^2 - 48z + 144 = 0$, $x^2 - 6x + y^2 - 4y + z^2 - 16z + 61 = 0$, $(x-3)^2 + (y-2)^2 + (z-8)^2 + 61 = 9 + 4 + 64$, $(x-3)^2 + (y-2)^2 + (z-8)^2 = 16$. Thus, the locus is the sphere with center $(3, 2, 8)$ and radius 4.

40.25 Find the vector $\overrightarrow{PQ}$ from $P = (1, -2, 4)$ to $Q = (3, 4, 3)$.

▐ In general, the vector $\overrightarrow{PQ}$ from $P(x_1, y_1, z_1)$ to $Q(x_2, y_2, z_2)$ is $\overrightarrow{PQ} = (x_2 - x_1, y_2 - y_1, z_2 - z_1)$. In this case, $\overrightarrow{PQ} = (2, 6, -1)$.

40.26 Find the length of the vector $\overrightarrow{PQ}$ of Problem 40.25.

▐ The length $|\mathbf{A}|$ of a vector $\mathbf{A} = (u, v, w)$ is $\sqrt{u^2 + v^2 + w^2}$. Since $\overrightarrow{PQ} = (2, 6, -1)$, the length $|\overrightarrow{PQ}| = \sqrt{4 + 36 + 1} = \sqrt{41}$.

40.27 Find the direction cosines of the vector $\overrightarrow{PQ}$ of Problem 40.25.

▐ In general, if $\mathbf{A} = (a, b, c)$, then the direction cosines of $\mathbf{A}$ are $\cos \alpha = a/|\mathbf{A}|$, $\cos \beta = b/|\mathbf{A}|$, $\cos \gamma = c/|\mathbf{A}|$, where α, β, and γ are the direction angles between $\mathbf{A}$ and the positive x-axis, y-axis, and z-axis, respectively. These angles are between 0 and π (inclusive). Thus, $\mathbf{A} = (|\mathbf{A}| \cos \alpha, |\mathbf{A}| \cos \beta, |\mathbf{A}| \cos \gamma) = |\mathbf{A}| (\cos \alpha, \cos \beta, \cos \gamma)$. In this case, $\overrightarrow{PQ} = (2, 6, -1)$ and $|\overrightarrow{PQ}| = \sqrt{41}$. So, $\cos \alpha = 2/\sqrt{41}$, $\cos \beta = 6/\sqrt{41}$, $\cos \gamma = -1/\sqrt{41}$. Thus, α and β are acute and γ is obtuse.

40.28 Find the direction cosines of $A = 3i + 12j + 4k$. [Recall that $i = (1, 0, 0)$, $j = (0, 1, 0)$, and $k = (0, 0, 1)$.]

▮ $|A| = \sqrt{9 + 144 + 16} = \sqrt{169} = 13$. Hence, $\cos \alpha = \frac{3}{13}$, $\cos \beta = \frac{12}{13}$, $\cos \gamma = \frac{4}{13}$.

40.29 Let vector $A = (a, b, c)$ have direction cosines $\cos \alpha$, $\cos \beta$, $\cos \gamma$. Show that $\cos^2 \alpha + \cos^2 \beta + \cos^2 \gamma = 1$.

▮ $\cos^2 \alpha + \cos^2 \beta + \cos^2 \gamma = \dfrac{a^2}{|A|^2} + \dfrac{b^2}{|A|^2} + \dfrac{c^2}{|A|^2} = \dfrac{a^2 + b^2 + c^2}{|A|^2} = \dfrac{|A|^2}{|A|^2} = 1$.

40.30 Given vectors $A = (3, 2, -1)$, $B = (5, 3, 0)$, and $C = (-2, 4, 1)$, calculate $A + B$, $A - C$, and $\frac{1}{2}A$.

▮ Vector addition and subtraction and multiplication by a scalar are all done componentwise. Thus, $A + B = (3 + 5, 2 + 3, -1 + 0) = (8, 5, -1)$, $A - C = (3 - (-2), 2 - 4, -1 - 1) = (5, -2, -2)$, and $\frac{1}{2}A = (\frac{1}{2}(3), \frac{1}{2}(2), \frac{1}{2}(-1)) = (\frac{3}{2}, 1, -\frac{1}{2})$.

40.31 Find the unit vector u in the same direction as $A = (-2, 3, 6)$.

▮ $|A| = \sqrt{4 + 9 + 36} = \sqrt{49} = 7$. So, $u = \frac{1}{7}A = (-\frac{2}{7}, \frac{3}{7}, \frac{6}{7})$. Note that $u = (\cos \alpha, \cos \beta, \cos \gamma)$, the vector composed of the direction cosines of A.

40.32 Assume the direction angles of a vector A are equal. What are these angles?

▮ We have $\alpha = \beta = \gamma$. By Problem 40.29, $\cos^2 \alpha + \cos^2 \beta + \cos^2 \gamma = 1$. So, $3\cos^2 \alpha = 1$, $\cos \alpha = \pm 1/\sqrt{3}$. Therefore, either $\alpha = \beta = \gamma = \cos^{-1}(1/\sqrt{3})$ or $\alpha = \beta = \gamma = \pi - \cos^{-1}(1/\sqrt{3})$.

40.33 Find the vector projection of $A = (1, 2, 4)$ on $B = (4, -2, 4)$.

▮ As in the planar case, the scalar projection of A on B is $A \cdot B/|B|$ (see Fig. 40-3). As the unit vector in the direction of B is $B/|B|$, the vector projection is $\dfrac{A \cdot B}{|B|^2} B = \dfrac{A \cdot B}{B \cdot B} B$. For the data, $A \cdot B = 1(4) + 2(-2) + 4(4) = 16$ and $B \cdot B = 4^2 + (-2)^2 + 4^2 = 36$. So, the vector projection of A on B is $\frac{16}{36}B = \frac{4}{9}B = (\frac{16}{9}, -\frac{8}{9}, \frac{16}{9})$.

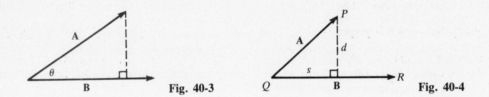

Fig. 40-3   Fig. 40-4

40.34 Find the distance from the point $P(1, -1, 2)$ to the line connecting the point $Q(3, 1, 4)$ to $R(1, -3, 0)$.

▮ Let $A = \overrightarrow{QP} = (-2, -2, -2)$, $B = \overrightarrow{QR} = (-2, -4, -4)$. Then (see Fig. 40-4) the scalar projection of A on B is $s = A \cdot B/|B| = (4 + 8 + 8)/\sqrt{4 + 16 + 16} = \frac{10}{3}$ and the Pythagorean theorem gives $d^2 = |A|^2 - s^2 = (4 + 4 + 4) - \frac{100}{9} = \frac{8}{9}$ or $d = 2\sqrt{2}/3$.

40.35 Find the angle θ between the vectors $A = (1, 2, 3)$ and $B = (2, -3, -1)$.

▮ $A \cdot B = |A|\,|B| \cos \theta$, $1(2) + 2(-3) + 3(-1) = \sqrt{1 + 4 + 9}\,\sqrt{4 + 9 + 1}\,\cos \theta$, $-7 = 14\cos \theta$, $\cos \theta = -\frac{1}{2}$, $\theta = 120° = 2\pi/3$.

40.36 Show that the triangle with vertices $P(4, 3, 6)$, $Q(-2, 0, 8)$, $R(1, 5, 0)$ is a right triangle and find its area.

▮ $\overrightarrow{PQ} = (-6, -3, 2)$ and $\overrightarrow{PR} = (-3, 2, -6)$. So, $\overrightarrow{PQ} \cdot \overrightarrow{PR} = 18 - 6 - 12 = 0$. Therefore, $\overrightarrow{PQ}$ is perpendicular to $\overrightarrow{PR}$. $|\overrightarrow{PQ}| = \sqrt{36 + 9 + 4} = 7$, $|\overrightarrow{PR}| = \sqrt{9 + 4 + 36} = 7$. Hence, the area is $\frac{1}{2}(7)(7) = \frac{49}{2}$.

40.37 For a unit cube, find the angle θ between a diagonal $\overrightarrow{OP}$ and the diagonal $\overrightarrow{OQ}$ of an adjacent face. (See Fig. 40-5.)

▮ $\overrightarrow{OP} = (1, 1, 1)$, $\overrightarrow{OQ} = (1, 1, 0)$. $\overrightarrow{OP} \cdot \overrightarrow{OQ} = |\overrightarrow{OP}|\,|\overrightarrow{OQ}| \cos \theta$, $2 = \sqrt{3}\,\sqrt{2}\cos \theta$, $\cos \theta = \sqrt{6}/3$. Hence, $\theta \approx 35°16'$.

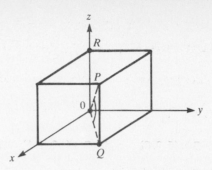

Fig. 40-5

40.38 For a unit cube, find the angle ψ between a diagonal $\overrightarrow{OP}$ and an adjacent edge $\overrightarrow{OR}$.   (See Fig. 40-5.)

▮ $\overrightarrow{OP} = (1, 1, 1)$, $\overrightarrow{OR} = (0, 0, 1)$. $\overrightarrow{OP} \cdot \overrightarrow{OR} = |\overrightarrow{OP}|\,|\overrightarrow{OR}|\cos\psi$, $1 = \sqrt{3}\cos\psi = \sqrt{3}/3$, $\psi \approx 54°\ 44'$.   [By geometry, $\psi = 90° - \theta$.]

40.39 Find a value of c for which $\mathbf{A} = 3\mathbf{i} - 2\mathbf{j} + 5\mathbf{k}$ and $\mathbf{B} = 2\mathbf{i} + 4\mathbf{j} + c\mathbf{k}$ will be perpendicular.

▮   We must have $0 = \mathbf{A} \cdot \mathbf{B} = 3(2) + (-2)(4) + 5c$, $5c = 2$, $c = \frac{2}{5}$.

40.40 Write the formula for the cross product $\mathbf{A} \times \mathbf{B}$,   where $\mathbf{A} = (a_1, a_2, a_3)$ and $\mathbf{B} = (b_1, b_2, b_3)$.

▮ $\mathbf{A} \times \mathbf{B} = (a_2 b_3 - a_3 b_2,\ a_3 b_1 - a_1 b_3,\ a_1 b_2 - a_2 b_1)$.   In pseudo-determinant form,

$$\mathbf{A} \times \mathbf{B} = \begin{vmatrix} \mathbf{i} & \mathbf{j} & \mathbf{k} \\ a_1 & a_2 & a_3 \\ b_1 & b_2 & b_3 \end{vmatrix}$$

If we expand along the first row, we obtain

$$\begin{vmatrix} a_2 & a_3 \\ b_2 & b_3 \end{vmatrix}\mathbf{i} - \begin{vmatrix} a_1 & a_3 \\ b_1 & b_3 \end{vmatrix}\mathbf{j} + \begin{vmatrix} a_1 & a_2 \\ b_1 & b_2 \end{vmatrix}\mathbf{k} = (a_2 b_3 - a_3 b_2)\mathbf{i} + (a_3 b_1 - a_1 b_3)\mathbf{j} + (a_1 b_2 - a_2 b_1)\mathbf{k}$$

40.41 Verify that $\mathbf{A} \times \mathbf{B}$ is perpendicular to both $\mathbf{A}$ and $\mathbf{B}$, when $\mathbf{A} = (1, 3, -1)$ and $\mathbf{B} = (2, 0, 1)$.

▮ $\mathbf{A} \times \mathbf{B} = (3(1) - (-1)0,\ (-1)2 - 1(1),\ 1(0) - 3(2)) = (3, -3, -6)$, $\mathbf{A} \cdot (\mathbf{A} \times \mathbf{B}) = (1, 3, -1) \cdot (3, -3, -6) = 3 - 9 + 6 = 0$, $\mathbf{B} \cdot (\mathbf{A} \times \mathbf{B}) = (2, 0, 1) \cdot (3, -3, -6) = 6 - 6 = 0$.   Therefore, $\mathbf{A} \perp (\mathbf{A} \times \mathbf{B})$ and $\mathbf{B} \perp (\mathbf{A} \times \mathbf{B})$.

40.42 Find a vector $\mathbf{N}$ that is perpendicular to the plane of the three points $P(1, -1, 4)$, $Q(2, 0, 1)$, and $R(0, 2, 3)$.

▮ $\overrightarrow{PQ} = (1, 1, -3)$, $\overrightarrow{PR} = (-1, 3, -1)$.   We can take $\mathbf{N} = \overrightarrow{PQ} \times \overrightarrow{PR} = (1(-1) - (-3)3,\ (-3)(-1) - 1(-1),\ 1(3) - 1(-1)) = (8, 4, 4)$.

40.43 Find the area of $\triangle PQR$ of Problem 40.42.

▮   Recall that $|\overrightarrow{PQ} \times \overrightarrow{PR}| = |\overrightarrow{PQ}|\,|\overrightarrow{PR}|\sin\theta$ is the area of the parallelogram formed by $\overrightarrow{PQ}$ and $\overrightarrow{PR}$.   So, the area is $|\mathbf{N}| = |(8, 4, 4)| = \sqrt{64 + 16 + 16} = \sqrt{96} = 4\sqrt{6}$.   The area of $\triangle PQR$ is half the area of the parallelogram, that is, $2\sqrt{6}$.

40.44 Find the distance d from the origin to the plane of Problem 40.42.

▮   Let X denote any point on the plane.   The distance d is the magnitude of the scalar projection of $\overrightarrow{OX}$ on the vector $\mathbf{N} = (8, 4, 4)$.   Choosing $X = P(1, -1, 4)$,   we have

$$d = \frac{|\mathbf{N} \cdot \overrightarrow{OP}|}{|\mathbf{N}|} = \frac{|(8, 4, 4) \cdot (1 - 1, 4)|}{4\sqrt{6}} = \frac{20}{4\sqrt{6}} = \frac{5}{\sqrt{6}} = \frac{5\sqrt{6}}{6}$$

40.45 Find the volume of the parallelepiped formed by the vectors $\overrightarrow{PQ}$ and $\overrightarrow{PR}$ of Problem 40.42 and the vector $\overrightarrow{PS}$, where $S = (3, 5, 7)$.

▮   The volume of the parallelepiped determined by noncoplanar vectors $\mathbf{A}$, $\mathbf{B}$, $\mathbf{C}$ is $|\mathbf{A} \cdot (\mathbf{B} \times \mathbf{C})|$.   Hence, in this case, the volume is $|\overrightarrow{PS} \cdot \mathbf{N}| = (2, 4, 10) \cdot (8, 4, 4) = 16 + 16 + 40 = 72$.

40.46 If $\mathbf{B} \times \mathbf{C} = \mathbf{0}$, what can be concluded about $\mathbf{B}$ and $\mathbf{C}$?

▮ $|\mathbf{B} \times \mathbf{C}| = |\mathbf{B}|\,|\mathbf{C}|\,\sin\theta$, where θ is the angle between $\mathbf{B}$ and $\mathbf{C}$. So, if $\mathbf{B} \times \mathbf{C} = \mathbf{0}$, either $\mathbf{B} = \mathbf{0}$ or $\mathbf{C} = \mathbf{0}$ or $\sin\theta = 0$. Note that $\sin\theta = 0$ is equivalent to $\mathbf{B}$ and $\mathbf{C}$ being parallel; in other words, linearly dependent.

40.47 If $\mathbf{A} \cdot (\mathbf{B} \times \mathbf{C}) = 0$, what can be concluded about the configuration of $\mathbf{A}$, $\mathbf{B}$, and $\mathbf{C}$?

▮ One possibility is that $\mathbf{B} \times \mathbf{C} = \mathbf{0}$, which, by Problem 40.46, means that $\mathbf{B} = \mathbf{0}$ or $\mathbf{C} = \mathbf{0}$ or $\mathbf{B}$ and $\mathbf{C}$ are parallel. Another possibility is that $\mathbf{A} = \mathbf{0}$. If $\mathbf{A} \neq \mathbf{0}$ and $\mathbf{B} \times \mathbf{C} \neq \mathbf{0}$, then $\mathbf{A} \cdot (\mathbf{B} \times \mathbf{C}) = 0$ means that $\mathbf{A} \perp (\mathbf{B} \times \mathbf{C})$, which is equivalent to $\mathbf{A}$ lying in the plane determined by $\mathbf{B}$ and $\mathbf{C}$ (and therefore not codetermining a parallelepiped with $\mathbf{B}$ and $\mathbf{C}$).

40.48 Verify the identity $|\mathbf{A} \times \mathbf{B}|^2 = |\mathbf{A}|^2\,|\mathbf{B}|^2 - (\mathbf{A} \cdot \mathbf{B})^2$.

▮ Let θ be the angle between $\mathbf{A}$ and $\mathbf{B}$. Then, $|\mathbf{A}|^2|\mathbf{B}|^2 - (\mathbf{A} \cdot \mathbf{B})^2 = |\mathbf{A}|^2|\mathbf{B}|^2 - |\mathbf{A}|^2|\mathbf{B}|^2 \cos^2\theta = |\mathbf{A}|^2|\mathbf{B}|^2(1 - \cos^2\theta) = |\mathbf{A}|^2|\mathbf{B}|^2 \sin^2\theta = |\mathbf{A} \times \mathbf{B}|^2$.

40.49 Establish the formula

$$\mathbf{A} \cdot (\mathbf{B} \times \mathbf{C}) = \begin{vmatrix} a_1 & a_2 & a_3 \\ b_1 & b_2 & b_3 \\ c_1 & c_2 & c_3 \end{vmatrix}$$

where $\mathbf{A} = (a_1, a_2, a_3)$, $\mathbf{B} = (b_1, b_2, b_3)$, $\mathbf{C} = (c_1, c_2, c_3)$.

▮ By Problem 40.40, the cofactors of the first row are the respective components of $\mathbf{B} \times \mathbf{C}$.

40.50 If $\mathbf{A} \times \mathbf{B} = \mathbf{A} \times \mathbf{C}$ and $\mathbf{A} \neq \mathbf{0}$, does it follow that $\mathbf{B} = \mathbf{C}$?

▮ No. $\mathbf{A} \times \mathbf{B} = \mathbf{A} \times \mathbf{C}$ is equivalent to $\mathbf{A} \times (\mathbf{B} - \mathbf{C}) = \mathbf{0}$. The last equation holds when $\mathbf{A}$ is parallel to $\mathbf{B} - \mathbf{C}$, and this can happen when $\mathbf{B} \neq \mathbf{C}$.

40.51 If $\mathbf{A}, \mathbf{B}, \mathbf{C}$ are mutually perpendicular, show that $\mathbf{A} \times (\mathbf{B} \times \mathbf{C}) = \mathbf{0}$.

▮ $\mathbf{B} \times \mathbf{C}$ is a vector perpendicular to the plane of $\mathbf{B}$ and $\mathbf{C}$. By hypothesis, $\mathbf{A}$ is also such a vector and, therefore, $\mathbf{A}$ and $\mathbf{B} \times \mathbf{C}$ are parallel. Hence, $\mathbf{A} \times (\mathbf{B} \times \mathbf{C}) = \mathbf{0}$.

40.52 Verify that $\mathbf{B} \times \mathbf{A} = -(\mathbf{A} \times \mathbf{B})$.

▮ Let $\mathbf{A} = (a_1, a_2, a_3)$ and $\mathbf{B} = (b_1, b_2, b_3)$. Then $\mathbf{A} \times \mathbf{B} = (a_2 b_3 - a_3 b_2, \ a_3 b_1 - a_1 b_3, \ a_1 b_2 - a_2 b_1)$ and $\mathbf{B} \times \mathbf{A} = (b_2 a_3 - b_3 a_2, \ b_3 a_1 - b_1 a_3, \ b_1 a_2 - b_2 a_1) = -(\mathbf{A} \times \mathbf{B})$.

40.53 Verify that $\mathbf{A} \times \mathbf{A} = \mathbf{0}$.

▮ By Problem 40.52, $\mathbf{A} \times \mathbf{A} = -(\mathbf{A} \times \mathbf{A})$.

40.54 Verify that $\mathbf{i} \times \mathbf{j} = \mathbf{k}$, $\mathbf{j} \times \mathbf{k} = \mathbf{i}$, and $\mathbf{k} \times \mathbf{i} = \mathbf{j}$.

▮ $\mathbf{i} \times \mathbf{j} = (1, 0, 0) \times (0, 1, 0) = (0(0) - 0(1), \ 0(0) - 1(0), \ 1(1) - 0(0)) = (0, 0, 1) = \mathbf{k}$. $\mathbf{j} \times \mathbf{k} = (0, 1, 0) \times (0, 0, 1) = (1(1) - 0(0), \ 0(0) - 0(1), \ 0(0) - 1(0)) = (1, 0, 0) = \mathbf{i}$. Finally, $\mathbf{k} \times \mathbf{i} = (0, 0, 1) \times (1, 0, 0) = (0(0) - 1(0), \ 1(1) - 0(0), \ 0(0) - 0(1)) = (0, 1, 0) = \mathbf{j}$.

40.55 Show that $\mathbf{A} \cdot (\mathbf{B} \times \mathbf{C}) = (\mathbf{A} \times \mathbf{B}) \cdot \mathbf{C}$ (*exchange of dot and cross*).

▮ $(\mathbf{A} \times \mathbf{B}) \cdot \mathbf{C} = \mathbf{C} \cdot (\mathbf{A} \times \mathbf{B}) = \mathbf{A} \cdot (\mathbf{B} \times \mathbf{C})$. The first equality follows from the definition of the dot product; the second is established by making two row interchanges in the determinant of Problem 40.49.

40.56 Find the volume of the parallelepiped whose edges are $\overrightarrow{OA}$, $\overrightarrow{OB}$, $\overrightarrow{OC}$, where $O = (0, 0, 0)$, $A = (1, 2, 3)$, $B = (1, 1, 2)$, $C = (2, 1, 1)$.

▮ The volume is $|\overrightarrow{OA} \cdot (\overrightarrow{OB} \times \overrightarrow{OC})| = |(1, 2, 3) \cdot ((1, 1, 2) \times (2, 1, 1))| = |(1, 2, 3) \cdot (-1, 3, -1)| = |{-1} + 6 - 3| = 2$.

40.57   Show that $(\mathbf{A} + \mathbf{B}) \cdot ((\mathbf{B} + \mathbf{C}) \times (\mathbf{C} + \mathbf{A})) = 2\mathbf{A} \cdot (\mathbf{B} \times \mathbf{C})$.

▌ $(\mathbf{A} + \mathbf{B}) \cdot ((\mathbf{B} + \mathbf{C}) \times (\mathbf{C} + \mathbf{A})) = (\mathbf{A} + \mathbf{B}) \cdot ((\mathbf{B} \times \mathbf{C}) + (\mathbf{B} \times \mathbf{A}) + (\mathbf{C} \times \mathbf{C}) + (\mathbf{C} \times \mathbf{A})) = (\mathbf{A} + \mathbf{B}) \cdot ((\mathbf{B} \times \mathbf{C}) + (\mathbf{B} \times \mathbf{A}) + (\mathbf{C} \times \mathbf{A})) = \mathbf{A} \cdot (\mathbf{B} \times \mathbf{C}) + \mathbf{A} \cdot (\mathbf{B} \times \mathbf{A}) + \mathbf{A} \cdot (\mathbf{C} \times \mathbf{A}) + \mathbf{B} \cdot (\mathbf{B} \times \mathbf{C}) + \mathbf{B} \cdot (\mathbf{B} \times \mathbf{A}) + \mathbf{B} \cdot (\mathbf{C} \times \mathbf{A}) = \mathbf{A} \cdot (\mathbf{B} \times \mathbf{C}) + \mathbf{B} \cdot (\mathbf{C} \times \mathbf{A})$,   since, for all $\mathbf{D}$ and $\mathbf{E}$, $\mathbf{D} \cdot (\mathbf{D} \times \mathbf{E}) = 0$ and $\mathbf{D} \cdot (\mathbf{E} \times \mathbf{D}) = 0$.   Hence, we obtain $2\mathbf{A} \cdot (\mathbf{B} \times \mathbf{C})$,   since $\mathbf{B} \cdot (\mathbf{C} \times \mathbf{A}) = (\mathbf{B} \times \mathbf{C}) \cdot \mathbf{A}$ by Problem 40.55.

40.58   Prove $(\mathbf{A} \times \mathbf{B}) \cdot (\mathbf{C} \times \mathbf{D}) = (\mathbf{A} \cdot \mathbf{C})(\mathbf{B} \cdot \mathbf{D}) - (\mathbf{A} \cdot \mathbf{D})(\mathbf{B} \cdot \mathbf{C})$.

▌ $\mathbf{A} \times \mathbf{B} = (a_2b_3 - a_3b_2, \ a_3b_1 - a_1b_3, \ a_1b_2 - a_2b_1)$ and $\mathbf{C} \times \mathbf{D} = (c_2d_3 - c_3d_2, \ c_3d_1 - c_1d_3, \ c_1d_2 - c_2d_1)$. So, $(\mathbf{A} \times \mathbf{B}) \cdot (\mathbf{C} \times \mathbf{D}) = (a_2b_3 - a_3b_2)(c_2d_3 - c_3d_2) + (a_3b_1 - a_1b_3)(c_3d_1 - c_1d_3) + (a_1b_2 - a_2b_1)(c_1d_2 - c_2d_1) = a_2b_3c_2d_3 - a_2b_3c_3d_2 - a_3b_2c_2d_3 + a_3b_2c_3d_2 + a_3b_1c_3d_1 - a_3b_1c_1d_3 - a_1b_3c_3d_1 + a_1b_3c_1d_3 + a_1b_2c_1d_2 - a_1b_2c_2d_1 - a_2b_1c_1d_2 + a_2b_1c_2d_1$. On the other hand, $(\mathbf{A} \cdot \mathbf{C})(\mathbf{B} \cdot \mathbf{D}) - (\mathbf{A} \cdot \mathbf{D})(\mathbf{B} \cdot \mathbf{C}) = (a_1c_1 + a_2c_2 + a_3c_3) \times (b_1d_1 + b_2d_2 + b_3d_3) - (a_1d_1 + a_2d_2 + a_3d_3)(b_1c_1 + b_2c_2 + b_3c_3) = a_1b_1c_1d_1 + a_1b_2c_1d_2 + a_1b_3c_1d_3 + a_2b_2c_2d_2 + a_2b_3c_2d_3 + a_3b_1c_3d_1 + a_3b_2c_3d_2 + a_3b_3c_3d_3 - a_1b_1c_1d_1 - a_1b_2c_2d_1 - a_1b_3c_3d_1 - a_2b_1c_1d_2 - a_2b_2c_2d_2 - a_2b_3c_3d_2 - a_3b_1c_1d_3 - a_3b_2c_2d_3 - a_3b_3c_3d_3$.   Hence, the two sides are equal.

40.59   Prove $(\mathbf{A} - \mathbf{B}) \times (\mathbf{A} + \mathbf{B}) = 2(\mathbf{A} \times \mathbf{B})$.

▌ $(\mathbf{A} - \mathbf{B}) \times (\mathbf{A} + \mathbf{B}) = (\mathbf{A} \times \mathbf{A}) + (\mathbf{A} \times \mathbf{B}) - (\mathbf{B} \times \mathbf{A}) - (\mathbf{B} \times \mathbf{B})$.   Since $\mathbf{A} \times \mathbf{A} = \mathbf{B} \times \mathbf{B} = 0$ and $\mathbf{B} \times \mathbf{A} = -(\mathbf{A} \times \mathbf{B})$,   the result is $2(\mathbf{A} \times \mathbf{B})$.

40.60   Show that $\mathbf{C} \times (\mathbf{A} \times \mathbf{B})$ is a linear combination of $\mathbf{A}$ and $\mathbf{B}$; namely, $\mathbf{C} \times (\mathbf{A} \times \mathbf{B}) = (\mathbf{C} \cdot \mathbf{B})\mathbf{A} - (\mathbf{C} \cdot \mathbf{A})\mathbf{B}$.

▌ $\mathbf{A} \times \mathbf{B} = (a_2b_3 - a_3b_2, \ a_3b_1 - a_1b_3, \ a_1b_2 - a_2b_1)$. So, $\mathbf{C} \times (\mathbf{A} \times \mathbf{B}) = (c_2(a_1b_2 - a_2b_1) - c_3(a_3b_1 - a_1b_3), c_3(a_2b_3 - a_3b_2) - c_1(a_1b_2 - a_2b_1), \quad c_1(a_3b_1 - a_1b_3) - c_2(a_2b_3 - a_3b_2)) = (c_1b_1 + c_2b_2 + c_3b_3)(a_1, a_2, a_3) - (a_1c_1 + a_2c_2 + a_3c_3)(b_1, b_2, b_3) = (\mathbf{C} \cdot \mathbf{B})\mathbf{A} - (\mathbf{C} \cdot \mathbf{A})\mathbf{B}$.

40.61   Show that $\mathbf{A} \times (\mathbf{A} \times \mathbf{B}) = (\mathbf{A} \cdot \mathbf{B})\mathbf{A} - (\mathbf{A} \cdot \mathbf{A})\mathbf{B}$.

▌ In Problem 40.60, let $\mathbf{C} = \mathbf{A}$.

40.62   Show that $(\mathbf{A} \times \mathbf{B}) \times (\mathbf{C} \times \mathbf{D}) = ((\mathbf{A} \times \mathbf{B}) \cdot \mathbf{D})\mathbf{C} - ((\mathbf{A} \times \mathbf{B}) \cdot \mathbf{C})\mathbf{D}$.

▌ In Problem 40.60, substitute $\mathbf{C}$ for $\mathbf{A}$, $\mathbf{D}$ for $\mathbf{B}$, and $\mathbf{A} \times \mathbf{B}$ for $\mathbf{C}$.

40.63   Show that the points $(0, 0, 0)$, (a_1, a_2, a_3), (b_1, b_2, b_3), and (c_1, c_2, c_3) are coplanar if and only if

$$\begin{vmatrix} a_1 & a_2 & a_3 \\ b_1 & b_2 & b_3 \\ c_1 & c_2 & c_3 \end{vmatrix} = 0$$

▌ Let $\mathbf{A} = (a_1, a_2, a_3)$, $\mathbf{B} = (b_1, b_2, b_3)$, $\mathbf{C} = (c_1, c_2, c_3)$.   By Problems 40.45 and 40.49, the determinant above is equal to $\mathbf{A} \cdot (\mathbf{B} \times \mathbf{C})$,   which is equal in magnitude to the nonzero volume of the parallelepiped formed by $\mathbf{A}, \mathbf{B}, \mathbf{C}$, when the given points are not coplanar.   If those points are coplanar, either $\mathbf{B} \times \mathbf{C} = 0$,   and, therefore, $\mathbf{A} \cdot (\mathbf{B} \times \mathbf{C}) = 0$;   or $\mathbf{B} \times \mathbf{C} \neq 0$ and $\mathbf{A}$ is perpendicular to $\mathbf{B} \times \mathbf{C}$ (because $\mathbf{A}$ is in the plane of $\mathbf{B}$ and $\mathbf{C}$), so that again $\mathbf{A} \cdot (\mathbf{B} \times \mathbf{C}) = 0$.

40.64   Show that $\mathbf{A} \cdot (\mathbf{B} \times \mathbf{C}) = \mathbf{B} \cdot (\mathbf{C} \times \mathbf{A})$.

▌ This follows at once from Problem 40.55 (exchange dot and cross on the right side).

40.65   (Reciprocal crystal lattices).   Assume $\mathbf{V}_1, \mathbf{V}_2, \mathbf{V}_3$ are noncoplanar vectors.   Let

$$\mathbf{K}_1 = \frac{\mathbf{V}_2 \times \mathbf{V}_3}{\mathbf{V}_1 \cdot (\mathbf{V}_2 \times \mathbf{V}_3)} \qquad \mathbf{K}_2 = \frac{\mathbf{V}_3 \times \mathbf{V}_1}{\mathbf{V}_1 \cdot (\mathbf{V}_2 \times \mathbf{V}_3)} \qquad \mathbf{K}_3 = \frac{\mathbf{V}_1 \times \mathbf{V}_2}{\mathbf{V}_1 \cdot (\mathbf{V}_2 \times \mathbf{V}_3)}$$

Show that $\mathbf{K}_1 \cdot (\mathbf{K}_2 \times \mathbf{K}_3) = 1/[\mathbf{V}_1 \cdot (\mathbf{V}_2 \times \mathbf{V}_3)]$.

▌ Let $c = \mathbf{V}_1 \cdot (\mathbf{V}_2 \times \mathbf{V}_3)$.   Then

$$\mathbf{K}_2 \times \mathbf{K}_3 = \frac{\mathbf{V}_3 \times \mathbf{V}_1}{c} \times \frac{\mathbf{V}_1 \times \mathbf{V}_2}{c} = \frac{1}{c^2} \{[(\mathbf{V}_3 \times \mathbf{V}_1) \cdot \mathbf{V}_2]\mathbf{V}_1 - [(\mathbf{V}_3 \times \mathbf{V}_1) \cdot \mathbf{V}_1]\mathbf{V}_2\}$$

by Problem 40.62. By virtue of the fact that $\mathbf{V}_1$ is perpendicular to $\mathbf{V}_3 \times \mathbf{V}_1$, we have $(\mathbf{V}_3 \times \mathbf{V}_1) \cdot \mathbf{V}_1 = 0$, and, therefore, $\mathbf{K}_2 \times \mathbf{K}_3 = (1/c^2)[(\mathbf{V}_3 \times \mathbf{V}_1) \cdot \mathbf{V}_2]\mathbf{V}_1$.   By Problem 40.64, $(\mathbf{V}_3 \times \mathbf{V}_1) \cdot \mathbf{V}_2 = \mathbf{V}_2 \cdot (\mathbf{V}_3 \times \mathbf{V}_1) = \mathbf{V}_1 \cdot (\mathbf{V}_2 \times \mathbf{V}_3) = c$. Hence, $\mathbf{K}_2 \times \mathbf{K}_3 = (1/c)\mathbf{V}_1$. Therefore,

$$\mathbf{K}_1 \cdot (\mathbf{K}_2 \times \mathbf{K}_3) = \frac{\mathbf{V}_2 \times \mathbf{V}_3}{c} \cdot \frac{\mathbf{V}_1}{c} = \frac{1}{c^2}\, \mathbf{V}_1 \cdot (\mathbf{V}_2 \times \mathbf{V}_3) = \frac{c}{c^2} = \frac{1}{c}$$

40.66   Show that $\mathbf{K}_i \perp \mathbf{V}_j$ for $i \ne j$. (For the notation, see Problem 40.65.)

▌ $\mathbf{K}_1 \cdot \mathbf{V}_2 = \dfrac{\mathbf{V}_2 \times \mathbf{V}_3}{c} \cdot \mathbf{V}_2 = 0$ since $\mathbf{V}_2$ is perpendicular to $\mathbf{V}_2 \times \mathbf{V}_3$. $\mathbf{K}_1 \cdot \mathbf{V}_3 = \dfrac{\mathbf{V}_2 \times \mathbf{V}_3}{c} \cdot \mathbf{V}_3 = 0$ since $\mathbf{V}_3$ is perpendicular to $\mathbf{V}_2 \times \mathbf{V}_3$. The other cases are handled similarly.

40.67   Show that $\mathbf{K}_i \cdot \mathbf{V}_i = 1$ for $i = 1, 2, 3$. (For the notation, see Problem 40.65.)

▌ $\mathbf{K}_1 \cdot \mathbf{V}_1 = \dfrac{\mathbf{V}_2 \times \mathbf{V}_3}{c} \cdot \mathbf{V}_1 = \dfrac{\mathbf{V}_1 \cdot (\mathbf{V}_2 \times \mathbf{V}_3)}{c} = \dfrac{c}{c} = 1$.   The other cases are similar.

40.68   If $\mathbf{A} = (2, -3, 1)$ is normal to one plane $\mathscr{P}_1$ and $\mathbf{B} = (-1, 4, -2)$ is normal to another plane $\mathscr{P}_2$, show that $\mathscr{P}_1$ and $\mathscr{P}_2$ intersect and find a vector parallel to the line $\mathscr{L}$ in which they intersect.

▌ If $\mathscr{P}_1$ and $\mathscr{P}_2$ were parallel, $\mathscr{A}$ and $\mathscr{B}$ would have to be parallel. But $\mathbf{A}$ and $\mathbf{B}$ are not parallel (since $\mathbf{A}$ is not a scalar multiple of $\mathbf{B}$). Therefore, $\mathscr{P}_1$ and $\mathscr{P}_2$ intersect. Since $\mathbf{A}$ is perpendicular to plane $\mathscr{P}_1$, $\mathbf{A}$ is perpendicular to the line $\mathscr{L}$ in $\mathscr{P}_1$. Likewise, $\mathbf{B}$ is perpendicular to $\mathscr{L}$. Hence, $\mathscr{L}$ is parallel to $\mathbf{A} \times \mathbf{B}$.

40.69   Find the vector representation, the parametric equations, and the rectangular equations for the line through the points $P(1, -2, 5)$ and $Q(3, 4, 6)$.

▌ If R is any point on the line, then $\overrightarrow{PR} = t\overrightarrow{PQ}$ for some scalar t (see Fig. 40-6). Hence, $\overrightarrow{OR} = \overrightarrow{OP} + \overrightarrow{PR} = \overrightarrow{OP} + t\overrightarrow{PQ}$. Thus, $(1, 2, -5) + t(2, 6, 1)$ is a vector function that generates the line. A parametric form is $x = 1 + 2t$, $y = 2 + 6t$, $z = -5 + t$. If we eliminate t from these equations, we obtain the rectangular equations $\dfrac{x-1}{2} = \dfrac{y-2}{6} = \dfrac{z+5}{1}$.

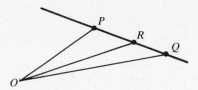

Fig. 40-6

40.70   Find the points at which the line of Problem 40.69 cuts the coordinate planes.

▌ By Problem 40.69, the parametric equations are $x = 1 + 2t$, $y = 2 + 6t$, $z = -5 + t$. To find the intersection with the xy-plane, set $z = 0$. Then $t = 5$, $x = 11$, $y = 32$. So, the intersection point is $(11, 32, 0)$. To find the intersection with the xz-plane, set $y = 0$. Then $t = -\frac{1}{3}$, $x = \frac{1}{3}$, $z = -\frac{16}{3}$. So, the intersection is $(\frac{1}{3}, 0, -\frac{16}{3})$. To find the intersection with the yz-plane, set $x = 0$. Then $t = -\frac{1}{2}$, $y = -1$, $z = -\frac{11}{2}$. So, the intersection is $(0, -1, -\frac{11}{2})$.

40.71   Find the vector representation, parametric equations, and rectangular equations for the line through the points $P(3, 2, 1)$ and $Q(-1, 2, 4)$.

▌ $\overrightarrow{PQ} = (-4, 0, 3)$. So, the line is generated by the vector function $(3, 2, 1) + t(-4, 0, 3)$. The parametric equations are $x = 3 - 4t$, $y = 2$, $z = 1 + 3t$. The rectangular equations are $\dfrac{x-3}{-4} = \dfrac{z-1}{3}$, $y = 2$.

40.72   Write equations for the line through the point $(1, 2, -6)$ and parallel to the vector $(4, 1, 3)$.

▌ The line is generated by the vector function $(1, 2, -6) + t(4, 1, 3)$. Parametric equations are $x = 1 + 4t$, $y = 2 + t$, $z = -6 + 3t$. A set of rectangular equations is $\dfrac{x-1}{4} = \dfrac{y-2}{1} = \dfrac{z+6}{3}$.

40.73 Write parametric equations for the line through $(-1, 4, 2)$ and parallel to the line $\dfrac{x-2}{4} = \dfrac{y}{5} = \dfrac{z+2}{3}$.

▌ The given line is parallel to the vector $(4, 5, 3)$. Hence, the desired equations are $x = -1 + 4t$, $y = 4 + 5t$, $z = 2 + 3t$.

40.74 Show that the lines $\mathscr{L}_1$: $x = x_0 + at$, $y = y_0 + bt$, $z = z_0 + ct$ and $\mathscr{L}_2$: $x = x_1 + At$, $y = y_1 + Bt$, $z = z_1 + Ct$ are parallel if and only if a, b, c are proportional to A, B, C.

▌ $\mathscr{L}_1$ is parallel to the vector (a, b, c). $\mathscr{L}_2$ is parallel to the vector (A, B, C). Therefore, $\mathscr{L}_1$ is parallel to $\mathscr{L}_2$ if and only if (a, b, c) is parallel to (A, B, C), that is, if and only if $(a, b, c) = \lambda(A, B, C)$ for some λ. The latter condition means that a, b, c are proportional to A, B, C.

40.75 Show that the lines $\mathscr{L}_1$: $x = x_0 + at$, $y = y_0 + bt$, $z = z_0 + ct$ and $\mathscr{L}_2$: $x = x_0 + At$, $y = y_0 + Bt$, $z = z_0 + Ct$ are perpendicular if and only if $aA + bB + cC = 0$.

▌ $\mathscr{L}_1$ is parallel to the vector (a, b, c). $\mathscr{L}_2$ is parallel to the vector (A, B, C). $\mathscr{L}_1$ and $\mathscr{L}_2$ have a common point (x_0, y_0, z_0). Hence, $\mathscr{L}_1$ is perpendicular to $\mathscr{L}_2$ if and only if $(a, b, c) \perp (A, B, C)$, which is equivalent to $(a, b, c) \cdot (A, B, C) = 0$, or $aA + bB + cC = 0$.

40.76 Show that the lines $x = 1 + t$, $y = 2t$, $z = 1 + 3t$ and $x = 3s$, $y = 2s$, $z = 2 + s$ intersect, and find their point of intersection.

▌ Assume (x, y, z) is a point of intersection. Then $2t = y = 2s$. So, $t = s$. From $1 + t = x = 3s$, we then have $1 + s = 3s$, $2s = 1$, $s = \frac{1}{2}$. So, $s = t = \frac{1}{2}$. Note that $1 + 3t = 1 + 3(\frac{1}{2}) = \frac{5}{2}$ and $2 + s = 2 + \frac{1}{2} = \frac{5}{2}$. Thus, the equations for z are compatible. The intersection point is $(\frac{3}{2}, 1, \frac{5}{2})$.

40.77 By the methods of calculus, find the point $P_1(x, y, z)$ on the line $x = 3 + t$, $y = 2 + t$, $z = 1 + t$ that is closest to the point $P_0(1, 2, 1)$, and verify that $\overrightarrow{P_0 P_1}$ is perpendicular to the line.

▌ The distance from $P_0(1, 2, 1)$ to (x, y, z) is $\sqrt{(x-1)^2 + (y-2)^2 + (z-1)^2} = \sqrt{(t+2)^2 + t^2 + t^2} = \sqrt{3t^2 + 4t + 4}$. It suffices to minimize $3t^2 + 4t + 4$. Set $D_t(3t^2 + 4t + 4) = 6t + 4 = 0$. So, $t = -\frac{2}{3}$. Hence, $P_1 = (\frac{7}{3}, \frac{4}{3}, \frac{1}{3})$. Note that $\overrightarrow{P_0 P_1}$ is $(\frac{4}{3}, -\frac{2}{3}, -\frac{2}{3})$, and $\overrightarrow{P_0 P_1} \cdot (1, 1, 1) = 0$. Hence, $\overrightarrow{P_0 P_1}$ is perpendicular to the line. The distance from P_1 to the line is $\sqrt{3(\frac{4}{9}) - \frac{8}{3} + 4} = \sqrt{\frac{8}{3}}$.

40.78 Describe a method based on vectors for finding the distance from a point P_0 to a line $x = x_0 + at$, $y = y_0 + bt$, $z = z_0 + ct$ that does not contain P.

▌ Choose any two points R and Q on the line (i.e., choose any two t-values). Let $\mathbf{A} = \overrightarrow{P_0 R} \times \overrightarrow{P_0 Q}$. $\mathbf{A}$ is perpendicular to the plane containing P_0 and the line (see Fig. 40-7). Now, we want the line $P_0 P_1$ from P_0 to the nearest point P_1 on the line. Clearly, $\mathbf{A} \times \overrightarrow{RQ}$ is parallel to $P_0 P_1$, so that we can write an equation for the line $P_0 P_1$. The intersection of $P_0 P_1$ with the given line yields the point P_1, and the distance $d = \overline{P_0 P_1}$.

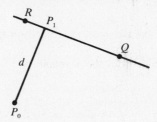

Fig. 40-7

40.79 Apply the method of Problem 40.78 to Problem 40.77.

▌ The line is $x = 3 + t$, $y = 2 + t$, $z = 1 + t$. P_0 is $(1, 2, 1)$. Let $R = (3, 2, 1)$ and $Q = (4, 3, 2)$. $\overrightarrow{RQ}$ is $(1, 1, 1)$, $\overrightarrow{P_0 R}$ is $(2, 0, 0)$, and $\overrightarrow{P_0 Q} = (3, 1, 1)$. Then $\mathbf{A} = \overrightarrow{P_0 R} \times \overrightarrow{P_0 Q} = (0, -2, 2)$, and $\mathbf{A} \times \overrightarrow{RQ} = (0, -2, 2) \times (1, 1, 1) = (-4, 2, 2)$. So, the line $P_0 P_1$ is $x = 1 - 4s$, $y = 2 + 2s$, $z = 1 + 2s$. To get the intersection of line $P_0 P_1$ with the given line, equate the x-coordinates: $3 + t = 1 - 4s$. Then $t = -2 - 4s$. Substitute in the y-coordinates: $2 + (-2 - 4s) = 2 + 2s$. So, $s = -\frac{1}{3}$. Hence, $t = -\frac{2}{3}$. Note that the z-coordinates then agree: $1 - \frac{2}{3} = 1 + 2(-\frac{1}{3})$. The intersection point P_1 is $x = \frac{7}{3}$, $y = \frac{4}{3}$, $z = \frac{1}{3}$. The distance between P_0 and the given line is $\overline{P_0 P_1} = \sqrt{(\frac{4}{3})^2 + (-\frac{2}{3})^2 + (-\frac{2}{3})^2} = \sqrt{\frac{8}{3}}$. These are the same results that were found in Problem 40.77.

40.80   Find an equation of the plane containing the point $P_1(3, -2, 5)$ and perpendicular to the vector $N = (4, 2, -7)$.

▮   For any point $P(x, y, z)$, P is in the plane if and only if $\overrightarrow{P_1P} \perp N$, that is, if and only if $\overrightarrow{P_1P} \cdot N = 4(x - 3) + 2(y + 2) - 7(z - 5) = 0$, which can also be written as $4x + 2y - 7z = -27$.

40.81   Find an equation of the plane containing the point $P_1(4, 3, -2)$ and perpendicular to the vector $(5, -4, 6)$.

▮   We know (from Problem 40.80) that the plane has an equation of the form $5x - 4y + 6z = d$. Since P_1 lies in the plane, $5(4) - 4(3) + 6(-2) = d$. So, $d = -4$ and the plane has the equation $5x - 4y + 6z = -4$.

40.82   Find an equation for the plane through the points $P(1, 3, 5)$, $Q(-1, 2, 4)$, and $R(4, 4, 0)$.

▮ $N = \overrightarrow{PQ} \times \overrightarrow{PR}$ is normal to the plane. $N = (-2, -1, -1) \times (3, 1, -5) = (6, -13, 1)$. So, the equation has the form $6x - 13y + z = d$. Since R is in the plane, $6(4) - 13(4) + 0 = d$, $d = -28$. Thus, the equation is $6x - 13y + z = -28$.

40.83   Find an equation for the plane through the points $(3, 2, -1)$, $(1, -1, 3)$, and $(3, -2, 4)$.

▮   We shall use a method different from the one used in Problem 40.82. The equation of the plane has the form $ax + by + cz = d$. Substitute the values corresponding to the three given points: (1) $3a + 2b - c = d$; (2) $a - b + 3c = d$; (3) $3a - 2b + 4c = d$. Eliminating a from (1) and (2), we get (4) $5b - 10c = -2d$. Eliminating a from (2) and (3), we get (5) $b - 5c = -2d$. Eliminating d from (4) and (5), we get $5b - 10c = b - 5c$, $4b = 5c$, $b = \frac{5}{4}c$. From (5), $-2d = \frac{5}{4}c - 5c$, $d = \frac{15}{8}c$. From (2), $a = d + b - 3c = \frac{15}{8}c + \frac{5}{4}c - 3c = c/8$. So, the equation of the plane is $(c/8)x + \frac{5}{4}cy + cz = \frac{15}{8}c$. Multiplying by $8/c$ yields $x + 10y + 8z = 15$.

40.84   Find the cosine of the angle θ between the planes $4x + 4y - 2z = 9$ and $2x + y + z = -3$.

▮ θ is the angle between the normal vectors $(4, 4, -2)$ and $(2, 1, 1)$. So,

$$\cos \theta = \frac{(4, 4, -2) \cdot (2, 1, 1)}{|(4, 4, -2)| \, |(2, 1, 1)|} = \frac{10}{6\sqrt{6}} = \frac{5\sqrt{6}}{18}$$

Of course, there is another angle between the planes, the supplement of the angle whose cosine was just found. The cosine of the other angle is $-\cos \theta = -5\sqrt{6}/18$.

40.85   Find parametric equations for the line $\mathscr{L}$ that is the intersection of the planes in Problem 40.84.

▮   The line $\mathscr{L}$ is perpendicular to the normal vectors of the planes, and is, therefore, parallel to their cross product $(4, 4, -2) \times (2, 1, 1) = (6, -8, -4)$. We also need a point on the intersection. To find one, set $x = 0$ and solve the two resulting equations, $4y - 2z = 9$, $y + z = -3$. Multiplying the second equation by 2 and adding, we get $6y = 3$, $y = \frac{1}{2}$. So, $z = -\frac{7}{2}$. Thus, the point is $(0, \frac{1}{2}, -\frac{7}{2})$ and we obtain the line $x = 6t$, $y = \frac{1}{2} - 8t$, $z = -\frac{7}{2} - 4t$.

40.86   Find an equation of the plane containing the point $P(1, 3, 1)$ and the line $\mathscr{L}$: $x = t$, $y = t$, $z = t + 2$.

▮   The point $Q(0, 0, 2)$ is on the given line $\mathscr{L}$ (set $t = 0$). The vector $\overrightarrow{QP} = (1, 3, -1)$ is parallel to the plane. Since the vector $(1, 1, 1)$ is parallel to $\mathscr{L}$, the cross product $(1, 3, -1) \times (1, 1, 1) = (4, -2, -2)$ is normal to the plane. So, an equation of the plane is $4x - 2y - 2z = d$. Since the point $Q(0, 0, 2)$ is in the plane, $-4 = d$. Thus, the plane has the equation $4x - 2y - 2z = -4$, or $2x - y - z = -2$.

40.87   (a) Express vectorially the distance from the origin to the intersection of two planes. (b) Check the result of (a) geometrically.

▮   (a) Figure 40-8(a) indicates two planes, $\mathscr{P}_1$ and $\mathscr{P}_2$, with respective normals N_1 and N_2, and the common point A. The line of intersection, $\mathscr{L}$, has the vector equation $\overrightarrow{OX} = \overrightarrow{OA} + tN$ (cf. Problem 40.85). If P is the point of $\mathscr{L}$ closest to O, then $\overrightarrow{OP} \perp N$, or $0 = \overrightarrow{OP} \cdot N = (\overrightarrow{OA} + t_P N) \cdot N = \overrightarrow{OA} \cdot N + t_P N \cdot N$. Hence,

$$t_P = -(\overrightarrow{OA} \cdot N)/(N \cdot N) \quad \text{and} \quad \overrightarrow{OP} = \overrightarrow{OA} - \frac{\overrightarrow{OA} \cdot N}{N \cdot N} N = \frac{N \times (\overrightarrow{OA} \times N)}{N \cdot N}$$

where the last step follows from Problem 40.60. Finally, $d = |\overrightarrow{OP}|$.

(b) Since $\overrightarrow{OA} \times \mathbf{N}$ and $\mathbf{N}$ are orthogonal,

$$d = \frac{1}{|\mathbf{N}|^2} (|\mathbf{N}| \, |\overrightarrow{OA} \times \mathbf{N}|) = \frac{1}{|\mathbf{N}|^2} (|\mathbf{N}| \, |\overrightarrow{OA}| \, |\mathbf{N}| \sin \phi) = |\overrightarrow{OA}| \sin \phi$$

where ϕ is the angle between $\overrightarrow{OA}$ and $\mathbf{N}$.   But this is geometrically obvious; see Fig. 40-8(b).

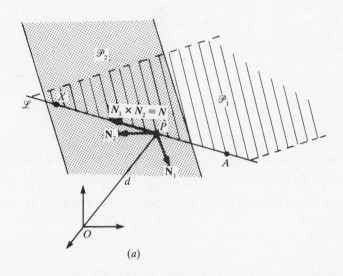
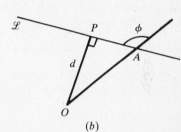

(a)   (b)   **Fig. 40-8**

40.88   Show that the distance D from the point $P(x_1, y_1, z_1)$ to the plane $ax + by + cz + d = 0$ is given by $D = |ax_1 + by_1 + cz_1 + d| / \sqrt{a^2 + b^2 + c^2}$.

▌ Let $Q(x_2, y_2, z_2)$ be any point on the given plane.   Then the distance D is the magnitude of the scalar projection of $\overrightarrow{PQ} = (x_2 - x_1, \ y_2 - y_1, \ z_2 - z_1)$ on the normal vector $\mathbf{N}(a, b, c)$ to the plane.   So, D is

$$\frac{|\mathbf{N} \cdot \overrightarrow{PQ}|}{|\mathbf{N}|} = \frac{|(a, b, c) \cdot (x_2 - x_1, \ y_2 - y_1, \ z_2 - z_1)|}{\sqrt{a^2 + b^2 + c^2}} = \frac{|a(x_2 - x_1) + b(y_2 - y_1) + c(z_2 - z_1)|}{\sqrt{a^2 + b^2 + c^2}}$$

$$= \frac{|ax_2 + by_2 + cz_2 - ax_1 - by_1 - cz_1|}{\sqrt{a^2 + b^2 + c^2}}$$

Since $Q(x_2, y_2, z_2)$ is a point of the plane, $ax_2 + by_2 + cz_2 + d = 0$.   Hence,

$$D = \frac{|-d - ax_1 - by_1 - cz_1|}{\sqrt{a^2 + b^2 + c^2}} = \frac{|ax_1 + by_1 + cz_1 + d|}{\sqrt{a^2 + b^2 + c^2}}$$

40.89   Find the distance D from the point $(3, -5, 2)$ to the plane $8x - 2y + z = 5$.

▌ By the formula of Problem 40.88, $D = \dfrac{|8(3) - 2(-5) + 2 - 5|}{\sqrt{(8)^2 + (2)^2 + (1)^2}} = \dfrac{31}{\sqrt{69}}$.

40.90   Show that the planes $ax + by + cz + d_1 = 0$ and $ax + by + cz + d_2 = 0$ are parallel.

▌ The planes have the same normal vector $\mathbf{N}(a, b, c)$.   Since they are both perpendicular to $\mathbf{N}$, they must be parallel.

40.91   Show that the distance between the parallel planes $\mathcal{P}_1$: $ax + by + cz + d_1 = 0$ and $\mathcal{P}_2$: $ax + by + cz + d_2 = 0$ is $|d_2 - d_1| / \sqrt{a^2 + b^2 + c^2}$.

▌ Let (x_1, y_1, z_1) be a point of plane $\mathcal{P}_1$.   Hence, $ax_1 + by_1 + cz_1 + d_1 = 0$.   The distance between the planes is equal to the distance between the point (x_1, y_1, z_1) and plane $\mathcal{P}_2$, which is, by Problem 40.88,

$$\frac{|ax_1 + by_1 + cz_1 + d|}{\sqrt{a^2 + b^2 + c^2}} = \frac{|-d_1 + d_2|}{\sqrt{a^2 + b^2 + c^2}} = \frac{|d_2 - d_1|}{\sqrt{a^2 + b^2 + c^2}}$$

40.92 Find the distance between the planes $x - 2y + 2z = 1$ and $2x - 4y + 4z = 3$.

▌ The second plane also has the equation $x - 2y + 2z = \frac{3}{2}$ and is, therefore, parallel to the first plane. By Problem 40.91, the distance between the planes is $|\frac{3}{2} - 1|/\sqrt{1^2 + 2^2 + 2^2} = \frac{1}{2}/\sqrt{9} = \frac{1}{6}$.

40.93 Find an equation for the plane $\mathscr{P}$ that is parallel to the plane $\mathscr{P}_1$: $x - 2y + 2z = 1$ and passes through the point $(1, -1, 2)$.

▌ Since $\mathscr{P}$ and $\mathscr{P}_1$ are parallel, the normal vector $(1, -2, 2)$ of $\mathscr{P}_1$ is also a normal vector of $\mathscr{P}$. Hence, an equation of $\mathscr{P}$ is $x - 2y + 2z = d$. Since the point $(1, -1, 2)$ lies on $\mathscr{P}$, $1 - 2(-1) + 2(2) = d$, $d = 7$. So, an equation of $\mathscr{P}$ is $x - 2y + 2z = 7$.

40.94 Consider the sphere of radius 3 and center at the origin. Find the coordinates of the point P where the plane tangent to this sphere at $(1, 2, 2)$ cuts the x-axis.

▌ The radius vector $(1, 2, 2)$ is perpendicular to the tangent plane at $(1, 2, 2)$. Hence, the tangent plane has an equation of the form $x + 2y + 2z = d$. Since $(1, 2, 2)$ is in the plane, $1 + 2(2) + 2(2) = d$, $d = 9$. So, the plane has the equation $x + 2y + 2z = 9$. When $y = 0$ and $z = 0$, $x = 9$. Hence, the point P is $(9, 0, 0)$.

40.95 Find the value of k for which the planes $3x - 4y + 2z + 9 = 0$ and $3x + 4y - kz + 7 = 0$ are perpendicular.

▌ The planes are perpendicular if and only if their normal vectors $(3, -4, 2)$ and $(3, 4, -k)$ are perpendicular, which is equivalent to $(3, -4, 2) \cdot (3, 4, -k) = 0$, $9 - 16 - 2k = 0$, $k = -\frac{7}{2}$.

40.96 Check that the planes $x - 2y + 2z = 1$ and $3x - y - z = 2$ intersect and find their line of intersection.

▌ Since the normal vectors $(1, -2, 2)$ and $(3, -1, -1)$ are not parallel, the planes are not parallel and must intersect. Their line of intersection $\mathscr{L}$ is parallel to the cross product $(1, -2, 2) \times (3, -1, -1) = (4, 7, 5)$. To find a point on $\mathscr{L}$, set $x = 0$ in the equations of the planes: $-2y + 2z = 1$, $-y - z = 2$. Multiply the second equation by 2 and add: $-4y = 5$, $y = -\frac{5}{4}$, $z = -\frac{3}{4}$. So, the point $(0, -\frac{5}{4}, -\frac{3}{4})$ is on $\mathscr{L}$, and $\mathscr{L}$ has the equations $x = 4t$, $y = -\frac{5}{4} + 7t$, $z = -\frac{3}{4} + 5t$.

40.97 Show that the two sets of equations

$$\frac{x-4}{3} = \frac{y-6}{4} = \frac{z+9}{-12} \quad \text{and} \quad \frac{x-1}{-6} = \frac{y-2}{-8} = \frac{z-3}{24}$$

represent the same straight line.

▌ The point $(4, 6, -9)$ lies on the first line, and substitution in the second system of equations yields the equalities $(4 - 1)/(-6) = (6 - 2)/(-8) = (-9 - 3)/24$. So, the two lines have a point in common. But the denominators of their equations represent vectors parallel to the lines, in the first case, $(3, 4, -12)$ and, in the second case, $(-6, -8, 24)$. Since $(-6, -8, 24) = -2(3, 4, -12)$, the vectors are parallel. Hence, the two lines must be identical.

40.98 Find an equation of a plane containing the intersection of the planes $3x - 2y + 4z = 5$ and $2x + 4y - z = 7$ and passing through the point $(2, 1, 2)$.

▌ We look for a suitable constant k so that the plane $(3x - 2y + 4z - 5) + k(2x + 4y - z - 7) = 0$ contains the given point. Thus, $(3 + 2k)x + (-2 + 4k)y + (4 - k)z - (5 + 7k) = 0$ must be satisfied by $(2, 1, 2)$: $(3 + 2k)2 + (-2 + 4k) + (4 - k)2 - (5 + 7k) = 0$, $-k + 7 = 0$, $k = 7$. So, the desired plane is $17x + 26y - 3z - 54 = 0$.

40.99 Find the coordinates of the point P at which the line $\dfrac{x+8}{9} = \dfrac{y-10}{-4} = \dfrac{z-9}{-2}$ cuts the plane $3x + 4y + 5z = 76$.

▌ Write the equations of the line in parametric form: $x = -8 + 9t$, $y = 10 - 4t$, $z = 9 - 2t$. Substitute in the equation for the plane: $3(-8 + 9t) + 4(10 - 4t) + 5(9 - 2t) = 76$. Then, $t + 61 = 76$, $t = 15$. Thus, the point P is $(127, -50, -21)$.

40.100 Show that the line $\dfrac{x-3}{5} = \dfrac{y+1}{5} = \dfrac{z+4}{7}$ lies in the plane $3x + 4y - 5z = 25$.

▌ The equations of the line in parametric form are $x = 3 + 5t$, $y = -1 + 5t$, $z = -4 + 7t$. Substitute in the equation of the plane: $3(3 + 5t) + 4(-1 + 5t) - 5(-4 + 7t) = 25$, $0(t) + 25 = 25$, which holds identically in t. Hence, all points of the line satisfy the equation of the plane.

40.101   Show that the line $\mathscr{L}$ of intersection of the planes $x + y - z = 0$ and $x - y - 5z + 7 = 0$ is parallel to the line $\mathscr{M}$ determined by $\dfrac{x+3}{3} = \dfrac{y-1}{-2} = \dfrac{z-5}{1}$.

    ▌   The line $\mathscr{L}$ must be perpendicular to the normal vectors of the planes, $(1, 1, -1)$ and $(1, -1, -5)$.   Note that $\mathscr{M}$ is parallel to the vector $(3, -2, 1)$, which is perpendicular to both $(1, 1, -1)$ and $(1, -1, -5)$.   Hence, $\mathscr{M}$ is parallel to $\mathscr{L}$.

40.102   Show that the second-degree equation $(2x + y - z - 3)^2 + (x + 2y - 3z + 5)^2 = 0$ represents a straight line in space.

    ▌ $a^2 + b^2 = 0$ if and only if $a = 0$ and $b = 0$.   So, the given equation is equivalent to $2x + y - z - 3 = 0$ and $x + 2y - 3z + 5 = 0$,   a pair of intersecting planes that determine a straight line.

40.103   Find $\cos\theta$, where θ is the angle between the planes $2x - y + 2z = 3$ and $3x + 2y - 6z = 7$.

    ▌ θ is equal to the angle between the normal vectors to the planes: $(2, -1, 2)$ and $(3, 2, -6)$.   So,

$$\cos\theta = \frac{(2, -1, 2) \cdot (3, 2, -6)}{\sqrt{4+1+4}\,\sqrt{9+4+36}} = \frac{-8}{3(7)} = -\frac{8}{21}$$

40.104   Find an equation of the line through $P(4, 2, -1)$ and perpendicular to the plane $6x - 3y + z = 5$.

    ▌   The line is parallel to the normal vector to the plane, $(6, -3, 2)$.   Hence, the line has the parametric equations $x = 4 + 6t$, $y = 2 - 3t$, $z = -1 + 2t$.

40.105   Find equations of the line through $P(4, 2, -1)$ and parallel to the intersection $\mathscr{L}$ of the planes $x - y + 2z + 4 = 0$ and $2x + 3y + 6z - 12 = 0$.

    ▌   The line $\mathscr{L}$ is perpendicular to the normal vectors of the planes, $(1, -1, 2)$ and $(2, 3, 6)$, and is, therefore, parallel to their cross product $(1, -1, 2) \times (2, 3, 6) = (-12, -2, 5)$.   Hence, the required equations are $x = 4 - 12t$, $y = 2 - 2t$, $z = -1 + 5t$.

40.106   Find an equation of the plane through $P(1, 2, 3)$ and parallel to the vectors $(2, 1, -1)$ and $(3, 6, -2)$.

    ▌   A normal vector to the plane is $(2, 1, -1) \times (3, 6, -2) = (4, 1, 9)$.   So, the plane has an equation of the form $4x + y + 9z = d$.   Since $(1, 2, 3)$ lies in the plane, $4 + 2 + 27 = d$, $d = 33$.   So, the equation is $4x + y + 9z = 33$.

40.107   Find an equation of the plane through $(2, -3, 2)$ and the line $\mathscr{L}$ determined by the planes $6x + 4y + 3z + 5 = 0$ and $2x + y + z - 2 = 0$.

    ▌   Consider the plane $(6x + 4y + 3z + 5) + k(2x + y + z - 2) = 0$.   This plane passes through $\mathscr{L}$.   We want the point $(2, -3, 2)$ to be on the plane.   So, $(12 - 12 + 6 + 5) + k(4 - 3 + 2 - 2) = 0$, $11 + k = 0$, $k = -11$. Therefore,   we   get $6x + 4y + 3z + 5 - 11(2x + y + z - 2) = 0$, $-16x - 7y - 8z + 27 = 0$, $16x + 7y + 8z - 27 = 0$.

40.108   Find an equation of the plane through $P_0(2, -1, -1)$ and $P_1(1, 2, 3)$ and perpendicular to the plane $2x + 3y - 5z - 6 = 0$.

    ▌ $\overrightarrow{P_0P_1} = (-1, 3, 4)$ is parallel to the plane.   Also, the normal vector $(2, 3, -5)$ to the plane $2x + 3y - 5z - 6 = 0$ is parallel to the sought plane.   Hence, a normal vector to the required plane is $(-1, 3, 4) \times (2, 3, -5) = (-27, 3, -9)$.   Thus, an equation of that plane has the form $-27x + 3y - 9z = d$.   Since $(1, 2, 3)$ lies in the plane, $-27 + 6 - 27 = d$, $d = -48$.   Thus, we get $-27x + 3y - 9z = -48$,   or $9x - y + 3z = 16$.

40.109   Let $A(1, 2, 3)$, $B(2, -1, 5)$, and $C(4, 1, 3)$ be consecutive vertices of a parallelogram $ABCD$ (Fig. 40-9).   Find the coordinates of D.

    ▌ $\overrightarrow{AD} = \overrightarrow{BC} = (2, 2, -2)$. $\overrightarrow{OD} = \overrightarrow{OA} + \overrightarrow{AD} = (1, 2, 3) + (2, 2, -2) = (3, 4, 1)$.   Hence, D has coordinates $(3, 4, 1)$.

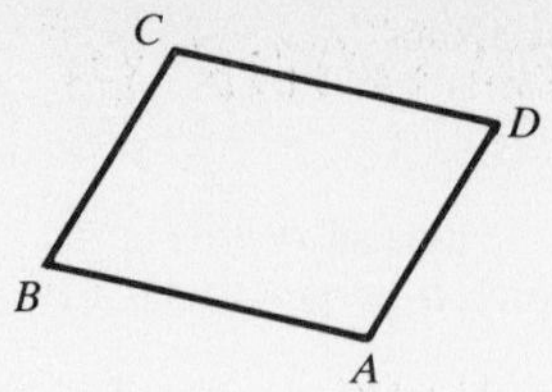

Fig. 40-9

40.110 Find the area of the parallelogram $ABCD$ of Problem 40.109.

▎ The area is $|\overrightarrow{AB} \times \overrightarrow{AD}| = |(1, -3, 2) \times (2, 2, -2)| = |(2, 6, 8)| = 2|(1, 3, 4)| = 2\sqrt{1 + 9 + 16} = 2\sqrt{26}$.

40.111 Find the area of the orthogonal projection of the parallelogram of Problem 40.109 onto the xy-plane.

▎ Let $A'B'C'D'$ be the projection. $A' = (1, 2, 0)$, $B' = (2, -1, 0)$, $C' = (4, 1, 0)$, $D' = (3, 4, 0)$. The desired area is $|\overrightarrow{A'B'} \times A'D'| = |(1, -3, 0) \times (2, 2, 0)| = |(0, 0, 8)| = 8$.

40.112 Find the smaller angle of intersection of the planes $5x - 14y + 2z - 8 = 0$ and $10x - 11y + 2z + 15 = 0$.

▎ The desired angle θ is the smaller angle between the normal vectors to the planes, $(5, -14, 2)$ and $(10, -11, 2)$. Now,

$$\cos \theta = \frac{(5, -14, 2) \cdot (10, -11, 2)}{\sqrt{25 + 196 + 4} \sqrt{100 + 121 + 4}} = \frac{208}{15(15)} = \frac{208}{225} \approx 0.9244$$

Then, $\theta \approx 22° 25'$.

40.113 Find a method for determining the distance between two nonintersecting, nonparallel lines

$\mathscr{L}_1$:
$$\frac{x - x_1}{a_1} = \frac{y - y_1}{b_1} = \frac{z - z_1}{c_1}$$

$\mathscr{L}_2$:
$$\frac{x - x_2}{a_2} = \frac{y - y_2}{b_2} = \frac{z - z_2}{c_2}$$

▎ The plane $\mathscr{P}$ through $\mathscr{L}_1$ parallel to $\mathscr{L}_2$ has as a normal vector $\mathbf{N} = (a_1, b_1, c_1) \times (a_2, b_2, c_2)$. The distance between $\mathscr{L}_1$ and $\mathscr{L}_2$ is equal to the distance from a point on $\mathscr{L}_2$ to the plane $\mathscr{P}$. That distance is equal to the magnitude of the scalar projection of the vector $\overrightarrow{P_1 P_2}$ [from $P_1 = (x_1, y_1, z_1)$ on $\mathscr{L}_1$ to $P_2 = (x_2, y_2, z_2)$ on $\mathscr{L}_2$] on the normal vector $\mathbf{N}$. Thus, we obtain $|\overrightarrow{P_1 P_2} \cdot \mathbf{N}| / |\mathbf{N}|$.

40.114 Find the distance d between the lines

$$\mathscr{L}_1: \quad \frac{x + 2}{2} = \frac{y - 3}{-1} = \frac{z + 3}{-2} \quad \text{and} \quad L_2: \quad \frac{x + 1}{4} = \frac{y - 2}{-3} = \frac{z}{-5}$$

▎ Use the method of Problem 40.113. $\mathbf{N} = (2, -1, -2) \times (4, -3, -5) = (-1, 2, -2)$. The point $P_1(-2, 3, -3)$ lies on $\mathscr{L}_1$, and the point $P_2(-1, 2, 0)$ lies on $\mathscr{L}_2$. $\overrightarrow{P_1 P_2} = (1, -1, 3)$. Hence,

$$d = \frac{|\overrightarrow{P_1 P_2} \cdot \mathbf{N}|}{|\mathbf{N}|} = \frac{|(1, -1, 3) \cdot (-1, 2, -2)|}{\sqrt{1 + 4 + 4}} = \frac{|-1 - 2 - 6|}{3} = \frac{9}{3} = 3$$

CHAPTER 41

Functions of Several Variables

MULTIVARIATE FUNCTIONS AND THEIR GRAPHS

41.1 Sketch the cylinder $\dfrac{x^2}{16} + \dfrac{y^2}{9} = 1$.

▮ See Fig. 41-1. The surface is generated by taking the ellipse $\dfrac{x^2}{16} + \dfrac{y^2}{9} = 1$ in the xy-plane and moving it parallel to the z-axis (z is the missing variable).

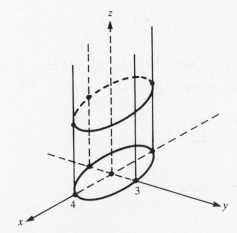

Fig. 41-1

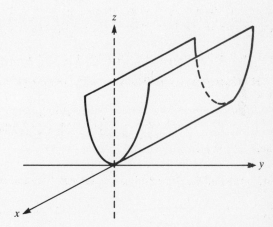

Fig. 41-2

41.2 Describe and sketch the cylinder $z = y^2$.

▮ See Fig. 41-2. Take the parabola in the yz-plane $z = y^2$ and move it parallel to the x-axis (x is the missing variable).

41.3 Describe and sketch the graph of $2x + 3y = 6$.

▮ See Fig. 41-3. The graph is a plane, obtained by taking the line $2x + 3y = 6$, lying in the xy-plane, and moving it parallel to the z-axis.

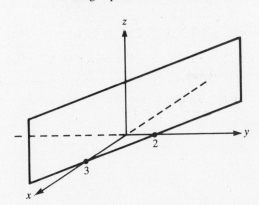

Fig. 41-3

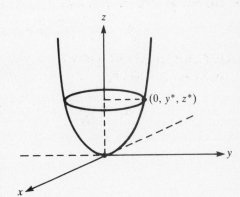

Fig. 41-4

41.4 Write the equation for the surface obtained by revolving the curve $z = y^2$ (in the yz-plane) about the z-axis.

▮ When the point $(0, y^*, z^*)$ on $z = y^2$ is rotated about the z-axis (see Fig. 41-4), consider any resulting point (x, y, z). Clearly, $z = z^*$ and $\sqrt{x^2 + y^2} = y^*$. Hence, $x^2 + y^2 = (y^*)^2 = z^* = z$. So, the resulting points satisfy the equation $z = x^2 + y^2$.

361

41.5 Write an equation for the surface obtained by rotating a curve $f(y, z) = 0$ (in the yz-plane) about the z-axis.

   ▮ This is a generalization of Problem 41.4. A point $(0, y^*, z^*)$ on the curve yields points (x, y, z), where $z = z^*$ and $\sqrt{x^2 + y^2} = y^*$. Hence, the point (x, y, z) satisfies the equation $f(\sqrt{x^2 + y^2}, z) = 0$.

41.6 Write an equation for the surface obtained by rotating the curve $\dfrac{y^2}{4} + \dfrac{z^2}{25} = 1$ (in the yz-plane) about the z-axis.

   ▮ By Problem 41.5, the equation is obtained by replacing y by $\sqrt{x^2 + y^2}$ in the original equation. So, we get

$$\frac{(\sqrt{x^2 + y^2})^2}{4} + \frac{z^2}{25} = 1 \qquad \frac{x^2 + y^2}{4} + \frac{z^2}{25} = 1 \qquad \frac{x^2}{4} + \frac{y^2}{4} + \frac{z^2}{25} = 1$$

an ellipsoid.

41.7 Write an equation of the surface obtained by rotating the hyperbola $\dfrac{x^2}{9} - \dfrac{y^2}{4} = 1$ (in the xy-plane) about the x-axis.

   ▮ By analogy with Problem 41.5, we replace y by $\sqrt{y^2 + z^2}$, obtaining $\dfrac{x^2}{9} - \dfrac{y^2 + z^2}{4} = 1$, $\dfrac{x^2}{9} - \dfrac{y^2}{4} - \dfrac{z^2}{4} = 1$. (This surface is called a hyperboloid of two sheets.)

41.8 Write an equation of the surface obtained by rotating the line $z = 2y$ (in the yz-plane) about the z-axis.

   ▮ By Problem 41.5, an equation is $z = 2\sqrt{x^2 + y^2}$, $z^2 = 4(x^2 + y^2)$. This is a cone (with both nappes) having the z-axis as axis of symmetry (see Fig. 41-5).

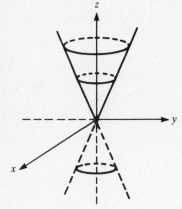

Fig. 41-5

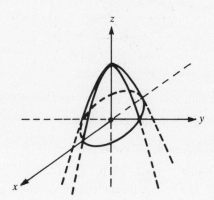

Fig. 41-6

41.9 Write an equation of the surface obtained by rotating the parabola $z = 4 - x^2$ (in the xz-plane) about the z-axis. (See Fig. 41-6).

   ▮ By Problem 41.5, an equation is $z = 4 - (\sqrt{x^2 + y^2})^2$, $z = 4 - (x^2 + y^2)$. This is a circular paraboloid.

41.10 Describe and sketch the surface obtained by rotating the curve $z = |y|$ (in the yz-plane) about the z-axis.

   ▮ By Problem 41.5, an equation is $z = |\sqrt{x^2 + y^2}|$, which is equivalent to $z^2 = x^2 + y^2$ for $z \geq 0$. This is a right circular cone (with a 90° apex angle); see Fig. 41-7.

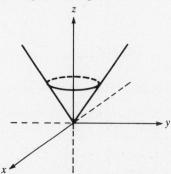

Fig. 41-7

41.11 Of what curve in the yz-plane is the surface $x^2 + y^2 - z^2 = 1$ a surface of revolution?

▌ It is equivalent to $(\sqrt{x^2 + y^2})^2 - z^2 = 1$, which is obtained from the hyperbola $y^2 - z^2 = 1$ by rotation around the z-axis.

41.12 From what curve in the xy-plane is the surface $9x^2 + 25y^2 + 9z^2 = 225$ obtained by rotation about the y-axis?

▌ It is equivalent to $9(\sqrt{x^2 + z^2})^2 + 25y^2 = 225$, which is obtained from $9x^2 + 25y^2 = 225$ by rotation about the y-axis. The latter curve is the ellipse $\dfrac{x^2}{25} + \dfrac{y^2}{9} = 1$.

41.13 Describe and sketch the graph of $\dfrac{x^2}{a^2} + \dfrac{y^2}{b^2} + \dfrac{z^2}{c^2} = 1$, where $a, b, c > 0$.

▌ See Fig. 41-8. Each section parallel to one of the coordinate planes is an ellipse (or a point or nothing). The surface is bounded, $|x| \le a$, $|y| \le b$, $|z| \le c$. This surface is called an ellipsoid. When $a = b = c$, the surface is a sphere.

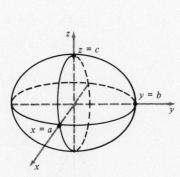

Fig. 41-8

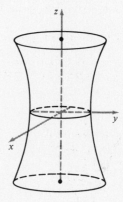

Fig. 41-9

41.14 Describe and sketch the graph of $\dfrac{x^2}{a^2} + \dfrac{y^2}{b^2} - \dfrac{z^2}{c^2} = 1$, where $a, b, c > 0$.

▌ See Fig. 41-9. Each section $z = k$ parallel to the xy-plane cuts out an ellipse. The ellipses get bigger as $|z|$ increases. (For $z = 0$, the xy-plane, the section is the ellipse $\dfrac{x^2}{a^2} + \dfrac{y^2}{b^2} = 1$.) For sections made by planes $x = k$, we obtain hyperbolas, and, similarly for sections made by planes $y = k$. The surface is called a hyperboloid of one sheet.

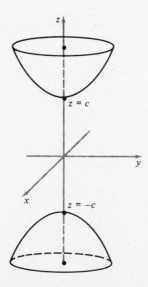

Fig. 41-10

41.15 Describe and sketch the graph of $\dfrac{z^2}{c^2} - \dfrac{x^2}{a^2} - \dfrac{y^2}{b^2} = 1$, where $a, b, c > 0$.

▮ See Fig. 41-10. Note that $\dfrac{z^2}{c^2} = 1 + \dfrac{x^2}{a^2} + \dfrac{y^2}{b^2} \ge 1$. Hence, $|z| \ge c$. The sections by planes $z = k$, with $|k| > c$, are ellipses. When $|z| = c$, we obtain a point $(0, 0, \pm c)$. The sections determined by planes $x = k$ or $y = k$ are hyperbolas. The surface is called a hyperboloid of two sheets.

41.16 Describe and sketch the graph of $\dfrac{x^2}{a^2} + \dfrac{y^2}{b^2} = \dfrac{z^2}{c^2}$, where $a, b, c > 0$.

▮ See Fig. 41-11. This is an elliptic cone. The horizontal cross sections $z = c \ne 0$ are ellipses. The horizontal cross section $z = 0$ is a point, the origin. The surface intersects the xz-plane $(y = 0)$ in a pair of lines, $z = \pm \dfrac{c}{a} x$, and intersects the yz-plane $(x = 0)$ in a pair of lines $z = \pm \dfrac{c}{b} y$. The other cross sections, determined by $x = k$ or $y = k$, are hyperbolas.

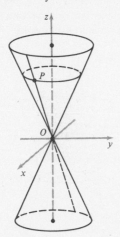

Fig. 41-11

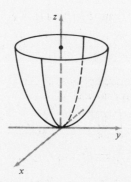

Fig. 41-12

41.17 Describe and sketch the graph of the function $f(x, y) = x^2 + y^2$.

▮ See Fig. 41-12. This is the graph of $z = x^2 + y^2$. Note that $z \ge 0$. When $z = 0$, $x^2 + y^2 = 0$, and, therefore, $x = y = 0$. So, the intersection with the xy-plane is the origin. Sections made by planes $z = k > 0$ are circles with centers on the z-axis. Sections made by planes $x = k$ or $y = k$ are parabolas. The surface is called a circular paraboloid.

41.18 Describe and sketch the graph of the function $f(x, y) = 2x + 5y - 10$.

▮ This is the graph of $z = 2x + 5y - 10$, or $2x + 5y - z = 10$, a plane having $(2, 5, -1)$ as a normal vector.

41.19 Describe and sketch the graph of $z = y^2 - x^2$.

▮ See Fig. 41-13. This is called a saddle surface, with the "seat" at the origin. The plane sections $z = c > 0$ are hyperbolas with principal axis the y-axis. For $z = c < 0$, the plane sections are hyperbolas with principal axis the x-axis. The section made by $z = 0$ is $y^2 - x^2 = 0$, $(y - x)(y + x) = 0$, the pair of lines $y = x$ and $y = -x$. The sections made by planes $x = c$ or $y = c$ are parabolas.

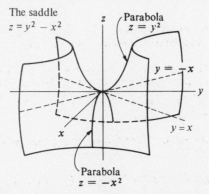

Fig. 41-13

41.20 Find the points at which the line $\dfrac{x-6}{3} = \dfrac{y+2}{-6} = \dfrac{z-2}{-4}$ intersects the ellipsoid $\dfrac{x^2}{81} + \dfrac{y^2}{36} + \dfrac{z^2}{9} = 1$.

❚ The line can be written in parametric form as $x = 6 + 3t$, $y = -2 - 6t$, $z = 2 + 4t$. Hence, substituting in the equation of the ellipsoid, we get $\dfrac{(6+3t)^2}{81} + \dfrac{(2+6t)^2}{36} + \dfrac{(2+4t)^2}{9} = 1$, $4(6+3t)^2 + 9(2+6t)^2 + 36(2 + 4t)^2 = 324$, $26(36t^2) + 26(36t) + 4(81) = 324$, $t^2 + t = 0$, $t(t+1) = 0$, $t = 0$ or $t = -1$. So, the points are $(6, -2, 2)$ and $(3, 4, -2)$.

41.21 Show that the plane $2x - y - 2z = 10$ intersects the paraboloid $2z = \dfrac{x^2}{9} + \dfrac{y^2}{4}$ at a single point, and find the point.

❚ Solve the equations simultaneously. $2x - y - 10 = \dfrac{x^2}{9} + \dfrac{y^2}{4}$, $4x^2 + 9y^2 - 72x + 36y = -360$, $4(x-9)^2 + 9(y+2)^2 = -360 + 324 + 36 = 0$. Hence, the only solution is $x = 9$, $y = -2$. Then $z = 5$. Thus, the only point of intersection is $(9, -2, 5)$, where the plane is tangent to the paraboloid.

41.22 Find the volume of the ellipsoid $\dfrac{x^2}{a^2} + \dfrac{y^2}{b^2} + \dfrac{z^2}{c^2} = 1$.

❚ The plane $z = k$ cuts the ellipsoid in an ellipse

$$\frac{x^2}{a^2} + \frac{y^2}{b^2} = 1 - \frac{k^2}{c^2} = \frac{c^2 - k^2}{c^2} \qquad \frac{x^2}{(a\sqrt{c^2 - k^2}/c)^2} + \frac{y^2}{(b\sqrt{c^2 - k^2}/c)^2} = 1$$

By Problem 20.72, the area of this ellipse is $\pi(a\sqrt{c^2 - k^2}/c)(b\sqrt{c^2 - k^2}/c) = \pi ab(c^2 - k^2)/c^2$. Hence, by the cross-section formula for volume,

$$V = 2 \int_0^c \frac{\pi ab}{c^2} (c^2 - k^2)\, dk = \frac{2\pi ab}{c^2} (c^2 k - \tfrac{1}{3}k^3) \Big]_0^c = \frac{2\pi ab}{c^2} (c^3 - \tfrac{1}{3}c^3) = \tfrac{4}{3}\pi abc$$

41.23 Identify the graph of $9x^2 - y^2 + 16z^2 = 144$.

❚ This equation is equivalent to $(x^2/16) - (y^2/144) + (z^2/9) = 1$. This is an elliptic hyperboloid of one sheet, with axis the y-axis. The cross sections $y = k$ are ellipses $(x^2/16) + (z^2/9) = 1 + (k^2/144)$. The cross sections $x = k$ are hyperbolas $(z^2/9) - (y^2/144) = 1 - (k^2/16)$, and the cross sections $z = k$ are hyperbolas $(x^2/16) - (y^2/144) = 1 - (k^2/9)$.

41.24 Identify the graph of $25x^2 - y^2 - z^2 = 25$.

❚ This is equivalent to $x^2 - (y^2/25) - (z^2/25) = 1$, which is a circular hyperboloid of two sheets. Each cross section $x = k$ is a circle $y^2 + z^2 = 25(k^2 - 1)$. (We must have $|k| \geq 1$.) Each cross section $y = k$ is a hyperbola $x^2 - (z^2/25) = 1 + (k^2/25)$, and each cross section $z = k$ is a hyperbola $x^2 - (y^2/25) = 1 + (k^2/25)$. The axis of the hyperboloid is the x-axis.

41.25 Identify the graph of $x^2 + 4z^2 = 2y$.

❚ This is equivalent to $(x^2/4) + z^2 = y/2$. This is an elliptic paraboloid, with the y-axis as its axis. For $y = k > 0$, the cross section is an ellipse $(x^2/4) + z^2 = k/2$. ($y = 0$ yields the origin, and the cross sections $y = k < 0$ are empty.) The cross section $x = k$ is a parabola $y/2 = z^2 + (k^2/4)$. The cross section $z = k$ is a parabola $y/2 = (x^2/4) + k^2$.

41.26 Show that, if a curve $\mathscr{C}$ in the xy-plane has the equation $f(x, y) = 0$, then an equation of the cylinder generated by a line intersecting $\mathscr{C}$ and moving parallel to the vector $\mathbf{A} = (a, b, 1)$ is $f(x - az, y - bz) = 0$.

❚ The line $\mathscr{L}$ through a point $(x_0, y_0, 0)$ on $\mathscr{C}$ and parallel to $\mathbf{A}$ has parametric equations $x = x_0 + at$, $y = y_0 + bt$, $z = t$. Then, $x_0 = x - az$ and $y_0 = y - bz$. Hence, $f(x - az, y - bz) = 0$. Conversely, if $f(x - az, y - bz) = 0$, then, setting $x_0 = x - az$, $y_0 = y - bz$, we see that $(x_0, y_0, 0)$ is on $\mathscr{C}$ and (x, y, z) is on the corresponding line $\mathscr{L}$.

41.27 Find an equation of the cylinder generated by a line through the curve $y^2 = x - y$ in the xy-plane that moves parallel to the vector $\mathbf{A} = (2, 2, 1)$.

❚ By Problem 41.26, an equation is $(y - 2z)^2 = (x - 2z) - (y - 2z)$, $(y - 2z)^2 = x - y$.

41.28 The two equations $x^2 + 3y^2 - z^2 + 3x = 0$ and $2x^2 + 6y^2 - 2z^2 - 4y = 3$ together determine the curve in which the corresponding surfaces intersect. Show that this curve lies in a plane.

▐ Eliminate z by multiplying the first equation by 2 and subtracting the result from the second equation: $6x + 4y = -3$. Hence, all points of the curve lie in the plane $6x + 4y = -3$.

41.29 Show that the hyperboloid of one sheet $x^2 + y^2 - z^2 = 1$ is a ruled surface, that is, each of its points lies on a line that is entirely included in the surface. (See Fig. 41-14.)

▐ Note that the circle $\mathscr{C}$, $x^2 + y^2 = 1$ in the xy-plane, lies in the surface. Now consider any point (x, y, z) on the surface. Then, $x^2 + y^2 - z^2 = 1$. Let $x_0 = \dfrac{x - zy}{1 + z^2}$ and $y_0 = \dfrac{xz + y}{1 + z^2}$. Hence,

$$x_0^2 + y_0^2 = \frac{x^2z^2 + 2xyz + y^2 + x^2 - 2xyz + z^2y^2}{(1 + z^2)^2} = \frac{(x^2 + y^2)(1 + z^2)}{(1 + z^2)^2} = 1$$

So, $(x_0, y_0, 0)$ is on the surface and lies on the circle $\mathscr{C}$. The line $\mathscr{L}$: $x = x_0 + y_0 t$, $y = y_0 - x_0 t$, $z = t$ contains the point $(x_0, y_0, 0)$ and lies entirely on the given surface, since $(x_0 + y_0 t)^2 + (y_0 - x_0 t)^2 - t^2 = 1$. The original point (x, y, z) appears on $\mathscr{L}$ for the parameter value $t = z$, since

$$x_0 + y_0 t = \frac{x - zy}{1 + z^2} + \frac{xz + y}{1 + z^2} z = \frac{x - zy + xz^2 + yz}{1 + z^2} = x$$

and

$$y_0 - x_0 t = \frac{xz + y}{1 + z^2} - \frac{x - zy}{1 + z^2} z = \frac{xz + y - xz + z^2 y}{1 + z^2} = y$$

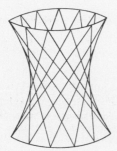

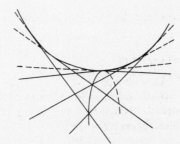

Fig. 41-14   Fig. 41-15

41.30 Show that the hyperbolic paraboloid $z = y^2 - x^2$ is a ruled surface. (See the definition in Problem 41.29.)

▐ Consider any point (x_0, y_0, z_0) on the surface. The following line L: $x = x_0 + t$, $y = y_0 + t$, $z = z_0 + 2(y_0 - x_0)t$ contains the given point (when $t = 0$). Note that $\mathscr{L}$ lies on the paraboloid: for (x, y, z) on $\mathscr{L}$, $y^2 - x^2 = (y_0 + t)^2 - (x_0 + t)^2 = y_0^2 + 2y_0 t - x_0^2 - 2x_0 t = (y_0^2 - x_0^2) + 2(y_0 - x_0)t = z_0 + 2(y_0 - x_0)t = z$. (See Fig. 41-15.)

41.31 Describe the graph of the function $f(x, y) = \sqrt{a^2 - x^2 - y^2}$, where $a > 0$.

▐ This is the graph of $z = \sqrt{a^2 - x^2 - y^2}$, which, for $z \geq 0$, is equivalent to $z^2 = a^2 - x^2 - y^2$, $x^2 + y^2 + z^2 = a^2$, the equation of the sphere with center at the origin and radius a. Hence, the graph is the upper half of that sphere (including the circle $x^2 + y^2 = a^2$ in the xy-plane).

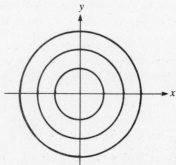

Fig. 41-16

41.32 Describe the level curves (contour map) of $f(x, y) = x^2 + y^2$.

▮ In general, the level curves of a function $f(x, y)$ are the family of curves $f(x, y) = k$, $z = 0$. Here, the level curves are the circles $x^2 + y^2 = r^2$ (write $k = r^2 > 0$) of radius r with center at the origin (Fig. 41-16). There is one through every point of the xy-plane except the origin (unless we consider the origin as the level curve $x^2 + y^2 = 0$).

41.33 Describe the level curves of $f(x, y) = y - x$.

▮ The level curves form the family of parallel lines $y - x = k$ with slope 1 (Fig. 41-17). There is one level curve through every point.

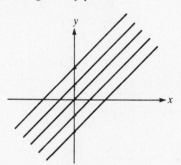

Fig. 41-17

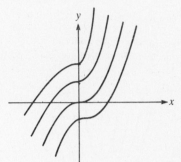

Fig. 41-18

41.34 Describe the level curves of $f(x, y) = y - x^3$.

▮ The level curves form the family of cubic curves $y = x^3 + k$ (Fig. 41-18). There is one level curve through every point.

41.35 Describe the level curves of $f(x, y) = y/x^2$.

See Fig. 41-19. The level curves form the family of curves $y = kx^2$. When $k > 0$, we get a parabola that opens upward; when $k < 0$, the parabola opens downward. When $k = 0$, we get the x-axis. In each case, the level curve is "punctured" at the origin, since y/x^2 is undefined when $x = 0$. There is a level curve through every point not on the y-axis.

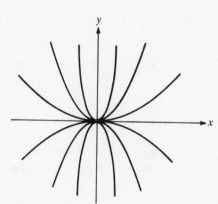

Fig. 41-19

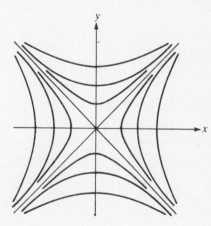

Fig. 41-20

41.36 Describe the level curves of $f(x, y) = y^2 - x^2$.

▮ See Fig. 41-20. The level curves, $y^2 - x^2 = k$ are two sets of rectangular hyperbolas (corresponding to $k > 0$ and $k < 0$), plus two straight lines ($k = 0$) to which all the hyperbolas are asymptotic. There is a single level curve through every point (x, y) except $(0, 0)$, which lies on two.

41.37 Describe the level curves of $f(x, y) = 4x^2 + 9y^2$.

▮ See Fig. 41-21. The level curves form a family of ellipses $4x^2 + 9y^2 = k^2$, or $\dfrac{x^2}{(k/2)^2} + \dfrac{y^2}{(k/3)^2} = 1$. There is a level curve through every point (considering the origin to be a level curve consisting of a single point).

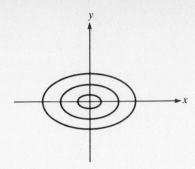

Fig. 41-21

41.38 Describe the level curves of $f(x, y) = y - x^2$.

❚ This is a family of parabolas $y = x^2 + k$, all opening upwards and having the y-axis as axis of symmetry. There is one level curve through every point.

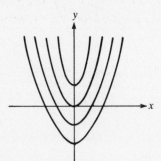

Fig. 41-22

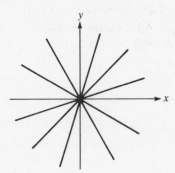

Fig. 41-23

41.39 Describe the level curves of $f(x, y) = y/x$.

❚ This is the family of all "punctured" lines $y = kx$ through the origin, but with the origin excluded. There is a level curve through every point not on the y-axis.

41.40 Describe the level curves of $f(x, y) = 1/xy$.

❚ These are the rectangular hyperbolas $xy = k$ $(k > 0)$ and $xy = k$ $(k < 0)$; see Fig. 41-24. There is a level curve through every point not on a coordinate axis.

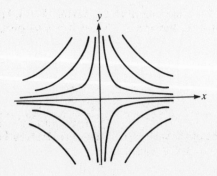

Fig. 41-24

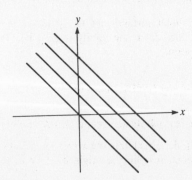

Fig. 41-25

41.41 Describe the level curves of $f(x, y) = \sqrt{y + x}$.

❚ The level curves are $y + x = k \geq 0$. These are the parallel lines of slope -1 and nonnegative y-intercept (Fig. 41-25). There is a level curve through every point on or above the line $y = -x$.

41.42 Describe the level curves of $f(x, y) = \sqrt{1 - y - x^2}$.

❚ The level curves are the curves $1 - y - x^2 = k \geq 0$, or $y = -x^2 + 1 - k = -x^2 + c$, where $c \leq 1$. This is a family of parabolas with the y-axis as axis of symmetry, opening downward, and with vertex at $(0, c)$; see Fig. 41-26. There is a level curve through every point on or below the parabola $y = -x^2 + 1$.

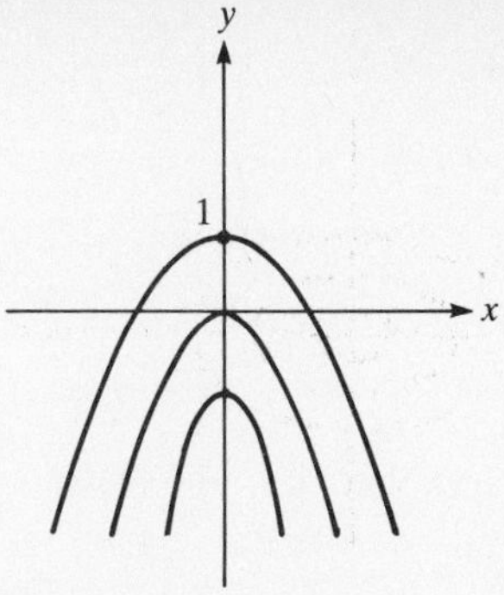

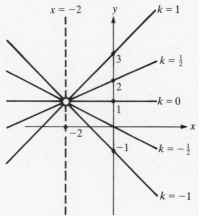

Fig. 41-26

41.43 Describe the level curves of $f(x, y) = \dfrac{y-1}{x+2}$.

▌ See Fig. 41-27. The level curves are $\dfrac{y-1}{x+2} = k$, or $y = kx + (2k+1)$. This is a family of punctured straight lines. All these lines pass through the point $(-2, 1)$, which is excluded from them. There is a level curve through every point not on the vertical line $x = -2$.

Fig. 41-27

41.44 Describe the level surfaces of $f(x, y, z) = 3x - 2y + z$.

▌ The level surfaces are the planes $3x - 2y + z = k$. Since they all have the vector $(3, -2, 1)$ as normal vector, they form the family of parallel planes perpendicular to that vector. There is a level surface through every point.

41.45 Describe the level surfaces of $f(x, y, z) = \dfrac{x^2}{25} + \dfrac{y^2}{9} + z^2$.

▌ The level surfaces form a family of ellipsoids $\dfrac{x^2}{25} + \dfrac{y^2}{9} + z^2 = k > 0$, or $\dfrac{x^2}{(5\sqrt{k})^2} + \dfrac{y^2}{(3\sqrt{k})^2} + \dfrac{z^2}{(\sqrt{k})^2} = 1$, in which the axes along the x-axis, y-axis, and z-axis are in the proportion $5:3:1$. There is a level surface through every point (except the origin if one does not count a point as a level surface).

41.46 Describe the level surfaces of $f(x, y, z) = 3x^2 + 5y^2 - z^2$.

▌ The level surfaces $3x^2 + 5y^2 - z^2 = k > 0$ are hyperboloids of one sheet around the z-axis. The level surfaces $3x^2 + 5y^2 - z^2 = k \le 0$ are hyperboloids of two sheets around the z-axis. There is a level surface through every point.

41.47 Describe the level surfaces of $f(x, y, z) = \dfrac{1}{x^2 + y^2 + z^2}$.

▌ The level surfaces are $x^2 + y^2 + z^2 = k^2 > 0$, concentric spheres with center at the origin. Every point except the origin lies on a level surface.

41.48 What is signified by $\lim\limits_{(x,\,y)\to(a,\,b)} f(x, y) = L$?

▯ Recall that $|(u, v)| = \sqrt{u^2 + v^2}$. Then $\lim\limits_{(x,\,y)\to(a,\,b)} f(x, y) = L$ means that, for every $\varepsilon > 0$, there exists $\delta > 0$ such that $|f(x, y) - L| < \varepsilon$ whenever $|(x, y) - (a, b)| < \delta$.

41.49 Prove that $\lim\limits_{(x,\,y)\to(1,\,2)} 2x - 3y = -4$.

▯ Let $\varepsilon > 0$. We must find $\delta > 0$ such that $|(x, y) - (1, 2)| < \delta$ implies $|(2x - 3y) - (-4)| < \varepsilon$; that is, $\sqrt{(x-1)^2 + (y-2)^2} < \delta$ implies $|2x - 3y + 4| < \varepsilon$. Let $\delta = \varepsilon/5$. Assume $\sqrt{(x-1)^2 + (y-2)^2} < \delta$. Then $|x - 1| < \delta$ and $|y - 2| < \delta$. So, $|2x - 3y + 4| = |2(x - 1) - 3(y - 2)| \leq 2|x - 1| + 3|y - 2| < 2\delta + 3\delta = 5\delta < \varepsilon$.

41.50 Find $\lim\limits_{(x,\,y)\to(0,\,0)} \dfrac{2xy^2}{x^2 + y^2}$, if it exists.

▯ Note that, if $y = mx$ and $(x, y) \to (0, 0)$, then $\dfrac{2xy^2}{x^2 + y^2} = \dfrac{2xm^2x^2}{x^2 + m^2x^2} = \dfrac{2m^2}{1 + m^2} x \to 0$. However, this is not enough to ensure that $2xy^2/(x^2 + y^2) \to 0$ no matter how $(x, y) \to (0, 0)$. Take any $\varepsilon > 0$. Note that $|(x, y) - (0, 0)| = \sqrt{x^2 + y^2}$. Observe also that $|x| \leq \sqrt{x^2 + y^2}$ and $y^2 \leq x^2 + y^2$. So,

$$\left| \frac{2xy^2}{x^2 + y^2} \right| \leq \frac{2\sqrt{x^2 + y^2}(x^2 + y^2)}{x^2 + y^2} = 2\sqrt{x^2 + y^2}$$

Hence, if we choose $\delta = \varepsilon/2$ and if $\sqrt{x^2 + y^2} < \delta$, then $|2xy^2/(x^2 + y^2)| \leq 2\sqrt{x^2 + y^2} < 2\delta = \varepsilon$. Thus,

$$\lim_{(x,\,y)\to(0,\,0)} \frac{2xy^2}{x^2 + y^2} = 0.$$

41.51 Find $\lim\limits_{(x,\,y)\to(0,\,0)} \dfrac{2xy^2}{x^2 + y^4}$, if it exists.

▯ Let $y = mx$. Then

$$\frac{2xy^2}{x^2 + y^4} = \frac{2x(m^2x^2)}{x^2 + m^4x^4} = \frac{2m^2x}{1 + m^4x^2} \to \frac{0}{1} = 0$$

as $(x, y) \to (0, 0)$. However, if we let $x = y^2$, then $\dfrac{2xy^2}{x^2 + y^4} = \dfrac{2y^4}{y^4 + y^4} = 1$. Hence, as $(x, y) \to (0, 0)$ along the parabola $x = y^2$, $\dfrac{2xy^2}{x^2 + y^4} \to 1$. Therefore, $\lim\limits_{(x,\,y)\to(0,\,0)} \dfrac{2xy^2}{x^2 + y^4}$ does not exist.

41.52 Find $\lim\limits_{(x,\,y)\to(0,\,0)} \dfrac{x^2 - y^2}{x^2 + y^2}$, if it exists.

▯ Let $y = mx$. Then

$$\frac{x^2 - y^2}{x^2 + y^2} = \frac{x^2 - m^2x^2}{x^2 + m^2x^2} = \frac{1 - m^2}{1 + m^2}$$

Since $(1 - m^2)/(1 + m^2)$ depends on m, $(x^2 - y^2)/(x^2 + y^2)$ approaches different numbers as $(x, y) \to (0, 0)$ along different lines. Hence, $\lim\limits_{(x,\,y)\to(0,\,0)} \dfrac{x^2 - y^2}{x^2 + y^2}$ does not exist.

41.53 Find $\lim\limits_{(x,\,y)\to(0,\,0)} \dfrac{xy}{\sqrt{x^2 + y^2}}$, if it exists.

❚ Let $y = mx$. Then

$$\frac{xy}{\sqrt{x^2 + y^2}} = \frac{mx^2}{\sqrt{x^2 + m^2x^2}} = \frac{mx^2}{|x|\sqrt{1 + m^2}} = \frac{m|x|}{\sqrt{1 + m^2}} \to 0 \quad \text{as} \quad (x, y) \to (0, 0)$$

We can show that $\lim\limits_{(x, y) \to (0, 0)} \dfrac{xy}{\sqrt{x^2 + y^2}} = 0$. For, $|x| \le \sqrt{x^2 + y^2}$. Hence,

$$\left| \frac{xy}{\sqrt{x^2 + y^2}} \right| \le |y| \quad \text{and} \quad \frac{xy}{\sqrt{x^2 + y^2}} \to 0 \quad \text{as} \quad (x, y) \to (0, 0)$$

41.54 Find $\lim\limits_{(x, y) \to (0, 0)} \dfrac{x^2}{x^2 + y^2}$, if it exists.

❚ Let $y = mx$. Then $\dfrac{x^2}{x^2 + y^2} = \dfrac{x^2}{x^2 + m^2x^2} = \dfrac{1}{1 + m^2}$. Hence, as $(x, y) \to (0, 0)$ along the line $y = mx$, $x^2/(x^2 + y^2) \to 1/(1 + m^2)$. Therefore, $\lim\limits_{(x, y) \to (0, 0)} \dfrac{x^2}{x^2 + y^2}$ does not exist, since many different values of m yield different values of $1/(1 + m^2)$.

41.55 Find $\lim\limits_{(x, y) \to (0, 0)} \dfrac{x}{x^2 + y^2}$, if it exists.

❚ Let $y = mx$. Then

$$\left| \frac{x}{x^2 + y^2} \right| = \left| \frac{x}{x^2 + m^2x^2} \right| = \frac{1}{|x|(1 + m^2)} \to +\infty \quad \text{as} \quad (x, y) \to (0, 0)$$

Hence, $\lim\limits_{(x, y) \to (0, 0)} \dfrac{x}{x^2 + y^2}$ does not exist.

41.56 Find $\lim\limits_{(x, y) \to (0, 0)} \dfrac{x^2y^2}{x^3 + y^3}$, if it exists.

❚ Let $y = mx$. Then

$$\frac{x^2y^2}{x^3 + y^3} = \frac{x^2(m^2x^2)}{x^3 + m^3x^3} = \frac{xm^2}{1 + m^3} \to 0 \quad \text{as} \quad (x, y) \to (0, 0)$$

Now let $y = -xe^x$. Then

$$\frac{x^2y^2}{x^3 + y^3} = \frac{-x^4e^{2x}}{x^3 - x^3e^{3x}} = \frac{xe^{2x}}{e^{3x} - 1}$$

By L'Hôpital's rule,

$$\lim_{x \to 0} \frac{xe^{2x}}{e^{3x} - 1} = \lim_{x \to 0} \frac{e^{2x}(2x + 1)}{3e^{3x}} = \frac{1}{3}$$

Hence, the limit does not exist.

41.57 Find $\lim\limits_{(x, y) \to (0, 0)} \dfrac{x^3 + y^3}{x^2 + y}$, if it exists.

❚ Let $y = mx$. Then

$$\frac{x^3 + y^3}{x^2 + y} = \frac{x^3 + m^3x^3}{x^2 + mx} = \frac{x^2(1 + m^3)}{x + m} \to 0 \quad \text{as} \quad (x, y) \to (0, 0)$$

[This is also true when $m = 0$. In that case, $y = 0$ and $(x^3 + y^3)/(x^2 + y) = x \to 0$]. However, let $(x, y) \to (0, 0)$ along $y = -x^2e^x$. Then

$$\frac{x^3 + y^3}{x^2 + y} = \frac{x^3 - x^6e^{3x}}{x^2 - x^2e^x} = \frac{x - x^4e^{3x}}{1 - e^x}$$

By L'Hôpital's rule,

$$\lim_{x \to 0} \frac{x - x^4 e^{3x}}{1 - e^x} = \lim_{x \to 0} \frac{1 - 3x^4 e^{3x} - 4x^3 e^{3x}}{-e^x} = -1.$$

Hence, the limit cannot exist.

41.58 Find $\displaystyle\lim_{(x,\, y) \to (0,\, 0)} \frac{x^2 y^2}{x^4 + y^4}$, if it exists.

▮ Let $y = mx$. Then

$$\frac{x^2 y^2}{x^4 + y^4} = \frac{x^2(m^2 x^2)}{x^4 + m^4 x^4} = \frac{m^2}{1 + m^4}$$

Therefore, the limit does not exist. (For example, we get different limits for $m = 0$ and $m = 1$.)

41.59 Find $\displaystyle\lim_{(x,\, y) \to (0,\, 0)} \frac{x^2 y^2}{x^2 + y^2}$, if it exists.

▮ Note that $0 \le x^2 y^2/(x^2 + y^2) \le y^2 \to 0$ as $(x, y) \to (0, 0)$. Hence, the limit is 0.

41.60 Is it possible to define $\displaystyle f(x, y) = \frac{x^3 + y^3}{x^2 + y^2}$ at $(0, 0)$ so that $f(x, y)$ is continuous?

▮ Note that

$$\left| \frac{x^3 + y^3}{x^2 + y^2} \right| \le \frac{|x^3|}{x^2 + y^2} + \frac{|y^3|}{x^2 + y^2} = \frac{x^2}{x^2 + y^2} |x| + \frac{y^2}{x^2 + y^2} |y| \le |x| + |y| \to 0 \qquad \text{as} \quad (x, y) \to (0, 0)$$

So, $\displaystyle\lim_{(x,\, y) \to (0,\, 0)} \frac{x^3 + y^3}{x^2 + y^2} = 0$. Hence, if we define $f(0, 0) = 0$, then $f(x, y)$ will be continuous, since it is obvious that $f(x, y)$ is continuous at all points different from the origin.

41.61 Is it possible to define $\displaystyle f(x, y) = \frac{xy}{\sqrt{x^2 + y^2}}$ at the origin so that $f(x, y)$ is continuous?

▮ In general, $\displaystyle\left(\frac{xy}{\sqrt{x^2 + y^2}} \right)^2 = \frac{x^2 y^2}{x^2 + y^2} \le y^2 \to 0$ as $(x, y) \to (0, 0)$. Hence, $|xy/\sqrt{x^2 + y^2}| \to 0$. So, if we define $f(0, 0) = 0$, $f(x, y)$ will be continuous at the origin, and, therefore, everywhere.

41.62 Determine whether the function

$$f(x, y) = \begin{cases} \dfrac{xy}{x^2 + y^2} & \text{if } (x, y) \neq (0, 0) \\ 0 & \text{if } (x, y) = (0, 0) \end{cases}$$

is continuous at the origin.

▮ As $(x, y) \to (0, 0)$ along the line $y = mx$, $\displaystyle f(x, y) = \frac{mx^2}{(1 + m^2)x^2} = \frac{m}{1 + m^2}$. Hence, $\displaystyle\lim_{(x,\, y) \to (0,\, 0)} f(x, y)$ does not exist and, therefore, $f(x, y)$ is not continuous [and cannot be made continuous by redefining $f(0, 0)$].

41.63 Is it possible to extend $\displaystyle f(x, y) = \frac{x + y}{x^2 + y^2}$ to the origin so that the resulting function is continuous?

▮ Let $y = mx$. Then $\displaystyle\left| \frac{x + y}{x^2 + y^2} \right| = \left| \frac{x + mx}{x^2 + m^2 x^2} \right| = \frac{1 + m}{|x|(1 + m^2)} \to +\infty$ as $(x, y) \to (0, 0)$. Hence, continuous extension is impossible.

41.64 Find the domain of definition of the function $\displaystyle f(x, y) = \frac{x + 2y}{\sqrt{4 - x^2 - y^2}}$.

▮ The function is defined when the denominator is defined and $\neq 0$. The latter holds when and only when $4 - x^2 - y^2 > 0$, that is, when $x^2 + y^2 < 4$. So, the domain is the inside of the circle of radius 2 with center at the origin.

41.65   Find the domain of definition of the function $f(x, y) = \ln(16 - x^2 - y^2) + \ln(x^2 + y^2 - 1)$.

▌   This is defined when the arguments of the ln function are positive, that is, when $16 - x^2 - y^2 > 0$ and $x^2 + y^2 - 1 > 0$,   or, equivalently, $1 < x^2 + y^2 < 16$.   Thus, the domain consists of all points between the concentric circles around the origin of radii 1 and 4.

41.66   Find the domain of definition of $f(x, y) = e^x \ln(xy)$.

▌   The function is defined when and only when $xy > 0$,   that is, in the first and third quadrants (and not on the axes).

CYLINDRICAL AND SPHERICAL COORDINATES

41.67   Give the equations connecting rectangular and cylindrical coordinates of a point in space.

▌   For a point with rectangular coordinates (x, y, z), corresponding cylindrical coordinates are (r, θ, z), where $r^2 = x^2 + y^2$ and $\tan \theta = y/x$.   Conversely, $x = r \cos \theta$ and $y = r \sin \theta$.   Thus, (r, θ) are "polar" coordinates corresponding to (x, y).

41.68   Describe the surface with the cylindrical equation $r = k$.

▌   When $k \neq 0$,   this is the equation of a right circular cylinder with radius $|k|$ and the z-axis as axis of symmetry.   When $k = 0$,   the graph is just the z-axis.

41.69   Describe the surface with cylindrical equation $\theta = k$.

▌   This is a plane containing the z-axis and making an angle of k radians with the xz-plane.

41.70   Find cylindrical coordinates for the point with rectangular coordinates $(2, 2\sqrt{3}, 8)$.

▌ $r = \sqrt{2^2 + (2\sqrt{3})^2} = \sqrt{16} = 4$, $\tan \theta = 2\sqrt{3}/2 = \sqrt{3}$, $\theta = \pi/3$.   So, a set of cylindrical coordinates is $(4, \pi/3, 8)$.   Other cylindrical coordinates for the same point are $(4, (\pi/3) + 2\pi n, 8)$ for any integer n, as well as $(-4, (\pi/3) + (2n + 1)\pi, 8)$ for any integer n.

41.71   Find rectangular coordinates for the point with cylindrical coordinates $(5, \pi/6, 2)$.

▌ $x = r \cos \theta = 5 \cos(\pi/6) = 5(\sqrt{3}/2)$. $y = r \sin \theta = 5 \sin(\pi/6) = 5(\frac{1}{2}) = \frac{5}{2}$. $z = 2$.

41.72   Find cylindrical coordinates for the point with rectangular coordinates $(2, 2, 2)$.

▌ $r\sqrt{2^2 + 2^2} = \sqrt{8} = 2\sqrt{2}$.   Further, $\theta = \tan^{-1} \frac{2}{2} = \tan^{-1} 1 = \pi/4$.   So, a set of cylindrical coordinates is $(2\sqrt{2}, \pi/4, 2)$.   Other sets are $(2\sqrt{2}, (\pi/4) + 2\pi n, 2)$ for any integer n, and $(-2\sqrt{2}, (\pi/4) + (2n + 1)\pi, 2)$ for any integer n.

41.73   Find rectangular coordinates for the point with cylindrical coordinates $(1/\sqrt{3}, 7\pi/6, 4)$.

▌ $x = r \cos \theta = \dfrac{1}{\sqrt{3}} \cos \dfrac{7\pi}{6} = \dfrac{1}{\sqrt{3}} \left(-\dfrac{\sqrt{3}}{2}\right) = -\dfrac{1}{2}$ $y = r \sin \theta = \dfrac{1}{\sqrt{3}} \sin \dfrac{7\pi}{6} = \dfrac{1}{\sqrt{3}} \left(-\dfrac{1}{2}\right) = -\dfrac{1}{2\sqrt{3}} = -\dfrac{\sqrt{3}}{6}$

41.74   Show that, if the curve in the yz-plane with rectangular equation $f(y, z) = 0$ is rotated about the z-axis, the resulting surface has the cylindrical equation $f(r, z) = 0$.

▌   Since $r = \sqrt{x^2 + y^2}$,   this is a direct consequence of Problem 41.5.

41.75   Describe the surface with the cylindrical equation $z + r = 1$.

▌   By Problem 41.74, this is the surface that results from rotating the curve $z + y = 1$ about the z-axis.   That surface is a cone (with both nappes) having the z-axis as its axis of symmetry.

41.76   Describe the surface with cylindrical equation $z^2 + r^2 = 4$.

▌   Replacing r^2 by $x^2 + y^2$,   we obtain the equation $x^2 + y^2 + z^2 = 4$ of a sphere with center at the origin and radius 2.

41.77 Describe the surface having the cylindrical equation $r = z$.

▮ By Problem 41.74, this surface is the result of rotating the curve $y = z$ in the yz-plane about the z-axis. This is clearly a cone (with two nappes).

41.78 Find a cylindrical equation for the plane $2x - 3y + z = 4$.

▮ Replace x by $r \cos \theta$ and y by $r \sin \theta$: $2r \cos \theta - 3r \sin \theta + z = 4$, or $r(2 \cos \theta - 3 \sin \theta) = 4$.

41.79 Find a cylindrical equation for the ellipsoid $x^2 + y^2 + 4z^2 = 5$.

▮ Replace $x^2 + y^2$ by r^2, obtaining $r^2 + 4z^2 = 5$.

41.80 Find a rectangular equation corresponding to the cylindrical equation $z = r^2 \cos 2\theta$.

▮ $z = r^2 \cos 2\theta = r^2(\cos^2 \theta - \sin^2 \theta) = r^2 \cos^2 \theta - r^2 \sin^2 \theta = x^2 - y^2$. Thus, the surface is the hyperbolic paraboloid $z = x^2 - y^2$.

41.81 Find a cylindrical equation for the surface whose rectangular equation is $z^2(x^2 - y^2) = 4xy$.

▮ The corresponding equation is $z^2(r^2 \cos^2 \theta - r^2 \sin^2 \theta) = 4r \cos \theta \, (r \sin \theta)$, $z^2(\cos^2 \theta - \sin^2 \theta) = 4 \cos \theta \sin \theta$, $z^2 \cos 2\theta = 2 \sin 2\theta$, $z^2 = 2 \tan 2\theta$. (Note that, when we cancelled out r^2, the points on the z-axis were not lost. Any point on the z-axis satisfies $z^2 = 2 \tan 2\theta$ by a suitable choice of θ.)

41.82 Find a rectangular equation for the surface with cylindrical equation $\theta = \pi/3$.

▮ $\tan \theta = \tan (\pi/3) = \sqrt{3}$. So, $y/x = \sqrt{3}$, $y = \sqrt{3}x$, which is a plane through the z-axis.

41.83 Find a rectangular equation for the surface with the cylindrical equation $r = 2 \sin \theta$.

▮ $r^2 = 2r \sin \theta$, $x^2 + y^2 = 2y$, $x^2 + (y - 1)^2 = 1$. This is a right circular cylinder of radius 1 and having as axis of symmetry the line $x = 0$, $y = 1$.

41.84 Find the rectangular equation for the surface whose cylindrical equation is $r^2 \sin 2\theta = 2z$.

▮ $2r^2 \sin \theta \cos \theta = 2z$, $r \sin \theta \, (r \cos \theta) = z$, $xy = z$. This is a saddle surface (like that of Problem 41.19).

41.85 Write down the equations connecting spherical coordinates (ρ, ϕ, θ) with rectangular and cylindrical coordinates.

▮ See Fig. 41-28. $x = r \cos \theta = \rho \sin \phi \cos \theta$, $y = r \sin \theta = \rho \sin \phi \sin \theta$, $z = \rho \cos \phi$. $\rho^2 = r^2 + z^2 = x^2 + y^2 + z^2$.

$$\tan \phi = \frac{r}{z} = \frac{\sqrt{x^2 + y^2}}{z} \qquad \tan \theta = \frac{y}{x} \qquad r = \rho \sin \phi$$

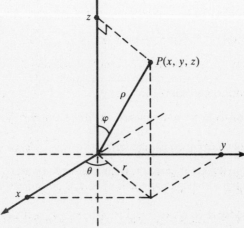

Fig. 41-28

41.86 Describe the surface with the equation $\rho = k$ in spherical coordinates.

▮ $\rho = k$ represents the sphere with center at the origin and radius k.

41.87 Describe the surface with the equation $\phi = k$ $(0 < k < \pi/2)$ in spherical coordinates.

▮ $\phi = k$ represents a one-napped cone with vertex at the origin whose generating lines make a fixed angle of k radians with the positive z-axis.

41.88 Find a set of spherical coordinates for the point whose rectangular coordinates are $(1, 1, \sqrt{6})$.

▮ $\rho^2 = 1^2 + 1^2 + (\sqrt{6})^2 = 8$.   Hence, $\rho = 2\sqrt{2}$. $\tan \phi = \sqrt{1^2 + 1^2}/\sqrt{6} = \sqrt{2}/\sqrt{6} = 1/\sqrt{3}$.   Therefore, $\phi = \pi/6$. $\tan \theta = \frac{1}{1} = 1$.   Hence, $\theta = \pi/4$.   So, the spherical coordinates are $(2\sqrt{2}, \pi/6, \pi/4)$.

41.89 Find a set of spherical coordinates for the point whose rectangular coordinates are $(0, -1, \sqrt{3})$.

▮ $\rho^2 = 0^2 + (-1)^2 + (\sqrt{3})^2 = 4$.   Hence, $\rho = 2$. $\tan \phi = \sqrt{0^2 + (-1)^2}/\sqrt{3} = 1/\sqrt{3}$.   Therefore, $\phi = \pi/6$. $\tan \theta = -1/0 = \infty$.   Hence, $\theta = \pi/2$.   So, a set of spherical coordinates is $(2, \pi/6, \pi/2)$.

41.90 Find the rectangular coordinates of the point with spherical coordinates $(3, \pi/2, \pi/2)$.

▮ Geometrically, it is easy to see that the point is $(0, 3, 0)$.   By calculation,

$$x = 3 \sin \frac{\pi}{2} \cos \frac{\pi}{2} = 3(1)(0) = 0 \qquad y = 3 \sin \frac{\pi}{2} \sin \frac{\pi}{2} = 3(1)(1) = 3 \qquad z = 3 \cos \frac{\pi}{2} = 3(0) = 0$$

41.91 Find the rectangular coordinates of the point with spherical coordinates $(4, 2\pi/3, \pi/3)$.

▮ $x = 4 \sin \frac{2\pi}{3} \cos \frac{\pi}{3} = 4\left(\frac{\sqrt{3}}{2}\right)\left(\frac{1}{2}\right) = \sqrt{3}$ $y = 4 \sin \frac{2\pi}{3} \sin \frac{\pi}{3} = 4\left(\frac{\sqrt{3}}{2}\right)\left(\frac{\sqrt{3}}{2}\right) = 3$ $z = 4 \cos \frac{2\pi}{3} = 4(\frac{1}{2}) = 2$

41.92 Find a spherical equation for the surface whose rectangular equation is $x^2 + y^2 + z^2 + 6z = 0$.

▮ $x^2 + y^2 + z^2 = \rho^2$ and $z = \rho \cos \phi$.   Thus, we get $\rho^2 + 6\rho \cos \phi = 0$.   Then $\rho = 0$ or $\rho + 6 \cos \phi = 0$.   Since the origin is the only solution of $\rho = 0$ and the origin also lies on $\rho + 6 \cos \phi = 0$,   then $\rho + 6 \cos \phi = 0$ is the desired equation.

41.93 Find a spherical equation for the surface whose rectangular equation is $x^2 + y^2 = 4$.

▮ Of course, the surface is the right circular cylinder with radius 2 and the z-axis as axis of symmetry.   Since $x^2 + y^2 = r^2 = \rho^2 \sin^2 \phi$,   we have $\rho^2 \sin^2 \phi = 4$.   This can be reduced to $\rho \sin \phi = 2$.   (The other possibility $\rho \sin \phi = -2$ yields no additional points.)

41.94 Find the graph of the spherical equation $\rho = 2a \sin \phi$,   where $a > 0$.

▮ Since θ is not present in the equation, the surface is obtained by rotating about the z-axis the intersection of the surface with the yz-plane.   In the latter plane, $\rho = 2a \sin \phi$ becomes $\sqrt{y^2 + z^2} = 2a \dfrac{y}{\sqrt{y^2 + z^2}}$, $y^2 + z^2 = 2ay$, $(y - a)^2 + z^2 = a^2$.   This is a circle with center $(a, 0)$ and radius a.   So, the resulting surface is obtained by rotating that circle about the z-axis.   The result can be thought of as a doughnut (torus) with no hole in the middle.

41.95 Describe the surface whose equation in spherical coordinates is $\rho \sin \phi = 3$.

▮ Since $r = \rho \sin \phi = 3$,   this is the right circular cylinder with radius 3 and the z-axis as axis of symmetry.

41.96 Describe the surface whose equation in spherical coordinates is $\rho \cos \phi = 3$.

▮ Since $z = \rho \cos \phi = 3$,   this is the plane that is parallel to, and three units above, the xy-plane.

41.97 Describe the surface whose equation in spherical coordinates is $\rho^2 \sin^2 \phi \cos 2\theta = 4$.

▮ $\rho^2 \sin^2 \phi \cos 2\theta = \rho^2 \sin^2 \phi (\cos^2 \theta - \sin^2 \theta) = \rho^2 \sin^2 \phi \cos^2 \theta - \rho^2 \sin^2 \phi \sin^2 \theta = x^2 - y^2$.   Hence, we have the cylindrical surface $x^2 - y^2 = 4$,   generated by the hyperbola $x^2 - y^2 = 4$ in the xy-plane.

41.98 Find an equation in spherical coordinates of the ellipsoid $x^2 + y^2 + 9z^2 = 9$.

▮ $x^2 + y^2 + 9z^2 = x^2 + y^2 + z^2 + 8z^2 = \rho^2 + 8\rho^2 \cos^2 \phi = 9$. Thus, we obtain the equation $\rho^2(1 + 8 \cos^2 \phi) = 9$.

CHAPTER 42

Partial Derivatives

42.1 If $f(x, y) = 4x^3 - 3x^2y^2 + 2x + 3y$, find the partial derivatives f_x and f_y.

▌ First, consider y constant. Then, differentiating with respect to x, we obtain $f_x = 12x^2 - 6y^2x + 2$. If we keep x fixed and differentiate with respect to y, we get $f_y = -6x^2y + 3$.

42.2 If $f(x, y) = x^5 \ln y$, find f_x and f_y.

▌ Differentiating with respect to x while keeping y fixed, we find that $f_x = 5x^4 \ln y$. Differentiating with respect to y while keeping x fixed, we get $f_y = x^5/y$.

42.3 For $f(x, y) = 3x^2 - 2x + 5$, find f_x and f_y.

▌ $f_x = 6x - 2$ and $f_y = 0$.

42.4 If $f(x, y) = \tan^{-1}(x + 2y)$, find f_x and f_y.

▌ $\qquad f_x = \dfrac{1}{1 + (x + 2y)^2} \cdot 1 = \dfrac{1}{1 + (x + 2y)^2} \qquad f_y = \dfrac{1}{1 + (x + 2y)^2} \cdot 2 = \dfrac{2}{1 + (x + 2y)^2}$

42.5 For $f(x, y) = \cos xy$, find f_x and f_y.

▌ $\qquad\qquad f_x = (-\sin xy)y = -y \sin xy \qquad f_y = (-\sin xy)x = -x \sin xy$

42.6 If $f(r, \theta) = r \cos \theta$, find $\partial f/\partial r$ and $\partial f/\partial \theta$.

▌ $\qquad\qquad\qquad \dfrac{\partial f}{\partial r} = \cos \theta \quad \text{and} \quad \dfrac{\partial f}{\partial \theta} = -r \sin \theta$

42.7 If $f(x, y) = \dfrac{x}{y^3} - \dfrac{y}{x^3}$, find f_x and f_y.

▌ $\qquad\qquad\qquad f_x = \dfrac{1}{y^3} + \dfrac{3y}{x^4} \qquad f_y = -\dfrac{3x}{y^4} - \dfrac{1}{x^3}$

42.8 Find the first partial derivatives of $f(x, y, z) = xy^2z^3$.

▌ $\qquad\qquad\qquad f_x = y^2z^3 \qquad f_y = 2xyz^3 \qquad f_z = 3xy^2z^2$

42.9 Find the first partial derivatives of $f(u, v, t) = e^{uv} \sin ut$.

▌ $\quad f_u = e^{uv}(\cos ut)t + ve^{uv} \sin ut = e^{uv}(t \cos ut + v \sin ut) \qquad f_v = ue^{uv} \sin ut \qquad f_t = ue^{uv} \cos ut$

42.10 Find the first partial derivatives of $f(x, y, z, u, v) = 2x + yz - ux + vy^2$.

▌ $\qquad\qquad f_x = 2 - u \qquad f_y = z + 2vy \qquad f_z = y \qquad f_u = -x \qquad f_v = y^2$

42.11 Find the first partial derivatives of $f(x, y, u, v) = \ln(x/y) - ve^{uy}$.

▌ Note that $f(x, y, u, v) = \ln x - \ln y - ve^{uy}$. Then,

$\qquad\qquad f_x = \dfrac{1}{x} \qquad f_y = -\dfrac{1}{y} - uve^{uy} \qquad f_u = -yve^{uy} \qquad f_v = -e^{uy}$

42.12 Give an example of a function $f(x, y)$ such that $f_x(0, 0) = f_y(0, 0) = 0$, but f is not continuous at $(0, 0)$. Hence, the existence of the first partial derivatives does not ensure continuity.

▮ Let
$$f(x, y) = \begin{cases} \dfrac{xy}{x^2 + y^2} & \text{if } (x, y) \neq (0, 0) \\ 0 & \text{if } (x, y) = (0, 0) \end{cases}$$

$$f_x(0, 0) = \lim_{\Delta x \to 0} \frac{f(\Delta x, 0) - f(0, 0)}{\Delta x} = \lim_{\Delta x \to 0} \frac{0 - 0}{\Delta x} = 0$$

$$f_y(0, 0) = \lim_{\Delta y \to 0} \frac{f(0, \Delta y) - f(0, 0)}{\Delta y} = \lim_{\Delta y \to 0} \frac{0 - 0}{\Delta y} = 0$$

If we approach $(0, 0)$ along the line $y = x$, $f(x, y) = x^2/2x^2 = \frac{1}{2}$. If we approach $(0, 0)$ along the line $y = 0$, $f(x, y) = 0$. Hence, $\lim_{(x, y) \to (0, 0)} f(x, y)$ does not exist and, therefore, f is not continuous at $(0, 0)$.

42.13 If z is implicitly defined as a function of x and y by $x^2 + y^2 - z^2 = 3$, find $\partial z/\partial x$ and $\partial z/\partial y$.

▮ By implicit differentiation with respect to x, $2x - 2z(\partial z/\partial x) = 0$, $x = z(\partial z/\partial x)$, $\partial z/\partial x = x/z$. By implicit differentiation with respect to y, $2y - 2z(\partial z/\partial y) = 0$, $y = z(\partial z/\partial y)$, $\partial z/\partial y = y/z$.

42.14 If z is implicitly defined as a function of x and y by $x \sin z - z^2 y = 1$, find $\partial z/\partial x$ and $\partial z/\partial y$.

▮ By implicit differentiation with respect to x, $x \cos z \, (\partial z/\partial x) + \sin z - 2yz(\partial z/\partial x) = 0$, $(\partial z/\partial x)(x \cos z - 2yz) = -\sin z$, $\partial z/\partial x = \sin z/(2yz - x \cos z)$. By implicit differentiation with respect to y, $x \cos z \, (\partial z/\partial y) - z^2 - 2yz(\partial z/\partial y) = 0$, $(\partial z/\partial y)(x \cos z - 2yz) = z^2$, $\partial z/\partial y = z^2/(x \cos z - 2yz)$.

42.15 If z is defined as a function of x and y by $xy - yz + xz = 0$, find $\partial z/\partial x$ and $\partial z/\partial y$.

▮ By implicit differentiation with respect to x,

$$y - y\frac{\partial z}{\partial x} + x\frac{\partial z}{\partial x} + z = 0 \qquad \frac{\partial z}{\partial x}(x - y) = -(y + z) \qquad \frac{\partial z}{\partial x} = \frac{y + z}{y - x}$$

By implicit differentiation with respect to y,

$$x - \left(y\frac{\partial z}{\partial y} + z\right) + x\frac{\partial z}{\partial y} = 0 \qquad \frac{\partial z}{\partial y}(x - y) = z - x \qquad \frac{\partial z}{\partial y} = \frac{z - x}{x - y}$$

42.16 If z is implicitly defined as a function of x and y by $x^2 + y^2 + z^2 = 1$, show that $x\dfrac{\partial z}{\partial x} + y\dfrac{\partial z}{\partial y} = z - \dfrac{1}{z}$.

▮ By implicit differentiation with respect to x, $2x + 2z(\partial z/\partial x) = 0$, $\partial z/\partial x = -x/z$. By implicit differentiation with respect to y, $2y + 2z(\partial z/\partial y) = 0$, $\partial z/\partial y = -y/z$. Thus,

$$x\frac{\partial z}{\partial x} + y\frac{\partial z}{\partial y} = x\left(-\frac{x}{z}\right) + y\left(-\frac{y}{z}\right) = -\frac{x^2 + y^2}{z} = \frac{z^2 - 1}{z} = z - \frac{1}{z}$$

42.17 If $z = \ln\sqrt{x^2 + y^2}$, show that $x\dfrac{\partial z}{\partial x} + y\dfrac{\partial z}{\partial y} = 1$.

▮ Because $z = \frac{1}{2}\ln(x^2 + y^2)$, we have $\partial z/\partial x = \frac{1}{2}[2x/(x^2 + y^2)] = x/(x^2 + y^2)$ and $\partial z/\partial y = \frac{1}{2}[2y/(x^2 + y^2)] = y/(x^2 + y^2)$. Therefore,

$$x\frac{\partial z}{\partial x} + y\frac{\partial z}{\partial y} = \frac{x^2}{x^2 + y^2} + \frac{y^2}{x^2 + y^2} = 1$$

42.18 If $x = e^{2r}\cos\theta$ and $y = e^{3r}\sin\theta$, find r_x, r_y, θ_x, and θ_y by implicit partial differentiation.

▮ Differentiate both equations implicitly with respect to x. $1 = 2e^{2r}(\cos\theta)r_x - e^{2r}(\sin\theta)\theta_x$, $0 = 3e^{3r}(\sin\theta)r_x + e^{3r}(\cos\theta)\theta_x$. From the latter, since $e^{3r} \neq 0$, $0 = 3(\sin\theta)r_x + (\cos\theta)\theta_x$. Now solve simultaneously for r_x and θ_x. $r_x = \cos\theta/[e^{2r}(2 + \sin^2\theta)]$, $\theta_x = -3\sin\theta/[e^{2r}(2 + \sin^2\theta)]$. Now differentiate the original equations for x and y implicitly with respect to y: $0 = 2e^{2r}(\cos\theta)r_y - e^{2r}(\sin\theta)\theta_y$, $1 = 3e^{3r}(\sin\theta)r_y + e^{3r}(\cos\theta)\theta_y$. From the first of these, since $e^{2r} \neq 0$, we get $0 = 2(\cos\theta)r_y - (\sin\theta)\theta_y$. Solving simultaneously for r_y and θ_y, we obtain $r_y = \sin\theta/[e^{3r}(2 + \sin^2\theta)]$ and $\theta_y = 2\cos\theta/[e^{3r}(2 + \sin^2\theta)]$.

42.19 If $z = e^{x/y} \sin(x/y) + e^{y/x} \cos(y/x)$, show that $x \dfrac{\partial z}{\partial x} + y \dfrac{\partial z}{\partial y} = 0$.

▮ It is easy to prove a more general result. Let $z = f(x/y)$, where f is an arbitrary differentiable function. Then

$$x \frac{\partial z}{\partial x} = xf'\left(\frac{x}{y}\right)\frac{1}{y} = \frac{x}{y} f'\left(\frac{x}{y}\right) \quad \text{and} \quad y \frac{\partial z}{\partial y} = yf'\left(\frac{x}{y}\right)\left(-\frac{x}{y^2}\right) = -\frac{x}{y} f'\left(\frac{x}{y}\right)$$

by addition, $x \dfrac{\partial z}{\partial x} + y \dfrac{\partial z}{\partial y} = 0$.

42.20 If $z = xe^{y/x}$, show that $x \dfrac{\partial z}{\partial x} + y \dfrac{\partial z}{\partial y} = z$.

▮ In general, if $z = xf(y/x)$, where f is differentiable, $x \dfrac{\partial z}{\partial x} + y \dfrac{\partial z}{\partial y} = x[xf'(y/x)(-y/x^2) + f(y/x)] + xyf'(y/x)(1/x) = xf(y/x) = z$.

42.21 If $z = \dfrac{y}{x+y}$, evaluate $x \dfrac{\partial z}{\partial x} + y \dfrac{\partial z}{\partial y}$.

▮ $z = \dfrac{1}{(x/y)+1}$, and Problem 42.19 applies.

42.22 If $z = \int_x^y e^{\sin t}\, dt$, find $\dfrac{\partial z}{\partial x}$ and $\dfrac{\partial z}{\partial y}$.

▮ $\dfrac{\partial z}{\partial x} = \dfrac{\partial}{\partial x}\left(-\int_y^x e^{\sin t}\, dt\right) = -e^{\sin x}$, since, in general, $\dfrac{d}{dx}\int_a^x f(t)\, dt = f(x)$. Similarly, $\dfrac{\partial z}{\partial y} = e^{\sin y}$.

42.23 The ideal gas law for a gas in a closed chamber is $pV = kT$, where p is pressure, V is volume, T is temperature, and k is a suitable constant. Show that $\dfrac{\partial p}{\partial V}\left(\dfrac{\partial T}{\partial p}\right)\left(\dfrac{\partial V}{\partial T}\right) = -1$.

▮ Think of p as a function of V and T, and differentiate with respect to V, keeping T fixed: $p + \dfrac{\partial p}{\partial V} V = 0$, $\partial p/\partial V = -p/V$. Similarly, think of T as a function of p and V and differentiate with respect to p: $V = k \dfrac{\partial T}{\partial p}$, $\partial T/\partial p = \dfrac{1}{k} V$. Finally, think of V as a function of p and T and differentiate with respect to T: $p \dfrac{\partial V}{\partial T} = k$, $\partial V/\partial T = k/p$. Hence, $\left(\dfrac{\partial p}{\partial V}\right)\left(\dfrac{\partial T}{\partial p}\right)\left(\dfrac{\partial V}{\partial T}\right) = -\left(\dfrac{p}{V}\right)\left(\dfrac{V}{k}\right)\left(\dfrac{k}{p}\right) = -1$.

42.24 If $f(x, y) = 3x^2 y - 2x^3 + 5y^2$, find $f_x(1, 2)$ and $f_y(1, 2)$.

▮ $f_x = 6xy - 6x^2$. Hence, $f_x(1, 2) = 12 - 6 = 6$. $f_y = 3x^2 + 10y$. Hence, $f_y(1, 2) = 3 + 20 = 23$.

42.25 If $f(x, y) = \cos 3x \sin 4y$, find $f_x(\pi/12, \pi/6)$ and $f_y(\pi/12, \pi/6)$.

▮ $f_x(x, y) = -3 \sin 3x \sin 4y$. Therefore,

$$f_x\left(\frac{\pi}{12}, \frac{\pi}{6}\right) = -3 \sin\frac{\pi}{4} \sin\frac{2\pi}{3} = -3\left(\frac{\sqrt{2}}{2}\right)\left(\frac{\sqrt{3}}{2}\right) = -\frac{3\sqrt{6}}{4} \qquad f_y(x, y) = 4 \cos 3x \cos 4y = 4\left(\frac{\sqrt{2}}{2}\right)\left(-\frac{1}{2}\right) = -\sqrt{2}$$

42.26 Find the slope of the tangent line to the curve that is the intersection of the surface $z = x^2 - y^2$ with the plane $x = 2$, at the point $(2, 1, 3)$.

▮ In the plane $x = 2$, x is constant. Hence, the slope of the tangent line to the curve is the derivative $\partial z/\partial y = -2y = -2(1) = -2$.

42.27 Find the slope of the tangent line to the curve that is the intersection of the sphere $x^2 + y^2 + z^2 = 1$ with the plane $y = \frac{1}{2}$, at the point $(\frac{1}{2}, \frac{1}{2}, \sqrt{2}/2)$.

▮ Since y is constant in the plane $y = \frac{1}{2}$, the slope of the tangent line to the curve is $\partial z/\partial x$. By implicit differentiation of the equation $x^2 + y^2 + z^2 = 1$, we get $2x + 2z(\partial z/\partial x) = 0$. Hence, at the point $(\frac{1}{2}, \frac{1}{2}, \sqrt{2}/2)$, $\partial z/\partial x = -x/z = -\frac{1}{2}/(\sqrt{2}/2) = -1/\sqrt{2} = -\sqrt{2}/2$.

42.28 Find the slopes of the tangent lines to the curves cut from the surface $z = 3x^2 + 4y^2 - 6$ by planes through the point $(1, 1, 1)$ and parallel to the xz- and yz-planes.

▮ The plane through $(1, 1, 1)$ and parallel to the xz-plane is $y = 1$. The slope of the tangent line to the resulting curve is $\partial z/\partial x = 6x = 6$. The plane through $(1, 1, 1)$ and parallel to the yz-plane is $x = 1$. The slope of the tangent line to the resulting curve is $\partial z/\partial y = 8y = 8$.

42.29 Find equations of the tangent line at the point $(-2, 1, 5)$ to the parabola that is the intersection of the surface $z = 2x^2 - 3y^2$ and the plane $y = 1$.

▮ The slope of the tangent line is $\partial z/\partial x = 4x = -8$. Hence, a vector in the direction of the tangent line is $(1, 0, -8)$. (This follows from the fact that there is no change in y and, for a change of 1 unit in x, there is a change of $\partial z/\partial x$ in z.) Therefore, a system of parametric equations for the tangent line is $x = 2 + t$, $y = 1$, $z = 5 - 8t$ (or, equivalently, we can use the pair of planes $y = 1$ and $8x + z = -11$).

42.30 Find equations of the tangent line at the point $(-2, 1, 5)$ to the hyperbola that is the intersection of the surface $z = 2x^2 - 3y^2$ and the plane $z = 5$.

▮ Think of y as a function of x and z. Then the slope of the tangent line to the curve is $\partial y/\partial x$. By implicit differentiation with respect to x, $0 = 4x - 6y(\partial y/\partial x)$. Hence, at $(-2, 1, 5)$, $0 = -8 - 6(\partial y/\partial x)$, $\partial y/\partial x = -\frac{4}{3}$. So, a vector in the direction of the tangent line is $(1, -\frac{4}{3}, 0)$, and parametric equations for the tangent line are $x = -2 + t$, $y = 1 - \frac{4}{3}t$, $z = 5$ (or, equivalently, the pair of planes $4x + 3y = -5$ and $z = 5$).

42.31 Show that the tangent lines of Problems 42.29 and 42.30 both lie in the plane $8x + 6y + z + 5 = 0$. [This is the *tangent plane* to the surface $z = 2x^2 - 3y^2$ at the point $(-2, 1, 5)$.]

▮ Both lines contain the point $(-2, 1, 5)$, which lies in the plane $\mathcal{P}$: $8x + 6y + z + 5 = 0$, since $8(-2) + 6(1) + 5 + 5 = 0$. A normal vector to the plane is $\mathbf{A} = (8, 6, 1)$. [$\mathbf{A}$ is a *surface normal* at $(-2, 1, 5)$.] The tangent line of Problem 42.29 is parallel to the vector $\mathbf{B} = (1, 0, -8)$, which is perpendicular to $\mathbf{A}$ [since $\mathbf{A} \cdot \mathbf{B} = 8(1) + 6(0) + 1(-8) = 0$]. Therefore, that tangent line lies in the plane $\mathcal{P}$. Likewise, the tangent line of Problem 42.30 is parallel to the vector $\mathbf{C} = (1, -\frac{4}{3}, 0)$, which is perpendicular to $\mathbf{A}$ [since $\mathbf{A} \cdot \mathbf{C} = 8(1) + 6(-\frac{4}{3}) + 1(0) = 0$]. Hence, that tangent line also lies in plane $\mathcal{P}$. (We have assumed here the fact that any line containing a point of a plane and perpendicular to a normal vector to the plane lies entirely in the plane.)

42.32 The plane $y = 3$ intersects the surface $z = 2x^2 + y^2$ in a curve. Find equations of the tangent line to this curve at the point $(2, 3, 17)$.

▮ The slope of the tangent line is the derivative $\partial z/\partial x = 4x = 8$. Hence, a pair of equations for the tangent line is $(z - 17)/(x - 2) = 8$, $y = 3$, or, equivalently, $z = 8x + 1$, $y = 3$.

42.33 The plane $x = 3$ intersects the surface $z = x^2/(y^2 - 3)$ in a curve. Find equations of the tangent line to this curve at $(3, 2, 9)$.

▮ The slope of the tangent line is $\dfrac{\partial z}{\partial y} = -\dfrac{x^2}{(y^2 - 3)^2}(2y) = -36$. Hence, a pair of equations for the tangent line is $(z - 9)/(y - 2) = -36$ and $x = 3$, or, equivalently, $z = -36y + 81$, $x = 3$. [Another method is to use the vector $(0, 1, -36)$, parallel to the line, to form the parametric equations $x = 3$, $y = 2 + t$, $z = 9 - 36t$.]

42.34 State a set of conditions under which the mixed partial derivatives $f_{xy}(x_0, y_0)$ and $f_{yx}(x_0, y_0)$ are equal.

▮ If (x_0, y_0) is inside an open disk throughout which f_{xy} and f_{yx} exist, and if f_{xy} and f_{yx} are continuous at (x_0, y_0), then $f_{xy}(x_0, y_0) = f_{yx}(x_0, y_0)$. Similar conditions ensure equality for $n \geq 3$ partial differentiations, regardless of the order in which the derivatives are taken.

42.35 For $f(x, y) = 3x^2y - 2xy + 5y^2$, verify that $f_{xy} = f_{yx}$.

▮ $f_x = 6xy - 2y$, $f_{xy} = 6x - 2$. $f_y = 3x^2 - 2x + 10y$, $f_{yx} = 6x - 2$.

42.36 For $f(x, y) = x^7 \ln y + \sin xy$, verify $f_{xy} = f_{yx}$.

▮
$$f_x = 7x^6 \ln y + y \cos xy \qquad f_{xy} = \frac{7x^6}{y} + \cos xy - xy \sin xy$$

$$f_y = \frac{x^7}{y} + x \cos xy \qquad f_{yx} = \frac{7x^6}{y} + \cos xy - yx \sin xy$$

42.37 For $f(x, y) = e^x \cos y$, verify that $f_{xy} = f_{yx}$.

❚ $\qquad f_x = e^x \cos y \qquad f_{xy} = -e^x \sin y \qquad f_y = -e^x \sin y \qquad f_{yx} = -e^x \sin y$

42.38 If $f(x, y) = 3x^2 - 2xy + 5y^3$, verify that $f_{xy} = f_{yx}$.

❚ $\qquad f_x = 6x - 2y \qquad f_{xy} = -2 \qquad f_y = -2x + 15y^2 \qquad f_{yx} = -2$

42.39 If $f(x, y) = x^2 \cos y + y^2 \sin x$, verify that $f_{xy} = f_{yx}$.

❚ $f_x = 2x \cos y + y^2 \cos x$, $f_{xy} = -2x \sin y + 2y \cos x$. $f_y = -x^2 \sin y + 2y \sin x$, $f_{yx} = -2x \sin y + 2y \cos x$.

42.40 For $f(x, y) = 3x^4 - 2x^3y^2 + 7y$, find $f_{xx}, f_{xy}, f_{yx},$ and f_{yy}.

❚ $f_x = 12x^3 - 6x^2y^2$, $f_y = -4x^3y + 7$, $f_{xx} = 36x^2 - 12xy^2$, $f_{yy} = -4x^3$, $f_{xy} = -12x^2y$, $f_{yx} = -12x^2y$.

42.41 If $f(x, y) = e^{xy}y^2 + \dfrac{x}{y}$, find $f_{xx}, f_{xy}, f_{yx},$ and f_{yy}.

❚ $\qquad f_x = ye^{xy}y^2 + \dfrac{1}{y} = e^{xy}y^3 + \dfrac{1}{y} \qquad f_y = 2e^{xy}y + xe^{xy}y^2 - \dfrac{x}{y^2} = e^{xy}(2y + xy^2) - \dfrac{x}{y^2}$

$\qquad f_{xx} = e^{xy}y^3(y) = e^{xy}y^4 \qquad f_{yy} = e^{xy}(2 + 2xy) + xe^{xy}(2y + xy^2) + \dfrac{2x}{y^3} = e^{xy}(2 + 4xy + x^2y^2) + \dfrac{2x}{y^3}$

$\qquad f_{xy} = 3e^{xy}y^2 + xe^{xy}y^3 - \dfrac{1}{y^2} \qquad f_{yx} = e^{xy}(y^2) + ye^{xy}(2y + xy^2) - \dfrac{1}{y^2} = 3e^{xy}y^2 + xe^{xy}y^3 - \dfrac{1}{y^2}$

42.42 If $f(x, y, z) = x^2y + y^2z - 2xz$, find $f_{xy}, f_{yx}, f_{xz}, f_{zx}, f_{yz}, f_{zy}$.

❚ $f_x = 2xy - 2z$, $f_y = x^2 + 2yz$, $f_z = y^2 - 2x$. $f_{xy} = 2x$, $f_{yx} = 2x$, $f_{xz} = -2$, $f_{zx} = -2$, $f_{yz} = 2y$, $f_{zy} = 2y$.

42.43 Give an example to show that the equation $f_{xy} = f_{yx}$ is not always valid.

❚ Let $$f(x, y) = \begin{cases} xy\left(\dfrac{x^2 - y^2}{x^2 + y^2}\right) & \text{if } (x, y) \neq (0, 0) \\ 0 & \text{if } (x, y) = (0, 0) \end{cases}$$

Then $\qquad f_x(0, y) = \lim\limits_{\Delta x \to 0} \dfrac{f(\Delta x, y) - f(0, y)}{\Delta x} = \lim\limits_{\Delta x \to 0} y\dfrac{(\Delta x)^2 - y^2}{(\Delta x)^2 + y^2} = -y$

and $\qquad f_y(x, 0) = \lim\limits_{\Delta y \to 0} \dfrac{f(x, \Delta y) - f(x, 0)}{\Delta y} = \lim\limits_{\Delta y \to 0} x\dfrac{x^2 - (\Delta y)^2}{x^2 + (\Delta y)^2} = x$

Consequently, $\qquad f_{xy}(0, 0) = \lim\limits_{\Delta y \to 0} \dfrac{f_x(0, \Delta y) - f_x(0, 0)}{\Delta y} = \lim\limits_{\Delta y \to 0} \dfrac{-\Delta y - 0}{\Delta y} = -1$

and $\qquad f_{yx}(0, 0) = \lim\limits_{\Delta x \to 0} \dfrac{f_y(\Delta x, 0) - f_y(0, 0)}{\Delta x} = \lim\limits_{\Delta x \to 0} \dfrac{\Delta x - 0}{\Delta x} = 1$

Thus $f_{xy}(0, 0) \neq f_{yx}(0, 0)$. (The conditions of Problem 42.34 are not met by this function.)

42.44 Is there a function $f(x, y)$ such that $f_x = e^x \cos y$ and $f_y = e^x \sin y$?

❚ Assume that there is such a function. Then f_{xy} and f_{yx} will be continuous everywhere. Hence, $f_{xy} = f_{yx}$. Thus, $-e^x \sin y = e^x \sin y$, or $\sin y = 0$ for all y, which is false. No such function exists.

42.45 If $f(x, y) = e^x y^2 - x^3 \ln y$, verify that $f_{xxy}, f_{xyx},$ and f_{yxx} all are equal.

❚ $f_x = e^x y^2 - 3x^2 \ln y$, $f_y = 2e^x y - x^3/y$. $f_{xx} = e^x y^2 - 6x \ln y$, $f_{xy} = 2e^x y - 3x^2/y$, $f_{yx} = 2e^x y - 3x^2/y$, $f_{xxy} = 2e^x y - 6x/y$, $f_{xyx} = 2e^x y - 6x/y$, $f_{yxx} = 2e^x y - 6x/y$.

42.46   If $f(x, y) = y \sin x - x \sin y$,   verify that f_{xyy}, f_{yxy}, and f_{yyx} are equal.

▮ $f_x = y \cos x - \sin y$, $f_y = \sin x - x \cos y$, $f_{xy} = \cos x - \cos y$, $f_{yx} = \cos x - \cos y$, $f_{yy} = x \sin y$,
$f_{xyy} = \sin y$, $f_{yxy} = \sin y$, $f_{yyx} = \sin y$.

42.47   If $z = \dfrac{xy}{x - y}$,   show that $x^2 \dfrac{\partial^2 z}{\partial x^2} + 2xy \dfrac{\partial^2 z}{\partial x \, \partial y} + y^2 \dfrac{\partial^2 z}{\partial y^2} = 0$.

▮ $\dfrac{\partial z}{\partial x} = \dfrac{(x - y)y - xy}{(x - y)^2} = -\dfrac{y^2}{(x - y)^2}$ $\dfrac{\partial z}{\partial y} = \dfrac{(x - y)x - xy(-1)}{(x - y)^2} = \dfrac{x^2}{(x - y)^2}$ $\dfrac{\partial^2 z}{\partial x^2} = \dfrac{2y^2}{(x - y)^3}$

$\dfrac{\partial^2 z}{\partial x \, \partial y} = \dfrac{(x - y)^2 (2x) - x^2 [2(x - y)]}{(x - y)^4} = \dfrac{(x - y)(2x) - 2x^2}{(x - y)^3} = -\dfrac{2xy}{(x - y)^3}$ $\dfrac{\partial^2 z}{\partial y^2} = \dfrac{-2x^2}{(x - y)^3} (-1) = \dfrac{2x^2}{(x - y)^3}$

Thus, $x^2 \dfrac{\partial^2 z}{\partial x^2} + 2xy \dfrac{\partial^2 z}{\partial x \, \partial y} + y^2 \dfrac{\partial^2 z}{\partial y^2} = \dfrac{2x^2 y^2}{(x - y)^3} - \dfrac{4x^2 y^2}{(x - y)^3} + \dfrac{2x^2 y^2}{(x - y)^3} = 0$

For a simpler solution, see Problem 42.78.

42.48   If $z = e^{ax} \sin ay$,   show that $\dfrac{\partial^2 z}{\partial x^2} + \dfrac{\partial^2 z}{\partial y^2} = 0$.

▮ $\dfrac{\partial z}{\partial x} = ae^{ax} \sin ay$, $\dfrac{\partial^2 z}{\partial x^2} = a^2 e^{ax} \sin ay$, $\dfrac{\partial z}{\partial y} = ae^{ax} \cos ay$, $\dfrac{\partial^2 z}{\partial y^2} = -a^2 e^{ax} \sin ay$.

42.49   If $z = e^{-t}(\sin x + \cos y)$,   show that $\dfrac{\partial^2 z}{\partial x^2} + \dfrac{\partial^2 z}{\partial y^2} = \dfrac{\partial z}{\partial t}$.

▮ $\partial z / \partial x = e^{-t} \cos x$, $\partial^2 z / \partial x^2 = -e^{-t} \sin x$, $\partial z / \partial y = -e^{-t} \sin y$, $\partial^2 z / \partial y^2 = -e^{-t} \cos y$, $\partial z / \partial t =$
$-e^{-t}(\sin x + \cos y) = -e^{-t} \sin x - e^{-t} \cos y = \dfrac{\partial^2 z}{\partial x^2} + \dfrac{\partial^2 z}{\partial y^2}$.

42.50   If $f(x, y) = g(x)h(y)$,   show that $f_{xy} = f_{yx}$.

▮ $f_x = g'(x)h(y)$, $f_{xy} = g'(x)h'(y)$. $f_y = g(x)h'(y)$, $f_{yx} = g'(x)h'(y)$.

42.51   If $z = g(x)h(y)$,   show that $z \dfrac{\partial^2 z}{\partial x \, \partial y} = \dfrac{\partial z}{\partial x} \dfrac{\partial z}{\partial y}$.

▮ $\partial z / \partial x = g'(x)h(y)$, $\partial z / \partial y = g(x)h'(y)$, $\partial^2 z / (\partial x \, \partial y) = g'(x)h'(y)$.   So, $z \dfrac{\partial^2 z}{\partial x \, \partial y} = g(x)h(y)g'(x)h'(y)$,
while $\dfrac{\partial z}{\partial x} \dfrac{\partial z}{\partial y} = g'(x)h(y)g(x)h'(y)$.

42.52   Verify that $f(x, y) = \ln (x^2 + y^2)$ satisfies *Laplace's equation*, $f_{xx} + f_{yy} = 0$.

▮ $f_x = \dfrac{2x}{x^2 + y^2}$, $f_{xx} = \dfrac{(x^2 + y^2)(2) - 2x(2x)}{(x^2 + y^2)^2} = \dfrac{2(y^2 - x^2)}{(x^2 + y^2)^2}$. $f_y = \dfrac{2y}{x^2 + y^2}$, $f_{yy} = \dfrac{(x^2 + y^2)(2) - 2y(2y)}{(x^2 + y^2)^2} =$
$\dfrac{2(x^2 - y^2)}{(x^2 + y^2)^2}$.   Hence, $f_{xx} + f_{yy} = 0$.

42.53   If $f(x, y) = \tan^{-1} (y/x)$, verify that $f_{xx} + f_{yy} = 0$.

▮ $f_x = \dfrac{1}{1 + y^2/x^2} \left(\dfrac{-y}{x^2} \right) = -\dfrac{y}{x^2 + y^2}$, $f_{xx} = \dfrac{y}{(x^2 + y^2)^2} (2x) = \dfrac{2xy}{(x^2 + y^2)^2}$. $f_y = \dfrac{1}{1 + y^2/x^2} \left(\dfrac{1}{x} \right) = \dfrac{x}{x^2 + y^2}$,
$f_{yy} = -\dfrac{2yx}{(x^2 + y^2)^2}$.   Therefore, $f_{xx} + f_{yy} = 0$.

42.54   If the *Cauchy–Riemann equations* $f_x = g_y$ and $g_x = -f_y$ hold, prove that, under suitable assumptions, f
and g satisfy Laplace's equation (see Problem 42.52).

▮   Since $f_x = g_y$,   we have $f_{xx} = g_{yx}$.   Since $g_x = -f_y$, $g_{xy} = -f_{yy}$.   Now, assuming that the second
mixed partial derivatives exist and are continuous, we have $g_{yx} = g_{xy}$,   and, therefore, $f_{xx} = -f_{yy}$.   Hence,
$f_{xx} + f_{yy} = 0$.   Likewise, $g_{xx} = -f_{yx} = -f_{xy} = -g_{yy}$,   and, therefore, $g_{xx} + g_{yy} = 0$.

42.55 Show that $f(x, t) = (x + at)^3$ satisfies the *wave equation*, $a^2 f_{xx} = f_{tt}$.

■ $f_x = 3(x + at)^2$, $f_{xx} = 6(x + at)$. $f_t = 3(x + at)^2(a) = 3a(x + at)^2$, $f_{tt} = 6a(x + at)(a) = 6a^2(x + at) = a^2 f_{xx}$.

42.56 Show that $f(x, t) = \sin(x + at)$ satisfies the wave equation $a^2 f_{xx} = f_{tt}$.

■ $f_x = \cos(x + at)$, $f_{xx} = -\sin(x + at)$, $f_t = a \cos(x + at)$, $f_{tt} = -a^2 \sin(x + at) = a^2 f_{xx}$.

42.57 Show that $f(x, t) = e^{x-at}$ satisfies the wave equation $a^2 f_{xx} = f_{tt}$.

■ $f_x = e^{x-at}$, $f_{xx} = e^{x-at}$. $f_t = -ae^{x-at}$, $f_{tt} = a^2 e^{x-at} = a^2 f_{xx}$.

42.58 Let $f(x, t) = u(x + at) + v(x - at)$, where u and v are assumed to have continuous second partial derivatives. In generalization of Problems 42.55–42.57, show that f satisfies the wave equation $a^2 f_{xx} = f_{tt}$.

■ $f_x = u'(x + at) + v'(x - at)$, $f_{xx} = u''(x + at) + v''(x - at)$. $f_t = au'(x + at) - av'(x - at)$, $f_{tt} = a^2 u''(x + at) + a^2 v''(x - at) = a^2 f_{xx}$.

42.59 Verify the valid formula $\dfrac{d}{dx} \displaystyle\int_a^b f(x, y)\,dy = \int_a^b \dfrac{\partial f}{\partial x}\,dy$ when $f(x, y) = \sin xy$, $a = 0$, $b = \pi$.

■ $\int_0^\pi \sin xy\,dy = -\dfrac{1}{x} \cos xy \Big]_0^\pi = -\dfrac{1}{x}(\cos \pi x - \cos 0) = -\dfrac{1}{x}(\cos \pi x - 1)$. So, $\dfrac{d}{dx} \displaystyle\int_0^\pi \sin xy\,dy = \dfrac{d}{dx}\left[-\dfrac{1}{x}(\cos \pi x - 1)\right] = -\dfrac{x(-\pi \sin \pi x) - (\cos \pi x - 1)}{x^2} = \dfrac{\cos \pi x + \pi x \sin \pi x - 1}{x^2}$ On the other hand, $\partial f/\partial x = y \cos xy$ and $\int_0^\pi y \cos xy\,dy$ can be integrated by parts:

$$\int_0^\pi y \cos xy\,dy = \dfrac{y}{x} \sin xy \Big]_0^\pi - \dfrac{1}{x} \int_0^\pi \sin xy\,dy = \left[\dfrac{y}{x} \sin xy + \dfrac{1}{x}\left(\dfrac{1}{x}\right) \cos xy\right]_0^\pi = \dfrac{xy \sin xy + \cos xy}{x^2}\bigg]_0^\pi$$

$$= \dfrac{\pi x \sin \pi x + \cos \pi x - 1}{x^2} = \dfrac{d}{dx} \int_a^b f(x, y)\,dy$$

42.60 Verify the valid formula $\dfrac{d}{dx} \displaystyle\int_a^b f(x, y)\,dy = \int_a^b \dfrac{\partial f}{\partial x}\,dy$, when $f(x, y) = x + y$, $a = 0$, $b = 1$.

■ $\int_0^1 (x + y)\,dy = (xy + \frac{1}{2}y^2) \big]_0^1 = x + \frac{1}{2}$. Hence, $\dfrac{d}{dx} \displaystyle\int_0^1 f(x, y)\,dy = \dfrac{d}{dx}(x + \frac{1}{2}) = 1$. On the other hand, $\dfrac{\partial f}{\partial x} = 1$, $\int_a^b \dfrac{\partial f}{\partial x}\,dy = \int_0^1 dy = y \big]_0^1 = 1$.

42.61 If $f(x, y) = \int_0^1 \cos(x + 2y + t)\,dt$, find f_x and f_y.

■ $f_x = \displaystyle\int_0^1 \dfrac{\partial}{\partial x} \cos(x + 2y + t)\,dt = -\int_0^1 \sin(x + 2y + t)\,dt = \cos(x + 2y + t)\Big]_0^1 = \cos(x + 2y + 1) - \cos(x + 2y)$.

$$f_y = \int_0^1 \dfrac{\partial}{\partial y} \cos(x + 2y + t)\,dt = -\int_0^1 2 \sin(x + 2y + t)\,dt = 2 \cos(x + 2y + t)\Big]_0^1$$

$$= 2[\cos(x + 2y + 1) - \cos(x + 2y)]$$

42.62 Let $f(x, y) = \int_0^y (x^2 + tx)\,dt$. Find f_x and f_y and verify that $f_{xy} = f_{yx}$.

■ $f_x = \displaystyle\int_0^y \dfrac{\partial}{\partial x}(x^2 + tx)\,dt = \int_0^y (2x + t)\,dt = (2xt + \frac{1}{2}t^2) \big]_0^y = 2xy + \frac{1}{2}y^2$. $f_y = x^2 + yx$. $f_{xy} = 2x + y$. $f_{yx} = 2x + y$.

42.63 Let $M(x, y)$ and $N(x, y)$ satisfy $\partial M/\partial y = \partial N/\partial x$ for all (x, y). Show the existence of a function $f(x, y)$ such that $\partial f/\partial x = M$ and $\partial f/\partial y = N$.

■ Let $f(u, v) = \int_0^v N(0, y)\,dy + \int_0^u M(x, v)\,dx$. Then $\dfrac{\partial f}{\partial u} = \dfrac{\partial}{\partial u} \int_0^u M(x, v)\,dx = M(u, v)$. Also,

$$\dfrac{\partial f}{\partial v} = N(0, v) + \int_0^u \dfrac{\partial}{\partial v} M(x, v)\,dx = N(0, v) + \int_0^u \dfrac{\partial N}{\partial x}(x, v)\,dx \quad \left(\text{since } \dfrac{\partial M}{\partial y} = \dfrac{\partial N}{\partial x}\right)$$

$$= N(0, v) + \{N(x, v)\,]_0^u\} = N(0, v) + [N(u, v) - N(0, v)] = N(u, v)$$

42.64 If $u = f(x_1, \ldots, x_n)$ and $x_1 = h_1(t), \ldots, x_n = h_n(t)$, state the chain rule for du/dt.

▮ $\dfrac{du}{dt} = f_{x_1} \dfrac{dx_1}{dt} + \cdots + f_{x_n} \dfrac{dx_n}{dt}$.

42.65 If $u = x^2 - 2y^2 + z^3$ and $x = \sin t$, $y = e^t$, $z = 3t$, find du/dt.

▮ By the formula of Problem 42.64,

$$\frac{du}{dt} = \frac{\partial u}{\partial x}\frac{dx}{dt} + \frac{\partial u}{\partial y}\frac{dy}{dt} + \frac{\partial u}{\partial z}\frac{dz}{dt} = 2x \cos t - 4y\, e^t + 3z^2(3) = 2\sin t \cos t - 4e^t(e^t) + 9(3t)^2 = \sin 2t - 4e^{2t} + 81t^2.$$

42.66 If $w = \Phi(x, y, z)$ and $x = f(u, v)$, $y = g(u, v)$, $z = h(u, v)$, state the chain rule for $\partial w/\partial u$ and $\partial w/\partial v$.

▮ $\dfrac{\partial w}{\partial u} = \dfrac{\partial \Phi}{\partial x}\dfrac{\partial x}{\partial u} + \dfrac{\partial \Phi}{\partial y}\dfrac{\partial y}{\partial u} + \dfrac{\partial \Phi}{\partial z}\dfrac{\partial z}{\partial u}$. $\dfrac{\partial w}{\partial v} = \dfrac{\partial \Phi}{\partial x}\dfrac{\partial x}{\partial v} + \dfrac{\partial \Phi}{\partial y}\dfrac{\partial y}{\partial v} + \dfrac{\partial \Phi}{\partial z}\dfrac{\partial z}{\partial v}$.

42.67 Let $z = r^2 + s^2 + t^2$ and $t = rsu$. Clarify the two possible values of $\partial z/\partial r$ and find both values.

▮ (1) If $z = f(r, s, t) = r^2 + s^2 + t^2$, then $\partial z/\partial r$ means $f_r(r, s, t) = 2r$. (2) If $z = r^2 + s^2 + t^2 = r^2 + s^2 + (rsu)^2 = r^2 + s^2 + r^2s^2u^2 = g(r, s, u)$, then $\partial z/\partial r$ means $g_r(r, s, u) = 2r + 2rs^2u^2$. To distinguish the two possible values, one often uses $\left(\dfrac{\partial z}{\partial r}\right)_{s,t}$ for the first value, and $\left(\dfrac{\partial z}{\partial r}\right)_{s,u}$ for the second value.

42.68 How fast is the volume V of a rectangular box changing when its length l is 10 feet and increasing at the rate of 2 ft/s, its width w is 5 feet and decreasing at the rate of 1 ft/s, and its height h is 3 feet and increasing at the rate of 2 ft/s?

▮ $V = lwh$. $\dfrac{dV}{dt} = \dfrac{\partial V}{\partial l}\dfrac{dl}{dt} + \dfrac{\partial V}{\partial w}\dfrac{dw}{dt} + \dfrac{\partial V}{\partial h}\dfrac{dh}{dt} = wh(2) + lh(-1) + lw(2) = 5(3)(2) + 10(3)(-1) + 10(5)(2) = 30 - 30 + 100 = 100$ ft^3/s.

42.69 If $w = x^2 + 3xy - 2y^2$, and $x = r\cos\theta$, $y = r\sin\theta$, find $\partial w/\partial r$ and $\partial w/\partial\theta$.

▮ $\dfrac{\partial w}{\partial r} = \dfrac{\partial w}{\partial x}\dfrac{\partial x}{\partial r} + \dfrac{\partial w}{\partial y}\dfrac{\partial y}{\partial r} = (2x + 3y)\cos\theta + (3x - 4y)\sin\theta$

$\qquad = (2r\cos\theta + 3r\sin\theta)\cos\theta + (3r\cos\theta - 4r\sin\theta)\sin\theta = r(2\cos^2\theta + 6\sin\theta\cos\theta - 4\sin^2\theta)$

$\dfrac{\partial w}{\partial\theta} = \dfrac{\partial w}{\partial x}\dfrac{\partial x}{\partial\theta} + \dfrac{\partial w}{\partial y}\dfrac{\partial y}{\partial\theta} = (2x + 3y)(-r\sin\theta) + (3x - 4y)(r\cos\theta)$

$\qquad = (2r\cos\theta + 3r\sin\theta)(-r\sin\theta) + (3r\cos\theta - 4r\sin\theta)(r\cos\theta)$

$\qquad = r^2(-2\cos\theta\sin\theta - 3\sin^2\theta + 3\cos^2\theta - 4\sin\theta\cos\theta) = r^2(3\cos^2\theta - 6\cos\theta\sin\theta - 3\sin^2\theta)$

$\qquad = 3r^2(\cos^2\theta - 2\cos\theta\sin\theta - \sin^2\theta) = 3r^2(\cos 2\theta - \sin 2\theta)$

42.70 Let $u = f(x, y)$, $x = r\cos\theta$, $y = r\sin\theta$. Show that $\left(\dfrac{\partial u}{\partial x}\right)^2 + \left(\dfrac{\partial u}{\partial y}\right)^2 = \left(\dfrac{\partial u}{\partial r}\right)^2 + \dfrac{1}{r^2}\left(\dfrac{\partial u}{\partial\theta}\right)^2$.

▮ By the chain rule, $\dfrac{\partial u}{\partial r} = \dfrac{\partial u}{\partial x}\cos\theta + \dfrac{\partial u}{\partial y}\sin\theta$, and $\dfrac{\partial u}{\partial\theta} = \dfrac{\partial u}{\partial x}(-r\sin\theta) + \dfrac{\partial u}{\partial y}(r\cos\theta) = r\left(-\dfrac{\partial u}{\partial x}\sin\theta + \dfrac{\partial u}{\partial r}\cos\theta\right)$. Thus,

$$\left(\frac{\partial u}{\partial r}\right)^2 + \frac{1}{r^2}\left(\frac{\partial u}{\partial\theta}\right)^2 = \left(\frac{\partial u}{\partial x}\right)^2\cos^2\theta + 2\cos\theta\sin\theta\,\frac{\partial u}{\partial x}\frac{\partial u}{\partial y} + \left(\frac{\partial u}{\partial y}\right)^2\sin^2\theta + \left(\frac{\partial u}{\partial x}\right)^2\sin^2\theta - 2\sin\theta\cos\theta\,\frac{\partial u}{\partial x}\frac{\partial u}{\partial y}$$

$$+ \left(\frac{\partial u}{\partial r}\right)^2\cos^2\theta = \left(\frac{\partial u}{\partial x}\right)^2 + \left(\frac{\partial u}{\partial y}\right)^2$$

42.71 Let $u = f(x, y)$, $x = r\cos\theta$, $y = r\sin\theta$. Show that $u_{rr} = f_{xx}\cos^2\theta + 2f_{xy}\cos\theta\sin\theta + f_{yy}\sin^2\theta$.

▮ $u_r = f_x\cos\theta + f_y\sin\theta$. Hence, $u_{rr} = \cos\theta\,(f_{xx}\cos\theta + f_{xy}\sin\theta) + \sin\theta\,(f_{yx}\cos\theta + f_{yy}\sin^2\theta) = f_{xx}\cos^2\theta + 2f_{xy}\cos\theta\sin\theta + f_{yy}\sin^2\theta$.

42.72 Let $u = f(x, y)$, $x = r\cos\theta$, $y = r\sin\theta$. Show that $f_{xx} + f_{yy} = u_{rr} + \dfrac{1}{r^2}u_{\theta\theta} + \dfrac{1}{r}u_r$.

$u_\theta = -f_x r \sin \theta + r f_y \cos \theta$. Hence, $u_{\theta\theta} = -r\{f_x \cos \theta + \sin \theta[f_{xx}(-r \sin \theta) + f_{xy}(r \cos \theta)]\} + r\{-f_y \sin \theta + \cos \theta[f_{yx}(-r \sin \theta) + f_{yy}(r \cos \theta)]\} = f_{xx} r^2 \sin \theta - 2f_{xy} r^2 \sin \theta \cos \theta + f_{yy} r^2 \cos^2 \theta - r(f_x \cos \theta + f_y \sin \theta)$. Using Problem 42.71, we get $u_{rr} + \dfrac{1}{r^2} u_{\theta\theta} + \dfrac{1}{r} u_r = f_{xx}(\cos^2 \theta + \sin^2 \theta) + f_{yy}(\sin^2 \theta + \cos^2 \theta) - \dfrac{1}{r}(f_x \cos \theta + f_y \sin \theta) + \dfrac{1}{r}(f_x \cos \theta + f_y \sin \theta) = f_{xx} + f_{yy}$.

42.73 Let $z = u^3 v^5$, where $u = x + y$ and $v = x - y$. Find $\dfrac{\partial z}{\partial y}$ (a) by the chain rule, (b) by substitution and explicit computation.

  ▌ (a) $\dfrac{\partial z}{\partial y} = \dfrac{\partial z}{\partial u}\dfrac{\partial u}{\partial y} + \dfrac{\partial z}{\partial v}\dfrac{\partial v}{\partial y} = 3u^2 v^5 - 5u^3 v^4 = u^2 v^4(3v - 5u) = u^2 v^4(-2x - 8y)$

  $= -2(x + 4y)(x + y)^2(x - y)^4$.

  (b) $z = (x + y)^3(x - y)^5$. Hence,

  $\dfrac{\partial z}{\partial y} = (x + y)^3[5(x - y)^4(-1)] + 3(x + y)^2(x - y)^5$

  $= (x + y)^2(x - y)^4[-5(x + y) + 3(x - y)]$

  $= (x + y)^2(x - y)^4(-2x - 8y) = -2(x + 4y)(x + y)^2(x - y)^4$.

42.74 Prove Euler's theorem: If $f(x, y)$ is homogeneous of degree n, then $xf_x + yf_y = nf$. [Recall that $f(x, y)$ is homogeneous of degree n if and only if $f(tx, ty) = t^n f(x, y)$ for all x, y and for all $t > 0$].

  ▌ Differentiate $f(tx, ty) = t^n f(x, y)$ with respect to t. By the chain rule, $f_x(tx, ty)\dfrac{\partial(tx)}{\partial t} + f_y(tx, ty)\dfrac{\partial(ty)}{\partial t} = nt^{n-1}f(x, y)$. Hence, $f_x(tx, ty)(x) + f_y(tx, ty)(y) = nt^{n-1}f(x, y)$. Let $t = 1$. Then, $xf_x(x, y) + yf_y(x, y) = nf(x, y)$. (A similar result holds for functions f of more than two variables.)

42.75 Verify Euler's theorem (Problem 42.74) for the function $f(x, y) = xy^2 + x^2 y - y^3$.

  ▌ $f(x, y)$ is homogeneous of degree 3, since $f(tx, ty) = (tx)(ty)^2 + (tx)^2(ty) - (ty)^3 = t^3(xy^2 + x^2 y - y^3) = t^3 f(x, y)$. So, we must show that $xf_x + yf_y = 3f$. $f_x = y^2 + 2xy$, $f_y = 2xy + x^2 - 3y^2$. Hence, $xf_x + yf_y = x(y^2 + 2xy) + y(2xy + x^2 - 3y^2) = xy^2 + 2x^2 y + 2xy^2 + x^2 y - 3y^3 = 3(xy^2 + x^2 y - y^3) = 3f(x, y)$.

42.76 Verify Euler's theorem (Problem 42.74) for the function $f(x, y) = \sqrt{x^2 + y^2}$.

  ▌ $f(x, y)$ is homogeneous of degree 1, since $f(tx, ty) = \sqrt{(tx)^2 + (ty)^2} = t\sqrt{x^2 + y^2} = t f(x, y)$ for $t > 0$. We must check that $xf_x + yf_y = \sqrt{x^2 + y^2}$. But, $f_x = x/\sqrt{x^2 + y^2}$ and $f_y = y/\sqrt{x^2 + y^2}$. Thus, $xf_x + yf_y = (x^2/\sqrt{x^2 + y^2}) + (y^2/\sqrt{x^2 + y^2}) = (x^2 + y^2)/\sqrt{x^2 + y^2} = \sqrt{x^2 + y^2}$.

42.77 Verify Euler's theorem (Problem 42.74) for the function $f(x, y, z) = 3xz^2 - 2xyz + y^2 z$.

  ▌ f is clearly homogeneous of degree 3. Now, $f_x = 3z^2 - 2yz$, $f_y = -2xz + 2zy$, $f_z = 6xz - 2xy + y^2$. Thus, $xf_x + yf_y + zf_z = x(3z^2 - 2yz) + y(-2xz + 2zy) + z(6xz - 2xy + y^2) = 3xz^2 - 2xyz - 2xyz + 2y^2 z + 6xz^2 - 2xyz + y^2 z = 9xz^2 - 6xyz + 3y^2 z = 3 f(x, y, z)$.

42.78 If $f(x, y)$ is homogeneous of degree n and has continuous second-order partial derivatives, prove $x^2 f_{xx} + 2xy f_{xy} + y^2 f_{yy} = n(n - 1)f$.

  ▌ By Problem 42.74, $f_x(tx, ty)(x) + f_y(tx, ty)(y) = nt^{n-1}f(x, y)$. Differentiate with respect to t: $x[f_{xx}(tx, ty)(x) + f_{xy}(tx, ty)(y)] + y[f_{yx}(tx, ty)(x) + f_{yy}(tx, ty)(y)] = n(n - 1)t^{n-2}f(x, y)$. Now let $t = 1$: $x(f_{xx} \cdot x + f_{xy} \cdot y) + y(f_{yx} \cdot x + f_{yy} \cdot y) = n(n - 1)f$, $x^2 f_{xx} + 2xy f_{xy} + y^2 f_{yy} = n(n - 1)f$, since $f_{xy} = f_{yx}$.

42.79 If $f(x, y)$ is homogeneous of degree n, show that f_x is homogeneous of degree $n - 1$.

  ▌ $f(tx, ty) = t^n f(x, y)$. Differentiate with respect to x: $f_x(tx, ty)(t) + f_y(tx, ty)(0) = t^n[f_x(x, y)(1) + f_y(x, y)(0)]$, $f_x(tx, ty)(t) = t^n f_x(x, y)$, $f_x(tx, ty) = t^{n-1}f_x(x, y)$.

42.80 Show that any function $f(x, y)$ that is homogeneous of degree n is separable in polar coordinates and has the form $f(x, y) = r^n \Theta(\theta)$.

▌ Choose new variables $u = \ln x$, $v = y/x$, and write

$$\frac{f(x, y)}{(\sqrt{x^2 + y^2})^n} \equiv \phi(u, v)$$

Replacing x and y by tx and ty $(t > 0)$, we have

$$\phi(u + \ln t, v) = \frac{f(tx, ty)}{[\sqrt{(tx)^2 + (ty)^2}]^n} = \frac{t^n f(x, y)}{t^n (\sqrt{x^2 + y^2})^n} = \phi(u, v)$$

Because $\ln t$ assumes all real values, the above equation can hold only if ϕ is independent of u; i.e., $\phi(u, v) = \psi(v)$. Hence, $f(x, y) = (\sqrt{x^2 + y^2})^n \psi(y/x) = r^n \psi(\tan \theta) = r^n \Theta(\theta)$.

42.81 Find a general solution $f(x, y)$ of the equation $a f_x = f_y$, where $a \neq 0$.

▌ Let $u = x + ay$, $v = x - ay$. Then x and y can be found in terms of u and v, and $f(x, y)$ can be considered a function $w = F(u, v)$. By the chain rule, $f_x = \frac{\partial w}{\partial x} = F_u \frac{\partial u}{\partial x} + F_v \frac{\partial v}{\partial x} = F_u + F_v$, $f_y = \frac{\partial w}{\partial y} = F_u \frac{\partial u}{\partial y} + F_v \frac{\partial v}{\partial y} = a F_u - a F_v$. Substituting in $a f_x = f_y$, we get $a F_u + a F_v = a F_u - a F_v$, and, therefore, $F_v = 0$. Thus, F is a function $g(u)$ of u alone. Hence, $w = g(u) = g(x + ay)$. So, $g(x + ay)$ is the general solution, where g is any continuously differentiable function.

42.82 If $z = 2x^2 - 3xy + 7y^2$, $x = \sin t$, $y = \cos t$, find dz/dt.

▌ By the chain rule, $\frac{dz}{dt} = \frac{\partial z}{\partial x} \frac{dx}{dt} + \frac{\partial z}{\partial y} \frac{dy}{dt} = (4x - 3y) \cos t + (-3x + 14y)(-\sin t) = (4 \sin t - 3 \cos t) \cos t - (14 \cos t - 3 \sin t) \sin t = 3 \sin^2 t - 10 \sin t \cos t - 3 \cos^2 t$.

42.83 If $z = \ln (x^2 + y^2)$, $x = e^{-t}$, $y = e^t$, find dz/dt.

▌ By the chain rule,

$$\frac{dz}{dt} = \frac{\partial z}{\partial x} \frac{dx}{dt} + \frac{\partial z}{\partial y} \frac{dy}{dt} = \frac{2x}{x^2 + y^2} (-e^{-t}) + \frac{2y}{x^2 + y^2} e^t = -\frac{2e^{-2t}}{e^{-2t} + e^{2t}} + \frac{2e^{2t}}{e^{-2t} + e^{2t}} = 2 \frac{e^{2t} - e^{-2t}}{e^{2t} + e^{-2t}} = 2 \tanh 2t$$

42.84 If $z = f(x, y) = x^4 + 3xy - y^2$ and $y = \sin x$, find dz/dx.

▌ By the chain rule, $\frac{dz}{dx} = f_x \frac{dx}{dx} + f_y \frac{dy}{dx} = (4x^3 + 3y) + (3x - 2y) \cos x = (4x^3 + 3 \sin x) + (3x - 2 \sin x) \cos x$.

42.85 If $z = f(x, y) = xy^2 + x^2y$ and $y = \ln x$, find dz/dx and dz/dy.

▌ First, think of z as a composite function of x. By the chain rule, $dz/dx = f_x + f_y (dy/dx) = y^2 + 2xy + (2xy + x^2)(1/x) = y^2 + 2xy + 2y + x = (\ln x)^2 + 2(x + 1) \ln x + x$. Next, think of z as a composite function of y (by virtue of $x = e^y$). Then, $\frac{dz}{dy} = f_x \frac{dx}{dy} + f_y = (y^2 + 2xy)x + (2xy + x^2) = x(y^2 + 2xy + 2y + x) = e^y(y^2 + 2ye^y + 2y + e^y)$.

42.86 The altitude h of a right circular cone is decreasing at the rate of 3 mm/s, while the radius r of the base is increasing at the rate of 2 mm/s. How fast is the volume V changing when the altitude and radius are 100 mm and 50 mm, respectively?

▌ $V = \frac{1}{3} \pi r^2 h$. So, $\frac{dV}{dt} = \frac{\partial V}{\partial r} \frac{dr}{dt} + \frac{\partial V}{\partial h} \frac{dh}{dt} = (\frac{2}{3} \pi rh)(2) + (\frac{1}{3} \pi r^2)(-3) = \frac{4}{3} \pi rh - \pi r^2 = \pi[\frac{4}{3}(5000) - 2500] = \frac{12{,}500\pi}{3}$ mm^3/s.

42.87 A point P is moving along the curve of intersection of the paraboloid $\frac{x^2}{16} - \frac{y^2}{9} = z$ and the cylinder $x^2 + y^2 = 5$. If x is increasing at the rate of 5 cm/s, how fast is z changing when $x = 2$ cm and $y = 1$ cm?

▌ Apply the chain rule to $z = \frac{x^2}{16} - \frac{y^2}{9}$. $\frac{dz}{dt} = \frac{\partial z}{\partial x} \frac{dx}{dt} + \frac{\partial z}{\partial y} \frac{dy}{dt} = \frac{x}{8} \frac{dx}{dt} - \frac{2y}{9} \frac{dy}{dt}$. From $x^2 + y^2 = 5$, $x \frac{dx}{dt} + y \frac{dy}{dt} = 0$. Since $dx/dt = 5$, $5x + y \frac{dy}{dt} = 0$. So, when $x = 2$ and $y = 1$, $dy/dt = -10$, and $dz/dt = \frac{2}{8}(5) - \frac{2}{9}(-10) = 10(\frac{1}{8} + \frac{2}{9}) = \frac{125}{36}$ cm/s.

42.88 Find du/dt given that $u = x^2y^3$, $x = 2t^3$, $y = 3t^2$.

▮ $\dfrac{du}{dt} = u_x \dfrac{dx}{dt} + u_y \dfrac{dy}{dt} = 2xy^3(6t^2) + 3x^2y^2(6t) = 6xy^2t(2yt + 3x) = 6(2t^3)(9t^4)(t)(6t^3 + 6t^3) = 1296t^{11}$.

42.89 If the radius r of a right circular cylinder is increasing at the rate of 3 in/s and the altitude h is increasing at the rate of 2 in/s, how fast is the surface area S changing when $r = 10$ inches and $h = 5$ inches?

▮ $S = 2\pi rh$. By the chain rule, $\dfrac{dS}{dt} = \dfrac{\partial S}{\partial r} \dfrac{dr}{dt} + \dfrac{\partial S}{\partial h} \dfrac{dh}{dt} = (2\pi h)(3) + (2\pi r)(2) = 2\pi(3h + 2r) = 2\pi(15 + 20) = 70\pi$ in^2/s.

42.90 If a point is moving on the curve of intersection of $x^2 + 3xy + 3y^2 = z^2$ and the plane $x - 2y + 4 = 0$, how fast is it moving when $x = 2$, if x is increasing at the rate of 3 units per second?

▮ From $x - 2y + 4 = 0$, $\dfrac{dx}{dt} - 2\dfrac{dy}{dt} = 0$. Since $dx/dt = 3$, $dy/dt = \frac{3}{2}$. From $x^2 + 3xy + 3y^2 = z^2$, by the chain rule, $(2x + 3y)(dx/dt) + (3x + 6y)(dy/dt) = 2z(dz/dt)$. Hence,

$$(2x + 3y)(3) + (3x + 6y)(\tfrac{3}{2}) = 2z\dfrac{dz}{dt} \qquad (*)$$

When $x = 2$, the original equations become $3y^2 + 6y + 4 = z^2$ and $-2y + 6 = 0$, yielding $y = 3$, $z = \pm 7$. Thus, by $(*)$, $39 + 36 = 2z(dz/dt)$, $dz/dt = \pm\frac{75}{14}$. Hence, the speed

$$\dfrac{ds}{dt} = \sqrt{\left(\dfrac{dx}{dt}\right)^2 + \left(\dfrac{dy}{dt}\right)^2 + \left(\dfrac{dz}{dt}\right)^2} = \sqrt{9 + \tfrac{9}{4} + 9(\tfrac{25}{14})^2} = 3\sqrt{1 + \tfrac{1}{4} + \tfrac{625}{196}} = 3\sqrt{\tfrac{870}{196}} = \tfrac{3}{14}\sqrt{870} \approx 6.3 \text{ units per second}$$

42.91 If $u = f(x, y)$ and $x = r \cosh s$, $y = \sinh s$, show that $\left(\dfrac{\partial u}{\partial x}\right)^2 - \left(\dfrac{\partial u}{\partial y}\right)^2 = \left(\dfrac{\partial u}{\partial r}\right)^2 - \dfrac{1}{r^2}\left(\dfrac{\partial u}{\partial s}\right)^2$.

▮ $\dfrac{\partial u}{\partial r} = \dfrac{\partial u}{\partial x}\dfrac{\partial x}{\partial r} + \dfrac{\partial u}{\partial y}\dfrac{\partial y}{\partial r} = \dfrac{\partial u}{\partial x}(\cosh s) + \dfrac{\partial u}{\partial y}(\sinh s)$. $\dfrac{\partial u}{\partial s} = \dfrac{\partial u}{\partial x}\dfrac{\partial x}{\partial s} + \dfrac{\partial u}{\partial y}\dfrac{\partial y}{\partial s} = \dfrac{\partial u}{\partial x}(r \sinh s) + \dfrac{\partial u}{\partial y}(r \cosh s)$. Hence,

$$\left(\dfrac{\partial u}{\partial r}\right)^2 - \dfrac{1}{r^2}\left(\dfrac{\partial u}{\partial s}\right)^2 = \left(\dfrac{\partial u}{\partial x}\right)^2 \cosh^2 s + 2\dfrac{\partial u}{\partial x}\dfrac{\partial u}{\partial y}\cosh s \sinh s + \left(\dfrac{\partial u}{\partial y}\right)^2 \sinh^2 s$$
$$- \left[\left(\dfrac{\partial u}{\partial x}\right)^2 \sinh^2 s + 2\dfrac{\partial u}{\partial x}\dfrac{\partial u}{\partial y}\sinh s \cosh s + \left(\dfrac{\partial u}{\partial y}\right)^2 \cosh^2 s\right]$$
$$= \left(\dfrac{\partial u}{\partial x}\right)^2 (\cosh^2 s - \sinh^2 s) + \left(\dfrac{\partial u}{\partial y}\right)^2 (\sinh^2 s - \cosh^2 s) = \left(\dfrac{\partial u}{\partial x}\right)^2 - \left(\dfrac{\partial u}{\partial y}\right)^2$$

42.92 If $z = H(u, v)$, and $u = f(x, y)$, $v = g(x, y)$ satisfy the Cauchy–Riemann equations $\partial u/\partial x = \partial v/\partial y$ and $\partial u/\partial y = -\partial v/\partial x$, show that $\dfrac{\partial^2 H}{\partial x^2} + \dfrac{\partial^2 H}{\partial y^2} = \left[\left(\dfrac{\partial u}{\partial x}\right)^2 + \left(\dfrac{\partial v}{\partial x}\right)^2\right]\left(\dfrac{\partial^2 H}{\partial u^2} + \dfrac{\partial^2 H}{\partial v^2}\right)$.

▮ $H_x = H_u u_x + H_v v_x$, $H_y = H_u u_y + H_v v_y$. $H_{xx} = (H_{uu} u_x + H_{uv} v_x)u_x + H_u u_{xx} + (H_{vu} u_x + H_{vv} v_x)v_x + H_v v_{xx}$, $H_{yy} = (H_{uu} u_y + H_{uv} v_y)u_y + H_u u_{yy} + (H_{vu} u_y + H_{vv} v_y)v_y + H_v v_{yy}$. By Problem 42.54, $u_{xx} = -u_{yy}$ and $v_{xx} = -v_{yy}$. Hence, $H_{xx} + H_{yy} = (H_{uu} u_x + H_{uv} v_x)u_x + (H_{vu} u_x + H_{vv} v_x)v_x + [H_{uu}(-v_x) + H_{uv} u_x](-v_x) + [H_{vu}(-v_x) + H_{vv} u_x]u_x = H_{uu} u_x^2 + H_{vv} v_x^2 + H_{uu} v_x^2 + H_{vv} u_x^2 = (u_x^2 + v_x^2)(H_{uu} + H_{vv})$.

42.93 If $g(u)$ is continuously differentiable, show that $w = g(x^2 - y^2)$ is a solution of $y\dfrac{\partial w}{\partial x} + x\dfrac{\partial w}{\partial y} = 0$.

▮ $\partial w/\partial x = g'(x^2 - y^2)(2x)$, $\partial w/\partial y = g'(x^2 - y^2)(-2y)$. Hence,

$$y\dfrac{\partial w}{\partial x} + x\dfrac{\partial w}{\partial y} = [y(2x) + x(-2y)]g'(x^2 - y^2) = 0.$$

42.94 If $w = f(x^2 - y^2, y^2 - x^2)$, show that $y\dfrac{\partial w}{\partial x} + x\dfrac{\partial w}{\partial y} = 0$.

▮ $\dfrac{\partial w}{\partial x} = f_1\dfrac{\partial(x^2 - y^2)}{\partial x} + f_2\dfrac{\partial(y^2 - x^2)}{\partial x} = f_1 \cdot (2x) + f_2 \cdot (-2x) = 2x(f_1 - f_2)$. (Here, f_1 is the partial derivative of f with respect to the first variable, and f_2 is the partial derivative of f with respect to the second variable.)

$$\frac{\partial w}{\partial y} = f_1 \frac{\partial(x^2 - y^2)}{\partial y} + f_2 \frac{\partial(y^2 - x^2)}{\partial y} = f_1 \cdot (-2y) + f_2 \cdot (2y) = 2y(f_2 - f_1)$$

Therefore, $y \dfrac{\partial w}{\partial x} + x \dfrac{\partial w}{\partial y} = 2xy(f_1 - f_2) + 2xy(f_2 - f_1) = 0$.

42.95 If $w = f\left(\dfrac{y - x}{xy}, \dfrac{z - y}{yz}\right)$, show that $x^2 \dfrac{\partial w}{\partial x} + y^2 \dfrac{\partial w}{\partial y} + z^2 \dfrac{\partial w}{\partial z} = 0$.

▮ $$\frac{\partial w}{\partial x} = f_1 \frac{\partial((y - x)/xy)}{\partial x} + f_2 \frac{\partial((z - y)/yz)}{\partial x} = f_1 \cdot \left(-\frac{1}{x^2}\right) + f_2 \cdot 0 = -\frac{1}{x^2} f_1$$

$$\frac{\partial w}{\partial y} = f_1 \frac{\partial((y - x)/xy)}{\partial y} + f_2 \frac{\partial((z - y)/yz)}{\partial y} = f_1 \cdot \left(\frac{1}{y^2}\right) - f_2 \cdot \left(\frac{1}{y^2}\right) = \frac{1}{y^2}(f_1 - f_2)$$

$$\frac{\partial w}{\partial z} = f_1 \frac{\partial((y - x)/xy)}{\partial z} + f_2 \frac{\partial((z - y)/yz)}{\partial z} = 0 + f_2 \cdot \left(\frac{1}{z^2}\right) = \frac{1}{z^2} f_2$$

Thus, $x^2 \dfrac{\partial w}{\partial x} + y^2 \dfrac{\partial w}{\partial y} + z^2 \dfrac{\partial w}{\partial z} = -f_1 + f_1 - f_2 + f_2 = 0$.

42.96 Prove Liebniz's formula: For differentiable functions $u(x)$ and $v(x)$,

$$\frac{d}{dx} \int_u^v f(x, y)\,dy = -f(x, u) \frac{du}{dx} + f(x, v) \frac{dv}{dx} + \int_u^v \frac{\partial f}{\partial x}\,dy.$$

▮ Let $w = \int_u^v f(x, y)\,dy$. By the chain rule, $\dfrac{dw}{dx} = \dfrac{\partial w}{\partial u} \dfrac{du}{dx} + \dfrac{\partial w}{\partial v} \dfrac{dv}{dx} + \dfrac{\partial w}{\partial x}$. Now, $\partial w/\partial u = -f(x, u)$ and $\partial w/\partial v = f(x, v)$, and (Problem 42.60) $\partial w/\partial x = \int_u^v \dfrac{\partial f}{\partial x}\,dy$, yielding Leibniz's formula.

42.97 Verify Leibniz's formula (Problem 42.96) for $u = x$, $v = x^2$, and $f(x, y) = x^3 y^2 + x^2 y^3$.

▮ $du/dx = 1$, $dv/dx = 2x$, $\dfrac{\partial f}{\partial x} = 3x^2 y^2 + 2xy^3$, and $\int_u^v \dfrac{\partial f}{\partial x}\,dy = \int_x^{x^2}(3x^2 y^2 + 2xy^3)\,dy = (x^2 y^3 + \frac{1}{2}xy^4)\,]_x^{x^2} = (x^8 + \frac{1}{2}x^9) - (x^5 + \frac{1}{2}x^5)$. So, $-f(x, u)\dfrac{du}{dx} + f(x, v)\dfrac{dv}{dx} + \int_u^v \dfrac{\partial f}{\partial x}\,dy = -(x^3 u^2 + x^2 u^3) + (x^3 v^2 + x^2 v^3)2x + x^8 + \frac{1}{2}x^9 - x^5 - \frac{1}{2}x^5 = -2x^5 + (x^7 + x^8)(2x) + x^8 + \frac{1}{2}x^9 - \frac{3}{2}x^5 = -\frac{7}{2}x^5 + 3x^8 + \frac{5}{2}x^9$. On the other hand, $\int_u^v f(x, y)\,dy = \int_x^{x^2}(x^3 y^2 + x^2 y^3)\,dy = (\frac{1}{3}x^3 y^3 + \frac{1}{4}x^2 y^4)\,]_x^{x^2} = (\frac{1}{3}x^9 + \frac{1}{4}x^{10}) - (\frac{1}{3}x^6 + \frac{1}{4}x^6) = \frac{1}{3}x^9 + \frac{1}{4}x^{10} - \frac{7}{12}x^6$. Hence, $\dfrac{d}{dx} \int_u^v f(x, y)\,dy = 3x^8 + \frac{5}{2}x^9 - \frac{7}{2}x^5$, verifying Leibniz's formula.

42.98 Assume $\partial f/\partial x = 0$ for all (x, y). Show that $f(x, y) = h(y)$ for some function h.

▮ For each y_0, $\dfrac{d}{dx} f(x, y_0) = \dfrac{\partial f}{\partial x}(x, y_0) = 0$. Hence, $f(x, y_0)$ is a constant, c. Let $h(y_0) = c$. Then $f(x, y) = h(y)$ for all x and y.

42.99 Assume $\partial f/\partial x = x$ for all (x, y). Show that $f(x, y) = \frac{1}{2}x^2 + h(y)$ for some function $h(y)$.

▮ Let $F(x, y) = f(x, y) - \frac{1}{2}x^2$. Then, $\dfrac{\partial F}{\partial x} = \dfrac{\partial f}{\partial x} - x = 0$. By Problem 42.98, $F(x, y) = h(y)$ for a suitable h.

42.100 Assume $\partial f/\partial x = G(x)$ for all (x, y). Then prove there are functions $g(x)$ and $h(y)$ such that $f(x, y) = g(x) + h(y)$.

▮ Let $F(x, y) = f(x, y) - \int_0^x G(t)\,dt$. Then $\dfrac{\partial F}{\partial x} = \dfrac{\partial f}{\partial x} - G(x) = 0$. By Problem 42.98, $F(x, y) = h(y)$ for some h. Thus, $f(x, y) = \int_0^x G(t)\,dt + h(y)$. Let $g(x) = \int_0^x G(t)\,dt$.

42.101 Assume $\partial f/\partial x = g(y)$ for all (x, y). Show that $f(x, y) = x\,g(y) + h(y)$ for some function h.

▮ Let $H(x, y) = f(x, y) - x\,g(y)$. Then, $\dfrac{\partial H}{\partial x} = \dfrac{\partial f}{\partial x} - g(y) = 0$. By Problem 42.98, $H(x, y) = h(y)$ for a suitable function h. Hence, $f(x, y) = x\,g(y) + h(y)$.

42.102 Find a general solution for $\partial^2 f/\partial x^2 = 0$.

❚ Let $K(x, y) = \partial f/\partial x$. Then $\partial K/\partial x = 0$. By Problem 42.98, $K(x, y) = g(y)$ for some function g. Hence, $\partial f/\partial x = g(y)$. By Problem 42.101, $f(x, y) = x g(y) + h(y)$ for a suitable function h. Conversely, any linear function of x (with coefficients depending on y) satisfies $f_{xx} = 0$.

42.103 Find a general solution of $\dfrac{\partial^2 f}{\partial x \, \partial y} = 0$.

❚ Let $L(x, y) = \partial f/\partial y$. Then $\partial L/\partial x = 0$. By Problem 42.98, $L(x, y) = g(y)$ for some g. So $\partial f/\partial y = g(y)$. By an analogue of Problem 42.100, there are functions $A(x)$ and $B(y)$ such that $f(x, y) = A(x) + B(y)$. Conversely, any such function $f(x, y) = A(x) + B(y)$, where A and B are twice differentiable, satisfies $\dfrac{\partial^2 f}{\partial x \, \partial y} = 0$.

42.104 Find a general solution of $\dfrac{\partial^2 f}{\partial x \, \partial y} = 1$.

❚ Note that $\dfrac{\partial^2 (xy)}{\partial x \, \partial y} = 1$. Let $C(x, y) = f(x, y) - xy$. Then $\dfrac{\partial^2 C}{\partial x \, \partial y} = 1 - 1 = 0$. By Problem 42.103, $C(x, y) = A(x) + B(y)$ for suitable $A(x)$ and $B(y)$. Then, $f(x, y) = A(x) + B(y) + xy$. This is the general solution of $\dfrac{\partial^2 f}{\partial x \, \partial y} = 1$.

42.105 Show that the tangent plane to a surface $z = f(x, y)$ at a point (x_0, y_0, z_0) has a normal vector $(f_x(x_0, y_0), f_y(x_0, y_0), -1)$.

❚ One vector in the tangent plane at (x_0, y_0, z_0) is $(1, 0, f_x)$, and another is $(0, 1, f_y)$. Hence, a normal vector is $(0, 1, f_y) \times (1, 0, f_x) = (f_x, f_y, -1)$.

42.106 Find an equation of the tangent plane to $z = x^2 + y^2$ at $(1, 2, 5)$.

❚ $\partial z/\partial x = 2x = 2$, $\partial z/\partial y = 2y = 4$. Hence, by Problem 42.105, a normal vector to the tangent plane is $(2, 4, -1)$. Therefore, an equation of that plane is $2(x - 1) + 4(y - 2) - (z - 5) = 0$, or, equivalently, $2x + 4y - z = 5$.

42.107 Find an equation of the tangent plane to $z = xy$ at $(2, \frac{1}{2}, 1)$.

❚ $\partial z/\partial x = y = \frac{1}{2}$, $\partial z/\partial y = x = 2$. Thus, a normal vector to the tangent plane is $(\frac{1}{2}, 2, -1)$, and an equation of that plane is $\frac{1}{2}(x - 2) + 2(y - \frac{1}{2}) - (z - 1) = 0$, or, equivalently, $x + 4y - 2z = 2$.

42.108 Find an equation of the tangent plane to the surface $z = 2x^2 - y^2$ at the point $(1, 1, 1)$.

❚ $\partial z/\partial x = 4x = 4$, $\partial z/\partial y = -2y = -2$. Hence, a normal vector to the tangent plane is $(4, -2, -1)$, and an equation of the plane is $4(x - 1) - 2(y - 1) - (z - 1) = 0$, or, equivalently, $4x - 2y - z = 1$.

42.109 If a surface has the equation $F(x, y, z) = 0$, show that a normal vector to the tangent plane at (x_0, y_0, z_0) is $(F_x(x_0, y_0, z_0), F_y(x_0, y_0, z_0), F_z(x_0, y_0, z_0))$.

❚ Assume that $F_z(x_0, y_0, z_0) \neq 0$ so that $F(x, y, z) = 0$ implicitly defines z as a function of x and y in a neighborhood of (x_0, y_0, z_0). Then, by Problem 42.105, a normal vector to the tangent plane is $\left(\dfrac{\partial z}{\partial x}, \dfrac{\partial z}{\partial y}, -1\right)$. Differentiate $F(x, y, z) = 0$ with respect to x: $F_x + F_y \dfrac{\partial y}{\partial x} + F_z \dfrac{\partial z}{\partial x} = 0$. Hence, $F_x + F_z \dfrac{\partial z}{\partial x} = 0$, since $\partial y/\partial x = 0$. Therefore, $\partial z/\partial x = -F_x/F_z$. Similarly, $\partial z/\partial y = -F_y/F_z$. Hence, a normal vector is $(-F_x/F_z, -F_y/F_z, -1)$. Multiplying by the scalar $-F_z$, we obtain another normal vector (F_x, F_y, F_z).

42.110 Find an equation of the tangent plane to the sphere $x^2 + y^2 + z^2 = 1$ at the point $(\frac{1}{2}, \frac{1}{2}, 1/\sqrt{2})$.

❚ By Problem 42.109, a normal vector to the tangent plane will be $(2x, 2y, 2z) = (1, 1, \sqrt{2})$. Hence, an equation for the tangent plane is $(x - \frac{1}{2}) + (y - \frac{1}{2}) + \sqrt{2}[z - (1/\sqrt{2})] = 0$, or, equivalently, $x + y + \sqrt{2}z = 2$.

42.111 Find an equation for the tangent plane to $z^3 + xyz - 2 = 0$ at $(1, 1, 1)$.

❚ By Problem 42.109, a normal vector to the tangent plane is $(yz, xz, 3z^2 + xy) = (1, 1, 4)$. Hence, the tangent plane is $(x - 1) + (y - 1) + 4(z - 1) = 0$, or, equivalently, $x + y + 4z = 6$.

42.112   Find an equation of the tangent plane to the ellipsoid $\dfrac{x^2}{a^2} + \dfrac{y^2}{b^2} + \dfrac{z^2}{c^2} = 1$ at a point (x_0, y_0, z_0).

▌ By Problem 42.109, a normal vector to the tangent plane is $\left(\dfrac{2x_0}{a^2}, \dfrac{2y_0}{b^2}, \dfrac{2z_0}{c^2}\right)$, or, better, the vector $\left(\dfrac{x_0}{a^2}, \dfrac{y_0}{b^2}, \dfrac{z_0}{c^2}\right)$. Hence, an equation of the tangent plane is $\dfrac{x_0}{a^2}(x - x_0) + \dfrac{y_0}{b^2}(y - y_0) + \dfrac{z_0}{c^2}(z - z_0) = 0$, or, equivalently $\dfrac{x_0}{a^2}x + \dfrac{y_0}{b^2}y + \dfrac{z_0}{c^2}z = 1$.

42.113   Let $a > 0$. The tangent plane to the surface $xyz = a$ at a point (x_0, y_0, z_0) in the first octant forms a tetrahedron with the coordinate planes.   Show that such tetrahedrons all have the same volume.

▌ By Problem 42.109, a normal vector to the tangent plane is $(y_0 z_0, x_0 z_0, x_0 y_0)$. Hence, that tangent plane has an equation $y_0 z_0(x - x_0) + x_0 z_0(y - y_0) + x_0 y_0(z - z_0) = 0$, or, equivalently, $y_0 z_0 x + x_0 z_0 y + x_0 y_0 z = 3x_0 y_0 z_0$. This plane cuts the x, y, and z-axes at $(3x_0, 0, 0)$, $(0, 3y_0, 0)$, $(0, 0, 3z_0)$, respectively (Fig. 42-1). Hence, the volume of the resulting tetrahedron is $\frac{1}{6}(3x_0)(3y_0)(3z_0) = \frac{9}{2}x_0 y_0 z_0 = \frac{9}{2}a$.

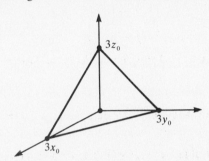

$3z_0$

$3y_0$

$3x_0$ **Fig. 42-1**

42.114   Find a vector tangent at the point $(2, 1, 4)$ to the curve of intersection of the cone $z^2 = 3x^2 + 4y^2$ and the plane $3x - 2y + z = 8$.

▌ A normal vector to $z^2 = 3x^2 + 4y^2$ at $(2, 1, 4)$ is $\mathbf{A} = (6x, 8y, -2z) = (12, 8, -8)$. A normal vector to the plane $3x - 2y + z = 8$ is $\mathbf{B} = (3, -2, 1)$. A vector parallel to the tangent line of the curve of the intersection will be perpendicular to both normal vectors and, therefore, will be parallel to their cross product $\mathbf{A} \times \mathbf{B} = (12, 8, -8) \times (3, -2, 1) = (-8, -12, -48)$. A simpler tangent vector would be $(2, 3, 12)$.

42.115   If a surface has an equation of the form $z = x\, f(x/y)$, show that all of its tangent planes have a common point.

▌ A normal vector to the surface at (x, y, z) is

$$(\partial z/\partial x, \partial z/\partial y, -1) = \left(x\, f'\!\left(\frac{x}{y}\right)\!\left(\frac{1}{y}\right) + f\!\left(\frac{x}{y}\right),\ x\, f'\!\left(\frac{x}{y}\right)\!\left(-\frac{x}{y^2}\right),\ -1\right).$$

Hence, the tangent plane at (x_0, y_0, z_0) has an equation

$$\left[\frac{x_0}{y_0}\, f'\!\left(\frac{x_0}{y_0}\right) + f\!\left(\frac{x_0}{y_0}\right)\right]x - \left[\frac{x_0^2}{y_0^2}\, f'\!\left(\frac{x_0}{y_0}\right)\right]y - z = 0$$

Thus, the plane goes through the origin.

42.116   Let normal lines be drawn at all points on the surface $z = ax^2 + by^2$ that are at a given height h above the xy-plane.   Find an equation of the curve in which these lines intersect the xy-plane.

▌ The normal vectors are $(2ax, 2by, -1)$. Hence, the normal line at (x_0, y_0, h) has parametric equations $x = x_0 + 2ax_0 t$, $y = y_0 + 2by_0 t$, $z = h - t$. This line hits the xy-plane when $z = 0$, that is, when $t = h$. Thus, the point of intersection is $x = x_0 + 2ax_0 h = x_0(1 + 2ah)$, $y = y_0 + 2by_0 h = y_0(1 + 2bh)$. Note that $h = ax_0^2 + by_0^2 = \dfrac{ax^2}{(1 + 2ah)^2} + \dfrac{by^2}{(1 + 2bh)^2}$. Hence, the desired equation is $\dfrac{ax^2}{(1 + 2ah)^2} + \dfrac{by^2}{(1 + 2bh)^2} = h$.

42.117   Find equations of the normal line to the surface $x^2 + 4y^2 = z^2$ at $(3, 2, 5)$.

▌ A normal vector is $(2x, 8y, -2z) = (6, 16, -10)$, by Problem 42.109 [with $f(x, y, z) = x^2 + 4y^2 - z^2$]. Hence, equations for the normal line are $\dfrac{x - 3}{6} = \dfrac{y - 2}{16} = \dfrac{z - 5}{-10}$,   or, in parametric form, $x = 3 + 6t$, $y = 2 + 16t$, $z = 5 - 10t$.

42.118 Give an expression for a tangent vector to a curve $\mathscr{C}$ that is the intersection of the surfaces $F(x, y, z) = 0$ and $G(x, y, z) = 0$.

▌ A normal vector to the surface $F(x, y, z) = 0$ is $\mathbf{A} = (F_x, F_y, F_z)$, and a normal vector to the surface $G(x, y, z) = 0$ is $\mathbf{B} = (G_x, G_y, G_z)$. Since the curve is perpendicular to both $\mathbf{A}$ and $\mathbf{B}$, a tangent vector would be given by

$$\mathbf{A} \times \mathbf{B} = \begin{vmatrix} \mathbf{i} & \mathbf{j} & \mathbf{k} \\ F_x & F_y & F_z \\ G_x & G_y & G_z \end{vmatrix}$$

42.119 Find equations of the tangent line to the curve that is the intersection of $x^2 + 2y^2 + 2z^2 = 5$ and $3x - 2y - z = 0$ at $(1, 1, 1)$.

▌ By Problem 42.118, a tangent vector is

$$\begin{vmatrix} \mathbf{i} & \mathbf{j} & \mathbf{k} \\ 2x & 4y & 4z \\ 3 & -2 & -1 \end{vmatrix} = \begin{vmatrix} \mathbf{i} & \mathbf{j} & \mathbf{k} \\ 2 & 4 & 4 \\ 3 & -2 & -1 \end{vmatrix} = (4, 14, -16)$$

or, more simply, $(2, 7, -8)$. Hence, equations for the tangent line are $x = 1 + 2t$, $y = 1 + 7t$, $z = 1 - 8t$.

42.120 Write an equation for the normal plane at (x_0, y_0, z_0) to the curve $\mathscr{C}$ of Problem 42.118.

▌ If (x, y, z) is a point of the normal plane, the vector $\mathbf{C} = (x - x_0, \; y - y_0, \; z - z_0)$ is orthogonal to the tangent vector $\mathbf{A} \times \mathbf{B}$ at (x_0, y_0, z_0). Thus, the normal plane is given by

$$0 = \mathbf{C} \cdot (\mathbf{A} \times \mathbf{B}) = \begin{vmatrix} x - x_0 & y - y_0 & z - z_0 \\ F_x & F_y & F_z \\ G_x & G_y & G_z \end{vmatrix}$$

where the derivatives are evaluated at (x_0, y_0, z_0).

42.121 Find an equation of the normal plane to the curve that is the intersection of $9x^2 + 4y^2 - 36z = 0$ and $3x + y + z - z^2 - 1 = 0$, at the point $(2, -3, 2)$.

▌ By Problem 42.120, an equation is

$$0 = \begin{vmatrix} x - 2 & y + 3 & z - 2 \\ 36 & -24 & -36 \\ 3 & 1 & -3 \end{vmatrix} = \begin{vmatrix} x + z - 4 & y + 3 & z - 2 \\ 0 & -24 & -36 \\ 0 & 1 & -3 \end{vmatrix} = 108(x + z - 4), \quad \text{or} \quad x + z = 4$$

42.122 Show that the surfaces $x^2 + y^2 + z^2 = 18$ and $xy = 9$ are tangent at $(3, 3, 0)$.

▌ We must show that the surfaces have the same tangent plane at $(3, 3, 0)$, or, equivalently, that they have parallel normal vectors at $(3, 3, 0)$. A normal vector to the sphere $x^2 + y^2 + z^2 = 18$ is $(2x, 2y, 2z) = (6, 6, 0)$. A normal vector to the cylindrical surface $xy = 9$ is $(y, x, 0) = (3, 3, 0)$. Since $(6, 6, 0)$ and $(3, 3, 0)$ are parallel, the surfaces are tangent.

42.123 Show that the surfaces $x^2 + y^2 + z^2 - 8x - 8y - 6z + 24 = 0$ and $x^2 + 3y^2 + 2z^2 = 9$ are tangent at $(2, 1, 1)$.

▌ The first surface has normal vector $(2x - 8, \; 2y - 8, \; 2z - 6) = (-4, -6, -4)$, and the second surface has normal vector $(2x, 6y, 4z) = (4, 6, 4)$. Since $(4, 6, 4)$ and $(-4, -6, -4)$ are parallel, the surfaces are tangent at $(2, 1, 1)$.

42.124 Show that the surfaces $x^2 + 2y^2 - 4z^2 = 8$ and $4x^2 - y^2 + 2z^2 = 14$ are perpendicular at the point $(2, 2, 1)$.

▌ It suffices to show that the tangent planes are perpendicular, or, equivalently, that the normal vectors are perpendicular. A normal vector to the first surface is $\mathbf{A} = (2x, 4y, -8z) = (4, 8, -8)$, and a normal vector to the second surface is $\mathbf{B} = (8x, -2y, 4z) = (16, -4, 4)$. Since $\mathbf{A} \cdot \mathbf{B} = 4(16) + 8(-4) + (-8)4 = 0$, $\mathbf{A}$ and $\mathbf{B}$ are perpendicular.

42.125 Show that the three surfaces $\mathscr{S}_1$: $14x^2 + 11y^2 + 8z^2 = 66$, $\mathscr{S}_2$: $3z^2 - 5x + y = 0$, $\mathscr{S}_3$: $xy + yz - 4zx = 0$ are mutually perpendicular at the point $(1, 2, 1)$.

❚ A normal vector to $\mathscr{S}_1$ is $\mathbf{A} = (28x, 22y, 16z) = (28, 44, 16)$. A normal vector to $\mathscr{S}_2$ is $\mathbf{B} = (-5, 1, 6z) = (-5, 1, 6)$. A normal vector to $\mathscr{S}_3$ is $\mathbf{C} = (y - 4z,\ \ x + z,\ \ y - 4x) = (-2, 2, -2)$. Since $\mathbf{A} \cdot \mathbf{B} = 0$, $\mathbf{A} \cdot \mathbf{C} = 0$, and $\mathbf{B} \cdot \mathbf{C} = 0$, the normal vectors are mutually perpendicular and, therefore, so are the surfaces.

42.126 Show that the sum of the intercepts of the tangent plane to the surface $x^{1/2} + y^{1/2} + z^{1/2} = a^{1/2}$ at any of its points is equal to a.

❚ A normal vector at (x_0, y_0, z_0) is $(1/2\sqrt{x_0}, 1/2\sqrt{y_0}, 1/2\sqrt{z_0})$, or, more simply, $(1/\sqrt{x_0}, 1/\sqrt{y_0}, 1/\sqrt{z_0})$. Hence, an equation of the tangent plane at (x_0, y_0, z_0) is $\dfrac{1}{\sqrt{x_0}}(x - x_0) + \dfrac{1}{\sqrt{y_0}}(y - y_0) + \dfrac{1}{\sqrt{z_0}}(z - z_0) = 0$, or, equivalently, $\dfrac{1}{\sqrt{x_0}}x + \dfrac{1}{\sqrt{y_0}}y + \dfrac{1}{\sqrt{z_0}}z = \sqrt{x_0} + \sqrt{y_0} + \sqrt{z_0} = \sqrt{a}$. Thus, the x-intercept is $\sqrt{a}\,\sqrt{x_0}$, the y-intercept is $\sqrt{a}\,\sqrt{y_0}$, and the z-intercept is $\sqrt{a}\,\sqrt{z_0}$. Therefore, the sum of the intercepts is $\sqrt{a}(\sqrt{x_0} + \sqrt{y_0} + \sqrt{z_0}) = \sqrt{a}\,\sqrt{a} = a$.

CHAPTER 43

Directional Derivatives and the Gradient. Extreme Values

43.1 Given a function $f(x, y)$ and a unit vector $\mathbf{u}$, define the *derivative of f in the direction* $\mathbf{u}$, at a point (x_0, y_0), and state its connection with the gradient $\nabla f = (f_x, f_y)$.

    ❚ The ray in the direction $\mathbf{u}$ starting at the point (x_0, y_0) is $(x_0, y_0) + t\mathbf{u}$ $(t \geq 0)$. The (directional) derivative, at (x_0, y_0), of the function f in the direction $\mathbf{u}$ is the rate of change, at (x_0, y_0), of f along that ray, that is, $\dfrac{d}{dt} f((x_0, y_0) + t\mathbf{u})$, evaluated at $t = 0$. It is equal to $\nabla f \cdot \mathbf{u}$, the scalar projection of the gradient on $\mathbf{u}$. (Similar definitions and results apply for functions f of three or more variables.)

43.2 Show that the gradient ∇f is the direction in which the derivative achieves its maximum, $|\nabla f|$, and $-\nabla f$ is the direction in which the derivative achieves its minimum, $-|\nabla f|$.

    ❚ The derivative in the direction of a unit vector $\mathbf{u}$ is $\nabla f \cdot \mathbf{u} = |\nabla f| \cos \theta$, where θ is the angle between ∇f and $\mathbf{u}$. Since $\cos \theta$ takes on its maximum value 1 when $\theta = 0$, the maximum value of $\nabla f \cdot \mathbf{u}$ is obtained when $\mathbf{u}$ is the unit vector in the direction of ∇f. That maximum value is $|\nabla f|$. Similarly, since $\cos \theta$ takes on its minimum value -1 when $\theta = \pi$, the minimum value of $\nabla f \cdot \mathbf{u}$ is attained when $\mathbf{u}$ has the direction of $-\nabla f$. That minimum value is $-|\nabla f|$.

43.3 Find the derivative of $f(x, y) = 2x^2 - 3xy + 5y^2$ at the point $(1, 2)$ in the direction of the unit vector $\mathbf{u}$ making an angle of $45°$ with the positive x-axis.

    ❚ $\mathbf{u} = (\cos 45°, \sin 45°) = (\sqrt{2}/2, \sqrt{2}/2)$. $\nabla f = (4x - 3y, -3x + 10y) = (-2, 17)$ at $(1, 2)$. Hence, the derivative is $\nabla f \cdot \mathbf{u} = (-2, 17) \cdot (\sqrt{2}/2, \sqrt{2}/2) = \dfrac{\sqrt{2}}{2} (15) = \dfrac{15\sqrt{2}}{2}$.

43.4 Find the derivative of $f(x, y) = x - \sin xy$ at $(1, \pi/2)$ in the direction of the unit vector $\mathbf{u} = \left(\cos \dfrac{\pi}{3}, \sin \dfrac{\pi}{3} \right) = (\frac{1}{2}, \sqrt{3}/2)$.

    ❚ $\nabla f = (1 - y \cos xy, -x \cos xy) = (1, 0)$. Hence, the derivative is $\nabla f \cdot (\frac{1}{2}, \sqrt{3}/2) = \frac{1}{2}(1, 0) \cdot (1, \sqrt{3}) = \frac{1}{2}$.

43.5 Find the derivative of $f(x, y) = xy^2$ at $(1, 3)$ in the direction toward $(4, 5)$.

    ❚ The indicated direction is that of the vector $(4, 5) - (1, 3) = (3, 2)$. The unit vector in that direction is $\mathbf{u} = (1/\sqrt{13})(3, 2)$. The gradient $\nabla f = (y^2, 2xy) = (9, 6)$. So, the derivative is $\nabla f \cdot \mathbf{u} = (9, 6) \cdot \dfrac{1}{\sqrt{13}}(3, 2) = \dfrac{1}{\sqrt{13}} 39 = 3\sqrt{13}$.

43.6 Find the derivative of $f(x, y, z) = x^2 y^2 z$ at $(2, 1, 4)$ in the direction of the vector $(1, 2, 2)$.

    ❚ The unit vector in the indicated direction is $\mathbf{u} = \frac{1}{3}(1, 2, 2)$. $\nabla f = (2xy^2z, 2x^2yz, x^2y^2) = (16, 32, 64) = 16(1, 2, 4)$. Hence, the derivative is $\nabla f \cdot \mathbf{u} = 16(1, 2, 4) \cdot \frac{1}{3}(1, 2, 2) = \frac{16}{3}(13) = \frac{208}{3}$.

43.7 Find the derivative of $f(x, y, z) = x^3 + y^3 z$ at $(-1, 2, 1)$ in the direction toward $(0, 3, 3)$.

    ❚ A vector in the indicated direction is $(0, 3, 3) - (-1, 2, 1) = (1, 1, 2)$. The unit vector in that direction is $\mathbf{u} = \dfrac{1}{\sqrt{6}}(1, 1, 2)$. $\nabla f = (3x^2, 3y^2z, y^3) = (3, 12, 8)$. Hence, the directional derivative is $\nabla f \cdot \mathbf{u} = (3, 12, 8) \cdot \dfrac{1}{\sqrt{6}}(1, 2, 2) = \dfrac{1}{\sqrt{6}}(31)$.

43.8 On a hill represented by $z = 8 - 4x^2 - 2y^2$, find the direction of the steepest grade at $(1, 1, 2)$.

    ❚ The steepest grade is the maximum directional derivative. By Problem 43.2, this occurs in the direction of $\nabla z = (-8x, -4y) = (-8, -4)$. Note that this is the direction of steepest *ascent* at $(1, 1, 2)$.

43.9 For the surface of Problem 43.8, find the direction of the level curve through $(1, 1)$ and show that it is perpendicular to the direction of steepest grade.

▮ The level curve $z = 2$ is $8 - 4x^2 - 2y^2 = 2$, or $2x^2 + y^2 = 3$. By implicit differentiation, $4x + 2y \dfrac{dy}{dx} = 0$, $dy/dx = -2x/y = -2$. Hence, a tangent vector is $(1, -2)$. A vector in the direction of steepest grade is $(-8, -4)$, by Problem 43.8. Because $(-8, -4) \cdot (1, -2) = 0$, the two directions are perpendicular.

43.10 Show that the sum of the squares of the directional derivatives of $z = f(x, y)$ at any point is constant for any two mutually perpendicular directions and is equal to the square of the gradient.

▮ Let **u** and **v** be two perpendicular unit vectors. The sum S of the squares of the derivatives in these directions is $S = (\nabla f \cdot \mathbf{u})^2 + (\nabla f \cdot \mathbf{v})^2 = |\nabla f|^2 \cos^2 \theta_1 + |\nabla f|^2 \cos^2 \theta_2 = |\nabla f|^2 (\cos^2 \theta_1 + \cos^2 \theta_2)$, where θ_1 and θ_2 are the angles between ∇f and **u**, and ∇f and **v**, respectively. Since $\mathbf{u} \perp \mathbf{v}$, we may assume that $\theta_1 - \theta_2 = \pm \pi/2$. So, $\cos \theta_1 = \cos [\theta_2 \pm (\pi/2)] = \pm \sin \theta_2$. Hence, $\cos^2 \theta_1 + \cos^2 \theta_2 = \sin^2 \theta_2 + \cos^2 \theta_2 = 1$. Thus, $S = |\nabla f|^2$.

43.11 Find the derivative of $z = x \ln y$ at the point $(1, 2)$ in the direction making an angle of $30°$ with the positive x-axis.

▮ The unit vector in the given direction is $\mathbf{u} = (\cos 30°, \sin 30°) = (\sqrt{3}/2, \tfrac{1}{2})$. $\nabla f = (\ln y, x/y) = (\ln 2, \tfrac{1}{2})$. So, the derivative is $\nabla f \cdot \mathbf{u} = (\ln 2, \tfrac{1}{2}) \cdot \tfrac{1}{2}(\sqrt{3}, 1) = \tfrac{1}{2}(\sqrt{3} \ln 2 + \tfrac{1}{2})$.

43.12 If the electric potential V at any point (x, y) is $V = \ln \sqrt{x^2 + y^2}$, find the rate of change of V at $(3, 4)$ in the direction toward the point $(2, 6)$.

▮ A vector in the given direction is $(-1, 2)$, and a corresponding unit vector is $\mathbf{u} = (1/\sqrt{5})(-1, 2)$, $\nabla V = \left(\dfrac{x}{x^2 + y^2}, \dfrac{y}{x^2 + y^2} \right) = (\tfrac{3}{25}, \tfrac{4}{25})$. Hence, the rate of change of V in the specified direction is $\nabla V \cdot \mathbf{u} = \tfrac{1}{25}(3, 4) \cdot \dfrac{1}{\sqrt{5}} (-1, 2) = \dfrac{1}{25\sqrt{5}} (5) = \dfrac{\sqrt{5}}{25}$.

43.13 If the temperature is given by $f(x, y, z) = 3x^2 - 5y^2 + 2z^2$ and you are located at $(\tfrac{1}{3}, \tfrac{1}{5}, \tfrac{1}{2})$ and want to get cool as soon as possible, in which direction should you set out?

▮ $\nabla f = (6x, -10y, 4z) = (2, -2, 2) = 2(1, -1, 1)$. The direction in which f decreases the most rapidly is that of $-\nabla f$; thus, you should move in the direction of the vector $(-1, 1, -1)$.

43.14 Prove that the gradient ∇F of a function $F(x, y, z)$ at a point $P(x_0, y_0, z_0)$ is perpendicular to the level surface of F going through P.

▮ Let $F(x_0, y_0, z_0) = k$. Then the level surface through P is $F(x, y, z) = k$. By Problem 42.109, a normal vector to that surface is (F_x, F_y, F_z), which is just ∇F.

43.15 In what direction should one initially travel, starting at the origin, to obtain the most rapid rate of increase of the function $f(x, y, z) = (3 - x + y)^2 + (4x - y + z + 2)^3$?

▮ The appropriate direction is that of $\nabla f = (f_x, f_y, f_z) = (2(3 - x + y)(-1) + 3(4x - y + z + 2)^2(4),\ 2(3 - x + y) + 3(4x - y + z + 2)^2(-1),\ 3(4x - y + z + 2)^2) = (42, -6, 12) = 6(7, -1, 2)$.

43.16 If $f(x, y, z) = x^3 + y^3 - z$, find the rate of change of f at the point $(1, 1, 2)$ along the line $\dfrac{x - 1}{3} = \dfrac{y - 1}{2} = \dfrac{z - 2}{-2}$ in the direction of decreasing x.

▮ A vector parallel to the given line is $(3, 2, -2)$. Since the first component, 3, is positive, the vector is pointing in the direction of increasing x. Hence, we want the opposite vector $(-3, -2, 2)$. A unit vector in that direction is $\mathbf{u} = \dfrac{1}{\sqrt{17}} (-3, -2, 2)$. Hence, the required rate of change is $\nabla f \cdot \mathbf{u}$. Since $\nabla f = (3x^2, 3y^2, -1) = (3, 3, -1)$, we get $(3, 3, -1) \cdot \dfrac{1}{\sqrt{17}} (-3, -2, 2) = \dfrac{1}{\sqrt{17}} (-9 - 6 - 2) = -\sqrt{17}$.

43.17 For the function f of Problem 43.16, find the rate of change of f at $(4, 1, 0)$ along the normal line to the plane $3(x - 4) - (y - 1) + 2z = 0$ in the direction of increasing x.

▌ A vector parallel to the normal line to the given plane is $(3, -1, 2)$ and it is pointing in the direction of increasing x. The unit vector in this direction is $\mathbf{u} = \dfrac{1}{\sqrt{14}} (3, -1, 2)$. Since $\nabla f = (8, 8, -8)$, the rate of change is $\nabla f \cdot \mathbf{u} = 8(1, 1, -1) \cdot \dfrac{1}{\sqrt{14}} (3, -1, 2) = \dfrac{8}{\sqrt{14}} (0) = 0$.

43.18 For functions $f(x, y, z)$ and $g(x, y, z)$, prove $\nabla(f + g) = \nabla f + \nabla g$.

▌ $\nabla f + \nabla g = (f_x, f_y, f_z) + (g_x, g_y, g_z) = ((f + g)_x, (f + g)_y, (f + g)_z) = \nabla(f + g)$.

43.19 Prove $\nabla(fg) = f \nabla g + g \nabla f$.

▌ $\nabla(fg) = (fg_x + f_x g, \ fg_y + f_y g, \ fg_z + f_z g) = (fg_x, fg_y, fg_z) + (f_x g, f_y g, f_z g) = f \nabla g + g \nabla f$.

43.20 Prove $\nabla(f^n) = nf^{n-1}\nabla f$.

▌ The proof is by induction on n. The result is clearly true when $n = 1$. Assume the result true for n. Then

$$\nabla(f^{n+1}) = \nabla(f^n \cdot f) = f^n \nabla f + \nabla(f^n)f \quad \text{(by Problem 43.19)}$$

$$= f^n \nabla f + nf^{n-1}\nabla f(f) \quad \text{(by the inductive hypothesis)}$$

$$= f^n \nabla f + nf^n \nabla f = (n + 1)f^n \nabla f$$

Thus, the identity also holds for $n + 1$.

43.21 Prove $\nabla\left(\dfrac{f}{g}\right) = \dfrac{g\nabla f - f\nabla g}{g^2}$.

▌ By Problem 43.19, $\nabla f = \nabla\left(g \cdot \dfrac{f}{g}\right) = g\nabla\left(\dfrac{f}{g}\right) + \dfrac{f}{g} \nabla g$. Now solve for $\nabla(f/g)$.

43.22 If $z = f(x, y)$ has a relative maximum or minimum at a point (x_0, y_0), show that $\nabla z = \mathbf{0}$ at (x_0, y_0).

▌ We wish to show that both partial derivatives of z vanish at (x_0, y_0). The plane $y = y_0$ intersects the surface $z = f(x, y)$ in a curve $z = f(x, y_0)$ that has a relative maximum (or minimum) at $x = x_0$. Hence, $\partial z/\partial x = 0$ at (x_0, y_0). Similarly, the plane $x = x_0$ intersects the surface $z = f(x, y)$ in a curve $z = f(x_0, y)$ that has a relative maximum (or minimum) at $y = y_0$. Hence, the derivative $\partial z/\partial y = 0$ at (x_0, y_0). Similar results hold for functions f of more than two variables.

43.23 Assume that $f(x, y)$ has continuous second partial derivatives in a disk containing the point (x_0, y_0) inside it. Assume also that (x_0, y_0) is a critical point of f, that is, $f_x = f_y = 0$ at (x_0, y_0). Let $\Delta \equiv f_{xx}f_{yy} - (f_{xy})^2$ (the *Hessian determinant*). State sufficient conditions for f to have a relative maximum or minimum at (x_0, y_0).

▌ **Case 1.** Assume $\Delta > 0$ at (x_0, y_0). (a) If $f_{xx} + f_{yy} < 0$ at (x_0, y_0), then f has a relative maximum at (x_0, y_0). (b) If $f_{xx} + f_{yy} > 0$ at (x_0, y_0), then f has a relative minimum at (x_0, y_0). **Case 2.** If $\Delta < 0$, f has neither a relative maximum nor a relative minimum at (x_0, y_0). **Case 3.** If $\Delta = 0$, no conclusions can be drawn.

43.24 Using the assumptions and notation of Problem 43.23, give examples to show that case 3, where $\Delta = 0$, allows no conclusions to be drawn.

▌ Each of the functions $f_1(x, y) = x^4 + y^4$, $f_2(x, y) = -(x^4 + y^4)$, and $f_3(x, y) = x^3 - y^3$ vanishes at $(0, 0)$ together with its first and second partials; hence $\Delta(0, 0) = 0$ for each function. But, at $(0, 0)$, f_1 has a relative minimum, f_2 has a relative maximum, and f_3 has neither $[f_3(x, 0) = x^3$ takes on both positive and negative values in any neighborhood of the origin].

43.25 Find the relative maxima and minima of the function $f(x, y) = 2x + 4y - x^2 - y^2 - 3$.

▌ $f_x = 2 - 2x$, $f_y = 4 - 2y$. Setting $f_x = 0$, $f_y = 0$, we have $x = 1$, $y = 2$. Thus, $(1, 2)$ is the only critical point. $f_{xy} = 0$, $f_{xx} = -2$, and $f_{yy} = -2$. So, $\Delta = f_{xx}f_{yy} - (f_{xy})^2 = 4 > 0$. Since $f_{xx} + f_{yy} = -4 < 0$, there is a relative maximum at $(1, 2)$, by Problem 43.23.

43.26 Find the relative maxima and minima of $f(x, y) = x^3 + y^3 - 3xy$.

▮ $f_x = 3x^2 - 3y$, $f_y = 3y^2 - 3x$. Setting $f_x = 0$, $f_y = 0$, we have $x^2 = y$ and $y^2 = x$. So, $y^4 = y$ and, therefore, either $y = 0$ or $y = 1$. So, the critical points are $(0, 0)$ and $(1, 1)$. $f_{xy} = -3$, $f_{xx} = 6x$, $f_{yy} = 6y$. At $(0, 0)$, $\Delta = f_{xx}f_{yy} - (f_{xy})^2 = -9 < 0$. Therefore, by Problem 43.23, there is neither a relative maximum nor a relative minimum at $(0, 0)$. At $(1, 1)$, $\Delta = 36 - 9 > 0$. Also, $f_{xx} + f_{yy} = 12 > 0$. Hence, by Problem 43.23, there is a relative minimum at $(1, 1)$.

43.27 Find all relative maxima and minima of $f(x, y) = x^2 + 2xy + 2y^2$.

▮ Since $f(x, y) = (x + y)^2 + y^2 \geq 0$, there is an absolute minimum 0 at $(0, 0)$. $f_x = 2x + 2y$, $f_y = 2x + 4y$. Setting $f_x = 0$, $f_y = 0$, we get $x + y = 0$, $x + 2y = 0$, and, therefore, $x = y = 0$. Thus, $(0, 0)$ is the only critical point—the only possible site of an extremum.

43.28 Find all relative maxima and minima of $f(x, y) = (x - y)(1 - xy)$.

▮ $f(x, y) = x - y - x^2y + xy^2$. Then, $f_x = 1 - 2xy + y^2$, $f_y = -1 - x^2 + 2xy$. Setting $f_x = 0$, $f_y = 0$, we have $2xy = 1 + y^2$, $2xy = 1 + x^2$. Therefore, $1 + y^2 = 1 + x^2$, $y^2 = x^2$, $y = \pm x$. Hence, $\pm 2x^2 = 1 + x^2$. $-2x^2 = 1 + x^2$ is impossible. So, $2x^2 = 1 + x^2$, $x^2 = 1$, $x = \pm 1$. Thus, the critical points are $(1, 1)$, $(1, -1), (-1, 1), (-1, -1)$. $f_{xy} = -2x + 2y$, $f_{xx} = -2y$, $f_{yy} = 2x$. Hence, $\Delta = -4xy - 4(y - x)^2$. For $(1, 1)$ and $(-1, -1)$, $\Delta = -4 < 0$. For $(1, -1)$ and $(-1, 1)$, $\Delta = -12 < 0$. Hence, by Problem 43.23, there are no relative maxima or minima.

43.29 Find all relative maxima and minima of $f(x, y) = 2x^2 + y^2 + 6xy + 10x - 6y + 5$.

▮ $f_x = 4x + 6y + 10$, $f_y = 2y + 6x - 6$. Setting $f_x = 0$ and $f_y = 0$, we get $2x + 3y + 5 = 0$ and $3x + y - 3 = 0$. Solving simultaneously, we have $x = 2$, $y = -3$. Further, $f_{xy} = 6$, $f_{xx} = 4$, $f_{yy} = 2$. So, $\Delta = f_{xx}f_{yy} - (f_{xy})^2 = -28 < 0$. By Problem 43.23, there is no relative maximum or minimum.

43.30 Find all relative maxima and minima of $f(x, y) = xy(2x + 4y + 1)$.

▮ $f(x, y) = 2x^2y + 4xy^2 + xy$. Then $f_x = 4xy + 4y^2 + y$, $f_y = 2x^2 + 8xy + x$. Setting $f_x = 0$, we get $y = 0$ or $4x + 4y + 1 = 0$. Setting $f_y = 0$, we get $x = 0$ or $2x + 8y + 1 = 0$. If $4x + 4y + 1 = 0$ and $2x + 8y + 1 = 0$, then $x = \frac{1}{6}$ and $y = -\frac{1}{12}$. Therefore, the critical points are $(0, 0)$, $(0, -\frac{1}{4})$, $(-\frac{1}{2}, 0)$, and $(-\frac{1}{6}, -\frac{1}{12})$. Now, $f_{xy} = 4x + 8y + 1$, $f_{yy} = 8x$, $f_{xx} = 4y$. Hence, $\Delta = f_{xx}f_{yy} - (f_{xy})^2 = 32xy - (4x + 8y + 1)^2$. Now use Problem 43.23. For $(0, 0)$, $(0, -\frac{1}{4})$, and $(-\frac{1}{2}, 0)$, $\Delta = -1 < 0$, and, therefore, there is no relative extremum. For $(-\frac{1}{6}, -\frac{1}{12})$, $\Delta = \frac{1}{3} > 0$. Moreover, $f_{xx} + f_{yy} = -\frac{5}{3} < 0$. So, there is a relative maximum at $(-\frac{1}{6}, -\frac{1}{12})$.

43.31 Find the shortest distance between the lines $\dfrac{x - 2}{4} = \dfrac{y + 1}{-7} = \dfrac{z + 1}{1}$ and $\dfrac{x - 2}{-2} = \dfrac{y - 1}{1} = \dfrac{z - 2}{-3}$.

▮ Write the first line in parametric form as $x = 2 + 4t$, $y = -1 - 7t$, $z = -1 + t$, and the second line as $x = 2 - 2s$, $y = 1 + s$, $z = 2 - 3s$. For any t and s, the distance between the corresponding points on the two lines is $\sqrt{(4t + 2s)^2 + (2 + s + 7t)^2 + (-3 + 3s + t)^2} = \sqrt{66t^2 + 14s^2 + 36st - 14s + 22t + 13}$. Minimizing this quantity is equivalent to minimizing its square, $f(s, t) = 66t^2 + 14s^2 + 36st - 14s + 22t + 13$. $f_s = 28s + 36t - 14$, $f_t = 132t + 36s + 22$. Solving $f_s = 0$, $f_t = 0$, we find $s = \frac{11}{10}$, $t = -\frac{7}{15}$. Also, $f_{st} = 36$, $f_{ss} = 28$, $f_{tt} = 132$, $\Delta = (28)(132) - (36)^2 = 2400 > 0$. $f_{ss} + f_{tt} = 160 > 0$. Thus, by Problem 43.23, we have a relative minimum and, by geometric intuition, we know it is an absolute minimum. Substitution of $s = \frac{11}{10}$, $t = -\frac{7}{15}$ in the distance formula above yields the distance $\sqrt{6}/6$ between the lines.

43.32 Find positive numbers x, y, z such that $x + y + z = 18$ and xyz is a maximum.

▮ $xyz = xy(18 - x - y) = 18xy - x^2y - xy^2$. Hence, we must maximize $f(x, y) = 18xy - x^2y - xy^2$ over the triangle $x > 0$, $y > 0$, $x + y < 18$ (Fig. 43-1). $f_x = 18y - 2xy - y^2$, $f_y = 18x - x^2 - 2xy$. Setting $f_x = 0$, $f_y = 0$, and then subtracting, we get $(y - x)(18 - x - y) = 0$. Since $z = 18 - x - y > 0$, $y - x = 0$, that is, $y = x$. Substituting in $18y - 2xy - y^2 = 0$, we find that $y = 6$. Hence, $x = 6$ and $z = 6$. $f_{xy} = 18 - 2x - 2y = -6$, $f_{xx} = -2y = -12$, $f_{yy} = -2x = -12$. So, $\Delta = 144 - 36 > 0$. Also, $f_{xx} + f_{yy} = -24 < 0$. Hence, $(6, 6)$ yields the relative maximum 6^3 for $f(x, y)$. That this is actually an absolute maximum follows from the facts that (i) the continuous function $f(x, y)$ must have an absolute maximum on the *closed* triangle of Fig. 43-1; (ii) on the boundary of the triangle, $f(x, y) = 0$.

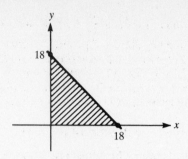

Fig. 43-1

43.33 Find positive numbers x, y, z such that $xyz = 64$ and $x + y + z$ is a minimum.

▮ Instead of following the solution to Problem 43.32, let us apply the theorem of the means (Problem 43.51):

$$\sqrt[3]{xyz} \le \frac{x + y + z}{3} \quad \text{or} \quad x + y + z \ge 3\sqrt[3]{xyz} = 3\sqrt[3]{64} = 12$$

Equality—and hence an absolute minimum—is obtained for $x = y = z = 4$.

43.34 Find positive numbers x, y, z such that $x + y + z = 20$ and xyz^2 is a maximum.

▮ $xyz^2 = (20 - y - z)yz^2 = 20yz^2 - y^2z^2 - yz^3$. We must maximize this function $f(y, z)$ under the conditions $y > 0$, $z > 0$, $y + z < 20$. $f_y = 20z^2 - 2yz^2 - z^3$, $f_z = 40yz - 2y^2z - 3yz^2$. Set $f_y = 0$, $f_z = 0$. Then, $z^2(20 - 2y - z) = 0$ and $yz(40 - 2y - 3z) = 0$. Since $y > 0$ and $z > 0$, $20 - 2y - z = 0$ and $40 - 2y - 3z = 0$. Solving these equations simultaneously, we obtain $y = 5$, $z = 10$. Hence, $x = 5$. This point $(5, 5, 10)$ yields an absolute maximum (by an argument similar to that given in Problem 43.32).

43.35 Find the maximum value of xy^2z^3 on the part of the plane $x + y + z = 12$ in the first octant.

▮ We must maximize $f(y, z) = (12 - y - z)y^2z^3 = 12y^2z^3 - y^3z^3 - y^2z^4$ subject to the conditions $y + z < 12$, $y > 0$, $z > 0$. $f_y = 24yz^3 - 3y^2z^3 - 2yz^4$, $f_z = 36y^2z^2 - 3y^3z^2 - 4y^2z^3$. Set $f_y = 0$, $f_z = 0$. Then $yz^3(24 - 3y - 2z) = 0$ and $y^2z^2(36 - 3y - 4z) = 0$. Since $y \ne 0$ and $z \ne 0$, $24 - 3y - 2z = 0$ and $36 - 3y - 4z = 0$. Subtracting, we get $12 - 2z = 0$, $z = 6$. Hence, $y = 4$ and $x = 2$. That this point yields the absolute maximum can be shown by an argument similar to that in Problem 43.32.

43.36 Using gradient methods, find the minimum value of the distance from the origin to the plane $Ax + By + Cz + D = 0$.

▮ The distance from the origin to a point $(x\ y, z)$ in the plane is $\sqrt{x^2 + y^2 + z^2}$. Minimizing this distance is equivalent to minimizing the square of the distance, $x^2 + y^2 + z^2$. At least one of A, B, and C is nonzero; we can rename the coordinates, if need be, so as to make $C \ne 0$. Then $z = (1/C)(D - Ax - By)$, and we must minimize $f(x, y) = x^2 + y^2 + (1/C^2)(D - Ax - By)^2$. $f_x = 2x + (2/C^2)(D - Ax - By)(-A)$, $f_y = 2y + (2/C^2)(D - Ax - By)(-B)$. Set $f_x = 0$ and $f_y = 0$. Then

$$x = \frac{A}{C^2}(D - Ax - By) \qquad y = \frac{B}{C^2}(D - Ax - By) \qquad (*)$$

Hence, $Bx = Ay$. Substitute in $(*)$: $C^2x = AD - A^2x - BAy = AD - Ax^2 - B^2x$, $(A^2 + B^2 + C^2)x = AD$, $x = AD/(A^2 + B^2 + C^2)$. Similarly, $y = BD/(A^2 + B^2 + C^2)$. So, $z = (1/C)(D - Ax - By) = CD/(A^2 + B^2 + C^2)$. Then the minimum distance $\sqrt{x^2 + y^2 + z^2}$ is

$$\sqrt{\frac{D^2(A^2 + B^2 + C^2)}{(A^2 + B^2 + C^2)^2}} = \frac{|D|}{\sqrt{A^2 + B^2 + C^2}}$$

(Compare Problem 40.88.)

43.37 The surface area of a rectangular box without a top is to be 108 cubic feet. Find the greatest possible volume.

▮ Let x, y, z be the length, width, and height. Then the surface area $S = xy + 2xz + 2yz = 108$. The volume $V = xyz$. Think of z as a function of x and y. Then V becomes a function of x and y. $V_x = y\left(x\dfrac{\partial z}{\partial x} + z\right)$, $V_y = x\left(y\dfrac{\partial z}{\partial y} + z\right)$. From $xy + 2xz + 2yz = 108$ by differentiation with respect to x, $y + 2\left(x\dfrac{\partial z}{\partial x} + z\right) + 2y\dfrac{\partial z}{\partial x} = 0$, and, therefore, $\partial z/\partial x = -(2z + y)/[2(x + y)]$. Similarly, $\partial z/\partial y = -(2z + x)/[2(x + y)]$. Set

$V_x = 0$ and $V_y = 0$. Since $y \neq 0$ and $x \neq 0$, $x \dfrac{\partial z}{\partial x} + z = 0$ and $y \dfrac{\partial z}{\partial y} + z = 0$. So, $\partial z/\partial x = -z/x$ and $\partial z/\partial y = -z/y$. Then, $-z/x = -(2z + y)/[2(x + y)]$ and $-z/y = -(2z + x)/[2(x + y)]$. Thus, $2xz + 2yz = 2xz + xy$ and $2xz + 2yz = 2yz + xy$, and, therefore, $z = x/2$ and $z = y/2$. Substitute in $xy + 2xz + 2yz = 108$. Then $4z^2 + 4z^2 + 4z^2 = 108$, $12z^2 = 108$, $z^2 = 9$, $z = 3$. Hence, $x = 6$, $y = 6$. Therefore, the maximum volume is 108 cubic feet. (This can be shown in the usual way to be a relative maximum. An involved argument is necessary to show that it is an absolute maximum.)

43.38 Find the point on the surface $z = xy - 1$ that is nearest the origin.

▎ It suffices to minimize $x^2 + y^2 + z^2$ for arbitrary points (x, y, z) on the surface. Hence, we must minimize $f(x, y) = x^2 + y^2 + (xy - 1)^2$. $f_x = 2x + 2(xy - 1)(y)$, $f_y = 2y + 2(xy - 1)(x)$. Set $f_x = 0$, $f_y = 0$. Then $x + (xy - 1)y = 0$ and $y + (xy - 1)x = 0$. Hence, $x^2 + xy(xy - 1) = 0$ and $y^2 + xy(xy - 1) = 0$. Therefore, $x^2 = y^2$, $x = \pm y$. Substitute in $x + (xy - 1)y = 0$, getting $x + x^3 \pm x = 0$. Hence, $x^3 = 0$ or $x(x^2 + 2) = 0$. In either case, $x = 0$, $y = 0$, $z = -1$. By geometric reasoning, there must be a minimum. Hence, it must be located at $(0, 0, -1)$.

43.39 Find an equation of the plane through $(1, 1, 2)$ that cuts off the least volume in the first octant.

▎ Let a, b, c be the intercepts of a plane through $(1, 1, 2)$. Then an equation of the plane is $\dfrac{x}{a} + \dfrac{y}{b} + \dfrac{z}{c} = 1$. The volume $V = \frac{1}{6}abc$. We can think of c as a function of a and b, and, therefore, of V as a function of a and b. Thus, $V_a = \frac{1}{6}b\left(a\dfrac{\partial c}{\partial a} + c\right)$, $V_b = \frac{1}{6}a\left(b\dfrac{\partial c}{\partial b} + c\right)$. Since the plane contains $(1, 1, 2)$, $\dfrac{1}{a} + \dfrac{1}{b} + \dfrac{2}{c} = 1$. Hence, by differentiation with respect to a, $-\dfrac{1}{a^2} - \dfrac{2}{c^2}\dfrac{\partial c}{\partial a} = 0$, and, by differentiation with respect to b, $-\dfrac{1}{b^2} - \dfrac{1}{c^2}\dfrac{\partial c}{\partial b} = 0$. Therefore, $\partial c/\partial a = -c^2/2a^2$ and $\partial c/\partial b = -c^2/2b^2$. Set $V_a = 0$ and $V_b = 0$. Then $a\dfrac{\partial c}{\partial a} + c = 0$, $b\dfrac{\partial c}{\partial b} + c = 0$, or $a[-(c^2/2a^2)] + c = 0$ and $b[-(c^2/2b^2)] + c = 0$, that is, $c = c^2/2a$ and $c = c^2/2b$. Therefore, $2a = 2b$, $b = a$. From $c = c^2/2a$, $c = 2a$. Substitute in $\dfrac{1}{a} + \dfrac{1}{b} + \dfrac{2}{c} = 1$. Then, $\dfrac{1}{a} + \dfrac{1}{a} + \dfrac{1}{a} = 1$, and so, $a = 3$. Hence, $b = 3$, $c = 6$. Thus the desired equation is $\dfrac{x}{3} + \dfrac{y}{3} + \dfrac{z}{6} = 1$, or $2x + 2y + z = 6$.

43.40 Determine the values of p and q so that the sum S of the squares of the vertical distances of the points $(0, 2)$, $(1, 3)$, and $(2, 5)$ from the line $y = px + q$ shall be a minimum.

▎ We must minimize $S = (q - 2)^2 + (p + q - 3)^2 + (2p + q - 5)^2$. $S_q = 2(q - 2) + 2(p + q - 3) + 2(2p + q - 5) = 6q + 6p - 20$. $S_p = 2(p + q - 3) + 2(2p + q - 5)(2) = 10p + 6q - 26$. Let $S_q = 0$, $S_p = 0$. Then $3q + 3p - 10 = 0$, $5p + 3q - 13 = 0$. So, $2p - 3 = 0$, $p = \frac{3}{2}$, $q = \frac{11}{6}$.

43.41 Show that a rectangular box (with top) of maximum volume V having prescribed surface area S is a cube.

▎ Let x, y, z be the length, width, and height, respectively. $S = 2xy + 2xz + 2yz$. We must maximize $V = xyz$. Think of z as a function of x and y; then so is V. Now,

$$V_x = y\left(x\dfrac{\partial z}{\partial x} + z\right) = yz + yx\dfrac{\partial z}{\partial x} \quad\text{and}\quad V_y = x\left(y\dfrac{\partial z}{\partial y} + z\right) = xz + xy\dfrac{\partial z}{\partial y}$$

Differentiate the surface area equation with respect to x and to y:

$$0 = 2y + 2\left(x\dfrac{\partial z}{\partial x} + z\right) + 2y\dfrac{\partial z}{\partial x} \tag{1}$$

$$0 = 2x + 2\left(y\dfrac{\partial z}{\partial y} + z\right) + 2x\dfrac{\partial z}{\partial y} \tag{2}$$

From (1), $\partial z/\partial x = -(y + z)/(x + y)$. From (2), $\partial z/\partial y = -(x + z)/(x + y)$. Then

$$V_x = yz - yx\left(\dfrac{y + z}{x + y}\right) \quad\text{and}\quad V_y = xz - xy\left(\dfrac{x + z}{x + y}\right)$$

Set $V_x = V_y = 0$. Then $z = x[(y + z)/(x + y)]$ and $z = y[(x + z)/(x + y)]$, $zx + zy = xy + xz$ and $xz + yz = yx + yz$, $zy = xy$ and $xz = yx$, $z = x$ and $z = y$. Thus, $x = y = z$ and the box is a cube.

43.42 Find the volume V of the largest rectangular box that may be inscribed in the ellipsoid $\dfrac{x^2}{a^2} + \dfrac{y^2}{b^2} + \dfrac{z^2}{c^2} = 1$.

❚ Let (x, y, z) be the vertex in the first octant. Then the sides must have lengths $2x, 2y, 2z$, and the volume $V = 8xyz$. Think of z as a function of x and y. Then V is a function of x and y. $V_x = 8y\left(x\dfrac{\partial z}{\partial x} + z\right)$, $V_y = 8x\left(y\dfrac{\partial z}{\partial y} + z\right)$. Differentiating the equation of the ellipsoid, we get $\dfrac{2x}{a^2} + \dfrac{2z}{c^2}\dfrac{\partial z}{\partial x} = 0$ and $\dfrac{2y}{b^2} + \dfrac{2z}{c^2}\dfrac{\partial z}{\partial y} = 0$. Thus, $\dfrac{\partial z}{\partial x} = -\dfrac{c^2}{z}\dfrac{x}{a^2}$ and $\dfrac{\partial z}{\partial y} = -\dfrac{c^2}{z}\dfrac{y}{b^2}$. Hence, $V_x = 8y\left(-\dfrac{c^2}{z}\dfrac{x^2}{a^2} + z\right)$ and $V_y = 8x\left(-\dfrac{c^2}{z}\dfrac{y^2}{b^2} + z\right)$. Set $V_x = V_y = 0$. Then $z = c^2x^2/za^2$, $z = c^2y^2/zb^2$. Therefore, $x^2/a^2 = x^2/c^2 = y^2/b^2$. Substitute in the equation of the ellipsoid: $3x^2 = a^2$, $3z^2 = c^2$, $3y^2 = b^2$. So, $x = a/\sqrt{3}$, $z = c/\sqrt{3}$, $y = b/\sqrt{3}$, and $V = 8abc/3\sqrt{3}$.

43.43 Divide 120 into three nonnegative parts so that the sum of the products taken two at a time is a maximum.

❚ Let $120 = x + y + (120 - x - y)$, with $x \geq 0$, $y \geq 0$. We must maximize $f(x, y) = xy + x(120 - x - y) + y(120 - x - y) = xy + 120x - x^2 - xy + 120y - xy - y^2 = 120x + 120y - x^2 - xy - y^2$. $f_x = 120 - 2x - y$, $f_y = 120 - x - 2y$. Set $f_x = f_y = 0$ and solve: $120 - 2x - y = 0$, $240 - 2x - 4y = 0$. Hence, $120 - 3y = 0$, $y = 40$, $x = 40$. Thus, the division is into three equal parts. Note that $f_{xy} = -1$, $f_{xx} = -2$, $f_{yy} = -2$, $\Delta = 4 - 1 = 3 > 0$, and $f_{xx} + f_{yy} = -4 > 0$. Hence, the critical point $(40, 40)$ yields a relative maximum $f(40, 40) = 4800$. To see that this is an absolute maximum, observe that the continuous function $f(x, y)$ must have an absolute maximum on the closed triangle in the first quadrant bounded above by the line $x + y = 120$ (Fig. 43-2). The only possibility for an absolute maximum inside the triangle occurs at a critical point, and $(40, 40)$ is the only such point. Consider the boundary of the triangle. On the lower leg, where $y = 0$, $f(x, 0) = x(120 - x) = 120x - x^2$, with $0 \leq x \leq 120$. The critical number of this function turns out to be 60. $f(60, 0) = 3600 < 4800 = f(40, 40)$. The endpoints $x = 0$ and $x = 120$ yield 0 as the value of f. Similarly, the other leg of the triangle, where $x = 0$, gives values of f less than $f(40, 40)$. Lastly, on the hypotenuse of the triangle, where $x + y = 120$, $f(x, y) = xy = x(120 - x)$, with $0 \leq x \leq 120$, and the same analysis as above shows that the values of f are less than $f(40, 40)$. Hence, $(40, 40)$ actually does give an absolute maximum.

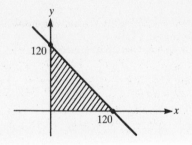

Fig. 43-2

43.44 Find the point in the plane $2x - y + 2x = 16$ nearest the origin.

❚ It suffices to minimize $x^2 + y^2 + z^2$ over all points (x, y, z) in the plane, or, equivalently, to minimize $f(x, y) = x^2 + y^2 + \left(\dfrac{16 - 2x + y}{2}\right)^2$. $f_x = 2x + \frac{1}{2}(16 - 2x + y)(-2)$, $f_y = 2y + \frac{1}{2}(16 - 2x + y)$. Set $f_x = f_y = 0$. $2x = 16 - 2x + y$, $-4y = 16 - 2x + y$. Then $y = 4x - 16$, $-5y = 16 - 2x$. Hence, $-5(4x - 16) = 16 - 2x$, $-20x + 80 = 16 - 2x$, $64 = 18x$, $x = \frac{32}{9}$. Then $y = 4(\frac{32}{9}) - 16 = -\frac{16}{9}$, $z = \dfrac{16 - 2x + y}{2} = \dfrac{1}{2}\left(\dfrac{144 - 64 - 16}{9}\right) = \dfrac{32}{9}$. Thus, the nearest point is $(\frac{32}{9}, -\frac{16}{9}, \frac{32}{9})$.

43.45 Find the relative extrema of $f(x, y) = x^4 + y^3 - 32x - 27y - 1$.

❚ $f_x = 4x^3 - 32$, $f_y = 3y^2 - 27$. Set $f_x = f_y = 0$. Then $x^3 = 8$, $y^2 = 9$, or $x = 2$, $y = \pm 3$. Thus, the critical points are $(2, 3)$ and $(2, -3)$. Applying Problem 43.23, $f_{xy} = 0$, $f_{xx} = 12x^2$, $f_{yy} = 6y$, $\Delta = 72x^2y$. At $(2, 3)$, $\Delta = 72(4)(3) > 0$, and $f_{xx} + f_{yy} = 48 + 18 > 0$. So, there is a relative minimum at $(2, 3)$. At $(2, -3)$, $\Delta = 72(4)(-3) < 0$. Hence, there is no relative extremum at $(2, -3)$.

43.46 Find the shortest distance between the line $x = 2 + 4s$, $y = 4 + s$, $z = 4 + 5s$ and the line $x = 1 - 2t$, $y = -3 + 3t$, $z = 2 + t$.

▮ It suffices to minimize the square of the distance: $f(s, t) = (1 + 4s + 2t)^2 + (7 + s - 3t)^2 + (2 + 5s - t)^2$. $f_s = 2(1 + 4s + 2t)(4) + 2(7 + s - 3t) + 2(2 + 5s - t)(5) = 42 + 84s$. $f_t = 2(1 + 4s + 2t)(2) + 2(7 + s - 3t)(-3) +$ $2(2 + 5s - t)(-1) = -42 + 28t$. Set $f_s = f_t = 0$. Then $s = -\frac{1}{2}$, $t = \frac{3}{2}$. Hence, $f(s, t) = 4 + 4 + 4 = 12$. Therefore, the distance between the lines is $\sqrt{12} = 2\sqrt{3}$.

43.47 Find the shortest distance from the point $(-1, 2, 4)$ to the plane $3x - 4y + 2z + 32 = 0$.

▮ It suffices to minimize the square of the distance between $(-1, 2, 4)$ and an arbitrary point (x, y, z) in the plane: $f(x, y, z) = (x + 1)^2 + (y - 2)^2 + (z - 4)^2$. Think of z as a function of x and y. $f_x = 2(x + 1) +$ $2(z - 4)(\partial z/\partial x)$, $f_y = 2(y - 2) + 2(z - 4)(\partial z/\partial y)$. From the equation of the plane, by differentiation with respect to x and y, respectively, we get $3 + 2(\partial z/\partial x) = 0$ and $-4 + 2(\partial z/\partial y) = 0$. Hence, $\partial z/\partial x = -\frac{3}{2}$ and $\partial z/\partial y = 2$. Set $f_x = f_y = 0$, whence

$$x + 1 + (z - 4)(-\tfrac{3}{2}) = 0 \qquad (1)$$

$$y - 2 + (z - 4)(2) = 0 \qquad (2)$$

Therefore, $\frac{2}{3}(x + 1) = (2 - y)/2$, which yields $4x + 4 = 6 - 3y$, $x = -\frac{3}{4}y + \frac{1}{2}$. From (2), $z = -\frac{1}{2}y + 5$. Substituting in the equation of the plane, we have $3(-\frac{3}{4}y + \frac{1}{2}) - 4y + (-y + 10) + 32 = 0$, $-\frac{29}{4}y + \frac{23}{2} +$ $32 = 0$, $y = 6$. Then $x = -\frac{3}{4}(6) + \frac{1}{2} = -4$, and $z = -\frac{1}{2}(6) + 5 = 2$. So, the closest point in the plane is $(-4, 6, 2)$. Hence, the shortest distance from $(-1, 2, 4)$ to the plane is $\sqrt{(-3)^2 + 4^2 + (-2)^2} = \sqrt{29}$. (This can be checked against the formula obtained in Problem 40.88.)

43.48 Find the shortest distance between the line $\mathcal{L}_1$ through $(1, 0, 1)$ parallel to the vector $(1, 2, 1)$ and the line $\mathcal{L}_2$ through $(2, 1, 4)$ and parallel to $(1, -1, 1)$.

▮ Parametric equations for $\mathcal{L}_1$ are $x = 1 + t$, $y = 2t$, $z = 1 + t$. Parametric equations for $\mathcal{L}_2$ are $x = 2 + s$, $y = 1 - s$, $z = 4 + s$. It suffices to minimize the square of the distance between arbitrary points on the lines: $f(s, t) = (1 + s - t)^2 + (1 - s - 2t)^2 + (3 + s - t)^2$. $f_s = 2(1 + s - t) + 2(1 - s - 2t)(-1) + 2(3 + s - t) = 6 + 6s$. $f_t = 2(1 + s - t)(-1) + 2(1 - s - 2t)(-2) + 2(3 + s - t)(-1) = -12 + 12t$. Set $f_s = f_t = 0$. Then $s = -1$, $t = 1$. So, the shortest distance is $(-1)^2 + (0)^2 + (1)^2 = 2$.

43.49 For which value(s) of k does $f(x, y) = x^2 + kxy + 4y^2$ have a relative minimum at $(0, 0)$?

▮ $f_x = 2x + ky$, $f_y = kx + 8y$. Set $f_x = f_y = 0$. Then $2x + ky = 0$, $kx + 8y = 0$. Multiply the first equation by $-k$, the second equation by 2, and add: $(-k^2 + 16)y = 0$. **Case 1.** $k^2 \neq 16$. Then $y = 0$. Hence, $x = 0$. So, $(0, 0)$ is a critical point. $f_{xy} = k$, $f_{xx} = 2$, $f_{yy} = 8$. Then $\Delta = f_{xx}f_{yy} - (f_{xy})^2 = 16 - k^2$. If $k^2 > 16$, then $\Delta < 0$ and there is no relative extremum. If $k^2 < 16$, $\Delta > 0$. Note that $f_{xx} + f_{yy} = 2 +$ $8 = 10 > 0$. Hence, there is a relative minimum at $(0, 0)$ when $k^2 < 16$. **Case 2.** $k^2 = 16$. $k = \pm 4$. Then $f(x, y) = x^2 \pm 4xy + 4y^2 = (x \pm 2y)^2 \geq 0$. Since $f(0, 0) = 0$, there is an absolute minimum at $(0, 0)$.

43.50 For all positive x, y, z such that $xyz^2 = 2500$, what is the smallest value of $x + y + z$?

▮ This problem is the dual of Problem 43.34; so we already know the solution: $\min (x + y + z) = 20$. Let us verify this solution by the gradient method. We want to minimize $F(x, y) = x + y + (50/\sqrt{xy})$ over $x > 0$, $y > 0$. Setting $F_x = 1 - (25/\sqrt{x^3 y}) = 0$, $F_y = 1 - (25/\sqrt{xy^3}) = 0$, and solving, we obtain $x = y = 5$. Hence, $z = \frac{50}{5} = 10$, and $x + y + z = 20$.

43.51 Prove that the geometric mean of three nonnegative numbers is not greater than their arithmetic mean; that is, $\sqrt[3]{abc} \leq (a + b + c)/3$ for $a, b, c \geq 0$.

▮ This is clear when $a = 0$ or $b = 0$ or $c = 0$. So, we may assume $a, b, c > 0$. Let $d = a + b + c$, and let $x = 18a/d$, $y = 18b/d$, $z = 18c/d$. Then x, y, and z are positive and $x + y + z = 18$. By Problem 43.32, $xyz \leq 6^3$. So, $\frac{a}{d} \frac{b}{d} \frac{c}{d} \leq \frac{6^3}{18^3} = \frac{1}{3^3}$, $abc \leq (d/3)^3$, $\sqrt[3]{abc} \leq d/3 = (a + b + c)/3$. In general, for nonnegative $a_1, \ldots, a_n$, we have $\sqrt[n]{a_1 \cdots a_n} \leq (a_1 + \cdots + a_n)/n$, with equality if and only if $a_1 = \cdots = a_n$.

43.52 Find the relative extrema of $f(x, y) = \frac{A}{x} + \frac{B}{y} + xy$, where $A \neq 0$ and $B \neq 0$.

▮ $f_x = -\dfrac{A}{x^2} + y$, $f_y = -\dfrac{B}{y^2} + x$. Set $f_x = f_y = 0$. Then $y = A/x^2$, $x = B/y^2$. Hence, $yx^2 = A$, $y(B/y^2)^2 = A$, $y^3 = B^2/A$, $y = \sqrt[3]{B^2/A}$. Similarly, $x = \sqrt[3]{A^2/B}$. Thus, the only critical point is $(\sqrt[3]{A^2/B}, \sqrt[3]{B^2/A})$. $f_{xy} = 1$, $f_{xx} = 2A/x^3$, $f_{yy} = 2B/y^3$. Hence, $\Delta = 4AB/x^3y^3 - 1 = 4 - 1 > 0$. In addition, $f_{xx} + f_{yy} = \dfrac{2A}{x^3} + \dfrac{2B}{y^3} = \dfrac{2B}{A} + \dfrac{2A}{B} = \dfrac{2(A^2 + B^2)}{AB}$. Therefore, at the critical point, there is a relative maximum when A and B are of opposite sign and a relative minimum when A and B have the same sign.

43.53 A closed rectangular box costs A cents per square foot for the top and bottom and B cents per square foot for the sides. If the volume V is fixed, what should the dimensions be to minimize the cost?

▮ Let l, w, h be the length, width, and height, respectively. The cost $C = 2Alw + 2B(lh + wh)$. $lwh = V$(constant). Hence, $C = 2Alw + 2B\left(\dfrac{V}{w} + \dfrac{V}{l}\right)$. Then $\partial C/\partial l = 2Aw - (2VB/l^2)$, $\partial C/\partial w = 2Al - (2VB/w^2)$. Set $\partial C/\partial l = \partial C/\partial w = 0$. Then $Awl^2 = VB$, $Aw^2l = VB$. So, $Awl^2 = Aw^2l$. Hence, $l = w$. Therefore, $Al^3 = VB$, $l = 2\sqrt[3]{VB/A}$, $w = 2\sqrt[3]{VB/A}$, $h = 8/lw = 2(\sqrt[3]{A/VB})^2$.

43.54 Find the absolute maximum and minimum of $f(x, y) = 4x^2 + 2xy - 3y^2$ on the unit square $0 \le x \le 1$, $0 \le y \le 1$.

▮ $f_x = 8x + 2y$, $f_y = 2x - 6y$. Set $f_x = f_y = 0$. Then $4x + y = 0$, $x - 3y = 0$. Solving, we get $x = y = 0$. Note that $f(0, 0) = 0$. Let us look at the boundary of the square (Fig. 43-3). (1) On the segment L_1, $x = 0$, $0 \le y \le 1$. Then $f(0, y) = -3y^2$. The maximum is 0, and the minimum is -3 at $(0, 1)$. (2) On the segment L_2, $x = 1$, $0 \le y \le 1$. $f(1, y) = 4 + 2y - 3y^2$. $\dfrac{df}{dy}(1, y) = 2 - 6y$. Thus, $\frac{1}{3}$ is a critical number. We must evaluate $f(1, y)$ for $y = 0$, $y = \frac{1}{3}$, and $y = 1$. $f(1, 0) = 4$, $f(1, \frac{1}{3}) = \frac{13}{3}$, $f(1, 1) = 3$. (3) On the segment L_3, $y = 0$, $0 \le x \le 1$. $f(x, 0) = 4x^2$. Hence, the maximum is 4 at $(1, 0)$, and the minimum is 0 at $(0, 0)$. (4) On the line segment L_4, $y = 1$, $0 \le x \le 1$. Then $f(x, 1) = 4x^2 + 2x - 3$. $\dfrac{df}{dx}(x, 1) = 8x + 2$. The critical number is $-\frac{1}{4}$, which does not lie in the interval $0 \le x \le 1$. Thus, we need only look at $f(x, 1)$ when $x = 0$ and $x = 1$. $f(0, 1) = -3$ and $f(1, 1) = 3$. Therefore, the absolute maximum is $\frac{13}{3}$ at $(1, \frac{1}{3})$, and the absolute minimum is -3 at $(0, 1)$.

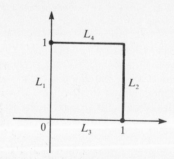

Fig. 43-3

43.55 Find the absolute extrema of $f(x, y) = \sin x + \sin y + \sin (x + y)$.

▮ By the periodicity of the sine function, we may restrict attention to the square $S: -\pi \le x \le \pi$, $-\pi \le y \le \pi$. Since f is continuous, f will have an absolute maximum and minimum on S (and, therefore, for all x and y). These extrema will occur either on the boundary or in the interior (where they will show up as critical points). $f_x = \cos x + \cos (x + y)$, $f_y = \cos y + \cos (x + y)$. Let $f_x = f_y = 0$. Then $\cos x = -\cos (x + y) = \cos y$. Hence, either $x = y$ or $x = -y$. **Case 1:** $x = y$. Then $\cos x = -\cos (x + y) = -\cos 2x = 1 - 2\cos^2 x$. So, $2\cos^2 x + \cos x - 1 = 0$, $(2\cos x - 1)(\cos x + 1) = 0$, $\cos x = \frac{1}{2}$ or $\cos x = -1$, $x = \pm\pi/3$ or $x = \pm\pi$. So, the critical points are $(\pi/3, \pi/3)$, $(-\pi/3, -\pi/3)$, (π, π), $(-\pi, -\pi)$. The latter two are on the boundary and need not be considered separately. Note that $f\left(\dfrac{\pi}{3}, \dfrac{\pi}{3}\right) = \dfrac{\sqrt{3}}{2} + \dfrac{\sqrt{3}}{2} + \dfrac{\sqrt{3}}{2} = \dfrac{3\sqrt{3}}{2}$ and $f\left(\dfrac{-\pi}{3}, -\dfrac{\pi}{3}\right) = -\dfrac{\sqrt{3}}{2} - \dfrac{\sqrt{3}}{2} - \dfrac{\sqrt{3}}{2} = \dfrac{-3\sqrt{3}}{2}$. **Case 2:** $x = -y$. Then $\cos x = -\cos (x + y) = -\cos 0 = -1$. Hence, $x = \pm\pi$. This yields the critical numbers $(\pi, -\pi)$ and $(-\pi, \pi)$. Since these are on the boundary, they need not be treated separately. Now consider the boundary of S (Fig. 43-4). (1) L_1: $f(\pi, y) = \sin \pi + \sin y + \sin (\pi + y) = \sin y - \sin y = 0$. (2) L_2: $f(-\pi, y) = f(\pi, y) = 0$. (3) L_3: $f(x, \pi) = \sin x + \sin \pi + \sin (x + \pi) = \sin x - \sin x = 0$. (4) L_4: $f(x, -\pi) = f(x, \pi) = 0$. Thus, f is 0 on the boundary. Hence, the absolute maximum is $3\sqrt{3}/2$ and the absolute minimum is $-3\sqrt{3}/2$.

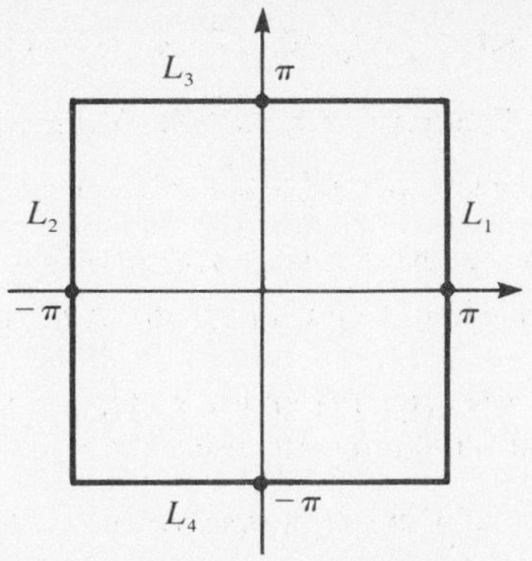

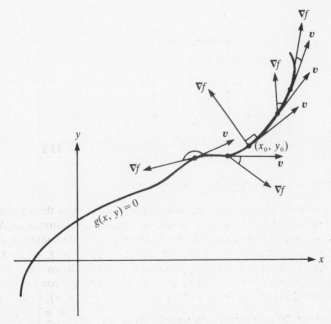

Fig. 43-4

43.56 State a theorem that justifies the *method of Lagrange multipliers*.

    ▮ Assume $f(x, y)$ and $g(x, y)$ have continuous partial derivatives in an open disk containing (x_0, y_0). If $g(x_0, y_0) = 0$ and f has an extreme value at (x_0, y_0) relative to all nearby points that satisfy the "constraint" $g(x, y) = 0$, and if $\nabla g(x_0, y_0) \neq 0$, then $\nabla f(x_0, y_0) = \lambda \nabla g(x_0, y_0)$ for some constant λ (called a *Lagrange multiplier*). Thus, the points at which extrema occur will be found among the common solutions of $g(x, y) = 0$ and $\nabla f = \lambda \nabla g$ [or, equivalently, $\nabla (f - \lambda g) = 0$]. Similar results hold for more than two variables. Moreover, for a function $f(x, y, z)$ and two constraints $g(x, y, z) = 0$ and $h(x, y, z) = 0$, one solves $\nabla f = \lambda \nabla g + \mu \nabla h$ together with the constraint equations.

43.57 With reference to Problem 43.56, interpret the condition $\nabla f(x_0, y_0) = \lambda \nabla g(x_0, y_0)$ in terms of the directional derivative.

    ▮ Figure 43-5 shows a portion of the curve $g(x, y) = 0$, together with its field of tangent vectors $\boldsymbol{v}$. If (x_0, y_0) is to be an extreme point for f along the curve, the derivative of f in the direction of $\boldsymbol{v}$ must vanish at (x_0, y_0). Thus, ∇f must be perpendicular to $\boldsymbol{v}$, and therefore parallel to the curve normal, at (x_0, y_0). But, by Problem 43.14, the curve normal can be taken to be ∇g; so $\nabla f = \lambda \nabla g$ at (x_0, y_0).

Fig. 43-5

43.58 Find the point(s) on the ellipsoid $4x^2 + 9y^2 + 36z^2 = 36$ nearest the origin.

    ▮ We must minimize $f(x, y, z) = x^2 + y^2 + z^2$ subject to the constraint $g(x, y, z) = 4x^2 + 9y^2 + 36z^2 - 36 = 0$. $\nabla f = (2x, 2y, 2z)$, $\nabla g = (8x, 18y, 72z)$. Let $\nabla f = \lambda \nabla g$. Then $(2x, 2y, 2z) = \lambda(8x, 18y, 72z)$, that is, $2x = \lambda(8x)$, $2y = \lambda(18y)$, $2z = \lambda(72z)$, which are equivalent to $x(4\lambda - 1) = 0$, $y(9\lambda - 1) = 0$, $z(36\lambda - 1) = 0$. Therefore, either $x = 0$ or $\lambda = \frac{1}{4}$; and either $y = 0$ or $\lambda = \frac{1}{9}$; and either $z = 0$ or $\lambda = \frac{1}{36}$.

Clearly, $x = 0$, $y = 0$, $z = 0$ does not satisfy the constraint equation $4x^2 + 9y^2 + 36z^2 = 36$. Hence, we must have either $\lambda = \frac{1}{4}$, $\lambda = \frac{1}{9}$, or $\lambda = \frac{1}{36}$. **Case 1.** $\lambda = \frac{1}{4}$. Then $y = 0$, $z = 0$. Hence, $4x^2 = 36$, $x^2 = 9$, $x = \pm 3$. Therefore, $\sqrt{x^2 + y^2 + z^2} = \sqrt{9} = 3$. **Case 2.** $\lambda = \frac{1}{9}$. Then $x = 0$, $z = 0$, $9y^2 = 36$, $y^2 = 4$, $y = \pm 2$. Therefore, $\sqrt{x^2 + y^2 + z^2} = \sqrt{4} = 2$. **Case 3.** $\lambda = \frac{1}{36}$. Then $x = 0$, $y = 0$, $36z^2 = 36$, $z^2 = 1$, $z = \pm 1$. Therefore, $\sqrt{x^2 + y^2 + z^2} = \sqrt{1} = 1$. Hence, the minimum distance from the origin is 1, achieved at $(0, 0, 1)$ and $(0, 0, -1)$.

43.59 Find the point(s) on the sphere $x^2 + y^2 + z^2 = 1$ furthest from the point $(2, 1, 2)$.

▌ We must maximize $(x - 2)^2 + (y - 1)^2 + (z - 2)^2$ subject to the constraint $g(x, y, z) = x^2 + y^2 + z^2 - 1 = 0$. $\nabla f = (2(x - 2),\ 2(y - 1),\ 2(z - 2))$. $\nabla g = (2x, 2y, 2z)$. Let $\nabla f = \lambda \nabla g$. Then $(2(x - 2),\ 2(y - 1),\ 2(z - 2)) = \lambda(2x, 2y, 2z)$, or $2(x - 2) = 2\lambda x$, $2(y - 1) = 2\lambda y$, $2(z - 2) = 2\lambda z$. These are equivalent to $x(1 - \lambda) = 2$, $y(1 - \lambda) = 1$, $z(1 - \lambda) = 2$. Hence, $1 - \lambda \neq 0$. Substitute $x = 2/(1 - \lambda)$, $y = 1/(1 - \lambda)$, $z = 2/(1 - \lambda)$ in the constraint equation: $\dfrac{4}{(1 - \lambda)^2} + \dfrac{1}{(1 - \lambda)^2} + \dfrac{4}{(1 - \lambda)^2} = 1$, $(1 - \lambda)^2 = 9$, $1 - \lambda = \pm 3$, $\lambda = -2$ or $\lambda = 4$. **Case 1:** $\lambda = -2$. Then $1 - \lambda = 3$, $x = \frac{2}{3}$, $y = \frac{1}{3}$, $z = \frac{2}{3}$, and $\sqrt{(x - 2)^2 + (y - 1)^2 + (z - 2)^2} = \sqrt{\frac{16}{9} + \frac{4}{9} + \frac{16}{9}} = \sqrt{\frac{36}{9}} = 2$. **Case 2:** $\lambda = 4$. Then $1 - \lambda = -3$, $x = -\frac{2}{3}$, $y = -\frac{1}{3}$, $z = -\frac{2}{3}$, and $\sqrt{(x - 2)^2 + (y - 1)^2 + (z - 2)^2} = \sqrt{\frac{64}{9} + \frac{16}{9} + \frac{64}{9}} = \sqrt{\frac{144}{9}} = 4$. So, the point on the sphere furthest from $(2, 1, 2)$ is $(-\frac{2}{3}, -\frac{1}{3}, -\frac{2}{3})$.

43.60 Find the point(s) on the cone $x^2 = y^2 + z^2$ nearest the point $(0, 1, 3)$.

▌ We must minimize $f(x, y, z) = x^2 + (y - 1)^2 + (z - 3)^2$ subject to the constraint $g(x, y, z) = y^2 + z^2 - x^2 = 0$. $\nabla f = (2x, 2(y - 1),\ 2(z - 3))$, $\nabla g = (-2x, 2y, 2z)$. Let $\nabla f = \lambda \nabla g$. Then $(2x, 2(y - 1),\ 2(z - 3)) = \lambda(-2x, 2y, 2z)$, or $2x = -2\lambda x$, $2(y - 1) = 2\lambda y$, $2(z - 3) = 2\lambda z$. These are equivalent to $x(\lambda + 1) = 0$, $y(1 - \lambda) = 1$, $z(1 - \lambda) = 3$. Hence, $1 - \lambda \neq 0$. Substitute $y = 1/(1 - \lambda)$, $z = 3/(1 - \lambda)$ in the constraint equation: $x^2 = \dfrac{1}{(1 - \lambda)^2} + \dfrac{9}{(1 - \lambda)^2} = \dfrac{10}{(1 - \lambda)^2}$. Hence, $x \neq 0$. Therefore, $x(\lambda + 1) = 0$ implies $\lambda + 1 = 0$, $\lambda = -1$. Then $y = 1(1 - \lambda) = \frac{1}{2}$, $z = 3/(1 - \lambda) = \frac{3}{2}$, and $x^2 = 10/(1 - \lambda)^2 = \frac{10}{4} = \frac{5}{2}$, $x = \pm\sqrt{\frac{5}{2}}$. Thus, $f(\sqrt{\frac{5}{2}}, \frac{1}{2}, \frac{3}{2}) = \frac{5}{2} + \frac{1}{4} + \frac{9}{4} = 5$, and $f(-\sqrt{\frac{5}{2}}, \frac{1}{2}, \frac{3}{2}) = 5$. Hence, the required points are $(\pm\sqrt{\frac{5}{2}}, \frac{1}{2}, \frac{3}{2})$.

43.61 Find the point(s) on the sphere $x^2 + y^2 + z^2 = 14$ where $3x - 2y + z$ attains its maximum value.

▌ We must maximize $f(x, y, z) = 3x - 2y + z$ subject to the constraint $g(x, y, z) = x^2 + y^2 + z^2 - 14 = 0$. $\nabla f = (3, -2, 1)$, $\nabla g = (2x, 2y, 2z)$. Let $\nabla f = \lambda \nabla g$, that is, $(3, -2, 1) = \lambda(2x, 2y, 2z)$. Then $3 = 2\lambda x$, $-2 = 2\lambda y$, $1 = 2\lambda z$. Hence, $\lambda \neq 0$ (since $2\lambda x = 3 \neq 0$). Substitute $x = 3/2\lambda$, $y = -1/\lambda$, $z = 1/2\lambda$ in the constraint equation: $\dfrac{9}{4\lambda^2} + \dfrac{1}{\lambda^2} + \dfrac{1}{4\lambda^2} = 14$, $14/4\lambda^2 = 14$, $\lambda^2 = \frac{1}{4}$, $\lambda = \pm\frac{1}{2}$. **Case 1:** $\lambda = \frac{1}{2}$. Then $x = 3$, $y = -2$, $z = 1$, and $3x - 2y + z = 9 + 4 + 1 = 14$. **Case 2:** $\lambda = -\frac{1}{2}$. Then $x = -3$, $y = 2$, $z = -1$, and $3x - 2y + z = -9 - 4 - 1 = -14$. Hence, the maximum is attained at $(3, -2, 1)$.

43.62 If a rectangular box has three faces in the coordinate planes and one vertex in the first octant on the paraboloid $z = 4 - x^2 - y^2$, find the maximum volume V of such a box.

▌ We must maximize $V = xyz$ subject to the constraint $g(x, y, z) = x^2 + y^2 + z - 4 = 0$. $\nabla V = (yz, xz, xy)$, $\nabla g = (2x, 2y, 1)$. Let $\nabla V = \lambda \nabla g$, that is, $(yz, xz, xy) = \lambda(2x, 2y, 1)$. This is equivalent to $yz = 2\lambda x$, $xz = 2\lambda y$, $xy = \lambda$. Since $\lambda = xy$, we have $yz = 2(xy)x$, $xz = 2(xy)y$, or $z = 2x^2$, $z = 2y^2$, $x^2 = y^2$. Substitute in the constraint equation: $x^2 + x^2 + 2x^2 = 4$, $4x^2 = 4$, $x^2 = 1$, $x = 1$, $y = 1$, $z = 2$. Hence, the maximum volume $V = xyz = 1(1)(2) = 2$.

43.63 Find the points on the curve of intersection of the ellipsoid $4x^2 + 4y^2 + z^2 = 1428$ and the plane $x + 4y - z = 0$ that are closest to the origin.

▌ We must minimize $x^2 + y^2 + z^2$ subject to the constraints $g(x, y, z) = 4x^2 + z^2 - 1428 = 0$ and $h(x, y, z) = x + 4y - z = 0$. $\nabla f = (2x, 2y, 2z)$, $\nabla g = (8x, 8y, 2z)$, $\nabla h = (1, 4, -1)$. Let $\nabla f = \lambda \nabla g + \mu \nabla h$, that is, $(2x, 2y, 2z) = \lambda(8x, 8y, 2z) + \mu(1, 4, -1)$. This is equivalent to

$$2x = 8\lambda x + \mu \tag{1}$$

$$2y = 8\lambda y + 4\mu \tag{2}$$

$$2z = 2\lambda z - \mu \tag{3}$$

From (1) and (2), $2y - 8\lambda y = 4(2x - 8\lambda x)$. Hence, $2y(1 - 4\lambda) = 8x(1 - 4\lambda)$. **Case 1.** $1 - 4\lambda \neq 0$. Then $y = 4x$. From the second constraint, $z = x + 4y = 17x$. From the first constraint, $4x^2 + 4y^2 + z^2 = 4x^2 + 64x^2 + 289x^2 = 1428$, $357x^2 = 1428$, $x^2 = 4$, $x = \pm 2$, $y = \pm 8$, $z = \pm 34$. The distance to the origin is $\sqrt{x^2 + y^2 + z^2} = \sqrt{4 + 64 + 1156} = \sqrt{1224} = 6\sqrt{34}$. **Case 2.** $1 - 4\lambda = 0$. Then $\lambda = \frac{1}{4}$. By (1), $2x = 8\lambda x + \mu = 2x + \mu$. Hence, $\mu = 0$. By (3), $z = \frac{1}{2}z$. Therefore, $z = 0$. By the second constraint, $x = -4y$. By the first constraint, $64y^2 + 4y^2 = 1428$, $68y^2 = 1428$, $y^2 = 21$, $y = \pm\sqrt{21}$, $x = \pm 4\sqrt{21}$. So, the distance to the origin is $\sqrt{16(21) + 21 + 0} = \sqrt{21}\sqrt{17}$. The minimum distance in Case 1 is $6\sqrt{34} = 6\sqrt{2}\sqrt{17}$, and, in Case 2, $\sqrt{21}\sqrt{17}$. Since $\sqrt{21} < 6\sqrt{2}$, the minimum distance is $\sqrt{21}\sqrt{17}$, attained at $(\pm 4\sqrt{21}, \pm\sqrt{21}, 0)$.

43.64 A wire L units long is cut into three pieces. The first piece is bent into a square, the second into a circle, and the third into an equilateral triangle. Find the manner of cutting up the wire that will produce a minimum total area of the square, circle, and triangle, and the manner that will produce the maximum total area.

▮ Let the first piece be $4x$, the second $2\pi y$, and the third $3z$. Then the total area is $A = x^2 + \pi y^2 + (\sqrt{3}/4)z^2$. The constraint is $g(x, y, z) = 4x + 2\pi y + 3z - L = 0$. $\nabla A = (2x, 2\pi y, (\sqrt{3}/2)z)$ and $\nabla g = (4, 2\pi, 3)$. Let $\nabla A = \lambda \nabla g$, that is, $(2x, 2\pi y, z\sqrt{3}/2) = \lambda(4, 2\pi, 3)$, or $2x = 4\lambda$, $2\pi y = 2\lambda\pi$, $z\sqrt{3}/2 = 3$. Hence, $x = 2\lambda$, $y = \lambda$, $z = 2\sqrt{3}\lambda$. Substitute in the constraint equation: $8\lambda + 2\pi\lambda + 6\sqrt{3}\lambda = L$. Hence, $\lambda = L/(8 + 2\pi + 6\sqrt{3}) = L/[2(4 + \pi + 3\sqrt{3})]$. Therefore, $A = 4\lambda^2 + \pi\lambda^2 + 12\lambda^2(\sqrt{3}/4) = (4 + \pi + 3\sqrt{3})\lambda^2 = L^2/[4(4 + \pi + 3\sqrt{3})]$. We must also consider the "boundary" values when $x = 0$ or $y = 0$ or $z = 0$. When $x = 0$, we get $2\lambda(\pi + 3\sqrt{3}) = L$, $\lambda = L/[2(\pi + 3\sqrt{3})]$, $A = \pi\lambda^2 + 12\lambda^2(\sqrt{3}/4) = \lambda^2(\pi + 3\sqrt{3}) = L^2/[4(\pi + 3\sqrt{3})]$. When $y = 0$, a similar calculation yields $A = L^2/[4(4 + 3\sqrt{3})]$, and, when $z = 0$, we get $A = L^2/[4(4 + \pi)]$. Thus, the maximum area $A = L^2/[4(4 + \pi)]$ occurs when $z = 0$, $x = L/(4 + \pi)$, $y = L/[2(4 + \pi)]$. The minimum area $A = L^2/[4(4 + \pi + 3\sqrt{3})]$ occurs when $x = L/(4 + \pi + 3\sqrt{3})$, $y = L/[2(4 + \pi + 3\sqrt{3})]$, $z = \sqrt{3}L/(4 + \pi + 3\sqrt{3})$.

43.65 Minimize xy for points on the circle $x^2 + y^2 = 1$.

▮ Sometimes it is better to avoid calculus. Since $(x + y)^2 = (x^2 + y^2) + 2xy = 1 + 2xy$, an absolute minimum occurs for $x = -y$; that is, for $(1/\sqrt{2}, -1/\sqrt{2})$ and $(-1/\sqrt{2}, 1/\sqrt{2})$.

43.66 Use Lagrange multipliers to maximize $x_1y_1 + \cdots + x_ny_n$ subject to the constraints $\sum_{i=1}^{n} x_i^2 = 1$ and $\sum_{i=1}^{n} y_i^2 = 1$.

▮ It is obvious that we can restrict our attention to nonnegative numbers. We must maximize $f(x_1, \ldots, x_n, y_1, \ldots, y_n) = x_1y_1 + \cdots + x_ny_n$ subject to the constraints $g(x_1, \ldots, x_n) = \left(\sum_{i=1}^{n} x_i^2\right) - 1 = 0$ and $h(y_1, \ldots, y_n) = \left(\sum_{i=1}^{n} y_i^2\right) - 1 = 0$. $\nabla f = (y_1, y_2, \ldots, y_n, x_1, \ldots, x_n)$, $\nabla g = (2x_1, \ldots, 2x_n, 0, \ldots, 0)$, $\nabla h = (0, \ldots, 0, 2y_1, \ldots, 2y_n)$. Let $\nabla f = \lambda \nabla g + \mu \nabla h$. Then $(y_1, \ldots, y_n, x_1, \ldots, x_n) = \lambda(2x_1, \ldots, 2x_n, 0, \ldots, 0) + \mu(0, \ldots, 0, 2y_1, \ldots, 2y_n)$. Hence, $y_1 = 2\lambda x_1, \ldots, y_n = 2\lambda x_n$, $x_1 = 2\mu y_1, \ldots, x_n = 2\mu y_n$.

Clearly, $\lambda \neq 0$. (Otherwise, $y_1 = \cdots = y_n = 0$, contradicting $\sum_{i=1}^{n} y_i^2 = 1$.) Similarly, $\mu \neq 0$. Now, $1 = \sum_{i=1}^{n} y_i^2 = \sum_{i=1}^{n} 4\lambda^2 x_i^2 = 4\lambda^2 \sum_{i=1}^{n} x_i^2 = 4\lambda^2$. Hence, $\lambda^2 = \frac{1}{4}$ and $\lambda = \pm\frac{1}{2}$. Since $y_i = 2\lambda x_i$ and $x_i \geq 0$ and $y_i \geq 0$, $\lambda = \frac{1}{2}$. Similarly, $\mu = \frac{1}{2}$. Hence, $y_i = x_i$. Therefore, $x_1y_1 + \cdots + x_ny_n = x_1^2 + \cdots + x_n^2 = 1$. Hence, the maximum value is 1. [This result becomes obvious when the problem is restated as: Maximize the dot product of two unit vectors.]

43.67 A solid is to consist of a right circular cylinder surmounted by a right circular cone. For a fixed surface area S (including the base), what should be the dimensions to maximize the volume V?

▮ Let r and h be the radius of the base and the height of the cylinder, respectively. Let 2α be the vertex angle of the cone. $S = \pi r^2 + 2\pi rh + \pi rs = \pi r^2 + 2\pi rh + \pi r^2 \csc \alpha$ (since $\sin \alpha = r/s$). $V = \pi r^2 h + \frac{1}{3}\pi r^2(r \cot \alpha) = \pi r^2 h + \frac{1}{3}\pi r^3 \cot \alpha$. We have to maximize $f(r, h, \alpha) = \pi rh^2 + \frac{1}{3}\pi r^3 \cot \alpha$ under the constraint $g(r, h, \alpha) = \pi r^2 + 2\pi rh + \pi r^2 \csc \alpha - S = 0$. $\nabla f = (2\pi rh + \pi r^2 \cot \alpha, \pi r^2, -\frac{1}{3}\pi r^3 \csc^2 \alpha)$. $\nabla g = (2\pi r + 2\pi h + 2\pi r \csc \alpha, 2\pi r, -\pi r^2 \csc \alpha \cot \alpha)$. Let $\nabla f = \lambda \nabla g$. Then,

$$2\pi rh + \pi r^2 \cot \alpha = 2\pi\lambda(r + h + r \csc \alpha) \tag{1}$$

$$\pi r^2 = 2\pi r\lambda \tag{2}$$

$$-\tfrac{1}{3}\pi r^3 \csc^2 \alpha = -\pi\lambda r^2 \csc \alpha \cot \alpha \tag{3}$$

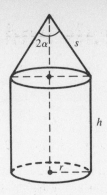

Fig. 43-6

Hence, by (2), $r = 2\lambda$. In addition, $2rh + r^2 \cot \alpha = 2\lambda(r + h + r \csc \alpha)$ follows from (1) and yields, with $r = 2\lambda$, $2rh + r^2 \cot \alpha = r(r + h + r \csc \alpha)$, and, therefore, $r^2(\cot \alpha - \csc \alpha) = r(r - h)$. Hence, $r(\cot \alpha - \csc \alpha) = r - h$. From (3) and $r = 2\lambda$, we get $\frac{1}{3} \csc^2 \alpha = \frac{1}{2} \csc \alpha \cot \alpha$, from which follows $\cos \alpha = \frac{2}{3}$. Then $\sin \alpha = \sqrt{5}/3$, $\csc \alpha = 3/\sqrt{5}$, $\cot \alpha = 2/\sqrt{5}$, $\cot \alpha - \csc \alpha = -1/\sqrt{5}$. Therefore, $-r/\sqrt{5} = r - h$. Then $h = r\left(1 + \dfrac{1}{\sqrt{5}}\right) = \dfrac{\sqrt{5} + 1}{\sqrt{5}} r$. (The value of r, and therefore of h, can be found in terms of S by substitution in the equation for S.)

43.68 Prove Cauchy's inequality: $\displaystyle\sum_{i=1}^{n} a_i b_i \le \left(\sum_{i=1}^{n} a_i^2\right)^{1/2} \left(\sum_{i=1}^{n} b_i^2\right)^{1/2}$.

▎ Let $A = \left(\displaystyle\sum_{i=1}^{n} a_i^2\right)^{1/2}$ and $B = \left(\displaystyle\sum_{i=1}^{n} b_i^2\right)^{1/2}$. Let $x_i = a_i/A$ and $y_i = b_i/B$. (If $A = 0$ or $B = 0$, the desired result is obvious.) Then $\displaystyle\sum_{i=1}^{n} x_i^2 = 1$ and $\displaystyle\sum_{i=1}^{n} y_i^2 = 1$. By Problem 43.66, $\displaystyle\sum_{i=1}^{n} x_i y_i \le 1$. Hence, $\displaystyle\sum_{i=1}^{n} \dfrac{a_i}{A} \dfrac{b_i}{B} \le 1$, $\displaystyle\sum_{i=1}^{n} a_i b_i \le AB$.

CHAPTER 44
Multiple Integrals and their Applications

44.1 Evaluate the iterated integral $I = \int_2^3 \int_1^5 (x + 2y)\, dx\, dy$.

▪ $\int_1^5 (x + 2y)\, dx = (\frac{1}{2}x^2 + 2yx)\,]_1^5 = (\frac{25}{2} + 10y) - (\frac{1}{2} + 2y) = 12 + 8y$.   Therefore, $I = \int_2^3 (12 + 8y)\, dy = (12y + 4y^2)\,]_2^3 = (36 + 36) - (24 + 16) = \boxed{32}$.

44.2 Evaluate the iterated integral $I = \int_0^1 \int_{x^2}^{x^3} (x^2 + y^2)\, dy\, dx$.

▪ $\int_{x^2}^{x^3} (x^2 + y^2)\, dy = (x^2 y + \frac{1}{3}y^3)\,]_{x^2}^{x^3} = (x^5 + \frac{1}{3}x^9) - (x^4 + \frac{1}{3}x^6)$.   Therefore, $I = \int_0^1 (x^5 + \frac{1}{3}x^9 - x^4 - \frac{1}{3}x^6)\, dx = \frac{1}{6}x^6 + \frac{1}{30}x^{10} - \frac{1}{5}x^5 - \frac{1}{21}x^7\,]_0^1 = \frac{1}{6} + \frac{1}{30} - \frac{1}{5} - \frac{1}{21} = \boxed{-\frac{1}{21}}$.

44.3 Evaluate the iterated integral $I = \int_{-\pi}^{\pi} \int_0^2 r \sin \theta\, dr\, d\theta$.

▪ $\int_0^2 r \sin \theta\, dr = \frac{1}{2}r^2 \sin \theta\,]_0^2 = 2 \sin \theta$.   Therefore, $I = \int_{-\pi}^{\pi} 2 \sin \theta\, d\theta = -2 \cos \theta\,]_{-\pi}^{\pi} = -2[\cos \pi - \cos(-\pi)] = -2(\cos \pi - \cos \pi) = 0$.

44.4 Evaluate the iterated integral $I = \int_0^{\pi/2} \int_0^{\cos \theta} \rho^2 \sin \theta\, d\rho\, d\theta$.

▪ $\int_0^{\cos \theta} \rho^2 \sin \theta\, d\rho = \frac{1}{3}\rho^3 \sin \theta\,]_0^{\cos \theta} = \frac{1}{3} \cos^3 \theta \sin \theta$.   Hence, $I = \int_0^{\pi/2} \frac{1}{3} \cos^3 \theta \sin \theta\, d\theta = -\frac{1}{12} \cos^4 \theta\,]_0^{\pi/2} = -\frac{1}{12}[\cos^4(\pi/2) - \cos^4 0] = -\frac{1}{12}(0 - 1) = \boxed{\frac{1}{12}}$.

44.5 Evaluate the iterated integral $I = \int_0^1 \int_0^z \int_0^y (x + y + z)\, dx\, dy\, dz$.

▪ $\int_0^y (x + y + z)\, dx = \frac{1}{2}x^2 + (y + z)x\,]_0^y = \frac{1}{2}y^2 + (y + z)z.^{.y}$ Hence, $\int_0^z \int_0^y (x + y + z)\, dx\, dy = \int_0^z (\frac{1}{2}y^2 + yz + z^2)\, dy = \frac{1}{6}y^3 + \frac{1}{2}y^2 z + z^2 y\,]_0^z = \frac{1}{6}z^3 + \frac{1}{2}z^3 + z^3 = \frac{5}{3}z^3$.   Therefore, $I = \int_0^1 \frac{5}{3}z^3\, dz = \frac{5}{12}z^4\,]_0^1 = \frac{5}{12}$. $\boxed{1/4}$

44.6 Evaluate $I = \int_0^{\ln 4} \int_0^{\ln 3} e^{x+y}\, dx\, dy$. $\left(\text{OR} = \int_0^{\ln 4} [e^{x+y}]_0^{\ln 3}\, dy = \int_0^{\ln 4} (e^{\ln 3 + y} - e^y)\, dy \right.$

$= [e^{\ln 3 + y} - e^y]_0^{\ln 4} = e^{\ln 3 + \ln 4} - e^{\ln 4} - e^{\ln 3} + 1 = 6$

▪ In this case the double integral may be replaced by a product: $I = (\int_0^{\ln 3} e^x\, dx)(\int_0^{\ln 4} e^y\, dy) = (3 - 1)(4 - 1) = \boxed{6}$.   (See Problem 44.71.)

44.7 Evaluate $\int_1^2 \int_0^y x\sqrt{y^2 - x^2}\, dx\, dy$.

▪ $\int_0^y x\sqrt{y^2 - x^2}\, dx = -\frac{1}{2} \cdot \frac{2}{3}(y^2 - x^2)^{3/2}\,]_0^y = -\frac{1}{3}(y^2 - x^2)^{3/2}\,]_0^y = -\frac{1}{3}[-(y^2)^{3/2}] = \frac{1}{3}y^3$.   Therefore, $I = \int_1^2 \frac{1}{3}y^3\, dy = \frac{1}{12}y^4\,]_1^2 = \frac{1}{12}(16 - 1) = \frac{15}{12} = \boxed{\frac{5}{4}}$.

44.8 Evaluate $I = \int_0^1 \int_y^1 e^{x^2}\, dx\, dy$.   cf: $\iint 1(6)$ EX

▪ $\int e^{x^2}\, dx$ cannot be evaluated in terms of standard functions. Therefore, we change the order of integration, using Fig. 44-1. $I = \int_0^1 \int_0^x e^{x^2}\, dy\, dx = \int_0^1 xe^{x^2}\, dx = \frac{1}{2}e^{x^2}\,]_0^1 = \boxed{\frac{1}{2}(e - 1)}$.

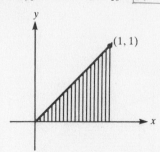

Fig. 44-1

44.9 Evaluate $\int_0^{\pi/2} \int_0^{\cos y} e^x \sin y\, dx\, dy$.

▪ $\int_0^{\cos y} e^x \sin y\, dx = (\sin y)(e^x)\,]_0^{\cos y} = (\sin y)(e^{\cos y} - 1)$.   Therefore, $I = \int_0^{\pi/2} [(\sin y)e^{\cos y} - \sin y]\, dy = (-e^{\cos y} + \cos y)\,]_0^{\pi/2} = (-e^0 + 0) - (-e + 1) = \boxed{e - 2}$.

44.10 Evaluate $I = \int_0^1 \int_{y^4}^{y^2} \sqrt{y/x}\, dx\, dy$.

$\int_{y^4}^{y^2} \sqrt{y/x}\, dx = 2\sqrt{y}\sqrt{x}\,]_{y^4}^{y^2} = 2\sqrt{y}(y - y^2) = 2(y^{3/2} - y^{5/2})$. Therefore, $I = 2\int_0^1 (y^{3/2} - y^{5/2})\, dy = 2(\frac{2}{5}y^{5/2} - \frac{2}{7}y^{7/2})\,]_0^1 = 2(\frac{2}{5} - \frac{2}{7}) = \boxed{\frac{8}{35}}$.

44.11 Evaluate $I = \iint_{\mathcal{R}} x\, dA$, where $\mathcal{R}$ is the region bounded by $y = x$ and $y = x^2$.

The curves $y = x$ and $y = x^2$ intersect at $(0,0)$ and $(1,1)$, and, for $0 < x < 1$, $y = x$ is above $y = x^2$ (see Fig. 44-2). $I = \int_0^1 \int_{x^2}^x x\, dy\, dx = \int_0^1 xy\,]_{x^2}^x\, dx = \int_0^1 (x^2 - x^3)\, dx = (\frac{1}{3}x^3 - \frac{1}{4}x^4)\,]_0^1 = \frac{1}{3} - \frac{1}{4} = \boxed{\frac{1}{12}}$.

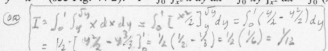

OR $I = \int_0^1 \int_y^{\sqrt{y}} x\, dx\, dy = \int_0^1 [\frac{x^2}{2}]_y^{\sqrt{y}}\, dy = \int_0^1 (\frac{y}{2} - \frac{y^2}{2})\, dy$
$= \frac{1}{2}\cdot[\frac{y^2}{2} - \frac{y^3}{3}]_0^1 = \frac{1}{2}(\frac{1}{2} - \frac{1}{3}) = \frac{1}{2}(\frac{1}{6}) = \frac{1}{12}$

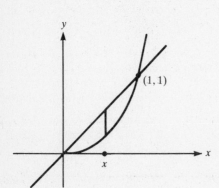

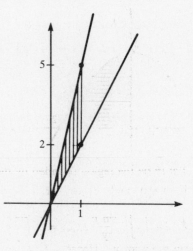

Fig. 44-2   Fig. 44-3

44.12 Evaluate $I = \iint_{\mathcal{R}} y^2\, dA$, where $\mathcal{R}$ is the region bounded by $y = 2x$, $y = 5x$, and $x = 1$.

The lines $y = 2x$ and $y = 5x$ intersect at the origin. For $0 < x \le 1$, the region runs from $y = 2x$ up to $y = 5x$ (Fig. 44-3). Hence, $I = \int_0^1 \int_{2x}^{5x} y^2\, dy\, dx = \int_0^1 \frac{1}{3}y^3\,]_{2x}^{5x}\, dx = \frac{1}{3}\int_0^1 (125x^3 - 8x^3)\, dx = \frac{1}{3}\int_0^1 117x^3\, dx = \frac{39}{4}x^4\,]_0^1 = \frac{39}{4}$.

44.13 Evaluate $I = \iint_{\mathcal{R}} (x - y)\, dA$, where $\mathcal{R}$ is the region above the x-axis bounded by $y^2 = 3x$ and $y^2 = 4 - x$ (see Fig. 44-4).

It is convenient to evaluate I by means of strips parallel to the x-axis. $I = \int_0^{\sqrt{3}} \int_{y^2/3}^{4-y^2} (x - y)\, dx\, dy = \int_0^{\sqrt{3}} (\frac{1}{2}x^2 - yx)\,]_{y^2/3}^{4-y^2}\, dy = \int_0^{\sqrt{3}} [\frac{1}{2}(4 - y^2)^2 - y(4 - y^2)] - [\frac{1}{2}(y^2/3)^2 - y^3/3]\, dy = \int_0^{\sqrt{3}} (8 - 4y^2 + \frac{1}{2}y^4 - 4y + y^3 - \frac{1}{18}y^4 - \frac{1}{3}y^3)\, dy = \int_0^{\sqrt{3}} (8 - 4y - 4y^2 + \frac{2}{3}y^3 + \frac{4}{9}y^4)\, dy = 8y - 2y^2 - \frac{4}{3}y^3 + \frac{1}{6}y^4 + \frac{4}{45}y^5\,]_0^{\sqrt{3}} = 8\sqrt{3} - 6 - 4\sqrt{3} + \frac{3}{2} + \frac{4}{5}\sqrt{3} = \frac{24}{5}\sqrt{3} - \frac{9}{2}$.

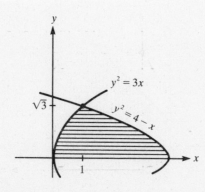

Fig. 44-4

44.14 Evaluate $I = \iint_{\mathcal{R}} \frac{1}{\sqrt{2y - y^2}}\, dA$, where $\mathcal{R}$ is the region in the first quadrant bounded by $x^2 = 4 - 2y$.

▊ The curve $x^2 = 4 - 2y$ is a parabola with vertex at $(0,2)$ and passing through the x-axis at $x = 2$ (Fig. 44-5). Hence, $I = \int_0^2 \int_0^{\sqrt{4-2y}} \dfrac{1}{\sqrt{2y - y^2}} \, dx \, dy = \int_0^2 \dfrac{x}{\sqrt{2y - y^2}} \Big]_0^{\sqrt{4-2y}} dy = \int_0^2 \dfrac{\sqrt{4-2y}}{\sqrt{2y - y^2}} \, dy = \int_0^2 \dfrac{\sqrt{2}}{\sqrt{y}} \dfrac{\sqrt{2-y}}{\sqrt{2-y}} \, dy = \sqrt{2} \int_0^2 y^{-1/2} \, dy = \sqrt{2} \cdot 2y^{1/2} \Big]_0^2 = 2\sqrt{2}(\sqrt{2} - 0) = 4.$ Note that, if we integrate using strips parallel to the y-axis, the integration is difficult.

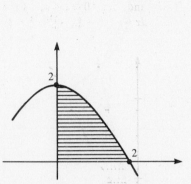

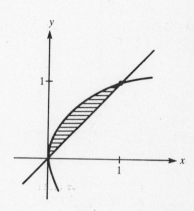

Fig. 44-5   Fig. 44-6

44.15 Let $\mathscr{R}$ be the region bounded by the curve $y = \sqrt{x}$ and the line $y = x$ (Fig. 44-6). Let $f(x, y) = \dfrac{\sin y}{y}$ if $y \neq 0$ and $f(x, 0) = 1$. Compute $I = \iint_{\mathscr{R}} f(x, y) \, dA.$

▊ $I = \int_0^1 \int_{y^2}^{y} \dfrac{\sin y}{y} \, dx \, dy = \int_0^1 \dfrac{\sin y}{y} x \Big]_{y^2}^{y} dy = \int_0^1 (\sin y - y \sin y) \, dy.$ Integration by parts yields $\int y \sin y \, dy = \sin y - y \cos y.$ Hence, $I = (-\cos y + y \cos y - \sin y) \Big]_0^1 = (-\sin 1) - (-1) = \boxed{1 - \sin 1.}$

$①$
H2D162
(7)(xviii)

44.16 Find the volume V under the plane $z = 3x + 4y$ and over the rectangle $\mathscr{R}$: $1 \leq x \leq 2$, $0 \leq y \leq 3$.

▊ $V = \iint_{\mathscr{R}} (3x + 4y) \, dA = \int_0^3 \int_1^2 (3x + 4y) \, dx \, dy = \int_0^3 (\frac{3}{2}x^2 + 4yx) \Big]_1^2 dy = \int_0^3 [(6 + 8y) - (\frac{3}{2} + 4y)] \, dy$

$= \int_0^3 (\frac{9}{2} + 4y) \, dy = (\frac{9}{2}y + 2y^2) \Big]_0^3 = \frac{27}{2} + 18 = \boxed{\frac{63}{2}.}$

44.17 Find the volume V in the first octant bounded by $z = y^2$, $x = 2$, and $y = 4$.

▊ $V = \iint_{\mathscr{R}} y^2 \, dA = \int_0^2 \int_0^4 y^2 \, dy \, dx = \int_0^2 \frac{1}{3}y^3 \Big]_0^4 dx = \int_0^2 \frac{64}{3} \, dx = \frac{64}{3} \cdot 2 = \boxed{\frac{128}{3}.}$

44.18 Find the volume V of the solid in the first octant bounded by $y = 0$, $z = 0$, $y = 3$, $z = x$, and $z + x = 4$ (Fig. 44-7).

▊ For given x and y, the z-value in the solid moves from $z = x$ to $z = -x + 4$. So $V = \int_0^3 \int_0^2 [(-x + 4) - x] \, dx \, dy = \int_0^3 \int_0^2 (4 - 2x) \, dx \, dy = \int_0^3 (4x - x^2) \Big]_0^2 dy = \int_0^3 (8 - 4) \, dy = 4 \int_0^3 dy = 4 \cdot 3 = 12.$

2

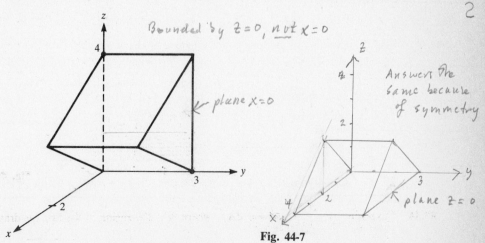

Fig. 44-7

Bounded by $z = 0$, not $x = 0$

plane $x = 0$

Answers the same because of symmetry

plane $z = 0$

$V = \int_0^3 \int_0^2 \int_z^{4-z} dx\, dz\, dy = \int_0^3 \int_0^2 [x]_z^{4-z} dz\, dy = \int_0^3 \int_0^2 (4 - z - z) \, dz\, dy$

$= \int_0^3 \int_0^2 (4 - 2z) \, dz\, dy = \int_0^3 [4z - z^2]_0^2 \, dy = \int_0^3 4 \, dy = [4y]_0^3 = \boxed{12}$

44.19 Find the volume V of the tetrahedron bounded by the coordinate planes and the plane $z = 6 - 2x + 3y$.

▮ As shown in Fig. 44-8, the solid lies above the triangle in the xy-plane bounded by $2x + 3y = 6$ and the x and y axes. $V = \int_0^3 \int_0^{2-2x/3} (6 - 2x - 3y) \, dy \, dx = \int_0^3 6y - 2xy - \frac{3}{2}y^2 \big]_0^{2-2x/3} \, dx = \int_0^3 (2 - \frac{2}{3}x)(6 - 2x - 3 + x) \, dx = \int_0^3 \frac{2}{3}(3 - x)(3 - x) \, dx = \frac{2}{3} \int_0^3 (3 - x)^2 \, dx = \frac{2}{3}(-\frac{1}{3})(3 - x)^3 \big]_0^3 = -\frac{2}{9}(-3^3) = 6$. (Check against the formula $V = \frac{1}{6}abc$.)

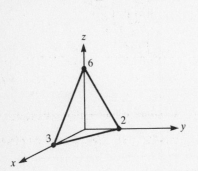

Fig. 44-8

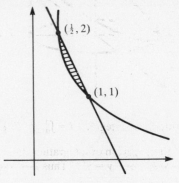

Fig. 44-9

44.20 Use a double integral to find the area of the region $\mathcal{R}$ bounded by $xy = 1$ and $2x + y = 3$.

▮ Figure 44-9 shows the region $\mathcal{R}$. $A = \iint_{\mathcal{R}} 1 \, dA = \int_{1/2}^1 \int_{1/x}^{3-2x} 1 \, dy \, dx = \int_{1/2}^1 (3 - 2x - 1/x) \, dx = 3x - x^2 - \ln x \big]_{1/2}^1 = (3 - 1 - 0) - (\frac{3}{2} - \frac{1}{4} - \ln \frac{1}{2}) = 2 - (\frac{5}{4} + \ln 2) = \boxed{\frac{3}{4} - \ln 2.}$

44.21 Find the volume V of the solid bounded by the right circular cylinder $x^2 + y^2 = 1$, the xy-plane, and the plane $x + z = 1$.

▮ As seen in Fig. 44-10, the base is the circle $x^2 + y^2 = 1$ in the xy-plane, the top is the plane $x + z = 1$. $V = \iint_{\mathcal{R}} (1 - x) \, dA = \int_{-1}^1 \int_{-\sqrt{1-x^2}}^{\sqrt{1-x^2}} (1 - x) \, dy \, dx = \int_{-1}^1 (1 - x)y \big]_{-\sqrt{1-x^2}}^{\sqrt{1-x^2}} \, dx = \int_{-1}^1 2(1 - x)\sqrt{1 - x^2} \, dx = 2 \int_{-1}^1 \sqrt{1 - x^2} \, dx - 2 \int_{-1}^1 x\sqrt{1 - x^2} \, dx = 2(\pi/2) + \frac{2}{3}(1 - x^2)^{3/2} \big]_{-1}^1 = \pi + \frac{2}{3}(0) = \pi$. (Note: We know that $\int_{-1}^1 \sqrt{1 - x^2} \, dx = \pi/2$, since the integral is the area of the unit semicircle.)

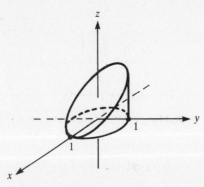

Fig. 44-10

44.22 Find the volume V of the solid bounded above by the plane $z = 3x + y + 6$, below by the xy-plane, and on the sides by $y = 0$ and $y = 4 - x^2$.

▮ Since $-2 \leq x \leq 2$ and $y \geq 0$, we have $z = 3x + y + 6 \geq 0$. Then $V = \int_{-2}^2 \int_0^{4-x^2} (3x + y + 6) \, dy \, dx = \int_{-2}^2 (3xy + \frac{1}{2}y^2 + 6y) \big]_0^{4-x^2} \, dx = \int_{-2}^2 [3x(4 - x^2) + \frac{1}{2}(4 - x^2)^2 + 6(4 - x^2)] \, dx = \int_{-2}^2 (32 + 12x - 10x^2 - 3x^3 + \frac{1}{2}x^4) \, dx = (32x + 6x^2 - \frac{10}{3}x^3 - \frac{3}{4}x^4 + \frac{1}{10}x^5) \big]_{-2}^2 = (64 + 24 - \frac{80}{3} - 12 + \frac{32}{10}) - (-64 + 24 + \frac{80}{3} - 12 - \frac{32}{10}) = \boxed{\frac{1216}{15}.}$

44.23 Find the volume of the wedge cut from the elliptical cylinder $9x^2 + 4y^2 = 36$ by the planes $z = 0$ and $z = y + 3$.

▮ On $9x^2 + 4y^2 = 36$, $-3 \leq y \leq 3$. Hence, $z = y + 3 \geq 0$. So the plane $z = y + 3$ will be above the plane $z = 0$ (see Fig. 44-11). Since the solid is symmetric with respect to the yz-plane, $V = 2 \int_{-3}^3 \int_0^{2\sqrt{9-y^2}/3} (y + 3) \, dx \, dy = 2 \int_{-3}^3 (y + 3)x \big]_0^{2\sqrt{9-y^2}/3} \, dy = 2 \int_{-3}^3 (y + 3) \cdot \frac{2}{3}\sqrt{9 - y^2} \, dy = \frac{4}{3} \int_{-3}^3 y\sqrt{9 - y^2} \, dy + 4 \int_{-3}^3 \sqrt{9 - y^2} \, dy = \frac{4}{3} \cdot (-\frac{1}{2}) \cdot \frac{2}{3}(9 - y^2)^{3/2} \big]_{-3}^3 + 4 \cdot \frac{1}{2}(9\pi) = -\frac{4}{9} \cdot 0 + 9\pi/8 = 9\pi/8$. [The integral $\int_{-3}^3 \sqrt{9 - y^2} \, dy$ represents the area of the upper semicircle of the circle $x^2 + y^2 = 9$. Hence, it is equal to $\frac{1}{2} \cdot \pi(3)^2 = \frac{1}{2} \cdot 9\pi$.]

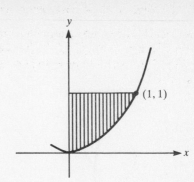

Fig. 44-11

Fig. 44-12

44.24 Express the integral $I = \int_0^1 \int_0^{\sqrt{y}} f(x, y) \, dx \, dy$ as an integral with the order of integration reversed.

▮ In the region of integration, the x-values for $0 \le y \le 1$ range from 0 to $\sqrt{y}$. Hence, the bounding curve is $x = \sqrt{y}$, or $y = x^2$. Thus (see Fig. 44-12), $I = \int_0^1 \int_{x^2}^1 f(x, y) \, dy \, dx$.

44.25 Express the integral $I = \int_0^4 \int_{x/2}^2 f(x, y) \, dy \, dx$ as an integral with the order of integration reversed.

▮ For $0 \le x \le 4$, the region of integration runs from $x/2$ to 2. Hence, the region of integration is the triangle indicated in Fig. 44-13. So, if we use strips parallel to the x-axis, $I = \int_0^2 \int_0^{2y} f(x, y) \, dx \, dy$.

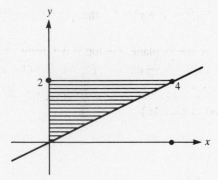

Fig. 44-13

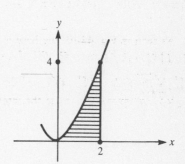

Fig. 44-14

44.26 Express $I = \int_0^2 \int_0^{x^2} f(x, y) \, dy \, dx$ as a double integral with the order of integration reversed.

▮ The region of integration is bounded by $y = 0$, $x = 2$, and $y = x^2$ (Fig. 44-14). $I = \int_0^4 \int_{\sqrt{y}}^2 f(x, y) \, dx \, dy$.

44.27 Express $I = \int_0^{\pi/2} \int_0^{\cos x} x^2 \, dy \, dx$ as double integral with the order of integration reversed and compute its value.

▮ The region of integration is bounded by $y = \cos x$, $y = 0$, and $x = 0$ (Fig. 44-15). So $I = \int_0^1 \int_0^{\cos^{-1} y} x^2 \, dx \, dy$. The original form is easier to calculate. $I = \int_0^{\pi/2} x^2 y \,]_0^{\cos x} \, dx = \int_0^{\pi/2} x^2 \cos x \, dx$. Two integrations by parts yields $\int x^2 \cos x \, dx = x^2 \sin x + 2x \cos x - 2 \sin x$. Hence, $I = (x^2 \sin x + 2x \cos x - 2 \sin x) \,]_0^{\pi/2} = \boxed{\pi^2/4 - 2.}$

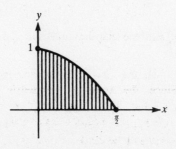

Fig. 44-15

44.28 Find $I = \iint_{\mathcal{R}} e^{x^3}\, dA$, where $\mathcal{R}$ is the region bounded by $y = x^2$, $x = 3$, and $y = 0$.

▮ Use strips parallel to the y-axis (see Fig. 44-16). $I = \int_0^3 \int_0^{x^2} e^{x^3}\, dy\, dx = \int_0^3 e^{x^3} y \,]_0^{x^2}\, dx = \int_0^3 e^{x^3} x^2\, dx = \tfrac{1}{3} e^{x^3}\,]_0^3 = \tfrac{1}{3}(e^{27} - e^0) = \boxed{\tfrac{1}{3}(e^{27} - 1).}$ Note that the integral with the variables in reverse order would have been impossible to calculate.

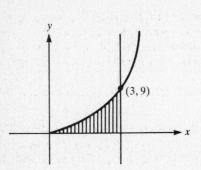

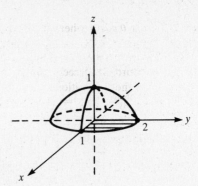

Fig. 44-16   Fig. 44-17

44.29 Find the volume cut from $4x^2 + y^2 + 4z = 4$ by the plane $z = 0$.

▮ The elliptical paraboloid $4x^2 + y^2 + 4z = 4$ has its vertex at $(0, 0, 1)$ and opens downward. It cuts the xy-plane in an ellipse, $4x^2 + y^2 = 4$, which is the boundary of the base $\mathcal{R}$ of the solid whose volume is to be computed (see Fig. 44-17). Because of symmetry, we need to integrate only over the first-quadrant portion of $\mathcal{R}$ and then multiply by 4. $V = 4 \cdot \tfrac{1}{4} \int_0^1 \int_0^{\sqrt{4-4x^2}} (4 - 4x^2 - y^2)\, dy\, dx = \int_0^1 [(4 - 4x^2)y - \tfrac{1}{3} y^3]\,]_0^{\sqrt{4-4x^2}}\, dx = \int_0^1 \sqrt{4 - 4x^2}[4 - 4x^2 - \tfrac{1}{3}(4 - 4x^2)]\, dx = \int_0^1 \tfrac{2}{3}(4 - 4x^2)^{3/2}\, dx = \tfrac{2}{3} \int_0^1 8(1 - x^2)^{3/2}\, dx = \tfrac{16}{3} \int_0^1 (1 - x^2)^{3/2}\, dx.$ Let

(1) use
$H2 \varnothing 1C2(9)$
(xvi)

$x = \sin \theta$, $dx = \cos \theta\, d\theta$. Then $V = \tfrac{16}{3} \int_0^{\pi/2} \cos^4 \theta\, d\theta = \dfrac{16}{3} \int_0^{\pi/2} \left(\dfrac{1 + \cos 2\theta}{2}\right)^2 d\theta = \tfrac{4}{3} \int_0^{\pi/2} (1 + 2\cos 2\theta + \cos^2 2\theta)\, d\theta = \tfrac{4}{3} \int_0^{\pi/2} \left(1 + 2\cos 2\theta + \dfrac{1 + \cos 4\theta}{2}\right) d\theta = \tfrac{4}{3}(\tfrac{3}{2}\theta + \sin 2\theta + \tfrac{1}{8}\sin 4\theta)\,]_0^{\pi/2} = \tfrac{4}{3}(3\pi/4) = \boxed{\pi.}$

44.30 Find the volume in the first octant bounded by $x^2 + z = 64$, $3x + 4y = 24$, $x = 0$, $y = 0$, and $z = 0$.

▮ See Fig. 44-18. The roof of the solid is given by $z = 64 - x^2$. The base $\mathcal{R}$ is the triangle in the first quadrant of the xy-plane bounded by the line $3x + 4y = 24$ and the coordinate axes. Hence, $V = \iint_{\mathcal{R}} (64 - x^2)\, dA = \int_0^8 \int_0^{(24-3x)/4} (64 - x^2)\, dy\, dx = \int_0^8 (64 - x^2)\,]_0^{(24-3x)/4}\, dx = \int_0^8 (64 - x^2) \cdot \tfrac{3}{4}(8 - x)\, dx = \tfrac{3}{4} \int_0^8 (512 - 64x - 8x^2 + x^3)\, dx = \tfrac{3}{4}(512x - 32x^2 - \tfrac{8}{3}x^3 + \tfrac{1}{4}x^4)\,]_0^8 = \tfrac{3}{4}(2^{12} - 2^{11} - \tfrac{1}{3} \cdot 2^{12} + \tfrac{1}{4} \cdot 2^{12}) = 3 \cdot 2^{10}(1 - \tfrac{1}{2} - \tfrac{1}{3} + \tfrac{1}{4}) = 3 \cdot 2^{10} \cdot \tfrac{5}{12} = 2^8 \cdot 5 = \boxed{1280.}$

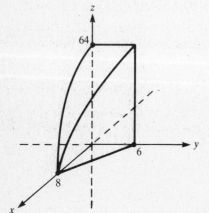

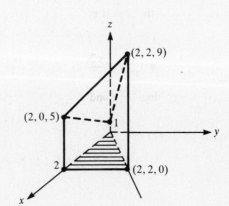

Fig. 44-18   Fig. 44-19

44.31 Find the volume in the first octant bounded by $2x + 2y - z + 1 = 0$, $y = x$, and $x = 2$.

▮ See Fig. 44-19. $V = \iint_{\mathcal{R}} (2x + 2y + 1)\, dA = \int_0^2 \int_0^x (2x + 2y + 1)\, dy\, dx = \int_0^2 (2xy + y^2 + y)\,]_0^x\, dx = \int_0^2 (3x^2 + x)\, dx = (x^3 + \tfrac{1}{2}x^2)\,]_0^2 = 8 + 2 = \boxed{10.}$

44.32 Find the volume of a wedge cut from the cylinder $4x^2 + y^2 = a^2$ by the planes $z = 0$ and $z = my$.

▮ The base is the semidisk $\mathcal{R}$ bounded by the ellipse $4x^2 + y^2 = a^2$, $y \ge 0$. Because of symmetry, we need only double the first-octant volume. Thus, $V = 2 \int_0^{a/2} \int_0^{\sqrt{a^2-4x^2}} my \, dy \, dx = m \int_0^{a/2} y^2 \,]_0^{\sqrt{a^2-4x^2}} dx = m \int_0^{a/2} (a^2 - 4x^2) \, dx = m(a^2 x - \frac{4}{3} x^3) \,]_0^{a/2} = m \cdot (a/2)[a^2 - \frac{4}{3}(a^2/4)] = (ma/2)(\frac{2}{3} a^2) = (m/3)a^3$.

44.33 Find $I = \iint_{\mathcal{R}} \sin \theta \, dA$, where $\mathcal{R}$ is the region outside the circle $r = 1$ and inside the cardioid $r = 1 + \cos \theta$ (see Fig. 44-20).

▮ For polar coordinates, recall that the factor r is introduced into the integrand via $dA = r \, dr \, d\theta$. By symmetry, we can restrict the integration to the first quadrant and double the result. $I = 2 \int_0^{\pi/2} \int_1^{1+\cos\theta} (\sin \theta) r \, dr \, d\theta = 2 \int_0^{\pi/2} \frac{1}{2} (\sin \theta) r^2 \,]_1^{1+\cos\theta} d\theta = \int_0^{\pi/2} [(1 + \cos \theta)^2 \sin \theta - \sin \theta] \, d\theta = [-(1+\cos\theta)^3/3 + \cos \theta] \,]_0^{\pi/2} = -\frac{1}{3} - (-\frac{8}{3} + 1) = \boxed{\frac{4}{3}}$.

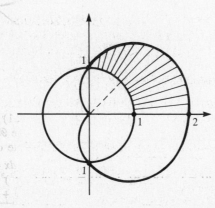

Fig. 44-20

44.34 Use cylindrical coordinates to calculate the volume of a sphere of radius a.

▮ In cylindrical coordinates, the sphere with center $(0, 0, 0)$ is $r^2 + z^2 = a^2$. Calculate the volume in the first octant and multiply it by 8. The base is the quarter disk $0 \le r \le a$, $0 \le \theta \le \pi/2$. $V = 8 \int_0^{\pi/2} \int_0^a \sqrt{a^2 - r^2} \, r \, dr \, d\theta = 8 \int_0^{\pi/2} -\frac{1}{2} \cdot \frac{2}{3} (a^2 - r^2)^{3/2} \,]_0^a \, d\theta = 8 \int_0^{\pi/2} \frac{1}{3} a^3 \, d\theta = (8a^3/3) \int_0^{\pi/2} d\theta = (8a^3/3)(\pi/2) = \boxed{\frac{4}{3} \pi a^3}$, the standard formula.

44.35 Use polar coordinates to evaluate $I = \int_0^1 \int_0^{\sqrt{1-x^2}} (x^2 + y^2)^{5/2} \, dy \, dx$.

▮ The region of integration is the part of the unit disk in the first quadrant: $0 \le \theta \le \pi/2$, $0 \le r \le 1$. Hence, $I = \int_0^{\pi/2} \int_0^1 r^5 \cdot r \, dr \, d\theta = \int_0^{\pi/2} \int_0^1 r^6 \, dr \, d\theta = \int_0^{\pi/2} \frac{1}{7} r^7 \,]_0^1 \, d\theta = \frac{1}{7} \int_0^{\pi/2} d\theta = \frac{1}{7}(\pi/2) = \boxed{\pi/14}$.

44.36 Find the area of the region enclosed by the cardioid $r = 1 + \cos \theta$.

▮ $A = \iint_{\mathcal{R}} 1 \, dA \stackrel{?}{=} \int_0^{2\pi} \int_0^{1+\cos\theta} r \, dr \, d\theta = \int_0^{2\pi} \frac{1}{2} r^2 \,]_0^{1+\cos\theta} \, d\theta = \frac{1}{2} \int_0^{2\pi} (1 + 2\cos\theta + \cos^2\theta) \, d\theta = \frac{1}{2} \int_0^{2\pi} \left(1 + 2\cos\theta + \frac{1+\cos 2\theta}{2} \right) d\theta = \frac{1}{2}(\frac{3}{2}\theta + 2\sin\theta + \frac{1}{4}\sin 2\theta) \,]_0^{2\pi} = \frac{1}{2}(\frac{3}{2} \cdot 2\pi) = \boxed{\frac{3\pi}{2}}$.   (or) $2\int_0^{\pi} \int_0^{1+\cos\theta} r \, dr \, d\theta$

44.37 Use polar coordinates to find the area of the region inside the circle $x^2 + y^2 = 9$ and to the right of the line $x = \frac{3}{2}$.

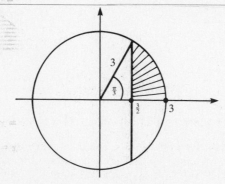

Fig. 44-21

2

▮ It suffices to double the area in the first quadrant. $x^2 + y^2 = 9$ becomes $r = 3$ in polar coordinates, and $x = \frac{3}{2}$ is equivalent to $r \cos \theta = \frac{3}{2}$. From Fig. 44-21, $0 \le \theta \le \pi/3$. So the area is $2 \int_0^{\pi/3} \int_{3/(2 \cos \theta)}^3 r \, dr \, d\theta =$ $\int_0^{\pi/3} r^2 \big]_{(3 \sec \theta)/2}^3 \, d\theta = \int_0^{\pi/3} (9 - \frac{9}{4} \sec^2 \theta) \, d\theta = (9\theta - \frac{9}{4} \tan \theta) \big]_0^{\pi/3} = 3\pi - (9\sqrt{3}/8).$ $\quad (\tan(\pi/3) = \sqrt{3})$

44.38 Describe the planar region $\mathcal{R}$ whose area is given by the iterated integral $I = \int_\pi^{2\pi} \int_1^{1 - \sin \theta} r \, dr \, d\theta.$ *see HSPT1 (2)*

▮ $r = 1 - \sin \theta$ is a cardioid, and $r = 1$ is the unit circle. Between $\theta = \pi$ and $\theta = 2\pi$, $1 < 1 - \sin \theta$ and the cardioid is outside the circle. Therefore, $\mathcal{R}$ is the region outside the unit circle and inside the cardioid (Fig. 44-22).

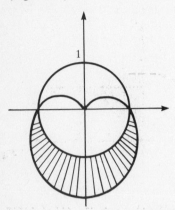

Fig. 44-22

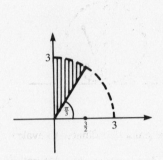

Fig. 44-23

44.39 Evaluate the integral $I = \int_0^{3/2} \int_{\sqrt{3}x}^{\sqrt{9 - x^2}} 2xy \, dy \, dx$ using (*a*) rectangular coordinates and (*b*) polar coordinates.

▮ (*a*) $I = \int_0^{3/2} xy^2 \big]_{\sqrt{3}x}^{\sqrt{9 - x^2}} dx = \int_0^{3/2} [x(9 - x^2) - 3x^3] \, dx = \int_0^{3/2} (9x - 4x^3) \, dx = (\frac{9}{2}x^2 - x^4) \big]_0^{3/2} = \frac{9}{4}(\frac{9}{2} - \frac{9}{4}) =$ $(\frac{9}{4})^2 = \boxed{\frac{81}{16}}.$ (*b*) As indicated in Fig. 44-23, the region of integration lies under the semicircle $y = \sqrt{9 - x^2}$ (or $r = 3$) and above the line $y = \sqrt{3}x$ (or $\theta = \pi/3$). Hence, $I = 2 \int_{\pi/3}^{\pi/2} \int_0^3 r \cos \theta \cdot r \cos \theta \cdot r \, dr \, d\theta =$ $2 \int_{\pi/3}^{\pi/2} \int_0^3 r^3 \cos \theta \sin \theta \, dr \, d\theta = \frac{1}{2} \int_{\pi/3}^{\pi/2} r^4 \cos \theta \sin \theta \big]_0^3 \, d\theta = \frac{81}{2} \int_{\pi/3}^{\pi/2} \cos \theta \sin \theta \, d\theta = \frac{81}{2} \cdot \frac{1}{2} \sin^2 \theta \big]_{\pi/3}^{\pi/2} = \frac{81}{4}(1 - \frac{3}{4}) =$ $\boxed{\frac{81}{16}}.$

44.40 Find the volume of the solid cut out from the sphere $x^2 + y^2 + z^2 \le 4$ by the cylinder $x^2 + y^2 = 1$ (see Fig. 44-24).

▮ It suffices to multiply by 8 the volume of the solid in the first octant. Use cylindrical coordinates. The sphere is $r^2 + z^2 = 4$ and the cylinder is $r = 1$. Thus, we have $V = 8 \int_0^{\pi/2} \int_0^1 \sqrt{4 - r^2} \, r \, dr \, d\theta =$ $8 \int_0^{\pi/2} [-\frac{1}{2} \cdot \frac{2}{3}(4 - r^2)^{3/2}] \big]_0^1 \, d\theta = -\frac{8}{3} \int_0^{\pi/2} [(3)^{3/2} - 8] \, d\theta = \frac{8}{3}(8 - 3\sqrt{3}) \int_0^{\pi/2} d\theta = \boxed{\frac{4\pi}{3}(8 - 3\sqrt{3})}.$

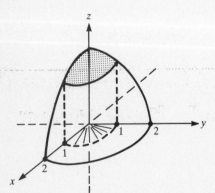

Fig. 44-24

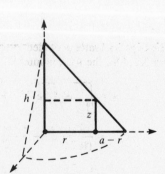

Fig. 44-25

44.41 Use integration in cylindrical coordinates to find the volume of a right circular cone of radius a and height h.

▮ The base is the disk of radius a, given by $r \le a$. For given r, the corresponding value of z on the cone is determined by $z/(a - r) = h/a$ (obtained by similar triangles; see Fig. 44-25.) Then $V = \int_0^{2\pi} \int_0^a (h/a)(a - r)r \, dr \, d\theta = \int_0^{2\pi} [\frac{1}{2}hr^2 - (h/3a)r^3] \big]_0^a \, d\theta = \int_0^{2\pi} (\frac{1}{2}ha^2 - \frac{1}{3}ha^2) \, d\theta = \frac{1}{6}ha^2 \int_0^{2\pi} d\theta = \frac{1}{6}ha^2(2\pi) = \boxed{\frac{1}{3}\pi a^2 h}$, the standard formula.

44.42 Find the average distance from points in the unit disk to a fixed point on the boundary.

▮ For the unit circle $r = 2 \sin \theta$, with the pole as the fixed point (Fig. 44-26), the distance of an interior point to the pole is r. Thus, $\bar{r} = \dfrac{1}{\text{area of } \mathcal{R}} \cdot \displaystyle\iint_{\mathcal{R}} r \, dA = \dfrac{1}{\pi} \int_0^\pi \int_0^{2 \sin \theta} r \cdot r \, dr \, d\theta = \dfrac{1}{\pi} \int_0^\pi \dfrac{1}{3} r^3 \Big]_0^{2 \sin \theta} d\theta =$

$\dfrac{8}{3\pi} \displaystyle\int_0^\pi \sin^3 \theta \, d\theta = \dfrac{8}{3\pi} \int_0^\pi (\sin \theta - \cos^2 \theta \sin \theta) \, d\theta = \dfrac{8}{3\pi} \left(-\cos \theta + \dfrac{1}{3} \cos^3 \theta \right) \Big]_0^\pi = \dfrac{8}{3\pi} [(1 - \tfrac{1}{3}) - (-1 + \tfrac{1}{3})] =$
$\dfrac{8}{3\pi} \left(\dfrac{4}{3} \right) = \dfrac{32}{9\pi}$.

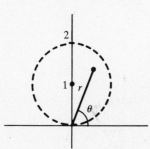

Fig. 44-26

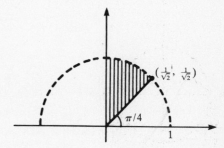

Fig. 44-27

44.43 Use polar coordinates to evaluate $I = \int_0^{1/\sqrt{2}} \int_x^{\sqrt{1-x^2}} \sqrt{x^2 + y^2} \, dy \, dx$.

▮ The region of integration is indicated in Fig. 44-27. $I = \int_{\pi/4}^{\pi/2} \int_0^1 r \cdot r \, dr \, d\theta = \left(\int_0^1 r^2 \, dr \right)\left(\int_{\pi/4}^{\pi/2} d\theta \right) =$
$\left(\dfrac{1}{3} \right)\left(\dfrac{\pi}{4} \right) = \boxed{\dfrac{\pi}{12}}$.

44.44 Use polar coordinates to evaluate $I = \int_0^2 \int_0^{\sqrt{2x - x^2}} x \, dy \, dx$.

▮ $y = \sqrt{2x - x^2}$ is equivalent to $y^2 = 2x - x^2$, $y \ge 0$, or the upper half of the circle $(x - 1)^2 + y^2 = 1$ (see Fig. 44-28). In polar coordinates, $x^2 + y^2 = 2x$ is equivalent to $r^2 = 2r \cos \theta$, or $r = 2 \cos \theta$. Thus, $I = \int_0^{\pi/2} \int_0^{2 \cos \theta} (r \cos \theta) r \, dr \, d\theta = \int_0^{\pi/2} \int_0^{2 \cos \theta} r^2 \cos \theta \, dr \, d\theta = \int_0^{\pi/2} \dfrac{1}{3} r^3 \cos \theta \Big]_0^{2 \cos \theta} d\theta = \dfrac{8}{3} \int_0^{\pi/2} \cos^4 \theta \, d\theta = \boxed{\dfrac{\pi}{2}}$ (by Problem 44.29).

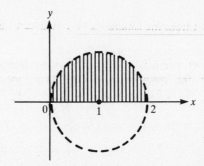

Fig. 44-28

44.45 If the depth of water provided by a water sprinkler in a given unit of time is 2^{-r} feet at a distance of r feet from the sprinkler, find the total volume of water within a distance of a feet from the sprinkler after one unit of time.

▮ $V = \displaystyle\int_0^{2\pi} d\theta \int_0^a e^{-r \ln 2} r \, dr = 2\pi \left[\dfrac{e^{-r \ln 2}}{(\ln 2)^2} (-r \ln 2 - 1) \right]_0^a = \dfrac{2\pi}{(\ln 2)^2} [1 - e^{-a \ln 2}(1 + a \ln 2)]$ See SCR p 520

$= \dfrac{2\pi}{(\ln 2)^2} \left(1 - \dfrac{1 + a \ln 2}{2^a} \right)$.

44.46 Evaluate $I = \displaystyle\int_{\sqrt{2}/2}^1 \int_{\sqrt{1 - x^2}}^x \dfrac{1}{\sqrt{x^2 + y^2}} \, dy \, dx$.

▮ The region of integration (Fig. 44-29) consists of all points in the first quadrant above the circle $x^2 + y^2 = 1$ and under the line $y = x$. Transform to polar coordinates, noting that $x = 1$ is equivalent to $r = \sec \theta$.
$I = \displaystyle\int_0^{\pi/4} \int_1^{\sec \theta} \dfrac{1}{r} \cdot r \, dr \, d\theta = \int_0^{\pi/4} \int_1^{\sec \theta} dr \, d\theta = \int_0^{\pi/4} (\sec \theta - 1) \, d\theta = (\ln |\sec \theta + \tan \theta| - \theta) \Big]_0^{\pi/4} = \ln (\sqrt{2} + 1) - \pi/4$.

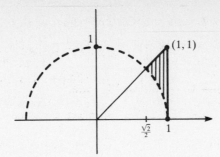

Fig. 44-29

44.47 | Show that $\int_0^\infty e^{-x^2} dx = \frac{\sqrt{\pi}}{2}$.

cf:
$1 \int\int 2(10)$

▌ Let $c = \int_0^\infty e^{-x^2} dx$. Then $c = \int_0^\infty e^{-y^2} dy$. Hence, $c^2 = \int_0^\infty e^{-x^2} dx \cdot \int_0^\infty e^{-y^2} dy = \int_0^\infty e^{-y^2} (\int_0^\infty e^{-x^2} dx) dy = \int_0^\infty \int_0^\infty e^{-x^2} e^{-y^2} dx \, dy = \int_0^\infty \int_0^\infty e^{-(x^2+y^2)} dx \, dy$. The region of integration is the entire first quadrant. Change to polar coordinates. $c^2 = \int_0^\infty \int_0^{\pi/2} e^{-r^2} r \, d\theta \, dr = \int_0^\infty e^{-r^2} \cdot r\theta \,]_0^{\pi/2} dr = \frac{\pi}{2} \int_0^\infty e^{-r^2} r \, dr = (\pi/2)(-\frac{1}{2})e^{-r^2}]_0^\infty = -(\pi/4)(\lim_{r\to+\infty} e^{-r^2} - e^0) = -(\pi/4)(0-1) = \pi/4$. Since $c^2 = \pi/4$, $c = \sqrt{\pi}/2$. (The rather loose reasoning in this computation can be made rigorous.)

44.48 | Evaluate $I = \int_0^\infty \frac{e^{-x}}{\sqrt{x}} dx$.

▌ Let $u = \sqrt{x}$, $x = u^2$, $dx = 2u \, du$. Then $I = \int_0^\infty \frac{e^{-u^2}}{u} \cdot 2u \, du = 2 \int_0^\infty e^{-u^2} du = 2 \cdot \frac{\sqrt{\pi}}{2} = \boxed{\sqrt{\pi},}$ by Problem 44.47.

44.49 | Evaluate $I = \int_0^\infty x^2 e^{-x^2} dx$.

▌ Consider $I_w = \int_0^w x^2 e^{-x^2} dx$. Use integration by parts. Let $u = x$, $dv = xe^{-x^2} dx$, $du = dx$, $v = -\frac{1}{2}e^{-x^2}$. Then $I_w = -\frac{1}{2}xe^{-x^2}]_0^w + \frac{1}{2}\int_0^w e^{-x^2} dx$. Hence, $I = \lim_{w\to+\infty} I_w = \lim_{w\to+\infty} (-\frac{1}{2}we^{-w^2}) + \frac{1}{2}\int_0^\infty e^{-x^2} dx = 0 + \frac{1}{2}\sqrt{\pi}/2 = \boxed{\sqrt{\pi}/4,}$ by Problem 44.47.

$1\int\int 6$ **44.50** | Use spherical coordinates to find the volume of a sphere of radius a.

(11)

▌ In spherical coordinates a sphere of radius a is characterized by $0 \le \rho \le a$, $0 \le \theta \le 2\pi$, $0 \le \phi \le \pi$. Recall that the volume element is given by $dV = \rho^2 \sin\phi \, d\rho \, d\theta \, d\phi$. $V = \int_0^\pi \int_0^{2\pi} \int_0^a \rho^2 \sin\phi \, d\rho \, d\theta \, d\phi = \int_0^\pi \int_0^{2\pi} \frac{1}{3}\rho^3 \sin\phi \,]_0^a d\theta \, d\phi = \int_0^\pi \int_0^{2\pi} \frac{1}{3}a^3 \sin\phi \, d\theta \, d\phi = \int_0^\pi \frac{1}{3}a^3 \sin\phi \cdot \theta \,]_0^{2\pi} d\phi = \int_0^\pi \frac{1}{3}a^3 \sin\phi \cdot 2\pi \, d\phi = \frac{2}{3}\pi a^3 \int_0^\pi \sin\phi \, d\phi = -\frac{2}{3}\pi a^3 \cos\phi \,]_0^\pi = -\frac{2}{3}\pi a^3(-1-1) = \frac{4}{3}\pi a^3$.

44.51 | Use spherical coordinates to find the volume of a right circular cone of height h and radius of base b.

$1\int\int 6$
(12)

▌ For the orientation shown in Fig. 44-30, the points of the cone satisfy $0 \le \theta \le 2\pi$, $0 \le \phi \le \tan^{-1}(b/h)$, $0 \le \rho \le h\sec\phi$. Thus, $V = \int_0^{2\pi} \int_0^{\tan^{-1}(b/h)} \int_0^{h\sec\phi} \rho^2 \sin\phi \, d\rho \, d\phi \, d\theta = \int_0^{2\pi} \int_0^{\tan^{-1}(b/h)} \frac{1}{3}\rho^3 \sin\phi \,]_0^{h\sec\phi} d\phi \, d\theta = \int_0^{2\pi} \int_0^{\tan^{-1}(b/h)} \frac{1}{3}h^3 \frac{\sin\phi}{\cos^3\phi} d\phi \, d\theta = \frac{1}{3}h^3 \int_0^{2\pi} \frac{1}{2\cos^2\theta} \,]_0^{\tan^{-1}(b/h)} d\theta = \frac{1}{3}h^3 \int_0^{2\pi} \frac{1}{2}\left(\frac{s^2}{h^2} - 1\right) d\theta = \frac{h^3}{6h^2} \int_0^{2\pi} (s^2 - h^2) \, d\theta = \frac{hb^2}{6} \int_0^{2\pi} d\theta = \frac{hb^2}{6} \cdot 2\pi = \frac{1}{3}\pi b^2 h$.

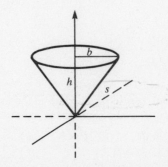

Fig. 44-30

44.52 Find the average distance $\bar{\rho}$ from the center of a ball $\mathscr{B}$ of radius a to all other points of the ball.

cf: 1554
(9)

▌ $\bar{\rho} = \dfrac{1}{\text{volume } V} \cdot \displaystyle\iiint_{\mathscr{B}} \rho\, dV = \dfrac{1}{\frac{4}{3}\pi a^3} \int_0^{2\pi} \int_0^{\pi} \int_0^{a} \rho \cdot \rho^2 \sin\phi\, d\rho\, d\phi\, d\theta = \dfrac{3}{4\pi a^3} \int_0^{2\pi} \int_0^{\pi} \dfrac{1}{4} \rho^4 \sin\phi \Big]_0^{a} d\phi\, d\theta =$

$\dfrac{3}{4\pi a^3} \displaystyle\int_0^{2\pi} \int_0^{\pi} \dfrac{1}{4} a^4 \sin\phi\, d\phi\, d\theta = \dfrac{3a}{16\pi} \int_0^{2\pi} -\cos\phi \Big]_0^{\pi} d\theta = \dfrac{3a}{16\pi} \int_0^{2\pi} -(-1-1)\, d\theta = \dfrac{3a}{8\pi} \int_0^{2\pi} d\theta = \dfrac{3a}{8\pi} \cdot 2\pi = \boxed{\dfrac{3}{4}a}.$ *

44.53 Find the area S of the part of the plane $x + 2y + z = 4$ which lies inside the cylinder $x^2 + y^2 = 1$.

1553
(5)EX

▌ Recall (Fig. 44-31) the relation $dA = dS \cos\theta = dS\, \dfrac{1}{\sqrt{1 + z_x^2 + z_y^2}}$ between an element of area dS of a surface $z = f(x, y)$ and its projection dA in the xy-plane. Thus, the formula for S is $\displaystyle\iint_{\mathscr{R}} \sqrt{1 + \left(\dfrac{\partial z}{\partial x}\right)^2 + \left(\dfrac{\partial z}{\partial y}\right)^2}\, dA$, where, here, $\mathscr{R}$ is the disk $x^2 + y^2 \le 1$. $1 + \dfrac{\partial z}{\partial x} = 0$, $\dfrac{\partial z}{\partial x} = -1$. $2 + \dfrac{\partial z}{\partial y} = 0$, $\dfrac{\partial z}{\partial y} = -2$. Hence, $S = \displaystyle\iint_{\mathscr{R}} \sqrt{6}\, dA = \sqrt{6}(\text{area } \mathscr{R}) = \boxed{\sqrt{6}\,\pi}.$

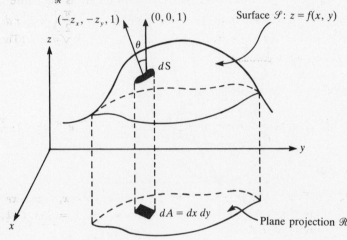

$(-z_x, -z_y, 1)$ $\quad (0,0,1)$ $\qquad$ Surface $\mathscr{S}: z = f(x, y)$

dS

$dA = dx\, dy$ $\qquad$ Plane projection $\mathscr{R}$

Fig. 44-31

44.54 Find the surface area S of the part of the sphere $x^2 + y^2 + z^2 = 36$ inside the cylinder $x^2 + y^2 = 6y$ and above the xy-plane.

1553
(6)EX

▌ $x^2 + y^2 = 6y$ is equivalent to $x^2 + (y-3)^2 = 9$. So the cylinder has axis $x = 0$, $y = 3$, and radius 3. The base $\mathscr{R}$ is the circle $x^2 + (y-3)^2 = 9$. (See Fig. 44-32.) $S = \displaystyle\iint_{\mathscr{R}} \sqrt{1 + \left(\dfrac{\partial z}{\partial x}\right)^2 + \left(\dfrac{\partial z}{\partial y}\right)^2}\, dA$. $2x +$

$2z\dfrac{\partial z}{\partial x} = 0$. Hence, $\dfrac{\partial z}{\partial x} = -\dfrac{x}{z}$. Similarly, $\dfrac{\partial z}{\partial y} = -\dfrac{y}{z}$. Thus, $1 + \left(\dfrac{\partial z}{\partial x}\right)^2 + \left(\dfrac{\partial z}{\partial y}\right)^2 = 1 + \dfrac{x^2}{z^2} + \dfrac{y^2}{z^2} = \dfrac{x^2 + y^2 + z^2}{z^2} = \dfrac{36}{z^2}$. Therefore, $S = \displaystyle\iint_{\mathscr{R}} \dfrac{6}{z}\, dA$. Now use polar coordinates. The circle $x^2 + y^2 = 6y$ is equivalent to $r = 6\sin\theta$. $x^2 + y^2 + z^2 = 36$ is equivalent to $r^2 + z^2 = 36$, or $z^2 = 36 - r^2$. Hence, $S = \displaystyle\int_0^{\pi} \int_0^{6\sin\theta} \dfrac{6}{\sqrt{36 - r^2}}\, r\, dr\, d\theta = \int_0^{\pi} -6\sqrt{36 - r^2} \Big]_0^{6\sin\theta} d\theta = -6 \int_0^{\pi} (6|\cos\theta| - 6)\, d\theta = -12\int_0^{\pi/2} (6\cos\theta - 6)\, d\theta$, since $|\cos(\pi - \theta)| = |\cos\theta|$. Hence, $S = -72(\sin\theta - \theta) \Big]_0^{\pi/2} = 72(\pi/2 - 1)$.

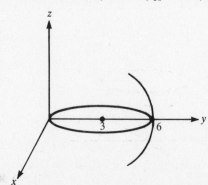

Fig. 44-32

44.55 Find the surface area S of the part of the sphere $x^2 + y^2 + z^2 = 4z$ inside the paraboloid $z = x^2 + y^2$.

$\int\int 3(8)$

▮ The region $\mathcal{R}$ under the spherical cap (Fig. 44-33) is obtained by finding the intersection of $x^2 + y^2 + z^2 = 4z$ and $z = x^2 + y^2$. This gives $z(z-3) = 0$. Hence, the paraboloid cuts the sphere when $z = 3$, and $\mathcal{R}$ is the disk $x^2 + y^2 \le 3$. $S = \iint\limits_{\mathcal{R}} \sqrt{1 + \left(\dfrac{\partial z}{\partial x}\right)^2 + \left(\dfrac{\partial z}{\partial y}\right)^2}\, dA$. $2x + 2z\dfrac{\partial z}{\partial x} = 4\dfrac{\partial z}{\partial x}$, $\dfrac{\partial z}{\partial x} = -\dfrac{x}{z-2}$. Similarly, $\dfrac{\partial z}{\partial y} = -\dfrac{y}{z-2}$. Hence,

$$1 + \left(\frac{\partial z}{\partial x}\right)^2 + \left(\frac{\partial z}{\partial y}\right)^2 = 1 + \frac{x^2}{(z-2)^2} + \frac{y^2}{(z-2)^2} = \frac{(z-2)^2 + x^2 + y^2}{(z-2)^2} = \frac{(x^2 + y^2 + z^2) - 4z + 4}{(z-2)^2} = \frac{4}{(z-2)^2}$$

Therefore,

$$S = \iint\limits_{\mathcal{R}} \frac{2}{z-2}\, dA = \int_0^{2\pi}\int_0^{\sqrt{3}} \frac{2}{\sqrt{4-r^2}}\, r\, dr\, d\theta = -\int_0^{2\pi} 2\sqrt{4-r^2}\, \Big]_0^{\sqrt{3}}\, d\theta = -2\int_0^{2\pi}(1-2)\, d\theta = 2\cdot 2\pi = 4\pi$$

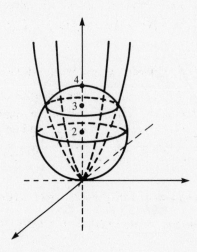

Fig. 44-33

44.56 Find the surface area of the part of the sphere $x^2 + y^2 + z^2 = 25$ between the planes $z = 2$ and $z = 4$.

$\int\int 3(9)$

▮ The surface lies above the ring-shaped region $\mathcal{R}$: $3 \le r \le \sqrt{21}$. $S = \iint\limits_{\mathcal{R}} \sqrt{1 + \left(\dfrac{\partial z}{\partial x}\right)^2 + \left(\dfrac{\partial z}{\partial y}\right)^2}\, dA = \iint\limits_{\mathcal{R}} \sqrt{1 + \dfrac{x^2}{z^2} + \dfrac{y^2}{z^2}}\, dA = \iint\limits_{\mathcal{R}} \sqrt{\dfrac{25}{z^2}}\, dA = \iint\limits_{\mathcal{R}} \dfrac{5}{z}\, dA = \int_0^{2\pi}\int_3^{\sqrt{21}} \dfrac{5}{\sqrt{25-r^2}}\, r\, dr\, d\theta$, since $r^2 + z^2 = 25$. Hence, $S = \int_0^{2\pi} -5\sqrt{25-r^2}\, \Big]_3^{\sqrt{21}}\, d\theta = -5\int_0^{2\pi}(2-4)\, d\theta = 10\int_0^{2\pi} d\theta = 20\pi$.

44.57 Find the surface area of a sphere of radius a.

$\int\int 3(10)$

▮ Consider the upper hemisphere of the sphere $x^2 + y^2 + z^2 = a^2$. It projects down onto the disk $\mathcal{R}$ of radius a whose center is at the origin. Hence, the surface area of the entire sphere is $2\iint\limits_{\mathcal{R}} \sqrt{1 + \left(\dfrac{\partial z}{\partial x}\right)^2 + \left(\dfrac{\partial z}{\partial y}\right)^2}\, dA = 2\iint\limits_{\mathcal{R}} \sqrt{1 + \dfrac{x^2}{z^2} + \dfrac{y^2}{z^2}}\, dA = 2\iint\limits_{\mathcal{R}} \sqrt{\dfrac{x^2 + y^2 + z^2}{z^2}}\, dA = 2\iint\limits_{\mathcal{R}} \dfrac{a}{z}\, dA = 2\int_0^{2\pi}\int_0^{a} \dfrac{a}{\sqrt{a^2-r^2}}\, r\, dr\, d\theta = 2a\int_0^{2\pi}(-\sqrt{a^2-r^2})\,]_0^a\, d\theta = -2a\int_0^{2\pi}(-a)\, d\theta = 2a^2\int_0^{2\pi} d\theta = 2a^2\cdot 2\pi = 4\pi a^2$.

44.58 Find the surface area of a cone of height h and radius of base b.

$\int\int 3(11)$

▮ Consider the cone $z = \dfrac{h}{b}\sqrt{x^2 + y^2}$, or $b^2 z^2 = h^2 x^2 + h^2 y^2$ (see Fig. 44-30). The portion of the cone under $z = h$ projects onto the interior $\mathcal{R}$ of the circle $r = b$ in the xy-plane. Then

$$S = \iint\limits_{\mathcal{R}} \sqrt{1 + \left(\frac{\partial z}{\partial x}\right)^2 + \left(\frac{\partial z}{\partial y}\right)^2} \, dA = \iint\limits_{\mathcal{R}} \sqrt{1 + \left(\frac{h^2 x}{b^2 z}\right)^2 + \left(\frac{h^2 y}{b^2 z}\right)^2} \, dA = \iint\limits_{\mathcal{R}} \sqrt{\frac{b^4 z^2 + h^4 x^2 + h^4 y^2}{b^4 z^2}} \, dA$$

$$= \iint\limits_{\mathcal{R}} \sqrt{\frac{h^2(b^2 + h^2)(x^2 + y^2)}{b^4 z^2}} \, dA = \iint\limits_{\mathcal{R}} \frac{h}{b^2} \sqrt{b^2 + h^2} \frac{\sqrt{x^2 + y^2}}{z} \, dA = \iint\limits_{\mathcal{R}} \frac{h}{b^2} \sqrt{b^2 + h^2} \cdot \frac{b}{h} \, dA$$

$$= \frac{1}{b} \sqrt{b^2 + h^2} \iint\limits_{\mathcal{R}} dA = \frac{\sqrt{b^2 + h^2}}{b} \, (\pi b^2) = \pi b \sqrt{b^2 + h^2} = \pi b s$$

where $s = \sqrt{b^2 + h^2}$ is the slant height of the cone.

44.59 Use a triple integral to find the volume V inside $x^2 + y^2 = 9$, above $z = 0$, and below $x + z = 4$.

▮ $V = \int_{-3}^{3} \int_{-\sqrt{9-x^2}}^{\sqrt{9-x^2}} \int_{0}^{4-x} 1 \, dz \, dy \, dx$. The bounds on z correspond to the requirement that the solid is above $z = 0$ and below $x + z = 4$. The bounds on y come from the equation $x^2 + y^2 = 9$. $V = \int_{-3}^{3} \int_{-\sqrt{9-x^2}}^{\sqrt{9-x^2}} z \,]_0^{4-x} \, dy \, dx = \int_{-3}^{3} \int_{-\sqrt{9-x^2}}^{\sqrt{9-x^2}} (4-x) \, dy \, dx = \int_{-3}^{3} (4-x)y \,]_{-\sqrt{9-x^2}}^{\sqrt{9-x^2}} \, dx = \int_{-3}^{\sqrt{3}} 2(4-x)\sqrt{9-x^2} \, dx = 8\int_{-3}^{3} \sqrt{9-x^2} \, dx + \int_{-3}^{3} (-2x)\sqrt{9-x^2} \, dx$. Now, $\int_{-3}^{3} \sqrt{9-x^2} \, dx$ gives the area of the upper half of the disk $x^2 + y^2 \le 9$ of radius 3, namely, $\frac{1}{2}(9\pi) = \frac{9}{2}\pi$. Also, $\int_{-3}^{3} (-2x)\sqrt{9-x^2} \, dx = \frac{2}{3} (9-x^2)^{3/2} \,]_{-3}^{3} = 0$. Therefore, $V = 8(\frac{9}{2}\pi) = 36\pi$.

44.60 Use a triple integral to find the volume V inside $x^2 + y^2 = 4x$, above $z = 0$, and below $x^2 + y^2 = 4z$.

▮ $x^2 + y^2 = 4x$ is equivalent to $(x-2)^2 + y^2 = 4$, the cylinder of radius 2 with axis $x = 2$, $y = 0$. Hence, $V = \int_0^4 \int_{-\sqrt{4x-x^2}}^{\sqrt{4x-x^2}} \int_0^{(x^2+y^2)/4} 1 \, dz \, dy \, dx$. This is difficult to compute; so let us switch to cylindrical coordinates. The circle $x^2 + y^2 = 4x$ becomes $r = 4\cos\theta$, and we get $V = \int_0^\pi \int_0^{4\cos\theta} \int_0^{r^2/4} r \, dz \, dr \, d\theta = \int_0^\pi \int_0^{4\cos\theta} rz \,]_0^{r^2/4} \, dr \, d\theta = \int_0^\pi \int_0^{4\cos\theta} \frac{1}{4} r^3 \, dr \, d\theta = \int_0^\pi \frac{1}{16} r^4 \,]_0^{4\cos\theta} \, d\theta = \int_0^\pi 16\cos^4\theta \, d\theta = 32 \int_0^{\pi/2} \cos^4\theta \, d\theta = 6\pi$ (Problem 44.29).

44.61 Use a triple integral to find the volume inside the cylinder $r = 4$, above $z = 0$, and below $2z = y$.

▮ The solid is wedge-shaped. The base is the half-disk $0 \le \theta \le \pi$, $0 \le r \le 4$. The height is $\frac{y}{2} = \frac{r\sin\theta}{2}$. Then

$$V = \int_0^\pi \int_0^4 \int_0^{(r\sin\theta)/2} r \, dz \, dr \, d\theta = \int_0^\pi \int_0^4 rz \,]_0^{(r\sin\theta)/2} \, dr \, d\theta = \int_0^\pi \int_0^4 \frac{1}{2} r^2 \sin\theta \, dr \, d\theta = \frac{1}{6} \int_0^\pi r^3 \sin\theta \,]_0^4 \, d\theta$$

$$= \frac{1}{6} \int_0^\pi 64 \sin\theta \, d\theta = \frac{32}{3} (-\cos\theta) \,]_0^\pi = -\frac{32}{3} (-1-1) = \frac{64}{3}$$

44.62 Use a triple integral to find the volume cut from the cone $\phi = \pi/4$ by the sphere $\rho = 2a\cos\phi$.

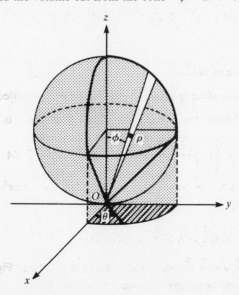

Fig. 44-34

❙ Refer to Fig. 44-34. Note that $\rho = 2a \cos \phi$ has the equivalent forms $\rho^2 = 2a\rho \cos \phi$, $x^2 + y^2 + z^2 = 2az$, $x^2 + y^2 + (z-a)^2 = a^2$. Thus, it is the sphere with center at $(0, 0, a)$ and radius a. Then, $V = \int_0^{2\pi} \int_0^{\pi/4} \int_0^{2a \cos \phi} \rho^2 \sin \phi \, d\rho \, d\phi \, d\theta = \int_0^{2\pi} \int_0^{\pi/4} \frac{1}{3} \rho^3 \sin \phi \,]_0^{2a \cos \phi} d\phi \, d\theta = \int_0^{2\pi} \int_0^{\pi/4} \frac{8a^3}{3} \sin \phi \cos^3 \phi \, d\phi \, d\theta = \frac{8a^3}{3} \int_0^{2\pi} \left(-\frac{\cos^4 \phi}{4} \right) \,]_0^{\pi/4} d\theta = -\frac{2a^3}{3} \int_0^{2\pi} \left(\frac{1}{4} - 1 \right) d\theta = \frac{1}{2} a^3 \int_0^{2\pi} d\theta = \frac{1}{2} a^3 \cdot 2\pi = \pi a^3$.

44.63 Find the volume within the cylinder $r = 4 \cos \theta$ bounded above by the sphere $r^2 + z^2 = 16$ and below by the plane $z = 0$.

❙ Integrate first with respect to z from $z = 0$ to $z = \sqrt{16 - r^2}$, then with respect to r from $r = 0$ to $r = 4 \cos \theta$, and then with respect to θ from $\theta = 0$ to $\theta = \pi$. (See Fig. 44-35.) $V = \int_0^\pi \int_0^{4 \cos \theta} \int_0^{\sqrt{16-r^2}} r \, dz \, dr \, d\theta = \int_0^\pi \int_0^{4 \cos \theta} rz \,]_0^{\sqrt{16-r^2}} dr \, d\theta = \int_0^\pi \int_0^{4 \cos \theta} r \sqrt{16 - r^2} \, dr \, d\theta = \int_0^\pi [-\frac{1}{2} \cdot \frac{2}{3} (16 - r^2)^{3/2}]]_0^{4 \cos \theta} d\theta = -\frac{1}{3} \int_0^\pi (64 \sin^3 \theta - 64) \, d\theta = -\frac{64}{3} \int_0^\pi (\sin \theta - \cos^2 \theta \sin \theta - 1) \, d\theta = -\frac{64}{3} (-\cos \theta + \frac{1}{3} \cos^3 \theta - \theta)]_0^\pi = -\frac{64}{3} [(1 - \frac{1}{3} - \pi) - (-1 + \frac{1}{3})] = -\frac{64}{3} (\frac{4}{3} - \pi) = \frac{64}{9} (3\pi - 4)$.

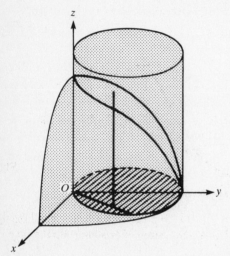

Fig. 44-35

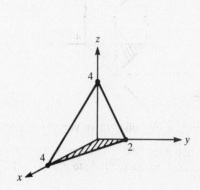

Fig. 44-36

44.64 Evaluate $I = \iiint_{\mathcal{R}} x \, dV$, where $\mathcal{R}$ is the tetrahedron bounded by the coordinate planes and the plane $x + 2y + z = 4$ (see Fig. 44-36).

❙ y can be integrated from 0 to 2. In the base triangle, for a given y, x runs from 0 to $4 - 2y$. For given x and y, z varies from 0 to $4 - x - 2y$. Hence, $I = \int_0^2 \int_0^{4-2y} \int_0^{4-x-2y} x \, dz \, dx \, dy = \int_0^2 \int_0^{4-2y} xz \,]_0^{4-x-2y} dx \, dy = \int_0^2 \int_0^{4-2y} x(4 - x - 2y) \, dx \, dy = \int_0^2 \int_0^{4-2y} (4x - x^2 - 2xy) \, dx \, dy = \int_0^2 (2x^2 - \frac{1}{3} x^3 - yx^2)]_0^{4-2y} dy = \int_0^2 (4 - 2y)^2 [2 - \frac{1}{3}(4 - 2y) - y] \, dy = \int_0^2 (4 - 2y)^2 (\frac{2}{3} - \frac{1}{3} y) \, dy = \frac{1}{3} \int_0^2 (2 - y)^3 \, dy = \frac{4}{3} \cdot (-\frac{1}{4})(2 - y)^4]_0^2 = -\frac{1}{3}(-16) = \frac{16}{3}$.

44.65 Evaluate $I = \iiint_{\mathcal{R}} (x^2 + y^2 + z^2) \, dV$, where $\mathcal{R}$ is the ball of radius a with center at the origin.

❙ Use spherical coordinates. $I = \int_0^{2\pi} \int_0^\pi \int_0^a \rho^2 \cdot \rho^2 \sin \phi \, d\rho \, d\phi \, d\theta = \int_0^{2\pi} \int_0^\pi \frac{1}{5} \rho^5 \sin \phi \,]_0^a d\phi \, d\theta = \frac{1}{5} a^5 \int_0^{2\pi} \int_0^\pi \sin \phi \, d\phi \, d\theta = \frac{1}{5} a^5 \int_0^{2\pi} (-\cos \phi)]_0^\pi d\theta = \frac{1}{5} a^5 \int_0^{2\pi} [1 - (-1)] \, d\theta = \frac{2}{5} a^5 \int_0^{2\pi} d\theta = \frac{2}{5} a^5 \cdot 2\pi = \frac{4}{5} \pi a^5$.

44.66 Use a triple integral to find the volume V of the solid inside the cylinder $x^2 + y^2 = 25$ and between the planes $z = 2$ and $x + z = 8$.

❙ The projection on the xy-plane is the circle $x^2 + y^2 = 25$. Use cylindrical coordinates.

$$V = \int_0^{2\pi} \int_0^5 \int_2^{8 - (r \cos \theta)} r \, dz \, dr \, d\theta = \int_0^{2\pi} \int_0^5 rz \,]_2^{8 - (r \cos \theta)} dr \, d\theta = \int_0^{2\pi} \int_0^5 r(6 - r \cos \theta) \, dr \, d\theta$$

$$= \int_0^{2\pi} (3r^2 - \frac{1}{3} r^3 \cos \theta)]_0^5 \, d\theta = \int_0^{2\pi} 25(3 - \frac{5}{3} \cos \theta) \, d\theta = 25(3\theta - \frac{5}{3} \sin \theta)]_0^{2\pi} = 25(6\pi) = 150\pi.$$

44.67 Find the volume of the solid enclosed by the paraboloids $z = x^2 + y^2$ (upward-opening) and $z = 36 - 3x^2 - 8y^2$ (downward-opening).

▌ The projection of the intersection of the surfaces is the ellipse $4x^2 + 9y^2 = 36$. By symmetry, we can calculate the integral with respect to x and y in the first quadrant and then multiply it by 4.

$$V = 4 \int_0^3 \int_0^{2\sqrt{9-x^2}/3} \int_{x^2+y^2}^{36-3x^2-8y^2} dz\, dy\, dx = 4 \int_0^3 \int_0^{2\sqrt{9-x^2}/3} [(36 - 3x^2 - 8y^2) - (x^2 + y^2)]\, dy\, dx$$

$$= 4 \int_0^3 \int_0^{2\sqrt{9-x^2}/3} (36 - 4x^2 - 9y^2)\, dy\, dx = 4 \int_0^3 (36y - 4x^2 y - 3y^3)\,]_0^{2\sqrt{9-x^2}/3}\, dx$$

$$= \frac{8}{3} \int_0^3 \sqrt{9 - x^2}\, [36 - 4x^2 - 4(9 - x^2)]\, dx = \frac{32}{3} \int_0^3 \sqrt{9 - x^2}\, \left[\frac{2}{3}(9 - x^2) \right] dx = \frac{64}{9} \int_0^3 (9 - x^2)^{3/2}\, dx$$

Let $x = 3 \sin \theta$, $dx = 3 \cos \theta\, d\theta$. Then $V = \dfrac{64}{9} \displaystyle\int_0^{\pi/2} 27 \cos^3 \theta \cdot 3 \cos \theta\, d\theta = 576 \int_0^{\pi/2} \cos^4 \theta\, d\theta = 108\pi$ (from Problem 44.29).

44.68 Describe the solid whose volume is given by the integral $\int_0^2 \int_{2x}^4 \int_0^1 dz\, dy\, dx$.

▌ See Fig. 44-37. The solid lies under the plane $z = 1$ and above the region in the first quadrant of the xy-plane bounded by the lines $y = 2x$ and $y = 4$.

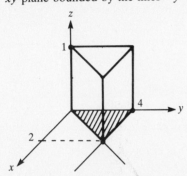

Fig. 44-37

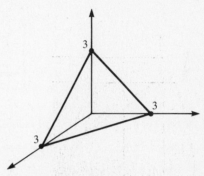

Fig. 44-38

44.69 Describe the solid whose volume is given by the integral $\int_0^3 \int_0^{3-x} \int_0^{3-x-y} dz\, dy\, dx$.

▌ See Fig. 44-38. The solid lies under the plane $z = 3 - x - y$ and above the triangle in the xy-plane bounded by the coordinate axes and the line $x + y = 3$. It is a tetrahedron of volume $\frac{1}{6}(3)^3 = \frac{9}{2}$.

44.70 Describe the solid whose volume is given by the integral $\int_0^5 \int_0^{\sqrt{25-x^2}} \int_0^3 dz\, dy\, dx$, and compute the volume.

▌ The solid is the part of the solid right circular cylinder $x^2 + y^2 \le 25$ lying in the first octant between $z = 0$ and $z = 3$ (see Fig. 44-39). Its volume is $\frac{1}{4}(\pi r^2 h) = \frac{1}{4}[\pi(5)^2 \cdot 3] = \frac{75}{4}\pi$.

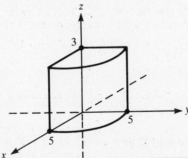

Fig. 44-39

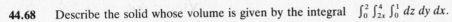

44.71 If $\mathcal{R}$ is a rectangular box $x_1 \le x \le x_2$, $y_1 \le y \le y_2$, $z_1 \le z \le z_2$, show that $I = \iiint_{\mathcal{R}} f(x)g(y)h(z)\, dV = \underbrace{(\int_{x_1}^{x_2} f(x)\, dx)}_{a} \underbrace{(\int_{y_1}^{y_2} g(y)\, dy)}_{b} \underbrace{(\int_{z_1}^{z_2} h(z)\, dz)}_{c}$.

▌ $I = \int_{z_1}^{z_2} \int_{y_1}^{y_2} \int_{x_1}^{x_2} f(x)g(y)h(z)\, dx\, dy\, dz = \int_{z_1}^{z_2} \int_{y_1}^{y_2} g(y)\, h(z)[\int_{x_1}^{x_2} f(x)\, dx]\, dy\, dz = a \int_{z_1}^{z_2} h(z)[\int_{y_1}^{y_2} g(y)\, dy]\, dz = ab \int_{z_1}^{z_2} h(z)\, dz = abc$.

44.72 Evaluate $I = \iiint_{\mathcal{R}} x^3 y e^z\, dV$, where $\mathcal{R}$ is the rectangular box: $1 \le x \le 2$, $0 \le y \le 1$, $0 \le z \le \ln 2$.

▌ By Problem 44.71, $I = \int_1^2 x^3\, dx \cdot \int_0^1 y\, dy \cdot \int_0^{\ln 2} e^z\, dz = \frac{1}{4}x^4\,]_1^2 \cdot \frac{1}{2}y^2\,]_0^1 \cdot e^z\,]_0^{\ln 2} = (4 - \frac{1}{4}) \cdot \frac{1}{2} \cdot (2 - 1) = \frac{15}{8}$.

44.73 Evaluate $J = \iiint_{\mathcal{R}} x^2 \, dV$ for the ball $\mathcal{R}$ of Problem 44.65.

▌ By spherical symmetry, $I \equiv \iiint_{\mathcal{R}} x^2 \, dV + \iiint_{\mathcal{R}} y^2 \, dV + \iiint_{\mathcal{R}} z^2 \, dV = J + J + J$; so $J = \frac{1}{3}I = \dfrac{4\pi a^5}{15}$.

44.74 Find the mass of a plate in the form of a right triangle $\mathcal{R}$ with legs a and b, if the density (mass per unit area) is numerically equal to the sum of the distances from the legs.

▌ Let the right angle be at the origin and the legs a and b be along the positive x and y axes, respectively (Fig. 44-40). The density $\delta(x, y) = x + y$. Hence, the mass $M = \iint (x + y) \, dA = \int_0^a \int_0^{b - (b/a)x} (x + y) \, dy \, dx = \int_0^a (xy + \frac{1}{2} y^2) \big]_0^{b-(b/a)x} dx = \dfrac{b}{a} \int_0^a (a - x)[x + \frac{1}{2}(b/a)(a - x)] \, dx = \dfrac{b}{a} \int_0^a [ax - x^2 + \frac{1}{2}(b/a)(a - x)^2] \, dx = (b/a)[\frac{1}{2}ax^2 - \frac{1}{3}x^3 - \frac{1}{2}(b/a) \cdot \frac{1}{3}(a - x)^3] \big]_0^a = (b/a)\{(\frac{1}{2}a^3 - \frac{1}{3}a^3) - [-\frac{1}{2}(b/a) \cdot \frac{1}{3}a^3]\} = (b/a)(\frac{1}{6}a^3 + \frac{1}{6}ba^2) = \frac{1}{6}ba(a + b)$.

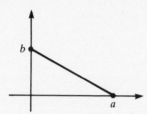

Fig. 44-40

44.75 Find the mass of a circular plate $\mathcal{R}$ of radius a whose density is numerically equal to the distance from the center.

▌ Let the circle be $r = a$. Then $M = \iint_{\mathcal{R}} r \, dA = \int_0^{2\pi} \int_0^a r \cdot r \, dr \, d\theta = \int_0^{2\pi} \frac{1}{3}r^3 \big]_0^a \, d\theta = \int_0^{2\pi} \frac{1}{3}a^3 \, d\theta = \frac{1}{3}a^3 \cdot 2\pi = \frac{2}{3}\pi a^3$.

44.76 Find the mass of a solid right circular cylinder $\mathcal{R}$ of height h and radius of base b, if the density (mass per unit volume) is numerically equal to the square of the distance from the axis of the cylinder.

▌ $M = \iiint_{\mathcal{R}} r^2 \, dV = \int_0^{2\pi} \int_0^b \int_0^h r^2 \cdot r \, dz \, dr \, d\theta = \int_0^{2\pi} \int_0^b r^3 h \, dr \, d\theta = \int_0^{2\pi} \frac{1}{4}hr^4 \big]_0^b \, d\theta = \int_0^{2\pi} \frac{1}{4}hb^4 \, d\theta = \frac{1}{4}hb^4 \cdot 2\pi = \frac{1}{2}\pi hb^4$.

44.77 Find the mass of a ball $\mathcal{B}$ of radius a whose density is numerically equal to the distance from a fixed diametral plane.

▌ Let the ball be the inside of the sphere $x^2 + y^2 + z^2 = a^2$, and let the fixed diametral plane be $z = 0$. Then $M = \iiint |z| \, dV$. Use the upper hemisphere and double the result. In spherical coordinates,

$$M = 2 \int_0^{2\pi} \int_0^{2\pi} \int_0^a z \cdot \rho^2 \sin \phi \, d\rho \, d\phi \, d\theta = 2 \int_0^{2\pi} \int_0^{\pi/2} \int_0^a \rho \cos \phi \cdot \rho^2 \sin \phi \, d\rho \, d\phi \, d\theta$$

$$= 2 \int_0^{2\pi} \int_0^{\pi/2} \frac{1}{4}\rho^4 \cos \phi \sin \phi \big]_0^a \, d\phi \, d\theta = \frac{1}{2}a^4 \int_0^{2\pi} \frac{1}{2} \sin^2 \phi \big]_0^{\pi/2} \, d\theta = \frac{1}{4}a^4 \int_0^{2\pi} d\theta = \frac{1}{4}a^4 \cdot 2\pi = \frac{1}{2}\pi a^4$$

44.78 Find the mass of a solid right circular cone $\mathcal{C}$ of height h and radius of base b whose density is numerically equal to the distance from its axis.

▌ Let $\alpha = \tan^{-1}(b/h)$. Then the cone is $\phi = \alpha$. Use spherical coordinates. $M = \iiint_{\mathcal{C}} r \, dV = \int_0^{2\pi} \int_0^\alpha \int_0^{h \sec \phi} \rho \sin \phi \cdot \rho^2 \sin \phi \, d\rho \, d\phi \, d\theta = \int_0^{2\pi} \int_0^\alpha \frac{1}{4}\rho^4 \sin^2 \phi \big]_0^{h \sec \phi} \, d\phi \, d\theta = \frac{1}{4}h^4 \int_0^{2\pi} \int_0^\alpha \sec^4 \phi \sin^2 \phi \, d\phi \, d\theta = \frac{1}{4}h^4 \int_0^{2\pi} \int_0^\alpha \sec^2 \phi \tan^2 \phi \, d\phi \, d\theta = \frac{1}{4}h^4 \int_0^{2\pi} \frac{1}{3} \tan^3 \phi \big]_0^\alpha \, d\theta = \frac{1}{4}h^4 \cdot \frac{1}{3} \tan^3 \alpha \int_0^{2\pi} d\theta = \frac{1}{12}h^4(b/h)^3 \cdot 2\pi = \frac{1}{6}\pi hb^3$.

44.79 Find the mass of a spherical surface $\mathcal{S}$ whose density is equal to the distance from a fixed diametral plane.

▌ Let $\mathcal{S}$ be $x^2 + y^2 + z^2 = a^2$, and let the fixed diametral plane be the xy-plane. Then $M = \iint_{\mathcal{C}} |z| \dfrac{dS}{dA} \, dA$, where $\mathcal{C}$ is the disk $x^2 + y^2 = a^2$. Then, if we double the mass of the upper hemisphere, $M = 2 \iint_{\mathcal{C}} z \sqrt{1 + \left(\dfrac{\partial z}{\partial x}\right)^2 + \left(\dfrac{\partial z}{\partial y}\right)^2} \, dA$. $\dfrac{\partial z}{\partial x} = -\dfrac{x}{z}$ and $\dfrac{\partial z}{\partial y} = -\dfrac{y}{z}$. Then $1 + \left(\dfrac{\partial z}{\partial x}\right)^2 + \left(\dfrac{\partial z}{\partial y}\right)^2 = 1 + \dfrac{x^2}{z^2} + \dfrac{y^2}{z^2} = \dfrac{x^2 + y^2 + z^2}{z^2} = \dfrac{a^2}{z^2}$. Hence, $M = 2 \iint_{\mathcal{C}} a \, dA = 2a \iint_{\mathcal{C}} dA = 2a(\text{area of a circle of radius } a) = 2a(\pi a^2) = 2\pi a^3$.

44.80 Find the center of mass $(\bar{x}, \bar{y})$ of the plate cut from the parabola $y^2 = 8x$ by its latus rectum $x = 2$ if the density is numerically equal to the distance from the latus rectum.

▌ See Fig. 44-41. By symmetry, $\bar{y} = 0$. The mass $M = \iint (2 - x)\, dA = \int_{-4}^{4} \int_{y^2/8}^{2} (2 - x)\, dx\, dy =$
$\int_{-4}^{4} \left(2x - \frac{1}{2} x^2\right)\Big]_{y^2/8}^{2}\, dy = \int_{-4}^{4} \left[2 - \left(\frac{y^2}{4} - \frac{y^4}{128}\right)\right] dy = \left(2y - \frac{1}{12} y^3 + \frac{1}{128} \frac{y^5}{5}\right)\Big]_{-4}^{4} = 8\left(2 - \frac{16}{12} + \frac{1}{128} \cdot \frac{256}{5}\right) =$
$\frac{128}{15}$. The moment about the y-axis is given by $M_y = \int_{-4}^{4} \int_{y^2/8}^{2} x(2 - x)\, dx\, dy = \int_{-4}^{4} \left(x^2 - \frac{1}{3} x^3\right)\Big]_{y^2/8}^{2}\, dy =$
$\int_{-4}^{4} \left[\frac{4}{3}\left(\frac{y^4}{64} - \frac{y^6}{3 \cdot 512}\right)\right] dy = \left(\frac{4y}{3} - \frac{1}{64} \frac{y^5}{5} + \frac{1}{3 \cdot 512} \frac{y^7}{7}\right)\Big]_{-4}^{4} = 8\left(\frac{4}{3} - \frac{4}{5} + \frac{8}{21}\right) = \frac{256}{35}$. Hence, $\bar{x} = \frac{M_y}{M} =$
$\frac{\frac{256}{35}}{\frac{128}{15}} = \frac{6}{7}$. Thus, the center of mass is $\left(\frac{6}{7}, 0\right)$.

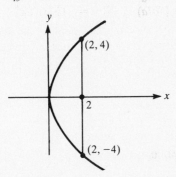

Fig. 44-41

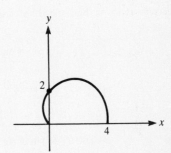

Fig. 44-42

44.81 Find the center of mass of a plate in the form of the upper half of the cardioid $r = 2(1 + \cos \theta)$ if the density is numerically equal to the distance from the pole.

▌ See Fig. 44-42. The mass $M = \iint_{\mathscr{R}} r\, dA = \int_0^\pi \int_0^{2(1+\cos \theta)} r \cdot r\, dr\, d\theta = \int_0^\pi \frac{1}{3} r^3 \Big]_0^{2(1+\cos \theta)}\, d\theta = \frac{8}{3} \int_0^\pi (1 +$
$\cos \theta)^3\, d\theta = \frac{8}{3} \int_0^\pi \left[1 + 3\cos \theta + 3\left(\frac{1 + \cos 2\theta}{2}\right) + (\cos \theta - \sin^2 \theta \cos \theta)\right] d\theta = \frac{8}{3}\left(\frac{5}{2}\theta + 4\sin \theta + \frac{3}{4} \sin 2\theta - \right.$
$\left. \frac{1}{3} \sin^3 \theta\right)\Big]_0^\pi = \frac{8}{3}\left(\frac{5}{2} \cdot \pi\right) = \frac{20\pi}{3}$. The moment about the x-axis is $M_x = \int_0^\pi \int_0^{2(1+\cos \theta)} yr^2\, dr\, d\theta =$
$\int_0^\pi \int_0^{2(1+\cos \theta)} r^3 \sin \theta\, dr\, d\theta = \int_0^\pi \frac{1}{4} r^4 \sin \theta \Big]_0^{2(1+\cos \theta)}\, d\theta = 4 \int_0^\pi (1 + \cos \theta)^4 \sin \theta\, d\theta = -4\frac{(1 + \cos \theta)^5}{5}\Big]_0^\pi =$
$-\frac{4}{5}(-32) = \frac{128}{5}$. Hence, $\bar{y} = \frac{M_x}{M} = \frac{128/5}{20\pi/3} = \frac{96}{25\pi}$. The moment about the y-axis is $M_y =$
$\int_0^\pi \int_0^{2(1+\cos \theta)} xr^2\, dr\, d\theta = \int_0^\pi \int_0^{2(1+\cos \theta)} r^3 \cos \theta\, dr\, d\theta = \int_0^\pi \frac{1}{4} r^4 \cos \theta \Big]_0^{2(1+\cos \theta)}\, d\theta = 4 \int_0^\pi (1 + \cos \theta)^4 \cos \theta\, d\theta =$
$4 \int_0^\pi \left[\cos \theta + 2(1 + \cos 2\theta) + 6(\cos \theta - \sin^2 \theta \cos \theta) + \left(1 + 2\cos 2\theta + \frac{1 + \cos 4\theta}{2}\right) + (\cos \theta - 2\sin^2 \theta \cos \theta + \right.$
$\left. \sin^4 \theta \cos \theta)\right] d\theta = 4(\frac{7}{2}\theta + 8\sin \theta + 2\sin 2\theta + \frac{1}{8} \sin 4\theta - \frac{8}{3} \sin^3 \theta + \frac{1}{5} \sin^5 \theta)\Big]_0^\pi = 4(\frac{7}{2}\pi) = 14\pi$. Hence, $\bar{x} =$
$\frac{M_y}{M} = \frac{14\pi}{20\pi/3} = \frac{21}{10}$. So the center of mass is $\left(\frac{21}{10}, \frac{96}{25\pi}\right)$.

44.82 Find the center of mass of the first-quadrant part of the disk of radius a with center at the origin, if the density function is y.

▌ The mass $M = \int_0^a \int_0^{\sqrt{a^2 - y^2}} y\, dx\, dy = \int_0^a y\sqrt{a^2 - y^2}\, dy = -\frac{1}{2} \cdot \frac{2}{3}(a^2 - y^2)^{3/2}\Big]_0^a = -\frac{1}{3}(-a^3) = \frac{1}{3}a^3$. The moment about the x-axis is

$$M_x = \iint_{\mathscr{R}} y \cdot y\, dA = \int_0^{\pi/2} \int_0^a y^2 \cdot r\, dr\, d\theta = \int_0^{\pi/2} \int_0^a r^3 \sin \theta\, dr\, d\theta = \int_0^{\pi/2} \frac{1}{4} a^4 \sin^2 \theta\, d\theta$$

$$= \frac{1}{4} a^4 \int_0^{\pi/2} \frac{1 - \cos 2\theta}{2}\, d\theta = \frac{1}{8} a^4 \left(\theta - \frac{1}{2} \sin 2\theta\right)\Big]_0^{\pi/2} = \frac{a^4}{8}\left(\frac{\pi}{2}\right) = \frac{\pi a^4}{16}$$

Hence, $\bar{y} = \frac{M_x}{M} = \frac{\pi a^4/16}{a^3/3} = \frac{3\pi a}{16}$. The moment about the y-axis is $M_y = \int_0^a \int_0^{\sqrt{a^2 - y^2}} xy\, dx\, dy =$
$\int_0^a \frac{1}{2} yx^2 \Big]_0^{\sqrt{a^2 - y^2}}\, dy = \frac{1}{2} \int_0^a y(a^2 - y^2)\, dy = \frac{1}{2}\left(\frac{a^2}{2} y^2 - \frac{y^4}{4}\right)\Big]_0^a = \frac{1}{2}\left(\frac{a^4}{2} - \frac{a^4}{4}\right) = \frac{a^4}{8}$. Therefore, $\bar{x} = \frac{M_y}{M} =$
$\frac{a^4/8}{a^3/3} = \frac{3a}{8}$. Hence, the center of mass is $\left(\frac{3a}{8}, \frac{3\pi a}{16}\right)$.

44.83 Find the center of mass of the cube of edge a with three faces on the coordinate planes, if the density is numerically equal to the sum of the distances from the coordinate planes.

▮ The mass $M = \int_0^a \int_0^a \int_0^a (x + y + z)\, dx\, dy\, dz = 3(\int_0^a x\, dx)(\int_0^a dy)(\int_0^a dz) = \frac{3}{2} a^4$. The moment about the xz-plane is $M_{xz} = \int_0^a \int_0^a \int_0^a y(x + y + z)\, dx\, dy\, dz = \int_0^a \int_0^a y\left[\frac{1}{2} x^2 + (y + z)x\right]\Big|_0^a dy\, dz = \int_0^a \int_0^a y\left[\frac{a^2}{2} + a(y + z)\right] dy\, dz = \int_0^a \left[\frac{a^2}{4} y^2 + a\left(\frac{1}{3} y^3 + \frac{1}{2} zy^2\right)\right]\Big|_0^a dz = \int_0^a \left(\frac{a^4}{4} + \frac{a^4}{3} + \frac{a^3}{2} z\right) dz = \frac{7}{12} a^4 z + \frac{a^3}{4} z^2\Big|_0^a = \frac{7}{12} a^5 + \frac{1}{4} a^5 = \frac{5}{6} a^5$. So $\bar{y} = \frac{M_{xz}}{M} = \frac{\frac{5}{6} a^5}{\frac{3}{2} a^4} = \frac{5}{9} a$. By symmetry, $\bar{x} = \frac{5}{9} a$, $\bar{z} = \frac{5}{9} a$.

44.84 Find the center of mass of the first octant of the ball of radius a, $x^2 + y^2 + z^2 \le a^2$, if the density is numerically equal to z.

▮ The mass is one-eighth of that of the ball in Problem 44.77, or $\pi a^4/16$. The moment about the xz-plane is

$$M_{xz} = \int_0^{\pi/2} \int_0^{\pi/2} \int_0^a \underbrace{\rho \sin\phi \sin\theta}_{y} \cdot \underbrace{\rho \cos\phi}_{z} \cdot \rho^2 \sin\phi\, d\rho\, d\phi\, d\theta = \int_0^{\pi/2} \int_0^{\pi/2} \rho^5 \sin^2\phi \cos\phi \sin\theta\, \Big|_0^a d\phi\, d\theta$$

$$= \frac{a^5}{5} \int_0^{\pi/2} \frac{1}{3} \sin^3\phi \sin\theta\, \Big|_0^{\pi/2} d\theta = \frac{a^5}{5} \int_0^{\pi/2} \frac{1}{3} \sin\theta\, d\theta = \frac{a^5}{15} (-\cos\theta)\Big|_0^{\pi/2} = \frac{a^5}{15}.$$

Hence, $\bar{y} = \frac{M_{xz}}{M} = \frac{a^5/15}{\pi a^4/16} = \frac{16a}{15\pi}$. By symmetry, $\bar{x} = \frac{16a}{15\pi}$. The moment about the xy-plane is $M_{xy} = \int_0^{\pi/2} \int_0^{\pi/2} \int_0^a \rho^2 \cos^2\theta \cdot \rho^2 \sin\phi\, d\rho\, d\phi\, d\theta = \int_0^{\pi/2} \int_0^{\pi/2} \frac{1}{5} a^5 \cos^2\phi \sin\theta\, d\phi\, d\theta = \frac{a^5}{5} \int_0^{\pi/2} \left(-\frac{\cos^3\phi}{3}\right)\Big|_0^{\pi/2} d\theta = \frac{a^5}{5} \int_0^{\pi/2} \frac{1}{3} d\theta = \frac{a^5}{15} \cdot \frac{\pi}{2} = \frac{\pi a^5}{30}$. Hence $\bar{z} = \frac{M_{xy}}{M} = \frac{\pi a^5/30}{\pi a^4/16} = \frac{8a}{15}$. Thus, the center of mass is $\left(\frac{16a}{15\pi}, \frac{16a}{15\pi}, \frac{8a}{15}\right)$.

44.85 Find the center of mass of a solid right circular cone $\mathscr{C}$ of height h and radius of base b, if the density is equal to the distance from the base.

▮ Let the cone have the base $x^2 + y^2 \le b$ in the xy-plane and vertex at $(0, 0, h)$; its equation is then $z = h - (h/b)r$. By symmetry, $\bar{x} = \bar{y} = 0$. The mass $M = \iiint z\, dV$. Use cylindrical coordinates. Then

$$M = \int_0^{2\pi} \int_0^b \int_0^{h - (h/b)r} zr\, dz\, dr\, d\theta = \int_0^{2\pi} \int_0^b \frac{1}{2} rz^2\, \Big|_0^{h-(h/b)r} dr\, d\theta = \frac{h^2}{2b^2} \int_0^{2\pi} \int_0^b r(b - r)^2\, dr\, d\theta = \frac{h^2}{2b^2} \int_0^{2\pi} \left(\frac{1}{2} b^2 r^2 - \frac{2b}{3} r^3 + \frac{1}{4} r^4\right)\Big|_0^b d\theta = \frac{h^2}{2b^2} \int_0^{2\pi} \left(\frac{1}{2} b^4 - \frac{2}{3} b^4 + \frac{1}{4} b^4\right) d\theta = \frac{h^2 b^2}{24} \cdot 2\pi = \frac{\pi h^2 b^2}{12}.$$

The moment about the xy-plane is $M_{xy} = \int_0^{2\pi} \int_0^b \int_0^{h - (h/b)r} z^2 r\, dz\, dr\, d\theta = \int_0^{2\pi} \int_0^b \frac{1}{3} rz^3\, \Big|_0^{h-(h/b)r} dr\, d\theta = \frac{h^3}{3b^3} \int_0^{2\pi} \int_0^b r(b - r)^3\, dr\, d\theta = \frac{h^3}{3b^3} \int_0^{2\pi} \left(\frac{b^3}{2} r^2 - b^2 r^3 + \frac{3}{4} br^4 - \frac{1}{5} r^5\right)\Big|_0^b d\theta = \frac{h^3}{3b^3} \int_0^{2\pi} \left(\frac{b^5}{2} - b^5 + \frac{3}{4} b^5 - \frac{1}{5} b^5\right) d\theta = \frac{h^3 b^2}{60} 2\pi = \frac{\pi h^3 b^2}{30}$. Hence, $\bar{z} = \frac{M_{xy}}{M} = \frac{\pi h^3 b^2/30}{\pi h^2 b^2/12} = \frac{2}{5} h$. Thus, the center of mass is $\left(0, 0, \frac{2}{5} h\right)$.

44.86 Find the moments of inertia of the triangle bounded by $3x + 4y = 24$, $x = 0$, and $y = 0$, and having density 1.

▮ The moment of inertia with respect to the x-axis is $I_x = \int_0^8 \int_0^{6-(3/4)x} y^2\, dy\, dx = \int_0^8 \frac{1}{3} y^3\, \Big|_0^{6-(3/4)x} dx = \frac{9}{64} \int_0^8 (8 - x)^3\, dx = \frac{9}{64} (-1) \frac{(8 - x)^4}{4}\Big|_0^8 = \frac{9}{64} \cdot \frac{8^4}{4} = 144$. The moment of inertia with respect to the y-axis is $I_y = \int_0^8 \int_0^{6-(3/4)x} x^2\, dy\, dx = \int_0^8 x^2 \cdot \frac{3}{4} (8 - x)\, dx = \frac{3}{4} \left(\frac{8}{3} x^3 - \frac{x^4}{4}\right)\Big|_0^8 = \frac{3}{4} \cdot \frac{1}{12} (8)^4 = 256$.

44.87 Find the moment of inertia of a square plate of side a with respect to a side, if the density is numerically equal to the distance from an extremity of that side.

▮ Let the square be $0 \le x \le a$, $0 \le y \le a$, and let the density at (x, y) be the distance $\sqrt{x^2 + y^2}$ from the origin. We want to find the moment of inertia I_x about the x-axis: $I_x = \int_0^a \int_0^a y^2 \sqrt{x^2 + y^2}\, dy\, dx$. Now, by the symmetry of the situation, the moment of inertia about the y-axis, $I_y = \int_0^a \int_0^a x^2 \sqrt{x^2 + y^2}\, dy\, dx$, must be equal to I_x. This allows us to write

$$I_x = \frac{1}{2} (I_x + I_y) = \frac{1}{2} \int_0^a \int_0^a (x^2 + y^2)^{3/2}\, dy\, dx = \int_0^x \int_0^x (x^2 + y^2)^{3/2}\, dy\, dx$$

where symmetry was again invoked in the last step.   For an easy evaluation, change to polar coordinates:

$$I_x = \int_0^{\pi/4} \int_0^{a\sec\theta} r^4\, dr\, d\theta = \frac{a^5}{5}\int_0^{\pi/4}\sec^5\theta\, d\theta = \frac{a^5}{5}\left[\frac14\sec^3\theta\tan\theta + \frac38\sec\theta\tan\theta + \frac38\ln\tan\left(\frac{\theta}{2}+\frac{\pi}{4}\right)\right]_0^{\pi/4}$$

$$= \frac{a^5}{5}\left[\frac{\sqrt2}{2} + \frac{3\sqrt2}{8} + \frac38\ln(1+\sqrt2) - 0 - 0 - 0\right] = \frac{a^5}{40}\left[7\sqrt2 + \ln(7+5\sqrt2)\right]$$

44.88   Find the moment of inertia of a cube of edge a with respect to an edge if the density is numerically equal to the square of the distance from one extremity of that edge.

▌ Consider the cube $0\le x\le a,\; 0\le y\le a,\; 0\le z\le a$.   Let the density be $x^2+y^2+z^2$, the square of the distance from the origin.   Let us calculate the moment of inertia around the x-axis. $I_x = \int_0^a\int_0^a\int_0^a(y^2+z^2)(x^2+y^2+z^2)\,dx\,dy\,dz$.   [The distance from (x,y,z) to the x-axis is $\sqrt{y^2+z^2}$.]   Then $I_x = \int_0^a\int_0^a(y^2+z^2)[\frac13 x^3 + (y^2+z^2)x]_0^a\,dy\,dz = \int_0^a\int_0^a(y^2+z^2)[\frac13 a^3 + (y^2+z^2)a]\,dy\,dz = \int_0^a[\frac13 a^3(\frac13 y^3 + z^2 y) + a(\frac15 y^5 + \frac23 y^3 z^2 + z^4 y)]_0^a\,dz = \int_0^a[\frac13 a^3(\frac13 a^3 + az^2) + a(\frac15 a^5 + \frac23 a^3 z^2 + az^4)]\,dz = a^2\int_0^a[\frac13 a^2(\frac13 a^2 + z^2) + (\frac14 a^4 + \frac29 a^2 z^2 + z^4)]\,dz = a^2[\frac13 a^2(\frac13 a^2 z + \frac13 z^3) + (\frac14 a^4 z + \frac29 a^2 z^3 + \frac15 z^5)]_0^a = a^2[\frac13 a^2(\frac23 a^3) + (\frac14 a^5 + \frac29 a^5 + \frac15 a^5)] = a^7(\frac29 + \frac14 + \frac29 + \frac15) = \frac{161}{180}a^7$.

44.89   Find the centroid of the region outside the circle $r=1$ and inside the cardioid $r=1+\cos\theta$.

▌ Refer to Fig. 44-20.   Clearly, $\bar y = 0$, and $\bar x$ is the same for the given region as for the half lying above the polar axis.   For the latter, the area $A = \int_0^{\pi/2}\int_1^{1+\cos\theta} r\, dr\, d\theta = \int_0^{\pi/2}\frac12 r^2\Big]_1^{1+\cos\theta} d\theta = \frac12\int_0^{\pi/2}(2\cos\theta + 1+\cos^2\theta)\,d\theta = \frac12\left(2\sin\theta + \frac{\theta}{2} + \frac{\sin2\theta}{4}\right)\Big]_0^{\pi/2} = \frac12\left(2 + \frac{\pi}{4}\right) = \frac{\pi+8}{8}$.   The moment about the y-axis is $M_y = \int_0^{\pi/2}\int_1^{1+\cos\theta} xr\, dr\, d\theta = \int_0^{\pi/2}\int_1^{1+\cos\theta} r^2\cos\theta\, dr\, d\theta = \int_0^{\pi/2}\frac13 r^3\cos\theta\Big]_1^{1+\cos\theta} d\theta = \frac13\int_0^{\pi/2}[(1+\cos\theta)^3 - 1]\cos\theta\, d\theta = \frac13\int_0^{\pi/2}(3\cos^2\theta + 3\cos^3\theta + \cos^4\theta)\,d\theta = \frac13\int_0^{\pi/2}\left[\frac32(1+\cos2\theta) + 3(\cos\theta - \sin^2\theta\cos\theta) + \frac14\left(1 + 2\cos2\theta + \frac{1+\cos4\theta}{2}\right)\right]_0^{\pi/2} = \frac13\left(\frac{15}{8}\theta + 3\sin\theta + \sin2\theta + \frac{\sin4\theta}{32} - \sin^3\theta\right)\Big]_0^{\pi/2} = \frac13\left(\frac{15\pi}{16} + 3 - 1\right) = \frac{5\pi}{16} + \frac23 = \frac{15\pi+32}{48}$.   Hence, $\bar x = \frac{M_y}{A} = \frac{(15\pi+32)/48}{(\pi+8)/8} = \frac{15\pi+32}{6(\pi+8)}$.

44.90   Find the centroid $(\bar x, \bar y)$ of the region in the first quadrant bounded by $y^2 = 6x$, $y=0$, and $x=6$ (Fig. 44-43).

▌ The area $A = \int_0^6\int_{y^2/6}^6 dx\, dy = \int_0^6\left(6 - \frac{y^2}{6}\right) dy = \left(6y - \frac{1}{18}y^3\right)\Big]_0^6 = 6(6-2) = 24$.   The moment about the x-axis is $M_x = \int_0^6\int_{y^2/6}^6 y\, dx\, dy = \int_0^6 y\left(6 - \frac{y^2}{6}\right) dy = \left(3y^2 - \frac{1}{24}y^4\right)\Big]_0^6 = 36\left(3 - \frac32\right) = 54$.   Hence, $\bar y = \frac{M_x}{A} = \frac{54}{24} = \frac94$.   The moment about the y-axis is $M_y = \int_0^6\int_{y^2/6}^6 x\, dx\, dy = \int_0^6\frac12 x^2\Big]_{y^2/6}^6 dy = \frac12\int_0^6\left(36 - \frac{y^4}{36}\right) dy = \frac12\left(36y - \frac{1}{180}y^5\right)\Big]_0^6 = 3\left(36 - \frac{36}{5}\right) = \frac{12\cdot36}{5}$.   Hence, $\bar x = \frac{M_y}{A} = \frac{12(36)/5}{24} = \frac{18}{5}$.   So the centroid is $\left(\frac{18}{5}, \frac94\right)$.

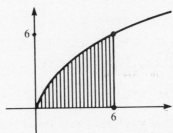

Fig. 44-43

44.91   Find the centroid of the solid under $z^2 = xy$ and above the triangle bounded by $y=x$, $y=0$, and $x=4$.

▌ The volume $V = \int_0^4\int_0^x\int_0^{\sqrt{xy}} dz\, dy\, dx = \int_0^4\int_0^x\sqrt{xy}\, dy\, dx = \int_0^4\frac23 x^{1/2}y^{3/2}\Big]_0^x dx = \int_0^4\frac23 x^2\, dx = \frac29 x^3\Big]_0^4 = \frac{128}{9}$.   The moment about the yz-plane is $M_{yz} = \int_0^4\int_0^x\int_0^{\sqrt{xy}} x\, dz\, dy\, dx = \int_0^4\int_0^x xz\Big]_0^{\sqrt{xy}} dy\, dx = \int_0^4\int_0^x x^{3/2} y^{1/2}\, dy\, dx = \int_0^4\frac23 x^{3/2} y^{3/2}\Big]_0^x dx = \int_0^4\frac23 x^3\, dx = \frac16 x^4\Big]_0^4 = \frac{128}{3}$.   Hence, $\bar x = \frac{M_{yz}}{V} = \frac{\frac{128}{3}}{\frac{128}{9}} = 3$.   The moment about the

xz-plane is $\quad M_{xz} = \int_0^4 \int_0^x \int_0^{\sqrt{xy}} y \, dz \, dy \, dx = \int_0^4 \int_0^x x^{1/2} y^{3/2} \, dy \, dx = \int_0^4 \frac{2}{5} x^{1/2} y^{5/2} \Big]_0^x \, dx = \int_0^4 \frac{2}{5} x^3 \, dx = \frac{1}{10} x^4 \Big]_0^4 = \frac{128}{5}$. $\quad$ Hence, $\quad \bar{y} = \dfrac{M_{xz}}{V} = \dfrac{\frac{128}{5}}{\frac{128}{9}} = \dfrac{9}{5}$. $\quad$ The moment about the xy-plane is $\quad M_{xy} = \int_0^4 \int_0^x \int_0^{\sqrt{xy}} z \, dz \, dy \, dx = \int_0^4 \int_0^x \frac{1}{2} z^2 \Big]_0^{\sqrt{xy}} \, dy \, dx = \int_0^4 \int_0^x \frac{1}{2} xy \, dy \, dx = \int_0^4 \frac{1}{4} xy^2 \Big]_0^x \, dx = \int_0^4 \frac{1}{4} x^3 \, dx = \frac{1}{16} x^4 \Big]_0^4 = 16$. $\quad$ Hence, $\quad \bar{z} = \dfrac{M_{xy}}{V} = \dfrac{16}{\frac{128}{9}} = \dfrac{9}{8}$. $\quad$ Thus, the centroid is $\left(3, \dfrac{9}{5}, \dfrac{9}{8}\right)$.

44.92 Find the centroid of the upper half $\mathcal{H}$ of the solid ball of radius a with center at the origin.

❙ We know $\quad V = \frac{1}{2}\left(\frac{4}{3}\pi a^3\right) = \frac{2}{3}\pi a^3$. $\quad$ By symmetry, $\quad \bar{x} = \bar{y} = 0$. $\quad$ The moment about the xy-plane is $M_{xy} = \iiint_{\mathcal{H}} z \, dV$. $\quad$ Use spherical coordinates.

$$M_{xy} = \int_0^{2\pi} \int_0^{\pi/2} \int_0^a \rho \cos\phi \cdot \rho^2 \sin\phi \, d\rho \, d\phi \, d\theta = \int_0^{2\pi} \int_0^{\pi/2} \frac{1}{4} \rho^4 \cos\phi \sin\phi \Big]_0^a \, d\phi \, d\theta$$

$$= \frac{a^4}{4} \int_0^{2\pi} \int_0^{\pi/2} \cos\phi \sin\phi \, d\phi \, d\theta = \frac{a^4}{4} \int_0^{2\pi} \frac{1}{2} \sin^2\phi \Big]_0^{\pi/2} \, d\theta = \frac{a^4}{8} \int_0^{2\pi} d\theta = \frac{a^4}{8} \cdot 2\pi = \frac{\pi a^4}{4}$$

Hence, $\quad \bar{z} = \dfrac{M_{xy}}{V} = \dfrac{\pi a^4/4}{2\pi a^3/3} = \dfrac{3}{8} a$.

CHAPTER 45

Vector Functions in Space. Divergence and Curl. Line Integrals

45.1 For the space curve $\mathbf{R}(t) = (t, t^2, t^3)$, find a tangent vector and the tangent line at the point $(1, 1, 1)$.

▌ A tangent vector is given by the derivative $\mathbf{R}'(t) = (1, 2t, 3t^2)$. At $(1, 1, 1)$, $t = 1$. Hence, a tangent vector is $(1, 2, 3)$. Parametric equations for the tangent line are $x = 1 + u$, $y = 1 + 2u$, $z = 1 + 3u$. As a vector function, the tangent line can be represented by $(1, 1, 1) + u(1, 2, 3)$.

45.2 Find the speed of a particle tracing the curve of Problem 45.1 at time $t = 1$. (The parameter t is usually, but not necessarily, interpreted as the time.)

▌ If s is the arc length, $\dfrac{ds}{dt} = \sqrt{\left(\dfrac{dx}{dt}\right)^2 + \left(\dfrac{dy}{dt}\right)^2 + \left(\dfrac{dz}{dt}\right)^2}$ is the speed. In general, if $\mathbf{R}(t) = (x(t), y(t), z(t))$, $\mathbf{R}'(t) = \left(\dfrac{dx}{dt}, \dfrac{dy}{dt}, \dfrac{dz}{dt}\right)$, and $\dfrac{ds}{dt} = |\mathbf{R}'(t)|$. For this particular case, $\dfrac{ds}{dt} = |(1, 2t, 3t^2)| = \sqrt{1 + 4t^2 + 9t^4}$. When $t = 1$, $\dfrac{ds}{dt} = \sqrt{14}$.

45.3 Find the normal plane to the curve of Problem 45.1 at $t = 1$.

▌ The tangent vector at $t = 1$ is $(1, 2, 3)$. That vector is perpendicular to the normal plane. Since the normal plane contains the point $(1, 1, 1)$, its equation is $1(x - 1) + 2(y - 1) + 3(z - 1) = 0$, or equivalently, $x + 2y + 3z = 6$.

45.4 Find a tangent vector, the tangent line, the speed, and the normal plane to the helical curve $\mathbf{R}(t) = (a \cos 2\pi t, a \sin 2\pi t, bt)$ at $t = 1$.

▌ A tangent vector is $\mathbf{R}'(t) = (-2\pi a \sin 2\pi t, 2\pi a \cos 2\pi t, b) = (0, 2\pi a, b)$. The tangent line is $(a, 0, b) + u(0, 2\pi a, b)$. The speed is $|\mathbf{R}'(t)| = \sqrt{4\pi^2 a^2 + b^2}$. The normal plane has the equation $(0)(x - a) + (2\pi a)(y - 0) + (b)(z - b) = 0$, or equivalently, $2\pi a y + bz = b^2$.

45.5 Prove that the angle θ between a tangent vector and the positive z-axis is the same for all points of the helix of Problem 45.4.

▌ $\mathbf{R}'(t) = (-2\pi a \sin 2\pi t, 2\pi a \cos 2\pi t, b)$ has a constant z-component; thus, θ is constant.

45.6 Find a tangent vector, the tangent line, the speed, and the normal plane to the curve $\mathbf{R}(t) = (t \cos t, t \sin t, t)$ at $t = \pi/2$.

▌ $\mathbf{R}'(t) = (\cos t - t \sin t, \sin t + t \cos t, 1) = (-\pi/2, 1, 1)$ is a tangent vector when $t = \pi/2$. The tangent line is traced out by $(0, \pi/2, \pi/2) + u(-\pi/2, 1, 1)$, that is, $x = (-\pi/2)u$, $y = \pi/2 + u$, $z = \pi/2 + u$. The speed at $t = \pi/2$ is $ds/dt = |\mathbf{R}'(t)| = \sqrt{\pi^2/4 + 2} = \frac{1}{2}\sqrt{\pi^2 + 8}$. An equation of the normal plane is $(-\pi/2)(x - 0) + (y - \pi/2) + (z - \pi/2) = 0$, or equivalently, $(-\pi/2)x + y + z = \pi$.

45.7 If $\mathbf{G}(t) = (t, t^3, \ln t)$ and $\mathbf{F}(t) = t^2 \mathbf{G}(t)$, find $\mathbf{F}'(t)$.

▌ In general, $\dfrac{d}{dt}[g(t)\mathbf{G}(t)] = g(t)\mathbf{G}'(t) + g'(t)\mathbf{G}(t)$. (The proof in Problem 34.53 is valid for arbitrary vector functions.) Hence, $\mathbf{F}'(t) = t^2(1, 3t^2, 1/t) + 2t(t, t^3, \ln t) = (3t^2, 5t^4, t + 2t \ln t)$.

45.8 If $\mathbf{G}(t) = (e^t, \cos t, t)$, find $\dfrac{d}{dt}[(\sin t)\mathbf{G}(t)]$.

▌ By the formula of Problem 45.7, $\dfrac{d}{dt}[(\sin t)\mathbf{G}(t)] = (\sin t)\mathbf{G}'(t) + \left(\dfrac{d}{dt}\sin t\right)\mathbf{G}(t) = (\sin t)(e^t, -\sin t, 1) + (\cos t)(e^t, \cos t, t) = [e^t(\sin t + \cos t), \cos^2 t - \sin^2 t, \sin t + t \cos t]$.

45.9 If $\mathbf{F}(t) = (\sin t, \cos t, t)$ and $\mathbf{G}(t) = (t, 1, \ln t)$, find $\dfrac{d}{dt}[\mathbf{F}(t) \cdot \mathbf{G}(t)]$.

▮ The product formula $\dfrac{d}{dt}[\mathbf{F}(t) \cdot \mathbf{G}(t)] = \mathbf{F}(t) \cdot \mathbf{G}'(t) + \mathbf{F}'(t) \cdot \mathbf{G}(t)$ of Problem 34.56 holds for arbitrary vector functions. So $\dfrac{d}{dt}[\mathbf{F}(t) \cdot \mathbf{G}(t)] = (\sin t, \cos t, t) \cdot (1, 0, 1/t) + (\cos t, -\sin t, 1) \cdot (t, 1, \ln t) = \sin t + 1 + t \cos t - \sin t + \ln t = 1 + t \cos t + \ln t$.

45.10 If $\mathbf{G}(3) = (1, 1, 2)$ and $\mathbf{G}'(3) = (3, -2, 5)$, find $\dfrac{d}{dt}[t^2 \mathbf{G}(t)]$ at $t = 3$.

▮ $\dfrac{d}{dt}[t^2 \mathbf{G}(t)] = t^2 \mathbf{G}'(t) + 2t\mathbf{G}(t)$. Hence, at $t = 3$, $\dfrac{d}{dt}[t^2 \mathbf{G}(t)] = 9(3, -2, 5) + 6(1, 1, 2) = (33, -12, 57)$.

45.11 Let $\mathbf{F}(t) = \mathbf{G}(t) \times \mathbf{H}(t)$. Prove $\mathbf{F}' = (\mathbf{G} \times \mathbf{H}') + (\mathbf{G}' \times \mathbf{H})$.

▮ $\mathbf{F}'(t) = \lim\limits_{\Delta t \to 0} \dfrac{\mathbf{F}(t + \Delta t) - \mathbf{F}(t)}{\Delta t} = \lim\limits_{\Delta t \to 0} \dfrac{[\mathbf{G}(t + \Delta t) \times \mathbf{H}(t + \Delta t)] - [\mathbf{G}(t) \times \mathbf{H}(t)]}{\Delta t}$

$= \lim\limits_{\Delta t \to 0} \dfrac{\{\mathbf{G}(t + \Delta t) \times [\mathbf{H}(t + \Delta t) - \mathbf{H}(t)]\} + \{[\mathbf{G}(t + \Delta t) - \mathbf{G}(t)] \times \mathbf{H}(t)\}}{\Delta t}$

$= \lim\limits_{\Delta t \to 0} \left[\mathbf{G}(t + \Delta t) \times \dfrac{\mathbf{H}(t + \Delta t) - \mathbf{H}(t)}{\Delta t} + \dfrac{\mathbf{G}(t + \Delta t) - \mathbf{G}(t)}{\Delta t} \times \mathbf{H}(t) \right] = [\mathbf{G}(t) \times \mathbf{H}'(t)] + [\mathbf{G}'(t) \times \mathbf{H}(t)]$.

45.12 Prove that $\dfrac{d}{dt}[\mathbf{R}(t) \times \mathbf{R}'(t)] = \mathbf{R}(t) \times \mathbf{R}''(t)$.

▮ By Problem 45.11, $\dfrac{d}{dt}[\mathbf{R}(t) \times \mathbf{R}'(t)] = [\mathbf{R}(t) \times \mathbf{R}''(t)] + [\mathbf{R}'(t) \times \mathbf{R}'(t)]$. But $\mathbf{A} \times \mathbf{A} = \mathbf{0}$ for all $\mathbf{A}$. Hence, the term $\mathbf{R}'(t) \times \mathbf{R}'(t)$ vanishes.

45.13 If $\mathbf{G}'(t)$ is perpendicular to $\mathbf{G}(t)$ for all t, show that $|\mathbf{G}(t)|$ is constant, that is, the point $\mathbf{G}(t)$ lies on the surface of a sphere with center at the origin.

▮ Since $\mathbf{G}'(t)$ is perpendicular to $\mathbf{G}(t)$, $\mathbf{G}'(t) \cdot \mathbf{G}(t) = 0$. Then $\dfrac{d}{dt}[|\mathbf{G}(t)|^2] = \dfrac{d}{dt}[\mathbf{G}(t) \cdot \mathbf{G}(t)] = \mathbf{G}(t) \cdot \mathbf{G}'(t) + \mathbf{G}'(t) \cdot \mathbf{G}(t) = 0 + 0 = 0$. Therefore, $|\mathbf{G}(t)|^2$ is constant, and hence $|\mathbf{G}(t)|$ is constant.

45.14 Derive the converse of Problem 45.13: If $|\mathbf{G}(t)|$ is constant, then $\mathbf{G}(t) \cdot \mathbf{G}'(t) = 0$.

▮ Let $|\mathbf{G}(t)| = c$ for all t. Then $\mathbf{G}(t) \cdot \mathbf{G}(t) = |\mathbf{G}(t)|^2 = c^2$. So $\dfrac{d}{dt}[\mathbf{G}(t) \cdot \mathbf{G}(t)] = 0$. Hence, $\mathbf{G}(t) \cdot \mathbf{G}'(t) + \mathbf{G}'(t) \cdot \mathbf{G}(t) = 0$. Thus, $2\mathbf{G}(t) \cdot \mathbf{G}'(t) = 0$, and, therefore, $\mathbf{G}(t) \cdot \mathbf{G}'(t) = 0$.

45.15 Assume that $\mathbf{R}(t) \neq \mathbf{0}$ and $\mathbf{R}'(t) \neq \mathbf{0}$ for all t. If $\mathbf{R}(t)$ is closest to the origin at $t = t_0$, show that the position vector $\mathbf{R}(t)$ is perpendicular to the velocity vector $\mathbf{R}'(t)$ at $t = t_0$. [Recall that, when t is the time, $\mathbf{R}'(t)$ is called the velocity vector.]

▮ $|\mathbf{R}(t)|^2$ has a relative minimum at $t = 0$. Hence, $\dfrac{d}{dt}[|\mathbf{R}(t)|^2] = 0$ at $t = t_0$. But $\dfrac{d}{dt}[|\mathbf{R}(t)|^2] = \dfrac{d}{dt}[\mathbf{R}(t) \cdot \mathbf{R}(t)] = \mathbf{R}(t) \cdot \mathbf{R}'(t) + \mathbf{R}'(t) \cdot \mathbf{R}(t) = 2\mathbf{R}(t) \cdot \mathbf{R}'(t)$. Therefore, $\mathbf{R}(t) \cdot \mathbf{R}'(t) = 0$ at $t = t_0$. Since $\mathbf{R}(t) \neq \mathbf{0}$ and $\mathbf{R}'(t) \neq \mathbf{0}$, $\mathbf{R}(t) \perp \mathbf{R}'(t)$ at $t = t_0$.

45.16 Find the principal unit normal vector $\mathbf{N}$ to the curve $\mathbf{R}(t) = (t, t^2, t^3)$ when $t = 1$.

▮ The unit tangent vector is $\mathbf{T} = \dfrac{\mathbf{R}'(t)}{|\mathbf{R}'(t)|} = \dfrac{(1, 2t, 3t^2)}{\sqrt{1 + 4t^2 + 9t^4}}$. The principal unit normal vector is $\mathbf{N} = \dfrac{\mathbf{T}'(t)}{|\mathbf{T}'(t)|}$.

Now, $\mathbf{T}'(t) = \dfrac{d}{dt}\left[\dfrac{\mathbf{R}'(t)}{|\mathbf{R}'(t)|}\right] = \dfrac{1}{\sqrt{1 + 4t^2 + 9t^4}} \mathbf{R}''(t) + \dfrac{d}{dt}\left(\dfrac{1}{\sqrt{1 + 4t^2 + 9t^4}}\right)\mathbf{R}'(t) = \dfrac{1}{\sqrt{1 + 4t^2 + 9t^4}}(0, 2, 6t) - \dfrac{4t + 18t^3}{(1 + 4t^2 + 9t^4)^{3/2}}(1, 2t, 3t^2)$. When $t = 1$, $\mathbf{T}'(t) = \dfrac{2(0, 1, 3)}{\sqrt{14}} - \dfrac{22}{14\sqrt{14}}(1, 2, 3) = \dfrac{1}{7\sqrt{14}}[14(0, 1, 3) - 11(1, 2, 3)] = \dfrac{1}{7\sqrt{14}}(-11, -8, 9)$. Therefore, $|\mathbf{T}'(t)| = \dfrac{1}{7\sqrt{14}}\sqrt{121 + 64 + 81} = \dfrac{\sqrt{266}}{7\sqrt{14}} = \dfrac{\sqrt{19}}{7}$. Thus, $\mathbf{N} = \dfrac{1}{\sqrt{266}}(-11, -8, 9)$.

45.17 Find a formula for the binormal vector $\mathbf{B} = \mathbf{T} \times \mathbf{N}$ in the case of the helix $\mathbf{R}(t) = (3 \cos 2t, 3 \sin 2t, 8t)$.

▮ $\mathbf{R}'(t) = (-6 \sin 2t, 6 \cos 2t, 8)$. $|\mathbf{R}(t)| = \sqrt{36 + 64} = 10$. Hence, $\mathbf{T} = \frac{1}{5}(-3 \sin 2t, 3 \cos 2t, 4)$. Therefore, $\mathbf{T}' = \frac{1}{5}(-6 \cos 2t, -6 \sin 2t, 0) = -\frac{6}{5}(\cos 2t, \sin 2t, 0)$, and $|\mathbf{T}'| = \frac{6}{5}$. Thus, $\mathbf{N} = \dfrac{\mathbf{T}'}{|\mathbf{T}'|} = -(\cos 2t, \sin 2t, 0)$ and $\mathbf{B} = \mathbf{T} \times \mathbf{N} = \frac{1}{5}(-3 \sin 2t, 3 \cos 2t, 4) \times -(\cos 2t, \sin 2t, 0) = -\frac{1}{5}(-4 \sin 2t, -4 \cos 2t, -3) = \frac{1}{5}(4 \sin 2t, 4 \cos 2t, 3)$.

45.18 Find an equation of the *osculating plane* to the curve $\mathbf{R}(t) = (2t - t^2, t^2, 2t + t^2)$ at $t = 1$.

▮ The osculating plane is the plane determined by $\mathbf{T}$ and $\mathbf{N}$; hence, the binormal vector $\mathbf{B} = \mathbf{T} \times \mathbf{N}$ is a normal vector to the osculating plane. $\mathbf{R}'(t) = (2 - 2t, 2t, 2 + 2t)$. $|\mathbf{R}'(t)| = \sqrt{4(1 - t)^2 + 4t^2 + 4(1 + t)^2} = 2\sqrt{3t^2 + 2}$. Thus, $\mathbf{T}(t) = \dfrac{1}{2\sqrt{3t^2 + 2}}(2 - 2t, 2t, 2 + 2t)$. Hence, $\mathbf{T}'(t) = \dfrac{1}{2\sqrt{3t^2 + 2}}(-2, 2, 2) - \dfrac{3}{2}\dfrac{t}{(3t^2 + 2)^{3/2}}(2 - 2t, 2t, 2 + 2t)$. At $t = 1$, $\mathbf{R}' = (0, 2, 4)$ and $\mathbf{T}' = \dfrac{1}{2\sqrt{5}}(-2, 2, 2) - \dfrac{3}{2}\dfrac{1}{5\sqrt{5}}(0, 2, 4) = \dfrac{1}{10\sqrt{5}}[5(-2, 2, 2) - 3(0, 2, 4)] = \dfrac{1}{10\sqrt{5}}(-10, 4, -2) = \dfrac{1}{5\sqrt{5}}(-5, 2, -1)$. Since $\mathbf{T}$ is parallel to $\mathbf{R}'$ and $\mathbf{N}$ is parallel to $\mathbf{T}'$, a normal vector to the osculating plane is given by $(0, 2, 4) \times (-5, 2, -1) = (-10, -20, 10) = -10(1, 2, -1)$. Therefore, an equation of the osculating plane at $(1, 1, 3)$ is $(x - 1) + 2(y - 1) - (z - 3) = 0$, or equivalently, $x + 2y - z = 0$.

45.19 Show that a normal vector to the osculating plane of a curve $\mathbf{R}(t)$ is given by $\mathbf{R}' \times \mathbf{R}''$.

▮ The osculating plane is the plane determined by $\mathbf{T}$ and $\mathbf{N}$. Let $\mathbf{D} = \mathbf{R}' \times \mathbf{R}''$. Now, $\mathbf{T} = \dfrac{\mathbf{R}'}{|\mathbf{R}'|}$. Hence, $\mathbf{T}' = \dfrac{1}{|\mathbf{R}'|}\mathbf{R}'' + \dfrac{d}{dt}\left(\dfrac{1}{|\mathbf{R}'|}\right)\mathbf{R}'$. Since $\mathbf{D} \cdot \mathbf{R}' = 0$ and $\mathbf{D} \cdot \mathbf{R}'' = 0$, it follows that $\mathbf{D} \cdot \mathbf{T}' = 0$. Thus, $\mathbf{D}$ is perpendicular to $\mathbf{R}'$ and to $\mathbf{T}'$, and, therefore, to $\mathbf{T}$ and to $\mathbf{N}$. Hence, $\mathbf{D}$ is a normal vector to the osculating plane.

45.20 Find an equation of the osculating plane of the curve $\mathbf{R}(t) = (3t^2, 1 - t, t^3)$ at $(12, -1, 8)$ corresponding to $t = 2$.

▮ $\mathbf{R}'(t) = (6t, -1, 3t^2) = (12, -1, 12)$, and $\mathbf{R}''(t) = (6, 0, 6t) = (6, 0, 12)$. Hence, $\mathbf{R}' \times \mathbf{R}'' = (12, -1, 12) \times (6, 0, 12) = (-12, -72, 72) = -12(1, 6, -6)$. By Problem 45.19, $-12(1, 6, -6)$, or more simply $(1, 6, -6)$, is a normal vector to the osculating plane. Hence, an equation of that plane is $x - 12 + 6(y + 1) - 6(z - 8) = 0$, or equivalently, $x + 6y - 6z + 42 = 0$.

45.21 Find the curvature κ of the curve $\mathbf{R}(t) = (\sin t, \cos t, \frac{1}{2}t^2)$ at $t = 0$.

▮ $\kappa = \left|\dfrac{d\mathbf{T}}{ds}\right|$. First, $\mathbf{R}' = (\cos t, -\sin t, t)$, $\dfrac{ds}{dt} = |\mathbf{R}'(t)| = \sqrt{1 + t^2}$, and $\mathbf{T} = \dfrac{1}{\sqrt{1 + t^2}}(\cos t, -\sin t, t)$. Hence, $\dfrac{d\mathbf{T}}{dt} = \dfrac{1}{\sqrt{1 + t^2}}(-\sin t, -\cos t, 1) - \dfrac{t}{(1 + t^2)^{3/2}}(\cos t, -\sin t, t)$. At $t = 0$, $\dfrac{ds}{dt} = 1$ and $\dfrac{d\mathbf{T}}{dt} = (0, -1, 1)$; therefore, $\dfrac{d\mathbf{T}}{ds} = \dfrac{d\mathbf{T}/dt}{ds/dt} = (0, -1, 1)$ and $\kappa = \left|\dfrac{d\mathbf{T}}{ds}\right| = \sqrt{2}$.

45.22 Prove the following formula for the curvature: $\kappa = \dfrac{|\mathbf{R}' \times \mathbf{R}''|}{|\mathbf{R}'|^3}$.

▮ $\mathbf{R}' = |\mathbf{R}'|\mathbf{T}$. Hence, $\mathbf{R}'' = \left[\dfrac{d}{dt}(|\mathbf{R}'|)\right]\mathbf{T} + |\mathbf{R}'|(\mathbf{T}')$. Note that $\mathbf{T}' = \dfrac{ds}{dt}\left(\dfrac{d\mathbf{T}}{ds}\right) = |\mathbf{R}'|\kappa\mathbf{N}$. Hence, $\mathbf{R}'' = \left[\dfrac{d}{dt}(|\mathbf{R}'|)\right]\mathbf{T} + \kappa|\mathbf{R}'|^2\mathbf{N}$. Therefore, $\mathbf{R}' \times \mathbf{R}'' = |\mathbf{R}'|\mathbf{T} \times \left\{\left[\dfrac{d}{dt}(|\mathbf{R}'|)\right]\mathbf{T} + \kappa|\mathbf{R}'|^2\mathbf{N}\right\} = |\mathbf{R}'|\left[\dfrac{d}{dt}(|\mathbf{R}'|)\right](\mathbf{T} \times \mathbf{T}) + \kappa|\mathbf{R}'|^3(\mathbf{T} \times \mathbf{N}) = \kappa|\mathbf{R}'|^3(\mathbf{T} \times \mathbf{N})$. Since $|\mathbf{T} \times \mathbf{N}| = 1$, it follows that $|\mathbf{R}' \times \mathbf{R}''| = \kappa|\mathbf{R}'|^3$.

45.23 Find the curvature of $\mathbf{R} = (e^t \sin t, e^t \cos t, e^t)$ at $t = 0$.

▮ $\mathbf{R} = e^t(\sin t, \cos t, 1)$. Hence, $\mathbf{R}' = e^t(\cos t, -\sin t, 0) + e^t(\sin t, \cos t, 1) = e^t(\cos t + \sin t, \cos t - \sin t, 1) = e^t(2 \cos t, -2 \sin t, 1)$. At $t = 0$, $\mathbf{R}' = (1, 1, 1)$ and $\mathbf{R}'' = (2, 0, 1)$. Hence, $\mathbf{R}' \times \mathbf{R}'' = (1, 1, -2)$, $|\mathbf{R}' \times \mathbf{R}''| = \sqrt{6}$, and $|\mathbf{R}'| = \sqrt{3}$. By Problem 45.22, $\kappa = \dfrac{|\mathbf{R}' \times \mathbf{R}''|}{|\mathbf{R}'|^3} = \dfrac{\sqrt{6}}{(\sqrt{3})^3} = \dfrac{\sqrt{2}}{3}$.

45.24 Show that the gradient ∇f of the scalar function $f(x, y, z) = \dfrac{1}{\sqrt{x^2 + y^2 + z^2}}$ satisfies $\nabla f = \dfrac{\mathbf{P}}{|\mathbf{P}|^3}$, where $\mathbf{P} = (x, y, z)$.

▌ $\nabla f = (f_x, f_y, f_z) = \left(-\dfrac{x}{(x^2 + y^2 + z^2)^{3/2}}, \ -\dfrac{y}{(x^2 + y^2 + z^2)^{3/2}}, \ -\dfrac{z}{(x^2 + y^2 + z^2)^{3/2}} \right) = -\dfrac{1}{|\mathbf{P}|^3} \mathbf{P}$.

45.25 Compute the divergence, div $\mathbf{F}$, for the vector field $\mathbf{F} = (xy, yz, xz)$.

▌ By definition, if $\mathbf{F}(x, y, z) = (f(x, y, z), g(x, y, z), h(x, y, z))$, then $\text{div } \mathbf{F} = \dfrac{\partial f}{\partial x} + \dfrac{\partial g}{\partial y} + \dfrac{\partial h}{\partial z}$. In this case, $\text{div } \mathbf{F} = y + z + x$.

45.26 Compute div $\mathbf{F}$ for $\mathbf{F} = \dfrac{(x, y, z)}{|(x, y, z)|^3} = \dfrac{\mathbf{P}}{|\mathbf{P}|^3}$, where $\mathbf{P} = (x, y, z)$ and, therefore, $|\mathbf{P}| = \sqrt{x^2 + y^2 + z^2}$.

▌ $\mathbf{F}(x, y, z) = \left(\dfrac{x}{(\sqrt{x^2 + y^2 + z^2})^3}, \ \dfrac{y}{(\sqrt{x^2 + y^2 + z^2})^3}, \ \dfrac{z}{(\sqrt{x^2 + y^2 + z^2})^3} \right)$. $\dfrac{\partial}{\partial x} \left(\dfrac{x}{|\mathbf{P}|^3} \right) = \dfrac{|\mathbf{P}|^3 - x(\partial/\partial x)(|\mathbf{P}|)^3}{|\mathbf{P}|^6} =$

$\dfrac{|\mathbf{P}|^3 - 3x|\mathbf{P}|^2(\partial |\mathbf{P}|/\partial x)}{|\mathbf{P}|^6} = \dfrac{|\mathbf{P}| - 3x(\partial |\mathbf{P}|/\partial x)}{|\mathbf{P}|^4}$. But $\dfrac{\partial |\mathbf{P}|}{\partial x} = \dfrac{\partial}{\partial x} (\sqrt{x^2 + y^2 + z^2}) = \dfrac{x}{\sqrt{x^2 + y^2 + z^2}} = \dfrac{x}{|\mathbf{P}|}$.

Therefore, $\dfrac{\partial}{\partial x} \left(\dfrac{x}{|\mathbf{P}|^3} \right) = \dfrac{|\mathbf{P}| - 3x(x/|\mathbf{P}|)}{|\mathbf{P}|^4} = \dfrac{1}{|\mathbf{P}|^3} - \dfrac{3x^2}{|\mathbf{P}|^5}$. Similarly, $\dfrac{\partial}{\partial y} \left(\dfrac{y}{|\mathbf{P}|^3} \right) = \dfrac{1}{|\mathbf{P}|^3} - \dfrac{3y^2}{|\mathbf{P}|^5}$ and $\dfrac{\partial}{\partial z} \left(\dfrac{z}{|\mathbf{P}|^3} \right) =$

$\dfrac{1}{|\mathbf{P}|^3} - \dfrac{3z^2}{|\mathbf{P}|^5}$. Hence, $\text{div } \mathbf{F} = \left(\dfrac{1}{|\mathbf{P}|^3} - \dfrac{3x^2}{|\mathbf{P}|^5} \right) + \left(\dfrac{1}{|\mathbf{P}|^3} - \dfrac{3y^2}{|\mathbf{P}|^5} \right) + \left(\dfrac{1}{|\mathbf{P}|^3} - \dfrac{3z^2}{|\mathbf{P}|^5} \right) = \dfrac{3}{|\mathbf{P}|^3} - \dfrac{3(x^2 + y^2 + z^2)}{|\mathbf{P}|^5} = \dfrac{3}{|\mathbf{P}|^3} -$

$\dfrac{3|\mathbf{P}|^2}{|\mathbf{P}|^5} = \dfrac{3}{|\mathbf{P}|^3} - \dfrac{3}{|\mathbf{P}|^3} = 0$. [In view of Problem 45.24, we have shown that $1/|\mathbf{P}|$ satisfies Laplace's equation at all points except $(0, 0, 0)$.]

45.27 Let $\mathbf{F}(x, y, z) = (xyz^2, x^2yz, -xyz)$. Find div $\mathbf{F}$.

▌ $\text{div } \mathbf{F} = \dfrac{\partial}{\partial x} (xyz^2) + \dfrac{\partial}{\partial y} (x^2yz) + \dfrac{\partial}{\partial z} (-xyz) = yz^2 + x^2z - xy$.

45.28 For any vector field $\mathbf{F}(x, y, z) = (f(x, y, z), g(x, y, z), h(x, y, z))$, define curl $\mathbf{F}$.

▌ $\text{curl } \mathbf{F} = \left(\dfrac{\partial h}{\partial y} - \dfrac{\partial g}{\partial z}, \ \dfrac{\partial f}{\partial z} - \dfrac{\partial h}{\partial x}, \ \dfrac{\partial g}{\partial x} - \dfrac{\partial f}{\partial y} \right)$, or, more vividly,

$$\text{curl } \mathbf{F} = \nabla \times \mathbf{F} = \begin{vmatrix} \mathbf{i} & \mathbf{j} & \mathbf{k} \\ \dfrac{\partial}{\partial x} & \dfrac{\partial}{\partial y} & \dfrac{\partial}{\partial z} \\ f & g & h \end{vmatrix}$$

where $\nabla = \left(\dfrac{\partial}{\partial x}, \dfrac{\partial}{\partial y}, \dfrac{\partial}{\partial z} \right)$.

45.29 Find curl $\mathbf{F}$ when $\mathbf{F} = (yz, xz, xy)$.

▌ $\text{curl } \mathbf{F} = \left(\dfrac{\partial(xy)}{\partial y} - \dfrac{\partial(xz)}{\partial z}, \ \dfrac{\partial(yz)}{\partial z} - \dfrac{\partial(xy)}{\partial x}, \ \dfrac{\partial(xz)}{\partial x} - \dfrac{\partial(yz)}{\partial y} \right) = (x - x, \ y - y, \ z - z) = (0, 0, 0) = \mathbf{0}$.

45.30 Compute div $\mathbf{P}$ and curl $\mathbf{P}$, for $\mathbf{P} = (x, y, z)$.

▌ $\text{div } \mathbf{P} = \dfrac{\partial x}{\partial x} + \dfrac{\partial y}{\partial y} + \dfrac{\partial z}{\partial z} = 1 + 1 + 1 = 3$. $\quad$ $\text{curl } \mathbf{P} = \left(\dfrac{\partial z}{\partial y} - \dfrac{\partial y}{\partial z}, \quad \dfrac{\partial x}{\partial z} - \dfrac{\partial z}{\partial x}, \quad \dfrac{\partial y}{\partial x} - \dfrac{\partial x}{\partial y} \right) = (0 - 0, \ 0 - 0, \ 0 - 0) = \mathbf{0}$.

45.31 Compute div $\mathbf{F}$ and curl $\mathbf{F}$ for $\mathbf{F} = (\cos 2x, \sin 2y, \tan z)$.

▌ $\text{div } \mathbf{F} = \dfrac{\partial}{\partial x} (\cos 2x) + \dfrac{\partial}{\partial y} (\sin 2y) + \dfrac{\partial}{\partial z} (\tan z) = -2 \sin 2x + 2 \cos 2y + \sec^2 z$. $\quad$ $\text{curl } \mathbf{F} = \left(\dfrac{\partial(\tan z)}{\partial y} - \right.$

$\dfrac{\partial(\sin 2y)}{\partial z}, \ \dfrac{\partial(\cos 2x)}{\partial z} - \dfrac{\partial(\tan z)}{\partial x}, \ \left. \dfrac{\partial(\sin 2y)}{\partial x} - \dfrac{\partial(\cos 2x)}{\partial y} \right) = (0 - 0, \ 0 - 0, \ 0 - 0) = \mathbf{0}$.

45.32 Show that, for any scalar function $f(x, y, z)$, div $\nabla f = f_{xx} + f_{yy} + f_{zz}$. (The latter sum, called the *Laplacian* of f, is often notated as $\nabla^2 f$.)

▌ $\nabla f = (f_x, f_y, f_z)$. Hence, div $\nabla f = \dfrac{\partial}{\partial x}(f_x) + \dfrac{\partial}{\partial y}(f_y) + \dfrac{\partial}{\partial z}(f_z) = f_{xx} + f_{yy} + f_{zz}$.

45.33 For any scalar function $f(x, y, z)$ with continuous mixed second partial derivatives, show that curl $\nabla f = \mathbf{0}$.

▌ $\nabla f = (f_x, f_y, f_z)$. Then curl $\nabla f = (f_{zy} - f_{yz}, \quad f_{xz} - f_{zx}, \quad f_{yx} - f_{xy}) = (0, 0, 0) = \mathbf{0}$. Here, we used the equality of the mixed second partial derivatives.

45.34 For a vector field $\mathbf{F}(x, y, z) = (f(x, y, z), g(x, y, z), h(x, y, z))$, where f, g, and h have continuous mixed second partial derivatives, show that div curl $\mathbf{F} = 0$.

▌ curl $\mathbf{F} = \left(\dfrac{\partial h}{\partial y} - \dfrac{\partial g}{\partial z}, \dfrac{\partial f}{\partial z} - \dfrac{\partial h}{\partial x}, \dfrac{\partial g}{\partial x} - \dfrac{\partial f}{\partial y}\right)$. So div curl $\mathbf{F} = \dfrac{\partial}{\partial x}\left(\dfrac{\partial h}{\partial y} - \dfrac{\partial g}{\partial z}\right) + \dfrac{\partial}{\partial y}\left(\dfrac{\partial f}{\partial z} - \dfrac{\partial h}{\partial x}\right) + \dfrac{\partial}{\partial z}\left(\dfrac{\partial g}{\partial x} - \dfrac{\partial f}{\partial y}\right) = h_{yx} - g_{zx} + f_{zy} - h_{xy} + g_{xz} - f_{yz} = (h_{yx} - h_{xy}) + (g_{xz} - g_{zx}) + (f_{zy} - f_{yz}) = 0 + 0 + 0 = 0$.

45.35 For a scalar field f and a vector field $\mathbf{F}$, prove the product rule div $(f\mathbf{F}) = f$ div $\mathbf{F} + \nabla f \cdot \mathbf{F}$.

▌ Let $\mathbf{F} = (\phi(x, y, z), \psi(x, y, z), \eta(x, y, z))$. Then div $(f\mathbf{F}) = $ div $(f\phi, f\psi, f\eta) = f\phi_x + f_x\phi + f\psi_y + f_y\psi + f\eta_z + f_z\eta = f(\phi_x + \psi_y + \eta_z) + (f_x, f_y, f_z) \cdot (\phi, \psi, \eta) = f$ div $\mathbf{F} + \nabla f \cdot \mathbf{F}$.

45.36 For a scalar field f and a vector field $\mathbf{F} = (\phi, \psi, \eta)$, prove the product rule curl $(f\mathbf{F}) = f$ curl $\mathbf{F} + \nabla f \times \mathbf{F}$.

▌ $f\mathbf{F} = (f\phi, f\psi, f\eta)$. Hence, curl $(f\mathbf{F}) = ((f\eta)_y - (f\psi)_z, \quad (f\phi)_z - (f\eta)_x, \quad (f\psi)_x - (f\phi)_y) = (f\eta_y + f_y\eta - f\psi_z - f_z\psi, \quad f\phi_z + f_z\phi - f\eta_x - f_x\eta, \quad f\psi_x + f_x\psi - f\phi_y - f_y\phi) = (f\eta_y - f\psi_z, \quad f\phi_z - f\eta_x, \quad f\psi_x - f\phi_y) + (f_y\eta - f_z\psi, \quad f_z\phi - f_x\eta, \quad f_x\psi - f_y\phi) = f$ curl $(\phi, \psi, \eta) + (f_x, f_y, f_z) \times (\phi, \psi, \eta) = f$ curl $\mathbf{F} + \nabla f \times \mathbf{F}$.

45.37 For vector fields $\mathbf{F} = (f, g, h)$ and $\mathbf{G} = (\phi, \psi, \eta)$, prove div $(\mathbf{F} \times \mathbf{G}) =$ curl $\mathbf{F} \cdot \mathbf{G} - \mathbf{F} \cdot$ curl $\mathbf{G}$.

▌ $\mathbf{F} \times \mathbf{G} = (g\eta - h\psi, \quad h\phi - f\eta, \quad f\psi - g\phi)$. Hence, div $(\mathbf{F} \times \mathbf{G}) = (g\eta - h\psi)_x + (h\phi - f\eta)_y + (f\psi - g\phi)_z = g\eta_x + g_x\eta - h\psi_x - h_x\psi + h\phi_y + h_y\phi - f\eta_y - f_y\eta + f\psi_z + f_z\psi - g\phi_z - g_z\phi = \phi(h_y - g_z) + \psi(f_z - h_x) + \eta(g_x - f_y) + f(\psi_z - \eta_y) + g(\eta_x - \phi_z) + h(\phi_y - \psi_x) = \mathbf{G} \cdot$ curl $\mathbf{F} - \mathbf{F} \cdot$ curl $\mathbf{G}$.

45.38 Let $\mathbf{P} = (x, y, z)$. For any function $f(u)$, show that curl $(f(|\mathbf{P}|)\mathbf{P}) = \mathbf{0}$.

▌ curl $(f(|\mathbf{P}|)\mathbf{P}) = $ curl $(xf(|\mathbf{P}|), yf(|\mathbf{P}|), zf(|\mathbf{P}|)) = \left(\dfrac{\partial}{\partial y}[zf(|\mathbf{P}|)] - \dfrac{\partial}{\partial z}[yf(|\mathbf{P}|)], \quad \dfrac{\partial}{\partial z}[xf(|\mathbf{P}|)] - \dfrac{\partial}{\partial x}[zf(|\mathbf{P}|)], \quad \dfrac{\partial}{\partial x}[yf(|\mathbf{P}|)] - \dfrac{\partial}{\partial y}[xf(|\mathbf{P}|)]\right)$. Consider the first component:

$$zf'(|\mathbf{P}|) \cdot \dfrac{\partial}{\partial y}\sqrt{x^2 + y^2 + z^2} - yf'(|\mathbf{P}|)\dfrac{\partial}{\partial z}\sqrt{x^2 + y^2 + z^2} = f'(|\mathbf{P}|)\left(z \cdot \dfrac{y}{\sqrt{x^2 + y^2 + z^2}} - y \cdot \dfrac{z}{\sqrt{x^2 + y^2 + z^2}}\right) = 0$$

Similarly, the second and third components also are 0.

45.39 Prove that div $(\nabla f \times \nabla g) = 0$ for any scalar functions $f(x, y, z)$ and $g(x, y, z)$ with continuous mixed second partial derivatives.

▌ By Problems 45.33 and 45.37, div $(\nabla f \times \nabla g) = \mathbf{0} \cdot \nabla g - \nabla f \cdot \mathbf{0} = 0$.

45.40 Use a line integral to find the mass of a wire running along the parabola $\mathscr{C}$: $y = x^2$ from $(0, 0)$ to $(1, 1)$, if the density (mass per unit length) of the wire at any point (x, y) is numerically equal to x.

▌ $m = \int_{\mathscr{C}} x \, ds = \int_0^1 x\sqrt{1 + \left(\dfrac{dy}{dx}\right)^2} \, dx = \int_0^1 x\sqrt{1 + 4x^2} \, dx = \frac{1}{8} \cdot \frac{2}{3}(1 + 4x^2)^{3/2}\big]_0^1 = \frac{1}{12}[(5)^{3/2} - 1] = \frac{1}{12}(5\sqrt{5} - 1)$

45.41 Evaluate the line integral $\int_{\mathscr{C}} \mathbf{F} \cdot \mathbf{T} \, ds$ for $\mathbf{F} = (y, 0, zy)$, where $\mathscr{C}$ is the helix $(\cos t, \sin t, 3t)$, $0 \le t \le 2\pi$. (Here, $\mathbf{T}$ denotes the unit tangent vector along the curve.)

▌ If s denotes the arc length, $\mathbf{T} = \left(\dfrac{dx}{ds}, \dfrac{dy}{ds}, \dfrac{dz}{ds}\right)$, then $\mathbf{F} \cdot \mathbf{T} = (y, 0, zy) \cdot \left(\dfrac{dx}{ds}, \dfrac{dy}{ds}, \dfrac{dz}{ds}\right) = y\dfrac{dx}{ds} + zy\dfrac{dz}{ds}$.

Hence, $\int_{\mathscr{C}} \mathbf{F} \cdot \mathbf{T} \, ds = \int_{\mathscr{C}} \left(y\dfrac{dx}{ds} + zy\dfrac{dz}{ds}\right) = \int_{\mathscr{C}} y \, dx + zy \, dz = \int_0^{2\pi}\left(y\dfrac{dx}{dt} + zy\dfrac{dz}{dt}\right) dt = \int_0^{2\pi}[(\sin t)(-\sin t) + $

$(3t \sin t)(3)] \, dt = \int_0^{2\pi}(9t \sin t - \sin^2 t) \, dt = \left[9(\sin t - t \cos t) - \dfrac{1}{2}\left(t - \dfrac{\sin 2t}{2}\right)\right]_0^{2\pi} = 9(-2\pi) - \frac{1}{2}(2\pi) = -19\pi$.

45.42 Evaluate $L = \int_{\mathscr{C}} y\, dx + z\, dy - x\, dz$, where $\mathscr{C}$ is the line segment from $(0, 1, -1)$ to $(1, 2, 1)$.

❚ Parametric equations for $\mathscr{C}$ are $x = t$, $y = 1 + t$, $z = -1 + 2t$ for $0 \le t \le 1$. Hence, $L = \int_0^1 \left(y\, \dfrac{dx}{dt} + z\, \dfrac{dy}{dt} - x\, \dfrac{dz}{dt} \right) dt = \int_0^1 [1 + t + (-1 + 2t) - 2t]\, dt = \int_0^1 t\, dt = \frac{1}{2} t^2 \big]_0^1 = \frac{1}{2}$.

45.43 Find the work W done by a force $\mathbf{F} = (xy, yz, xz)$ acting on an object moving along the curve $\mathbf{R}(t) = (t, t^2, t^3)$ for $0 \le t \le 1$.

❚ $W = \int_{\mathscr{C}} \mathbf{F} \cdot \mathbf{T}\, ds = \int_0^1 \mathbf{F} \cdot \mathbf{R}'(t)\, dt = \int_0^1 (t^3)(1) + (t^5)(2t) + (t^4)(3t^2)\, dt = \frac{1}{4} + \frac{5}{7} = \frac{27}{28}$.

45.44 Let $\mathscr{C}$ be any curve in the xy-plane. Show that $\int_{\mathscr{C}} y\, dx + x\, dy$ depends only on the endpoints of $\mathscr{C}$.

❚ $\int_{\mathscr{C}} y\, dx + x\, dy = \int_a^b \left(y\, \dfrac{dx}{dt} + x\, \dfrac{dy}{dt} \right) dt = \int_a^b \dfrac{d}{dt}(xy)\, dt = xy\, \Big]_a^b = x(b)y(b) - x(a)y(a)$.

45.45 Prove that any gradient field ∇f in a region is conservative. More precisely, if $\mathscr{C}$ is any curve in the region from a point P to a point Q, then $\int_{\mathscr{C}} \nabla f \cdot \mathbf{T}\, ds = f(Q) - f(P)$.

❚ If $\mathscr{C}$ is given by $(x(t), y(t), z(t))$ for $a \le t \le b$, $\int_{\mathscr{C}} \nabla f \cdot \mathbf{T}\, ds = \int_{\mathscr{C}} f_x\, dx + f_y\, dy + f_z\, dz = \int_a^b \left(f_x\, \dfrac{dx}{dt} + f_y\, \dfrac{dy}{dt} + f_z\, \dfrac{dz}{dt} \right) dt = \int_a^b \dfrac{d}{dt}\, [f(x(t), y(t), z(t))]\, dt = f(x(t), y(t), z(t))\, \big]_a^b = f(Q) - f(P)$. (Note: The converse also holds if any two points in the region can be connected by a continuous curve lying in the region.)

45.46 If $\mathscr{C}$ is the circle $x^2 + y^2 = 49$, $\mathbf{F}(x, y) = (x^2, y^2)$, and $\mathbf{n}$ denotes the exterior unit normal vector, compute $\int_{\mathscr{C}} \mathbf{F} \cdot \mathbf{n}\, ds$.

❚ $\int_{\mathscr{C}} \mathbf{F} \cdot \mathbf{n}\, ds = \int_{\mathscr{C}} (x^2, y^2) \cdot \left(\dfrac{dy}{ds}, -\dfrac{dx}{ds} \right) ds$, since $\mathbf{n} = \left(\dfrac{dy}{ds}, -\dfrac{dx}{ds} \right)$. Hence, we obtain $\int_{\mathscr{C}} x^2\, dy - y^2\, dx = \int_0^{2\pi} \left(x^2\, \dfrac{dy}{dt} - y^2\, \dfrac{dx}{dt} \right) dt$, where $\mathscr{C}$ is parametrized as $x = 7 \cos t$, $y = 7 \sin t$, $0 \le t \le 2\pi$. Thus, we have $\int_0^{2\pi} [49 \cos^2 t \cdot 7 \cos t - 49 \sin^2 t(-7 \sin t)]\, dt = (7)^3 \int_0^{2\pi} (\cos^3 t + \sin^3 t)\, dt = (7)^3 \int_0^{2\pi} [(\cos t - \sin^2 t \cos t) + (\sin t - \cos^2 t \sin t)]\, dt = (7)^3 (\sin t - \frac{1}{3} \sin^3 t - \cos t + \frac{1}{3} \cos^3 t)]_0^{2\pi} = (7)^3 [(-1 + \frac{1}{3}) - (-1 + \frac{1}{3})] = 0$.

45.47 State Green's theorem in the plane.

❚ If $\mathscr{C}$ is a closed plane curve bounding a region $\mathscr{R}$ and $\mathbf{n}$ denotes the exterior unit normal vector along $\mathscr{C}$, then $\int_{\mathscr{C}} \mathbf{F} \cdot \mathbf{n}\, ds = \iint_{\mathscr{R}} \text{div}\, \mathbf{F}\, dA$ for any vector field $\mathbf{F}$ on $\mathscr{R}$.

45.48 Verify Green's theorem (Problem 45.47) when $\mathbf{F} = (3x, 2y)$ and $\mathscr{R}$ is the disk of radius 1 with center at the origin.

❚ $\text{div}\, \mathbf{F} = \dfrac{\partial}{\partial x}(3x) + \dfrac{\partial}{\partial y}(2y) = 3 + 2 = 5$. Therefore, $\iint_{\mathscr{R}} \text{div}\, \mathbf{F}\, dA = 5 \iint_{\mathscr{R}} dA = 5\, (\text{area of } \mathscr{R}) = 5\pi$. $\mathscr{C}$ is the circle $x = \cos t$, $y = \sin t$, $0 \le t \le 2\pi$. Then, $\int_{\mathscr{C}} \mathbf{F} \cdot \mathbf{n}\, ds = \int_{\mathscr{C}} (3x, 2y) \cdot \left(\dfrac{dy}{ds}, -\dfrac{dx}{ds} \right) ds = \int_0^{2\pi} \left(3x\, \dfrac{dy}{dt} - 2y\, \dfrac{dx}{dt} \right) dt = \int_0^{2\pi} [3 \cos t \cdot \cos t - 2 \sin t\,(-\sin t)]\, dt = \int_0^{2\pi} (3 \cos^2 t + 2 \sin^2 t)\, dt = \int_0^{2\pi} (2 + \cos^2 t)\, dt = \int_0^{2\pi} \left(2 + \dfrac{1 + \cos 2t}{2} \right) dt = \left(\dfrac{5}{2} t + \dfrac{\sin 2t}{4} \right) \Big]_0^{2\pi} = \dfrac{5}{2} \cdot 2\pi = 5\pi$.

45.49 *Gauss' theorem,* $\iint_{\mathscr{C}} \mathbf{F} \cdot \mathbf{n}\, dS = \iiint_{\mathscr{R}} \text{div}\, \mathbf{F}\, dV$, where $\mathscr{C}$ is a closed surface bounding a three-dimensional region $\mathscr{R}$, is the analogue to Green's theorem in the plane. Verify Gauss' theorem for the surface $\mathscr{C}$: $x^2 + y^2 + z^2 = a^2$ and the vector field $\mathbf{F} = (x, y, z)$.

❚ We have $\text{div}\, \mathbf{F} = 1 + 1 + 1 = 3$ and, on $\mathscr{C}$, $\mathbf{F} = a\mathbf{n}$. Hence, the theorem reads

$$\iint_{\mathscr{C}} a\, dS = \iiint_{\mathscr{R}} 3\, dV \quad \text{or} \quad V = \tfrac{1}{3} aS$$

which is the correct equation $\frac{4}{3}\pi a^3 = \frac{1}{3} a(4\pi a^2)$.

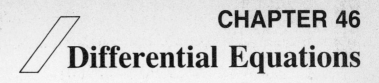

CHAPTER 46
Differential Equations

46.1 Solve $\dfrac{dy}{dx} = \dfrac{5x}{7y}$.

▮ The variables are separable: $7y\,dy = 5x\,dx$. Then, $\int 7y\,dy = \int 5x\,dx$, $7 \cdot \frac{1}{2}y^2 = 5 \cdot \frac{1}{2}x^2 + C_1$, $y^2 = \frac{5}{7}x^2 + C$. Here, as usual, C represents an arbitrary constant.

46.2 Solve $\dfrac{dy}{dx} = \dfrac{5y}{7x}$.

▮ The variables are separable. $7\int \dfrac{1}{y}\,dy = 5\int \dfrac{1}{x}\,dx$, $7\ln|y| = 5\ln|x| + C_1$, $\ln|y| = \frac{5}{7}\ln|x| + C_1$, $|y| = e^{5(\ln|x|+C_1)/7} = e^{C_1} \cdot e^{5(\ln|x|)/7}$, $|y| = C|x|^{5/7}$, where $C = e^{C_1}$. Note that any function of the form $y = Cx^{5/7}$ satisfies the given equation, where C is an arbitrary constant (not necessarily positive).

46.3 Solve $\dfrac{dy}{dx} = ky$.

▮ Separate the variables. $\int \dfrac{1}{y}\,dy = \int k\,dx$, $\ln|y| = kx + C_1$, $|y| = e^{kx+C_1} = e^{C_1} \cdot e^{kx} = Ce^{kx}$. Any function of the form $y = Ce^{kx}$ is a solution.

46.4 Solve $\dfrac{dy}{dx} = xy$.

▮ $\int \dfrac{dy}{y} = \int x\,dx$. $\ln|y| = \frac{1}{2}x^2 + C_1$, $|y| = e^{x^2/2+C_1} = e^{C_1} \cdot e^{x^2/2} = Ce^{x^2/2}$. Note that $y = Ce^{x^2/2}$ is a general solution.

46.5 Solve $\dfrac{dy}{dx} = ay(1-by)$.

▮ Separate the variables. $\int \dfrac{dy}{y(1-by)} = \int a\,dx$, $\int \left(\dfrac{1}{y} + \dfrac{b}{1-by}\right)dy = \int a\,dx$, $\ln|y| - \ln|1-by| = ax + C_1$, $\ln\left|\dfrac{y}{1-by}\right| = ax + C_1$, $\left|\dfrac{y}{1-by}\right| = e^{ax+C_1}$. Taking reciprocals, we get $\left|\dfrac{1-by}{y}\right| = e^{-ax-C_1} = e^{-C_1} \cdot e^{-ax} = Ce^{-ax}$. Hence, $\left|\dfrac{1}{y} - b\right| = Ce^{-ax}$. If we allow C to be negative (which means allowing C_1 to be complex), $\dfrac{1}{y} - b = Ce^{-ax}$, $\dfrac{1}{y} = b + Ce^{-ax}$, $y = \dfrac{1}{b + Ce^{-ax}}$.

46.6 Solve $\dfrac{dy}{dx} = \dfrac{x^2}{y^3}$.

▮ $\int y^3\,dy = \int x^2\,dx$, $\frac{1}{4}y^4 = \frac{1}{3}x^3 + C_1$, $y^4 = \frac{4}{3}x^3 + C$.

46.7 Solve $x - y^2\dfrac{dy}{dx} = 0$.

▮ $\dfrac{dy}{dx} = \dfrac{x}{y^2}$, $\int y^2\,dy = \int x\,dx$, $\frac{1}{3}y^3 = \frac{1}{2}x^2 + C_1$, $y^3 = \frac{3}{2}x^2 + C$.

46.8 Solve $x + y\dfrac{dy}{dx} = 2$.

▮ $y\dfrac{dy}{dx} = 2 - x$, $\int y\,dy = \int (2-x)\,dx$, $\frac{1}{2}y^2 = 2x - \frac{1}{2}x^2 + C_1$, $y^2 = 4x - x^2 + C$.

46.9 Solve $y\dfrac{dy}{dx} - (1+y)x^2 = 0$.

▮ $y\dfrac{dy}{dx} = (1+y)x^2$, $\int \dfrac{y}{1+y}\,dy = \int x^2\,dx$, $\int \left(1 - \dfrac{1}{1+y}\right)dy = \int x^2\,dx$, $y - \ln|1+y| = \frac{1}{3}x^3 + C$.

46.10 Solve $y \dfrac{dy}{dx} - (1+y^2)x^2 = 0$.

▮ $y \dfrac{dy}{dx} = (1+y^2)x^2$, $\displaystyle\int \dfrac{y}{1+y^2}\,dy = \int x^2\,dx$, $\tfrac{1}{2}\ln(1+y^2) = \tfrac{1}{3}x^3 + C_1$, $\ln(1+y^2) = \tfrac{2}{3}x^3 + C_2$, $1+y^2 = e^{2x^3/3+C_2} = e^{C_2}e^{2x^3/3} = Ce^{2x^3/3}$, $y^2 = Ce^{2x^3/3} - 1$.

46.11 Solve $\dfrac{dy}{dx} = \dfrac{1+y^2}{1+x^2}$.

▮ $\displaystyle\int \dfrac{dy}{1+y^2} = \int \dfrac{dx}{1+x^2}$, $\tan^{-1} y = \tan^{-1} x + C_1$, $y = \tan(\tan^{-1} x + C_1) = \dfrac{\tan(\tan^{-1} x) + \tan C_1}{1 - \tan(\tan^{-1} x)\tan C_1}$,

$y = \dfrac{x+C}{1-Cx}$, where $C = \tan C_1$.

46.12 Solve $(1+x)e^{3y}\dfrac{dy}{dx} = 1$.

▮ $\displaystyle\int e^{3y}\,dy = \int \dfrac{dx}{1+x}$, $\tfrac{1}{3}e^{3y} = \ln|1+x| + C_1$, $e^{3y} = 3\ln|1+x| + C$, $3y = \ln(3\ln|1+x| + C)$, $y = \tfrac{1}{3}\ln(3\ln|1+x| + C)$.

46.13 Solve $\dfrac{dy}{dx} = \sqrt{\dfrac{1-y^2}{1-x^2}}$.

▮ $\displaystyle\int \dfrac{dy}{\sqrt{1-y^2}} = \int \dfrac{dx}{\sqrt{1-x^2}}$, $\sin^{-1} y = \sin^{-1} x + C$,

$y = \sin(\sin^{-1} x + C) = \sin(\sin^{-1} x)\cos C + \cos(\sin^{-1} x)\sin C = x\cos C + \sqrt{1-x^2}\sin C$

46.14 Solve $\dfrac{dy}{dx} = 3x^2 y^2$, given $y(0) = \tfrac{1}{2}$.

▮ $\displaystyle\int \dfrac{dy}{y^2} = \int 3x^2\,dx$, $-\dfrac{1}{y} = x^3 + C$, $y = -\dfrac{1}{x^3 + C}$. Since $y(0) = \tfrac{1}{2}$, $\tfrac{1}{2} = -1/C$, $C = -2$. Hence, $y = -1/(x^3 - 2)$.

46.15 Solve $\dfrac{dy}{dx} = 1 + y + x^2 + yx^2$.

▮ $\dfrac{dy}{dx} = y(1+x^2) + 1 + x^2 = (y+1)(1+x^2)$. $\displaystyle\int \dfrac{dy}{y+1} = \int (1+x^2)\,dx$, $\ln|y+1| = x + \tfrac{1}{3}x^3 + C_1$, $|y+1| = e^{C_1}\cdot e^{x+x^3/3}$, $y = Ce^{x+x^3/3} - 1$.

46.16 Solve $x\,dx + y\,dy = xy(x\,dy - y\,dx)$.

▮ $x + y\dfrac{dy}{dx} = xy\left(x\dfrac{dy}{dx} - y\right)$, $\dfrac{dy}{dx}(y - x^2y) = -(x + xy^2)$, $\dfrac{dy}{dx}y(1 - x^2) = -x(1+y^2)$, $\displaystyle\int \dfrac{y\,dy}{1+y^2} = -\int \dfrac{x\,dx}{1-x^2}$, $\tfrac{1}{2}\ln(1+y^2) = \tfrac{1}{2}\ln|1-x^2| + \ln C_1$, $(1+y^2)^{1/2} = C_1(1-x^2)^{1/2}$, $1+y^2 = C(1-x^2)$.

46.17 Solve $y' = (x+y)^2$. $\left(\text{As usual, } y' \equiv \dfrac{dy}{dx}.\right)$

▮ The variables do not separate. Try the substitution $z = x + y$. Then, $z' = 1 + y'$. So $z' - 1 = z^2$, $\dfrac{dz}{dx} = z^2 + 1$. Now, the variables x and z are separable. $\displaystyle\int \dfrac{dz}{z^2+1} = \int dx$, $\tan^{-1} z = x + C$, $z = \tan(x+C)$, $x + y = \tan(x+C)$, $y = \tan(x+C) - x$.

46.18 Solve $\dfrac{dy}{dx} = \dfrac{x+2y}{3y-2x}$.

▮ The variables are not separable. However, on the right side, the numerator and denominator are homogeneous of the same degree (one) in x and y. In such a case, let $y = vx$. Then, $\dfrac{dy}{dx} = v + x\dfrac{dv}{dx}$. Hence, $v + x\dfrac{dv}{dx} = \dfrac{x+2vx}{3vx-2x} = \dfrac{1+2v}{3v-2}$. Thus, $x\dfrac{dv}{dx} = \dfrac{1+2v}{3v-2} - v = \dfrac{1+4v-3v^2}{3v-2}$. Now the variables x and

v are separable. $\int \dfrac{3v-2}{1+4v-3v^2}\,dv = \int \dfrac{dx}{x}$, $-\frac{1}{2}\ln|1+4v-3v^2| = \ln|x| + \ln C_1$, $\ln|1+4v-3v^2|^{-1/2} = \ln(C_1|x|)$, $1/\sqrt{1+4v-3v^2} = C_1|x|$, $\sqrt{1+4v-3v^2} = 1/(C_1|x|)$, $1+4v-3v^2 = 1/(C_2x^2)$, $1+4(y/x) - 3(y/x)^2 = 1/(C_2x^2)$, $x^2+4xy-3y^2 = C$.

46.19 Solve $\dfrac{dy}{dx} = -\dfrac{2x^2+y^2}{2xy+3y^2}$.

▮ The variables are not separable. But the numerator and denominator are homogeneous in x and y of degree 2. Let $y=vx$. Then $\dfrac{dy}{dx} = v + x\dfrac{dv}{dx}$. Therefore, $v + x\dfrac{dv}{dx} = -\dfrac{2x^2+x^2v^2}{2x(vx)+3(vx)^2} = -\dfrac{2+v^2}{2v+3v^2}$. Hence, $x\dfrac{dv}{dx} = -\left(\dfrac{2+v^2}{2v+3v^2} + v\right) = -\dfrac{2+3v^2+3v^3}{2v+3v^2}$. The variables v and x are separable. $\int \dfrac{(2v+3v^2)\,dv}{2+3v^2+3v^3} = -\int \dfrac{dx}{x}$, $\frac{1}{3}\ln|2+3v^2+3v^3| = -\ln|x| + \ln C_1$, $|2+3v^2+3v^3|^{1/3} = C_1/|x|$, $|2+3v^2+3v^3| = C_2/|x|^3$, $|2+3(y/x)^2+3(y/x)^3| = C_2/|x|^3$, $|2x^3+3xy^2+3y^3| = C_2$, $2x^3+3xy^2+3y^3 = C$.

46.20 Find the curve that passes through the point $(1,3)$ and whose tangent line at (x,y) has slope $-(1+y/x)$.

▮ $\dfrac{dy}{dx} = -\left(1+\dfrac{y}{x}\right) = -\dfrac{x+y}{x}$. Let $y=vx$. $v + x\dfrac{dv}{dx} = -\dfrac{x+vx}{x} = -(1+v)$. $x\dfrac{dv}{dx} = -(2v+1)$. Now x and v are separable. $\int \dfrac{dv}{2v+1} = -\int \dfrac{dx}{x}$, $\frac{1}{2}\ln|2v+1| = -\ln|x| + \ln C_1$, $|2v+1|^{1/2} = C_1/|x|$, $2v+1 = C/x^2$, $2(y/x)+1 = C/x^2$, $2xy+x^2 = C$. Since the curve passes through $(1,3)$, $6+1 = C$. Hence, $2xy+x^2 = 7$.

46.21 A boat starts off from one side of a river at a point B and heads toward a point A on the other side of the river directly across from B. The river has a uniform width of c feet and its current downstream is a constant a ft/s. At each moment, the boat is headed toward A with a speed through the water of b ft/s. Under what conditions on a, b, and c will the boat ever reach the opposite bank?

▮ See Fig. 46-1. Let A be the origin, let B lie on the positive x-axis, and let the y-axis point upstream. Then the components of the boat's velocity are $dx/dt = -b\cos\theta$ and $dy/dt = -a + b\sin\theta$, where θ is the angle from the positive x-axis to the boat. Hence, $\dfrac{dy}{dx} = \dfrac{-a+b\sin\theta}{-b\cos\theta} = \dfrac{-a+b(-y/\sqrt{x^2+y^2})}{-b(x/\sqrt{x^2+y^2})} = \dfrac{a\sqrt{x^2+y^2}+by}{bx}$. The numerator and denominator are homogeneous of order one in x and y. Let $y=zx$. Then, $z + x\dfrac{dz}{dx} = \dfrac{a\sqrt{x^2+z^2x^2}+bzx}{bx} = \dfrac{a\sqrt{1+z^2}+bz}{b}$. Thus, $x\dfrac{dz}{dx} = \dfrac{a\sqrt{1+z^2}}{b}$. $\int \dfrac{dz}{\sqrt{1+z^2}} = \dfrac{a}{b}\int \dfrac{dx}{x}$. By means of a trigonometric substitution on the left, we get $\ln|\sqrt{1+z^2}+z| = (a/b)\ln|x| + \ln K$, $|\sqrt{1+z^2}+z| = K|x|^{a/b}$, $\left|\dfrac{\sqrt{x^2+y^2}}{x} + \dfrac{y}{x}\right| = Kx^{a/b}$, $\sqrt{x^2+y^2}+y = Kx^{a/b+1}$. At $t=0$, $x=c$ and $y=0$. Hence, $c = Kc^{a/b+1}$, and, therefore, $K = c^{-a/b}$. Thus, $c^{a/b}(\sqrt{x^2+y^2}+y) = x^{a/b+1}$. Solve for $y = \dfrac{1}{2c^{a/b}}(x^{a/b+1} - c^{2a/b}x^{1-a/b})$.
Case 1. $1-a/b<0$. (In this case, $b<a$, that is, the current is faster than the speed of the boat.) Then, as $x\to 0$, $y\to -\infty$, and the boat never reaches the opposite bank. **Case 2.** $1-a/b=0$. (In this case, the current is the same as the speed of the boat.) Then, as $x\to 0$, $y\to -\frac{1}{2}c$, and the boat lands at a distance below A equal to half the width of the river. **Case 3.** $1-a/b>0$. (In this case, the boat moves faster than the current.) Then, as $x\to 0$, $y\to 0$, and the boat reaches point A.

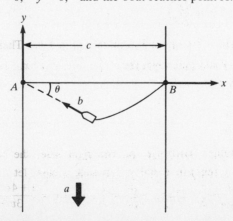

Fig. 46-1

46.22 Find the orthogonal trajectories of the family of circles tangent to the y-axis at the origin.

▎ The family consists of all curves $(x - a)^2 + y^2 = a^2$, with $a \neq 0$, or

$$x^2 + y^2 = 2ax \qquad (1)$$

Differentiate: $2x + 2y \dfrac{dy}{dx} = 2a$, $x + y \dfrac{dy}{dx} = a$. Substitute this value of a in (1), obtaining $x^2 + y^2 = 2x\left(x + y \dfrac{dy}{dx}\right)$. To find the orthogonal trajectories, replace $\dfrac{dy}{dx}$ by its negative reciprocal, which will be the slope of the orthogonal curves:

$$x^2 + y^2 = 2x\left(x - \frac{y}{y'}\right) \qquad (2)$$

Solve for y': $\dfrac{y}{y'} = x - \dfrac{x^2 + y^2}{2x} = \dfrac{x^2 - y^2}{2x}$, $y' = \dfrac{2xy}{x^2 - y^2}$. The numerator and denominator on the right are homogeneous. Let $y = vx$. Then $v + x \dfrac{dv}{dx} = \dfrac{2x^2 v}{x^2 - x^2 v^2} = \dfrac{2v}{1 - v^2}$, $x \dfrac{dv}{dx} = \dfrac{2v}{1 - v^2} - v = \dfrac{v + v^3}{1 - v^2}$. $\displaystyle\int \dfrac{1 - v^2}{v + v^3}\, dv = \int \dfrac{dx}{x}$. By a partial fractions decomposition, $\displaystyle\int \dfrac{dv}{v} - 2\int \dfrac{v}{1 + v^2}\, dv = \int \dfrac{dx}{x}$, $\ln|v| - \ln(1 + v^2) = \ln|x| + \ln C_1$, $\dfrac{v}{1 + v^2} = Cx$, $\dfrac{y/x}{1 + y^2/x^2} = Cx$, $\dfrac{y}{x^2 + y^2} = C$, $x^2 + y^2 = \dfrac{1}{C} y$, $x^2 + y^2 = 2Ky$, $x^2 + (y - K)^2 = K^2$. This is the family of all circles tangent to the x-axis at the origin.

46.23 Find the orthogonal trajectories of the family of hyperbolas $xy = c$.

▎ $xy' + y = 0$. Replace y' by $-1/y'$: $-x/y' + y = 0$, $\dfrac{dy}{dx} = \dfrac{x}{y}$, $\int y\, dy = \int x\, dx$, $\frac{1}{2} y^2 = \frac{1}{2} x^2 + C_1$, $y^2 = x^2 + C$, $y^2 - x^2 = C$. Hence, the orthogonal trajectories form another family of equilateral hyperbolas (and the pair of lines $y = \pm x$).

46.24 Find the orthogonal trajectories of the family of all straight lines through the origin $y = mx$.

▎ $\dfrac{dy}{dx} = m = \dfrac{y}{x}$. Replace $\dfrac{dy}{dx}$ by its negative reciprocal: $-\dfrac{1}{dy/dx} = \dfrac{y}{x}$, $\dfrac{dy}{dx} = -\dfrac{x}{y}$, $\int y\, dy = -\int x\, dx$, $\frac{1}{2} y^2 = -\frac{1}{2} x^2 + C_1$, $y^2 = -x^2 + C$, $x^2 + y^2 = C$. Thus, we have obtained the family of all circles with center at the origin.

46.25 Determine whether the equation $(y - x^3)\, dx + (x + y^3)\, dy = 0$ is *exact*. If it is, solve the equation.

▎ $M\, dx + N\, dy = 0$ is exact if and only if $\dfrac{\partial M}{\partial y} = \dfrac{\partial N}{\partial x}$. In this case, $\dfrac{\partial M}{\partial y} = \dfrac{\partial}{\partial y}(y - x^3) = 1$ and $\dfrac{\partial N}{\partial x} = \dfrac{\partial}{\partial x}(x + y^3) = 1$. Hence, the equation is exact. To solve, we first find $g(y) = \displaystyle\int \left(N - \dfrac{\partial}{\partial y} \int M\, dx\right) dy = \int \left\{x + y^3 - \dfrac{\partial}{\partial y}\left[\int (y - x^3)\, dx\right]\right\} dy = \int \left[x + y^3 - \dfrac{\partial}{\partial y}\left(yx - \dfrac{1}{4} x^4\right)\right] dy = \int (x + y^3 - x)\, dy = \int y^3\, dy = \frac{1}{4} y^4$. Then set $f(x, y) = \int M\, dx + g(y) = yx - \frac{1}{4} x^4 + \frac{1}{4} y^4$. The left side of the original equation is equal to df. Hence, the solution is $yx - \frac{1}{4} x^4 + \frac{1}{4} y^4 = C_1$, or $4yx - x^4 + y^4 = C$.

46.26 Determine whether the equation $(\cos y + y \cos x)\, dx + (\sin x - x \sin y)\, dy = 0$ is exact. If it is, solve the equation.

▎ $\dfrac{\partial}{\partial y}(\cos y + y \cos x) = -\sin y + \cos x$. $\dfrac{\partial}{\partial x}(\sin x - x \sin y) = \cos x - \sin y$. Hence, the equation is exact. As in Problem 46.25, let $g(y) = \displaystyle\int \left\{\sin x - x \sin y - \dfrac{\partial}{\partial y}\left[\int (\cos y + y \cos x)\, dx\right]\right\} dy = \int \left[\sin x - x \sin y - \dfrac{\partial}{\partial y}(x \cos y + y \sin x)\right] dy = \int [\sin x - x \sin y - (-x \sin y + \sin x)]\, dy = \int 0\, dy = K$. Set $f(x, y) = \int (\cos y + y \cos x)\, dx + K = x \cos y + y \sin x + K$. Hence, the solution is $x \cos y + y \sin x = C$.

46.27 Explain the method of solving $\dfrac{dy}{dx} + p(x)y = q(x)$ by means of an *integrating factor*.

▎ Define the integrating factor $\rho = \exp[\int p(x)\, dx]$. Hence, $\dfrac{d\rho}{dx} = \exp[\int p(x)\, dx] \cdot p(x) = \rho \cdot p(x)$. Thus, $\dfrac{d}{dx}(\rho y) = \rho \dfrac{dy}{dx} + \dfrac{d\rho}{dx} y = \rho \dfrac{dy}{dx} + \rho \cdot p(x) \cdot y$. If we multiply the original differential equation by ρ, $\rho \dfrac{dy}{dx} + \rho \cdot p(x) \cdot y = \rho \cdot q(x)$. Hence, $\dfrac{d}{dx}(\rho y) = \rho q(x)$. So $\rho y = \int \rho q(x)\, dx + C$, and $y = \dfrac{1}{\rho}\left[\int \rho q(x)\, dx + C\right]$.

46.28 Solve $\dfrac{dy}{dx} + y = e^x$.

▮ Use the method of Problem 46.27. $\rho = \exp(\int 1\,dx) = e^x$, and $\int \rho q(x)\,dx = \int e^x \cdot e^x\,dx = \int e^{2x}\,dx = \frac{1}{2}e^{2x}$. So $y = \dfrac{1}{e^x}\left(\frac{1}{2}e^{2x} + C\right) = \frac{1}{2}e^x + Ce^{-x}$.

46.29 Solve $\dfrac{dy}{dx} + \dfrac{1}{3}\,y = 1$.

▮ Use an integrating factor. Let $\rho = \exp\left(\int \frac{1}{3}\,dx\right) = e^{x/3}$. Then $\int \rho q(x)\,dx = \int e^{x/3}\,dx = 3e^{x/3}$. Therefore, $y = e^{-x/3}(3e^{x/3} + C) = 3 + Ce^{-x/3}$.

46.30 Solve $x\dfrac{dy}{dx} + y = x$.

▮ Divide by x: $\dfrac{dy}{dx} + \dfrac{1}{x}\,y = 1$. Use an integrating factor: $\rho = \exp\left(\int \frac{1}{x}\,dx\right) = e^{\ln x} = x$. Then $\int \rho q(x)\,dx = \int x\,dx = \frac{1}{2}x^2$, and $y = (1/x)(\frac{1}{2}x^2 + C) = \frac{1}{2}x + C/x$.

46.31 Solve $\dfrac{dy}{dx} - y = \sin x$.

▮ Use an integrating factor. $\rho = \exp[\int (-1)\,dx] = e^{-x}$, $\int \rho q(x)\,dx = \int e^{-x}\sin x\,dx = -\frac{1}{2}e^{-x}(\sin x + \cos x)$. (The last result is found by integration by parts.) Hence, $y = e^x[-\frac{1}{2}e^{-x}(\sin x + \cos x) + C] = -\frac{1}{2}(\sin x + \cos x) + Ce^x$.

46.32 Find a curve in the xy-plane that passes through the origin and for which the tangent at (x, y) has slope $x^2 + y$.

▮ $\dfrac{dy}{dx} = x^2 + y$, $\dfrac{dy}{dx} - y = x^2$. Use an integrating factor. $\rho = \exp[\int (-1)\,dx] = e^{-x}$. $\int \rho q(x)\,dx = \int e^{-x}x^2\,dx = -e^{-x}(x^2 + 2x + 2)$. (The last equation follows by integration by parts.) Hence, $y = e^x[-e^{-x}(x^2 + 2x + 2) + C]$, $y = -(x^2 + 2x + 2) + Ce^x$. Since the curve passes through $(0, 0)$, $0 = -2 + C$, $C = 2$. Thus, $y = -(x^2 + 2x + 2) + 2e^x$.

46.33 Find a curve that passes through $(0, 2)$ and has at each point (x, y) a tangent line with slope $y - 2e^{-x}$.

▮ $\dfrac{dy}{dx} = y - 2e^{-x}$, $\dfrac{dy}{dx} - y = -2e^{-x}$. Let $\rho = \exp[\int(-1)\,dx] = e^{-x}$. Then $\int \rho q(x)\,dx = -2\int e^{-2x}\,dx = e^{-2x}$. Therefore, $y = e^x(e^{-2x} + C) = e^{-x} + Ce^x$. Since the curve passes through $(0, 1)$, $2 = 1 + C$, $C = 1$. Hence, $y = e^{-x} + e^x = 2\cosh x$.

46.34 Find a curve that passes through $(2, 1)$ and has at each point (x, y) a normal line with slope $\dfrac{2xy}{y^2 - x^2}$.

▮ The slope of the tangent line is the negative reciprocal of the slope of the normal line. Hence, $\dfrac{dy}{dx} = \dfrac{x^2 - y^2}{2xy}$. Let $y = vx$. Then $v + x\dfrac{dv}{dx} = \dfrac{x^2 - x^2v^2}{2x(vx)} = \dfrac{1 - v^2}{2v}$. Thus, $x\dfrac{dv}{dx} = \dfrac{1 - v^2}{2v} - v = \dfrac{1 - 3v^2}{2v}$. Now x and v are separable. $2\int \dfrac{v}{1 - 3v^2}\,dv = \int \dfrac{dx}{x}$, $-\frac{1}{3}\ln|1 - 3v^2| = \ln|x| + \ln C_1$, $|1 - 3v^2|^{-1/3} = C_1|x|$, $1 - 3v^2 = C_2/x^3$, $1 - 3(y^2/x^2) = c_2/x^3$, $x^3 - 3xy^2 = C_2$. Since the curve passes through $(2, 1)$, $8 - 6 = C_2$. Thus, $x^3 - 3xy^2 = 2$.

46.35 Solve $y' + \dfrac{2}{x}\,y = 6x^2$.

▮ Use an integrating factor. Let $\rho = \exp\left(\int \frac{2}{x}\,dx\right) = e^{2\ln x} = x^2$. Hence, $\int \rho q(x)\,dx = \int x^2 \cdot 6x^2\,dx = 6\int x^4\,dx = \frac{6}{5}x^5$. Then $y = (1/x^2)(\frac{6}{5}x^5 + C) = \frac{6}{5}x^3 + C/x^2$.

46.36 Solve $\tan x\dfrac{dy}{dx} + y = \sec x$.

▮ $\dfrac{dy}{dx} + (\cot x)y = \csc x$. Use an integrating factor. $\rho = \exp(\int \cot x\,dx) = e^{\ln(\sin x)} = \sin x$. Then $\int \rho q(x)\,dx = \int \sin x \cdot \csc x\,dx = \int 1\,dx = x$. Therefore, $y = (x + C)\csc x$.

46.37 Find a curve in the xy-plane passing through the point $(1, 1)$ and for which the normal line at any point (x, y) has slope $-xy^2$.

▮ The tangent line has slope $\dfrac{1}{xy^2}$. So $\dfrac{dy}{dx} = \dfrac{1}{xy^2}$. Hence, $\int y^2\, dy = \int \dfrac{1}{x}\, dx$, $\frac{1}{3}y^3 = \ln|x| + C_1$, $y^3 = 3\ln|x| + C$. Since the curve passes through $(1, 1)$, $1 = 3\ln 1 + C$, $1 = 3\cdot 0 + C$, $C = 1$. Therefore, $y^3 = 3\ln|x| + 1$, or $x = \pm e^{(y^3-1)/3}$.

46.38 Find a curve in the xy-plane that passes through $(5, -1)$ such that every normal line to the curve passes through the point $(1, 2)$.

▮ The normal line at (x, y) has slope $\dfrac{y-2}{x-1}$. Hence, the tangent line has slope $\dfrac{1-x}{y-2}$. Thus, $\dfrac{dy}{dx} = \dfrac{1-x}{y-2}$. So $\int (y-2)\, dy = \int (1-x)\, dx$. Therefore, $\frac{1}{2}y^2 - 2y = x - \frac{1}{2}x^2 + C$, $x^2 - 2x + y^2 - 4y = C_1$, $(x-1)^2 + (y-2)^2 = C_2$. Since the curve passes through $(5, -1)$, $16 + 9 = C_2$. Hence, $(x-1)^2 + (y-2)^2 = 25$. Thus, the curve is the circle with center at $(1, 2)$ and radius 5.

46.39 Find the curve that passes through $(0, 4)$ such that every tangent line to the curve passes through $(1, 2)$.

▮ The slope of the tangent line at (x, y) is $\dfrac{y-2}{x-1}$. Hence, $\dfrac{dy}{dx} = \dfrac{y-2}{x-1}$. Thus, $\int \dfrac{dy}{y-2} = \int \dfrac{dx}{x-1}$, $\ln|y-2| = \ln|x-1| + C$, $\ln\left|\dfrac{y-2}{x-1}\right| = C$, $\dfrac{y-2}{x-1} = C_1$, $y - 2 = C_1(x-1)$. Since the curve passes through $(0, 4)$, $2 = C_1(-1)$, $C_1 = 2$. Hence, $y - 2 = -2(x-1)$, $y = -2x + 4$. Thus, the curve is the straight line through $(0, 4)$ and $(1, 2)$.

46.40 Find a procedure for solving a *Bernoulli equation*

$$y' + p(x)y = q(x)y^n \qquad (n \neq -1) \tag{1}$$

▮ Let $v = y^{1-n}$. Then $v' = (1-n)y^{-n}y'$. From (1), $(1-n)y^{-n}y' + (1-n)p(x)y^{1-n} = (1-n)q(x)$. Hence,

$$v' + (1-n)p(x)v = (1-n)q(x) \tag{2}$$

Now we can use an integrating factor. Let $\rho = \exp\int (1-n)p(x)\, dx$. Then $v = \dfrac{1}{\rho}\left[\int (1-n)\rho q(x)\, dx + C\right]$, and $y = v^{1/(1-n)}$.

46.41 Solve $xy' - 2y = 4x^3y^{1/2}$.

▮ The given equation is equivalent to

$$y' - \frac{2}{x}y = 4x^2y^{1/2} \tag{1}$$

Use the method of Problem 46.40, with $n = \frac{1}{2}$. We get the equation

$$v' - \frac{1}{x}v = 2x^2 \tag{2}$$

Let $\rho = \exp\left(\int -\dfrac{1}{x}\, dx\right) = e^{-\ln x} = \dfrac{1}{x}$. Hence, $v = x(\int 2x\, dx + C) = x(x^2 + C)$. Then, $y = v^2 = x^2(x^2 + C)^2$.

46.42 Solve the Bernoulli equation $y' - (1/x + 2x^4)y = x^3y^2$.

▮ Use the method of Problem 46.40. Let $v = 1/y$, $v' = -y'/y^2$. Then, $v' + (1/x + 2x^4)v = -x^3$. Let $\rho = \exp\left[\int \left(\dfrac{1}{x} + 2x\right)dx\right] = e^{\ln x + 2x^5/5} = xe^{2x^5/5}$. Then, $v = -\dfrac{1}{x}e^{-2x^5/5}(\int x^4 e^{2x^5/5}\, dx) = -\dfrac{e^{-2x^5/5}}{x}\left(\frac{1}{2}e^{2x^5/5} + C\right) = -\left(\dfrac{1 + Ke^{-2x^5/5}}{2x}\right)$. Therefore, $y = -\dfrac{2x}{1 + Ke^{-2x^5/5}}$.

46.43 Consider the *Riccati equation*

$$y' = P(x) + Q(x)y + R(x)y^2 \tag{1}$$

If y_p is a particular solution of (1) and z is the general solution of the Bernoulli equation

$$z' - (Q + 2Ry_p)z = Rz^2 \tag{2}$$

show that $y = y_p + z$ is a general solution of the Riccati equation (1).

▌ **1.** Assume $y = y_p + z$, where z is a solution of the Bernoulli equation (2). Then, $y' = y_p' + z' = (P + Qy_p + Ry_p^2) + [Rz^2 + (Q + 2Ry_p)z] = P + Q(y_p + z) + R(y_p^2 + 2y_p z + z^2) = P + Qy + R(y_p + z)^2 = P + Qy + y^2$. Hence, y is a solution of (1). **2.** Assume y is a solution of (1). Let $z = y - y_p$. Then $z' = y' - y_p' = (P + Qy + Ry^2) - (P + Qy_p + Ry_p^2) = Q(y - y_p) + R(y^2 - y_p^2) = (y - y_p)[Q + R(y + y_p)] = z[Q + R(z + 2y_p)] = z(Q + 2Ry_p + Rz) = Rz^2 + (Q + 2Ry_p)z$. Thus, z is a solution of the Bernoulli equation (2).

46.44 Solve the equation

$$y' = -x^5 + \frac{y}{x} + x^3 y^2 \tag{1}$$

▌ This is a Riccati equation (see Problem 46.43). Note that $y_p = x$ is a particular solution. Consider the corresponding Bernoulli equation $z' - (1/x + 2x^4)z = x^3 z^2$. By Problem 46.40, its general solution is $z = -\dfrac{2x}{1 + Ke^{-2x^5/5}}$. Hence, by Problem 46.43, a general solution of (1) is $y = x - \dfrac{2x}{1 + Ke^{-2x^5/5}}$.

46.45 Solve $xy'' + y' = x$.

▌ This is equivalent to $y'' + (1/x)y' = 1$. Let $z = y'$. Then $z' + (1/x)z = 1$ is a first-order equation of the form $dz/dx + p(x)z = q(x)$. Use an integrating factor. Let $\rho = \exp\left(\int \dfrac{1}{x}\, dx\right) = e^{\ln x} = x$. Then, $\int \rho q(x)\, dx = \int x\, dx = \frac{1}{2}x^2$. Hence, $z = (1/x)(\frac{1}{2}x^2 + C) = \frac{1}{2}x + C/x$. Therefore, $dy/dx = \frac{1}{2}x + C/x$, and $y = \frac{1}{4}x^2 + C \ln|x| + K$.

46.46 Solve $y'' - (y')^2 = 0$.

▌ Let $z = y'$. Then $z' - z^2 = 0$ is a first-order equation. $dz/dx = z^2$. The variables x and z are separable. $\int \dfrac{dz}{z^2} = \int dx$, $-1/z = x + C$, $z = -1/(x + C)$. Hence, $dy/dx = -1/(x + C)$, $y = -\ln|x + C| + K$.

46.47 Solve $y'' + a^2/y^3 = 0$.

▌ Let $z = y'$. Then $y'' = \dfrac{dz}{dx} = \dfrac{dz}{dy} \cdot \dfrac{dy}{dx} = \dfrac{dz}{dy} \cdot z$. Hence, $z\dfrac{dz}{dy} + \dfrac{a^2}{y^3} = 0$, $z\dfrac{dz}{dy} = -\dfrac{a^2}{y^3}$. y and z are separable. $\int z\, dz = -a^2 \int \dfrac{dy}{y^3}$, $\frac{1}{2}z^2 = \frac{1}{2}\dfrac{a^2}{y^2} + C_1$, $z^2 = \dfrac{a^2}{y^2} + C = \dfrac{a^2 + Cy^2}{y^2}$, $\dfrac{dy}{dx} = \dfrac{\sqrt{a^2 + Cy^2}}{y}$, $\int \dfrac{y\, dy}{\sqrt{a^2 + Cy^2}} = \int dx$, $\frac{1}{C}\sqrt{a^2 + Cy^2} = x + K$, $a^2 + Cy^2 = C^2(x + K)^2$.

46.48 Describe the solutions of a homogeneous linear differential equation of second order with constant coefficients:

$$y'' + by' + cy = 0 \tag{1}$$

▌ Consider the auxiliary quadratic equation

$$u^2 + bu + c = 0 \tag{2}$$

There are three cases. **Case 1.** $b^2 - 4c > 0$. Hence, (2) has two distinct real roots r_1 and r_2. Then, the general solution of (1) is $C_1 e^{r_1 x} + C_2 e^{r_2 x}$. **Case 2.** $b^2 - 4c = 0$. Hence, (2) has a double root r. Then the general solution of (1) is $(C_1 + C_2 x)e^{rx}$, or $C_1 e^{rx} + C_2 xe^{rx}$. **Case 3.** $b^2 - 4c < 0$. Hence, (2) has two distinct conjugate complex roots $r_1 = p + iq$ and $r_2 = p - iq$. Then the general solution of (1) is $(C_1 \cos qx + C_2 \sin qx)e^{px}$.

46.49 Find the general solution $y'' + 4y' + 3y = 0$.

▌ Solve the auxiliary equation $u^2 + 4u + 3 = 0$. $(u + 3)(u + 1) = 0$, $u = -3$ or -1. Hence, by Problem 46.48, the general solution is $C_1 e^{-x} + C_2 e^{-3x}$.

46.50 Find the general solution of $y'' - 5y' + 4y = 0$.

▌ Solve the auxiliary equation $u^2 - 5u + 4 = 0$. $(u - 4)(u - 1) = 0$, $u = 4$ or 1. By Problem 46.48, the general solution is $C_1 e^x + C_2 e^{4x}$.

46.51 Find the general solution of $y'' + y' - 6y = 0$.

▮ Solve the auxiliary equation $u^2 + u - 6 = 0$. $(u + 3)(u - 2) = 0$, $u = -3$ or $u = 2$. By Problem 46.48, the general solution is $C_1 e^{2x} + C_2 e^{-3x}$.

46.52 Find the general solution of $y'' - 4y' + 4y = 0$.

▮ Solve the auxiliary equation $u^2 - 4u + 4 = 0$. $(u - 2)^2 = 0$, $u = 2$. By Problem 46.48, the general solution is $(C_1 + C_2 x)e^{2x}$.

46.53 Find the general solution of $y'' + 2y' + 2y = 0$.

▮ Solve the auxiliary equation $u^2 + 2u + 2 = 0$. $u = \dfrac{-2 \pm \sqrt{-4}}{2} = -1 \pm i$. By Problem 46.48, the general solution is $(C_1 \cos x + C_2 \sin x)e^x$.

46.54 Find the general solution of $y'' + y' + y = 0$.

▮ Solve the auxiliary equation $u^2 + u + 1 = 0$, $u = \dfrac{-1 \pm \sqrt{-3}}{2} = -\dfrac{1}{2} \pm \dfrac{\sqrt{3}}{2} i$. By Problem 46.48, the general solution is $\left(C_1 \cos \dfrac{\sqrt{3}}{2} x + C_2 \sin \dfrac{\sqrt{3}}{2} x\right)e^{-x/2}$.

46.55 Find the general solution of $\dfrac{d^2 y}{dx^2} = -a^2 y$, where $a > 0$.

▮ This is equivalent to $y'' + a^2 y = 0$. Solve the auxiliary equation $u^2 + a^2 = 0$, $u = \pm ai$. By Problem 46.48, the general solution is $C_1 \cos ax + C_2 \sin ax$.

46.56 Given a linear differential equation of nth order with constant coefficients,

$$y^{(n)} + a_{n-1} y^{(n-1)} + \cdots + a_1 y' + a_0 y = f(x) \tag{1}$$

if y_p is a particular solution of (1) and if y_g is a general solution of the corresponding homogeneous equation

$$y^{(n)} + a_{n-1} y^{(n-1)} + \cdots + a_1 y' + a_0 y = 0 \tag{2}$$

show that $y_p + y_g$ is a general solution of (1).

▮ Clearly, $y_p + y_g$ is a solution of (1), since $(y_p + y_g)^{(k)} = y_p^{(k)} + y_g^{(k)}$. If y is a solution of (1), it is obvious, by the same reasoning, that $y - y_p$ is a solution of (2). Hence, $y - y_p$ has the form y_g. So $y = y_p + y_g$.

46.57 Find a general solution of

$$y'' + y' + y = x^2 \tag{1}$$

▮ By Problem 46.54, a general solution of $y'' + y' + y = 0$ is $\left(C_1 \cos \dfrac{\sqrt{3}}{2} x + C_2 \sin \dfrac{\sqrt{3}}{2} x\right)e^{-x/2}$. We must find a particular solution of (1). Noting that the right-hand side is a polynomial, we guess that a polynomial will work. Hence, it must have degree 2. Let us set $y = a_2 x^2 + a_1 x + a_0$. Then $y' = 2a_2 x + a_1$ and $y'' = 2a_2$. Hence, we must have $2a_2 + 2a_2 x + a_1 + a_2 x^2 + a_1 x + a_0 = x^2$, or $a_2 x^2 + (2a_2 + a_1)x + (2a_2 + a_1 + a_0) = x^2$. Hence, $a_2 = 1$, $2a_2 + a_1 = 0$, $2a_2 + a_1 + a_0 = 0$. Thus, $a_1 = -2a_2 = -2$, and $a_0 = 2a_2 - a_1 = -2 + 2 = 0$. Thus, $x^2 - 2x$ is a particular solution. Hence, a general solution of (1) is $x^2 - 2x + \left(C_1 \cos \dfrac{\sqrt{3}}{2} x + C_2 \sin \dfrac{\sqrt{3}}{2} x\right)e^{-x/2}$.

46.58 Find a general solution of

$$y'' + 5y' + 4y = 1 + x + x^2 \tag{1}$$

▮ A general solution of $y'' + 5y' + 4y = 0$ is obtained from the auxiliary equation $u^2 + 5u + 4 = (u + 4)(u + 1) = 0$. The roots are -1 and -4. Hence, the general solution is $C_1 e^{-x} + C_2 e^{-4x}$. We guess that a particular solution of (1) is of the form $y = Ax^2 + Bx + C$. Then, $y' = 2Ax + B$, and $y'' = 2A$. Therefore, $1 + x + x^2 = y'' + 5y' + 4y = 4Ax^2 + (10A + 4B)x + (2A + 5B + 4C)$. Thus, $4A = 1$, $A = \frac{1}{4}$. Also, $1 = 10A + 4B$, $1 = 10(\frac{1}{4}) + 4B$, $B = -\frac{3}{8}$. Finally, $1 = 2A + 5B + 4C$, $C = \frac{1}{4}(1 - 2A - 5B) = \frac{1}{4}(1 - \frac{1}{2} + \frac{15}{8}) = \frac{19}{32}$. So a particular solution is $\frac{1}{4}x^2 - \frac{3}{8}x + \frac{19}{32} = \frac{1}{32}(8x^2 - 12x + 19)$. The general solution is $\frac{1}{32}(8x^2 - 12x + 19) + C_1 e^{-x} + C_2 e^{-4x}$.

46.59 Solve $y'' - 5y' + 4y = \sin 3x$.

▌ A general solution of the homogeneous equation $y'' - 5y' + 4y = 0$ was found in Problem 46.50; it is $C_1 e^x + C_2 e^{4x}$. Let us guess that a particular solution of the given equation is $y = A \sin 3x + B \cos 3x$. (An alternative to guessing is provided by Problem 46.65.) Then $y' = 3A \cos 3x - 3B \sin 3x$ and $y'' = -9A \sin 3x - 9B \cos 3x$. So $\sin 3x = (-9A \sin 3x - 9B \cos 3x) - 5(3A \cos 3x - 3B \sin 3x) + 4(A \sin 3x + B \cos 3x)$. Equating coefficients of $\sin 3x$ and $\cos 3x$, we get $-9A + 15B + 4A = 1$ and $-9B - 15A + 4B = 0$, or, more simply, $-5A + 15B = 1$ and $-15A - 5B = 0$. Hence, solving simultaneously, we get $-50A = 1$, $A = -\frac{1}{50}$, and $B = \frac{3}{50}$. Therefore, $y = \frac{1}{50}(3 \cos 3x - \sin 3x)$. The general solution is $\frac{1}{50}(3 \cos 3x - \sin 3x) + C_1 e^x + C_2 e^{4x}$.

46.60 Solve $y'' + 2y' + 2y = e^{3x}$.

▌ The general solution of the corresponding homogeneous equation was found in Problem 46.53: $y_g = (C_1 \cos x + C_2 \sin x)e^{-x}$. Let us try $y = Ae^{3x}$ as a particular solution of the given equation. Then, $y' = 3Ae^{3x}$ and $y'' = 9Ae^{3x}$. So $9Ae^{3x} + 6Ae^{3x} + 2Ae^{3x} = e^{3x}$. Hence, $15A = 1$, $A = \frac{1}{15}$. Thus, a particular solution is $y = \frac{1}{15}e^{3x}$, and the general solution is $\frac{1}{15}e^{3x} + (C_1 \cos x + C_2 \sin x)e^{-x}$.

46.61 Solve $y'' + y' - 12y = xe^x$.

▌ A general solution of the corresponding homogeneous equation is $C_1 e^{3x} + C_2 e^{-4x}$, since 3 and -4 are the roots of the auxiliary equation $u^2 + u - 12 = 0$. Let us try $y = (A + Bx)e^x$ as a particular solution of the given equation. Then $y' = (A + B + Bx)e^x$, and $y'' = (A + 2B + Bx)e^x$. Thus, $(A + 2B + Bx)e^x + (A + B + Bx)e^x - 12(A + Bx)e^x = xe^x$. Hence, $(A + 2B + Bx) + (A + B + Bx) - 12(A + Bx) = x$. Equating the coefficients of x as well as the constant coefficients, we get $-10B = 1$ and $-10A + 3B = 0$. Therefore, $B = -\frac{1}{10}$ and $A = -\frac{3}{100}$. Thus, $y = -\frac{1}{100}(3 + 10x)e^x$, and the general solution is $-\frac{1}{100}(3 + 10x)e^x + C_1 e^{3x} + C_2 e^{-4x}$.

46.62 Find a general solution of

$$y'' + 9y = e^x \cos 2x \qquad (1)$$

▌ A general solution of the corresponding homogeneous equation was found in Problem 46.55: $y_g = C_1 \cos 3x + C_2 \sin 3x$. Let us try $y = Ae^x \cos 2x + Be^x \sin 2x$ as a particular solution of (1). Then $y' = A(e^x \cos 2x - 2e^x \sin 2x) + B(e^x \sin 2x + 2e^x \cos 2x) = (A + 2B)e^x \cos 2x + (B - 2A)e^x \sin 2x$, and $y'' = (4B - 3A)e^x \cos 2x - (3B + 4A)e^x \sin 2x$. Thus, $e^x \cos 2x = (4B - 3A)e^x \cos 2x - (3B + 4A)e^x \sin 2x + 9(Ae^x \cos 2x + Be^x \sin 2x)$. Equating coefficients of $e^x \cos 2x$ and $e^x \sin 2x$, we get $4B - 3A + 9A = 1$ and $-(3B + 4A) + 9B = 0$, which, when simplified, become $4B + 6A = 1$ and $6B - 4A = 0$. Solving simultaneously yields $B = \frac{1}{13}$ and $A = \frac{3}{26}$. Hence, $y = \frac{1}{26}(3e^x \cos 2x + 2e^x \sin 2x)$. The general solution is $\frac{1}{26}(3e^x \cos 2x + 2e^x \sin 2x) + C_1 \cos 3x + C_2 \sin 3x$.

46.63 Find a general solution of $y'' - y' - 6y = 2(\sin 4x + \cos 4x)$.

▌ A general solution of $y'' - y' - 6y = 0$ is obtained by solved the auxiliary equation $u^2 - u - 6 = 0$. Since $u^2 - u - 6 = (u - 3)(u + 2)$, the roots are $u = 3$ and $u = -2$. Hence, the general solution is $C_1 e^{3x} + C_2 e^{-2x}$. Now we try $y = A \sin 4x + B \cos 4x$ as a particular solution of the given equation. Then $y' = 4A \cos 4x - 4B \sin 4x$, and $y'' = -16A \sin 4x - 16B \cos 4x$. Thus, $y'' - y' - 6y = (4B - 22A) \sin 4x - (4A + 22B) \cos 4x = 2 \sin 4x + 2 \cos 4x$. Hence, $4B - 22A = 2$ and $-22B - 4A = 2$. Solving simultaneously, we get $A = -\frac{13}{125}$ and $B = -\frac{18}{125}$. So $y = -\frac{1}{125}(13 \sin 4x + 18 \cos 4x)$, and the general solution is $-\frac{1}{125}(13 \sin 4x + 18 \cos 4x) + C_1 e^{3x} + C_2 e^{-2x}$.

46.64 Solve $y'' + Ay = B$, where $A > 0$.

▌ The general solution of the corresponding homogeneous equation is $y_g = C_1 \cos \sqrt{A}x + C_2 \sin \sqrt{A}x$, by Problem 46.55. A particular solution is $y_p = B/A$. Hence, the general solution is $B/A + C_1 \cos \sqrt{A}x + C_2 \sin \sqrt{A}x$.

46.65 Solve by the method of variation of parameters,

$$y'' + y = \tan x \qquad (1)$$

▌ The general solution of $y'' + y = 0$ is obtained by solving the auxiliary equation $u^2 + 1 = 0$. The roots are $\pm i$. Hence, the general solution consists of all linear combinations of $y_1 = \cos x$ and $y_2 = \sin x$. Calculate the *Wronskian* of y_1 and y_2:

$$W(y_1, y_2) = \begin{vmatrix} y_1 & y_2 \\ y_1' & y_2' \end{vmatrix} = \begin{vmatrix} \cos x & \sin x \\ -\sin x & \cos x \end{vmatrix} = 1$$

and use it together with the inhomogenous term $(\tan x)$ to construct

$$v_1 = -\int \frac{y_2 \tan x}{W(y_1, y_2)} \, dx = -\int \sin x \tan x \, dx = -\int (\sec x - \cos x) \, dx = \sin x - \ln |\sec x + \tan x|$$

and

$$v_2 = \int \frac{y_1 \tan x}{W(y_1, y_2)} \, dx = \int \cos x \tan x \, dx = \int \sin x \, dx = -\cos x$$

Then a particular solution of (1) is $y_p = y_1 v_1 + y_2 v_2 = \cos x(\sin x - \ln |\sec x + \tan x|) + \sin x(-\cos x) = -\cos x \ln |\sec x + \tan x|$. Therefore, the general solution of (1) is $-\cos x \ln |\sec x + \tan x| + C_1 \cos x + C_2 \sin x$.

46.66 Solve by variation of parameters

$$y'' + 2y' + y = e^{-x} \ln x \tag{1}$$

▮ The general solution of $y'' + 2y' + y = 0$ is obtained from the auxiliary equation $u^2 + 2u + 1 = 0$. The latter has -1 as a repeated root. Hence, the general solution consists of all linear combinations of $y_1 = e^{-x}$ and $y_2 = xe^{-x}$. The Wronskian

$$W(y_1, y_2) = \begin{vmatrix} y_1 & y_2 \\ y_1' & y_2' \end{vmatrix} = \begin{vmatrix} e^{-x} & xe^{-x} \\ -e^{-x} & e^{-x}(1-x) \end{vmatrix} = e^{-2x}$$

By means of an integration by parts, $v_1 = -\int \dfrac{y_2 e^{-x} \ln x}{W(y_1, y_2)} \, dx = -\int \dfrac{xe^{-x} \cdot e^{-x} \ln x}{e^{-2x}} \, dx = -\int x \ln x \, dx = \frac{1}{4}x^2(1 - 2 \ln x)$. Similarly $v_2 = \int \dfrac{y_1 e^{-x} \ln x}{W(y_1, y_2)} \, dx = \int \dfrac{e^{-x} \cdot e^{-x} \ln x}{e^{-2x}} \, dx = \int \ln x \, dx = x(\ln x - 1)$. Therefore, a particular solution of (1) is $y = y_1 v_1 + y_2 v_2 = \frac{1}{4}x^2 e^{-x}(1 - 2 \ln x) + x^2 e^{-x}(\ln x - 1) = \frac{1}{4}x^2 e^{-x}(2 \ln x - 3)$. The general solution is $\frac{1}{4}x^2 e^{-x}(2 \ln x - 3) + e^{-x}(C_1 + C_2 x)$.

46.67 Solve

$$y'' - 3y' + 2y = xe^{3x} + 1 \tag{1}$$

▮ The general solution of $y'' - 3y' + 2y = 0$ is obtained from the auxiliary equation $u^2 - 3u + 2 = (u - 2)(u - 1) = 0$. The roots are 1 and 2. Hence, the general solution consists of all linear combinations of $y_1 = e^x$ and $y_2 = e^{2x}$. The Wronskian

$$W(y_1, y_2) = \begin{vmatrix} y_1 & y_2 \\ y_1' & y_2' \end{vmatrix} = \begin{vmatrix} e^x & e^{2x} \\ e^x & 2e^{2x} \end{vmatrix} = e^{3x}$$

Then $v_1 = -\int \dfrac{y_2(xe^{3x} + 1)}{W(y_1, y_2)} \, dx = -\int \dfrac{e^{2x}(xe^{3x} + 1)}{e^{3x}} \, dx = -\int (xe^{2x} + e^{-x}) \, dx = -[\frac{1}{4}e^{2x}(2x - 1) + e^{-x}] = e^{-x} - \frac{1}{4}e^{2x}(2x - 1)$. $v_2 = \int \dfrac{y_1(xe^{3x} + 1)}{W(y_1, y_2)} \, dx = \int \dfrac{e^x(xe^{3x} + 1)}{e^{3x}} \, dx = \int (xe^x + e^{-2x}) \, dx = e^x(x - 1) - \frac{1}{2}e^{-2x}$. So a particular solution of (1) is $y = y_1 v_1 + y_2 v_2 = e^x[e^{-x} - \frac{1}{4}e^{2x}(2x - 1)] + e^{2x}[e^x(x - 1) - \frac{1}{2}e^{-2x}] = 1 - \frac{1}{4}e^{3x}(2x - 1) + e^{3x}(x - 1) - \frac{1}{2} = \frac{1}{2} + \frac{1}{4}e^{3x}(2x - 3)$. Hence, the general solution is $\frac{1}{2} + \frac{1}{4}e^{3x}(2x - 3) + C_1 e^x + C_2 e^{2x}$.

46.68 Solve

$$y'' + y = \sec x \tag{1}$$

▮ As in Problem 46.65, the general solution of $y'' + y = 0$ consists of all linear combinations of $y_1 = \cos x$ and $y_2 = \sin x$, and $W(y_1, y_2) = 1$. Then $v_1 = -\int \dfrac{y_2 \sec x}{W(y_1, y_2)} \, dx = -\int \sin x \sec x \, dx = -\int \tan x \, dx = -\ln |\sec x|$, and $v_2 = \int \dfrac{y_1 \sec x}{W(y_1, y_2)} \, dx = \int \cos x \sec x \, dx = \int 1 \, dx = x$. So a particular solution of (1) is $y = y_1 v_1 + y_2 v_2 = -\cos x \ln |\sec x| + x \sin x$. The general solution of (1) is $-\cos x \ln |\sec x| + x \sin x + C_1 \cos x + C_2 \sin x$.

46.69 Find a general solution

$$y'' + y = x \sin x \tag{1}$$

▐ As in Problem 46.65, the general solution of $y'' + y = 0$ consists of all linear combinations of $y_1 = \cos x$ and $y_2 = \sin x$, and $W(y_1, y_2) = 1$. Then $v_1 = -\int \dfrac{y_2 \, x \sin x}{W(y_1, y_2)} \, dx = -\int x \sin^2 x \, dx = \frac{1}{8}(\cos 2x + 2x \sin 2x - 2x^2)$, by an integration by parts. Moreover, $v_2 = \int \dfrac{y_1 x \sin x}{W(y_1, y_2)} \, dx = \int x \sin x \cos x \, dx = \frac{1}{8}(\sin 2x - 2x \cos 2x)$, by another integration by parts. Therefore, a particular solution is $y = y_1 v_1 + y_2 v_2 = \frac{1}{8}\cos x (\cos 2x + 2x \sin 2x - 2x^2) + \frac{1}{8}\sin x(\sin 2x - 2x \cos 2x) = \frac{1}{8}(\cos x \cos 2x + \sin x \sin 2x) + \frac{1}{4}x(\cos x \sin 2x - \sin x \cos 2x) - \frac{1}{4}x^2 \cos x = \frac{1}{8}\cos x + \frac{1}{4}x \sin x - \frac{1}{4}x^2 \cos x = \frac{1}{8}(\cos x + 2x \sin x - 2x^2 \cos x)$. The general solution is $\frac{1}{8}(\cos x + 2x \sin x - 2x^2 \cos x) + C_1 \cos x + C_2 \sin x = (x/4)(\sin x - x \cos x) + D_1 \cos x + D_2 \sin x$.

46.70 Solve the *predator-prey system*

$$\frac{dy}{dt} = K(x - \bar{x}) \qquad (K, \bar{x} > 0) \tag{1}$$

$$\frac{dx}{dt} = -L(y - \bar{y}) \qquad (L, \bar{y} > 0) \tag{2}$$

▐ Differentiate (2) with respect to t: $\dfrac{d^2x}{dt^2} = -L\dfrac{dy}{dt} = -LK(x - \bar{x})$. Hence, $\dfrac{d^2x}{dt^2} + LKx = LK\bar{x}$. By Problem 46.64, the general solution of this equation is $x = \bar{x} + C_1 \cos(\sqrt{LK}\,t) + C_2 \sin(\sqrt{LK}\,t)$. Then $\dfrac{dx}{dt} = \sqrt{LK}\,[C_2 \cos(\sqrt{LK}\,t) - C_1 \sin(\sqrt{LK}\,t)]$. By (2), $y = \bar{y} + \sqrt{\dfrac{K}{L}}\,[C_1 \sin(\sqrt{LK}\,t) - C_2 \cos(\sqrt{LK}\,t)]$.

46.71 For any function y of x, if $x = e^t$, then y also can be considered a function of t. Show that $x\dfrac{dy}{dx} = \dfrac{dy}{dt}$ and $x^2\dfrac{d^2y}{dx^2} = \dfrac{d^2y}{dt^2} - \dfrac{dy}{dt}$.

▐ $\dfrac{dy}{dt} = \dfrac{dy}{dx}\dfrac{dx}{dt} = \dfrac{dy}{dx}e^t = \dfrac{dy}{dx}x$. Therefore, $\dfrac{d^2y}{dt^2} = \dfrac{d}{dx}\left(\dfrac{dy}{dt}\right)\cdot\dfrac{dx}{dt} = \dfrac{d}{dx}\left(x\dfrac{dy}{dx}\right)\cdot e^t = \left(x\dfrac{d^2y}{dx^2} + \dfrac{dy}{dx}\right)\cdot x$. Hence, $\dfrac{d^2y}{dt^2} - \dfrac{dy}{dt} = \left(x^2\dfrac{d^2y}{dx^2} + x\dfrac{dy}{dx}\right) - x\dfrac{dy}{dx} = x^2\dfrac{d^2y}{dx^2}$.

46.72 Solve

$$x^2\frac{d^2y}{dx^2} + 3x\frac{dy}{dx} + y = 0 \tag{1}$$

▐ Let $x = e^t$. By Problem 46.71, $x\dfrac{dy}{dx} = \dfrac{dy}{dt}$ and $x^2\dfrac{d^2y}{dx^2} = \dfrac{d^2y}{dt^2} - \dfrac{dy}{dt}$. Then (1) becomes $\dfrac{d^2y}{dt^2} - \dfrac{dy}{dt} + 3\dfrac{dy}{dt} + y = 0$, or, more simply,

$$\frac{d^2y}{dt^2} + 2\frac{dy}{dt} + y = 0 \tag{2}$$

The auxiliary equation is $u^2 + 2u + 1 = 0$, with the repeated root $u = -1$. By Problem 46.48, the general solution of (2) is $y = (C_1 + C_2 t)e^{-t}$. Since $x = e^t$, $t = \ln x$. Hence, $y = \dfrac{C_1 + C_2 \ln x}{x}$.

46.73 Solve the linear third-order equation with constant coefficients $y''' + 2y'' - y' - 2y = 0$.

▐ The auxiliary equation is $u^3 + 2u^2 - u - 2 = 0$. 1 is a root. Division by $u - 1$ yields $u^2 + 3u + 2 = (u + 1)(u + 2)$. Thus, the three roots are $1, -1$, and -2. Hence, by an extension of Problem 46.48, the general solution is $C_1 e^x + C_2 e^{-x} + C_3 e^{-2x}$. (The method of variation of parameters also extends to inhomogeneous equations of higher order.)

46.74 Solve the linear third-order equation with constant coefficients $y''' - 7y'' + 16y' - 12y = 0$.

▐ The auxiliary equation is $u^3 - 7u^2 + 16u - 12 = 0$. 2 is a root. Division by $u - 2$ yields $u^2 - 5u + 6 = (u - 2)(u - 3)$. Thus, 2 is a repeated root and 3 is a root. So, by an extension of Problem 46.48, the general solution is $C_1 e^{3x} + (C_2 + C_3 x)e^{2x}$.

46.75 Solve the linear third-order equation with constant coefficients $y''' + 2y'' - 2y' - y = 0$.

▌ The auxiliary equation is $u^3 + 2u^2 - 2u - 1 = 0$. 1 is a root. Division by $u - 1$ yields $u^2 + 3u + 1$, with the roots $\dfrac{-3 \pm \sqrt{5}i}{2}$. By an extension of Problem 46.48, the general solution is $C_1 e^x + e^{-3x/2}[C_2 \cos(\sqrt{5}x/2) + C_3 \sin(\sqrt{5}x/2)]$.

46.76 Solve $y^{(4)} - 4y''' + 9y'' - 10y' + 6y = 0$.

▌ The auxiliary equation is $u^4 - 4u^3 + 9u^2 - 10u + 6 = 0$. This can be factored into $(u^2 - 2u + 3)(u^2 - 2u + 2)$. The first factor has roots $1 \pm \sqrt{2}i$, and second factor has roots $1 \pm i$. Hence, by an extension of Problem 46.48, the general solution is $e^x[C_1 \cos(\sqrt{2}x) + C_2 \sin(\sqrt{2}x)] + e^x(C_3 \cos x + C_4 \sin x)$.

46.77 Solve $y^{(4)} + 4y''' + 10y'' + 12y' + 9y = 0$.

▌ The auxiliary equation is $u^4 + 4u^3 + 10u^2 + 12u + 9 = 0$. This factors into $(u^2 + 2u + 3)^2$, with double roots $-1 \pm \sqrt{2}i$. Then, instead of $e^{-x}[C_1 \cos(\sqrt{2}x) + C_2 \sin(\sqrt{2}x)]$, the general solution is $e^{-x}[(C_1 + C_2 x) \cos(\sqrt{2}x) + (C_3 + C_4 x) \sin(\sqrt{2}x)]$.

46.78 Find a linear differential equation with constant coefficients satisfied by e^{2x}, e^{-x}, e^{3x}, and e^{5x}.

▌ The auxiliary equation should have $2, -1, 3$, and 5 as roots: $(u - 2)(u + 1)(u - 3)(u - 5) = u^4 - 9u^3 + 21u^2 + u - 30$. Hence, the required equation is $y^{(4)} - 9y''' + 21y'' + y' - 30y = 0$.

46.79 Find a linear differential equation with constant coefficients satisfied by e^{-2x} and $e^x \cos 2x$.

▌ e^{-x} stems from a root -2 of the auxiliary equation. $e^x \cos 2x$ comes from a complex root $1 + 2i$. Hence, we also take $1 - 2i$ as a root. Thus, we get the auxiliary equation $(u + 2)[u - (1 + 2i)][u - (1 - 2i)] = (u + 2)(u^2 - 2u + 5) = u^3 + u + 10 = 0$. Hence, the required equation is $y''' + y' + 10y = 0$.

46.80 Find a general solution of $(1 - x^2)y'' - 2xy' + 2y = 0$.

▌ This is a special case of *Legendre's equation*, $(1 - x^2)y'' - 2xy' + p(p + 1)y = 0$. Note that x is a particular solution. A second linear independent solution can be found as follows: If y_1 is one nontrivial solution of $y'' + P(x)y' + Q(x)y = 0$, then another linearly independent solution y_2 can be found by letting $v = \displaystyle\int \frac{1}{y_1^2} \exp\left(-\int P\,dx\right) dx$. In our case, $y_1 = x$ and $v = \displaystyle\int \frac{1}{x^2} \exp\left(-\int \frac{-2x}{1 - x^2}\,dx\right) dx = \int \frac{1}{x^2} e^{-\ln(1-x^2)}\,dx = \int \frac{1}{x^2} \cdot \frac{1}{1 - x^2}\,dx = \frac{1}{2} \ln\left|\frac{1 + x}{1 - x}\right| - \frac{1}{x}$. So $y_2 = vy_1 = vx = \frac{1}{2} x \ln\left|\frac{1 + x}{1 - x}\right| - 1$. The general solution consists of all linear combinations of y_1 and y_2: $C_1 x + C_2\left(\frac{1}{2}x \ln\left|\frac{1 + x}{1 - x}\right| - 1\right)$.

46.81 Solve $x^2 y'' + xy' + (x^2 - \frac{1}{4})y = 0$ (a special case of Bessels' equation) for $x > 0$.

▌ First check that $y_1 = (\sin x)/\sqrt{x}$ is a solution for $x > 0$. Then, by the method of Problem 46.80, $v = \displaystyle\int \frac{x}{\sin^2 x} \exp\left(-\int \frac{1}{x}\,dx\right) dx = \int \frac{x}{\sin^2 x} e^{-\ln x}\,dx = \int \frac{x}{\sin^2 x} \cdot \frac{1}{x}\,dx = \int \csc^2 x\,dx = -\cot x$. Let $y_2 = vy_1 = -\cot x \cdot \frac{\sin x}{\sqrt{x}} = -\frac{\cos x}{\sqrt{x}}$. Hence, the general solution is $C_1 \frac{\sin x}{\sqrt{x}} + C_2 \frac{\cos x}{\sqrt{x}} = \frac{C_1 \sin x + C_2 \cos x}{\sqrt{x}}$.